BIOPROCESS ENGINEERING PRINCIPLES

THIRD EDITION

PAULINE M. DORAN

KATE MORRISSEY

ROSS P. CARLSON

ELSEVIER

AMSTERDAM • BOSTON • HEIDELBERG • LONDON
NEW YORK • OXFORD • PARIS • SAN DIEGO
SAN FRANCISCO • SINGAPORE • SYDNEY • TOKYO
Academic Press is an imprint of Elsevier

Academic Press is an imprint of Elsevier
125 London Wall, London EC2Y 5AS, United Kingdom
525 B Street, Suite 1650, San Diego, CA 92101, United States
50 Hampshire Street, 5th Floor, Cambridge, MA 02139, United States

ISBN: 978-0-128-22191-4

For Information on all Academic Press publications
visit our website at https://www.elsevier.com/books-and-journals

Publisher: Peter Linsley
Acquisitions Editor: Stephen Merken
Senior Editorial Project Manager: Helena Beauchamp
Publishing Services Manager: Shereen Jameel
Project Manager: Maria Shalini
Senior Designer: Vicky Pearson Esser

Typeset by MPS Limited, Chennai, India

Printed in India

Last digit is the print number: 9 8 7 6 5 4 3 2 1

Working together
to grow libraries in
developing countries

www.elsevier.com • www.bookaid.org

Contents

Preface

Scope of Textbook

Bioprocessing is essential for the global food, chemical, and pharmaceutical industries. It is also a very interdisciplinary field. Therefore, the principal objective of the third edition of *Bioprocess Engineering Principles* is to provide a foundation for engineering students to contribute to this exciting, biology-based field. The textbook is designed to integrate with a classic engineering education and covers methods, theories, and analyses relevant to bioprocess challenges spanning enzymes to bioreactors and onto downstream technologies. The content is reinforced with numerous examples and a wide range of homework problems. Additionally, the textbook will serve as a valuable resource for not only students but also researchers, technicians, industrial engineers, and instructors.

Our ability to harness the catalytic potential of cells and enzymes is closely tied to advances in biochemistry, microbiology, immunology, and cell physiology. Knowledge in these areas has expanded rapidly in recent decades; tools of modern biotechnology such as recombinant DNA, gene probes, cell fusion, CRISPR-Cas9, and tissue culture offer new opportunities to develop novel products or to improve existing bioprocessing methods. While new products and processes can be conceived and developed in the laboratory, bringing modern biotechnology to industrial realization requires engineering expertise. Biological systems can be complex and difficult to control; nevertheless, they obey the laws of chemistry and physics and are therefore amenable to engineering analyses. Engineering input is essential for many aspects of bioprocessing, including the design and operation of bioreactors, sterilizers, and equipment for product recovery; the development of process automation and control; and the efficient and safe layout of bioprocess factories. As society continues to harness the potential of applied biology, there will be an increased demand for scientists and engineers trained in multidisciplinary bioprocess theory who can translate new discoveries from the laboratory into industrial-scale production. As a bioprocess engineer, you will work at the cutting-edge interface of biology and engineering. This textbook is designed to prepare you for that rewarding challenge.

The third edition of *Bioprocess Engineering Principles* has been edited for a target audience of engineering students. Edits include a streamlining of topics that are commonly covered in an engineering education to avoid redundancies. The streamlining of some material is countered by an expansion of topics that will provide an engineer with a foundation in basic biology. The new Chapter 1 contains an updated timeline detailing important developments in the field of bioprocessing including such timely developments as vaccine production including the COVID-19 vaccine. Chapter 2 discusses fundamentals of microbiology and biochemistry including topics like organism naming conventions, the branches on the tree of life, the central dogma of molecular biology, and the major macromolecular components of cells. Furthermore, new examples have been added to the textbook including microalgae growth and over two dozen new figures supplement the text. Finally, the textbook also provides online teaching resources including example syllabi, exams, assessments, and unique problem sets to support the work of instructors. Please visit https://educate.elsevier.com/book/details/9780128221914 to register for access (instructors only).

Acknowledgments

Coauthors Morrissey and Carlson are honored to contribute to the third edition of this textbook. They want to acknowledge the help of chapter reviewers Drs. Jixun Zhan, Jason Berberich, and Nanette Boyle for their expert feedback, knowledgeable suggestions, and enthusiastic punctuation. The coauthors would also like to thank the patient and understanding editors Mr. Stephen Merken, Ms. Helena Beauchamp, and Ms. Maria Shalini who provided essential guidance during the entire process. Finally, Morrissey and Carlson want to acknowledge the vision of author Dr. Pauline M. Doran who created and developed the previous editions of *Bioprocessing Engineering Principles*. Dr. Doran sought to close the gap between the biologists and engineers contributing to the field of bioprocessing; this textbook would not be what it is today without her many talents and her investment of time.

Introduction: Bioprocessing

CHAPTER 1

The Field of Bioprocess Engineering

1.1 Steps in Bioprocess Development

Bioprocess operations make use of whole organisms, cells, or components from cells such as enzymes, to produce new products or degrade harmful wastes. Biotechnology and bioprocessing are not new fields. The history of biotechnology begins with food production and food preservation in the ancient world. Plants and animals were first domesticated at least 10,000 years ago. These whole organisms were used as catalysts to convert resources like sunlight and CO_2 or forage into products like sugars or protein which were readily digested by humans. These early agricultural technologies lead to the use of microorganisms such as yeast and bacteria for brewing and breadmaking as well as yogurt and cheese production. Table 1.1 provides a timeline of some significant developments in biotechnology, starting with these initial bioproducts, and details important inventions, processes, and bioproducts from this point in history onward. Fermentation of grains and fruits into alcoholic beverages was performed in the ancient world at least as early as 3000 BCE. Other food-preserving fermentations practiced for thousands of years include the transformation of milk into cheeses or yogurt and the fermentation of soybeans. However, these bioprocesses were used without an understanding of the root biology. It was not until 1857 that French scientist Louis Pasteur demonstrated that alcoholic fermentation was caused by living cells, namely, yeasts. In the ensuing decades, the intentional manipulation of microbial fermentations to obtain food products, solvents, beverages, and later, substances having therapeutic value such as antibiotics gave rise to a large bioprocess industry [1–3]. Biotechnology has emerged as an enabling technology defined as "any technique that uses living organisms (or parts of organisms) to make or modify products, to improve plants or animals, or to develop microorganisms for specific uses" [4]. A major change in biotechnology occurred in the mid-20th century with discovery of the molecular basis of biology —DNA [5]. Biotechnology has helped to catalyze the growth of the pharmaceutical, food, agricultural processing, and specialty product sectors of the global economy [6,7]. The scope of biotechnology is broad and deep as it encompasses the use of chemicals to modify the behavior of biological systems and the genetic modification of organisms to confer new traits including the insertion of foreign DNA into people to compensate for genes whose absence or mutations cause life-threatening conditions.

Table 1.1: Timeline of Important Bioproduct and Bioprocess Development

Date	Event
8500 BCE	Animal husbandry, the domestication of livestock, affords early civilizations food at hand [9]
8000 BCE	Domestication of sheep leads to cheesemaking [10]
6000 BCE	First indication of yeast fermentation in breadmaking [11]
3000 BCE	Molded soybean curd used to treat skin infections in China [11]
2500 BCE	Beer fermentation in Egypt and the malting process of barely [11]
	First recorded use of insecticides by the Sumerians using sulfur compounds [12]
1663	First reference of a cell described as a basic unit of life by Hooke [13]
1675	First microscopic observations of protozoa and bacteria [14]
1857	Louis Pasteur's work with yeast and alcohol fermentation proving that yeast are living cells. Recognized as the birth of microbiology [11]
1900	Rudolf Diesel, the inventor of the diesel engine, presents his engine by running it off of peanut oil at the World Exhibition of Paris, demonstrating one of the first uses of a biofuel [11]
1902	*Bacillus thuringiensis* is first isolated from silkworm culture in Ishiwata, used in pesticides [11]
1900-1920	Ethanol, glycerol, acetone, and butanol are produced commercially [11]
1923	Citric acid fermentation using *Aspergillus niger* by Charles Pfizer [11]
1928/1943	Alexander Fleming discovers penicillin from *Penicillium notatum* (1928), and later, a submerged culture of *Penicillium chrysogenum* facilitates large-scale production of penicillin (1943) [11]
1953	The structure and function of DNA is determined, and xylose isomerase is discovered [11]
1957-1960	Commercial production of natural amino acids via fermentation [11]
1961	First commercial production of monosodium glutamate (MSG), which was identified in 1908 as a flavor enhancer [11]
1967	Clinton Corn Processing distributes first enzymatically produced fructose syrup [11]
1973	One of the first reports of a genetic engineering technique is made by Cohen and Boyer [15]
1977	The US Department of Energy was created, facilitating the advancement of biofuels [16]
1978	Human insulin produced in *Escherichia coli* [17]
1977-1982	Ethanol fermentation processes are adapted for fuel-grade ethanol [11]
1980	US Supreme Court rules that life forms are patentable
1981	The first commercial hepatitis B vaccine is approved by the FDA [18]
1983-1985	Process for industrial drying of fuel alcohol using a corn-based adsorbent demonstrated on an industrial scale [11]
	Transgenic pig, rabbit, and sheep are generated by microinjection of foreign DNA into egg nuclei [19]
	Polymerase chain reaction (PCR) is developed at Cetus [20]
	HIV genome is sequenced by Chiron [21]
1991	Human Genome Project is started
	First attempt at human gene therapy [11]
1990-1998	Mergers, selling, and acquisition of a number of biotechnology companies occurs throughout a now billion-dollar industry [11]
2000	Adult stem cells recognized as having potential to generate cells for other organs [22]
2001	First consensus human genome sequence published by Sanger, Arber, and Wu [23]
	Mechanism discovered for how the immune system generates regulatory T-cells [24]
	Human embryo cloned to generate stem cells [25]
	Structure and function of ribosomes is deciphered, which facilitates improving and designing new antibiotics [26]
	E15 (15% ethanol fuel) is officially allowed to be sold publicly by the US Environmental Protection Agency [16]

Table 1.1: Timeline of Important Bioproduct and Bioprocess Development—cont'd

Date	Event
2003	China approves the world's first commercial gene therapy [27,28]
2005-2007	Clustered Regularly Interspaced Short Palindromic Repeats (CRISPR), a family of DNA sequences found in bacteria, is demonstrated as a gene editing technique and to provide cell immunity against viruses [29–31]
2008	The structure of telomerase, an enzyme that conserves the ends of chromosomes, is decoded [32]
2013	CRISPR-Cas is used to edit the genomes of humans, zebrafish, rats, *Saccharomyces cerevisiae*, and plant genomes (rice, wheat, tobacco, sorghum) [33]
2018	Genetically engineered viruses that kill cancer cells open new immunotherapy avenues [34]
2020	The COVID-19 pandemic catalyzes the most rapid vaccine development in history, with messenger RNA (mRNA) vaccines at the forefront [35]

Figure 1.1: Samsung BioLogics state-of-the-art bioreactor hall.
From Praxis Samsung, A publication by Bioengineering AG, 8636 Wald, Switzerland.

The science of genetic engineering is finding applications in enhancing microbial and plant technologies to address global concerns such as the mitigation of greenhouse gases and the production of renewable chemicals such as polymers and fuels [8].

The engineering fundamentals required to translate the discoveries of biotechnology into tangible commercial products define the discipline of bioprocess engineering [6]. Bioprocess engineering translates biotechnology into unit operations, biochemical processes, equipment, and facilities for manufacturing bioproducts, such as the facility pictured in Figure 1.1. Industrial bioprocessing entails the design and scaleup of bioreactors that generate large quantities of transformed microbes or cells and their products, as well as technologies for recovery, separation, and purification of these products. This book presents basic principles of life sciences and engineering for the practice of biotechnology manufacturing [5,36–38].

1.2 Example Bioprocesses

The following sections highlight examples of bioprocesses that have shaped the bioprocess engineering field. Table 1.1 illustrated a timeline of the development of specific critical bio-products and bioprocessing milestones, whereas Table 1.2 expands on the timeline presented in Table 1.1 and provides examples of bioprocesses categories with corresponding microorganisms used. These examples highlight various production strategies as well as differences in production scale. Figure 1.2 shows a bioprocess facility with bioreactors ranging from 200 to 5000 L.

1.2.1 Antibiotic Production: The Story of Penicillin

In 1928, British scientist Alexander Fleming was trying to isolate the bacterium, *Staphylococcus aureus*, which causes boils. One of his agar medium plates became contaminated with a foreign organism. Fleming noticed that no bacteria grew near the invading species, which he hypothesized was due to an antibacterial agent produced by the contaminant. He recovered the organism and found it was a common mold of the *Penicillium* genus (later identified as *Penicillium notatum*). Fleming grew up cultures of the organism and used extraction methods to obtain a small quantity of secreted compound. He subsequently demonstrated that the chemical had powerful antimicrobial properties and named the product penicillin. Fleming preserved the culture, but the discovery had virtually no impact on the medical field for over a decade.

World War II provided the impetus to resurrect the discovery. Howard Florey, Ernst Chain, and Norman Heatley played key roles in producing sufficient biomass to test the effectiveness of penicillin. Assays were developed to measure the kinetics of the bioprocess, to monitor the amount of penicillin made, to standardize growth protocols, and to devise a novel product extraction process. After months of effort, this team produced enough penicillin to treat laboratory animals and humans. The penicillin worked and brought a patient to the point of recovery, but unfortunately, without a large supply of the antibiotic, sustained treatment of the patient was not possible. To generate meaningful amounts of penicillin, a large-scale bioprocess needed to be developed, which would require engineers, microbial physiologists, and other life scientists. The industrial disruptions of the war further complicated the situation. Florey and his associates approached pharmaceutical firms in the United States to produce penicillin, since the United States was not involved in the war at that time. Many companies, government laboratories, and universities became involved in this effort, notably Merck, Pfizer, Squibb, and the USDA Northern Regional Research Laboratory in Peoria, Illinois.

Initially, the large-scale production of penicillin was planned to be produced via chemical synthesis. At that time, US companies had achieved a great deal of success with the chemical synthesis of other drugs, which provided a great deal of control over the production of drugs. However, the chemical synthesis of penicillin proved to be exceedingly difficult.

Table 1.2: Examples of Products From Bioprocessing

Product	Typical organism used
Biomass	
Agricultural inoculants for nitrogen fixation	*Rhizobium leguminosarum*
Bakers' yeast	*Saccharomyces cerevisiae*
Cheese starter cultures	*Lactococcus* spp.
Inoculants for silage production	*Lactobacillus plantarum*
Single-cell protein	*Candida utilis* or *Pseudomonas methylotrophus*
Yogurt starter cultures	*Streptococcus thermophilus* and *Lactobacillus bulgaricus*
Bulk organics	
Acetone/butanol	*Clostridium acetobutylicum*
Ethanol (nonbeverage)	*S. cerevisiae*
Glycerol	*S. cerevisiae*
Organic acids	
Citric acid	*Aspergillus niger*
Gluconic acid	*A. niger*
Itaconic acid	*Aspergillus itaconicus*
Lactic acid	*Lactobacillus delbrueckii*
Amino acids	
L-Arginine	*Brevibacterium flavum*
L-Glutamic acid	*Corynebacterium glutamicum*
L-Lysine	*B. flavum*
L-Phenylalanine	*C. glutamicum*
Others	*Corynebacterium* spp.
Nucleic acid–related compounds	
5′-Guanosine monophosphate (5′-GMP)	*Bacillus subtilis*
5′-Inosine monophosphate (5′-IMP)	*Brevibacterium ammoniagenes*
Enzymes	
α-Amylase	*Bacillus amyloliquefaciens*
Glucoamylase	*A. niger*
Glucose isomerase	*Bacillus coagulans*
Pectinases	*A. niger*
Proteases	*Bacillus* spp.
Rennin	*Mucor miehei* or recombinant yeast
Vitamins	
Cyanocobalamin (B_{12})	*Propionibacterium shermanii* or *Pseudomonas denitrificans*
Riboflavin (B_2)	*Eremothecium ashbyii*
Extracellular polysaccharides	
Dextran	*Leuconostoc mesenteroides*
Xanthan gum	*Xanthomonas campestris*
Other	*Polianthes tuberosa* (plant cell culture)

(Continued)

Table 1.2: Examples of Products From Bioprocessing—cont'd

Product	Typical organism used
Poly-β-hydroxyalkanoate polyesters	
Poly-β-hydroxybutyrate	*Alcaligenes eutrophus*
Antibiotics	
Cephalosporins	*Cephalosporium acremonium*
Penicillins	*Penicillium chrysogenum*
Aminoglycoside antibiotics (e.g., streptomycin)	*Streptomyces griseus*
Ansamycins (e.g., rifamycin)	*Nocardia mediterranei*
Aromatic antibiotics (e.g., griseofulvin)	*Penicillium griseofulvum*
Macrolide antibiotics (e.g., erythromycin)	*Streptomyces erythreus*
Nucleoside antibiotics (e.g., puromycin)	*Streptomyces alboniger*
Polyene macrolide antibiotics (e.g., candidin)	*Streptomyces viridoflavus*
Polypeptide antibiotics (e.g., gramicidin)	*Bacillus brevis*
Tetracyclines (e.g., 7-chlortetracycline)	*Streptomyces aureofaciens*
Alkaloids	
Ergot alkaloids	*Claviceps paspali*
Taxol	*Taxus brevifolia* (plant cell culture)
Saponins	
Ginseng saponins	*Panax ginseng* (plant cell culture)
Pigments	
β-Carotene	*Blakeslea trispora*
Plant growth regulators	
Gibberellins	*Gibberella fujikuroi*
Insecticides	
Bacterial spores	*Bacillus thuringiensis*
Fungal spores	*Hirsutella thompsonii*
Microbial transformations	
D-Sorbitol to L-sorbose (in vitamin C production)	*Acetobacter suboxydans*
Steroids	*Rhizopus arrhizus*
Vaccines	
Diphtheria	*Corynebacterium diphtheriae*
Hepatitis B	Surface antigen expressed in recombinant *S. cerevisiae*
Mumps	Attenuated viruses grown in chick embryo cell cultures
Pertussis (whooping cough)	*Bordetella pertussis*
Poliomyelitis virus	Attenuated viruses grown in monkey kidney or human diploid cells
Rubella	Attenuated viruses grown in baby hamster kidney cells
Tetanus	*Clostridium tetani*

Table 1.2: Examples of Products From Bioprocessing—cont'd

Product	Typical organism used
Therapeutic proteins	
Erythropoietin	Recombinant mammalian cells
Factor VIII	Recombinant mammalian cells
Follicle-stimulating hormone	Recombinant mammalian cells
Granulocyte-macrophage colony-stimulating factor	Recombinant *Escherichia coli*
Growth hormones	Recombinant *E. coli*
Hirudin	Recombinant *S. cerevisiae*
Insulin and insulin analogues	Recombinant *E. coli*
Interferons	Recombinant *E.coli*
Interleukins	Recombinant *E.coli*
Platelet-derived growth factor	Recombinant *S. cerevisiae*
Tissue plasminogen activator	Recombinant *E.coli* or recombinant mammalian cells
Monoclonal antibodies	
Various, including Fab and Fab₂ fragments	Hybridoma cells
Therapeutic tissues and cells	
Cartilage cells	Human (patient) chondrocytes
Skin	Human skin cells

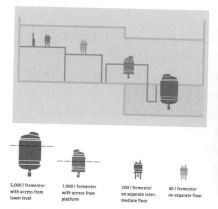

Figure 1.2: A bioprocess inoculation "train" where smaller bioreactors are used to inoculate larger bioreactors for production runs. *From Praxis Samsung, A publication by Bioengineering AG, 8636 Wald, Switzerland.*

In 1943, the War Production Board appointed A. L. Elder to coordinate the large-scale production of penicillin and the bioprocess route was selected. However, the low rate of penicillin production per unit volume would necessitate very large bioreactors and the low concentration (titer) made product recovery and purification difficult. Scientists at the Northern Regional Research Laboratory developed a corn steep liquor–lactose-based medium, which increased culture productivity approximately tenfold. Additionally, a worldwide search for better strains of *Penicillium* led to the isolation of a *Penicillium chrysogenum* strain. This strain, isolated from a moldy cantaloupe at a Peoria fruit market, proved superior to hundreds of other isolates and its progeny have been used in almost all commercial penicillin bioprocesses since.

The penicillin manufacturing process was another major technological hurdle. One method grew the mold on the surface of moist bran. This bran method was discarded because of difficulties in temperature control, sterilization, and equipment scaling. Engineers generally favored a submerged culture grown in a bioreactor. The submerged process also presented challenges as the technique needed to account for both mold physiology, namely the requirement for O_2, and mass transfer challenges associated with mixing viscous solutions. Large volumes of clean, sterile air were required. Large agitators were required, and the mechanical seal for the agitator shaft had to be designed to prevent contamination with other microorganisms [11]. There were similar hurdles in product recovery and purification. The fragile nature of the penicillin molecule required the development of special techniques, including pH shifts and rapid liquid-liquid extractions.

The engineers and biologists met the challenge. Pfizer completed the first commercial plant to produce penicillin using submerged cultivation [39]. The plant had fourteen 7000-gallon tanks. By the end of World War II, the United States had the capacity to produce enough penicillin for 100,000 patients per year.

It is important to reiterate that this accomplishment required a high degree of multidisciplinary work. For example, Merck realized that people who understood both engineering and biology were not available and therefore assigned a chemical engineer and microbiologist to work together on each aspect of the problem. Progress with penicillin production has continued. From 1939 to present day, the yield of penicillin has increased from $0.001\,g\,L^{-1}$ to over $50\,g\,L^{-1}$ of culture broth. Progress has involved better understanding of mold physiology, metabolic pathways, penicillin structure, methods of mutation and selection of mold genetics, process control, and bioreactor design. With the large-scale industrial production of penicillin, the modern concept of a bioprocess engineer was born.

Figure 1.3 details a process flow diagram for the bioproduction of bacitracin, another important antibiotic used to prevent skin infections.

1.2.2 Amino Acid Fermentation

"Amino acid fermentation" is the overproduction and secretion of amino acids by microorganisms and it is based on progress in academic research and industrial development. Unlike biopharmaceuticals, whose emergence as a major biotechnology business sector has been

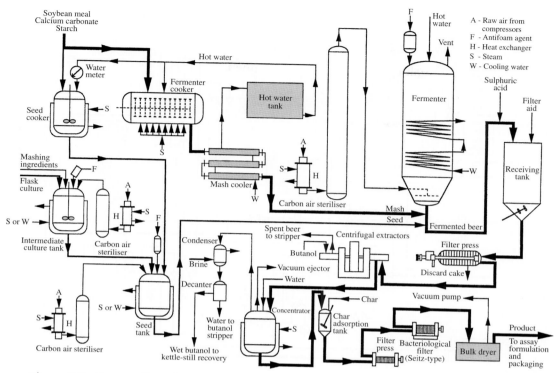

Figure 1.3: Process flow sheet showing the major operations for production of bacitracin. *Reprinted with permission from G.C. Inskeep, R.E. Bennett, J.F. Dudley, M.W. Shepard, Bacitracin: Product of biochemical engineering, Ind. Eng. Chem. 43 (1951) 1488–1498. Copyright 1951, American Chemical Society.*

relatively recent, the production of amino acids has a long history initiated with the isolation of asparagine from asparagus juice in 1806. Other amino acids were isolated from a variety of natural substances in subsequent years. In 1908, Japanese researcher Kikunae Ikeda discovered glutamate as an umami substance, leading to the commercialization of MSG (monosodium glutamate). Although glutamate was originally extracted from the hydrolysate of wheat or soybean, a microbial production method was invented in which glutamate was overproduced by strains of *Corynebacterium* grown on sugars. This discovery helped initiate the development of an amino acid production industry and also helped form the basis of modern biotechnology [11,40].

Today, microbial amino acid production is big business not only for human consumption but also as supplements for animal feed. Supplementing animal feed improves the health and weight gain of livestock. The global market for feed amino acids has been estimated at $6–7 billion in 2021. Major industrially produced amino acids include lysine, glutamate, threonine, and methionine. For example, approximately 2.5 million tons of lysine are sold annually.

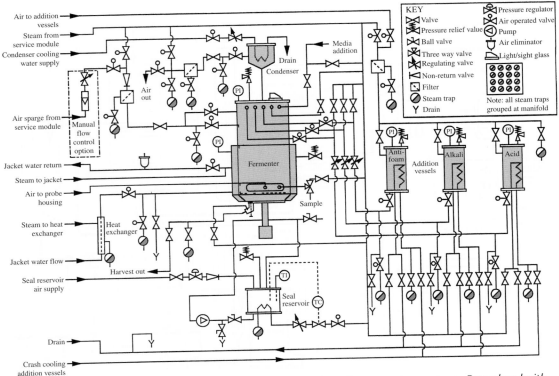

Figure 1.4: Detailed equipment diagram for a pilot-plant fermentation system. *Reproduced with permission from LH Engineering Ltd. Copyright 1983.*

Figure 1.4 illustrates a classic fermentation process flow diagram for upstream production of a bioproduct.

1.2.3 Biofuels and Modern Fuel Markets

Biofuels have been used since humans learned to control fire. Wood was the first biofuel used by ancient people for cooking and heating. While the use of wood as a biofuel is ancient, the modern concept of "biofuels" has roots in the early 1900s and is based in the automotive industry. Rudolf Diesel, the inventor of the diesel engine, and Henry Ford, the designer of the Model T, both utilized biofuels in their products. Diesel originally designed his engine to run on peanut oil and the Model T was designed to use hemp-derived biofuel. However, with the industrial development of large, naturally occurring deposits of crude oil, use of the more expensive biofuels was reduced [11].

Supply chain perturbations associated with World War II resulted in high demand for alternative fuel sources including biofuels. It was during this period that additional bioprocesses were developed, such as the use of gasoline-based fuels that were augmented with grain or potato-based alcohol. A serious fuel crisis occurred once again between 1973 and 1979. The

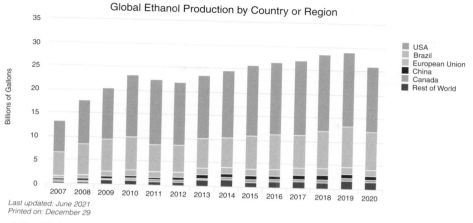

Figure 1.5: Global ethanol production by country or region. *From the U.S. Department of Energy's Alternative Fuels Data Center.* Data from http://afdc.energy.gov/data/10331

Organization of the Petroleum Exporting Countries (OPEC) reduced fuel exports which negatively impacted non-OPEC nations. The shortage of fuel attracted investment from academic institutions, industries, and governments and resulted in technological advances in biofuels like bioethanol and biodiesel. A combination of rising petroleum prices, increasing levels of greenhouse gases in the atmosphere, desire for sustainable fuels, and interest in rural development led to a new era of growth in biofuel biotechnology at the end of the 20th century.

The field of biofuels and renewable fuels continues to grow with advancements in bioprocess engineering [11]. The global market for bioethanol was approximately 25 billion gallons in 2020 (Figure 1.5). The United States is the world's largest producer, with production of 13.9 billion gallons in 2020. Presently, the United States and Brazil are the world's largest producers of bioethanol from corn and cane sugar, respectively [41]. Transitioning bioethanol production to nonfood feedstocks such as cellulose has been and will continue to be a major focal point for investment into sustainable bioprocesses.

Figure 1.6 illustrates trends in US biodiesel production, exports, and consumption from 2001 to 2020. Exports of biodiesel peaked in 2008 largely due to an unintended effect of a biodiesel tax credit in the European Union. Once this tax was eliminated, the exports dropped. The increased production and consumption seen from 2011 on is primarily driven by the Renewable Fuel Standard implemented in the US. These trends highlight important connections between global fuel markets, bioprocessing trends, and government policies [42].

1.2.4 2,3-Butanediol, the Industrial Range of a Single Molecule

Advances in bioprocessing and biotechnology have opened the doors to a range of innovations that can come from a single microorganism or molecule. An interesting example is

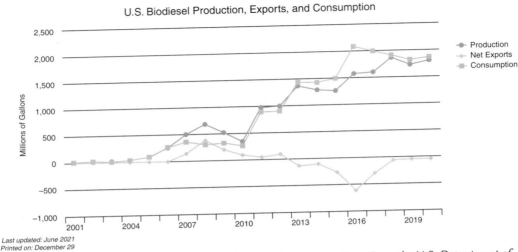

Figure 1.6: US biodiesel production exports, and consumption. *From the U.S. Department of Energy's Alternative Fuels Data Center.* From https://afdc.energy.gov/data/10325

2,3-butanediol. This compound ((CH$_3$CHOH)$_2$) is classified as a vicinal-diol or glycol and is used as a precursor to bioplastics, pesticides, and fuels [43–45]. 2,3-Butanediol has three stereoisomers, namely, two enantiomers (2 R,3R- and 2 S,3S-butanediol) and one meso compound (2 R,3S-butanediol). The different stereoisomers have different properties leading to the molecules used for different bioproducts. For example, the meso isomer of 2,3-butanediol is used to produce a polyurethane called "Vulkollan," an ultra-high-performance elastomer [43]. The (2 R,3 R)-stereoisomer of 2,3-butanediol is produced via fermentation by root-associated bacteria, including *Bacillus polymyxa* [44,46]. This isomer is effective in controlling plant viral diseases and increasing plant yields. [44]. Figure 1.7 details the downstream processing of 2,3-butanediol based on fermentation of 1000 bushels of wheat per day by *Aerobacillus polymyxa*. The bioprocess coproduces ethanol.

2,3-Butanediol produced through fermentation has also been researched as a precursor for 1,3-butadiene, which was used during World War II for synthetic rubber production [47,48]. 2,3-Butanediol has also been proposed as a rocket fuel that could be produced on Mars using a combination of light, CO$_2$, and a coculture of cyanobacteria and *E. coli*. [45]. The versatility and broad industrial potential of biologically produced molecules hints at a bright future for the field of bioprocessing and for students with solid foundations in biology and engineering.

1.2.5 Systems Biology

The microorganisms used in bioprocesses are complex systems composed of components including genes, RNA, proteins, and metabolites. These components are connected via

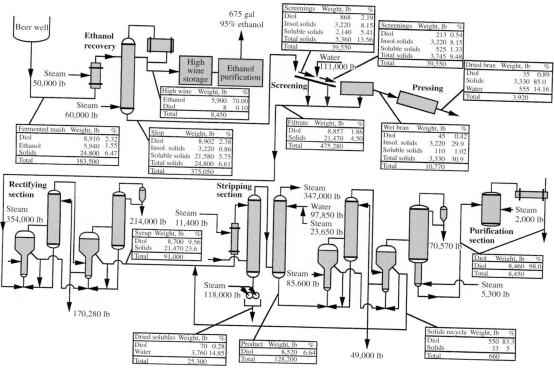

Figure 1.7: Quantitative flow sheet for the downstream processing of 2,3-butanediol based on fermentation of 1000 bushels of wheat per day by *Aerobacillus polymyxa*. From J.A. Wheat, J.D. Leslie, R.V. Tomkins, H.E. Mitton, D.S. Scott, G.A. Ledingham, Production and properties of 2,3-butanediol, XXVIII: Pilot plant recovery of levo-2,3-butanediol from whole wheat mashes fermented by *Aerobacillus polymyxa*, Can. J. Res. 26 F (1948) 469–496.

networks such as gene regulation systems and biochemical pathways, which ultimately make life possible (Figure 1.8). Systems often possess emergent properties which are defined as properties that are greater than the sum of the parts. Emergent properties are a sum of the parts and their interactions. Life is an emergent property. Engineers and scientists have made remarkable progress in compiling and decoding both the components and the interactions essential to life. The complexity of these systems often necessitates the use of computational tools to organize, analyze, and control the processes.

As you progress through this textbook, please think of all the hierarchical and simultaneous systems-based processes that make biology possible. The structure and regulation of these systems are the result of natural selection, a process where organisms possessing networks resulting in higher fitness pass their genes to successive generations at a higher frequency than organisms using networks which results in less fitness. Ecology is often credited as being the first biological science to truly embrace and apply the concept of systems biology.

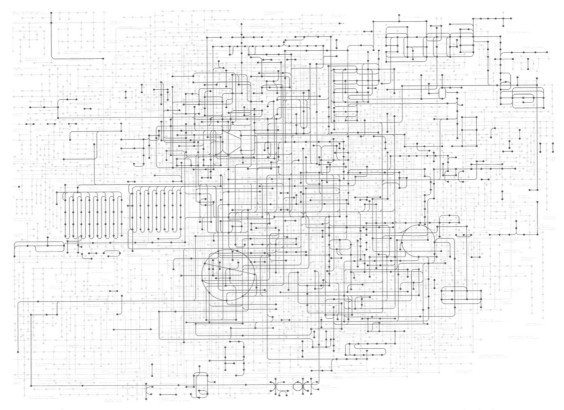

Figure 1.8: Graphical representation of the *E. coli* metabolic network where dark dots are metabolites and dark lines represent metabolic reactions occurring in this bacterium. Mass and energy flow through these networks converting substrates into new cells and useful products. *Diagram from KEGG (genome.jp).* http://www.genome.jp/pathway/map01100

Macroscale relationships can be simpler to observe and quantify. For example, the relationship between grass, rabbits, and foxes is one example of a biological system that is often converted into a mathematical formulation and studied in ecology courses and reactor kinetics courses.

Systems biology has expanded rapidly in recent decades. The outcomes of this field will continue to influence our societies by enhancing food production, cleaning the environment, producing more effective medicines, and hundreds of additional products. An introduction to some common computational systems biology approaches is presented in later chapters of the textbook.

1.2.6 Vaccines

Vaccines are treatments that prime an immune system to recognize and combat foreign microorganisms, viruses, or other biological compounds. The origins of this field lay in the distant past, with humans struggling to mitigate the effects of infectious diseases like smallpox. Smallpox is a virus-based pathogen that has killed millions of people and influenced the course of human history. For example, the fall of the Roman Empire is believed to have been influenced by a plague of smallpox infections [49–51].

Ancient peoples did not understand the biological bases of disease, but their keen observations led to basic preventative strategies. For example, healthy people were intentionally exposed to pathogens to protect them from future uncontrolled exposures. It is suggested that this occurred as long as 2000 years ago. Modern vaccine use is often described as starting in the CE 1700s. Various Europeans are noted as applying smallpox inoculation strategies, originally used in the Eastern Mediterranean, to build natural immunity. The approach used material from sick patients such as fluid from active smallpox pustules which was introduced under the skin using a needle. During the late 1700s, people realized that farmers who contracted a related, but less severe, virus-based disease, cowpox, became resilient to the more dangerous smallpox. This knowledge was used to develop the first vaccines for smallpox. English physician Edward Jenner was aware of this observation and tested the theory. Jenner inoculated a young boy, James Phipps, with material from an active cowpox patient using a needle. James got sick from the cowpox but recovered. Two months later, Jenner then inoculated James with material from an active smallpox patient. James did not develop smallpox and is considered the first person successfully vaccinated against smallpox. While ultimately successful, the experimentation on humans represents a major bioethics violation if we apply the norms of modern society. Although at the time, such practices were not regulated.

The smallpox vaccine is an example of a prophylactic vaccine that prevents disease. In the 1800s, French scientist Louis Pasteur continued pioneering vaccination strategies. He is credited with developing vaccines against cholera in farm animals and a postexposure vaccine for rabies that was effective with humans. The following centuries have resulted in vaccines for a wide range of human health challenges including yellow fever, pertussis, influenza, polio, measles, mumps, rubella, tetanus, and hepatitis B, to name a few major accomplishments.

Infectious diseases are still a major challenge around the world. The seasonal flu is responsible for 7000–20,000 deaths a year in the United States and the global novel coronavirus 2019 (SARS-CoV-2) pandemic has contributed to tens of millions of deaths globally. Vaccines can help prevent deaths from many infectious diseases and vaccines are major bioproducts.

Vaccines are commonly made from (1) the whole virus or bacterium, although they are typically attenuated or deactivated prior to use; (2) a specific component of the agent that triggers an immune response, like a surface protein or toxoid; or (3) just the genetic material or a derivative of the genetic material from the problematic agent, for example, the DNA or RNA from a virus (Figure 1.9).

Use whole virus or
bacterium, typically
attenuated or killed

Use parts that trigger
immune system like
spike proteins or
toxoids

Use genetic material
coding components
like spike protein

Figure 1.9: Three primary approaches to creating vaccines against viruses or bacteria. *Graphic from World Health Organization.* http://www.who.int/news-room/feature-stories/detail/the-race-for-a-covid-19-vaccine-explained

Production of a vaccine depends on factors like what agent is being used to trigger the immune response. This will play a major role in developing the bioprocess. For example, common flu vaccines often use inactivated viruses that are produced on large scales using methods like inoculating millions of chicken eggs. Viruses require a host to reproduce and chicken eggs can serve as bioreactors for virus production. While chicken eggs have been used for decades to produce vaccines, they can be problematic when huge numbers of eggs are needed quickly or when commercial chicken farms are impacted by outbreaks of infectious diseases like avian flu. Viruses can also be produced in eukaryotic cell cultures using bioreactor technologies. The use of cell culture permits rapid scale-up when many doses of a vaccine are needed quickly. Cell culture–based virus production will likely be a dominant strategy in coming decades.

The COVID-19 pandemic led to the rapid development and use of mRNA-based vaccines, although mRNA-based vaccines had been studied for decades prior to COVID-19. These vaccines introduce genetic information from the pathogen into the host, where the information is translated into a foreign protein. With COVID-19, the genetic information for the spike protein on the outside of the virus was used to develop vaccines. mRNA can be rapidly synthesized using modern nucleotide synthesis technologies and then incorporated within synthetic lipid droplets. When introduced into the host, the lipid droplets fuse with the host's cells transporting the mRNA within the cell and the mRNA is then translated into the target protein by the host's machinery. The target protein then triggers the host's immune system into developing antibodies to the protein, which provides protection if the host is exposed to the active virus. The mRNA from the vaccine has a limited lifetime in the host and eventually degrades. mRNA vaccines have a very promising future. Many therapeutic vaccines are being tested to elicit immune responses that are proposed to treat challenges like cancer.

1.3 Summary of Chapter 1

At the end of Chapter 1, you should:

- Be familiar with the history of critical bioproducts and bioprocesses
- Be familiar with specific examples of bioprocesses that have shaped the bioprocess engineering field
- Understand the complexity of bioprocesses through process flow diagram figures
- Be able to communicate the scale of modern renewable fuel markets
- Understand the impact and versatility of the field of bioprocessing

Problems and Topics of Discussion

The following are topic ideas or problems to help students understand the industry of bioprocessing. Use such ideas within the classroom, as part of assigned work or projects.

1.1. **Contemporary bioproducts**. Please research contemporary bioproducts using available resources including the internet. List four bioproducts you find interesting. For each product, provide a two- to four-sentence description of (i) what it is, (ii) what it is used for, and (iii) what are its societal impacts. The write-up should use complete sentences with proper grammar. Also include references for where you found the information.

1.2. **Know your industry**. Research the major corporations in an industrial field relevant to bioprocess engineering. For example, identify the top 10 global companies in the pharmaceutical industry and/or identify the top 10 global chemical companies. This information is widely available through many industry publications; however, the lists may vary from one source to another. Why do you think the lists are different? Is it easy to define an industry? Does everyone use the same definitions? For example, is a company involved in petroleum processing considered a chemical company?

1.3. **Critical events in the field of bioprocessing**. Helping students create links between the history of a field and how it has progressed can aid in understanding the "why" of course topics. What major events in the field of bioprocessing have shaped its scientific advances and contributions? For example, communicate in 1–2 paragraphs how the discovery of penicillin shaped antibiotic production. An additional example could be: how have historic events, world conflicts, and resource management influenced the progress of the bioprocessing field? Include references for where you found the information.

References

[1]   A.J. Hacking, Economic Aspects of Biotechnology, Cambridge University Press (1986)6–8, 18–20, 39–72.

[2]   S. Aiba, A.E. Humphrey, N.F. Millis, Biochemical Engineering, second ed., Academic Press, New York, 1973,1–11.

[3]   R.M. Evans, The Chemistry of Antibiotics Used in Medicine, Pergamon Press, Oxford, 1965,

[4]   Office of Technology Assessment, US Congress, Biotechnology in A Global Economy, B. Brown (Ed.), OTA-BA-494, US Government Printing Office, Washington, DC, 1991.

[5]   J. Houghton, S. Weatherwax, J. Ferrell, Breaking the Biological Barriers to Cellulosic Ethanol: A Joint Research Agenda, Office of Science and Office of Energy Efficiency and Renewable Energy, US Department of Energy, June 2006.

[6]   National Research Council (NRC) Committee on Bioprocess Engineering, Putting Biotechnology to Work: Bioprocess Engineering, National Academy of Sciences, Washington, DC, 1992,

[7]   National Research Council (NRC) Committee on Opportunities in Biotechnology for Future Army Applications, Opportunities in Biotechnology for Future Army Applications, National Academy of Sciences, Washington, DC, 2001.

[8]   S. Kim, B.E. Dale, Life cycle assessment of fuel ethanol derived from corn grain via dry milling, Bioresour. Technol. 99, (2008) 5250–5260.

[9]   A. Caliebe, A. Nebel, C. Makarewicz, M. Krawczak, B. Krause-Kyora, Insights into early pig domestication provided by ancient DNA analysis, Sci. Rep. 7 (2017) 44550.

[10] Y. Hatziminaoglou, J. Boyazoglu, The goat in ancient civilisations: from the Fertile Crescent to the Aegean Sea, Small Rumin. Res. 51 (2004) 123–129.

[11] N.S. Mosier, M.R. Ladisch, Modern Biotechnology: Connecting Innovations in Microbiology and Biochemistry to Engineering Fundamentals, Wiley (2009),

[12] J. Unsworth, History of pesticide use, Agrochemicals (201010). Retrieved December 29, 2021, from http://agrochemicals.iupac.org/index.php?option=com_sobi2&sobi2Task=sobi2Details&catid=3&sobi2Id=31

[13] P. Mazzarello, A unifying concept: the history of cell theory, Nat. Cell. Biol. 1 (1999) E13–E15.

[14] H. Gest, The discovery of microorganisms by Robert Hooke and Antoni van Leeuwenhoek, Fellows of The Royal Society, Notes Rec. R. Soc. Lond. 58 (2004) 187–201.

[15] S.N. Cohen, A.C.Y. Chang, H.W. Boyer, et al., Construction of biologically functional bacterial plasmids in vitro, Proc. Natl. Acad. Sci. 70 (1973) 3240–3244.

[16] S. Zhang, Timeline, All About Biofuels (2015). Retrieved December 29, 2021, from https://allaboutbiofuels.wixsite.com/biofuels/timeline

[17] D.V. Goeddel, D.G. Kleid, F. Bolivar, et al., Expression in *Escherichia coli* of chemically synthesized genes for human insulin, Proc. Natl. Acad. Sci. 76 (1979) 106–110.

[18] T.H. Maugh, II, FDA approves hepatitis B vaccine, Science. 214 (1981) 1113.

[19] R.E. Hammer, V.G. Pursel, C.E. Rexroad Jr, et al., Production of transgenic rabbits, sheep and pigs by microinjection, Nature. 315 (1985) 680–683.

[20] J.M.S. Bartlett, D. Stirling, A short history of the polymerase chain reaction, in: J.M.S. Bartlett D. Stirling(Eds.), PCR Protocols. Methods in Molecular Biology™, vol. 226, Humana Press (2003),

[21] G. Grandi, R. Zagursky, The impact of genomics in vaccine discovery: achievements and lessons, Expert Rev. Vaccines. 3 (2004) 621–623.

[22] D.L. Clarke, C.B. Johansson, J. Wilbertz, et al. Generalized potential of adult neural stem cells, Science. 288 (2000) 1660–1663.

[23] J.C. Venter, M.D. Adams, E.W. Myers, et al., The sequence of the Human Genome, Science. 291 (2001) 1304–1351.

[24] M. Jordan, A. Boesteanu, A. Reed, et al., Thymic selection of CD4+CD25+ regulatory T cells induced by an agonist self-peptide· Nat, Immunol 2 (2001) 301–306.

[25] Judith A. Johnson, Human Cloning, *Development* 2 (2001) 25–31.

[26] J. Harms, F. Schluenzen, R. Zarivach, et al., "High resolution structure of the large ribosomal subunit from a mesophilic eubacterium", Cell. 107(5) (2001) 679–688.

[27] W.W. Zhang, L. Li, D. Li, et al., "The first approved gene therapy product for cancer Ad-p53 (Gendicine): 12 years in the clinic", *Human gene therapy* 29.2 (2018) 160–179.

[28] "Gene therapy: An evolving story", in Lara V Marks, ed, *Engineering Health: How biotechnology changed medicine*, (Royal Society of Chemistry, October 2017).

[29] A. Bolotin, B. Quinquis, A. Sorokin, S.D. Ehrlich, Clustered regularly interspaced short palindrome repeats (CRISPRs) have spacers of extrachromosomal origin, Microbiology 151 (2005) 2551–2561.

[30] F.J.M. Mojica, D. ez-Villase, J.S. C.S., Garc a-Mart nez, E. Soria, Intervening Sequences of Regularly Spaced Prokaryotic Repeats Derive from Foreign Genetic Elements, J Mol Evol 60 (2005) 174–182.

[31] C. Pourcel, G. Salvignol, G. Vergnaud, CRISPR elements in Yersinia pestis acquire new repeats by preferential uptake of bacteriophage DNA, and provide additional tools for evolutionary studies, Microbiology 151 (2005) 653–663.

[32] Emmanuel Skordalakes, "Telomerase and the benefits of healthy living", *The Lancet Oncology* 9 (11) (2008) 1023–1024.

[33] Eric S. Lander, "The heroes of CRISPR", *Cell* 164 (1-2) (2016) 18–28.

[34] E.M. Scott, M.R. Duffy, J.D. Freedman, et al., "Solid tumor immunotherapy with T cell engager-armed oncolytic viruses." *Macromolecular bioscience* 18 (1) (2018) 1700187.

[35] Namit Chaudhary, Drew Weissman, Kathryn A. Whitehead., "mRNA vaccines for infectious diseases: principles, delivery and clinical translation.", Nature Reviews Drug Discovery 20 (11) (2021) 817–838.

[36] M.R. Ladisch, in: G.D. Considine (Ed.), Van Nostrand's Scientific Encyclopedia, 1, fifth ed., John Wiley & Sons, 2002, pp. 434–459.

[37] NABC Report 19, Agricultural Biofuels: Technology, Sustainability, and Profitability, A. Eaglesham, R.W.F. Hardy (Ed.), Cornell University Library, 2007, pp. 3–11.

[38] L.R. Lynd, M.S. Laser, D. Brandsby, et al., How biotech can transform biofuels, Nature Biotechnol. 26 (2008) 169–172.

[39] María Jesús Santesmases, *The circulation of penicillin in Spain: health, wealth and authority*, Springer (2017),

[40] S. Kinoshita, "Glutamic Acid Bacteria", in: A.L. Demain, N.A. Solomon, (Eds.), Biology of Industrial Microorganisms, Benjamin/Cummings, Menlo Park, CA, 1985, pp. 115–142.

[41] Alternative Fuels Data Center. (2021). *Maps and Data—Global Ethanol Production by County or Region*. Alternative Fuels Data Center: Maps and Data. Retrieved December 29, 2021, from https://afdc.energy.gov/data/

[42] Alternative Fuels Data Center. (2021). *Maps and Data—U.S. Biodiesel Production, Exports, and Consumption*. Alternative Fuels Data Center: Maps and Data. Retrieved December 29, 2021, from https://afdc.energy.gov/data/

[43] Heinz Gräfje, Wolfgang Körnig, Hans-Martin Weitz, Wolfgang Reiß, Guido Steffan, Herbert Diehl, Horst Bosche, Kurt Schneider, Heinz Kieczka, "Butanediols, Butenediol, and Butynediol": *Ullmann's Encyclopedia of Industrial Chemistry*, Wiley-VCH, Weinheim, 2000,

[44] H.G. Kong, T.S. Shin, T.H. Kim, C.-M. Ryu, Stereoisomers of the bacterial volatile compound 2,3-butanediol differently elicit systemic defense responses of pepper against multiple viruses in the field, Front. Plant Sci. 9 (2018) 90.

[45] N.S. Kruyer, M.J. Realff, W. Sun, et al., Designing the bioproduction of Martian rocket propellant via a biotechnology-enabled in situ resource utilization strategy, *Nat Commun* 12 (2021) 6166.

[46] C. De Mas, N.B. Jansen, G.T. Tsao, "Production of optically active 2,3-butanediol by Bacillus polymyxa", *Biotechnol. Bioeng.* 31 (4) (1988) 366–377.

[47] "Fermentation Derived 2,3-Butanediol", by Marcio Voloch, et al., Section 3 in *Comprehensive Biotechnology*, Vol 2, Pergamon Press Ltd, England, 1986, 933.

[48] Jian-Ying Dai, Pan Zhao, Cheng Xiao-Long, Xiu Zhi-Long, Enhanced production of 2,3-butanediol from sugarcane molasses", *Applied Biochemistry and Biotechnology* 175 (6) (2015) 3014–3024.

[49] S. Plotkin, History of vaccination, PNAS 111 (34) (2014) 12283–12287.

[50] I.J. Amanna, M.K. Slifka, Successful Vaccines, in: L. Hangartner D. Burton(Eds.), Vaccination Strategies Against Highly Variable Pathogens. Current Topics in Microbiology and Immunology, vol 428, Springer, Cham, 2018. https://doi.org/10.1007/82_2018_102

[51] www.who.int/news-room/spotlight/history-of-vaccination/a-brief-history-of-vaccination

CHAPTER 2

Microbiology and Biochemistry for Engineers

2.1 Cells and Organisms

Bioprocess engineering involves using cells or cellular parts like enzymes to transform less valuable chemicals into value-added chemicals or to transform harmful chemicals into less harmful ones. To be a successful bioprocess engineer, it is essential to understand the basic properties of the catalysts that perform these transformations. This chapter covers foundational properties of microbiology and biochemistry because they are the basis of the catalysts used throughout the bioprocess industry. More detailed treatments of all the topics can be found in a range of life science courses including microbiology, biochemistry, microbial physiology, and microbial ecology.

It all begins with a cell. A cell is the smallest quantum of life. A living thing, whether comprised of one cell, like a bacterium, or trillions of cells, like a human, is called an organism. Thus, cells are the basic building blocks of all organisms.

2.1.1 Nomenclature and Phylogeny

Life is ubiquitous on planet Earth; it can be found on most of the surface and can be found kilometers into the subsurface; it can be found on the bottoms of the oceans and can be found on the tops of clouds. Life takes on countless forms, shapes, and sizes. A systematic classification of life is necessary to organize our collective understanding of the relationships between organisms as well as for effective communication between scientists, engineers, and the public. Taxonomy and phylogeny are complementary scientific fields that name, define, classify, and organize organisms based on their traits, especially their genetic relatedness.

There are three domains of life including bacteria, archaea, and eukaryota (Figure 2.1). A prominent scientific hypothesis proposes that a last universal common ancestor (LUCA) gave rise to the three domains of life (Figure 2.1) and thus served as the base of the "tree of life." Bioinformatics has facilitated the analysis and visualization of all life on Earth. For example, the website onezoom.org maps the currently understood relationships between 2.2 million living species. It should be noted that viruses, such as those responsible for the COVID-19 pandemic, and prions, molecules responsible for medical conditions like mad cow disease,

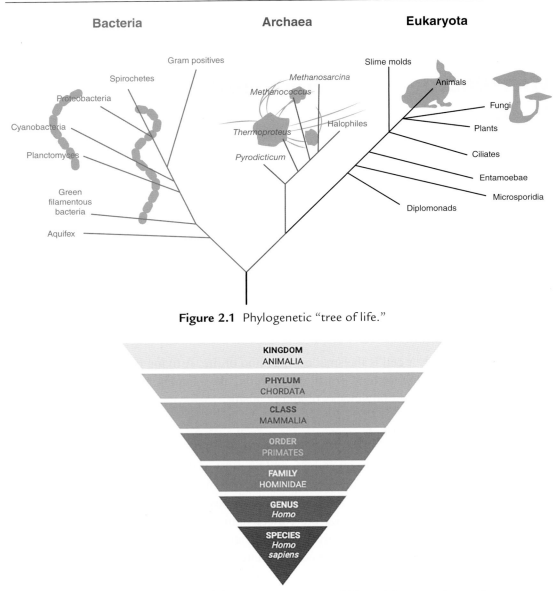

Figure 2.1 Phylogenetic "tree of life."

Figure 2.2 An example of the Linnaean system of binomial nomenclature for a human. *Created with BioRender.com.*

are not included in the "tree of life"; these classes of biological entities play critical roles in the global biome but are not free-living organisms.

Below the classification level of domain, there are successive categories of more related organisms. The genus and species levels are commonly used to name organisms based on a system known as the Latin dual name or binomial nomenclature (Figure 2.2). This system was established in the 1750s by Swedish botanist Carolus Linnaeus [1]. A genus is a group of related species, while a species includes organisms that are alike in reproduction processes. For example, the common bioprocess bacterium *Escherichia coli* belongs to the genus *Escherichia* and the species *coli*. In technical writing, it is convention to give the full binomial name when the organism is first mentioned, but in subsequent text, the name is abridged to the first letter of the genus followed by the species. In this case, *Escherichia coli* becomes *E. coli*. Also note that the binomial name is italicized every time it is used. While organisms that belong to the same species share major genetic content, there are subtle and often important variations within species that are often categorized as strains. *E. coli* strain K-12 MG1655 used in a bioprocess may express different physiological properties such as substrate preferences and growth rate from an *E. coli* strain found in the gut of mammals even though both belong to the same genus and species.

2.1.2 Prokaryotes and Eukaryotes

The domains bacteria and archaea are collectively known as prokaryotes while the domain eukaryota are called eukaryotes. A major difference between prokaryotic and eukaryotic cells is the absence or presence, respectively, of membrane-encased, subcellular structures known as organelles. The quintessential example of an organelle is the nucleus which contains genetic material for eukaryotic cells. Prokaryotes do not typically have distinct organelles, have a relatively simple structural organization, and typically possess a single circular chromosome. Eukaryotes often have more complex internal structures including numerous specialized organelles such as the mitochondria, endoplasmic reticulum, and Golgi apparatus where specialized chemistries occur. Most eukaryotes have more than one linear chromosome (DNA molecule) in the nucleus as well as additional chromosomes in some organelles such as the mitochondria. This text focuses on generalized rules of biology. Exceptions exist to most biological rules and are the topic of more specialized biology courses and texts.

The size, shape, and nutrition sources can vary substantially among the domains of life. Most prokaryotes vary from 0.5 to 3 micrometers (µm) in radius and different species can have different shapes, such as spherical or coccus (e.g., *Staphylococci*), cylindrical or bacillus (*E. coli*), or spiral or spirillum (*Rhodospirillum*). Compared to eukaryotes, prokaryotes typically grow at a rapid rate, with doubling times ranging from 30 min to several hours. Various nutrients can be used by prokaryotes as a carbon source, with examples including carbohydrates, hydrocarbons, proteins, and CO_2. Additionally, prokaryotes can obtain energy from numerous sources such as organic compounds like carbohydrates, inorganic compounds like reduced iron, or electromagnetic radiation like sunlight. A basic comparison of prokaryotes and eukaryotes is presented in Table 2.1 and microscopic images and schematic illustrations of common cells are provided in Figures 2.3 and 2.4 [2,3].

Bacteria

The domain bacteria can be divided into different classifications based on criteria like cell wall organization. The bacterial cell wall performs critical functions necessary for life

Table 2.1 Comparison of Common Prokaryote and Eukaryote Characteristics

Characteristic	Prokaryotes	Eukaryotes
Genome		
Nuclear membrane	None	Present
# of DNA molecules	1	>1
DNA in organelles	None	Present
DNA as chromosomes	None	Present
Mitotic and meiotic nucleus division	None	Present
Organelles		
Nucleus	None	Present
Mitochondria	None	Present
Endoplasmic reticulum	None	Present
Golgi apparatus	None	Present
Photosynthetic apparatus	Chlorosomes	Chloroplasts
Flagella	Single protein, simple structure	Complex structure with microtubules

including separating the internal components from the extracellular environment, retaining important cellular compounds like the chromosome and proteins, excluding unwanted molecules like toxic compounds, maintaining cell shape, and providing surface area for the cell to interact with the environment through mechanisms like nutrient transporters.

A bacterial classification system based on cell wall and membrane structure was developed by Hans Christian Gram in 1884 and now bacterial cells are categorized as being gram positive or gram negative. The staining procedure involves fixing cells with heat, straining the cells with crystal violet, followed by an iodine treatment, and an ethanol wash. Lastly, safranin is used as a counterstain such that gram-positive cells appear purple, while gram-negative cells appear pink. A diagram of a commonly studied, gram-negative bacterium *E. coli* is shown in Figure 2.5. Note that "bacteria" is the plural form of the word and "bacterium" is the singular form of the word.

Both gram-positive and gram-negative bacterial cell walls contain a polymer called peptidoglycan. Peptidoglycan is a complex macromolecule comprised of polysaccharides and amino acids and forms a mesh-like structure. Gram-positive bacteria contain a thick, rigid, and strong cell wall with multiple layers of peptidoglycan (Figure 2.6). As peptidoglycan is relatively porous, substances can pass through the gram-positive cell wall with ease. Molecules too large for simple diffusion can sometimes be brought into the cell using transporter or porin proteins.

Gram-negative bacteria possess a second plasma membrane located outside the peptidoglycan layer, known as the outer membrane (Figure 2.7). The outer membrane contains approximately 30% protein, 50% lipids, and 20% carbohydrates. The presence of lipopolysaccharides contributes to the net negative charge of the cell and helps stabilize the outer membrane while

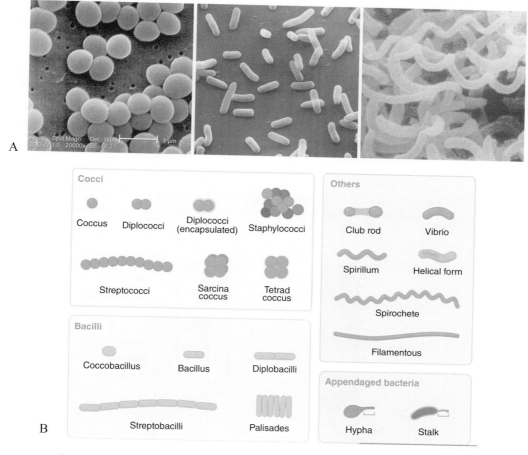

Figure 2.3 (A) Microscopic image panel (from left to right) of *Staphylococci,
E. coli, and Rhodospirillum.* (B) Illustrations of general microbial cell shape. *Adapted from images
by David M. Phillips (2014) and Science Materials (2016) and Created with BioRender.com.*

providing protection from certain substances. As gram-negative bacteria do possess a more
complex cell wall structure, this does present an obstacle when it comes to nutrient transport.
Therefore, the outer membrane is separated from the inner membrane by the periplasm, a
cellular structure that has many important functions including storing cellular energy, as will
be discussed later.

Other distinctions within the domain bacteria are based on the biochemical potential to use
different nutrients and energy sources. For example, cyanobacteria contain a suite of specialized molecules and enzymes that can capture energy from sunlight, use that captured energy
to split water, producing reducing equivalents and O_2, and then use the reducing equivalents to fix CO_2 into sugars and other complex organic molecules. This type of metabolism

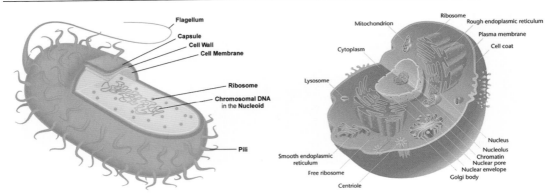

Figure 2.4 Schematic of a typical prokaryotic cell (left) and eukaryotic cell (right). The images are not to scale. Prokaryotes are generally smaller than eukaryotes. *Created with BioRender.com.*

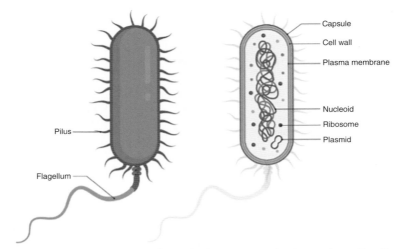

Figure 2.5 Schematic of a typical gram-negative bacterium, *E. coli*, which stains pink. *Created with BioRender.com.*

is classified as oxygenic photosynthesis and is the primary basis for the O_2 in the Earth's atmosphere. Anoxygenic photosynthetic bacteria, such as purple and green bacteria, are also able to capture energy from sunlight but are not able to split water and form O_2. Instead, these bacteria contain light-gathering pigments called bacteriochlorophyll, which pump protons into the periplasm that can be used to power cellular functions.

Archaea

Archaea and bacteria are both prokaryotes and can have similar sizes and shapes; however, they are very different on a genetic level. In fact, they are so different that they are categorized as different domains of life. Archaea are small (0.5–3 μm radius), mostly unicellular, and lack a nucleus or other organelles. Archaea have different cellular wall organization as

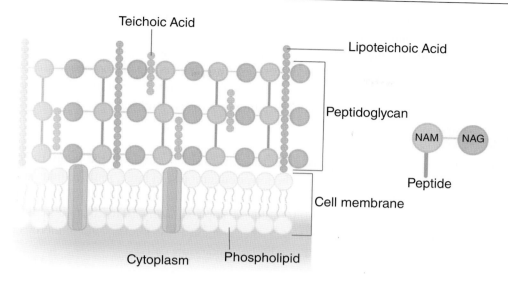

Figure 2.6 Gram-positive bacteria cell wall schematic. *Created with BioRender.com.*

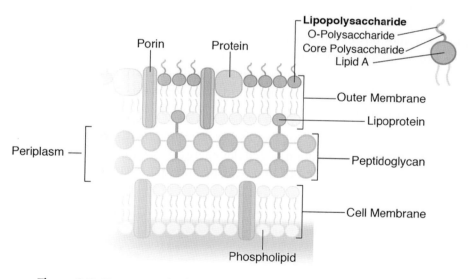

Figure 2.7 Gram-negative bacteria cell wall. *Created with BioRender.com.*

compared to bacteria; archaea do not use peptidoglycan, nor do they possess lipid bilayers, instead utilizing tetraether molecules to construct cellular membranes. Some archaea can grow in environments that are extreme for humans such as highly saline pools or the acidic hot springs of Yellowstone National Park or ocean hydrothermal vents. Some archaeal species have set the current understood limits of life including growth at temperatures exceeding 120°C, proton concentrations corresponding to pH = −0.02, or water activity of 0.6 [2]. To live at these conditions, the cells possess cellular structures and enzymes that retain function at these extreme conditions which can be very useful in harsh industrial processes. However, other archaea like methanogens are commonly found in mesophilic conditions such as the digestive system of birds and mammals including bioprocess students.

Eukaryotes

Eukaryotes consist of a wide range of organisms, from small, single-celled fungi, microalgae, and protozoa, which are known as the lower eukaryotes, to large, multicellular animals and plants, which are known as higher eukaryotes. Eukaryotes are generally 5 to 10 times larger than prokaryotes. Plant and animal cells are approximately 10 to 20 μm in radius, while eukaryotic cells like yeast are ~5 μm. Eukaryotes contain a true nucleus and numerous organelles. Figure 2.8 presents diagrams of two higher eukaryotic cells.

The cell wall and cell membrane structure of eukaryotes can sometimes be similar to bacteria. The plasma membrane is comprised of proteins and phospholipids that form a bilayer structure where hydrophobic proteins are embedded. One significant difference with bacteria is the presence of sterols in the cytoplasmic membrane of eukaryotes. Sterols strengthen the membrane structure, making it less flexible. The composition of eukaryotic cell walls can vary, with some eukaryotes possessing a peptidoglycan layer and some containing polysaccharides and cellulose (e.g., microalgae). Plant cell walls are composed of cellulose fibers embedded in pectin aggregates, which add strength. Animal cells do not have a cell wall, only a cytoplasmic membrane, and thus are extremely sensitive to shear; this complicates the design of large-scale bioreactors utilizing animal cells as catalysts [3].

The different organelles have different chemistries reflecting their specialized roles. Mitochondria are the primary location of cellular energy generation in aerobic eukaryotic cells and are responsible for respiration and oxidative phosphorylation. The mitochondria

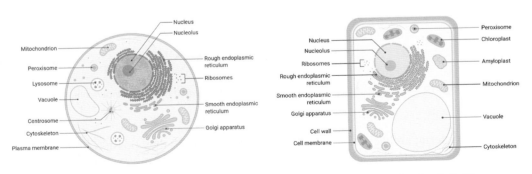

Figure 2.8 Schematics of two eukaryotic cells including an animal cell (left) and a plant cell (right). *Created with BioRender.com.*

possess their own reduced chromosome which contains genes for specialized mitochondrial functions. The rough endoplasmic reticulum is a complex membrane system and contains ribosomes on the inner membrane surfaces. This is the site of protein synthesis and protein modification after synthesis. The smooth endoplasmic reticulum is involved with lipid synthesis. Lysosomes are small membrane-bound organelles that contain and release digestive enzymes, contributing to the catabolism of nutrients. Peroxisomes are similar to lysosomes; however, peroxisomes perform oxidative reactions that produce hydrogen peroxide and also play roles in lipid catabolism. Golgi bodies are organelles composed of membrane aggregates and are responsible for modifying proteins with sugars in a process known as glycosylation. Vacuoles occupy a significant fraction of some eukaryotic cells (up to 90% of the volume in plant cells) and are responsible for food digestion, osmotic pressure regulation, and metabolite storage. Chloroplasts are green, chlorophyll-containing organelles responsible for photosynthesis in photosynthetic eukaryotes such as microalgae and plants. Chloroplasts, like mitochondria, contain their own DNA and some specific protein synthesis machinery.

Fungi

Fungi are eukaryotes that have colonized most surface ecosystems. Fungi possess great metabolic diversity, with different species able to produce a wide spectrum of antibiotics, enzymes, hormones, lipids, vitamins, and pigments, as well as the ability to degrade many types of recalcitrant materials such as lignocellulose and humic substances. Fungi can be divided into two major groups: yeasts and molds (Figure 2.9). Yeasts are unicellular microorganisms, 2 to 5 μm in radius, and typically have spherical, cylindrical, or oval shapes. Yeasts can reproduce by asexual or sexual means. Asexual reproduction occurs through either budding or fission. In budding, a small bud cell forms on the mother cell which grows

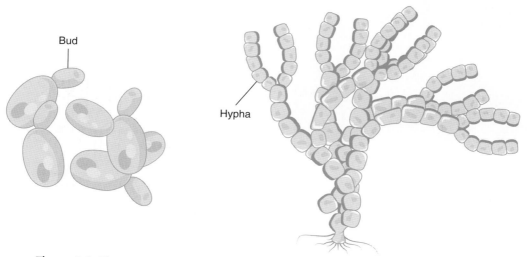

Figure 2.9 Illustrations of yeast (left) and mold (right). *Created with BioRender.com.*

and ultimately separates into a smaller daughter cell. Reproduction via fission is similar to that in bacteria: the cell grows to a certain size and then divides into two equal-sized cells. Classification of yeast is based on these modes of reproduction and the nutritional requirements. The yeast *Saccharomyces cerevisiae* has been used by humans for thousands of years. Its catalytic properties are still widely used for industrial processes like alcohol production (e.g., wine, beer, fuel ethanol) and baking (baker's yeast). Alcohol production occurs in the absence of O_2, a process known as fermentation, while baker's yeast is produced in bioreactors under oxic conditions.

Molds are filamentous fungi that form a complex mycelial structure comprised of branched filaments known as hyphae. The hyphae form an interconnected cytosol where cellular contents can be transported and shared at remarkable distances (1+ mm). Mold cells range in size from 10 μm up to 1 mm and can grow in submerged culture or can grow on solid nutrient surfaces. Some mycelia can grow into the air, where they can form asexual spores called conidia. Mold cultivations contribute to the bioprocess industry with the production of citric acid (*Aspergillus niger*) and many antibiotics, including penicillin (*Penicillium chrysogenum*). Filamentous fungi are also used as alternative protein sources with commercial names like Quorn, Meati Foods, and Nature's Fynd. The filamentous fungi used in Quorn products are grown semi-continuously in very large (50 m tall, >150 m³) airlift reactors (Figure 2.10)

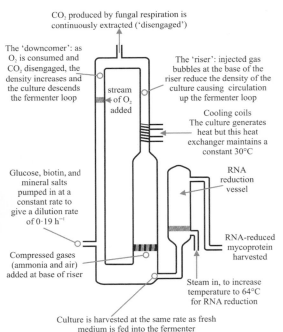

CO₂ produced by fungal respiration is continuously extracted ('disengaged')

The 'downcomer': as O₂ is consumed and CO₂ disengaged, the density increases and the culture descends the fermenter loop

stream of O₂ added

The 'riser': injected gas bubbles at the base of the riser reduce the density of the culture causing circulation up the fermenter loop

Cooling coils
The culture generates heat but this heat exchanger maintains a constant 30°C

RNA reduction vessel

Glucose, biotin, and mineral salts pumped in at a constant rate to give a dilution rate of 0.19 h^{-1}

RNA-reduced mycoprotein harvested

Compressed gases (ammonia and air) added at base of riser

Steam in, to increase temperature to 64°C for RNA reduction

Culture is harvested at the same rate as fresh medium is fed into the fermenter

Figure 2.10 Photograph of buildings housing Quorn airlift bioreactors (left) and a detailed schematic representation of the Quorn airlift reactor used for the production of mycoprotein in continuous flow culture (right). *Adapted from Trinci (1991, 1992, 1994) and Whittaker et al. (2020).*

which facilitates efficient production and mixing of the viscous cell suspension [4]. Filamentous fungi require special growth considerations due to the viscous nature of the culture broth, which we will discuss in Chapter 6.

Microalgae

Microalgae are photosynthetic eukaryotes, typically growing as unicellular organisms with cell radii on order of ~5 to 15 μm (Figure 2.11). Some microalgae, known as diatoms, contain silica or calcium carbonate in their elaborate cell walls known as frustules (Figure 2.12). Deposits of ancient diatoms are sometimes mined, and the cell walls used as filter aids in industry. Microalgal species including *Chlorella*, *Scenedesmus*, *Spirulina*, and *Dunaliella*, are used for wastewater treatment with simultaneous single-cell protein production, reducing unwanted nitrogen and phosphorous concentrations in the wastewater. Algal biofuels and high-protein food products are just some of the sustainable bioproducts investigated with these eukaryotes.

Protozoa

Protozoa are unicellular, motile, relatively large (upto 50–150 μm) eukaryotic cells that lack cell walls. Protozoa obtain food by ingesting other small organisms, such as bacteria, and are typically uninucleate, reproducing through sexual or asexual means. They are classified by their amoeboid motion, a crawling-like movement where the cytoplasm flows forward to form a pseudopodium (false foot), and the rest of the cell flows toward this anchor. Protozoa have demonstrated a beneficial role in wastewater treatment through bacterial removal [3].

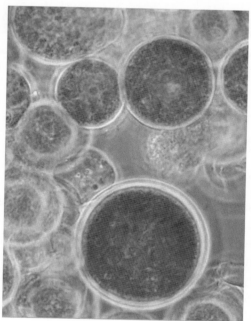

Figure 2.11 Microscopic image of microalgae species, *Chlorella vulgaris.* Credit: Flickr, t2ll2t.

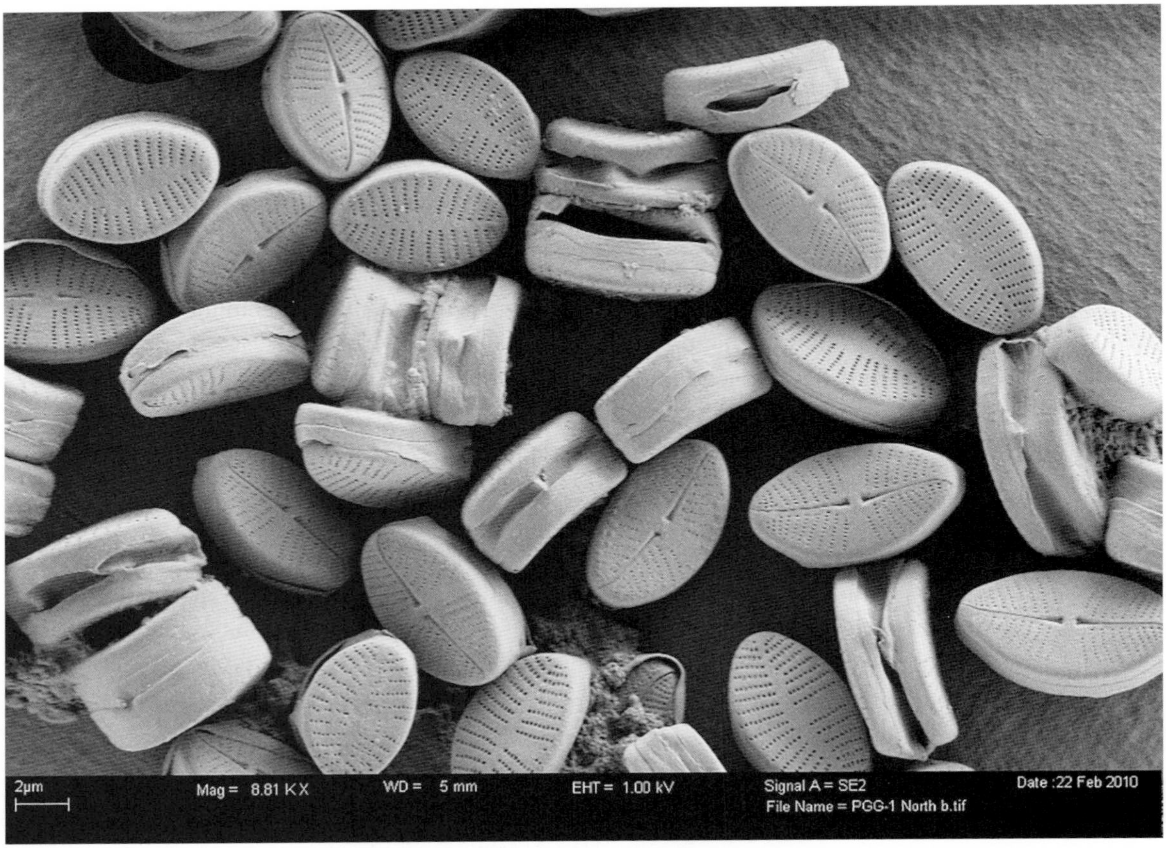

2μm   Mag = 8.81 KX   WD = 5 mm   EHT = 1.00 kV   Signal A = SE2   Date :22 Feb 2010
File Name = PGG-1 North b.tif

Figure 2.12 Microscopic image of diatoms. *Reprinted with permission from Karen Moll, Ross P. Carlson, and Brent M. Peyton, Montana State University (2010).*

Protozoa can also be problematic in microalgae production facilities as they can eat the phototrophs reducing system productivity.

2.1.3 Viruses and Prions

Viruses and prions are discussed here together, not because of any phylogenetic relationship to each other, but because both are nonliving biological entities that can play important roles in ecosystems. Viruses are typically extremely small (30–200 nm), obligate parasites of other cells such as bacterial, yeast, plant, or animal cells. Viruses are not free living, cannot capture or store energy, and are functionally active only when inside their host. Viruses contain genetic material necessary for their replication. The genetic material can be either double- or single-stranded deoxyribonucleic acid (DNA) or ribonucleic acid (RNA). In contrast, all documented, free-living cells store genetic information as double-stranded DNA. The viral genetic material is retained by a self-assembling protein coat called a capsid. It is hypothesized that all life is susceptible to infection by viruses. In fact, it has been proposed that some viruses are preyed upon by other viruses.

Viruses that infect bacteria are called bacteriophage. Many bacteriophages have a hexagonal head, tail, and tail fibers, as shown in Figure 2.13. Bacteriophages attach to the cell wall of a host cell with tail fibers, alter the cell wall of the host cell, and inject the viral nuclear material into the host cell. Once there, the bacteriophage nucleic acids reproduce inside the host along with the phage-specific components necessary to assemble new phage particles. When the host cell lyses, the phage particles are released into the local environment and can infect new cells. This mode of virus reproduction is called the lytic cycle. Viral genetic material can also integrate into the host chromosome, where it is reproduced along with the chromosome of the host until some future event triggers the viral sequence and a new lytic cycle begins.

Viruses cause numerous diseases including smallpox, common colds, HIV, and COVID-19. Antiviral agents and vaccines are critical targets for scientific development. In some cases, an "inactivated" virus is used as a vaccine; in others, empty shell virus particles (capsid without nucleic acid) can be used as vaccines without concern of viral replication, as all genetic material has been removed. mRNA vaccines use lipid nanoparticles to enable synthetic mRNA molecules, coding for a viral protein, to enter a cell. Once inside the cell, the mRNA results in the production of a viral protein, eliciting an adaptive immune response. Viruses are also problematic in bioprocessing. For example, if a bacteriophage were to infect a process for making yogurt, this could result in the loss of the entire culture and many thousands of liters of product.

Phages are adept at processing and moving genetic material and have been adapted as tools for genetic engineering. The KEIO *E. coli* gene deletion library is readily used to create mutant *E. coli* strains with the help of P1 phage to move modified DNA from host to host. Furthermore, viruses have been implemented in gene therapy, where viral

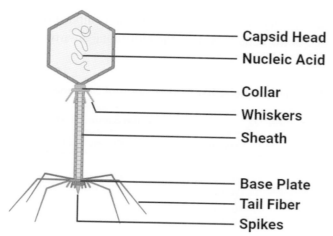

Figure 2.13 Schematic of a bacteriophage, a virus with a bacterial host. *Created with BioRender.com.*

genetic material is replaced with a desired gene to be inserted into the patient cells to correct or augment deficiencies.

Prions are infectious, misfolded proteins that can cause diseases including neurodegenerative disorders like "mad cow disease," chronic wasting disease, or Creutzfeldt–Jakob disease. The misfolded proteins are believed to corrupt normal variants of the same protein producing more infectious prion particles. Unlike living organisms or special biological entities like viruses, prions do not store genetic information as DNA or RNA. Prions are relatively new to science and there is still much to be learned.

2.2 Cellular Composition

2.2.1 Biomass Macromolecular Components

Cells are common catalysts in bioprocesses. It is therefore important to understand their basic physiochemical properties. Typical microbial cells are 60 to 80% water on a mass basis and they are denser than water having a live density of 1.05 to 1.1 g cm^{-3}. Live cells will sink in an aqueous solution, an important property utilized for harvesting cells with separation technologies like centrifuges. The elemental composition of a typical bacterium on a cell dry weight (cdw) basis is 50% carbon, 20% oxygen, 14% nitrogen, 8% hydrogen, 3% phosphorus, and 1% sulfur, with small amounts of K^+, Na^+, Ca^{2+}, Mg^{2+}, Cl^-, Fe^{2+}, and Cu^{2+}. The contents of a dry cell are composed primarily of high-molecular-weight, polymeric compounds including proteins, nucleic acids, polysaccharides, and lipids (Table 2.2). Many of the proteins are enzymes essential for cell growth and performing useful bioprocess chemistries. The nucleic acids, which store and process the genetic code (DNA and mRNA, respectively) as well as participate in protein synthesis machinery (rRNA), range from 10% to 30% of the cellular dry weight. The lipid content of most cells varies between 5% and 15% of the cellular dry weight, with some exceptions being strains of high lipid accumulating oleaginous yeast and microalgae which are used in bioprocesses as precursors for biodiesel production. These strains can be up to 50–80% lipid on a dry cell basis. In general, the composition of cells varies depending on the type and age of the cells as well as the nutritional environment in which the cells

Table 2.2 Macromolecular Composition of Microorganisms [3]

Organism	Composition (% Dry Weight)		
	Protein	Nucleic Acid	Lipid
Viruses	50–90	5–50	<1
Bacteria/archaea	40–70	13–34	10–15
Filamentous fungi	10–45	1–3	2–7
Yeast	40–50	4–10	1–6
Unicellular algae	10–60	1–5	4–80

are grown. In addition to biopolymers, cells contain other metabolites including inorganic salts (e.g., NH_4^+, PO_3^{4-}, SO_4^{2-}), metabolic intermediates (e.g., citrate, oxoglutarate, acetate), vitamins, and cofactors.

Proteins

Proteins are polymers built from amino acid monomers. Amino acids contain at least one carboxyl group and one alpha-amino group, while their R groups, or side chains, have different moieties. A moiety is a part of the chemical structure of a molecule that could include a substructure, such as a functional group. The primary structure of a protein is determined by the sequence of amino acids. The secondary and tertiary structures of a protein are the result of side group interactions which create complex 3-dimensional structures within a single protein and across multisubunit complexes. Proteins can be binned into five categories: structural proteins (glycoproteins, collagen, keratin), catalytic proteins (enzymes), transport proteins (hemoglobin, serum albumin), regulatory proteins (hormones, insulin, growth hormone), and protective proteins (antibodies, thrombin). Table 2.3 details each of these protein categories with additional descriptions, examples, and protein structure images. Catalytic proteins known as enzymes will be a major focus of the examples in this textbook.

Lipids, TAGS, and PHAs

Lipids are hydrophobic compounds that are insoluble in water, but soluble in nonpolar solvents such as chloroform or ether. Lipids serve as building blocks for molecules essential for membranes and can serve as carbon and energy storage molecules. Cells can alter the composition of lipids used to construct their membranes to acclimate to environmental stresses such as high or low temperature or the presence of chemicals such as ethanol. Fatty acids are a major component of most lipids and are synthesized by metabolic pathways which start with two carbon acetate units. Fatty acids have a hydrophilic end with a carboxyl group and a hydrophobic hydrocarbon tail. The carboxyl end of fatty acids can be bound to different chemical groups. Phospholipids are a family of lipids that are key components in membranes. Membranes are ordered with the hydrophobic tails of the phospholipids toward the core of the membrane and the hydrophilic heads on the outside of the lipid bilayer membrane (Figure 2.14). As seen in Figure 2.14a, the amphiphilic phospholipid contains two moieties. The hydrophobic moiety is the fatty acid tails, whereas the hydrophilic moiety is the glycerol head with a phosphate attached.

Triacylglycerols (TAGs) are a class of carbon and energy storage lipids. TAGs are produced and accumulated by organisms such as oleaginous yeast, microalgae, or oil seed plants like canola. TAGs are of industrial interest as precursors for biodiesel production based on inexpensive reactants like sunlight and CO_2. Another class of lipids of biotechnological importance is the polyhydroxyalkanoates (PHAs). PHAs are a group of naturally occurring,

Table 2.3 Protein Category Descriptions and Examples

Protein Function	Description	Example	Protein Image and Structure
Structural Protein	These proteins provide structure and support for cells. On a larger scale, they also allow the body to move.	Actin	Actin Single actin subunit Actin filament consisting of multiple subunits
Catalytic Protein (Enzyme)	Enzymes carry out the majority of the chemical reactions that take place in cells. They additionally assist with the formation of new molecules by reading the genetic information stored in DNA.	Phenylalanine hydroxylase	Phenylalanine hydroxylase Single phenylalanine hydroxylase subunit Phenylalanine hydroxylase protein consisting of four subunits
Transport Protein	These proteins bind and carry atoms and small molecules within cells and throughout the body.	Ferritin	Ferritin Single ferritin subunit Ferritin protein consisting of 24 subunits Cross section
Regulatory Protein (Hormones)	Types of hormones act as messenger proteins and transmit signals to coordinate biological processes between different cells, tissues, and organs.	Growth hormone	Growth hormone Growth hormone Growth hormone bound to receptor
Protective Protein (Antibody)	Antibodies bind to specific foreign particles, such as viruses and bacteria, to help protect the body.	Immunoglobulin G (IgG)	Immunoglobulin G (IgG) Foreign particle binding site Foreign particle binding site

Protein images from https://medlineplus.gov/ and https://medlineplus.gov/genetics/understanding/howgeneswork/protein.

biodegradable polyesters that can possess material properties similar to petroleum-based plastics. PHAs are also used for biomedical applications, including drug delivery and tissue engineering scaffolds, based on their biocompatibility and biodegradability [5].

Steroids are also classified as lipids. Steroids can be produced in bioprocesses and include cholesterol, present in the membranes of animal tissues, and cortisone, an antiinflammatory used to treat rheumatoid arthritis, injuries, and certain skin diseases.

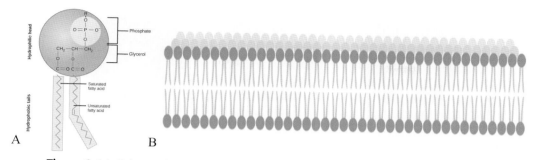

Figure 2.14 Schematic of an amphiphilic phospholipid (A) and a lipid bilayer (B). *Created with BioRender.com.*

RNA and DNA

There are two major classes of nucleic acids in cells: RNA and DNA. These macromolecules are built from monomers known as nucleotides. A nucleotide is comprised of a nitrogen-containing base, a ribose (RNA) or deoxyribose (DNA) sugar, and a polyphosphate group. RNA has many cellular functions including serving as an intermediate in genetic information conversion from DNA to proteins via messenger RNA (mRNA) and in the actual synthesis of protein via ribosomal RNA (rRNA). DNA forms the basis of the chromosome(s) which stores genetic information as a conserved sequence of nucleotides. The nucleotides are built on the following bases which can be found in oxy and deoxy variants: adenine (A), guanine (G), thymine (T), cytosine (C), and uracil (U). RNA uses oxy variants of A, G, C, and U, while DNA is comprised of the deoxy versions of A, T, G, and C.

Central Dogma of Molecular Biology

Life uses biopolymers to store information. This information has constraints on the direction it can flow. This framework is often called the "central dogma of molecular biology" and was originally proposed by Francis Crick, a scientist who played a central role in decoding the structure of DNA. The theory states that information flows in one direction, from DNA to RNA to protein (Figure 2.15). The information flow is based on the biological processes of transcription, which transfers information from DNA to RNA, and the process of translation, which transfers information from RNA to proteins.

The central dogma is a very useful theory for organizing biological data but was devised decades ago based on a limited knowledge of molecular processes. Like many aspects of biology, there are exceptions to the central dogma. For example, some viruses, which are biological entities but not living organisms, can store biological information as double- or single-stranded RNA that is then passed to proteins. These are known as RNA viruses. Additionally, prions—infectious proteins which are biological entities but not living

Figure 2.15 The central dogma of molecular biology defines constraints on the direction of biological information flow. *Graphic credit: National Cancer Institute.*

organisms—are able to pass biological information from protein to protein. Prions are responsible for medical maladies like mad cow disease and Creutzfeldt–Jakob disease.

While generally true, the central dogma of molecular biology is a great tool for organizing important concepts in biology. The shortcomings of the central dogma also highlight the exciting breadth of the biosphere and how relatively simple components like RNA and proteins can sometimes have functions that exceed their original descriptions.

2.3 Basic Cellular Metabolism

Organisms are classified based on the source of their metabolic energy and carbon. The primary energy sources for life are electromagnetic radiation such as sunlight and chemicals. Organisms that get their energy from light are classified as phototrophs, while organisms that get energy from chemicals are chemotrophs. Chemotrophs can be further subdivided based on the type of chemical energy. Chemoorganotrophs get energy from organic chemicals like sugars, while chemolithotrophs obtain energy from inorganic chemicals like reduced iron or sulfur compounds. The source of carbon is also a critical aspect of metabolism as all life is comprised of carbon-based macromolecules. Autotrophs utilize inorganic forms of carbon such as CO_2, while heterotrophs utilize organic (reduced) forms of carbon. Metabolism needs to combine strategies for the collection of energy and carbon sources for an organism to grow. For example, common bioprocess hosts like *E. coli* or *S. cerevisiae*, as well as bioprocess students, are chemoorganoheterotrophs that obtain energy from organic chemicals and utilize organic carbon sources such as glucose, maltose, and starch. Plants and many microalgae and cyanobacteria are photoautotrophs using light as an energy source and fixing inorganic carbon like CO_2. Organisms have developed many strategies to utilize the myriad of energy sources and carbon sources found in different environments. These strategies form the basis of many microbial physiology, metabolism, and ecology courses. The primary metabolisms discussed in this text are chemoorganoheterotrophy and photoautotrophy.

The energy required for life is based on the oxidation and reduction of chemicals. An energy source is oxidized when electrons are removed from it and a chemical is reduced when it gains electrons. The movement of electrons between different chemicals in a cell can be organized as either a catabolic or anabolic process, both being essential to life.

2.3.1 Catabolic and Anabolic Functions

Catabolism is the biochemical process of transforming relatively large and reduced substrates into smaller, often more oxidized products through combinations of breaking chemical bonds and oxidation reactions (e.g., glucose to CO_2 and H_2O). Catabolic processes concurrently harvest energy from the substrate for use in cellular processes. Anabolism is the biochemical process of synthesizing more complex compounds from simpler metabolites in processes that consume cellular energy and, in many cases, also reducing equivalents.

The key metabolite for storing and releasing cellular energy is adenosine triphosphate or ATP. The energy is stored and released from the phosphodiester bonds linking the three phosphate groups. ATP can be viewed as a type of "energy currency" which can store excess energy from thermodynamically favorable reactions and donate the energy to less favorable reactions permitting them to occur. ATP is produced via two major categories of biochemical reactions: substrate-level phosphorylation and oxidation phosphorylation. Substrate-level phosphorylation occurs in conjunction with the breaking and rearranging of chemical bonds, with some of the released energy being captured in the phosphodiester bond of ATP. Oxidative phosphorylation involves the conversion of energy stored as a chemiosmotic gradient of protons in the periplasm (proton motive force) into phosphodiester bonds based on the catalytic properties of the membrane-bound, ATP synthase enzyme. The energy used to create the chemiosmotic gradient of protons comes from the controlled, step-wise release of energy from electrons, which is used to pump protons against a concentration gradient. The cascade of enzymes that perform these chemistries is known as the electron transport chain.

Electrons, often referred to as reducing equivalents, can be conveyed from one metabolite to another using soluble electron carrier molecules like nicotinamide adenine dinucleotide (NADH) or nicotinamide adenine dinucleotide phosphate (NADPH) or membrane-soluble molecules like quinol. Electron carriers transition between a reduced form (carrying electrons) and an oxidized form (once the electrons have moved on) without being consumed in the reaction.

Metabolism is often arranged as a number of substrate-specific, biochemical pathways that channel carbon and electrons to a central metabolic hub where the molecules are oxidized, reduced, or rearranged before being funneled to metabolic products. Three important metabolic pathways in the central metabolism hub are (1) the Embden–Meyerhof–Parnas (EMP) pathway, also known as glycolysis ("sugar-lysis"), which breaks down and oxidizes glucose into pyruvate; (2) the tricarboxylic acid cycle (also known as the citric acid cycle or Krebs cycle), which breaks down and oxidizes pyruvate or acetyl-CoA into CO_2 and H_2O via a number of intermediates; and (3) the pentose–phosphate pathway (PPP), which can break down, oxidize, and/or rearrange a glucose-6-phosphate backbone into a variety of C_3, C_4, C_5, C_6, and C_7 metabolites, some of which can enter glycolysis for further oxidation. The three pathways serve as a central receiving hub and can perform chemistries required for both catabolic and anabolic roles. These three pathways are connected to many other reactions and may be partitioned between different cellular compartments like the cytosol or mitochondria in eukaryotes. Figure 2.16 details these major metabolic pathways along with other central metabolism reactions commonly found in *S. cerevisiae* [6].

As will be discussed in Chapter 4, metabolic pathways must conserve carbon and electrons. The conservation of mass is a major constraint on metabolism, dictating what catabolic and anabolic reactions are possible for a given environment. Electrons from oxidized substrates and carried on metabolites like NADH have two possible fates during catabolic processes. The electrons can flow through biochemical pathways to external electron acceptors like O_2 or NO_3^-, which are reduced to produce water or NO_2^-, respectively, in a biochemical strategy known as respiration. Respiration is often coupled with oxidative phosphorylation to generate ATP. Alternatively if no external electron acceptor is available, the electrons from the substrate can flow to an intracellular metabolite like pyruvate which is reduced to form lactic acid, which is then secreted as a metabolic byproduct. This metabolic strategy is known as fermentation and generates ATP based on substrate-level phosphorylation. It should be noted that many classic bioprocess studies use the term "fermentation" as a synonym for a bioprocess even if O_2 or another external electron acceptor is present. The same studies often use the term "fermenter" interchangeably with bioreactor, again even if an external electron acceptor is present. This terminology is avoided here to avoid miscommunication with microbiologists.

Many organisms can switch their metabolism between respiration and fermentation strategies as a function of environment to maximize the efficiency of energy metabolism. These organisms are called facultative or facultative anaerobes. For example, *S. cerevisiae* (baker's yeast) can oxidize glucose to CO_2 and water via respiration in the presence of O_2 but can also ferment glucose to ethanol and CO_2 in anoxic environments. The two metabolisms have very different efficiencies in producing ATP from glucose. Glucose can be catabolized to produce ~30 ATP in the presence of O_2 but can only be used to produce ~2 ATP in the absence of O_2. If O_2 is the final electron acceptor, the metabolism is called *aerobic respiration*. If another external electron acceptor like NO_3^- or fumarate is used in conjunction with the electron transport chain, the metabolism is called *anaerobic respiration*.

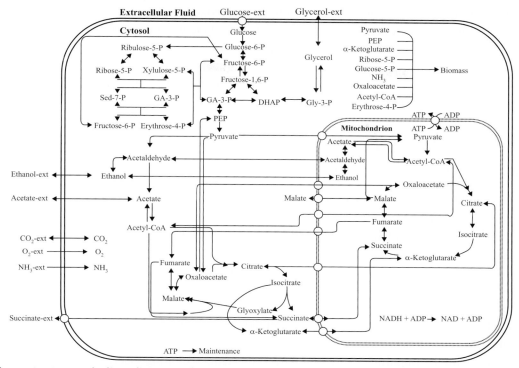

Figure 2.16 Metabolism diagram of major metabolic pathways and central metabolism reactions commonly found in *S. cerevisiae*. *Reprinted with permission from Carlson et al. (2002) [6].*

2.4 Cellular Nutrients and Growth Medium

Bioprocesses often produce cells through growth. To grow, cells require a food solution known as growth medium (singular) or growth media (plural) to enable both anabolic and catabolic processes. The nutrients in a growth medium can be classified as either macronutrients or micronutrients based on their abundance. A convenient definition for each category is that a macronutrient is present at an initial medium concentration of $>10^{-4}$ M, while a micronutrient is present in the medium at a concentration $<10^{-4}$ M. Nutrient sources of carbon, nitrogen, oxygen, hydrogen, sulfur, and phosphorus are typical macronutrients. Typical micronutrients include trace elements such as Mo^{2+}, Zn^{2+}, Cu^{2+}, Mn^{2+}, Ca^{2+}, Fe^{+2}, and Na^+, as well as vitamins, growth hormones, and select metabolic precursors [7]. The nutrients present in growth media are often called *substrates* in convention with microbiological scientists.

The macro- and micronutrients can be supplied using one of two major types of growth media, defined or complex. Defined media contain known amounts of all compounds including all nutrients required for cell growth and often a buffer to maintain pH. Complex media contain components which are not completely characterized, or which may vary between batches. Common components in complex medium include yeast extract, peptone, tryptic soy, molasses, brain-heart infusion, or corn steep liquor. Complex media often produce higher cell growth rates as compared to defined media. However, defined media advantages can include reduced costs, more reproducible results, more operator control of the bioprocess, and simplified separation processes.

2.5 Summary of Chapter 2

At the end of Chapter 2, you should:

* Be able to name cells using proper nomenclature
* Understand the differences between prokaryotes and eukaryotes and be able to determine which types of cells fall into these categories (i.e., bacteria, archaea, fungi, microalgae, and protozoa)
* Understand why neither a virus nor a prion is a living entity
* Be able to identify the both the major elemental components and the critical macromolecular components of a cell
* Understand the basic types of cellular metabolism including the roles of catabolic and anabolic reactions
* Understand the importance of essential cell nutrients including macro- and micronutrients in bioprocessing

References

[1] R.M. Baldini, C.E. Jarvis, Typification of some Linnaean names in Phalaris (Gramineae), Taxon. 40.3 (1991): 475–485.

[2] N. Merino, H.S. Aronson, D.P. Bojanova, J. Feyhl-Buska, M.L. Wong, S. Zhang, D. Giovannelli, Living at the extremes: extremophiles and the limits of life in a planetary context, Front. Microbiol. 10:780 (2019).

[3] M.L. Shuler, F. Kargi, "How cells grow." Bioprocess engineering basic concepts (2002) 168–170. Pearson College Division. ISBN-10: 0130819085.

[4] J.A. Whittaker, R.I. Johnson, T.J.A. Finnigan, S.V. Avery, P.S. Dyer, The biotechnology of quorn mycoprotein: past, present and future challenges, in: H. Nevalainen (Ed.), Grand Challenges in Fungal Biotechnology, Grand Challenges in Biology and Biotechnology. Springer, 2020.

[5] V. Amstutz, N. Hanik, J. Pott, et al., Tailored biosynthesis of polyhydroxyalkanoates in chemostat cultures, Methods Enzymol. 627 (2019) 99–123.

[6] R. Carlson, D. Fell, F. Srienc, Metabolic pathway analysis of a recombinant yeast for rational strain development, Biotechnol. Bioeng. 79.2 (2002) 121–134.

[7] B.L. Haines, E.L. Dunn, Coastal marshes, in: B.F. Chabot, H.A. Mooney (Eds.), Physiological Ecology of North American Plant Communities. Chapman & Hall Co, 1985, pp. 323–347.

CHAPTER 3

Introduction to Engineering Calculations and Units

Calculations used in bioprocess engineering require a systematic approach with well-defined methods and rules. Conventions and definitions that form the backbone of engineering analysis are presented in this chapter. You will use many of these repeatedly as you progress through this text and your career. In laying the foundations for calculations and problem-solving, this chapter will be a useful reference to review from time to time.

The first step in quantitative analysis of systems is to express the system properties using mathematical language. This chapter begins by considering how physical, chemical, and biological processes are characterized mathematically. The nature of physical variables, dimensions, and units is discussed, and formalized procedures for unit conversions are outlined. You have already encountered many of the concepts used in measurement, such as concentration, density, pressure, temperature, and so on; rules for quantifying these variables are summarized here in preparation for future chapters where they are applied to solve problems. The occurrence of reactions in biological systems is of particular importance; terminology involved in stoichiometric analysis is considered in this chapter. Finally, as equations representing biological processes often involve terms for the physical and chemical properties of materials, references for handbooks containing this information are provided.

Worked examples and problems are used to illustrate and reinforce the material described in the text. Although the terminology and engineering concepts in these examples may be unfamiliar, solutions to each problem can be obtained using techniques explained within this chapter. The emphasis in this chapter is on the use of basic mathematical principles irrespective of the particular application. A checklist is provided at the end of the chapter to assess your understanding of the material.

3.1 Physical Variables, Dimensions, and Units

Engineering calculations involve manipulation of numbers. Most of these numbers represent the magnitudes of measurable *physical variables*, such as mass, length, time, velocity, area, viscosity, temperature, density, and so on.

The seven quantities listed in Table 3.1 have been chosen by international agreements as a basis for measurement of physical variables. Two further supplementary units are used to express angular quantities. These base quantities are called *dimensions*, and it is from these that the dimensions of other physical variables are derived. For example, the dimensions of velocity, which is defined as distance or length traveled per unit time, are LT^{-1}; the dimensions of force, being mass $\times$ acceleration, are LMT^{-2}. A list of useful derived dimensional and nondimensional quantities is given in Table 3.2. Physical variables can be classified into two groups: *substantial variables* and *natural variables*.

3.1.1 Substantial Variables

Examples of substantial variables are mass, length, volume, viscosity, and temperature. Expression of the magnitude of substantial variables requires a precise physical standard against which the measurement is made. These standards are called *units*. You are already familiar with many units: meter, foot, and mile are units of length, for example, and hour and second are units of time. Statements about the magnitude of substantial variables must contain two parts: the number and the unit used for measurement. Reporting the speed of a moving car as 20 has no meaning unless information about the units, say km h^{-1} or ft s^{-1}, is also included.

As numbers representing substantial variables are multiplied, subtracted, divided, or added, their units must also be combined. The values of two or more substantial variables may be added or subtracted only if their units are the same. For example

$$5.0 \text{ kg} + 2.2 \text{ kg} = 7.2 \text{ kg}$$

On the other hand, the values and units of *any* substantial variables can be combined by multiplication or division; for example:

$$\frac{1500 \text{ km}}{12.5 \text{ h}} = 120 \text{ km h}^{-1}$$

Table 3.1: Base Quantities

Base quantity	Dimensional symbol	Base SI unit	Unit symbol
Length	L	meter	m
Mass	M	kilogram	kg
Time	T	second	s
Electric current	I	ampere	A
Temperature	Θ	kelvin	K
Amount of substance	N	gram-mole	mol or gmol
Luminous intensity	J	candela	cd
Supplementary units			
Plane angle	—	radian	rad
Solid angle	—	steradian	sr

Table 3.2: Examples of Derived Dimensional and Dimensionless Quantities

Quantity	Dimensions	Quantity	Dimensions
Acceleration	LT^{-2}	Momentum	LMT^{-1}
Angular velocity	T^{-1}	Osmotic pressure	$L^{-1}MT^{-2}$
Area	L^2	Partition coefficient	1
Atomic weight	1	Period	T
("relative atomic mass")		Power	L^2MT^{-3}
Concentration	$L^{-3}N$	Pressure	$L^{-1}MT^{-2}$
Conductivity	$L^{-3}M^{-1}T^3I^2$	Rotational frequency	T^{-1}
Density	$L^{-3}M$	Shear rate	T^{-1}
Diffusion coefficient	L^2T^{-1}	Shear strain	1
Distribution coefficient	1	Shear stress	$L^{-1}MT^{-2}$
Effectiveness factor	1	Specific death constant	T^{-1}
Efficiency	1	Specific gravity	1
Energy	L^2MT^{-2}	Specific growth rate	T^{-1}
Enthalpy	L^2MT^{-2}	Specific heat capacity	$L^2T^{-2}\Theta^{-1}$
Entropy	$L^2MT^{-2}\Theta^{-1}$	Specific interfacial area	L^{-1}
Equilibrium constant	1	Specific latent heat	L^2T^{-2}
Force	LMT^{-2}	Specific production rate	T^{-1}
Fouling factor	$MT^{-3}\Theta^{-1}$	Specific volume	L^3M^{-1}
Frequency	T^{-1}	Stress	$L^{-1}MT^{-2}$
Friction coefficient	1	Surface tension	MT^{-2}
Gas hold-up	1	Thermal conductivity	$LMT^{-3}\Theta^{-1}$
Half-life	T	Thermal resistance	$L^{-2}M^{-1}T^3\Theta$
Heat	L^2MT^{-2}	Torque	L^2MT^{-2}
Heat flux	MT^{-3}	Velocity	LT^{-1}
Heat transfer coefficient	$MT^{-3}\Theta^{-1}$	Viscosity (dynamic)	$L^{-1}MT^{-1}$
Ideal gas constant	$L^2MT^{-2}\Theta^{-1}N^{-1}$	Viscosity (kinematic)	L^2T^{-1}
Illuminance	$L^{-2}J$	Void fraction	1
Maintenance coefficient	T^{-1}	Volume	L^3
Mass flux	$L^{-2}MT^{-1}$	Weight	LMT^{-2}
Mass transfer coefficient	LT^{-1}	Work	L^2MT^{-2}
Molar mass	MN^{-1}	Yield coefficient	1
Molecular weight	1		
("relative molecular mass")			

Note: Dimensional symbols are defined in Table 3.1. Dimensionless quantities have dimension 1.

The way units are carried along during calculations has important consequences. Not only is proper treatment of units essential if the final answer is to have the correct units, but units and dimensions can also be used as a guide when deducing how physical variables are related in scientific theories and equations.

3.1.2 Natural Variables

The second group of physical variables are the natural variables. Specification of the magnitude of these variables does not require units or any other standard of measurement. Natural variables are also referred to as *dimensionless variables, dimensionless groups*, or *dimensionless numbers*. The simplest natural variables are ratios of substantial variables. For example, the aspect ratio of a cylinder is its length divided by its diameter; the result is a dimensionless number.

Other natural variables are not as obvious as this and involve combinations of substantial variables that do not have the same dimensions. Engineers make frequent use of dimensionless numbers for succinct representation of physical phenomena. For instance, a common dimensionless group in fluid mechanics is the Reynolds number, *Re*. For flow in a pipe, the Reynolds number is given by the equation:

$$Re = \frac{Du\rho}{\mu} \tag{3.1}$$

where D is the pipe diameter, u is fluid velocity, ρ is fluid density, and μ is fluid viscosity. When the dimensions of these variables are combined according to Eq. (3.1), the dimensions of the numerator exactly cancel those of the denominator. Other dimensionless variables relevant to bioprocess engineering are the Schmidt number, Prandtl number, Sherwood number, Peclet number, Nusselt number, Grashof number, power number, and many others. Definitions and applications of these natural variables are given in later chapters of this book.

Rotation is described based on the number of radians or revolutions:

$$\text{number of radians} = \frac{\text{length of arc}}{\text{radius}} = \frac{\text{length of arc}}{r} \tag{3.2}$$

$$\text{number of revolutions} = \frac{\text{length of arc}}{\text{circumference}} = \frac{\text{length of arc}}{2\pi r} \tag{3.3}$$

where r is radius. One revolution is equal to 2π radians. Radians and revolutions are nondimensional because the dimensions of length for arc, radius, and circumference in Eqs. (3.2) and (3.3) cancel. Consequently, rotational speed (e.g., number of revolutions per second) and angular velocity (e.g., number of radians per second) have dimensions T^{-1}. Degrees, which are subdivisions of a revolution, are converted into revolutions or radians before application in most engineering calculations. Frequency (e.g., number of vibrations per second) is another variable that has dimensions T^{-1}.

3.1.3 Dimensional Homogeneity in Equations

Rules about dimensions determine how equations are formulated. "Properly constructed" equations representing general relationships between physical variables must be dimensionally homogeneous. For dimensional homogeneity, the dimensions of terms that are added or subtracted must be the same, and the dimensions of the right side of the equation must be the same as those of the left side. As a simple example, consider the Margules equation for evaluating fluid viscosity from experimental measurements:

$$\mu = \frac{M}{4\pi h \Omega}\left(\frac{1}{R_o^2} - \frac{1}{R_i^2}\right) \tag{3.4}$$

The terms and dimensions in this equation are listed in Table 3.3. Numbers such as 4 have no dimensions; the symbol π represents the number 3.1415926536, which is also dimensionless. As discussed in Section 3.1.2, the number of radians per second represented by Ω has dimensions T^{-1}, so appropriate units would be, for example, s^{-1}. A quick check shows that Eq. (3.4) is dimensionally homogeneous since both sides of the equation have dimensions $L^{-1}MT^{-1}$ and all terms added or subtracted have the same dimensions. Note that when a term such as R_o is raised to a power such as 2, the units and dimensions of R_o must also be raised to that power.

For dimensional homogeneity, the argument of any transcendental function, such as a logarithmic, trigonometric, or exponential function, must be dimensionless. The following examples illustrate this principle.

1. An expression for cell growth is:

$$\ln\frac{x}{x_0} = \mu t \tag{3.5}$$

where x is cell concentration at time t, x_0 is initial cell concentration, and μ is the specific growth rate. The argument of the logarithm, the ratio of cell concentrations, is dimensionless.

2. The displacement y due to the action of a progressive wave with amplitude A, frequency $\omega/2\pi$, and velocity v is given by the equation:

$$y = A\sin\left[\omega\left(t - \frac{x}{v}\right)\right] \tag{3.6}$$

Table 3.3: Terms and Dimensions in Eq. (3.4)

Term	Dimensions	SI units
μ (dynamic viscosity)	$L^{-1}MT^{-1}$	pascal second (Pa s)
M (torque)	L^2MT^{-2}	newton meter (N m)
h (cylinder height)	L	meter (m)
Ω (angular velocity)	T^{-1}	radians per second (s^{-1})
R_o (outer radius)	L	meter (m)
R_i (inner radius)	L	meter (m)

where t is time and x is distance from the origin. The argument of the sine function,

$$\omega\left(t - \frac{x}{v}\right)$$, is dimensionless.

3.   The relationship between α, the mutation rate of *Escherichia coli*, and temperature T can be described using an Arrhenius-type equation:

$$\alpha = \alpha_0 e^{-E/RT} \tag{3.7}$$

where α_0 is the mutation reaction constant, E is the specific activation energy, and R is the ideal gas constant (see Section 3.5). The dimensions of RT are the same as those of E, so the exponent is as it should be: dimensionless.

The dimensional homogeneity of equations can sometimes be masked by mathematical manipulation. As an example, Eq. (2.5) might be written:

$$\ln x = \ln x_0 + \mu t \tag{3.8}$$

Inspection of this equation shows that rearrangement of the terms to group $\ln x$ and $\ln x_0$ together recovers dimensional homogeneity by providing a dimensionless argument for the logarithm.

Integration and differentiation of terms affect dimensionality. Integration of a function with respect to x increases the dimensions of that function by the dimensions of x. Conversely, differentiation with respect to x results in the dimensions being reduced by the dimensions of x. For example, if C is the concentration of a particular compound expressed as mass per unit volume and x is distance, dC/dx has dimensions $L^{-4}M$, whereas d^2C/dx^2 has dimensions $L^{-5}M$. On the other hand, if μ is the specific growth rate of an organism with dimensions T^{-1} and t is time, then $\int \mu dt$ is dimensionless.

3.1.4 Equations Without Dimensional Homogeneity

It is sometimes convenient to present the equation in a nonhomogeneous form for repetitive calculations or when an equation is derived from observation rather than from theoretical principles. Such equations are called *equations in numerics* or *empirical equations*. In empirical equations, the units associated with each variable must be stated explicitly. An example is Richards' correlation for the dimensionless gas hold-up ε in a stirred fermenter:

$$\left(\frac{P}{V}\right)^{0.4} u^{1/2} = 30\varepsilon + 1.33 \tag{3.9}$$

where P is power in units of horsepower, V is ungassed liquid volume in units of ft^3, u is linear gas velocity in units of ft s^{-1}, and ε is fractional gas hold-up, a dimensionless variable. The dimensions of each side of Eq. (3.9) are certainly not the same. Only the units specified can be used for direct application of Eq. (3.9).

3.2 Units

Several systems of units for expressing the magnitude of physical variables have been devised through the ages. The metric system of units originated from the National Assembly of France in 1790. In 1960 this system was rationalized, and the SI or Système International d'Unités was adopted as the international standard. Unit names and their abbreviations have been standardized; according to SI convention, unit abbreviations are the same for both singular and plural and are not followed by a period. SI prefixes used to indicate multiples and submultiples of units are listed in Table 3.4. Despite widespread use of SI units, no single system of units has universal application. Engineers in the United States continue to apply British or imperial units. In addition, many physical property data collected before 1960 are published in lists and tables using nonstandard units.

Familiarity with both metric and nonmetric units is necessary. Some units used in engineering, such as the slug (1 slug = 14.5939 kg), dram (1 dram = 1.77185 g), stoke (a unit of kinematic viscosity), poundal (a unit of force), and erg (a unit of energy), are probably not known to you. Although no longer commonly applied, these are legitimate units that may appear in engineering reports and tables of data.

It is often necessary to convert units in calculations. Units are changed using *conversion factors*. Some conversion factors, such as 1 in. = 2.54 cm and 2.20 lb = 1 kg, you probably already know. Tables of common conversion factors are given in the Appendix at the back of this book. Unit conversions are not only necessary to convert imperial units to metric; some physical variables have several metric units in common use. For example, viscosity may be reported as centipoise or $kg\ h^{-1}\ m^{-1}$; pressure may be given in standard atmospheres, pascals, or millimeters of mercury. Conversion of units seems simple enough; however, difficulties can arise when several variables are being converted into a single equation. Accordingly, an organized mathematical approach is needed.

Table 3.4: SI Prefixes

Factor	Prefix	Symbol	Factor	Prefix	Symbol
10^{-1}	deci*	d	10^{18}	exa	E
10^{-2}	centi*	c	10^{15}	peta	P
10^{-3}	milli	m	10^{12}	tera	T
10^{-6}	micro	μ	10^{9}	giga	G
10^{-9}	nano	n	10^{6}	mega	M
10^{-12}	pico	p	10^{3}	kilo	k
10^{-15}	femto	f	10^{2}	hecto*	h
10^{-18}	atto	a	10^{1}	deka*	da

*Used for areas and volumes.

From J.V. Drazil, *Quantities and Units of Measurement*, Mansell, 1983.

For each conversion factor, a *unity bracket* can be derived. The value of the unity bracket, as the name suggests, is unity. As an example, the conversion factor:

$$1 \text{ lb} = 453.6 \text{ g} \tag{3.10}$$

can be converted by dividing both sides of the equation by 1 lb to give a unity bracket denoted by vertical bars (| |):

$$1 = \left| \frac{453.6 \text{ g}}{1 \text{ lb}} \right| \tag{3.11}$$

Similarly, division of both sides of Eq. (3.10) by 453.6 g gives another unity bracket:

$$\left| \frac{1 \text{ lb}}{453.6 \text{ g}} \right| = 1 \tag{3.12}$$

To calculate how many pounds are in 200 g, we can multiply 200 g by the unity bracket in Eq. (3.12) or divide 200 g by the unity bracket in Eq. (3.11). This is permissible since the value of both unity brackets is unity, and multiplication or division by 1 does not change the value of 200 g. Using the option of multiplying by Eq. (3.12):

$$200 \text{ g} = 200 \text{ g} * \left| \frac{1 \text{ lb}}{453.6 \text{ g}} \right| \tag{3.13}$$

On the right side, canceling the old units leaves the desired unit, lb. Dividing the numbers gives:

$$200 \text{ g} = 0.441 \text{ lb} \tag{3.14}$$

A more complicated calculation involving a complete equation is given here in Example 3.1.

3.3 Force and Weight

According to Newton's law, the force exerted on a body in motion is proportional to its mass multiplied by the acceleration. The dimensions of force are LMT^{-2} (Table 3.2); the *natural units* of force in the SI system are kg m s^{-2}. Analogously, g cm s^{-2} and lb ft s^{-2} are natural units of force in the metric and British systems, respectively.

Force occurs frequently in engineering calculations, and *derived units* are used more commonly than natural units. In SI, the derived unit for force is the *newton*, abbreviated as N:

$$1 \text{ N} = 1 \text{ kg m s}^{-2} \tag{3.15}$$

In the British or imperial system, the derived unit for force is the *pound-force*, which is denoted lb_f. One pound-force is defined as (1 lb mass) × (gravitational acceleration at sea level and 45° latitude). In different systems of units, gravitational acceleration g at sea level and 45° latitude is:

EXAMPLE 3.1 Unit Conversion

Air is pumped through an orifice immersed in liquid. The size of the bubbles leaving the orifice depends on the diameter of the orifice and the properties of the liquid. The equation representing this situation is:

$$\frac{g(\rho_L - \rho_G)D_b^3}{\sigma D_o} = 6$$

where g = gravitational acceleration = 32.174 ft s^{-2}; ρ_L = liquid density = 1 g cm^{-3}; ρ_G = gas density = 0.081 lb ft^{-3}; D_b = bubble diameter; σ = gas–liquid surface tension = 70.8 dyn cm^{-1}; and D_o = orifice diameter = 1 mm.
Calculate the bubble diameter D_b.

Solution

Convert the data to a consistent set of units, for example, g, cm, s. From Appendix A, the conversion factors required are:

- 1 ft = 0.3048 m
- 1 lb = 453.6 g
- 1 dyn cm^{-1} = 1 g s^{-2}

Also:

- 1 m = 100 cm
- 10 mm = 1 cm

Converting units:

$$g = 32.174 \frac{\text{ft}}{\text{s}^2} \cdot \left|\frac{0.3048 \text{ m}}{1 \text{ ft}}\right| \cdot \left|\frac{100 \text{ cm}}{1 \text{ m}}\right| = 980.7 \text{ cm s}^{-2}$$

$$\rho_G = 0.081 \frac{\text{lb}}{\text{ft}^3} \cdot \left|\frac{453.6 \text{ g}}{1 \text{ lb}}\right| \cdot \left|\frac{1 \text{ ft}}{0.3048 \text{ m}}\right|^3 \cdot \left|\frac{1 \text{ m}}{100 \text{ cm}}\right|^3 = 1.30 \times 10^{-3} \text{ g cm}^{-3}$$

$$\sigma = 70.8 \text{ dyn cm}^{-1} \cdot \left|\frac{1 \text{ g s}^{-2}}{1 \text{ dyn cm}^{-1}}\right| = 70.8 \text{ g s}^{-2}$$

$$D_o = 1 \text{ mm} \cdot \left|\frac{1 \text{ cm}}{10 \text{ mm}}\right| = 0.1 \text{ cm}$$

Rearranging the equation to give an expression for D_b^3 :

$$D_b^3 = \frac{6\sigma D_o}{g(\rho_L - \rho_G)}$$

Substituting values gives:

$$D_b^3 = \frac{6(70.8 \text{ g s}^{-2})(0.1 \text{ cm})}{980.7 \text{ cm s}^{-2} (1 \text{ g cm}^{-3} - 1.30 \times 10^{-3} \text{ g cm}^{-3})} = 4.34 \times 10^{-2} \text{ cm}^3$$

Taking the cube root:

$$D_b = 0.35 \text{ cm}$$

Note that unity brackets are squared or cubed when appropriate, for example, when converting ft^3 to cm^3. This is permissible since the value of the unity bracket is 1, and 1^2 or 1^3 is still 1.

$$g = 9.8066 \text{ m s}^{-2} \tag{3.16}$$

$$g = 980.66 \text{ cm s}^{-2} \tag{3.17}$$

$$g = 32.174 \text{ ft s}^{-2} \tag{3.18}$$

Therefore:

$$1 \text{ lb}_f = 32.174 \text{ lb}_m \text{ft s}^{-2} \tag{3.19}$$

Note that pound-mass, which is usually represented as lb, has been shown here using the abbreviation lb_m to distinguish it from lb_f. Use of the pound in the imperial system for reporting both mass and force can be a source of confusion and requires care.

To convert force from a defined unit to a natural unit, a special dimensionless unity bracket called g_c is used. The form of g_c depends on the units being converted. From Eqs. (3.15) and (3.19):

$$g_c = 1 = \left| \frac{1 \text{ N}}{1 \text{ kg m s}^{-2}} \right| = \left| \frac{1 \text{ lb}_f}{32.174 \text{ lb}_m \text{ ft s}^{-2}} \right| \tag{3.20}$$

Application of g_c is illustrated in Example 3.2.

Weight is the force with which a body is attracted by gravity to the center of the Earth. Therefore, the weight of an object will change depending on its location, whereas its mass will not. Weight changes according to the value of the gravitational acceleration g, which varies by about 0.5% over the Earth's surface. Using Newton's law and depending on the exact value of g, the weight of a mass of 1 kg is about 9.8 newtons; the weight of a mass of 1 lb is about 1 lbf. Note that although the value of g changes with position on the Earth's surface (or in the universe), the value of g_c within a given system of units does not. g_c is a factor for converting units, not a physical variable.

EXAMPLE 3.2 Use of g_c

Calculate the kinetic energy of 250 lb_m of liquid flowing through a pipe at a speed of 35 ft s^{-1}. Express your answer in units of ft lb_f.

Solution

Kinetic energy is given by the equation:

$$\text{kinetic energy} = E_k = \frac{1}{2}Mv^2$$

where M is mass and v is velocity. Using the values given:

$$E_k = \frac{1}{2}(250\ lb_m)\left(35\frac{ft}{s}\right)^2 = 1.531 \times 10^5\ \frac{lb_m\,ft^2}{s^2}$$

Multiplying by g_c from Eq. (3.20) gives:

$$E_k = 1.531 \times 10^5\ \frac{lb_m\,ft^2}{s^2} \cdot \left|\frac{1\,lb_f}{32.174\ lb_m\,ft\ s^{-2}}\right|$$

Calculating and canceling units gives the answer:

$$E_k = 4760\ ft\ lb_f$$

3.4 Measurement Conventions

Familiarity with common physical variables and methods for expressing their magnitude is necessary for engineering analysis of bioprocesses. This section covers some useful definitions and engineering conventions that will be applied throughout the text.

3.4.1 Density

Density is a substantial variable defined as mass per unit volume. Its dimensions are $L^{-3}M$, and the usual symbol is ρ. Units for density are, for example, g cm^{-3}, kg m^{-3}, and lb ft^{-3}. If the density of acetone is 0.792 g cm^{-3}, the mass of 150 cm^3 acetone can be calculated as follows:

$$150\ cm^3 \left(\frac{0.792\ g}{cm^3}\right) = 119\ g$$

Densities of solids and liquids vary slightly with temperature. The density of water at 4°C is 1.0000 g cm^{-3}, or 62.4 lb ft^{-3}. The density of solutions is a function of both concentration and temperature. Gas densities are highly dependent on temperature and pressure.

3.4.2 Specific Gravity

Specific gravity, also known as "relative density," is a dimensionless variable. It is the ratio of two densities: that of the substance in question and that of a specified reference material. For liquids and solids, the reference material is usually water. For gases, air is commonly used as the reference, but other reference gases may also be specified.

Liquid densities vary with temperature. Accordingly, when reporting specific gravity, the temperatures of the substance and its reference material are specified. If the specific gravity of ethanol is given as, $0.789^{20°C}_{4°C}$ this means that the specific gravity is 0.789 for ethanol at 20°C referenced against water at 4°C. Since the density of water at 4°C is almost exactly $1.0000\,g\,cm^{-3}$, we can say immediately that the density of ethanol at 20°C is $0.789\,g\,cm^{-3}$.

3.4.3 Specific Volume

Specific volume is the inverse of density. The dimensions of specific volume are L^3M^{-1}.

3.4.4 Mole

In the SI system, a mole is "the amount of substance of a system which contains as many elementary entities as there are atoms in 0.012 kg of carbon-12" [1]. This means that a mole in the SI system is about 6.02×10^{23} molecules and is denoted by the term *gram-mole* or *gmol*. One thousand gmol is called a *kilogram-mole* or *kgmol*. In the American engineering system, the basic mole unit is the *pound-mole* or *lbmol*, which is $6.02 \times 10^{23} \times 453.6$ molecules. The gmol, kgmol, and lbmol therefore represent three different quantities. When molar quantities are specified simply as "moles," gmol is usually meant.

The number of moles in a given mass of material is calculated as follows:

$$\text{gram-moles} = \frac{\text{mass in grams}}{\text{molar mass in grams}} \tag{3.21}$$

$$\text{lb-moles} = \frac{\text{mass in lb}}{\text{molar mass in lb}} \tag{3.22}$$

Molar mass is the mass of one mole of substance and has dimensions MN^{-1}. Molar mass is routinely referred to as *molecular weight*, although the molecular weight of a compound is a dimensionless quantity calculated as the sum of the atomic weights of the elements constituting a molecule of that compound. The *atomic weight* of an element is its mass relative to carbon-12 having a mass of exactly 12; atomic weight is also dimensionless. The terms "molecular weight" and "atomic weight" are frequently used by engineers and chemists instead of the more correct terms "relative molecular mass" and "relative atomic mass."

3.4.5 Chemical Composition

Process streams usually consist of mixtures of components or solutions of one or more solutes. The following terms are used to define the composition of mixtures and solutions.

The *mole fraction* of component A in a mixture is defined as:

$$\text{mole fraction A} = \frac{\text{number of moles of A}}{\text{total number of moles}} \tag{3.23}$$

Mole percent is mole fraction × 100. In the absence of chemical reactions and loss of material from the system, the composition of a mixture expressed in mole fraction or mole percent does not vary with temperature.

The *mass fraction* of component A in a mixture is defined as:

$$\text{mass fraction A} = \frac{\text{mass of A}}{\text{total mass}} \tag{3.24}$$

Mass percent is mass fraction × 100; mass fraction and mass percent are also called *weight fraction* and *weight percent*, respectively. Another common expression for composition is weight-for-weight percent (% w/w). Although not so well defined, this is usually considered to be the same as weight percent. For example, a solution of sucrose in water with a concentration of 40% w/w contains 40 g sucrose per 100 g solution, 40 tons sucrose per 100 tons solution, 40 lb sucrose per 100 lb solution, and so on. In the absence of chemical reactions and loss of material from the system, mass and weight percent do not change with temperature.

The composition of liquids and solids is usually reported using mass percent, so this can be assumed even if not specified. For example, if an aqueous mixture is reported to contain 5% NaOH and 3% $MgSO_4$, it is conventional to assume that there are 5 g NaOH and 3 g $MgSO_4$ in every 100 g solution. Of course, mole or volume percent may be used for liquid and solid mixtures; however, this should be stated explicitly (e.g., 10 vol%, 50 mol%).

The *volume fraction* of component A in a mixture is:

$$\text{volume fraction A} = \frac{\text{volume of A}}{\text{total volume}} \tag{3.25}$$

Volume percent is volume fraction × 100. Although not as clearly defined as volume percent, volume-for-volume percent (% v/v) is usually interpreted in the same way as volume percent; for example, an aqueous sulfuric acid mixture containing 30 cm^3 acid in 100 cm^3 solution is referred to as a 30% v/v solution. Weight-for-volume percent (% w/v) is also often used; a codeine concentration of 0.15% w/v generally means 0.15 g codeine per 100 mL solution.

Compositions of gases are commonly given in volume percent; if percentage figures are given without specification, volume percent is assumed. According to the *International Critical Tables*, [2] the composition of air is 20.99% O_2, 78.03% N_2, 0.94% argon, and 0.03% carbon dioxide; small amounts of H_2, helium, neon, krypton, and xenon make up the remaining 0.01%. For most purposes, all inerts are lumped together with nitrogen and the composition of air is taken as approximately 21% O_2 and 79% N_2. This means that any sample of air will contain about 21% O_2 *by volume*. At low pressure, gas volume is directly proportional to number of moles; therefore, the composition of air as stated can also be interpreted as 21 mole% O_2. Because temperature changes at low pressure produce the same relative change in the partial volumes of the constituent gases as in the total volume, the volumetric composition of gas mixtures is not altered by variation in temperature. Temperature changes affect the component gases equally, so the overall composition is unchanged.

There are many other choices for expressing the concentration of a component in solutions and mixtures:

1. Moles per unit volume (e.g., gmol L^{-1}, lbmol ft^{-3}).
2. Mass per unit volume (e.g., kg m^{-3}, g L^{-1}, lb ft^{-3}).
3. Parts per million, ppm. This is used for very dilute solutions. Usually, ppm is a mass fraction for solids and liquids and a mole fraction for gases. For example, an aqueous solution of 20 ppm manganese contains 20 g manganese per 10^6 g solution. A sulfur dioxide concentration of 80 ppm in air means 80 gmol SO_2 per 10^6 gmol gas mixture. At low pressures this is equivalent to 80 L SO_2 per 10^6 L gas mixture.
4. Molarity, gmol L^{-1}. A molar concentration is abbreviated 1 M.
5. Molality, gmol per 1000 g solvent.
6. Normality, mole equivalents L^{-1}. A normal concentration is abbreviated 1 N and contains one equivalent gram-weight of solute per liter of solution. For an acid or base, an equivalent gram-weight is the weight of solute in grams that will produce or react with 1 gmol hydrogen ions. Accordingly, a 1 N solution of HCl is the same as a 1 M solution; on the other hand, a 1 N H_2SO_4 or 1 N $Ca(OH)_2$ solution is 0.5 M.
7. Formality, formula gram-weight L^{-1}. If the molecular weight of a solute is not clearly defined, formality may be used to express concentration. A formal solution contains one formula gram-weight of solute per liter of solution. If the formula gram-weight and molecular gram-weight are the same, molarity and formality are the same.

In several industries, concentration is expressed in an indirect way using specific gravity. The density and specific gravity of solutions are directly dependent on the concentration of solute for a given solute and solvent. Specific gravity is conveniently measured using a hydrometer, which may be calibrated using special scales. The *Baumé scale*, originally developed in France to measure levels of salt in brine, is in common use. One Baumé scale is used for liquids lighter than water; another is used for liquids heavier than water. For liquids heavier than water such as sugar solutions:

$$\text{degrees Baumé}\left(\, ^{\circ}\text{Bé} \, \right) = 145 - \frac{145}{G} \tag{3.26}$$

where G is specific gravity. Unfortunately, the reference temperature for the Baumé and other gravity scales is not standardized worldwide. If the Baumé hydrometer were calibrated at 60 °F (15.6°C), G in Eq. (2.26) would be the specific gravity at 60 °F relative to water at

60 °F; however, another common reference temperature is 20°C (68 °F). The Baumé scale is used widely in the wine and food industries as a measure of sugar concentration. For example, readings of °Bé from grape juice help determine when grapes should be harvested for wine making. The Baumé scale gives only an approximate indication of sugar levels; there is always some contribution to specific gravity from soluble compounds other than sugar.

Degrees Brix (°Brix), or *degrees Balling*, is another hydrometer scale used extensively in the sugar industry. Brix scales calibrated at 15.6°C and 20°C are in common use. With the 20°C scale, each degree Brix indicates 1 g of sucrose per 100 g liquid.

3.4.6 Temperature

Temperature is a measure of the thermal energy of a body at thermal equilibrium. As a dimension, it is denoted by Θ. Temperature is commonly measured in degrees *Celsius* (centigrade) or *Fahrenheit*. The Celsius scale is most common in science; 0°C is taken as the ice point of water and 100°C the normal boiling point of water. The Fahrenheit scale is in everyday use in the United States; 32 °F represents the ice point and 212 °F the normal boiling point of water. Both Fahrenheit and Celsius scales are *relative temperature scales*, meaning that their zero points have been arbitrarily assigned.

Sometimes it is necessary to use *absolute temperatures*. Absolute temperature scales have as their zero point the lowest temperature believed possible. Absolute temperature is used in applications of the ideal gas law and many other laws of thermodynamics. A scale for absolute temperature with degree units the same as on the Celsius scale is known as the *Kelvin* scale; the absolute temperature scale using Fahrenheit degree units is the *Rankine* scale. Accordingly, a temperature difference of one degree on the Celsius scale corresponds to a temperature difference of one degree on the Kelvin scale; the case is similar to the Fahrenheit and Rankine scales. Units on the Kelvin scale used to be termed "degrees Kelvin" and abbreviated °K. It is modern practice, however, to name the unit simply "Kelvin"; the SI symbol for Kelvin is K. Units on the Rankine scale are denoted °R. $0\,°R = 0\,K = -459.67\,°F = -273.15°C$. Comparison of the four temperature scales is shown in Figure 3.1.

Equations for converting temperature units are as follows; T represents the temperature reading:

$$T(K) = T(°C) + 273.15 \tag{3.27}$$

$$T(°R) = T(°F) + 459.67 \tag{3.28}$$

$$T(°R) = 1.8\,T(K) \tag{3.29}$$

$$T(°F) = 1.8\,T(°C) + 32 \tag{3.30}$$

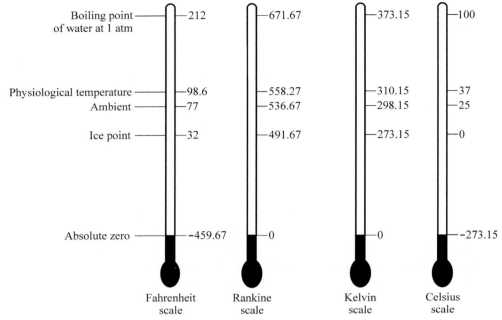

Figure 3.1 Comparison of temperature scales.

A temperature difference of 1 degree on the Kelvin–Celsius scale corresponds to a temperature difference of 1.8 degrees on the Rankine–Fahrenheit scale. This is readily deduced, for example, if we consider the difference between the freezing and boiling points of water, which is 100 degrees on the Kelvin–Celsius scale and $(212-32) = 180$ degrees on the Rankine–Fahrenheit scale.

The dimensions of several engineering parameters include temperature. Examples, such as specific heat capacity, heat transfer coefficient, and thermal conductivity, are listed in Table 3.2. These parameters are applied in engineering calculations to evaluate variations in the properties of materials caused by a *change* in temperature. For instance, the specific heat capacity is used to determine the change in enthalpy of a system resulting from a change in its temperature. Therefore, if the specific heat capacity of a certain material is known to be $0.56\,kcal\,kg^{-1}\,°C^{-1}$, this is the same as $0.56\,kcal\,kg^{-1}\,K^{-1}$, as any change in temperature measured in units of $°C$ is the same when measured on the Kelvin scale. Similarly, for engineering parameters quantified using the imperial system of units, $°F^{-1}$ can be substituted for $°R^{-1}$ and vice versa.

3.4.7 Pressure

Pressure is defined as force per unit area and has dimensions $L^{-1}MT^{-2}$. Units of pressure are numerous, including pounds per square inch (psi), millimeters of mercury (mmHg), standard atmospheres (atm), bar, newtons per square meter ($N\,m^{-2}$), and many others. The SI pressure unit, N m^{-2}, is called a pascal (Pa). Like temperature, pressure may be expressed using absolute or relative scales.

Absolute pressure is pressure relative to a complete vacuum. Because this reference pressure is independent of location, temperature, and weather, absolute pressure is a precise and invariant quantity. However, absolute pressure is not commonly measured. Most pressure-measuring devices sense the difference in pressure between the sample and the surrounding atmosphere at the time of measurement. These instruments give readings of *relative pressure*, also known as *gauge pressure*. Absolute pressure can be calculated from gauge pressure as follows:

$$\text{absolute pressure} = \text{gauge pressure} + \text{atmospheric pressure} \qquad (3.31)$$

As you know from listening to weather reports, atmospheric pressure varies with time and place and is measured using a *barometer*. Atmospheric pressure or *barometric pressure* should not be confused with the standard unit of pressure called the standard atmosphere (atm), defined as 1.013×10^5 N m^{-2}, 14.70 psi, or 760 mmHg at 0°C. Sometimes the units for pressure include information about whether the pressure is absolute or relative. Pounds per square inch is abbreviated as *psia* for absolute pressure or *psig* for gauge pressure. *Atma* denotes standard atmospheres of absolute pressure.

Vacuum pressure is another pressure term, used to indicate pressure below barometric pressure. A gauge pressure of -5 psig, or 5 psi below atmospheric, is the same as a vacuum of 5 psi. A perfect vacuum corresponds to an absolute pressure of zero.

3.5 Standard Conditions and Ideal Gases

A *standard state* of temperature and pressure has been defined and is used when specifying properties of gases, particularly molar volumes. Standard conditions are needed because the volume of a gas depends not only on the quantity present but also on the temperature and pressure. The most widely adopted standard state is 0°C and 1 atm.

Relationships between gas volume, pressure, and temperature were formulated in the 18th and 19th centuries. These correlations were developed under conditions of temperature and pressure such that the average distance between gas molecules was great enough to counteract the effect of intramolecular forces, and the volume of the molecules themselves could be neglected. A gas under these conditions became known as an *ideal gas*. This term now in common use refers to a gas that obeys certain simple physical laws, such as those of Boyle, Charles, and Dalton. Molar volumes for an ideal gas at standard conditions are:

$$1 \text{ gmol} = 22.4 \text{ L} \qquad (3.32)$$

$$1 \text{ kgmol} = 22.4 \text{ m}^3 \qquad (3.33)$$

$$1 \text{ lbmol} = 359 \text{ ft}^3 \qquad (3.34)$$

No real gas is an ideal gas at all temperatures and pressures. However, light gases such as hydrogen, oxygen, and air deviate negligibly from ideal behavior over a wide range of conditions. On the other hand, heavier gases such as sulfur dioxide and hydrocarbons can deviate considerably from ideal, particularly at high pressures. Vapors near the boiling point also deviate markedly from ideal. Nevertheless, for many applications in bioprocess engineering, gases can be considered ideal without much loss of accuracy.

Equations (3.32) through (3.34) can be verified using the *ideal gas law*:

$$pV = nRT \qquad (3.35)$$

where p is absolute pressure, V is volume, n is moles, T is absolute temperature, and R is the *ideal gas constant*. Equation (3.35) can be applied using various combinations of units for the physical variables, as long as the correct value and units of R are employed. A list of R values in different systems of units is given in Appendix B. Application of the ideal gas law is illustrated in Example 3.3.

EXAMPLE 3.3 Ideal Gas Law

Gas leaving a bioreactor at close to 1 atm pressure and 25°C has the following composition: 78.2% N_2, 19.2% O_2, 2.6% carbon dioxide.
   Calculate:

a. The mass composition of the bioreactor off-gas
b. The mass of CO_2 in each cubic meter of gas leaving the bioreactor

Solution
Molecular weights:

- $N_2 = 28$
- $O_2 = 32$
- Carbon dioxide = 44

a. As the gas is at low pressure, the percentages given for composition can be considered mole percentages. Therefore, using the molecular weights, 100 gmol off-gas contains:

$$78.2 \, \text{gmolN}_2 \cdot \left| \frac{28 \, \text{gN}_2}{1 \text{gmolN}_2} \right| = 2189.6 \, \text{gN}_2$$

$$19.2 \, \text{gmolO}_2 \cdot \left| \frac{32 \, \text{gO}_2}{1 \text{gmolO}_2} \right| = 614.4 \, \text{gO}_2$$

$$2.6 \, \text{gmolCO}_2 \cdot \left| \frac{44 \, \text{gCO}_2}{1 \text{gmolCO}_2} \right| = 114.4 \, \text{gCO}_2$$

Therefore, the total mass is $(2189.6 + 614.4 + 114.4)\,g = 2918.4\,g$. The mass composition can be calculated as follows:

$$\text{Mass percent } N_2 = \frac{2189.6\,g}{2918.4\,g} \times 100 = 75.0\%$$

$$\text{Mass percent } O_2 = \frac{614.4\,g}{2918.4\,g} \times 100 = 21.1\%$$

$$\text{Mass percent } CO_2 = \frac{114.4\,g}{2918.4\,g} \times 100 = 3.9\%$$

a. Therefore, the composition of the gas is 75.0 mass% N_2, 21.1 mass% O_2, and 3.9 mass% CO_2.

b. As the gas composition is given in volume percent, in each cubic meter of gas there must be 0.026 m^3 CO_2. The relationship between moles of gas and volume at 1 atm and 25°C is determined using Eq. (3.35) and an appropriate value of R from Appendix B:

$$(1\,atm)(0.026\,m^3) = n\left(0.000082057\,\frac{m^3 atm}{gmol\,K}\right)(298.15\,K)$$

Calculating the moles of CO_2 present:

$$n = 1.06\,gmol$$

Converting to mass of CO_2:

$$1.06\,gmol = 1.06\,gmol \cdot \left|\frac{44\,g}{1\,gmol}\right| = 46.8\,g$$

Therefore, each cubic meter of bioreactor off-gas contains 46.8 g CO_2.

3.6 Physical and Chemical Property Data

Information about the properties of materials is often required in engineering calculations. As measurement of physical and chemical properties is time consuming and expensive, handbooks containing this information are a tremendous resource. You may already be familiar with some handbooks of physical and chemical data, including:

- *International Critical Tables* [2]
- *CRC Handbook of Chemistry and Physics* [3]
- *Lange's Handbook of Chemistry* [4]

To these can be added:

- *Perry's Chemical Engineers' Handbook* [5]

And, for information about biological materials:

- *Biochemical Engineering and Biotechnology Handbook* [6]

A selection of physical and chemical property data is included in Appendix C.

3.7 Stoichiometry

In chemical or biochemical reactions, atoms and molecules rearrange to form new groups. Mass and molar relationships between the reactants consumed and products formed can be determined using stoichiometric calculations. This information is deduced from correctly written reaction equations and relevant atomic weights.

As an example, consider the principal reaction in alcohol fermentation: conversion of glucose to ethanol and carbon dioxide:

$$C_6H_{12}O_6 \rightarrow 2C_2H_6O + 2CO_2$$

(3.36)

This reaction equation states that one molecule of glucose breaks down to give two molecules of ethanol and two molecules of carbon dioxide. Another way of saying this is that one mole of glucose breaks down to give two moles of ethanol and two moles of carbon dioxide. Applying molecular weights, the equation also shows that reaction of 180 g glucose produces 92 g ethanol and 88 g carbon dioxide.

During chemical or biochemical reactions, the following two quantities are conserved:

1. *Total mass*, so that total mass of reactants = total mass of products
2. *Number of atoms of each element*, so that, for example, the number of C, H, and O atoms in the reactants = the number of C, H, and O atoms, respectively, in the products

Note that there is no corresponding law for conservation of moles: the number of moles of reactants is not necessarily equal to the number of moles of products.

By themselves, equations such as Eq. (3.36) suggest that all the reactants are converted into the products specified in the equation, and that the reaction proceeds to completion. This is often not the case for industrial reactions. Because the stoichiometry may not be known precisely, or to manipulate the reaction beneficially, reactants are not usually supplied in the exact proportions indicated by the reaction equation. Excess quantities of some reactants may be provided; this excess material is found in the product mixture once the reaction is stopped. In addition, reactants are often consumed in side reactions to make products not described by the principal reaction equation; these side products also form part of the final reaction mixture. In these circumstances, additional information is needed before the amounts of products formed or reactants consumed can be calculated. Several terms are used to describe partial and branched reactions.

1.   The *limiting reactant* or *limiting substrate* is the reactant present in the smallest *stoichiometric* amount. While other reactants may be present in smaller absolute quantities, at the time when the last molecule of the limiting reactant is consumed, residual amounts of all reactants except the limiting reactant will be present in the reaction mixture. As an illustration, for the L-glutamic acid reaction of Example 3.4, if 100 g glucose, 17 g NH_3, and 48 g O_2 are provided for conversion, glucose will be the limiting reactant even though a greater mass of it is available compared with the other substrates.

EXAMPLE 3.4 Stoichiometry of Amino Acid Synthesis

The overall reaction for microbial conversion of glucose to L-glutamic acid is:

$$\underset{\text{(glucose)}}{C_6H_{12}O_6} + NH_3 + 1.5O_2 \rightarrow \underset{\text{(glutamic acd)}}{C_5H_9NO_4} + CO_2 + 3H_2O$$

What mass of oxygen is required to produce 15 g L-glutamic acid?

Solution

Molecular weights:

- Oxygen = 32
- L-glutamic acid = 147

Because stoichiometric equations give relationships between moles, g glutamic acid is first converted to gmol using the unity bracket for molecular weight:

$$15 \text{ g glumatic acid} = 15 \text{ g glumatic acid} \cdot \left| \frac{1 \text{ gmol glutamic acid}}{147 \text{ g glutamic acid}} \right| = 0.102 \text{ gmol glutamic acid}$$

According to the reaction equation, production of 1 gmol of L-glutamic acid requires 1.5 gmol O_2. Therefore, production of 0.102 gmol L-glutamic acid requires $(0.102 \times 1.5) = 0.153$ gmol O_2. This can be expressed as mass of O_2 using the unity bracket for the molecular weight of O_2:

$$0.153 \text{ gmol } O_2 = 0.153 \text{ gmol } O_2 \cdot \left| \frac{32 \text{ g } O_2}{1 \text{ gmol } O_2} \right| = 4.9 \text{ g } O_2$$

Therefore, 4.9 g O_2 is required. More O_2 will be needed if microbial growth also occurs.

2. An *excess reactant* is a reactant present in an amount in excess of that required to combine with all of the limiting reactant. It follows that an excess reactant is one remaining in the reaction mixture once all the limiting reactant is consumed. The *percentage excess* is calculated using the amount of excess material relative to the quantity required for complete consumption of the limiting reactant:

$$\% \text{ excess} = \frac{\left(\begin{array}{c}\text{moles present} - \text{moles required to react} \\ \text{completely with the limiting reactant}\end{array}\right)}{\left(\begin{array}{c}\text{moles required to react} \\ \text{completely with the limiting reactant}\end{array}\right)} \times 100 \qquad (3.37)$$

or

$$\% \text{ excess} = \frac{\left(\begin{array}{c}\text{mass present} - \text{mass required to react} \\ \text{completely with the limiting reactant}\end{array}\right)}{\left(\begin{array}{c}\text{mass required to react} \\ \text{completely with the limiting reactant}\end{array}\right)} \times 100 \qquad (3.38)$$

The *required* amount of a reactant is the stoichiometric quantity needed for complete conversion of the limiting reactant. In the preceding glutamic acid example, the required amount of NH_3 for complete conversion of 100 g glucose is 9.4 g; therefore, if 17 g NH_3 is provided, the percent excess NH_3 is 80%. Even if only part of the reaction actually occurs, required and excess quantities are based on the entire amount of the limiting reactant.

Other reaction terms are not as well defined, with multiple definitions in common use:

1. *Conversion* is the fraction or percentage of a reactant converted into products.
2. *Degree of completion* is usually the fraction or percentage of the limiting reactant converted into products.
3. *Selectivity* is the amount of a particular product formed as a fraction of the amount that would have been formed if all the feed material had been converted to that product.
4. *Yield* is the *ratio* of mass or moles of product formed to the mass or moles of reactant consumed. If more than one product or reactant is involved in the reaction, the particular compounds referred to must be stated; for example, the yield of glutamic acid from glucose was 0.6 g glutamic acid (g glucose)$^{-1}$. Due to the complexity of metabolism and the frequent occurrence of side reactions, yield is an important term in bioprocess analysis. Application of the yield concept for cell and enzyme reactions is described in more detail in future chapters as well as in Example 3.5.

EXAMPLE 3.5 Incomplete Reaction and Yield

Depending on culture conditions, glucose can be catabolized by yeast to produce ethanol and carbon dioxide or can be diverted into other biosynthetic reactions. An inoculum of yeast is added to a solution containing $10\,g\,L^{-1}$ glucose. After some time, only $1\,g\,L^{-1}$ glucose remains, while the concentration of ethanol is $3.2\,g\,L^{-1}$. Determine:

a. The fractional conversion of glucose to ethanol
b. The yield of ethanol from glucose

Solution

a. To find the fractional conversion of glucose to ethanol, we must first determine how much glucose was directed into ethanol biosynthesis. Using a basis of $1\,L$, we can calculate the mass of glucose required for synthesis of $3.2\,g$ ethanol. First, g ethanol is converted to gmol using the unity bracket for molecular weight:

$$3.2\,g\ ethanol = 3.2\,g\ ethanol \cdot \left|\frac{1\,gmol\ ethanol}{46\,g\ ethanol}\right|$$

$$= 0.070\,gmol\ ethanol$$

According to Eq. (3.36), for ethanol fermentation, production of 1 gmol of ethanol requires 0.5 gmol glucose. Therefore, production of 0.070 gmol ethanol requires $(0.070 \times 0.5) = 0.035$ gmol glucose. This is converted to g using the molecular weight unity bracket for glucose:

$$0.035\,gmol\ glucose = 0.035\,gmol\ glucose \cdot \left|\frac{180\,g\ glucose}{1\ gmol\ glucose}\right|$$

$$= 6.3\,g\ glucose$$

Therefore, as $6.3\,g$ glucose was used for ethanol synthesis, based on the total amount of glucose provided per liter $(10\,g)$, the fractional conversion of glucose to ethanol was 0.63. Based on the amount of glucose actually consumed per liter $(9\,g)$, the fractional conversion to ethanol was 0.70.

b. Yield of ethanol from glucose is based on the total mass of glucose consumed. Since $9\,g$ glucose was consumed per liter to provide $3.2\,g\,L^{-1}$ ethanol, the yield of ethanol from glucose was $0.36\,g$ ethanol $(g\ glucose)^{-1}$. We can also conclude that, per liter, $(9 - 6.3) = 2.7\,g$ glucose was consumed but not used for ethanol synthesis.

3.8 *Methods for Checking and Estimating Results*

In this chapter, we have considered how to quantify variables and have begun to use different types of equations to solve simple problems. Applying equations to analyze practical situations involves calculations, which are usually performed with the aid of an electronic calculator. Each time you carry out a calculation, are you always happy and confident about the result? How can you tell if you have keyed in the wrong parameter values or made an error by pressing the function buttons of your calculator? As it is relatively easy to make mistakes in calculations, it is a good idea always to review your answers and check whether they are correct.

Professional engineers and scientists develop the habit of validating the outcomes of their mathematical analyses and calculations, preferably using independent means. Several approaches for checking and estimating results are available.

1. Ask yourself whether your answer is reasonable and makes sense. In some cases, judging whether a result is reasonable will depend on your specific technical knowledge and experience of the situation being examined. For example, you may find it difficult at this stage to know whether or not 2×10^{12} is a reasonable value for the Reynolds number in a stirred bioreactor. Nevertheless, you will already be able to judge the answers from other types of calculations. For instance, if you determine using design equations that the maximum cell concentration in a fermenter is 0.002 cells per liter, or that the cooling system provides a working fermentation temperature of 160°C, you should immediately suspect that you have made a mistake.

2. Simplify the calculation and obtain a rough or *order-of-magnitude* estimate of the answer. Instead of using exact numbers, round off the values to integers or powers of 10 and continue rounding off as you progress through the arithmetic. You can verify answers quickly using this method, often without needing a calculator. If the estimated answer is of the same order of magnitude as the result found using exact parameter values, you can be reasonably sure that the exact result is free from gross error. An order-of-magnitude calculation is illustrated in Example 3.6.

The following methods for checking calculated results can also be used.

Substitute the calculated answer back into the equations for checking. For example, if Re_i (Example 3.6) was determined as 1.10×10^5, this value could be used to back-calculate one of the other parameters in the equation such as ρ:

$$\rho = \frac{Re_i \mu}{N_i D_i^2}$$

If the value of ρ obtained in this way is $1015 \, kg \, m^{-3}$, we would know that our result for Re_i was free of accidental calculator error.

3. If none of the preceding approaches can be readily applied, a final option is to check your answer by repeating the calculation from the beginning. This strategy has the disadvantage of using a less independent method of checking, so there is a greater chance that you will make the same mistakes in the checking calculation as in the original. It is much

EXAMPLE 3.6 Order-of-Magnitude Calculation

The impeller Reynolds number Re_i for fluid flow in a stirred tank is defined as:

$$Re_i = \frac{N_i D_i^2 \rho}{\mu}$$

Re_i was calculated as 1.10×10^5 for the following parameter values: $N_i = 30.6\,\text{rpm}$, $D_i = 1.15\,\text{m}$, $\rho = 1015\,\text{kg m}^{-3}$, and $\mu = 6.23 \times 10^{-3}\,\text{kg m}^{-1}\,\text{s}^{-1}$. Check this answer using order-of-magnitude estimation.

Solution

The calculation using exact parameter values

$$Re_i = \frac{\left(30.6\,\text{min}^{-1}\right) \cdot \left|\dfrac{1\,\text{min}}{60\,\text{s}}\right| \cdot \left(1.15\,\text{m}\right)^2 \left(1015\,\text{kg m}^{-3}\right)}{6.23 \times 10^{-3}\,\text{kg m}^{-1}\text{s}^{-1}}$$

can be rounded off and approximated as:

$$Re_i = \frac{\left(30\,\text{min}^{-1}\right) \cdot \left|\dfrac{1\,\text{min}}{60\,\text{s}}\right| \cdot \left(1\,\text{m}^2\right)\left(1 \times 10^3\,\text{kg m}^{-3}\right)}{6 \times 10^{-3}\,\text{kg m}^{-1}\text{s}^{-1}}$$

Combining values and canceling units gives:

$$Re_i = \frac{0.5 \times 10^6}{6} \approx \frac{0.6 \times 10^6}{6} = 0.1 \times 10^6 = 10^5$$

This calculation is simple enough to be performed without using a calculator. As the rough answer of 10^5 is close to the original result of 1.10×10^5, we can conclude that no gross error was made in the original calculation. Note that units must still be considered and converted using unity brackets in order-of-magnitude calculations.

better, however, than leaving your answer completely unverified. If possible, you should use a different order of calculator keystrokes in the repeat calculation.

Note that methods 2 through 4 only address the issue of arithmetic or calculation mistakes. Your answer will still be wrong if the equation itself contains an error or if you are applying the wrong equation to solve the problem. You should check for this type of mistake separately. If you get an unreasonable result as described in method 1 but find you have made no calculation error, the equation is likely to be the cause.

Summary of Chapter 3

Having studied the contents of Chapter 3, you should:

- Understand dimensionality and be able to convert units with ease
- Understand the terms *mole, molecular weight, density, specific gravity, temperature,* and *pressure*; know various ways of expressing the *concentration* of solutions and mixtures; and be able to work simple problems involving these concepts
- Be able to apply the ideal gas law
- Understand reaction terms such as *limiting reactant, excess reactant, conversion, degree of completion, selectivity,* and *yield,* and be able to apply stoichiometric principles to reaction problems
- Know where to find physical and chemical property data in the literature
- Be able to perform order-of-magnitude calculations to estimate results and check the answers from calculations

Problems

3.1. Unit conversion
   a. Convert 1.5×10^{-6} centipoise to $kg\ s^{-1}\ cm^{-1}$.
   b. Convert 0.122 horsepower (British) to British thermal units per minute ($Btu\ min^{-1}$).
   c. Convert 10,000 rpm to s^{-1}.
   d. Convert $4335\ W\ m^{-2}\ °C^{-1}$ to $l\ atm\ min^{-1}\ ft^{-2}\ K^{-1}$.

3.2. Unit conversion
   a. Convert $345\ Btu\ lb^{-1}$ to $kcal\ g^{-1}$.
   b. Convert $670\ mmHg\ ft^3$ to metric horsepower h.
   c. Convert $0.554\ cal\ g^{-1}\ °C^{-1}$ to $kJ\ kg^{-1}\ K^{-1}$.
   d. Convert $10^3\ g\ L^{-1}$ to $kg\ m^{-3}$.

3.3. Unit conversion
   a. Convert $10^6\ \mu g\ mL^{-1}$ to $g\ m^{-3}$.
   b. Convert 3.2 centipoise to millipascal seconds (mPa s).
   c. Convert $150\ Btu\ h^{-1}\ ft^{-2}\ (°F\ ft^{-1})^{-1}$ to $W\ m^{-1}\ K^{-1}$.
   d. Convert 66 revolutions per hour to s^{-1}.

3.4. Unit conversion and calculation
   The mixing time t_m in a stirred fermenter can be estimated using the following equation:

$$t_m = 5.9D_T^{2/3}\left(\frac{\rho V_L}{P}\right)^{1/3}\left(\frac{D_T}{D_i}\right)^{1/3}$$

   Evaluate the mixing time in seconds for a vessel of diameter $D_T = 2.3\,m$ containing liquid volume $V_L = 10,000\,L$ stirred with an impeller of diameter $D_i = 45$ in. The liquid density $\rho = 65\,\text{lb ft}^{-3}$ and the power dissipated by the impeller $P = 0.70$ metric horsepower.

3.5. Unit conversion and dimensionless numbers

Using Eq. (3.1) for the Reynolds number, calculate Re for the two sets of data in the following table.

Parameter	Case 1	Case 2
D	2 mm	1 in.
u	$3\,\text{cm s}^{-1}$	$1\,\text{m s}^{-1}$
ρ	$25\,\text{lb ft}^{-3}$	$12.5\,\text{kg m}^{-3}$
μ	10^{-6} cP	$0.14 \times 10^{-4}\,\text{lb m s}^{-1}\,\text{ft}^{-1}$

3.6. Property data

Using appropriate handbooks, find values for:

a. The viscosity of ethanol at 40°C

b. The diffusivity of oxygen in water at 25°C and 1 atm

c. The thermal conductivity of Pyrex borosilicate glass at 37°C

d. The density of acetic acid at 20°C

e. The specific heat capacity of liquid water at 80°C

Make sure you reference the source of your information and explain any assumptions you make.

3.7. Dimensionless groups and property data

The rate at which oxygen is transported from gas phase to liquid phase is a very important parameter in fermenter design. A well-known correlation for transfer of gas is:

$$Sh = 0.31 Gr^{1/3} Sc^{1/3}$$

where Sh is the Sherwood number, Gr is the Grashof number, and Sc is the Schmidt number. These dimensionless numbers are defined as follows:

$$Sh = \frac{k_L D_b}{\mathcal{D}}$$

$$Gr = \frac{D_b^3 \rho_G (\rho_L - \rho_G) g}{\mu_L^2}$$

$$Sc = \frac{\mu_L}{\rho_L \mathcal{D}}$$

where k_L is the mass transfer coefficient, D_b is bubble diameter, $\mathcal{D}$ is the diffusivity of gas in the liquid, ρ_G is the density of the gas, ρ_L is the density of the liquid, μ_L is the viscosity of the liquid, and g is gravitational acceleration. A gas sparger in a fermenter operated at 28°C and 1 atm produces bubbles of about 2 mm diameter. Calculate the value of the mass transfer coefficient, k_L. Collect property data from, for example,

Perry's Chemical Engineers' Handbook, and assume that the culture broth has properties similar to those of water. (Do you think this is a reasonable assumption?) Report
the literature source for any property data used. State explicitly any other assumptions
you make.

3.8. Dimensionless numbers and dimensional homogeneity

The Colburn equation for heat transfer is:

$$\left(\frac{h}{C_p G}\right)\left(\frac{C_p \mu}{k}\right)^{2/3} = \frac{0.023}{\left(\frac{DG}{\mu}\right)^{0.2}}$$

where C_p is heat capacity, Btu lb^{-1} °F^{-1}; μ is viscosity, lb h^{-1} ft^{-1}; k is thermal conductivity, Btu h^{-1} ft^{-2} (°F ft^{-1})$^{-1}$; D is pipe diameter, ft; and G is mass velocity per unit area,
lb h^{-1} ft^{-2}. The Colburn equation is dimensionally consistent. What are the units and
dimensions of the heat transfer coefficient, h?

3.9. Dimensional homogeneity

The terminal eddy size in a fluid in turbulent flow is given by the Kolmogorov scale:

$$\lambda = \left(\frac{\nu^3}{\varepsilon}\right)^{1/4}$$

where λ is the length scale of the eddies (μm) and ε is the local rate of energy dissipation per unit mass of fluid (W kg^{-1}). Using the principle of dimensional homogeneity,
what are the dimensions of the term represented as ν?

3.10. Dimensional homogeneity and gc

Two students have reported different versions of the dimensionless power number N_P
used to relate fluid properties to the power required for stirring:

$$N_P = \frac{P_g}{\rho N_i^3 D_i^5}$$

and

$$N_P = \frac{P g_c}{\rho N_i^3 D_i^5}$$

where P is power, g is gravitational acceleration, ρ is fluid density, N_i is stirrer speed,
D_i is stirrer diameter, and g_c is the force unity bracket. Which equation is correct?

3.11. Mass and weight

a. The density of water is 62.4 lb$m$ ft^{-3}. What is the weight of 10 ft^3 of water:
b. At sea level and 45° latitude?
c. Somewhere above the Earth's surface where $g = 9.76$ m s^{-2}?

3.12. Molar units

If a bucket holds 20.0 lb NaOH, how many:

a. lbmol NaOH

b. gmol NaOH

c. kgmol NaOH

does it contain?

3.13. Density and specific gravity

a. The specific gravity of nitric acid is $1.5129_{4°C}^{20°C}$

(i) What is its density at 20°C in kg m^{-3}?

(ii) What is its molar specific volume?

b. The volumetric flow rate of carbon tetrachloride (CCl_4) in a pipe is 50 cm^3 min^{-1}. The density of CCl_4 is 1.6 g cm^{-3}.

(i) What is the mass flow rate of CCl_4?

(ii) What is the molar flow rate of CCl_4?

3.14. Molecular weight

Calculate the average molecular weight of air.

3.15. Mole fraction

A solution contains 30 wt% water, 25 wt% ethanol, 15 wt% methanol, 12 wt% glycerol, 10 wt% acetic acid, and 8 wt% benzaldehyde. What is the mole fraction of each component?

3.16. Solution preparation

You are asked to make up a 6% w/v solution of $MgSO_4 \cdot 7H_2O$ in water and are provided with $MgSO_4 \cdot 7H_2O$ crystals, water, a 500-mL beaker, a 250-mL measuring cylinder with 2-mL graduations, a balance that shows weight up to 100 g in increments of 0.1 g, and a stirring rod. Explain how you would prepare the solution in any quantity you wish, carefully explaining each step in the process.

3.17. Moles, molarity, and composition

a. How many gmol are there in 21.2 kg of isobutyl succinate ($C_{12}H_{22}O_4$)?

b. If sucrose ($C_{12}H_{22}O_{11}$) crystals flow into a hopper at a rate of 4.5 kg s^{-1}, how many gram-moles of sucrose are transferred in 30 min?

c. A solution of 75 mM tartaric acid ($C_4H_6O_6$) in water is used as a standard in HPLC analysis. How many grams of $C_4H_6O_6 \cdot H_2O$ are needed to make up 10 mL of solution?

d. An aqueous solution contains 60 µM salicylaldehyde ($C_7H_6O_2$) and 330 ppm dichloroacetic acid ($C_2H_2Cl_2O_2$). How many grams of each component are present in 250 mL?

3.18. Concentration

A holding tank with a capacity of 5000 L initially contains 1500 L of 25 mM NaCl solution.

a. What is the final concentration of NaCl if an additional 3000 L of 25 mM NaCl solution is pumped into the tank?

b. Instead of (a), if 3000 L of water is added to the tank, what is the final concentration of NaCl?

c. Instead of (a) or (b), if an additional 500 L of 25 mM NaCl solution plus 3000 L of water are added, what is the final concentration of NaCl:

    (i) Expressed as molarity?
    (ii) Expressed as % w/v?
    (iii) Expressed as g cm^{-3}?
    (iv) Expressed as ppm? Justify your answer.

3.19. Gas composition

A gas mixture containing 30% O_2, 5% carbon dioxide, 2% ammonia, and 63% N_2 is stored at 35 psia and room temperature (25°C) prior to injection into the headspace of an industrial-scale bioreactor used to culture normal fetal lung fibroblast cells. If the reactor is operated at 37°C and atmospheric pressure, explain how the change in temperature and pressure of the gas as it enters the vessel will affect its composition.

3.20. Specific gravity and composition

Broth harvested from a fermenter is treated for recovery and purification of a pharmaceutical compound. After filtration of the broth and partial evaporation of the aqueous filtrate, a solution containing 38.6% (w/w) pharmaceutical leaves the evaporator at a flow rate of 8.6 L min^{-1}. The specific gravity of the solution is 1.036. If the molecular weight of the drug is 1421, calculate:

a. The concentration of pharmaceutical in the solution in units of kg L^{-1}

b. The flow rate of pharmaceutical in gmol min^{-1}

3.21. Temperature scales

What is −40 °F in degrees centigrade? Degrees Rankine? Kelvin?

3.22. Pressure scales

a. The pressure gauge on an autoclave reads 15 psi. What is the absolute pressure in the chamber in psi? In atm?

b. A vacuum gauge reads 3 psi. What is the absolute pressure?

3.23. Gas leak

A steel cylinder containing compressed air is stored in a fermentation laboratory ready to provide aeration gas to a small-scale bioreactor. The capacity of the cylinder is 48 L, the absolute pressure is 0.35 MPa and the temperature is 22°C. One day in midsummer when the air conditioning breaks down, the temperature in the laboratory rises to 33°C and the valve at the top of the cylinder is accidentally left open. Estimate the proportion of air that will be lost. What assumptions will you make?

3.24. Gas supply

A small airlift bioreactor is used to culture suspended *Solanum aviculare* (kangaroo apple) plant cells. The reactor contains 1.5 L of culture broth and is operated at 25°C and atmospheric pressure. The air flow rate under these conditions is 0.8 vvm (1 vvm means 1 volume of gas per volume of liquid per minute). Air is supplied from a 48-L gas cylinder at 20°C. If, at 4 PM on Friday afternoon, the gauge pressure is 800 psi, is there enough air in the cylinder to operate the reactor over the weekend until 9 AM on Monday morning? At ambient temperature and for pressures up to 100 atm, the ideal gas law can be used to estimate the amount of air present to within 5%.

3.25. Stoichiometry and incomplete reaction

For production of penicillin ($C_{16}H_{18}O_4N_2S$) using *Penicillium* mold, glucose ($C_6H_{12}O_6$) is used as substrate and phenylacetic acid ($C_8H_8O_2$) is added as precursor. The stoichiometry for overall synthesis is:

$$1.67\,C_6H_{12}O_6 + 2\,NH_3 + 0.5\,O_2 + H_2SO_4 + C_8H_8O_2 \rightarrow C_{16}H_{18}O_4N_2S + 2\,CO_2 + 9\,H_2O$$

a. What is the maximum theoretical yield of penicillin from glucose?

b. When results from a particular penicillin fermentation were analyzed, it was found that 24% of the glucose had been used for growth, 70% for cell maintenance activities (such as membrane transport and macromolecule turnover), and only 6% for penicillin synthesis. Calculate the yield of penicillin from glucose under these conditions.

c. Batch fermentation under the conditions described in (b) is carried out in a 100-L tank. Initially, the tank is filled with nutrient medium containing 50 g L^{-1} glucose and 4 g L^{-1} phenylacetic acid. If the reaction is stopped when the glucose concentration is 5.5 g L^{-1}, determine:

   (i) Which is the limiting substrate if NH_3, O_2, and H_2SO_4 are provided in excess

   (ii) The total mass of glucose used for growth

   (iii) The amount of penicillin produced

   (iv) The final concentration of phenylacetic acid

3.26. Stoichiometry, yield, and the ideal gas law

Stoichiometric equations can be used to represent the growth of microorganisms provided a "molecular formula" for the cells is available. The molecular formula for biomass is obtained by measuring the amounts of C, N, H, O, and other elements in cells. For a particular bacterial strain, the molecular formula was determined to be $C_{4.4}H_{7.3}O_{1.2}N_{0.86}$. These bacteria are grown under aerobic conditions with hexadecane ($C_{16}H_{34}$) as substrate. The reaction equation describing growth is:

$$C_{16}H_{34} + 16.28\,O_2 + 1.42\,NH_3 \rightarrow 1.65\,C_{4.4}H_{7.3}O_{1.2}N_{0.86} + 8.74\,CO_2 + 13.11\,H_2O$$

   a. Is the stoichiometric equation balanced?

   b. Assuming 100% conversion, what is the yield of cells from hexadecane in g g^{-1}?

   c. Assuming 100% conversion, what is the yield of cells from oxygen in g g^{-1}?

   d. You have been put in charge of a small batch fermenter for growing the bacteria and aim to produce 2.5 kg of cells for inoculation of a pilot-scale reactor.

      (i) What minimum amount of hexadecane substrate must be contained in your culture medium?

      (ii) What must be the minimum concentration of hexadecane in the medium if the fermenter working volume is 3 m³?

      (iii) What minimum volume of air at 20°C and 1 atm pressure must be pumped into the fermenter during growth to produce the required amount of cells?

3.27. Stoichiometry and the ideal gas law

Roots of the *Begonia rex* plant are cultivated in an air-driven bioreactor in medium containing glucose ($C_6H_{12}O_6$) and two nitrogen sources, ammonia (NH_3) and nitrate (HNO_3). The root biomass can be represented stoichiometrically using the formula $CH_{1.63}O_{0.80}N_{0.13}$. A simplified reaction equation for growth of the roots is:

$$C_6H_{12}O_6 + 3.4O_2 + 0.15NH_3 + 0.18HNO_3 \rightarrow 2.5CH_{1.63}O_{0.80}N_{0.13} + 3.5CO_2 + 4.3H_2O$$

   a. Is the stoichiometric equation balanced?

   b. If the medium contains 30 g L^{-1} glucose, what minimum concentration of nitrate is required to achieve complete conversion of the sugar? Express your answer in units of gmol L^{-1} or M.

   c. If the bioreactor holds 50 L of medium and there is complete conversion of the glucose, what mass of roots will be generated?

   d. For the conditions described in (c), what minimum volume of air at 20°C and 1 atm pressure must be provided to the bioreactor during growth?

3.28. Stoichiometry, yield, and limiting substrate

Under anoxic conditions, biological denitrification of wastewater by activated sludge results in the conversion of nitrate to nitrogen gas. When acetate provides the carbon source, the reaction can be represented as follows:

$$5CH_3COOH + 8NO_3^- \rightarrow 4N_2 + 10CO_2 + 6H_2O + 8OH^-$$

   a. Is the stoichiometric equation balanced?

   b. In the absence of side reactions, what is the yield of nitrogen from acetate in g g^{-1}?

   c. A certain wastewater contains 6.0 mM acetic acid and 7 mM NaNO$_3$. If 25% of the acetate and 15% of the nitrate are consumed in other reactions (e.g., for growth of organisms in the sludge), which is the limiting substrate in the denitrification reaction?

   d. For the situation described in (c), what mass of gaseous nitrogen is produced from treatment of 5000 L of wastewater if the reaction is allowed to proceed until the limiting substrate is exhausted?

3.29. Order-of-magnitude calculation

The value of the sigma factor for a tubular-bowl centrifuge is approximated by the equation:

$$\Sigma = \frac{\pi \omega^2 b r^2}{2g}$$

An estimate of Σ corresponding to the following parameter values is required: $\omega = 12{,}000\,\text{rpm}$, $b = 1.25\,\text{m}$, $r = 0.37\,\text{m}$, and $g = 9.8066\,\text{m\,s}^{-2}$. Using an order-of-magnitude calculation, determine whether the value of Σ is closer to $100\,\text{m}^2$ or $1000\,\text{m}^2$.

3.30. Order-of-magnitude calculation

Two work-experience students at your startup company have been asked to evaluate the rate of reaction occurring in a transparent gel particle containing immobilized mouse melanoma cells. The equation for the reaction rate r^*_{As} is:

$$r^*_{As} = \frac{4}{3}\pi R^3 \left(\frac{v_{max} C_{As}}{K_m + C_{As}} \right)$$

where $R = 3.2\,\text{mm}$, $v_{max} = 0.12\,\text{gmol s}^{-1}\,\text{m}^{-3}$, $C_{As} = 41\,\text{gmol m}^{-3}$, and $K_m = 0.8\,\text{gmol m}^{-3}$. One student reports a reaction rate of $1.6 \times 10^{-8}\,\text{gmol s}^{-1}$; the other reports $1.6 \times 10^{-10}\,\text{gmol s}^{-1}$. You left your calculator on the bus this morning but must know quickly which student is correct. Use an order-of-magnitude calculation to identify the right answer.

References

[1] The International System of Units (SI), National Bureau of Standards Special Publication 330, US Government Printing Office, 1977. www.nist.gov/pml/owm/metric-si/si-units.

[2] National Research Council. International Critical Tables of Numerical Data, Physics, Chemistry and Technology. The National Academies Press, Washington, DC, 1930. https://doi.org/10.17226/20230.

[3] D.R. Lide, CRC Handbook of Chemistry and Physics, eighty-fourth ed., CRC Press, New York, 2003.

[4] J.G. Speight, Lange's Handbook of Chemistry, Sixteenth Edition (McGraw-Hill Education: New York, Chicago, San Francisco, Lisbon, London, Madrid, Mexico City, Milan, New Delhi, San Juan, Seoul, Singapore, Sydney, Toronto, 2005). https://www.accessengineeringlibrary.com/content/book/9781259586095.

[5] D.W. Green, R.H. Perry. Perry's Chemical Engineers' Handbook, Eighth Edition (McGraw-Hill: New York, Chicago, San Francisco, Lisbon, London, Madrid, Mexico City, Milan, New Delhi, San Juan, Seoul, Singapore, Sydney, Toronto, 2008, 1997, 1984, 1973, 1963, 1950, 1941, 1934). https://www.accessengineeringlibrary.com/content/book/9780071422949.

[6] B. Atkinson, F. Mavituna, Biochemical Engineering and Biotechnology Handbook, second ed., Macmillan, 1991.

Suggestions for Further Reading

Units and Dimensions

F. Cardarelli, Scientific Unit Conversion, Springer, 1997.

B.S. Massey, Measures in Science and Engineering, Ellis Horwood, 1986.

S.H. Qasim, SI Units in Engineering and Technology, Pergamon Press, 1977.

C. Wandmacher, A.I. Johnson, Metric Units in Engineering: Going SI, American Society of Civil Engineers, 1995.

Wildi, 1991 Wildi, T. (1991). *Units and Conversion Charts*. Institute of Electrical and Electronics Engineers.

Engineering Variables

Felder, R.M., Rousseau, R.W., 2005. Elementary Principles of Chemical Processes. 3rd ed. 2005, Wiley, Chapters 2 and 3.

Himmelblau, D.M., Riggs, J.B., 2004. Basic Principles and Calculations in Chemical Engineering. 7th ed., Chapters 1–5. Prentice Hall.

PART 2

Fundamental Concepts

CHAPTER 4

Mass Balances

Mass balances are fundamental to bioprocess engineering. The term *balance* implies that the masses entering and leaving a system must be equal for steady state processes. Mass balances are accounting procedures: the total mass entering must be accounted for at the end of the process, even if it undergoes heating, mixing, drying, cell growth, or any other operation except nuclear reactions. Sometimes it is not feasible to measure all masses or compositions of streams entering or leaving a system; unknown quantities can be calculated using the fundamental concept of mass balances. Mass balance problems have a constant theme: given the masses of some input and output streams, calculate the masses of the others.

Mass balances provide a very powerful and practical tool in engineering analyses. Many complex situations can be simplified by quantifying the movement of mass from reactants to products. Questions such as: What fraction of the substrate consumed is converted into salable products and what fraction is lost as carbon dioxide? What is the concentration of carbon dioxide in the bioreactor off-gas? How much reactant is needed to produce the target quantity of product? How much O_2 must be provided for the process to stay aerobic? Can be answered using mass balances. This chapter explains how the law of conservation of mass is applied to atoms, molecular species, and total mass and sets up formal techniques for solving mass balance problems with and without reaction. Aspects of metabolic stoichiometry are also discussed for calculation of nutrient and O_2 requirements during bioprocesses.

4.1 Thermodynamic Preliminaries

Thermodynamics is a fundamental branch of science focused on the properties of matter and energy. Thermodynamic principles are useful in setting up mass balances; some terms borrowed from thermodynamics are defined in the following sections.

4.1.1 System and Process

In thermodynamics, a *system* consists of any matter identified for investigation and is separated from the *surroundings*, which are the remainder of the *universe*, by a *system boundary* (Figure 4.1). The system boundary may be real and tangible, such as the walls of a bioreactor, or it may be

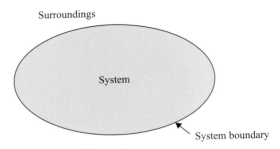

Figure 4.1 Thermodynamic system.

virtual or notional construct employed by the engineer to simplify calculations. If the system boundary does not permit mass to transit, the system is classified as a *closed* system with constant mass. Conversely, a system able to exchange mass with its surroundings is an *open* system.

A *process* causes changes in the system or surroundings. Several terms are commonly used to describe processes.

1. A *batch process* operates in a closed system. All materials are added to the system at the start of the process; the system is then closed, and the products removed only when the process is complete.
2. A *semi-batch process* allows either input or output of mass.
3. A *fed-batch process* allows input of material to the system but not output.
4. A *continuous process* allows matter to flow in and out of the system.

4.1.2 Steady State and Equilibrium

A process is said to be at *steady state* if all properties of a system, such as temperature, pressure, concentration, volume, mass, and so on, do not vary with time. Thus, if we monitor any variable of a steady-state system, its value will be unchanging with time.

Batch, fed-batch, and semi-batch processes are inherently *transient* or *unsteady-state*; system properties such as composition or total mass change with time. On the other hand, continuous processes may be either steady-state or transient. It is usual to run continuous processes as close to steady state as possible; however, unsteady-state conditions will exist during start-up and for some time after any change in operating conditions.

Steady state is a central theoretical concept in engineering analysis. It is not to be confused with another term, *equilibrium*, which is a thermodynamic concept. A system at equilibrium has no potential energy driving force for change. The system energy is minimized, and all opposing forces are exactly counter-balanced so that the properties of the system do not change with time.

There must be an overall change in the system to convert reactants into useful products. Therefore, equilibrium needs to be disturbed to maintain productivity. Continuously disturbing a system by adding raw materials and removing products will ensure an energetic driving force. Continuous, steady state processes constantly exchange mass with the surroundings; this disturbance drives the system away from equilibrium so that a net change in both the system and the universe can occur. Large-scale equilibrium does not often occur in engineering systems; steady states are more common.

4.2 Law of Conservation of Mass

Mass is conserved in ordinary chemical and physical processes. Consider the system of Figure 4.2, in which streams containing glucose enter and leave. The mass of glucose entering the system is M_i kg; the mass of glucose leaving is M_o kg. If M_i and M_o are different, there are four possible explanations:

1. The measurements of M_i and/or M_o are wrong.
2. The system has a leak allowing glucose to enter or escape undetected.
3. Glucose is consumed or generated by chemical reaction within the system.
4. Glucose accumulates within the system.

If we assume that the measurements are correct and there are no leaks, the difference between M_i and M_o must be due to consumption or generation of glucose by reaction and/or accumulation of glucose within the system. A mass balance for the system can be written in a general way to account for these possibilities:

$$
\begin{Bmatrix} \text{mass in} \\ \text{through} \\ \text{system} \\ \text{boundaries} \end{Bmatrix} - \begin{Bmatrix} \text{mass out} \\ \text{through} \\ \text{system} \\ \text{boundaries} \end{Bmatrix} + \begin{Bmatrix} \text{mass} \\ \text{generated} \\ \text{within} \\ \text{system} \end{Bmatrix} - \begin{Bmatrix} \text{mass} \\ \text{consumed} \\ \text{within} \\ \text{system} \end{Bmatrix} = \begin{Bmatrix} \text{mass} \\ \text{accumulated} \\ \text{within} \\ \text{system} \end{Bmatrix}
\tag{4.1}
$$

The accumulation term in the preceding equation can be either positive or negative; negative accumulation represents depletion of preexisting reserves. Equation (4.1) is known as the *general mass balance equation*. The mass referred to in the equation can be total mass, mass of a

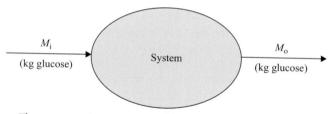

Figure 4.2 Flow sheet for a mass balance on glucose.

specific molecular or atomic species, or mass of a particular combination of compounds such as cells. Use of Eq. (4.1) is illustrated in Example 4.1.

4.2.1 Types of Mass Balance

The general mass balance Eq. (4.1) can be applied with equal utility to two different types of mass balance problems. For continuous processes, it is usual to collect information about the system referring to a particular instant in time. The amounts of mass entering and leaving are specified using flow rates: for example, molasses enters the system at a rate of $50 \, kg \, h^{-1}$; at the same instant in time, culture broth leaves at a rate of $20 \, kg \, h^{-1}$. These two quantities can be used directly in Eq. (4.1) as the input and output terms. A mass balance based on rates is called a *differential balance*.

An alternative approach is required for batch and semi-batch processes. Information about these systems is usually collected over a period of time rather than at a particular instant. For example, 100 kg substrate is added to the reactor; after 3 days of incubation, 45 kg product is recovered. Each term of the mass balance equation in this case is a quantity of mass, not a rate. This type of balance is called an *integral balance*.

In this chapter, we will be using differential balances for continuous processes operating at steady state, and integral balances for batch or semi-batch processes between their initial and final states. Calculation procedures for the two types of material balance are very similar.

EXAMPLE 4.1 General Mass Balance Equation

A continuous process is set up for treatment of wastewater. Each day, $10^5 \, kg$ cellulose and $10^3 \, kg$ bacteria enter in the feed stream, while $10^4 \, kg$ cellulose and $1.5 \times 10^4 \, kg$ bacteria leave in the effluent. The rate of cellulose digestion by the bacteria is $7 \times 10^4 \, kg \, day^{-1}$. The rate of bacterial growth is $2 \times 10^4 \, kg \, day^{-1}$; the rate of cell death by lysis is $5 \times 10^2 \, kg \, day^{-1}$. Write balances for cellulose and bacteria in the system.

Solution
Cellulose is not generated by the process, only consumed. Using a basis of 1 day, the cellulose balance in kg from Eq. (4.1) is:

$$\left(10^5 - 10^4 + 0 - 7 \times 10^4\right) = 2 \times 10^4 = \text{ accumulation}$$

Therefore, $2 \times 10^4 \, kg$ of cellulose accumulates in the system each day.
Performing the same balance for bacteria:

$$\left(10^3 - 1.5 \times 10^4 + 2 \times 10^4 - 5 \times 10^2\right) = 5.5 \times 10^3 = \text{ accumulation}$$

Therefore, $5.5 \times 10^3 \, kg$ of bacterial cells accumulate in the system each day.

4.2.2 Simplification of the General Mass Balance Equation

Equation (4.1) can be simplified in certain situations. If a continuous process is at steady state, the accumulation term on the right side of the equation must be zero. This follows from the definition of steady state: all properties of the system, including its mass, must be unchanging with time. A system at steady state cannot accumulate mass therefore Eq. (4.1) becomes:

$$\text{mass in} + \text{mass generated} = \text{mass out} + \text{mass consumed} \qquad (4.2)$$

Equation (4.2) is called the *general steady-state mass balance equation*. Equation (4.2) also applies over the entire duration of batch and fed-batch processes; "mass out" in this case is the total mass harvested from the system so that at the end of the process there is no accumulation.

If reaction does not occur in the system, or if the mass balance is applied to a substance that is neither a reactant nor a product of reaction, the generation and consumption terms in Eqs. (4.1) and (4.2) are zero. The generation and consumption terms must also be zero in balances applied to total mass because total mass can be neither created nor destroyed except by nuclear reaction. Similarly, generation and consumption of atomic species such as carbon, nitrogen, oxygen, and so on, cannot occur in normal chemical reactions. Therefore, at steady state, for balances on total mass or atomic species or when reaction does not occur, Eq. (4.2) can be further simplified to:

$$\text{mass in} = \text{mass out} \qquad (4.3)$$

Table 4.1 summarizes the types of material balance for which Eq. (4.3) is valid.

4.3 Procedure for Mass Balance Calculations

The first step in mass balance calculations is to understand the problem. Certain information is available about a process; the task is to calculate unknown quantities. Because it is sometimes difficult to sort through all the details provided, it is best to use standard procedures to translate process information into a form that can be used in calculations.

Table 4.1: Application of the Simplified Mass Balance, Eq. (4.3)

	At steady state, does mass in = mass out?	
Material	**Without reaction**	**With reaction**
Total mass	Yes	Yes
Total number of moles	Yes	No
Mass of a molecular species	Yes	No
Number of moles of a molecular species	Yes	No
Mass of an atomic species	Yes	Yes
Number of moles of an atomic species	Yes	Yes

Mass balances should be performed in an organized manner; this makes the solution easier to follow, check, and be used by others. In this chapter, a formalized series of steps is followed for each mass balance problem. These procedures may seem long-winded and unnecessary for easier problems; however, a standard method is helpful when you are first learning mass balance techniques. The same procedures are used in the next chapter for energy balances.

The following points are essential.

- *Draw a clear process flow diagram labeling all relevant information.* A simple box diagram showing all streams entering or leaving the system allows information about a process to be organized and summarized in a clear manner. All given quantitative information should be recorded on the diagram. Note that the variables of interest in mass balances are masses, mass flow rates, and mass compositions. If information about particular streams is given using volume or molar quantities, mass flow rates, and compositions should be calculated before labeling the flow sheet.
- *Select a set of units and state it clearly.* Calculations are easier when all quantities are expressed using consistent units. Units must also be indicated for all variables shown on process diagrams.
- *Select a basis for the calculation and state it clearly.* It is helpful to focus on a specific quantity of material entering or leaving the system when approaching mass balance problems. For continuous processes at steady state, we usually base the calculation on the amount of mass entering or leaving the system within a specified period of time. For batch or semi-batch processes, it is convenient to use either the total amount of mass fed to the system or the amount withdrawn at the end. Selection of a basis for calculation makes it easier to visualize the problem; the utility of the approach will be apparent in the worked examples of the next section.
- *State all assumptions.* To solve problems in this and the following chapters, you will need to apply some "engineering" judgment. Real-life situations are complex, and there will be times when one or more assumptions are required before you can proceed with calculations. Problems posed in this text may not give you all the necessary information to give you experience making assumptions. The details omitted can be assumed, provided your assumptions are reasonable. Engineers make assumptions all the time; knowing when an assumption is permissible and what constitutes a reasonable assumption is one of the marks of a skilled engineer. It is vitally important that you state the assumptions explicitly. Other scientists looking through your calculations need to know the conditions under which your results are applicable; they will also want to decide whether your assumptions are acceptable or whether they should be modified. Another assumption we must make in mass balance problems is that the system under investigation does not leak. When analyzing real systems, it is always a good idea to check for leaks before carrying out mass balances.

- *Identify which components of the system, if any, are involved in reaction.* This is necessary for determining which mass balance equation, (4.2) or (4.3), is appropriate. The simpler Eq. (4.3) can be applied to molecular species that are neither reactants nor products of reaction.

Example 4.2 demonstrates the procedure.

EXAMPLE 4.2 Setting up a Flow Sheet

Humid air enriched with O_2 is prepared for a gluconic acid bioprocess. The air is prepared in a special humidifying chamber. Liquid water enters the chamber at a rate of 1.5 L h^{-1} at the same time as dry air and 15 gmol min^{-1} of dry O_2 gas. All the water is evaporated. The outflowing gas is found to contain 1% (w/w) water. Draw and label the flow sheet for this process.

Solution

Let us choose units of g and min for this process; the information provided is first converted to mass flow rates in these units. The density of water is taken to be 10^3 g L^{-1}; therefore:

$$1.5\,\text{l h}^{-1} = \frac{1.5\,\text{l}}{\text{h}} \times \frac{10^3\,\text{g}}{1} \cdot \left| \frac{1\,\text{h}}{60\,\text{min}} \right| = 25\ \text{g min}^{-1}$$

As the molecular weight of O_2 is 32:

$$15\ \text{gmol min}^{-1} = \frac{15\ \text{gmol}}{\text{min}} \cdot \left| \frac{32\,\text{g}}{1\,\text{gmol}} \right| = 480\ \text{g min}^{-1}$$

Unknown flow rates are represented with symbols. As shown in Figure 4.3, the flow rate of dry air is denoted D g min^{-1} and the flow rate of humid, O_2-rich air is H g min^{-1}. The water content in the humid air is shown as 1 mass%.

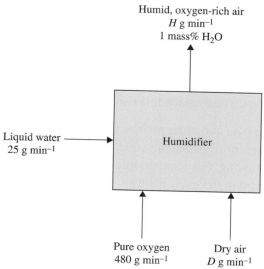

Figure 4.3 Flow sheet for O_2 enrichment and humidification of air.

4.4 Mass Balance Worked Examples

Procedures for performing mass balance calculations are outlined in this section. The methods shown will assist your problem-solving efforts by formalizing the mathematical approach. Mass balance calculations are divided into four steps: *assemble, analyze, calculate*, and *finalize*. Differential and integral mass balances with and without reaction are illustrated.

Note in Example 4.3 shown below, that the complete composition of the bioreactor slurry are not provided. Cell and kanamycin concentrations are given; however, the slurry most probably contains a variety of other components such as residual carbohydrates, minerals, vitamins, and amino acids, as well as additional cell growth products. These components are ignored in the mass balance; the liquid phase of the slurry is considered to be water only. This assumption is reasonable as the concentration of dissolved substances in culture broths is typically very small; water in spent broth usually accounts for more than 90% of the liquid phase.

Note also that the masses of some of the components in Example 4.3 were different by several orders of magnitude; for example, the mass of kanamycin in the filtrate was of the order 10^{-2} kg whereas the total mass of this stream was of the order 10^2 kg. Calculation of the mass of water by difference therefore involved subtracting a very small number from a large one and carrying more significant figures than warranted. This is an unavoidable feature of most mass balances for biological processes, which are characterized by dilute solutions, low product concentrations, and large amounts of water. However, although excess significant figures were carried in the mass balance table, the final answers were reported with due regard to data accuracy and appropriate significant figures.

Example 4.3 illustrated mass balance procedures for a simple steady-state process without reaction. An integral mass balance for a batch system without reaction is outlined in Example 4.4.

Mass balances on reactive systems are slightly more complicated than Examples 4.3 and 4.4. To solve problems with reaction, stoichiometric relationships must be used in conjunction with mass balance equations. These procedures are illustrated in Examples 4.5 and 4.6.

There are several points to note about the problem and calculation of Example 4.5. First, cell growth and its requirement for substrate were not considered because the cells used in this process were nongrowing. For many bioprocesses, cell growth and other metabolic activities must be considered in the mass balance. Use of nongrowing immobilized cells in Example 4.5 meant that the cells were not components of any stream flowing into or out of the process, nor were they generated in reaction. Therefore, cell mass did not have to be included in the calculation.

EXAMPLE 4.3 Continuous Filtration

A bioreactor slurry containing *Streptomyces kanamyceticus* cells is filtered using a continuous rotary vacuum filter. Slurry is fed to the filter at a rate of 120 kg h^{-1}; 1 kg slurry contains 60 g cell solids. To improve filtration rates, particles of diatomaceous earth filter aid are added at a rate of 10 kg h^{-1}. The concentration of kanamycin in the slurry is 0.05% by weight. Liquid filtrate is collected at a rate of 112 kg h^{-1}; the concentration of kanamycin in the filtrate is 0.045% (w/w). Filter cake containing cells and filter aid is removed continuously from the filter cloth.

(a) What percentage of water is the filter cake?
(b) If the concentration of kanamycin dissolved in the liquid within the filter cake is the same as that in the filtrate, how much kanamycin is absorbed per kg filter aid?

Solution

1. Assemble
   (i) Draw the flow sheet showing all data with units. This is shown in Figure 4.4.
   (ii) Define the system boundary by drawing on the flow sheet. The system boundary is shown in Figure 4.4.
2. Analyze
   (i) State any assumptions.
      —process is operating at steady state
      —system does not leak
      —filtrate contains no solids
      —cells do not absorb or release kanamycin during filtration
      —filter aid added is dry
      —the liquid phase of the slurry, excluding kanamycin, can be considered water
   (ii) Collect and state any extra data needed. No extra data are required.
   (iii) Select and state a basis. The calculation is based on 120 kg slurry entering the filter, or 1 hour.
   (iv) List the compounds, if any, that are involved in reaction. No compounds are involved in reaction.
   (v) Write down the appropriate general mass balance equation. The system is at steady state and no reaction occurs; therefore, Eq. (4.3) is appropriate:

$$\text{mass in} = \text{mass out}$$

3. Calculate
   (i) Set up a calculation table showing all components of all streams passing across the system boundaries. State the units used for the table. Enter all known quantities as shown in Figure 4.4, four streams cross the system boundaries: culture slurry, filter aid, filtrate, and filter cake. The components of these streams—cells, kanamycin, filter aid, and water—are represented in Table 4.2. The table is divided into two major sections: In and Out. Masses entering or leaving the system each hour are shown in the table; the units used are kg. Because filtrate and filter cake flow out of the system, there are no entries for these streams on the In side of the table. Conversely, there are no entries for the culture slurry and filter aid streams on the Out side of the table. The total mass of each stream is given in the last column on each side of the table. The total amount of each

component flowing in and out of the system are shown in the last row. With all known quantities entered, several masses remain unknown; these quantities are indicated by question marks.

(ii) Calculate unknown quantities; apply the mass balance equation. To complete Table 4.2, let us consider each row and column separately. In the row representing the culture slurry, the total mass of the stream is 120 kg and the masses of each component except water are known. The entry for water can therefore be determined as the difference between 120 kg and the sum of the known components: $(120 - 7.2 - 0.06 - 0)$ kg = 112.74 kg. This mass for water has been entered in Table 4.3. The row for the filter aid stream is already complete in Table 4.2: no cells or kanamycin are present in the diatomaceous earth entering the system; we have also assumed that the filter aid is dry. We can now fill in the final row of the In side of the table; numbers in this row are obtained by adding the values in each column. The total mass of cells input to the system in all streams is 7.2 kg, the total kanamycin entering is 0.06 kg, and so on. The total mass of all components fed into the system is the sum of the last column of the In side: $(120 + 10)$ kg = 130 kg. On the Out side, we can complete the row for filtrate. We have assumed there are no solids such as cells or filter aid in the filtrate; therefore, the mass of water in the filtrate is $(112 - 0.05)$ kg = 111.95. As yet, the entire composition and mass of the filter cake remain unknown. To complete the table, we must consider the mass balance equation relevant to this problem, Eq. (4.3). In the absence of reaction, this equation can be applied to total mass and to the masses of each component of the system.

Total mass balance

$$130 \text{ kg total mass in} = \text{total mass out}$$
$$\therefore \text{ total mass out} = 130 \text{ kg}$$

Cell balance

$$7.2 \text{ kg cells in} = \text{cells out}$$
$$\therefore \text{ cells out} = 7.2 \text{ kg}$$

Kanamycin balance

$$0.06 \text{ kg kanamycin in} = \text{kanamycin out}$$
$$\therefore \text{ kanamycin out} = 0.06 \text{ kg}$$

Filter aid balance

$$10 \text{ kg filter aid in} = \text{filter aid out}$$
$$\therefore \text{ filter aid out} = 10 \text{ kg}$$

Water balance

$$112.74 \text{ kg water in} = \text{water out}$$
$$\therefore \text{ water out} = 112.74 \text{ kg}$$

These results are entered in the last row of the Out side of Table 4.3. In the absence of reaction, this row is always identical to the final row of the In side. The component masses for the filter cake can now be filled in as the difference between numbers in the final row and the masses of each component in the filtrate. Take time to look over Table 4.3; you should understand how all the numbers shown were obtained.

(iii) Check that your results are reasonable and make sense. Mass balance calculations must be checked. Make sure that all columns and rows of Table 4.3 add up to the totals shown.

4. Finalize

(i). Answer the specific questions asked in the problem. The percentage of water in the filter cake can be calculated from the results in Table 4.3. Dividing the mass of water in the filter cake by the total mass of this stream, the percentage of water is:

$$\frac{0.79 \text{ kg}}{18 \text{ kg}} \times 100 = 4.39\%$$

(ii). Kanamycin is dissolved in the water to form the filtrate and the liquid phase is retained within the filter cake. If the concentration of kanamycin in the liquid phase is 0.045% (w/w), the mass of kanamycin in the filter-cake liquid is:

$$\frac{0.045}{100} \times (0.79 + 0.01) \text{ kg} = 3.6 \times 10^{-4} \text{ kg}$$

(a) However, we know from Table 4.3 that a total of 0.01 kg kanamycin is contained in the filter cake; therefore $(0.01 - 3.6 \times 10^{-4})$ kg $= 0.00964$ kg kanamycin is so far unaccounted for. Following our assumption that kanamycin is not adsorbed by the cells, 0.00964 kg kanamycin must be retained by the filter aid within the filter cake. As 10 kg filter aid is present, the kanamycin absorbed per kg filter aid is:

$$\frac{0.00964 \text{ kg kanamycin}}{10 \text{ kg filter aid}} = 9.64 \times 10^{-4} \text{ kg kanamycin (kg filter aid)}^{-1}$$

(iii). State the answers clearly and unambiguously, checking significant figures.
(a) The water content of the filter cake is 4.4%.
(b) The amount of kanamycin absorbed by the filter aid is 9.6×10^{-4} kg kanamycin (kg^{-1} filter aid).

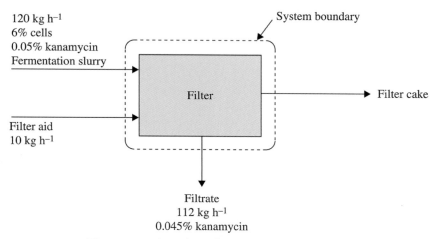

120 kg h^{-1}
6% cells
0.05% kanamycin
Fermentation slurry

System boundary

Filter

Filter cake

Filter aid
10 kg h^{-1}

Filtrate
112 kg h^{-1}
0.045% kanamycin

Figure 4.4 Flow sheet for continuous filtration.

Table 4.2: Mass Balance Table (kg)

Stream	In					Out				
	Cells	Kanamycin	Filter aid	Water	Total	Cells	Kanamycin	Filter aid	Water	Total
Fermentation slurry	7.2	0.06	0	?	120	–	–	–	–	–
Filter aid	0	0	10	0	10	–	–	–	–	–
Filtrate	–	–	–	–	–	0	0.05	0	?	112
Filter cake	–	–	–	–	–	?	?	?	?	?
Total	?	?	?	?	?	?	?	?	?	?

Table 4.3: Completed Mass Balance Table (kg)

Stream	In					Out				
	Cells	Kanamycin	Filter aid	Water	Total	Cells	Kanamycin	Filter aid	Water	Total
Bioreactor slurry	7.2	0.06	0	112.74	120	–	–	–	–	–
Filter aid	0	0	10	0	10	–	–	–	–	–
Filtrate	–	–	–	–	–	0	0.05	0	111.95	112
Filter cake	–	–	–	–	–	7.2	0.01	10	0.79	18
Total	7.2	0.06	10	112.74	130	7.2	0.06	10	112.74	130

EXAMPLE 4.4 Batch Mixing

Corn-steep liquor contains 2.5% invert sugars and 50% water on a mass basis; the rest can be considered solids. Beet molasses contains 50% sucrose, 1% invert sugars, and 18% water on a mass basis; the remainder is solids. A mixing tank contains 125 kg corn-steep liquor and 45 kg molasses; water is then added to produce a diluted sugar mixture containing 2% (w/w) invert sugars.

(a) How much water is required?
(b) What is the concentration of sucrose in the final mixture?

Solution

1. Assemble
   (i) Flow sheet
      The flow sheet for this batch process is shown in Figure 4.5. Unlike in Figure 4.4 where the streams represented continuously flowing inputs and outputs, the streams in Figure 4.5 represent masses added and removed at the beginning and end of the mixing process, respectively.
   (ii) System boundary
      The system boundary is indicated in Figure 4.5.

2. Analyze
   (i) Assumptions
       —no leaks
       —no inversion of sucrose to reducing sugars, or any other reaction
   (ii) Extra data
        No extra data are required
   (iii) Basis
        125 kg corn-steep liquor
   (iv) Compounds involved in reaction
        No compounds are involved in reaction
   (v) Mass balance equation
        The appropriate mass balance equation is Eq. (4.3):

$$\text{mass in} = \text{mass out}$$

3. Calculate
   (i) Calculation table
       Table 4.4 shows all given quantities in kg. Rows and columns on each side of the
       table have been completed as much as possible from the information provided. Two
       unknown quantities are given symbols: the mass of water added is denoted W, and the
       total mass of product mixture is denoted P
   (ii) Mass balance calculations
       Total mass balance

$$(170 + W) \text{ kg total mass in} = P \text{ kg total mass out}$$
$$\therefore 170 + W = P \tag{1}$$

   Invert sugars balance

$$3.575 \text{ kg invert sugars in} = (0.02P) \text{ kg invert sugars out}$$
$$\therefore 3.575 = 0.02 \, P$$
$$P = 178.75 \text{ kg}$$

   Using this result in (1):

$$W = 8.75 \text{ kg} \tag{2}$$

   Sucrose balance

$$22.5 \text{ kg sucrose in} = \text{sucrose out}$$
$$\therefore \text{Sucrose out} = 22.5 \text{ kg}$$

   Solids balance

$$73.325 \text{ kg solids in} = \text{solids out}$$
$$\therefore \text{Solids out} = 73.325 \text{ kg}$$

H₂O balance

$$(70.6 + W) \text{ kg in} = H_2O \text{ out}$$

Using the result from (2):

$$79.35 \text{ kg } H_2O \text{ in} = H_2O \text{ out}$$
$$\therefore H_2O \text{ out} = 79.35 \text{ kg}$$

These results allow the mass balance table to be completed, as shown in Table 4.5.
(iii) Check the results all columns and rows of Table 4.5 add up correctly.

4. Finalize

(i) The specific questions
The water required is 8.75 kg. The following is the sucrose concentration in the product mixture:

$$\frac{22.5}{178.75} \times 100 = 12.6\%$$

(ii) Answers
(a) 8.75 kg water is required.
(b) The product mixture contains 13% (w/w) sucrose.

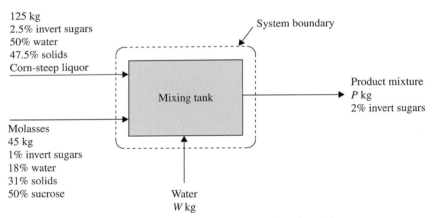

125 kg
2.5% invert sugars
50% water
47.5% solids
Corn-steep liquor

System boundary

Mixing tank

Product mixture
P kg
2% invert sugars

Molasses
45 kg
1% invert sugars
18% water
31% solids
50% sucrose

Water
W kg

Figure 4.5 Graphic of a flow sheet for the batch mixing process.

Example 4.5 illustrates the importance of phase separations. Unreacted O_2 and N_2 were assumed to leave the system as off-gas rather than as components of the liquid product stream. This assumption is reasonable due to the very poor solubility of O_2 and N_2 in aqueous liquids: although the product stream most likely contained some dissolved gases, the quantities would be relatively small. This assumption may need to be reviewed for gases with higher solubility, such as ammonia.

Table 4.4: Mass Balance Table (kg)

Stream	In					Out				
	Invert sugars	Sucrose	Solids	H_2O	Total	Invert sugars	Sucrose	Solids	H_2O	Total
Corn-steep liquor	3.125	0	59.375	62.5	125	—	—	—	—	—
Molasses	0.45	22.5	13.95	8.1	45	—	—	—	—	—
Water	0	0	0	W	W	—	—	—	—	—
Product mixture	—	—	—	—	—	0.02P	?	?	?	P
Total	3.575	22.5	73.325	70.6 + W	170 + W	0.02P	?	?	?	P

Table 4.5: Completed Mass Balance Table (kg)

Stream	In					Out				
	Invert sugars	Sucrose	Solids	H_2O	Total	Invert sugars	Sucrose	Solids	H_2O	Total
Corn-steep liquor	3.125	0	59.375	62.5	125	—	—	—	—	—
Molasses	0.45	22.5	13.95	8.1	45	—	—	—	—	—
Water	0	0	0	8.75	8.75	—	—	—	—	—
Product mixture	—	—	—	—	—	3.575	22.5	73.325	79.35	178.75
Total	3.575	22.5	73.325	79.35	178.75	3.575	22.5	73.325	79.35	178.75

EXAMPLE 4.5 Continuous Acetic Acid Bioprocess

Acetobacter aceti bacteria convert ethanol to acetic acid under aerobic conditions. A continuous bioprocess for vinegar production is proposed using nongrowing *A. aceti* cells immobilized on the surface of gelatin beads. Air is pumped into the bioreactor at a rate of 200 gmol h^{-1}. The production target is 2 kg h^{-1} acetic acid and the maximum acetic acid concentration tolerated by the cells is 12% (w/w).

(a) What minimum amount of ethanol is required?
(b) What minimum amount of water must be used to dilute the ethanol to avoid acid inhibition?
(c) What is the composition of the bioreactor off-gas?

Solution

1. Assemble
    (i) Flow sheet
         The flow sheet for this process is shown in Figure 4.6.

(ii) System boundary
The system boundary is shown in Figure 4.6.

(iii) Reaction equation
In the absence of cell growth, maintenance, or other metabolism of substrate, the reaction equation is:

$$C_2H_5OH + O_2 \rightarrow CH_3COOH + H_2O$$
$$\text{(ethanol)} \qquad\qquad \text{(acetic acid)}$$

2. **Analyze**

(i) Assumptions
—steady state
—no leaks
—inlet air is dry
—gas volume% = mole% (ideal gas assumption)
—no evaporation of ethanol, H_2O, or acetic acid
—complete conversion of ethanol
—ethanol is used by the cells for synthesis of acetic acid only; no side-reactions occur
—O_2 transfer is sufficiently rapid to meet the demands of the cells
—solubility of O_2 and N_2 in the liquid phase is negligible
—concentration of acetic acid in the product stream is 12% (w/w)

(ii) Extra data
Molecular weights:
a. Ethanol = 46
b. Acetic acid = 60
c. O_2 = 32
d. N_2 = 28
e. H_2O = 18
Composition of air (volume or mole): 21% O_2, 79% N_2

(iii) Basis
The calculation is based on 2 kg acetic acid leaving the system, or 1 hour.

(iv) Compounds involved in reaction
The compounds involved in reaction are ethanol, acetic acid, O_2, and H_2O. N_2 is not involved in reaction.

(v) Mass balance equations: For ethanol, acetic acid, O_2, and H_2O, the appropriate mass balance equation is Eq. (4.2):

$$\text{mass in} + \text{mass generated} = \text{mass out} + \text{mass consumed}$$

For total mass and N_2, the appropriate mass balance equation is Eq. (4.3):

$$\text{mass in} = \text{mass out}$$

3. **Calculate**

(i) Calculation table. The mass balance table listing the data provided is shown in Table 4.6; the units are kg. EtOH denotes ethanol; HAc is acetic acid. If 2 kg acetic acid represents 12 mass% of the product stream, the total mass of the product stream must be 2/0.12 = 16.67 kg. If we assume complete conversion of ethanol, the only components of the product stream are acetic acid and water; therefore, water must account for 88 mass% of the product stream $= 14.67$ kg. E and W denote the unknown quantities of ethanol and water in the feed stream, respectively; G represents the total mass of off-gas. The question marks in the table show which other quantities must be calculated. In order to represent what is known about the inlet air, some preliminary calculations are needed.

$$O_2 \text{ content} = (0.21)(200 \text{ gmol}) \cdot \left|\frac{32 \text{ g}}{\text{gmol}}\right| = 1344 \text{ g} = 1.344 \text{ kg}$$

$$N_2 \text{ content} = (0.79)(200 \text{ gmol}) \cdot \left|\frac{28 \text{ g}}{\text{gmol}}\right| = 4424 \text{ g} = 4.424 \text{ kg}$$

Therefore, the total mass of air in $= 5.768$ kg. The masses of O_2 and N_2 can now be entered in Table 4.6 as shown.

(ii) Mass balance and stoichiometry calculations

As N_2 is a tie component, its mass balance is straightforward.

N_2 balance

$$4.424 \text{ kg } N_2 \text{ in} = N_2 \text{ out}$$
$$\therefore N_2 \text{ out} = 4.424 \text{ kg}$$

To deduce the other unknowns, we must use stoichiometric analysis as well as mass balances.

HAc balance

$$0 \text{ kg HAc in} + \text{HAc generated} = 2 \text{ kg HAc out} + 0 \text{ kg HAc consumed}$$
$$\therefore \text{HAc generated} = 2 \text{ kg}$$

$$2 \text{ kg} = 2 \text{ kg} \cdot \left|\frac{1 \text{ kgmol}}{60 \text{ kg}}\right| = 3.333 \times 10^{-2} \text{ kgmol}$$

From reaction stoichiometry, we know that generation of 3.333×10^{-2} kg mol HAc requires 3.333×10^{-2} kg mol each of EtOH and O_2, and is accompanied by generation of 3.333×10^{-2} kg mol H_2O:

$$3.333 \times 10^{-2} \text{ kgmol} \cdot \left|\frac{46 \text{ kg}}{1 \text{ kgmol}}\right| = 1.533 \text{ kg EtOH is consumed}$$

$$3.333 \times 10^{-2} \text{ kgmol} \cdot \left|\frac{32 \text{ kg}}{1 \text{ kgmol}}\right| = 1.067 \text{ kg } O_2 \text{ is consumed}$$

$$3.333 \times 10^{-2} \text{ kgmol} \cdot \left|\frac{18 \text{ kg}}{1 \text{ kgmol}}\right| = 0.600 \text{ kg } H_2O \text{ is generated}$$

We can use this information to complete the mass balances for EtOH, O_2, and H_2O

EtOH balance

EtOH in $+ 0$ kg EtOH generated $= 0$ kg EtOH out $+ 1.533$ kg EtOH consumed

$$\therefore \text{ EtOH in} = 1.533 \text{ kg} = E$$

O_2 balance

1.344 kg O_2 in $+ 0$ kg O_2 generated $= O_2$ out $+ 1.067$ kg O_2 consumed

$$\therefore O_2 \text{ out} = 0.277 \text{ kg}$$

Therefore, summing the O_2 and N_2 components of the off-gas:

$$G = (0.277 + 4.424) \text{ kg} = 4.701 \text{ kg}$$

H_2O balance

W kg H_2O in $+ 0.600$ kg H_2O generated

$$= 14.67 \text{ kg } H_2O \text{ out} + 0 \text{ kg } H_2O \text{ consumed}$$
$$\therefore W = 14.07 \text{ kg}$$

These results allow us to complete the mass balance table, as shown in Table 4.7.

(iii) Check the results

All rows and columns of Table 4.7 add up correctly.

4. Finalize

(i) The specific questions

The ethanol required is 1.533 kg. The water required is 14.07 kg. The off-gas contains 0.277 kg O_2 and 4.424 kg N_2. As gas compositions are normally expressed using volume% or mole%, we convert these values to moles:

$$O_2 \text{ content} = 0.277 \text{ kg} \cdot \left| \frac{1 \text{ kgmol}}{32 \text{ kg}} \right| = 8.656 \times 10^{-3} \text{ kgmol}$$

$$N_2 \text{ content} = 4.424 \text{ kg} \cdot \left| \frac{1 \text{ kgmol}}{28 \text{ kg}} \right| = 0.1580 \text{ kgmol}$$

Therefore, the total molar quantity of off-gas is 0.1667 kg mol. The off-gas composition is:

$$\frac{8.656 \times 10^{-3} \text{ kgmol}}{0.1667 \text{ kgmol}} \times 100 = 5.19\% \text{ } O_2$$

$$\frac{0.1580 \text{ kgmol}}{0.1667 \text{ kgmol}} \times 100 = 94.8\% \text{ } N_2$$

(ii) Answers

Quantities are expressed in kg h^{-1} rather than kg to reflect the continuous nature of the process and the basis used for calculation.

(a) 1.5 kg h^{-1} ethanol is required.

(b) 14 kg h^{-1} water must be used to dilute the ethanol in the feed stream.

(c) The composition of the bioreactor off-gas is 5.2% O_2 and 95% N_2.

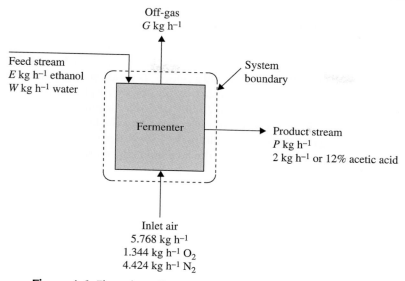

Figure 4.6 Flow sheet for continuous acetic acid production.

Table 4.6: Mass Balance Table (kg)

			In						Out				
Stream	EtOH	HAc	H_2O	O_2	N_2	Total	EtOH	HAc	H_2O	O_2	N_2	Total
Feed stream	E	0	W	0	0	$E + W$	–	–	–	–	–	–
Inlet air	0	0	0	1.344	4.424	5.768	–	–	–	–	–	–
Product stream	–	–	–	–	–	–	0	2	14.67	0	0	16.67
Off-gas	–	–	–	–	–	–	0	0	0	?	?	G
Total	E	0	W	1.344	4.424	$5.768 + E + W$	0	2	14.67	?	?	$16.67 + G$

Table 4.7: Completed Mass Balance Table (kg)

			In						Out				
Stream	EtOH	HAc	H_2O	O_2	N_2	Total	EtOH	HAc	H_2O	O_2	N_2	Total
Feed stream	1.533	0	14.07	0	0	15.603	–	–	–	–	–	–
Inlet air	0	0	0	1.344	4.424	5.768	–	–	–	–	–	–
Product stream	–	–	–	–	–	–	0	2	14.67	0	0	16.67
Off-gas	–	–	–	–	–	–	0	0	0	0.277	4.424	4.701
Total	1.533	0	14.07	1.344	4.424	21.371	0	2	14.67	0.277	4.424	21.371

EXAMPLE 4.6 Xanthan Gum Production

Xanthan gum is produced using *Xanthomonas campestris* in batch culture. Laboratory experiments have shown that for each gram of glucose utilized by the bacteria, 0.23 g O_2 and 0.01 g ammonia are consumed, while 0.75 g gum, 0.09 g cells, 0.27 g gaseous CO_2, and 0.13 g H_2O are formed. Other components of the system such as phosphate can be neglected. Medium containing glucose and ammonia dissolved in 20,000 L of water is pumped into a stirred bioreactor and inoculated with *X. campestris*. Air is sparged into the bioreactor; the total amount of off-gas recovered during the entire batch culture is 1250 kg. Because xanthan gum solutions have high viscosity and are difficult to handle, the final gum concentration should not be allowed to exceed 3.5 wt%.

a. How much glucose and ammonia are required?
b. What percentage of excess air is provided?

Solution

1. Assemble
   (i)   Flow sheet
         The flow sheet for this process is shown in Figure 4.7.
   (ii) System boundary
         The system boundary is shown in Figure 4.7.
   (iii) Reaction equation

$$1 \text{ g glucose} + 0.23 \text{ g } O_2 + 0.01 \text{ g } NH_3$$
$$\rightarrow 0.75 \text{ g gum} + 0.09 \text{ g cells} + 0.27 \text{ g } CO_2 + 0.13 \text{ g } H_2O$$

2. Analyze

   (i)   Assumptions
         —no leaks
         —inlet air and off-gas are dry
         —conversion of glucose and NH_3 is 100% complete
         —solubility of O_2 and N_2 in the liquid phase is negligible
         —CO_2 leaves in the off-gas
   (ii) Extra data
         Molecular weights:
         —$O_2 = 32$
         —$N_2 = 28$
         Density of water = 1 kg L^{-1}
         Composition of air: 21% O_2, 79% N_2
   (iii) Basis
         1250 kg off-gas
   (iv) Compounds involved in reaction. The compounds involved in reaction are glucose, O_2, NH_3, gum, cells, CO_2, and H_2O. N_2 is not involved in reaction.

(v) Mass balance equations

For glucose, O_2, NH_3, gum, cells, CO_2, and H_2O, the appropriate mass balance equation is Eq. (4.2):

$$\text{mass in} + \text{mass generated} = \text{mass out} + \text{mass consumed}$$

For total mass and N_2, the appropriate mass balance equation is Eq. (4.3):

$$\text{mass in} = \text{mass out}$$

3. Calculate

(i) Calculation table. Some preliminary calculations are required to start the mass balance table. First, using $1\,kg\,L^{-1}$ as the density of water, 20,000 L of water is equivalent to 20,000 kg. Let A be the unknown mass of air added. Air is composed of 21 mol% O_2 and 79 mol% N_2; we need to determine the composition of air as mass fractions. In 100 gmol air:

$$O_2 \text{ content} = 21\,\text{gmol} \cdot \left| \frac{32\,g}{1\,\text{gmol}} \right| = 672\,g$$

$$N_2 \text{ content} = 79\,\text{gmol} \cdot \left| \frac{28\,g}{1\,\text{gmol}} \right| = 2212\,g$$

If the total mass of air in 100 gmol is $(2212 + 672) = 2884\,g$, the composition of air is:

$$\frac{672\,g}{2884\,g} \times 100 = 23.3\,\text{mass}\%\,O_2$$

$$\frac{2212\,g}{2884\,g} \times 100 = 76.7\,\text{mass}\%\,N_2$$

Therefore, the mass of O_2 in the inlet air is $0.233\,A$; the mass of N_2 is $0.767\,A$. Let F denote the total mass of feed added; let P denote the total mass of product. We will perform the calculation to produce the maximum allowable gum concentration; therefore, the mass of gum in the product is $0.035P$. With the assumption of 100% conversion of glucose and NH_3, these compounds are not present in the product. Quantities known at the beginning of the problem are shown in Table 4.8.

(ii) Mass balance and stoichiometry calculations *Total mass balance*

$$(F + A) \text{ kg total mass in} = (1250 + P) \text{ kg total mass out}$$

$$\therefore F + A = 1250 + P \tag{1}$$

Gum balance

$$0 \text{ kg gum in} + \text{gum generated} = (0.035P) \text{ kg gum out} + 0 \text{ kg gum consumed}$$
$$\therefore \text{Gum generated} = (0.035P) \text{ kg}$$

From reaction stoichiometry, synthesis of $(0.035\,P)$ kg gum requires:

$$\frac{0.035P}{0.75}(1\,\text{kg}) = (0.0467\,P)\,\text{kg glucose}$$

$$\frac{0.035P}{0.75}(0.23\,\text{kg}) = (0.0107\,P)\,\text{kg } O_2$$

$$\frac{0.035P}{0.75}(0.01\,\text{kg}) = (0.00047\,P)\,\text{kg } NH_3$$

and generates:

$$\frac{0.035P}{0.75}(0.09\,\text{kg}) = (0.0042\,P)\,\text{kg cells}$$

$$\frac{0.035P}{0.75}(0.27\,\text{kg}) = (0.0126\,P)\,\text{kg } CO_2$$

$$\frac{0.035P}{0.75}(0.13\,\text{kg}) = (0.00607\,P)\,\text{kg } H_2O$$

O_2 balance

$$(0.233A)\,\text{kg } O_2 \text{ in} + 0\,\text{kg } O_2 \text{ generated} = O_2 \text{ out} + (0.0107P)\,\text{kg } O_2 \text{ consumed}$$

$$\therefore O_2 \text{ out} = (0.233A - 0.0107P)\,\text{kg} \tag{2}$$

N_2 balance
N_2 is a tie component.

$$(0.767A)\,\text{kg } N_2 \text{ in} = N_2 \text{ out}$$

$$\therefore N_2 \text{ out} = (0.767A)\,\text{kg} \tag{3}$$

CO_2 balance

$$0\,\text{kg } CO_2 \text{ in} + (0.0126P)\,\text{kg } CO_2 \text{ generated} = CO_2 \text{ out} + 0\,\text{kg } CO_2 \text{ consumed}$$

$$\therefore CO_2 \text{ out} = (0.0126P)\,\text{kg} \tag{4}$$

The total mass of gas out is 1250 kg. Therefore, adding the amounts of O_2, N_2, and CO_2 out from (2), (3), and (4):

$$1250 = (0.233A - 0.107P) + (0.767A) + (0.0126P)$$

$$1250 = A + 0.0019P$$

$$\therefore A = 1250 - 0.0019P \tag{5}$$

Glucose balance

glucose in $+ 0$ kg glucose generated
$= 0$ kg glucose out $+ (0.0467P)$ kg glucose consumed

$$\therefore \text{Glucose in} = (0.0467P) \text{ kg} \qquad (6)$$

NH_3 balance

NH_3 in $+ 0$ kg NH_3 generated
$= 0$ kg NH_3 out $+ (0.00047P)$ kg NH_3 consumed

$$\therefore NH_3 \text{ in} = (0.00047P) \text{ kg} \qquad (7)$$

We can now calculate the total mass of the feed, F:

$$F = \text{glucose in} + NH_3 \text{ in} + \text{water in}$$

From (6) and (7):

$$F = (0.0467P) \text{ kg} + (0.00047P) \text{ kg} + 20{,}000 \text{ kg}$$

$$F = (20{,}000 + 0.04717P) \text{ kg} \qquad (8)$$

We can now use (8) and (5) in (1):

$$(20{,}000 + 0.04717P) + (1250 - 0.0019P) = 1250 + P$$
$$20{,}000 = 0.95473P$$
$$\therefore P = 20{,}948.3 \text{ kg}$$

Substituting this result in (5) and (8):

$$A = 1210.2 \text{ kg}$$
$$F = 20{,}988.1 \text{ kg}$$

Also:

$$\text{Gum out} = 0.035P = 733.2 \text{ kg}$$

From Table 4.8:

$$O_2 \text{ in} = 282.0 \text{ kg}$$
$$N_2 \text{ in} = 928.2 \text{ kg}$$

Using the results for P, A, and F in (2), (3), (4), (6), and (7):

$$O_2 \text{ out} = 57.8 \text{ kg}$$
$$N_2 \text{ out} = 928.2 \text{ kg}$$
$$CO_2 \text{ out} = 263.9 \text{ kg}$$
$$\text{Glucose in} = 978.3 \text{ kg}$$
$$NH_3 \text{ in} = 9.8 \text{ kg}$$

Cell balance

$$0 \text{ kg cells in} + (0.0042P) \text{ kg cells generated}$$
$$= \text{cells out} + 0 \text{ kg cells consumed}$$
$$\therefore \text{Cells out} = (0.0042P) \text{ kg}$$
$$\text{Cells out} = 88.0 \text{ kg}$$

H$_2$O balance

$$20,000 \text{ kg } H_2O \text{ in} + (0.00607P) \text{ kg } H_2O \text{ generated}$$
$$= H_2O \text{ out} + 0 \text{ kg } H_2O \text{ consumed}$$
$$\therefore H_2O \text{ out} = 20,000 + (0.00607P) \text{ kg}$$
$$H_2O \text{ out} = 20,127.2 \text{ kg}$$

These entries are included in Table 4.9.

(iii) Check the results

All the columns and rows of Table 4.9 add up correctly to within round-off error.

4. Finalize

(i) The specific questions from the completed mass balance table, 978.3 kg glucose and 9.8 kg NH_3 are required. Calculation of the percentage of excess air is based on O_2, because O_2 is the reacting component of air. The percentage excess can be calculated using:

$$\% \text{ excess air} = \frac{\left(\begin{array}{c} \text{kg } O_2 \text{ present} - \text{kg } O_2 \text{ required to react} \\ \text{completely with the limiting substrate} \end{array} \right)}{\left(\begin{array}{c} \text{kg } O_2 \text{ required to react completely} \\ \text{with the limiting substrate} \end{array} \right)} \times 100$$

In this problem, both glucose and ammonia are limiting substrates. From stoichiometry and the mass balance table, the mass of O_2 required to react completely with 978.3 kg glucose and 9.8 kg NH_3 is:

$$\frac{978.3 \text{ kg}}{1 \text{ kg}}(0.23 \text{ kg}) = 225.0 \text{ kg } O_2$$

The mass provided is 282.0 kg; therefore:

$$\% \text{ excess air} = \frac{282.0 - 225.0}{225.0} \times 100 = 25.3\%$$

(ii) Answers
  (a) 980 kg glucose and 9.8 kg NH_3 are required.
  (b) 25% excess air is provided.

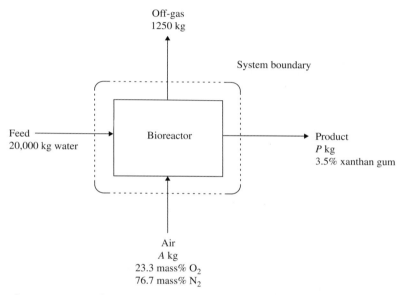

Off-gas
1250 kg

System boundary

Feed
20,000 kg water

Bioreactor

Product
P kg
3.5% xanthan gum

Air
A kg
23.3 mass% O_2
76.7 mass% N_2

Figure 4.7 Graphic of a flow sheet for bioproduction of xanthan gum.

In Example 4.5, N_2 did not react, nor were there more than one stream in and one stream out carrying N_2. A material that goes directly from one stream to another is called a *tie component*; the mass balance for a tie component is relatively simple. Tie components are useful because they can provide partial solutions to mass balance problems, making subsequent calculations easier. More than one tie component may be present in a process.

One of the listed assumptions in Example 4.5 is rapid O_2 transfer. Because cells use O_2 in dissolved form, O_2 must be transferred into the liquid phase from gas bubbles supplied to the bioreactor. The speed of this process depends on the culture conditions and operation of the bioreactor as described in more detail in Chapter 9. In mass balance problems, we assume that all O_2 required by the stoichiometric equation is immediately available to the cells.

Table 4.8: Mass Balance Table (kg)

Stream	In									Out								
	Glucose	O_2	N_2	CO_2	Gum	Cells	NH_3	H_2O	Total	Glucose	O_2	N_2	CO_2	Gum	Cells	NH_3	H_2O	Total
Feed	?	0	0	0	0	0	?	20,000	F	–	–	–	–	–	–	–	–	–
Air	0	0.233A	0.767A	0	0	0	0	0	A	–	–	–	–	–	–	–	–	–
Off-gas	–	–	–	–	–	–	–	–	–	0	?	?	?	0	0	0	0	1250
Product	–	–	–	–	–	–	–	–	–	0	0	0	0	0.035P	?	0	0	P
Total	?	0.233A	0.767A	0	0	0	?	20,000	$F+A$	0	?	?	?	0.035P	?	0	?	1250 + P

Table 4.9: Completed Mass Balance Table (kg)

Stream	In									Out								
	Glucose	O_2	N_2	CO_2	Gum	Cells	NH_3	H_2O	Total	Glucose	O_2	N_2	CO_2	Gum	Cells	NH_3	H_2O	Total
Feed	978.3	0	0	0	0	0	9.8	20,000	20,988.1	–	–	–	–	–	–	–	–	–
Air	0	282.0	928.2	0	0	0	0	0	1210.2	–	–	–	–	–	–	–	–	–
Off-gas	–	–	–	–	–	–	–	–	–	0	57.8	928.2	263.9	0	0	0	0	1250
Product	–	–	–	–	–	–	–	–	–	0	0	0	0	733.2	88.0	0	20,127.2	20,948.3
Total	978.3	282.0	928.2	0	0	0	9.8	20,000	22,198.3	0	57.8	928.2	263.9	733.2	88.0	0	20,127.2	22,198.3

Sometimes it is not possible to solve for unknown quantities in mass balances until near the end of the calculation. In such cases, symbols for various components rather than numerical values must be used in the balance equations. This is illustrated in the integral mass balance of Example 4.6, which analyses the batch culture of growing cells for production of xanthan gum.

4.5 Mass Balances With Recycle, Bypass, and Purge Streams

So far, we have performed mass balances on simple single-unit processes. However, steady-state systems incorporating recycle, bypass, and purge streams are common in bioprocess industries. Flow sheets illustrating these modes of operation are shown in Figure 4.8. Mass balance calculations for such systems can be more involved than those in Examples 4.3 through 4.6; several balances are required before all mass flows can be determined.

As an example, consider the system of Figure 4.9. Because cells are the catalysts in many bioprocesses, it is often advantageous to recycle them from spent broth. Cell recycle requires a separation device, such as a centrifuge or gravity settling tank, to provide a concentrated recycle stream. The flow sheet for cell recycle is shown in Figure 4.10; as indicated, at least four different system boundaries can be defined. System I represents the overall recycle process; only the fresh feed and final product streams cross this system boundary. In addition, separate mass balances can be performed over each process unit: the mixer, the bioreactor, and the settler. Other system boundaries could also be defined; for example, we could group the mixer and bioreactor, or settler and bioreactor, together. Mass balances with recycle involve carrying out individual mass balance calculations for each designated system. Depending on which quantities are known and what information is sought, analysis of more than one system may be required before the flow rates and compositions of all streams are known.

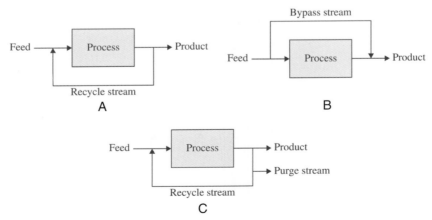

Figure 4.8 Flow sheet for processes with (A) recycle, (B) bypass, and (C) purge streams.

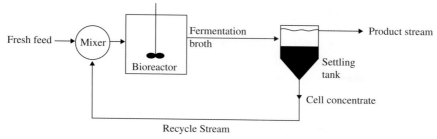

Figure 4.9 Bioreactor with cell recycle.

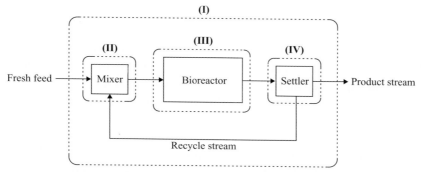

Figure 4.10 System boundaries for cell recycle system.

Mass balances with recycle, bypass, or purge streams usually involve longer calculations than for simple processes but are not more difficult conceptually. Examples of mass balance procedures for multiunit processes can be found in standard chemical engineering texts (e.g., [1–3]).

4.6 Stoichiometry of Cell Growth and Product Formation

So far in this chapter, the law of conservation of mass has been used to determine unknown quantities entering or leaving bioprocesses. For mass balances with reaction such as in Examples 4.5 and 4.6, the stoichiometry of conversion must be known before the mass balance can be solved. When cell growth occurs, cells are a product of reaction and must be represented in the reaction equation. A widely used term for cells in bioprocesses is *biomass*. In this section we discuss how reaction equations for biomass growth and product synthesis are formulated. Metabolic stoichiometry has many applications in bioprocessing. It can be used to compare theoretical and actual product yields, check the consistency of experimental bioreactor data, and formulate nutrient media.

4.6.1 Growth Stoichiometry and Elemental Balances

Cell growth obeys the law of conservation of matter despite its complexity and the thousands of intracellular reactions involved. All atoms of carbon, hydrogen, oxygen, nitrogen, and other elements consumed during growth are incorporated into new cells or excreted as products. Confining our attention to those compounds taken up or produced in significant quantity, we can write the following general equation for aerobic cell growth, assuming the only extracellular products formed are CO_2 and H_2O:

$$C_wH_xO_yN_z + aO_2 + bH_gO_hN_i \rightarrow cCH_\alpha O_\beta N_\delta + dCO_2 + eH_2O \qquad (4.4)$$

In Eq. (4.4):

- $C_wH_xO_yN_z$ is the chemical formula for the carbon source or substrate (e.g., for glucose $C_6H_{12}O_6$, $w = 6$, $x = 12$, $y = 6$, and $z = 0$). Once the identity of the substrate is known, $C_wH_xO_yN_z$ is fully specified and contains no unknown variables.

- $H_gO_hN_i$ is the chemical formula for the nitrogen source (e.g., for ammonia NH_3, $g = 3$, $h = 0$, and $i = 1$). Once the identity of the nitrogen source is known, $H_gO_hN_i$ contains no unknown variables.

- $CH_\alpha O_\beta N_\delta$ is the chemical "formula" for dry biomass. The formula quantifies the dry biomass composition and is based on one C atom: α, β, and δ are the numbers of H, O, and N atoms, respectively, present in the biomass per C atom. Microorganisms such as *Escherichia coli* contain a wide range of elements; however, 90% to 95% of the dry biomass can be accounted for by four major elements: C, H, O, and N as shown in Table 4.10. Compositions of several microbial species in terms of these four elements are listed in Table 4.11. Bacteria tend to have slightly higher nitrogen contents (11%–14%) than fungi (6.3%–9.0%) [1]. Cell composition also depends on the substrate utilized and the culturing conditions; hence, the different entries in Table 4.11 for the same organism. However, the results are remarkably similar for different cells and conditions; $CH_{1.8}O_{0.5}N_{0.2}$ can be used as a general formula for cell biomass when composition analysis is not available. The average "molecular weight" of cells based on C, H, O, and N content is therefore 24.6 g per C mole, although 5% to 10% residual ash is often added to account for those elements not included in the formula.

- a, b, c, d, and e are stoichiometric coefficients. Stoichiometric coefficients quantify the number of molecules and are often expressed with units of moles. Stoichiometric coefficients in this text do not represent the mass of the molecules. Because Eq. (4.4) is written using a basis of one mole of substrate, a moles of O_2 are consumed and d moles of CO_2 are formed, for example, per mole of substrate reacted. The total amount of biomass formed during growth is accounted for by the stoichiometric coefficient c.

Table 4.10: Elemental Composition of *Escherichia coli* Bacteria

Element	% Dry weight
C	50
O	20
N	14
H	8
P	3
S	1
K	1
Na	1
Ca	0.5
Mg	0.5
Cl	0.5
Fe	0.2
All others	0.3

From R.Y. Stanier, J.L. Ingraham, M.L. Wheelis, and P.R. Painter, 1986, *The Microbial World*, 5th ed., Prentice Hall, Upper Saddle River, NJ.

Table 4.11: Elemental Composition and Degree of Reduction for Selected Organisms

Organism	Elemental formula	Degree of reduction γ (relative to NH_3)
Bacteria		
Aerobacter aerogenes	$CH_{1.83}O_{0.55}N_{0.25}$	3.98
Escherichia coli	$CH_{1.77}O_{0.49}N_{0.24}$	4.07
Klebsiella aerogenes	$CH_{1.75}O_{0.43}N_{0.22}$	4.23
Klebsiella aerogenes	$CH_{1.73}O_{0.43}N_{0.24}$	4.15
Klebsiella aerogenes	$CH_{1.75}O_{0.47}N_{0.17}$	4.30
Klebsiella aerogenes	$CH_{1.73}O_{0.43}N_{0.24}$	4.15
Paracoccus denitrificans	$CH_{1.81}O_{0.51}N_{0.20}$	4.19
Paracoccus denitrificans	$CH_{1.51}O_{0.46}N_{0.19}$	3.96
Pseudomonas $C_{12}B$	$CH_{2.00}O_{0.52}N_{0.23}$	4.27
Fungi		
Candida utilis	$CH_{1.83}O_{0.54}N_{0.10}$	4.45
Candida utilis	$CH_{1.87}O_{0.56}N_{0.20}$	4.15
Candida utilis	$CH_{1.83}O_{0.46}N_{0.19}$	4.34
Candida utilis	$CH_{1.87}O_{0.56}N_{0.20}$	4.15
Saccharomyces cerevisiae	$CH_{1.64}O_{0.52}N_{0.16}$	4.12
Saccharomyces cerevisiae	$CH_{1.83}O_{0.56}N_{0.17}$	4.20
Saccharomyces cerevisiae	$CH_{1.81}O_{0.51}N_{0.17}$	4.28
Average	$CH_{1.79}O_{0.50}N_{0.20}$	4.19 (standard deviation = 3%)

From J.A. Roels, 1980, Application of macroscopic principles to microbial metabolism, *Biotechnol. Bioeng.* 22, 2457–2514.

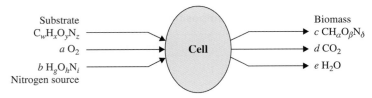

Figure 4.11 Conversion of substrate, O_2, and nitrogen for cell growth.

Equation (4.4) represents a macroscopic view of metabolism as illustrated in Figure 4.11; it ignores the detailed structure of the system and considers only those components that have net interchange with the environment. Equation (4.4) does not include a multitude of compounds such as ATP and NADH that are integral to metabolism and undergo exchange cycles within cells but are not subject to net exchange with the environment. Components such as vitamins and minerals taken up during metabolism could be included; however, since these materials are generally consumed in small quantities, we assume here that their contribution to the stoichiometry and energetics of reaction can be neglected. Other substrates and products can be added easily if appropriate. Despite its simplicity, the macroscopic approach provides a powerful tool for analysis.

Equation (4.4) is not complete unless the stoichiometric coefficients a, b, c, d, and e are known. Once a formula for biomass is obtained, these coefficients can be evaluated using normal procedures for balancing equations, that is, elemental balances and solution of simultaneous equations. The equations are:

Carbon balance:

$$1 \; mol \; substrate \cdot \frac{w \; Cmol}{1 \; mol \; sub.} = cmol \; biomass \cdot \frac{1 \; Cmol}{1 \; mol \; biom} + d \; mol \; CO_2 \cdot \frac{1 \; Cmol}{1 \; mol \; CO_2} \quad (4.5)$$

Hydrogen balance:

$$1 \; mol \; substrate \cdot \frac{x \; Hmol}{1 \; mol \; sub.} + b \; mol \; N \; substrate \cdot \frac{g \; Hmol}{1 \; mol \; Nsub}$$
$$= c \; mol \; biom \cdot \frac{\alpha \; Hmol}{1 \; mol \; biom} + e \; mol \; H_2O \cdot \frac{2 \; Hmol}{1 \; mol \; H_2O} \quad (4.6)$$

Oxygen balance:

$$1 \; mol \; substrate \cdot \frac{y \; Omol}{1 \; mol \; sub.} + a \; mol \; O_2 \cdot \frac{2 \; Omol}{1 \; molO_2} + b \; molN \; substrate \cdot \frac{h \; Omol}{1 \; mol \; Nsub.}$$
$$= c \; mol \; biomass \cdot \frac{\beta \; Omol}{1 \; mol \; biom} + d \; mol \; CO_2 \cdot \frac{2 \; Omol}{1 \; mol \; CO_2} + e \; mol \; H_2O \cdot \frac{1 \; Omol}{1 \; mol \; H_2O} \quad (4.7)$$

Nitrogen balance:

$$b \text{ mol N substrate} \cdot \frac{i \, Nmol}{1 \, mol \, N \, sub.} = cmol \text{ biomass} \cdot \frac{\delta \, Nmol}{1 \, mol \, biom} \tag{4.8}$$

The mass balance Eqs. (4.5) to (4.8) are sometimes written without units as shown below. It is important an engineer understands what the appropriate units are for each term and how the simplifications were made.

$$\text{C balance: } w = c + d \tag{4.5}$$

$$\text{H balance: } x + bg = c\alpha + 2e \tag{4.6}$$

$$\text{O balance: } y + 2a + bh = c\beta + 2d + e \tag{4.7}$$

$$\text{N balance: } z + bi = c\delta \tag{4.8}$$

Notice that we have five unknown coefficients (a, b, c, d, and e) but only four balance equations. This means that additional information is required before the equations can be solved uniquely. Usually, this information is obtained from experiments. A useful measurable parameter is the *respiratory quotient, RQ*:

$$RQ = \frac{\text{moles } CO_2 \text{ produced}}{\text{moles } O_2 \text{ consumed}} = \frac{d}{a} \tag{4.9}$$

When an experimental value of RQ is available, Eqs. (4.5) through (4.9) can be solved to determine the stoichiometric coefficients. The results, however, are sensitive to small errors in RQ, which must be measured very accurately. When Eq. (4.4) is completed, the quantities of substrate, nitrogen, and O_2 required for production of biomass can be determined directly.

Example 4.7 illustrates the use of mass balances to calculate stoichiometric relationships between substrates and products.

EXAMPLE 4.7 Stoichiometric Coefficients for Cell Growth

Production of single-cell protein from hexadecane is described by the following reaction equation:

$$C_{16}H_{34} + aO_2 + bNH_3 \rightarrow cCH_{1.66}O_{0.27}N_{0.20} + dCO_2 + eH_2O$$

Where $cCH_{1.66}O_{0.27}N_{0.20}$ represents the biomass (single-cell protein). If $RQ = 0.43$, determine the stoichiometric coefficients.

Solution

$$\text{C balance: } 16 = c + d \tag{1}$$

$$\text{H balance: } 34 + 3b = 1.66c + 2e \tag{2}$$

(Continued)

$$\text{O balance: } 2a = 0.27c + 2d + e \tag{3}$$

$$\text{N balance: } b = 0.20c \tag{4}$$

$$RQ : 0.43 = d/a \tag{5}$$

We must solve this set of simultaneous equations. A solution can be achieved in many different ways. The linear equations can be transformed into a classic linear algebra format of $Ax = b$ and solved using spread sheet applications or programs like MATLAB or Python. Here, for simplicity, we will use the manual solution. It is often a good idea to express each variable as a function of only one other variable. b is written simply as a function of c in (4): let us try expressing the other variables solely in terms of c. From (1):

$$d = 16 - c \tag{6}$$

From (5):

$$a = d/0.43 = 2.326d \tag{7}$$

Combining (6) and (7) gives an expression for a in terms of c only:

$$a = 2.326(16 - c) \tag{8}$$

$$a = 37.22 - 2.32c \tag{9}$$

Substituting (4) into (2) gives:

$$34 + 3(0.20c) = 1.66c + 2e$$

$$34 = 1.06c + 2e$$

$$e = 17 - 0.53c \tag{10}$$

Substituting (8), (6), and (10) into (3) gives:

$$2(37.22 - 2.326c) = 0.27c + 2(16 - c) + (17 - 0.53c)$$

$$25.44 = 2.39c$$

$$c = 10.64 \text{ Cmol biomass}$$

Using this result for c in (8), (4), (6), and (10) gives:

$$a = 12.48 \text{ moles of } O_2$$
$$b = 2.13 \text{ moles of } NH_3$$
$$d = 5.37 \text{ moles of } CO_2$$
$$e = 11.36 \text{ moles of } H_2O$$

Check that these values satisfy Eqs. (1) through (5).
The complete reaction equation, with units of moles, is:

$$C_{16}H_{34} + 12.48\,O_2 + 2.13\,NH_3 \rightarrow 10.64\,CH_{1.66}O_{0.27}N_{0.20} + 5.37\,CO_2 + 11.36\,H_2O$$

4.6.2 Electron Balances

Available electrons refer to the number of electrons available for transfer to O_2 and are defined here on a combustion basis with compounds oxidized to CO_2, H_2O, and nitrogen-containing compounds. The number of available electrons per atom of organic material is calculated from the valence of the elements: 4 for C, 1 for H, −2 for O, 5 for P, and 6 for S. The negative sign for O indicates it accepts two electrons while the positive values for the other elements indicate they donate electrons to O. The number of available electrons per atom of N depends on the reference state: −3 if ammonia is the reference, 0 for molecular nitrogen N_2, and 5 for nitrate. The N reference state for cell growth is usually chosen to be the same as the nitrogen source in the medium. In the following discussion, it will be assumed for convenience that ammonia is used as the nitrogen source; this can be changed easily if other nitrogen sources are used.

Degree of reduction, γ, is defined as the number of electrons available for transfer to O_2 per C mole of compound. For the generic substrate $C_w H_x O_y N_z$:

$$
\text{Number of available electrons} = \frac{4\,e^-mol}{1\,Cmol} * \frac{w\,Cmol}{1\,mol\ substrate} + \frac{1\,e^-mol}{1\,Hmol} * \frac{x\,Hmol}{1\,mol\ substrate}
$$
$$
+ \frac{-2\,e^-mol}{1\,Omol} * \frac{y\,Omol}{1\,mol\ substrate} + \frac{-3\,e^-mol}{1\,Nmol} * \frac{z\,Nmol}{1\,mol\ substrate}
$$
(4.10)

Which is sometimes written without units as:

$$
\text{Number of available electrons} = 4w + x - 2y - 3z
$$

Therefore, the degree of reduction of the substrate, γ_S (with units of $e^-mol\ Cmol^{-1}$), is:

$$
\gamma_s = \frac{4w + x - 2y - 3z}{w}
$$
(4.11)

Degrees of reduction relative to NH_3 and N_2 for several biological materials are given in Table C.1 in Appendix C. The number of available electrons and the degree of reduction of CO_2, H_2O, and NH_3 are zero. This means that the stoichiometric coefficients for these compounds do not appear in the electron balance, thus simplifying balance calculations.

Available electrons are conserved. The number of available electrons is conserved by virtue of each chemical element being conserved. Applying this principle to Eq. (4.4) with ammonia as the nitrogen source, and recognizing that CO_2, H_2O, and NH_3 have zero available electrons, the available electron balance is:

Number of availabe electrons in the substrate + number of availabe electrons in O_2
= number of availabe electrons in the biomass
(4.12)

This relationship can be written as:

$$w\gamma_s - 4a = c\gamma_B \tag{4.13}$$

where γ_S and γ_B are the degrees of reduction of substrate and biomass, respectively. Note that the available-electron balance is not independent of the complete set of elemental balances: if the stoichiometric equation is balanced in terms of each element including H and O, the electron balance is implicitly satisfied.

Electron balances are very useful and powerful calculations for predicting or analyzing the relationships between substrate and product stoichiometries. Life is based on the movement of energy using electrons. Therefore, electron balances constrain microbial metabolisms such as the anaerobic production of ethanol from sugars, the anaerobic production of lactic acid from sugars, or the growth of photoautotrophic cyanobacteria as illustrated in Examples 4.8 and 4.9.

EXAMPLE 4.8 Anaerobic Metabolism and Electron Balances

Ethanol and lactic acid are two common bioprocesses made by organisms like yeast or lactic acid bacteria. The two products have been used by humans for thousands to make beer or wine and to make yogurt, respectively. Modern applications of these products include fuel ethanol to power vehicles and polylactic acid as a degradable polymer with applications like petroleum-based plastics. Electron balances provide insight into the metabolic limits of biology. For example, is it possible to convert 1 glucose ($C_6H_{12}O_6$) molecule into 3 moles of ethanol (C_2H_6O)? Glucose contains 6 Cmoles and ethanol contains 2 Cmoles so a student unfamiliar with electron balances might think it is possible. It is not.

Find the maximum yield (mole product [mole glucose]$^{-1}$) of ethanol or lactic acid from glucose.

$$1 \text{ glucose} = a \text{ ethanol} + b \text{ CO}_2 + c \text{ H}_2\text{O}$$

$$1 \text{ glucose} = d \text{ lactic acid} + e \text{ CO}_2 + f \text{ H}_2\text{O}$$

Each reaction equation is normalized to 1 mole of glucose and has three additional unknowns (*a-c, d-f*). The problem statement, find the maximum product yield, can be answered with a single equation due to the useful nature of electron balances. Neither CO_2 nor H_2O convey electrons that can be oxidized by O_2, therefore the degree of reduction for both chemicals is 0. The degree of reduction for glucose, ethanol, and lactic acid are all found in the Appendix Table C1.

The electron balance for ethanol fermentation becomes:

$$1 * 4.00 \ e^- mol \ Cmol^{-1} - 1 * 6 \ Cmol = a * 6.00 \ e^- mol \ Cmol^{-1} * 2 \ Cmol$$

solving the equation for *a* results in a maximum of 2 moles ethanol produced per mole of glucose fermented.

The electron balance for lactic acid fermentation becomes:

$$1 * 4.00 \ e^- mol \ Cmol^{-1} - *6 \ Cmol = d * 4.00 \ e^- mol \ Cmol^{-1} * 3 \ Cmol$$

solving the equation for *d* results in a maximum of 2 moles lactic acid produced per mole of glucose fermented.

While glucose contains 6 moles of carbon, it is not possible to produce 3 moles of ethanol because there are not enough available electrons. Ethanol is more reduced than glucose. Some of the glucose carbon is oxidized so that some of the carbon can be reduced. The oxidized carbon is lost as CO_2. Glucose and lactic acid have the same degree of reduction; all the carbon and electrons from glucose are retained in lactic acid. Bioprocesses must purchase substrates which are the source of carbon and electrons. Retaining these valuable resources in the products is important for process economic viability.

EXAMPLE 4.9 Electron Balances, Photoautotrophic Metabolisms, and O_2 Production

The O_2 we breathe is a byproduct of oxygenic, photoautotrophic organisms. Electron balances provide insight into the generation of O_2 as a necessary step in conserving mass and electrons during photosynthesis. A cyanobacteria grows on sunlight fixing CO_2 and N_2 to produce biomass with O_2 produced as a byproduct. If 1 Cmole of biomass ($CH_{1.8}O_{0.5}N_{0.2}$) is produced, how much O_2 is coproduced?

$$a\, CO_2 + b\, N_2 + c\, H_2O + (\text{sunlight}) = 1\, \text{biomass} + d\, O_2$$

There are four unknowns in the growth equation. However, we only need to solve the electron balance to determine how much O_2 is produced making 1 Cmole of biomass. The reaction contains nitrogen species so we need to designate a nitrogen basis for the electron balance. We will select an N_2 basis to simplify the electron balance. Accordingly, the degree of reduction of N_2 is 0 e^-mol $(\text{mol})^{-1}$. The degree of reduction for CO_2 and H_2O are also both 0 e^-mol $(\text{mol})^{-1}$ and sunlight does not contain atoms or electrons. Therefore, the substrate side of an electron balance equation is equal to 0 e^-mol. The product side of the conservation equation has biomass, with a degree of reduction 4.8 e^-mol $(\text{Cmol biomass})^{-1}$, and O_2, which has a degree of reduction of -4 e^-mol $(\text{mol } O_2)^{-1}$. Our electron balance becomes:

$$0 = 1 * 4.8\, e^-\text{mol}\, Cmol^{-1} * 1\, Cmol + d * -4e^-\text{mol}$$

Solving the equation for *d* results in 1.2 mole O_2 produced per Cmole of biomass produced.

Look around your classroom. The organic materials including students, books, and petroleum-derived plastics were originally derived from similar chemical reactions as listed here. There are now moles O_2 circulating in the atmosphere as byproducts of these reactions.

4.6.3 Biomass Yield

Equation (4.13) is often used with carbon and nitrogen balances, Eqs. (4.5) and (4.8), and an experimental measurement, like *RQ*, for evaluation of stoichiometric coefficients. However, as one electron balance, two elemental balances, and one measured quantity are inadequate information for solution of five unknown coefficients, another experimental quantity is required. During cell growth, there is, as a general approximation, a linear relationship between the amount of biomass produced and the amount of substrate consumed. This relationship is expressed quantitatively using the *biomass yield*, $Y_{X/S}$:

$$Y_{X/S} = \frac{\text{g cells produced}}{\text{g substrate consumed}} \tag{4.14}$$

Many factors influence biomass yield, including medium composition, nature of the carbon and nitrogen sources, pH, and temperature. Choice of electron acceptor (e.g., O_2, nitrate, or sulfate) can also have a significant effect; biomass yields are typically greater in aerobic than in anaerobic cultures.

When $Y_{X/S}$ is constant throughout growth, its experimentally determined value can be used to evaluate the stoichiometric coefficient c in Eq. (4.4). Equation (4.14) expressed in terms of the stoichiometric relationship of Eq. (4.4) is:

$$Y_{X/S} = \frac{c \text{ (MW cells)}}{\text{MW substrate}} \tag{4.15}$$

where MW is molecular weight. Recall that the MW of cells calculated from the CHON elemental formula adjusted for residual ash. However, before applying measured values of $Y_{X/S}$ and Eq. (4.15) to evaluate c, we must be sure that the experimental culture system is well represented by the stoichiometric equation. For example, we must be sure that substrate is not used in other types of reactions not represented by the reaction equation. One complication with real cultures is that a fraction of substrate consumed is always used for *maintenance activities* such as maintenance of membrane potential and internal pH, turnover of cellular components, and cell motility. These metabolic functions require substrate but do not necessarily produce cell biomass, CO_2, and H_2O in the way described by Eq. (4.4). Maintenance requirements and the difference between observed and true yields are discussed further in Chapter 10. For the time being, we will assume that the available values for biomass yield reflect consumption of substrate only in the reaction represented by the stoichiometric equation.

4.6.4 Product Stoichiometry

Extracellular products are formed during growth in addition to biomass in many bioprocesses. When this occurs, the stoichiometric equation needs to be modified to reflect product synthesis. Consider the formation of an extracellular product $C_j H_k O_l N_m$ during growth. Equation (4.4) is extended to include product synthesis as follows:

$$C_w H_x O_y N_z + a\, O_2 + b\, H_g O_h N_i$$
$$\rightarrow c\, CH_\alpha O_\beta N_\delta + d\, CO_2 + e\, H_2O + f\, C_j H_k O_l N_m \tag{4.16}$$

where f is the stoichiometric coefficient for the product. Product synthesis introduces one additional unknown stoichiometric coefficient to the equation; thus, an additional relationship is required to solve the equations. This is usually provided as another experimentally determined yield, the *product yield from substrate, $Y_{P/S}$*:

$$Y_{P/S} = \frac{\text{g product formed}}{\text{g substrate consumed}} = \frac{f\ (\text{MW product})}{\text{MW substrate}} \qquad (4.17)$$

Equation (4.16) is appropriate only if product formation is linked directly with cell growth; accordingly, it cannot be applied for secondary metabolite production such as penicillin. An independent reaction equation must be used if product synthesis is distinct from growth.

4.6.5 Theoretical O_2 Demand

O_2 demand is an important parameter in bioprocessing because O_2 is often the limiting substrate in aerobic processes. O_2 demand is represented by the stoichiometric coefficient a in Eqs. (4.4) and (4.16). The requirement for O_2 is related directly to the available electrons balance. The electron balance is as follows when product synthesis occurs as represented by Eq. (4.16):

Number of available electrons in the substrate
+ number of available electrons in O_2
= number of available electrons in the biomass
+ number of available electrons in the product $\qquad$ (4.18)

This relationship can be expressed as:

$$w\gamma_s - 4a = c\gamma_B + fj\gamma_P \qquad (4.19)$$

where γ_P is the degree of reduction of the product and j is the number of Cmols $(\text{mol})^{-1}$ of product. Rearranging gives:

$$a = \frac{1}{4}\left(w\gamma_s - c\gamma_B - fj\gamma_P\right) \qquad (4.20)$$

Equation (4.20) is a very useful equation. It means that if we know which organism (γ_B), substrate (w and γ_s), and product (j and γ_P) are involved in cell culture, and the yields of biomass (c) and product (f), we can quickly calculate the O_2 demand. Of course, we could also determine a by solving for all the stoichiometric coefficients of Eq. (4.16) as described in Section 4.6.1. However, Eq. (4.20) allows more rapid evaluation and does not require that the quantities of NH_3, CO_2, and H_2O involved in the reaction be known.

4.6.6 Maximum Possible Yield

From Eq. (4.19), the fractional allocation of available electrons in the substrate can be written as:

$$1 = \frac{4a}{w\gamma_S} + \frac{c\gamma_B}{w\gamma_S} + \frac{f\,j\gamma_P}{w\gamma_S} \tag{4.21}$$

The first term on the right side is the fraction of available electrons transferred from the substrate to O_2, the second term is the fraction of available electrons transferred to the biomass, and the third term is the fraction of available electrons transferred to the product. This relationship can be used to obtain upper bounds for the yields of biomass and product from substrate.

Let us define ζ_B as the fraction of available electrons in the substrate transferred to biomass:

$$\zeta_B = \frac{c\gamma_B}{w\gamma_S} \tag{4.22}$$

If all available electrons were used for biomass synthesis, ζ_B would equal unity. The maximum value of the stoichiometric coefficient c is under these conditions:

$$c_{max} = \frac{w\gamma_S}{\gamma_B} \tag{4.23}$$

c_{max} can be converted to a biomass yield with mass units using Eq. (4.15). Therefore, even if we do not know the stoichiometry of growth, we can quickly calculate an upper limit for biomass yield from the molecular formulae for the substrate and product. If the composition of the cells is unknown, γ_B can be taken as 4.2 corresponding to the average biomass formula $CH_{1.8}O_{0.5}N_{0.2}$.

Maximum biomass yields for several substrates are listed in Table 4.12. The maximum biomass yield can be expressed in terms of mass ($Y_{X/S,max}$), or as the number of C atoms in the biomass per substrate C-atom consumed (c_{max}/w). These quantities are sometimes known as *thermodynamic maximum biomass yields*. Table 4.12 shows that substrates with high γ_S values and therefore high energy content can enable high maximum biomass yields.

Likewise, the maximum possible product yield in the absence of biomass synthesis can be determined from Eq. (4.21):

$$f_{max} = \frac{w\gamma_S}{j\gamma_P} \tag{4.24}$$

Equation (4.24) allows us to quickly calculate an upper limit for the product yield from the molecular formulae for the substrate and product.

Example 4.10 illustrates two important points. First, the chemical reaction equation for conversion of substrate without growth is a poor approximation of overall stoichiometry when

Table 4.12: Thermodynamic Maximum Biomass Yields

Substrate	Formula	γ_S	Thermodynamic maximum yield corresponding to $\zeta_B = 1$	
			Carbon yield (c_{max}/w)	Mass yield $Y_{X/S,max}$
Alkanes				
Methane	CH_4	8.0	1.9	2.9
Hexane (*n*)	C_6H_{14}	6.3	1.5	2.6
Hexadecane (*n*)	$C_{16}H_{34}$	6.1	1.5	2.5
Alcohols				
Methanol	CH_4O	6.0	1.4	1.1
Ethanol	C_2H_6O	6.0	1.4	1.5
Ethylene glycol	$C_2H_6O_2$	5.0	1.2	0.9
Glycerol	$C_3H_8O_3$	4.7	1.1	0.9
Carbohydrates				
Formaldehyde	CH_2O	4.0	0.95	0.8
Glucose	$C_6H_{12}O_6$	4.0	0.95	0.8
Sucrose	$C_{12}H_{22}O_{11}$	4.0	0.95	0.8
Starch	$(C_6H_{10}O_5)x$	4.0	0.95	0.9
Organic acids				
Formic acid	CH_2O_2	2.0	0.5	0.3
Acetic acid	$C_2H_4O_2$	4.0	0.95	0.8
Propionic acid	$C_3H_6O_2$	4.7	1.1	1.1
Lactic acid	$C_3H_6O_3$	4.0	0.95	0.8
Fumaric acid	$C_4H_4O_4$	3.0	0.7	0.6
Oxalic acid	$C_2H_2O_4$	1.0	0.24	0.1

From L.E. Erickson, I.G. Minkevich, and V.K. Eroshin, 1978, Application of mass and energy balance regularities in fermentation, *Biotechnol. Bioeng.* 20, 1595–1621.

EXAMPLE 4.10 Product Yield and O_2 Demand

The chemical reaction equation for respiration of glucose is:

$$C_6H_{12}O_6 + 6\,O_2 \rightarrow 6\,CO_2 + 6\,H_2O,$$

Candida utilis cells convert glucose to CO_2 and H_2O during growth. The cell composition is $CH_{1.84}O_{0.55}N_{0.2}$ plus 5% ash. The yield of biomass from substrate is 0.5 g biomass (g glucose)$^{-1}$. Ammonia is used as the nitrogen source.

(a) What is the O_2 demand with growth compared to that without?
(b) *C. utilis* is also able to grow using ethanol as substrate, producing cells of the same composition as above. On a mass basis, how does the maximum possible biomass yield from ethanol compare with the maximum possible yield from glucose?

Solution
Molecular weights:

—Glucose = $180 \, \mathrm{g \, mol^{-1}}$
—Ethanol = $46 \, \mathrm{g \, mol^{-1}}$

MW biomass is (25.44 + ash). Since ash accounts for 5% of the total weight, 95% of the total MW = 25.44. Therefore, MW biomass = 25.44/0.95 = 26.78. From Appendix Table C.1, γ for glucose is 4.00 e$^-$mol (Cmol glucose)$^{-1}$; γ for ethanol is 6.00 e$^-$mol (Cmol ethanol)$^{-1}$. γ = $(4 \times 1 + 1 \times 1.84 - 2 \times 0.55 - 3 \times 0.2) = 4.14$ e$^-$mol (Cmol biomass)$^{-1}$. For glucose $w = 6$; for ethanol $w = 2$.

(a) $Y_{X/S} = 0.5 \, \mathrm{g \, biomass \, (g \, glucose)^{-1}}$. Converting this mass yield to a molar yield:

$$Y_{XS} = \frac{0.5 \text{ g biomass}}{\text{g glucose}} \cdot \left| \frac{180 \text{ g biomass}}{1 \text{ gmol glucose}} \right| \cdot \left| \frac{1 \text{ gmol biomass}}{26.78 \text{ g biomass}} \right|$$

$$Y_{XS} = 3.36 \frac{\text{gmol biomass}}{\text{gmol glucose}} = c$$

O_2 demand is given by Eq. (4.20). In the absence of product formation:

$$a = \frac{1}{4}[6(4.00) - 3.36(4.14)] = 2.52$$

Therefore, the O_2 demand for glucose respiration with growth is 2.5 gmol O_2 per gmol glucose consumed. By comparison with the chemical reaction equation for respiration, this is only about 42% of that required in the absence of growth.

(b) The maximum possible biomass yield is given by Eq. (4.23). Using the data above, for glucose:

$$c_{max} = \frac{6(4.00)}{4.14} = 5.80$$

Converting this to a mass basis:

$$Y_{X/S \, max} = \frac{5.80 \text{ gmol biomass}}{\text{gmol glucose}} \cdot \left| \frac{1 \text{ gmol glucose}}{180 \text{ g glucose}} \right| \cdot \left| \frac{26.78 \text{ g biomass}}{1 \text{ gmol biomass}} \right|$$

$$Y_{X/S \, max} = 0.86 \frac{\text{g biomass}}{\text{g glucose}}$$

For ethanol, from Eq. (4.23):

$$c_{max} = \frac{2(6.00)}{4.14} = 2.90$$

and

$$Y_{X/S\,max} = \frac{2.90 \text{ gmol biomass}}{\text{gmol ethanol}} \cdot \left|\frac{1 \text{ gmol ethanol}}{46 \text{ g ethanol}}\right| \cdot \left|\frac{26.78 \text{ g biomass}}{1 \text{ gmol biomass}}\right|$$

$$Y_{X/S\,max} = 1.69 \frac{\text{g biomass}}{\text{g ethanol}}$$

Therefore, on a mass basis, the maximum possible amount of biomass produced per gram of ethanol consumed is roughly twice that per gram of glucose consumed. This result is consistent with the data for $Y_{X/S\,max}$ listed in Table 4.12.

cell growth occurs. When estimating yields and O_2 requirements for any process involving cell growth, the full stoichiometric equation including biomass should be used. Second, the chemical nature or oxidation state of the substrate has a major influence on biomass and product yields through the number of available electrons.

4.7 Unsteady-State Mass Balances

An unsteady-state or transient process is one that causes system properties to vary with time. Batch and semi-batch systems are inherently transient; continuous systems are unsteady during start-up and shut-down. Changing from one set of process conditions to another also creates an unsteady state, as does any fluctuation in input or control variables.

The principles of mass balances developed in this chapter can be applied to unsteady-state processes. Balance equations are used to determine the rate of change of system parameters; solution of these equations generally requires application of calculus. Questions such as: What is the concentration of product in the reactor as a function of time? Can be answered using unsteady-state mass balances. In this section we will consider some simple unsteady-state problems.

4.7.1 Unsteady-State Mass Balance Equations

When the mass of a system is not constant, we generally need to know how the mass varies as a function of time. To evaluate the *rate of change* of mass in the system, let us first return to the general mass balance equation:

$$\left\{\begin{matrix} \text{mass in} \\ \text{through} \\ \text{system} \\ \text{boundaries} \end{matrix}\right\} - \left\{\begin{matrix} \text{mass out} \\ \text{through} \\ \text{system} \\ \text{boundaries} \end{matrix}\right\} + \left\{\begin{matrix} \text{mass} \\ \text{generated} \\ \text{within} \\ \text{system} \end{matrix}\right\} - \left\{\begin{matrix} \text{mass} \\ \text{consumed} \\ \text{within} \\ \text{system} \end{matrix}\right\} = \left\{\begin{matrix} \text{mass} \\ \text{accumulated} \\ \text{within} \\ \text{system} \end{matrix}\right\} \quad (4.1)$$

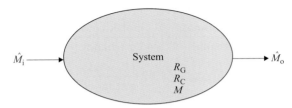

Figure 4.12 Flow system for an unsteady state mass balance.

Consider the flow system of Figure 4.12 in which reactions are taking place. Species A is involved in the process; M is the mass of A in the system. The "hat" symbol ^ denotes rate, so $\hat{M}_i$ is the mass flow rate of A entering the system, and $\hat{M}_o$ is the mass flow rate of A leaving. R_G is the mass rate of generation of species A by chemical reaction; R_C is the mass rate of consumption of A by reaction. The dimensions of $\hat{M}_i$, $\hat{M}_o$, R_G, and R_C are MT^{-1} and the units are, for example, g s^{-1}, kg h^{-1}, lb min^{-1}.

All variables, $\hat{M}_i$, $\hat{M}_o$, R_G, and R_C, may vary with time. However, let us focus on an infinitesimally small interval of time Δt between times t and $t + \Delta t$. Even though the system variables may be changing, if Δt is sufficiently small we can treat the flow rates $\hat{M}$ and rates of reaction R as if they were constant during this period. Under these circumstances, the terms of the general mass balance Eq. (4.1) may be written as follows.

- *Mass in.* During period Δt, the mass of species A transported into the system is $M_i \Delta t$. Note that the dimensions of $\hat{M}_i \Delta t$ are M and the units are, for example, g, kg, lb.
- *Mass out.* Similarly, the mass of species A transported out during time Δt is $\hat{M}_o \Delta t$.
- *Generation.* The mass of A generated during Δt is $R_G \Delta t$.
- *Consumption.* The mass of A consumed during Δt is $R_C \Delta t$.
- *Accumulation.* Let ΔM be the mass of A accumulated in the system during Δt. ΔM may be either positive (accumulation) or negative (depletion).

Entering these terms into the general mass balance Eq. (4.1) with the accumulation term on the left side gives:

$$\Delta M = \hat{M}_i \Delta t - \hat{M}_o \Delta t + R_G \Delta t - R_C \Delta t \tag{4.25}$$

We can divide both sides of Eq. (4.25) by Δt to give:

$$\frac{\Delta M}{\Delta t} = \hat{M}_i - \hat{M}_o + R_G - R_C \tag{4.26}$$

Equation (4.26) applies when Δt is infinitesimally small. If we take the limit as Δt approaches zero, that is, as t and $t + \Delta t$ become virtually the same, Eq. (4.26) represents the system at an instant rather than over an interval of time. Mathematical techniques for handling this type of situation are embodied in the rules of calculus. In calculus, the *derivative* of y with respect to x, dy/dx, is defined as:

$$\frac{dy}{dx} = \lim_{\Delta x \to 0} \frac{\Delta y}{\Delta x} \tag{4.27}$$

where $\lim_{\Delta x \to 0}$ represents the limit as Δx approaches zero. As Eq. (4.26) is valid for $\Delta t \to 0$, we can write it as:

$$\frac{dM}{dt} = \lim_{\Delta t \to 0} \frac{\Delta M}{\Delta t} = \hat{M}_i - \hat{M}_o + R_G - R_C \tag{4.28}$$

The derivative dM/dt represents the rate of change of mass with time measured at a particular instant. We have thus derived a differential equation for the rate of change of M as a function of the system variables, $\hat{M}_i$, $\hat{M}_o$, R_G, and R_C:

$$\frac{dM}{dt} = \hat{M}_i - \hat{M}_o + R_G - R_C \tag{4.29}$$

At steady state, there can be no change in the mass of the system, so the rate of change dM/dt must be zero. Therefore, at steady state, Eq. (4.29) reduces to a form of the familiar steady-state mass balance equation:

$$\text{mass in} + \text{mass generated} = \text{mass out} + \text{mass consumed} \tag{4.2}$$

Unsteady-state mass balance calculations begin with derivation of a differential equation to describe the process. Equation (4.29) was developed on a mass basis and contains parameters such as mass flow rate $\hat{M}$ and mass rate of reaction R. Another common form of the unsteady-state mass balance is based on volume. The reason for this variation is that reaction rates are usually expressed on a per-volume basis. For example, the rate of a first-order reaction is expressed in terms of the concentration of reactant:

$$r_C = k_1 C_A \tag{4.30}$$

where r_C is the *volumetric rate of consumption of A by reaction* (with units of, e.g., g cm^{-3} s^{-1}), k_1 is the first-order reaction rate constant, and C_A is the concentration of reactant A. This and other reaction rate equations are described in more detail in Chapter 10. When rate expressions are used in mass and energy balance problems, the relationship between mass and volume must enter the analysis. This is illustrated in Example 4.11.

EXAMPLE 4.11 Unsteady-State Material Balance for a Continuous Stirred Tank Reactor

A continuous stirred tank reactor (CSTR) is operated as shown in Figure 4.13. The volume of liquid in the tank is V. Feed enters with volumetric flow rate F_i; product leaves with volumetric flow rate F_o. The concentration of reactant A in the feed is C_{Ai}; the concentration of A in the exit stream is C_{Ao}. The density of the feed stream is ρ_i; the density of the product stream is ρ_o. The tank is well mixed. The concentration of A in the tank is C_A and the density of liquid in the tank is ρ. In the reactor, compound A undergoes reaction and is transformed into compound B. The volumetric rate of consumption of A by reaction is given by the expression $r_C = k_1 C_A$.

Using unsteady-state balances, derive differential equations for:

(a) Total mass
(b) The mass of component A

Solution
The general unsteady-state mass balance equation is Eq. (4.29):

$$\frac{dM}{dt} = \hat{M}_i - \hat{M}_b + R_C - R_C$$

(a) For the balance on total mass, R_G and R_C are zero; total mass cannot be generated or consumed by chemical reaction. From the definition of density, total mass can be expressed as the product of volume and density. Similarly, mass flow rate can be expressed as the product of volumetric flow rate and density.

$$\text{Total mass in the tank: } M = \rho V; \text{ therefore } \frac{dM}{dt} = \frac{d(\rho V)}{dt}$$

$$\text{Mass flow rate in: } \dot{M}_i = F_i \rho_i$$

$$\text{Mass flow rate out: } \dot{M}_s = F_s \rho_s$$

Substituting these terms into Eq. (4.29):

$$\frac{d(\rho V)}{dt} = F_i \rho_i - F_\omega \rho_\omega \tag{4.31}$$

Equation (4.31) is a differential equation representing an unsteady-state mass balance on total mass.

(b) Compound A is not generated in the reaction; therefore $R_G = 0$. The other terms of Eq. (4.29) can be expressed as follows:

$$\text{Mass of A in the tank: } M = VC_A; \text{ therefore } \frac{dM}{dt} = \frac{d(VC_A)}{dt}$$

$$\text{Mass flow rate of A in: } \dot{M}_i = F_i C_{Ai}$$

$$\text{Mass flow rate of A out: } \dot{M}_o = F_o C_{Ao}$$

$$\text{Rate of consumption of A: } R_C = k_1 C_A V$$

Substituting into Eq. (4.29):

$$\frac{d(VC_A)}{dt} = F_i C_{Ai} - F_o C_{Ao} - k_1 C_A V \tag{4.32}$$

Equation (4.32) is a differential equation representing an unsteady-state mass balance on A.

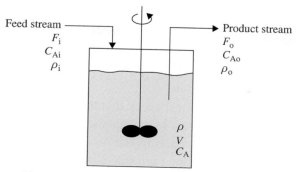

Figure 4.13 Continuous stirred tank reactor.

4.7.2 Solving Differential Equations

Unsteady-state mass balances are represented using differential equations. Once the differential equation for a system has been developed, the equation must be solved to obtain an expression for mass M as a function of time. Differential equations are solved by integration. Further details can be found in any elementary calculus textbook.

Before we proceed with solution techniques for unsteady-state mass balances, there are several general points to consider.

1. *A differential equation can be solved directly only if it contains no more than two variables.* For mass balance problems, the differential equation must have the form:

$$\frac{dM}{dt} = f(M,t) \tag{4.31}$$

where $f(M,t)$ represents some function of M and t. The function may contain constants, but no other variables besides M and t should appear in the expression.

2. *Solution of differential equations requires knowledge of boundary conditions.* Boundary conditions contain extra information about the system. The number of boundary conditions required depends on the *order* of the differential equation, which is equal to the order of the highest differential coefficient in the equation. For example, if the equation contains a second derivative (e.g., d^2x/dt^2), the equation is second order. All equations developed in this chapter have been first order because they involve only first-order derivatives of the form dx/dt. One boundary condition is required to solve a first-order differential equation; two boundary conditions are required for a second-order differential equation, and so on. Boundary conditions that apply at the beginning of the process when $t = 0$ are called *initial conditions.*

3. *Not all differential equations can be solved algebraically,* even if the equation contains only two variables and the boundary conditions are available. Solution of some differential equations requires application of numerical techniques, preferably using a computer. In this chapter we will be concerned mostly with simple equations that can be solved using elementary calculus.

The easiest way of solving differential equations is to *separate variables* so that each variable appears on only one side of the equation. For example, consider the simple differential equation:

$$\frac{dx}{dt} = a(b - x) \tag{4.32}$$

where a and b are constants. First, we must check that the equation contains only two variables x and t, and that all other parameters in the equation are constants. Once this is verified, the equation is separated so that x and t each appear on only one side of the equation. In the case of Eq. (4.30), this is done by dividing each side of the equation by $(b-x)$ and multiplying each side by dt:

$$\frac{dx}{(b - x)} = a\, dt \tag{4.33}$$

The equation is now ready for integration:

$$\int \frac{dx}{(b - x)} = \int a\, dt \tag{4.34}$$

Using classic integration rules results in:

$$-\ln(b - x) = at + K \tag{4.35}$$

Note that the constants of integration from both sides of the equation have been condensed into one constant K; this is valid because a constant $\pm$ a constant = a constant.

4.7.3 Solving Unsteady-State Mass Balances

Solution of unsteady-state mass balances is sometimes difficult unless certain simplifications are made. Because the aim here is to illustrate the application of unsteady-state balances without becoming too involved in integral calculus, the problems presented will be relatively simple. For most problems in this chapter, an analytical solution is possible.

The following restrictions are common in unsteady-state mass balance problems.

- The system is *well mixed* so that properties of the system do not vary with position. If properties within the system are the same at all points, this includes the point from which any product stream is drawn. Accordingly, when the system is well mixed, properties of the outlet stream are the same as those within the system.
- Expressions for reaction rate involve the concentration of only one reactive species. The mass balance equation for this species can then be derived and solved; if other chemical species appear in the kinetic expression, this introduces extra variables into the differential equation making solution more complex.

Example 4.12 illustrates solution of an unsteady-state mass balance without reaction.

EXAMPLE 4.12 Dilution of Salt Solution

To make 100 L of solution, 1.5 kg salt is dissolved in water. Pure water is pumped into a tank containing this solution at a rate of 5 L min^{-1}; salt solution overflows at the same rate. The tank is well mixed. How much salt is in the tank at the end of 15 min? Because the salt solution is dilute, assume that its density is constant and equal to that of water.

Solution

1. Flow sheet and system boundary.
   These are shown in Figure 4.14.
2. Define variables
   C_A = concentration of salt in the tank; V = volume of solution in the tank; ρ = density of salt solution and water.
3. Assumptions
   —no leaks
   —tank is well mixed
   —density of the salt solution is the same as that of water
4. Boundary conditions
   At the beginning of the process, the salt concentration is 1.5 kg in 100 L, or 0.015 kg L^{-1}. If we call this initial salt concentration C_{A0}, the initial condition is:

$$\text{at } t = 0, \quad C_A = C_{A0} = 0.015 \text{ kg L}^{-1} \tag{1}$$

   We also know that the initial volume of liquid in the tank is 100 L. Therefore, another initial condition is:

$$\text{at } t = 0, \quad V = V_0 = 100 L \tag{2}$$

5. Total mass balance
   The unsteady-state balance equation for total mass was derived in Example 4.11 as Eq. (1):

$$\frac{d(\rho V)}{dt} = F_i \rho_i - F_o \rho_o$$

   In this problem we are told that the volumetric flow rates of the inlet and outlet streams are equal; therefore $F_i = F_o$. In addition, the density of the system is constant so that $\rho_i = \rho_o = \rho$. Under these conditions, the terms on the right side of the equation above cancel to zero. On the left side, because ρ is constant, it can be taken outside of the differential. Therefore:

$$\rho \frac{dV}{dt} = 0$$

or

$$\frac{dV}{dt} = 0$$

   If the derivative of V with respect to t is zero, V must be a constant:

$$V = K$$

where K is the constant of integration. This result means that the volume of the tank is constant and independent of time. Initial condition (2) tells us that $V = 100\,L$ at $t = 0$; therefore, V must equal $100\,L$ at all times. Consequently, the constant of integration K is equal to $100\,L$, and the volume of liquid in the tank does not vary from $100\,L$.

6. Mass balance for salt

An unsteady-state mass balance equation for component A such as salt was derived in Example 4.11 as Eq. (2):

$$\frac{d(VC_A)}{dt} = F_i C_{Ai} - F_o C_{Ao} - k_1 C_A V$$

In the present problem there is no reaction, so k_1 is zero. Also, $F_i = F_o = F = 5\,L\,min^{-1}$. Because the tank is well mixed, the concentration of salt in the outlet stream is equal to that inside the tank, that is, $C_{Ao} = C_A$. In addition, since the inlet stream does not contain salt, $C_{Ai} = 0$. From the balance on total mass, we know that V is constant and therefore can be placed outside of the differential. Taking these factors into consideration, the equation above becomes:

$$V \frac{dC_A}{dt} = -FC_A$$

This differential equation contains only two variables C_A and t; F and V are constants. The variables are easy to separate by dividing both sides by VC_A and multiplying by dt:

$$\frac{dC_A}{C_A} = \frac{-F}{V} dt$$

The equation is now ready to integrate:

$$\int \frac{dC_A}{C_A} = \int \frac{-F}{V} dt$$

Using classic integration rules and combining the constants of integration:

$$\ln C_A = \frac{-F}{V} t + K \tag{3}$$

We have yet to determine the value of K. From initial condition (1), at $t = 0$, $C_A = C_{A0}$. Substituting this information into (3):

$$\ln C_{A0} = K$$

We have thus determined K. Substituting this value for K back into (3):

$$\ln C_A = \frac{-F}{V} t + \ln C_{A0} \tag{4}$$

This is the solution to the mass balance; it gives an expression for the concentration of salt in the tank as a function of time. Notice that if we had forgotten to add the constant of integration, the answer would not contain the term $\ln C_{A0}$. The equation would then say that at $t = 0$, $\ln C_A = 0$; that is, $C_A = 1$. We know this is not true; instead, at $t = 0$, $C_A = 0.015\,\text{kg}\,\text{L}^{-1}$, so the result without the boundary condition is incorrect. It is important to apply boundary conditions every time you integrate. The solution equation is usually rearranged to give an exponential expression. This is achieved by subtracting $\ln C_{A0}$ from both sides of (4):

$$\ln C_A - \ln C_{A0} = \frac{-F}{V}t$$

and noting that $(\ln C_A - \ln C_{A0})$ is the same as $\ln(C_A/C_{A0})$:

$$\ln\frac{C_A}{C_{A0}} = \frac{-F}{V}t$$

Taking the antilogarithm of both sides:

$$\frac{C_A}{C_{A0}} = e^{\frac{-F}{V}t}$$

or

$$C_A = C_{A0}e^{\frac{-F}{V}t}$$

We can check that this is the correct solution by taking the derivative of both sides with respect to t and making sure that the original differential equation is recovered. For $F = 5\,\text{L}\,\text{min}^{-1}$, $V = 100\,\text{L}$, and $C_{A0} = 0.015\,\text{g}\,\text{L}^{-1}$, at $t = 15\,\text{min}$:

$$C_A = \left(0.015\,\text{kg}\,\text{L}^{-1}\right)e^{\left(\frac{-5\,\text{L}\,\text{min}^{-1}}{100\,\text{L}}\right)(15\,\text{min})} = 7.09 \times 10^{-3}\,\text{kg}\,\text{L}^{-1}$$

The salt concentration after 15 min is $7.09 \times 10^{-3}\,\text{kg}\,\text{L}^{-1}$. Therefore:

$$\text{mass of salt} = C_A V = \left(7.09 \times 10^{-3}\,\text{kg}\,\text{L}^{-1}\right)(100\,\text{L}) = 0.71\,\text{kg}$$

7. Finalize

After 15 min, the mass of salt in the tank is 0.71 kg.

In Example 4.12 the density of the system was assumed constant. This simplified the mathematics of the problem so that ρ could be taken outside the differential and canceled from the total mass balance. The assumption of constant density is justified for dilute solutions because the density does not differ greatly from that of the solvent. The result of the total mass balance makes intuitive sense: for a tank with equal flow rates in and out and constant density, the volume of liquid inside the tank should remain constant.

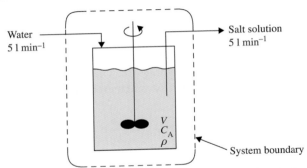

Figure 4.14 Well-mixed tank for dilution of salt solution.

The effect of reaction on the unsteady-state mass balance is illustrated in Example 4.13.

EXAMPLE 4.13 Flow Reactor

Rework Example 4.11 to include reaction. Assume that a reaction in the tank consumes salt at a rate given by the first-order equation:

$$r = k_i C_A$$

where k_1 is the first-order reaction constant and C_A is the concentration of salt in the tank. Derive an expression for C_A as a function of time. If $k_1 = 0.02 \, \text{min}^{-1}$, how long does it take for the concentration of salt to fall to a value $1/20$ at the initial level?

Solution

The flow sheet, boundary conditions, and assumptions for this problem are the same as in Example 4.11. The total mass balance is also the same; total mass in the system is unaffected by reaction.

1. Mass balance for salt: The unsteady-state mass balance equation for salt with first-order reaction is Eq. (2) from Example 4.11:

$$\frac{d(V C_A)}{dt} = F_i C_{Ai} - F_o C_{Ao} - k_1 C_A V$$

In this problem $F_i = F_o = F$, $C_{Ai} = 0$, and V is constant. Because the tank is well mixed, $C_{Ao} = C_A$. Therefore, the equation above becomes:

$$V \frac{dC_A}{dt} = -F C_A - k_i C_A V$$

This equation contains only two variables, C_A and t; F, V, and k_1 are constants. Separate variables by dividing both sides by $V C_A$ and multiplying by dt:

$$\frac{dC_A}{C_A} = \left(\frac{-F}{V} - k_1 \right) dt$$

Integrating both sides gives:

$$\ln C_A = \left(\frac{-F}{V} - k_\tau \right) t + K$$

where K is the constant of integration. K is determined from initial condition (1) in Example 4.11: at $t = 0$, $C_A = C_{A0}$. Substituting these values gives:

$$\ln C_{A0} = K$$

Substituting this value for K back into the answer:

$$\ln C_A = \left(\frac{-F}{V} - k_I \right) t + \ln C_{A0}$$

or

$$\ln \frac{C_A}{C_{A0}} = \left(\frac{-F}{V} - k_I \right) t$$

For $F = 5\,\text{L min}^{-1}$, $V = 100\,\text{L}$, $k_1 = 0.02\,\text{min}^{-1}$, and $C_A/C_{A0} = 1/20$, this equation becomes:

$$\ln \left(\frac{1}{20} \right) = \left(\frac{-5\,\text{L min}^{-1}}{100\,\text{L}} - 0.02\,\text{min}^{-1} \right) t$$

or

$$-3.00 = \left(0.07\,\text{min}^{-1} \right) t$$

Solving for t:

$$t = 42.8\,\text{min}$$

2. Finalize

The concentration of salt in the tank reaches $1/20$ of its initial level after 43 min.

4.8 Summary of Chapter 4

At the end of Chapter 4 you should:

- Understand the terms *system*, *surroundings*, *boundary*, and *process* in thermodynamics
- Be able to identify *open* and *closed systems*, and *batch*, *semibatch*, *fed-batch*, and *continuous processes*
- Understand the difference between *steady state* and *equilibrium*

- Be able to write appropriate equations for conservation of mass for processes with and without reaction
- Be able to solve simple mass balance problems with and without reaction
- Be able to apply stoichiometric principles for macroscopic analysis of cell growth and product formation
- Know what types of processes require unsteady-state analysis
- Be able to derive appropriate *differential equations* for unsteady-state mass balances
- Understand the need for *boundary conditions* to solve differential equations representing actual processes
- Be able to solve simple unsteady-state mass balances to obtain equations for system parameters as a function of time

Problems

4.1. Cell concentration using membranes

A battery of cylindrical hollow-fiber membranes is operated at steady state to concentrate a bacterial suspension harvested from a bioreactor. Culture broth is pumped at a rate of $350 \, kg \, min^{-1}$ through a stack of hollow-fiber membranes as shown in Figure 4P1.1. The broth contains 1% bacteria; the rest may be considered water. Buffer solution enters the annular space around the membrane tubes at a rate of $80 \, kg \, min^{-1}$; because broth in the membrane tubes is under pressure, water is forced across the membrane into the buffer. Cells in the broth are too large to pass through the membrane and pass out of the tubes as a concentrate. The aim of the membrane system is to produce a cell suspension containing 6% biomass.

a. What is the flow rate from the annular space?

b. What is the flow rate of cell suspension from the membrane tubes?

Assume that the cells are not active, that is, they do not grow. Assume further that the membrane does not allow any molecules other than water to pass from annulus to inner cylinder, or vice versa.

4.2. Raspberry coulis manufacture

A food company produces raspberry coulis as a topping for its best-selling line, double-strength chocolate mousse pie. Fresh raspberries comprising 5% seeds, 20% pulp solids, and 75% water on a mass basis are homogenized and placed in a stainless steel vat. Sugar is added to give a

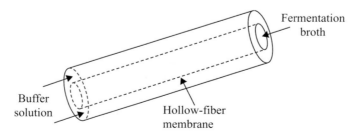

Figure 4P1.1 Hollow-fiber membrane for concentration of cells.

raspberry: sugar mass ratio of 3.5:1. The mixture is blended, strained to remove the seeds, then heated to reduce the water content to 35% (w/w). 500 kg of coulis is produced every day.

a. What mass of raspberries is required per week?

b. How much sugar is required per week?

c. What is the sugar content of the coulis?

4.3. Ethanol distillation

Liquid from a brewery bioreactor can be considered to contain 10% ethanol and 90% water on a mass basis. This fermentation product is pumped at a rate of $50{,}000 \, kg \, h^{-1}$ to a distillation column on the factory site. Under current operating conditions, a distillate of 45% ethanol and 55% water on a mass basis is produced from the top of the column at a rate of one-tenth that of the feed.

a. What is the composition of the waste "bottoms" from the still?

b. What is the rate of alcohol loss in the bottoms?

4.4. Polyethylene glycol–salt mixture

Aqueous two-phase extraction is used to purify a recombinant HIV–β-galactosidase fusion peptide produced in *Escherichia coli*. For optimum separation, 450 kg of a mixture of 19.7% w/w polyethylene glycol (PEG) and 17.7% w/w potassium phosphate salt in water is needed. Left over from previous pilot-plant trials is 100 kg of a mixture of 20% w/w PEG in water, and 150 kg of a mixture of 20% w/w PEG and 25% w/w salt in water. Also on hand is 200 kg of an aqueous stock solution of 50% w/w PEG, 200 kg of an aqueous stock solution of 40% w/w salt, and an unlimited supply of extra water. If all of both leftover mixtures must be used, how much of each stock solution and additional water is required?

4.5. Tetracycline crystallization

Tetracycline produced in *Streptomyces aureus* cultures is purified by crystallization. One hundred kg of a supersaturated solution containing 7.7 wt% tetracycline is cooled in a batch fluidized-bed crystallizer. Seed crystals of tetracycline are added at a concentration of 40 ppm to promote crystal growth. At the end of the crystallization process, the remaining solution contains 2.8% tetracycline.

a. What is the mass of the residual tetracycline solution?

b. What mass of tetracycline crystals is produced?

4.6. Flow rate calculation

A solution of 5% NaCl (w/w) in water is flowing in a stainless steel pipe. To estimate the flow rate, one of your colleagues starts pumping a 30% NaCl (w/w) tracer solution into the pipe at a rate of $80 \, mL \, s^{-1}$ while you measure the concentration of NaCl downstream after the solutions are well mixed. If the downstream NaCl concentration is 9.5% (w/w), use mass balance principles to determine the flow rate of the initial 5% NaCl solution.

4.7. Azeotropic distillation

Absolute or 100% ethanol is produced from a mixture of 95% (w/w) ethanol and 5% (w/w) water using the Keyes distillation process. A third component, benzene, is added to lower the volatility of the alcohol. Under these conditions, the overhead product is a constant-boiling mixture of 18.5% ethanol, 7.4% H_2O, and 74.1% benzene. The process is outlined in Figure 4P8.1. Use the following data to calculate the volume of benzene that should be fed to the still

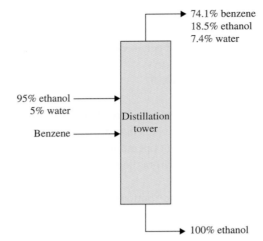

74.1% benzene
18.5% ethanol
7.4% water

95% ethanol
5% water

Distillation
tower

Benzene

100% ethanol

Figure 4P7.1 Flow sheet for Keyes distillation process.

in order to produce 250 L of absolute ethanol: ρ (100% alcohol) $= 0.785 \, \mathrm{g \, cm^{-3}}$; ρ (benzene) $= 0.872 \, \mathrm{g \, cm^{-3}}$.

4.8.   Cell culture using whey

Waste dairy whey is used to grow *Kluyveromyces fragilis* yeast in a continuous culture system operated at 30°C and 1 atm pressure. Medium containing 4% w/w lactose ($C_{12}H_{22}O_{11}$) and 0.15 % w/w NH_3 flows into a specially designed aerated bioreactor at a rate of 200 kg h^{-1}. The reactor is compartmentalized to facilitate gravity settling of the yeast; a suspension containing concentrated cells is drawn off continuously from the bottom of the reactor at a rate of 40 kg h^{-1}, while an aqueous waste stream containing 0.5 kg cells per 100 kg leaves from the top. All of the lactose provided is utilized by the culture. The biomass yield from lactose is known from preliminary experiments to be 0.25 g g^{-1}. The composition of the *K. fragilis* biomass is determined by elemental analysis to be $CH_{1.63}O_{0.54}N_{0.16}$+7.5% ash.

  a. What is the *RQ* for this culture?

  b. It is proposed to supply the reactor with air at a maximum flow rate of 305 L min^{-1}. Will this provide sufficient O_2? If not, what minimum level of O_2 enrichment of the air is needed if the total gas flow rate is 305 L min^{-1}?

  c. What is the concentration of residual NH_3 in the aqueous waste stream?

  d. What is the concentration of cells in the cell concentrate?

4.9. O_2 requirement for growth on glycerol

Klebsiella aerogenes is produced from glycerol in aerobic culture with ammonia as nitrogen source. The biomass contains 8% ash, 0.40 g biomass is produced for each g of glycerol consumed, and no major metabolic products are formed. What is the O_2 requirement for this culture in mass terms?

4.10. Product yield in anaerobic digestion

Anaerobic digestion of volatile acids by methane bacteria is represented by the equation:

$$CH_3COOH + NH_3 \rightarrow \text{biomass} + CO_2 + H_2O + CH_4$$

(acetic acid)   (methane)

The composition of methane bacteria is approximated by the empirical formula $CH_{1.4}O_{0.40}N_{0.20}$. For each kg of acetic acid consumed, 0.67 kg of CO_2 is evolved. How does the yield of methane under these conditions compare with the maximum possible yield?

4.11. Production of PHB

Poly-3-hydroxybutyrate (PHB) is a biodegradable thermoplastic accumulated intracellularly by many microorganisms under unfavorable growth conditions. *Azotobacter chroococcum* is being investigated for commercial PHB production using cheap soluble starch as the raw material and ammonia as the nitrogen source. Synthesis of PHB is observed to be growth-associated with maximum production occurring when the culture is provided with limited O_2. During steady-state continuous culture of *A. chroococcum*, the concentration of PHB in the cells is 44% w/w and the respiratory coefficient is 1.3. From elemental analysis, *A. chroococcum* biomass without PHB can be represented as $CH_2O_{0.5}N_{0.25}$. The monomeric unit for starch is $C_6H_{10}O_5$; $C_4H_6O_2$ is the monomeric unit for PHB.

a. Develop an empirical reaction equation for PHB production and cell growth. PHB can be considered a separate product of the culture even though it is not excreted from the biomass.

b. What is the yield of PHB-containing cells from starch in units of g g^{-1}?

c. If downstream recovery of PHB from the cells involves losses of about 35%, how many kg of starch are needed for production and recovery of 25 kg PHB?

4.12. Substrate requirements for continuous culture

The acetic acid bacterium *Acetobacter pasteurianus* is cultured under aerobic conditions in a continuous bioreactor using ethanol (C_2H_6O) as the substrate and ammonia as the nitrogen source. Under these conditions, acetate accumulation is completely suppressed so that biomass is the only major product. The rate of ethanol consumption is 150 g h^{-1} and the rate of biomass production is 45 g h^{-1}.

a. What is the rate of O_2 consumption during the culture?

b. Ammonia is fed to the culture at a rate of 20 g h^{-1}. At what rate does unreacted ammonia leave the reactor?

4.13. O_2 and sulfur requirements for bacterial culture

Filamentous *Saccharopolyspora erythraea* bacteria are grown in medium containing glucose as the carbon source, ammonia as the nitrogen source, and sulfate as the sulfur source. No major products other than biomass are produced during cell growth. Sulfur is included easily in the stoichiometric equation if we use sulfuric acid (H_2SO_4) to represent the sulfur source and define the valence of S as +6. The molecular formula for *S. erythraea* biomass is found to be $CH_{1.73}O_{0.52}N_{0.17}S_{0.0032}$ and the biomass yield from glucose is 0.29 g g^{-1}.

a. What are the O_2 requirements for this culture?

b. If the medium contains 20 g L^{-1} glucose and the culture proceeds until all the glucose is consumed, what minimum concentration of sulfate must also be included in the medium in units of gmol L^{-1}?

4.14.   Stoichiometry of single-cell protein synthesis

    a.   *Cellulomonas* bacteria used as single-cell protein for human or animal food are produced from glucose under anaerobic conditions. All carbon in the substrate is converted into biomass; ammonia is used as the nitrogen source. The molecular formula for the biomass is $CH_{1.56}O_{0.54}N_{0.16}$; the cells also contain 5% ash. How does the yield of biomass from substrate in mass and molar terms compare with the maximum possible biomass yield?

    b.   Another system for manufacture of single-cell protein is *Methylophilus methylotrophus*. This organism is produced aerobically from methanol with ammonia as nitrogen source. The molecular formula for the biomass is $CH_{1.68}O_{0.36}N_{0.22}$; these cells contain 6% ash.

        I.   How does the maximum yield of biomass compare with that found in (a)? What is the main reason for the difference?

        II.   If the actual yield of biomass from methanol is 42% of the thermodynamic maximum, what is the O_2 demand?

4.15.   Ethanol production by yeast and bacteria

    Both *Saccharomyces cerevisiae* yeast and *Zymomonas mobilis* bacteria produce ethanol from glucose under anaerobic conditions without external electron acceptors. The biomass yield from glucose is $0.11 \, g \, g^{-1}$ for yeast and $0.05 \, g \, g^{-1}$ for *Z. mobilis*. In both cases the nitrogen source is NH_3. Both cell compositions are represented by the formula $CH_{1.8}O_{0.5}N_{0.2}$.

    a.   What is the yield of ethanol from glucose in both cases?

    b.   How do the yields calculated in (a) compare with the thermodynamic maximum?

4.16. O_2 demand for production of recombinant protein

    Recombinant protein is produced by a genetically engineered strain of *Escherichia coli* during cell growth. The recombinant protein can be considered a product of cell culture even though it is not secreted from the cells; it is synthesized in addition to normal *E. coli* biomass. Ammonia is used as the nitrogen source for aerobic respiration of glucose. The recombinant protein has an overall formula of $CH_{1.55}O_{0.31}N_{0.25}$. The yield of biomass (excluding recombinant protein) from glucose is measured as $0.48 \, g \, g^{-1}$; the yield of recombinant protein from glucose is about 20% of that for cells.

    a.   How much ammonia is required?

    b.   What is the O_2 demand?

    c.   If the biomass yield remains at $0.48 \, g \, g^{-1}$, how much different are the ammonia and O_2 requirements for wild-type *E. coli* that is unable to synthesize recombinant protein?

4.17.   Effect of growth on O_2 demand

    The chemical reaction equation for conversion of ethanol (C_2H_6O) to acetic acid ($C_2H_4O_2$) is:

$$C_2H_6O + O_2 \rightarrow C_2H_4O_2 + H_2O$$

    Acetic acid is produced from ethanol during growth of *Acetobacter aceti*, which has the composition $CH_{1.8}O_{0.5}N_{0.2}$. The biomass yield from substrate is $0.14 \, g \, g^{-1}$; the product yield from substrate is $0.92 \, g \, g^{-1}$. Ammonia is used as the nitrogen source. How does growth in this culture affect the O_2 demand for acetic acid production?

4.18. Aerobic sugar metabolism

Candida stellata is a yeast frequently found in wine fermentations. Its sugar metabolism is being studied under aerobic conditions. In continuous culture with fructose ($C_6H_{12}O_6$) as the carbon source and ammonium phosphate as the nitrogen source, the yield of biomass from fructose is 0.025 g biomass (g biomass)$^{-1}$, the yield of ethanol (C_2H_6O) from fructose is 0.21 g ethanol (g fructose)$^{-1}$, and the yield of glycerol ($C_3H_8O_3$) from fructose is 0.07 g glycerol (g fructose)$^{-1}$.

a. If the rate of sugar consumption is 190 g h^{-1}, what is the rate of O_2 consumption in g h^{-1}?

b. What is the *RQ* for this culture?

4.19. Stoichiometry of animal cell growth

Analysis of the stoichiometry of animal cell growth can be complicated because of the large number of macronutrients involved (about 30 amino acids and vitamins plus other organic components and inorganic salts), and because the stoichiometry is sensitive to nutrient concentrations. Nevertheless, glucose ($C_6H_{12}O_6$) and glutamine ($C_5H_{10}O_3N_2$) can be considered the main carbon sources for animal cell growth; glutamine is also the primary nitrogen source. The major metabolic by-products are lactic acid ($C_3H_6O_3$) and ammonia (NH_3).

a. A simplified stoichiometric equation for growth of hybridoma cells is:

$$C_6H_{12}O_6 + pC_5H_{10}O_3N_2 + qO_2 + rCO_2$$
$$\rightarrow sCH_{1.82}O_{0.84}N_{0.25} + tC_3H_6O_3 + uNH_3 + vCO_2 + wH_2O$$

where $CH_{1.82}O_{0.84}N_{0.25}$ represents the biomass and p, q, r, s, t, u, v, and w are stoichiometric coefficients. In a test culture, for every g of glucose consumed, 0.42 g glutamine was taken up and 0.90 g lactic acid and 0.26 g cells were produced.

   I. What is the net carbon dioxide production per 100 g glucose?

   II. What is the O_2 demand?

   III. A typical animal cell culture medium contains 11 mM glucose and 2 mM glutamine. Which of these substrates is present in excess?

   IV. To prevent toxic effects on the cells, the concentrations of lactic acid and ammonia must remain below 1 g L^{-1} and 0.07 g L^{-1}, respectively. What maximum concentrations of glucose and glutamine should be provided in the medium used for batch culture of hybridoma cells?

b. The anabolic component of the stoichiometric equation in (a) can be represented using the equation:

$$C_6H_{12}O_6 + pC_5H_{10}O_3N_2 + rCO_2 \rightarrow sCH_{1.82}O_{0.84}N_{0.25}$$

   I. What proportion of the carbon in the biomass is derived from glucose, and what proportion is derived from glutamine?

   II. Considering both equations from (a) and (b), estimate the proportions of the carbon in glucose and glutamine, respectively, that are used for biomass production during culture of hybridoma cells.

References

[1]   J.-L. Cordier, B.M. Butsch, B. Birou, U. von Stockar, The relationship between elemental composition and heat of combustion of microbial biomass, Appl. Microbiol. Biotechnol. 25 (1987) 305–312.

[2]   R.M. Felder, R.W. Rousseau, Elementary Principles of Chemical Processes, third ed., Wiley, 2005. (Chapter 4)

[3]   D.M. Himmelblau, J.B. Riggs, Basic Principles and Calculations in Chemical Engineering, seventh ed., Prentice Hall, 2004. (Chapters 6–12)

[4]   R.K. Sinnott, Coulson and Richardson's Chemical Engineering, volume 6: Chemical Engineering Design, fourth ed., Elsevier, 2005. (Chapter 2)

Suggestions for Further Reading

Process Mass Balances

See references [2–4] through [2–4].

Metabolic Stoichiometry

See also reference [1].

B. Atkinson, F. Mavituna, Chapter 4, Biochemical engineering and biotechnology handbook, 2nd ed., Macmillan, 1991.

L.E. Erickson, I.G. Minkevich, V.K. Eroshin, Application of mass and energy balance regularities in fermentation, Biotechnol. Bioeng. 20 (1978) 1595–1621.

J.A. Roels, Energetics and kinetics in biotechnology (Chapter 3), Elsevier Biomedical Press, 1983.

W.M. van Gulik, H.J.G. ten Hoopen, J.J. Heijnen, Kinetics and stoichiometry of growth of plant cell cultures of *Catharanthus roseus* and *Nicotiana tabacum* in batch and continuous fermentors, Biotechnol. Bioeng. 40 (1992) 863–874.

A.-P. Zeng, W.-S. Hu, W.-D. Deckwer, Variation of stoichiometric ratios and their correlation for monitoring and control of animal cell cultures, Biotechnol. Prog. 14 (1998) 434–441.

CHAPTER 5

Energy Balances

Bioprocesses are not especially energy-intensive relative to many traditional chemical processes. Bioreactors and enzyme reactors are operated at temperatures and pressures close to ambient while energy input for downstream processing is often minimized to avoid damaging heat-labile products. Nevertheless, energy effects are important because cells and biological catalysts can be very sensitive to changes in temperature. In large-scale processes, heat released during cellular reactions can cause cell death or denaturation of enzymes if it is not removed quickly. For rational design of temperature-control facilities, energy flows in the system must be determined using energy balances. Energy effects are also important in other areas of bioprocessing such as steam sterilization of bioreactors and associated equipment.

The law of conservation of energy means that an energy accounting system can be set up to determine the amount of heating or cooling required to maintain optimum process temperatures. In this chapter, after the necessary thermodynamic concepts are explained, an energy conservation equation applicable to biological processes is derived. The calculation techniques outlined in Chapter 4 are then extended for solution of steady state and unsteady state energy balance problems.

5.1 Basic Energy Concepts

Energy takes three forms:

- Kinetic energy, E_k
- Potential energy, E_p
- Internal energy, U

Kinetic energy is the energy possessed by a moving system because of its velocity. *Potential energy* is due to the position of the system in a gravitational or electromagnetic field, or due to the conformation of the system relative to an equilibrium position (e.g., compression of a spring). *Internal energy* is the sum of all molecular, atomic, and subatomic energies of matter. Internal energy cannot be measured directly or known in absolute terms; we can only quantify change in internal energy.

Energy is transferred, in a closed system, as either heat or work. *Heat* is energy that flows across system boundaries because of a temperature difference between the system and its surroundings. *Work* is energy transferred as a result of any driving force other than temperature difference. There are two types of work: *shaft work* W_s, which is work done by a moving part within the system (e.g., an impeller mixing a bioreactor broth), and *flow work* W_f, the energy required to push matter into the system. In a flow-through process, fluid at the inlet has work done on it by fluid just outside of the system, while fluid at the outlet does work on the fluid in front to push the flow along. Flow work is given by the expression:

$$W_f = pV \qquad\qquad (5.1)$$

where p is pressure and V is volume. (Convince yourself that pV has the same dimensions as work and energy.)

5.1.1 Units

The SI unit for energy is the *joule* (J): $1\,J = 1$ newton meter (N m). Another unit is the *calorie* (cal), which is defined as the heat required to raise the temperature of 1 g of pure water by $1°C$ at 1 atm pressure. The quantity of heat according to this definition depends somewhat on the temperature of the water; because there has been no universal agreement on a reference temperature. The *international table calorie* (cal_{IT}) is fixed at 4.1868 J. In imperial units, the British thermal unit (Btu) is common; this is defined as the amount of energy required to raise the temperature of 1 lb of water by $1°F$ at 1 atm pressure. As with the calorie, a reference temperature is required for this definition; $60°F$ is common although other temperatures are sometimes used.

5.1.2 Intensive and Extensive Properties

Properties of matter fall into two categories: those whose magnitude depends on the quantity of matter present and those whose magnitude does not. Mass, volume, and energy are *extensive variables*. The values of extensive properties change if the size of the system is altered or if material is added or removed. Temperature, density, and mole fraction are examples of properties that are independent of the size of the system; these quantities are called *intensive variables*.

Extensive variables can be converted to *specific* quantities by dividing by the mass; for example, specific volume is total volume divided by mass. Because specific properties are independent of the mass of the system, they are intensive variables. In this chapter, for extensive properties denoted by an upper-case symbol, the specific property is given in lower-case notation. Therefore, if U is internal energy, u denotes specific internal energy with units of, for example, kJ g^{-1}. Although, strictly speaking, the term *specific* refers to the quantity per unit mass, we will use the same lower-case symbols for molar quantities (with units, e.g., kJ gmol^{-1}).

5.1.3 Enthalpy

Enthalpy is a property used commonly in energy balance calculations. It is defined as the combination of two energy terms:

$$H = U + pV \tag{5.2}$$

where H is enthalpy, U is internal energy, p is pressure, and V is volume. Specific enthalpy h is therefore:

$$h = u + pv \tag{5.3}$$

where u is specific internal energy and v is specific volume. Since internal energy cannot be measured or known in absolute terms, neither can enthalpy.

5.2 General Energy Balance Equations

The principle underlying all energy balance calculations is the law of conservation of energy, which states that energy can be neither created nor destroyed. We will derive equations used for solving energy balance problems.

The law of conservation of energy can be written as:

$$\left\{ \begin{array}{c} \text{energy in through} \\ \text{system boundaries} \end{array} \right\} - \left\{ \begin{array}{c} \text{energy out through} \\ \text{system boundaries} \end{array} \right\} = \left\{ \begin{array}{c} \text{energy accumulated} \\ \text{within the system} \end{array} \right\} \tag{5.4}$$

For practical application of this equation, consider the system depicted in Figure 5.1. Mass M_i enters the system while mass M_o leaves. Both masses have energy associated with them in the form of internal, kinetic, and potential energies; flow work is also being done. Energy leaves

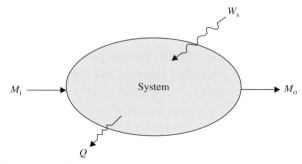

Figure 5.1 Flow system for energy balance calculations.

the system as heat Q; shaft work W_s is performed on the system by the surroundings. We will assume that the system is homogeneous without charge or surface energy effects.

To apply Eq. (5.4), we must identify the forms of energy involved in each term of the expression. If we group together the extensive properties and express them as specific variables multiplied by mass (M_i, M_o), Eq. (5.4) can be written:

$$M_i \left(u + e_k + e_p + pv \right)_i - M_o \left(u + e_k + e_p + pv \right)_o - Q + W_s = \Delta E \qquad (5.5)$$

where subscripts i and o refer to inlet and outlet conditions, respectively, and ΔE represents the total change or accumulation of energy in the system. u is specific internal energy, e_k is specific kinetic energy, e_p is specific potential energy, p is pressure, and v is specific volume. All energies associated with the masses crossing the system boundary are added together; the energy transfer terms Q and W_s are considered separately. Shaft work appears explicitly in Eq. (5.5) as W_s; flow work done by the inlet and outlet streams is represented as pv multiplied by mass.

The energy flows represented by Q and W_s can be directed into or out of the system; appropriate signs must be used to indicate the direction of flow. Because it is usual in bioprocesses that shaft work be done on the system by external sources, in this text we will adopt the convention that work is positive when energy flows *to* the system *from* the surroundings as shown in Figure 5.1. Conversely, work will be considered negative when the system supplies work to the surroundings. On the other hand, we will regard heat as positive when the surroundings receive energy from the system—that is, when the temperature of the system is higher than that of the surroundings. Therefore, when W_s and Q are positive quantities, W_s makes a positive contribution to the energy content of the system while Q causes a reduction. These effects are accounted for in Eq. (5.5). The opposite sign convention is sometimes used in thermodynamics texts; however, the choice of sign convention is arbitrary if used consistently.

Equation (5.5) refers to a process with only one input and one output stream. A more general equation is Eq. (5.6), which can be used for any number of separate mass flows:

$$\sum_{\substack{\text{input} \\ \text{streams}}} M \left(u + e_k + e_p + pv \right) - \sum_{\substack{\text{output} \\ \text{streams}}} M \left(u + e_k + e_p + pv \right) - Q + W_s = \Delta E \qquad (5.6)$$

The symbol $\sum$ means summation; the internal, kinetic, potential, and flow work energies associated with all output streams are added together and subtracted from the sum for all input streams. Equation (5.6) is a basic form of the *first law of thermodynamics*, a simple mathematical expression of the law of conservation of energy. The equation can be shortened by substituting enthalpy h for $u + pv$ as defined by Eq. (5.3):

$$\sum_{\substack{\text{input} \\ \text{streams}}} M\left(h + e_{\text{k}} + e_{\text{p}}\right) - \sum_{\substack{\text{output} \\ \text{streams}}} M\left(h + e_{\text{k}} + e_{\text{p}}\right) - Q + W_{\text{s}} = \Delta E \tag{5.7}$$

5.2.1 Special Cases

Equation (5.7) can be simplified considerably if the following assumptions are made:

- Kinetic energy is negligible
- Potential energy is negligible

These assumptions are acceptable for most bioprocesses, in which high-velocity motion and large changes in height or electromagnetic field do not generally occur. Thus, the energy balance equation becomes:

$$\sum_{\substack{\text{input} \\ \text{streams}}} (Mh) - \sum_{\substack{\text{output} \\ \text{streams}}} (Mh) - Q + W_{\text{s}} = \Delta E \tag{5.8}$$

Equation (5.8) can be simplified further in the following special cases:

- *Steady-state flow process.* At steady state, all properties of the system are invariant. Therefore, there can be no accumulation or change in the energy of the system: $\Delta E = 0$. The steady-state energy balance equation is:

$$\sum_{\substack{\text{input} \\ \text{streams}}} (Mh) - \sum_{\substack{\text{output} \\ \text{streams}}} (Mh) - Q + W_{\text{s}} = 0 \tag{5.9}$$

Equation (5.9) can also be applied over the entire duration of batch and fed-batch processes if there is no energy accumulation; "output streams" in this case refer to the harvesting of all mass in the system at the end of the process. Equation (5.9) is used frequently in bioprocess energy balances.

- *Adiabatic process.* A process in which no heat is transferred to or from the system is termed *adiabatic*; if the system has *adiabatic walls*, it cannot release or receive heat to or from the surroundings. Under these conditions $Q = 0$ and Eq. (5.8) becomes:

$$\sum_{\substack{\text{input} \\ \text{streams}}} (Mh) - \sum_{\substack{\text{output} \\ \text{streams}}} (Mh) + W_{\text{s}} = \Delta E \tag{5.10}$$

Equations (5.8), (5.9), and (5.10) are energy balance equations that allow us to predict, for example, how much heat must be removed from a bioreactor to maintain optimum conditions, or the effect of evaporation on cooling requirements. Furthermore, these approaches apply to bioprocessing units such as condensers, as shown in Figure 5.2. To apply the energy balance

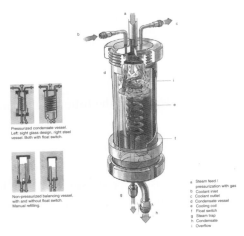

Figure 5.2 Illustration of an off-gas condenser (central) and different designs for condenser heat exchangers (left-side panels). *From L. Hasler, R. Butz, and P. Grad, 2006, Transparency, Form and Function: Fermenter Manufacturing-Art. DreiPunktVerlag. Germany. Courtesy of Bioengineering, Bioengineering AG, 8636 Wald, Switzerland.*

equations, we must know the specific enthalpy h of flow streams entering and leaving the system. Methods for calculating enthalpy are outlined in the following sections.

5.3 Enthalpy Calculation Procedures

5.3.1 Reference States

Specific enthalpy h appears explicitly in energy balance equations. What values of h do we use in these equations if enthalpy cannot be measured or known in absolute terms? Because energy balances are concerned with the *difference* in enthalpy between incoming and outgoing streams, we can overcome any difficulties by working in terms of enthalpy change. In many energy balance problems, changes in enthalpy are evaluated relative to a reference state that must be defined at the beginning of the calculation.

We will use various reference states in energy balance calculations to determine enthalpy change. Suppose, for example, that we want to calculate the change in enthalpy as a system moves from State 1 to State 2. If the enthalpies of States 1 and 2 are known relative to the same reference condition H_{ref}, ΔH is calculated as follows:

$$\text{State 1} \xrightarrow{\Delta H} \text{State 2}$$

$$\text{Enthalpy} = H_1 - H_{ref} \qquad \text{Enthalpy} = H_2 - H_{ref}$$

$$\Delta H = \left(H_2 - H_{ref} \right) - \left(H_1 - H_{ref} \right) = H_2 - H_1$$

ΔH is therefore independent of the reference state because H_{ref} cancels out in the calculation.

5.3.2 State Properties

The values of some variables depend only on the state of the system and not on how that state was reached. These variables are called *state properties* or *functions of state*; examples include temperature, pressure, density, and composition. On the other hand, work is a *path function* because the amount of work done depends on the way in which the final state is obtained from previous states.

Enthalpy is a state property. It means that the change in enthalpy of a system can be calculated by taking a series of hypothetical steps or *process path* leading from the initial state and eventually reaching the final state. Change in enthalpy is calculated for each step; the total enthalpy change for the process is then the sum of changes in the hypothetical path. This is true even though the process path used for calculation is not necessarily the same as that undergone by the actual system.

As an example, consider the enthalpy change for the process shown in Figure 5.3, in which hydrogen peroxide is converted to O_2 and water by catalase enzyme. The enthalpy change for the direct process at 35°C can be calculated using an alternative pathway in which hydrogen peroxide is first cooled to 25°C, O_2 and water are formed by reaction at 25°C, and the products are heated to 35°C. Because the initial and final states for both actual and hypothetical paths are the same, the total enthalpy change is also the same:

$$\Delta H = \Delta H_1 + \Delta H_2 + \Delta H_3$$

(5.11)

The reason for using hypothetical rather than actual pathways to calculate enthalpy change will become apparent later in the chapter.

5.4 Enthalpy Change in Nonreactive Processes

Change in enthalpy can occur as a result of:

1. Temperature change
2. Change of phase
3. Mixing or solution
4. Reaction

5.4.1 Change in Temperature

Heat transferred to raise or lower the temperature of a material is called *sensible heat*; change in the enthalpy of a system due to variation in temperature is called *sensible heat change*. Sensible heat change is determined using a property of matter called the *heat capacity at constant pressure*, or just *heat capacity*. We will use the symbol Cp for heat capacity; units for Cp are, for example, J gmol^{-1} K^{-1}, cal g^{-1} °C^{-1}, or Btu lb^{-1} °F^{-1}. The term *specific heat*

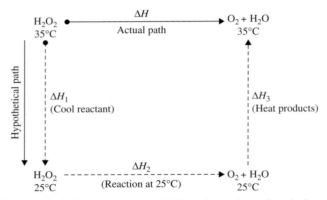

Figure 5.3 Hypothetical process path for calculation of enthalpy change.

capacity or *specific heat* is sometimes used when heat capacity is expressed on a per-unit-mass basis. Values for heat capacity must be known before enthalpy changes from heating or cooling can be determined.

Cp data and information about estimating heat capacities can be found in many reference sources such as *Perry's Chemical Engineers' Handbook*, [1] *CRC Handbook of Chemistry and Physics*, [2] and *International Critical Tables* [3].

There are several methods for calculating enthalpy change using *Cp* values. When *Cp* is approximately constant, the change in enthalpy of a substance at constant pressure due to a change in temperature ΔT is:

$$\Delta H = MC_p\Delta T = MC_p\left(T_2 - T_1\right) \tag{5.12}$$

where M is either mass or moles of the substance depending on the dimensions of *Cp*, T_1 is the initial temperature, and T_2 is the final temperature. The corresponding change in specific enthalpy is:

$$\Delta h = C_p\Delta T = C_p\left(T_2 - T_1\right) \tag{5.13}$$

Example 5.1 illustrates the calculation of sensible heat with a constant *Cp*.

Heat capacities for most substances vary with temperature. This means that when we calculate the enthalpy change due to a change in temperature ΔT, the value of *Cp* itself varies over the range of ΔT. Heat capacities are often tabulated as polynomial functions of temperature, such as:

$$C_p = a + bT + cT^2 + dT^3 \tag{5.14}$$

Coefficients a, b, c, and d for substances can be found in reference sources such as *Perry's Chemical Engineers' Handbook* [1] and *CRC Handbook of Chemistry and Physics* [2].

Sometimes we can assume that the heat capacity is constant; this will give results for sensible heat change that approximate the true value. Because the temperature range of interest in bioprocessing is often relatively small, assuming constant heat capacity for some materials does not introduce large errors.

5.4.2 Change of Phase

Phase changes, such as vaporization and melting, are accompanied by relatively large changes in internal energy and enthalpy. Heat transferred to or from a system causing change of phase at constant temperature and pressure is known as *latent heat*. Types of latent heat are:

- *Latent heat of vaporization* (Δh_v): the heat required to vaporize a liquid
- *Latent heat of fusion* (Δh_f): the heat required to melt a solid
- *Latent heat of sublimation* (Δh_s): the heat required to directly vaporize a solid

Condensation of vapor to liquid requires removal rather than addition of heat; the latent heat evolved in condensation is $-\Delta h_v$. Similarly, the latent heat evolved in freezing or solidification of liquid to solid is $-\Delta h_f$.

Latent heat is a property of substances and, like heat capacity, varies with temperature. Tabulated values of latent heats usually apply to substances at their normal boiling, melting, or sublimation point at 1 atm, and are called *standard heats of phase change*. Values may be found in *Perry's Chemical Engineers' Handbook* [1] and *CRC Handbook of Chemistry and Physics* [2].

EXAMPLE 5.1 Sensible Heat Change with Constant Cp

What is the enthalpy of 150 g formic acid at 70°C and 1 atm relative to 25°C and 1 atm?

Solution

Cp for formic acid in the temperature range of interest is 0.524 cal g^{-1} °C^{-1}. Substituting into (Eq. 5.12):

$$\Delta H = (150 \text{ g})(0.524 \text{ cal g}^{-1} \text{ °C}^{-1})(70 - 25)\text{°C}$$

$$\Delta H = 3537.0 \text{ cal}$$

or

$$\Delta H = 3.54 \text{ kcal}$$

Relative to $H = 0$ kcal at 25°C, the enthalpy of formic acid at 70°C is 3.54 kcal.

The change in enthalpy resulting from phase change is calculated directly from the latent heat. For example, the increase in enthalpy due to evaporation of liquid mass M at constant temperature is:

$$\Delta H = M \Delta h_v \tag{5.15}$$

The calculation is demonstrated in Example 5.2.

Phase changes often occur at a wide range of temperatures; for example, water can evaporate or boil at temperatures higher or lower than 100°C depending on system pressure. How can we determine ΔH when the latent heat at the actual temperature of the phase change is not listed in property tables? This problem is overcome by using a hypothetical process path as described in Section 5.3.2. Suppose a liquid is vaporized isothermally at 30°C, but tabulated values for the standard heat of vaporization refer to 60°C. We can consider a process whereby the liquid is heated from 30°C to 60°C, vaporized at 60°C, and the vapor cooled to 30°C as shown in Figure 5.4. The total enthalpy change for this hypothetical process is the same as if vaporization occurred directly at 30°C. ΔH_1 and ΔH_3 are sensible heat changes, which can be calculated using heat capacity values and the methods described in Section 5.4.1. ΔH_2 is the latent heat at standard conditions calculated using Δh_v data available from tables. Because enthalpy is a state property, ΔH for the actual path is the same as $\Delta H_1 + \Delta H_2 + \Delta H_3$.

5.4.3 Mixing and Solution

We have, so far, considered enthalpy changes for pure compounds. For an *ideal solution* or *ideal mixture* of multiple compounds, the thermodynamic properties of the mixture are a simple sum of contributions from the individual components weighted by the molar fractions.

EXAMPLE 5.2 Enthalpy of Condensation

Fifty grams of benzaldehyde vapor is condensed at 179°C. What is the enthalpy of the liquid relative to the vapor?

Solution
The molecular weight of benzaldehyde is 106.12 g gmol^{-1}, the normal boiling point is 179.0°C, and the standard heat of vaporization is 38.40 kJ gmol^{-1}. For condensation, the latent heat is -38.40 kJ gmol^{-1}. The enthalpy change is:

$$\Delta H = 50 \text{ g} \cdot \left| \frac{1 \text{ gmol}}{106.12 \text{ g}} \right| \cdot \left(-38.40 \text{ kJ gmol}^{-1} \right) = -18.09 \text{ kJ}$$

Therefore, the enthalpy of 50 g benzaldehyde liquid relative to the vapor at 179°C is -18.1 kJ. As heat is released during condensation, the enthalpy of the liquid is lower than the enthalpy of the vapor.

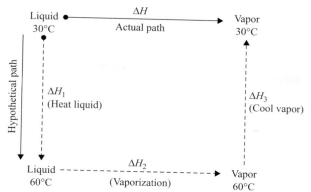

Figure 5.4 Process path for calculating latent heat change at a temperature other than the normal boiling point.

In *real solutions*, a net absorption or release of energy accompanies these processes, resulting in changes in the internal energy and enthalpy of the mixture. Dilution of sulfuric acid with water is a good example; in this case, energy is released and the solution temperature increases.

For real solutions, there is an additional energy term to consider in evaluating enthalpy: the *integral heat of mixing* or *integral heat of solution*, Δh_m. The integral heat of solution is defined as the change in enthalpy that occurs as one mole of solute is dissolved at constant temperature in a given quantity of solvent. The enthalpy of a nonideal mixture of two compounds A and B is:

$$H_{mixture} = H_A + H_B + \Delta H_m \qquad (5.16)$$

where H_A is the enthalpy of compound A, H_B is the enthalpy of compound B, and ΔH_m is the heat of mixing.

Heat of mixing is a property of the solution components and is dependent on the temperature and concentration of the mixture (illustrated in Example 5.3). Heats of solution for select aqueous systems are listed in *Perry's Chemical Engineers' Handbook* [1], *CRC Handbook of Chemistry and Physics* [2], and *Biochemical Engineering and Biotechnology Handbook* [4].

In biological systems, significant changes in enthalpy due to heats of mixing do not often occur. Most solutions in cell growth and enzyme processes are dilute aqueous mixtures; in energy balance calculations these solutions are usually considered ideal without much loss of accuracy.

EXAMPLE 5.3 Heat of Solution

Malonic acid and water are available separately at 25°C. If 15 g malonic acid is dissolved in 5 kg water, how much heat must be added for the solution to remain at 25°C? What is the solution enthalpy relative to the components?

Solution
The molecular weight of malonic acid is 104 g gmol^{-1}. Because the solution is very dilute (<0.3% w/w), we can use the integral heat of solution at infinite dilution. From handbooks, Δh_m at room temperature is 4.493 kcal gmol^{-1}. This value is positive; therefore the mixture enthalpy is greater than that of the components and heat is absorbed during solution. The heat required for the solution to remain at 25°C is:

$$\Delta H = 15\,g \cdot \left| \frac{1\,gmol}{104\,g} \right| \cdot \left(4.493\,kcal\,gmol^{-1}\right) = 0.648\,kcal$$

Relative to $H = 0$ kcal for water and malonic acid at 25°C, the enthalpy of the solution at 25°C is 0.65 kcal.

5.5 *Procedure for Energy Balance Calculations Without Reaction*

The methods described in Section 5.4 for evaluating enthalpy can be used to solve energy balance problems for systems in which no reactions occur. Many of the points described in Section 4.3 for mass balances also apply when establishing an energy balance.

1. A properly drawn and labeled *flow diagram* is essential to identify all inlet and outlet streams and their compositions. Temperatures, pressures, and phases of the material should also be indicated as appropriate.
2. The *units* selected for the energy balance should be stated; these units are also used when labeling the flow diagram.
3. A *basis* for the calculation is chosen and stated clearly (as in mass balance problems).
4. The *reference state* for $H = 0$ kJ gmol^{-1} is determined. In the absence of reaction, reference states for each molecular species in the system can be assigned arbitrarily.
5. State all *assumptions* used to solve the problem. Assumptions such as absence of leaks and steady-state operation for continuous processes are generally applicable.

Following on from (5), other assumptions commonly made for energy balances include:

- The system is homogeneous or well mixed. Under these conditions, product streams including gases leave the system at the system temperature.
- Heats of mixing are often neglected for mixtures containing compounds of similar molecular structure. Gas mixtures are always considered ideal.

- Sometimes shaft work can be neglected even though the system is stirred by mechanical means. This assumption may not apply when vigorous agitation is used or when the liquid being stirred is very viscous. When shaft work is not negligible, you will need to know how much mechanical energy is input through the stirrer.
- Evaporation in liquid systems may be considered negligible if the components are not particularly volatile or if the operating temperature is relatively low.
- Heat losses from the system to the surroundings are often ignored; this assumption is generally valid for large, insulated vessels when the operating temperature is close to ambient.

5.6 Energy Balance Worked Examples Without Reaction

The format described in Chapter 4 for mass balances can be used as a foundation for energy balance calculations as illustrated in the following Examples 5.4 and 5.5.

It is important to recognize that the final answers to energy balance problems do not depend on the choice of reference states for the components. Although values of h depend on the reference states, this dependence disappears when the energy balance equation is applied and the difference between the input and output enthalpies is determined. To prove this point, any of the examples in this chapter can be repeated using different reference conditions to obtain the same final answers.

5.7 Enthalpy Change Due to Reaction

Reactions in bioprocesses occur because of enzyme activity often related to cell metabolism. During reaction, relatively large changes in internal energy and enthalpy occur as bonds between atoms are broken or rearranged. *Heat of reaction* ΔH_{rxn} is the energy released or absorbed during reaction and is equal to the difference in enthalpy of reactants and products:

$$\Delta H_{rxn} = \sum_{products} Mh - \sum_{reactants} Mh \tag{5.17}$$

or

$$\Delta H_{rxn} = \sum_{products} nh - \sum_{reactants} nh \tag{5.18}$$

where $\sum$ denotes the sum, M is mass, n is number of moles, and h is specific enthalpy expressed on either a per-mass or per-mole basis. Note that M and n represent the mass and moles involved in the reaction, not the total amount present in the system. In an *exothermic reaction* the energy required to hold the atoms of product together is less than for the reactants; surplus energy is released as heat and ΔH_{rxn} is negative. On the other hand, energy is

EXAMPLE 5.4 Continuous Water Heater

Water at 25°C enters an open heating tank at a rate of $10\,\mathrm{kg\,h^{-1}}$. Liquid water leaves the tank at 88°C at a rate of $9\,\mathrm{kg\,h^{-1}}$; $1\,\mathrm{kg\,h^{-1}}$ water vapor is lost from the system through evaporation. At steady state, what is the rate of heat input to the system?

Solution

1. Assemble
   - (i)   Select units for the problem. kg, h, kJ, °C
   - (ii) Draw the flow sheet showing all data and units. The flow sheet is shown in Figure 5.4.
   - (iii) Define the system boundary by drawing on the flow sheet. The system boundary is indicated in Figure 5.4.
2. Analyze
   - (i)   State any assumptions.
     - process is operating at steady state
     - system does not leak
     - system is homogeneous
     - evaporation occurs at 88°C
     - vapor is saturated
     - shaft work is negligible
     - no heat losses
   - (ii) Select and state a basis.
     The calculation is based on 10 kg water entering the system, or 1 h.
   - (iii) Select a reference state.
     The reference state for water is: 0.01°C and 0.6112 kPa.
   - (iv) Collect any extra data needed from appropriate reference source.
     h (liquid water at 88°C) = $368.5\,\mathrm{kJ\,kg^{-1}}$
     h (saturated steam at 88°C) = $2656.9\,\mathrm{kJ\,kg^{-1}}$
     h (liquid water at 25°C) = $104.8\,\mathrm{kJ\,kg^{-1}}$
   - (v)   Determine which compounds are involved in reaction.
     No reaction occurs.
   - (vi) Write down the appropriate mass balance equation.
     The mass balance is already complete.
   - (vii) Write down the appropriate energy balance equation. At steady state, Eq. (5.9) applies:

$$\sum_{\substack{\text{input}\\\text{streams}}} (Mh) - \sum_{\substack{\text{output}\\\text{streams}}} (Mh) - Q + W_s = 0$$

3. Calculate.
   Identify the terms in the energy balance equation. For this problem $W_s = 0\,\mathrm{kJ}$. The energy balance equation becomes:

$$(Mh)_{\text{liq in}} - (Mh)_{\text{liq out}} - (Mh)_{\text{vap out}} - Q = 0$$

Substituting the information available:

$$(10\,\text{kg})(104.8\,\text{kJ}\,\text{kg}^{-1}) - (9\,\text{kg})(368.5\,\text{kJ}\,\text{kg}^{-1}) - (1\,\text{kg})(2656.9\,\text{kJ}\,\text{kg}^{-1}) - Q = 0$$
$$Q = -4925.4\,\text{kJ}$$

Q has a negative value. Thus, according to the sign convention outlined in Section 5.2, heat must be supplied to the system from the surroundings.

4. Finalize

Answer the specific questions asked in the problem; check the number of significant figures; state the answers clearly. The rate of heat input is $4.93 \times 10^3\,\text{kJ}\,\text{h}^{-1}$.

EXAMPLE 5.5 Cooling in Downstream Processing

In downstream processing of gluconic acid, concentrated cell broth containing 20% (w/w) gluconic acid is cooled prior to crystallization. The concentrated broth leaves an evaporator at a rate of 2000 kg h^{-1} and must be cooled from 90°C to 6°C. Cooling is achieved by heat exchange with 2700 kg h^{-1} water initially at 2°C. If the final temperature of the cooling water is 50°C, what is the rate of heat loss from the gluconic acid solution to the surroundings? Assume the heat capacity of gluconic acid is 0.35 cal g^{-1} °C^{-1}.

Solution

1. Assemble
    (i) Units
        kg, h, kJ, °C
    (ii) Flow sheet
        The flow sheet is shown in Figure 5.5.
    (iii) System boundary
        The system boundary indicated in Figure 5.5 separates the gluconic acid solution from the cooling water.
2. Analyze
    (i) Assumptions
        • steady state
        • no leaks
        • other components of the cell broth can be considered water
        • no shaft work
    (ii) Basis
        • 2000 kg feed, or 1 h
    (iii) Reference state
        • $H = 0\,\text{kJ}\,\text{gmol}^{-1}$ for gluconic acid at 90°C $H = 0\,\text{kJ}\,\text{gmol}^{-1}$ for water at its triple point

(iv) Extra data
*   The heat capacity of gluconic acid is $0.35 \, \mathrm{cal \, g^{-1} \, °C^{-1}}$; we will assume this C_p remains constant over the temperature range of interest. Converting units:

$$C_p(\text{gluconic acid}) = \frac{0.35 \, \mathrm{cal}}{\mathrm{g°C}} \cdot \left| \frac{4.187 \, \mathrm{J}}{1 \, \mathrm{cal}} \right| \cdot \left| \frac{1 \, \mathrm{kJ}}{1000 \, \mathrm{J}} \right| \cdot \left| \frac{1000 \, \mathrm{g}}{1 \, \mathrm{kg}} \right| = 1.47 \, \mathrm{kJ \, kg^{-1} \, °C^{-1}}$$

h (liquid water at 90°C) $= 37695 \, \mathrm{kJ \, kg^{-1}}$
h (liquid water at 6°C) $= 25.2 \, \mathrm{kJ \, kg^{-1}}$
h (liquid water at 2°C) $= 8.4 \, \mathrm{kJ \, kg^{-1}}$
h (liquid water at 50°C) $= 209.3 \, \mathrm{kJ \, kg^{-1}}$

(v) Compounds involved in reaction
No reaction occurs.

(vi) Mass balance equation
The mass balance equation for total mass, gluconic acid, and water is:

$$\text{mass in} = \text{mass out}$$

The mass flow rates are as shown in Figure 5.5.

(vii) Energy balance equation

$$\underset{\substack{\text{input} \\ \text{streams}}}{\sum (Mh)} - \underset{\substack{\text{output} \\ \text{streams}}}{\sum (Mh)} - Q + W_s = 0$$

3. Calculate
$W_s = 0 \, \mathrm{kJ \, gmol^{-1}}$. There are two heat flows out of the system: one to the cooling water (Q) and one representing loss to the surroundings (Q_{loss}). With symbols W = water and G = gluconic acid, the energy balance equation is:

$$\left(Mh\right)_{W_{in}} + (Mh)_{G_{in}} - (Mh)_{W_{out}} - \left(Mh\right)_{G_{out}} - Q_{\text{loss}} - Q = 0$$

$$(M)_{W_{in}} = (1600 \, \mathrm{kg})\left(376.9 \, \mathrm{kJ \, kg^{-1}}\right) = 6.03 \times 10^5 \, \mathrm{kJ}$$

$$\left(Mh\right)_{G_{in}} = 0 \, (\text{reference state})$$

$$(Mh)_{W_{out}} = (1600 \, \mathrm{kg})\left(1.47 \, \mathrm{kJ \, kg^{-1}}\right) = 4.03 \times 10^4 \, \mathrm{kJ}$$

$(Mh)_{G_{out}}$ at 6°C is calculated as a sensible heat change from 90°C using Eq. (5.12):

$$(Mh)_{G_{out}} = MC_y\left(T_2 - T_1\right) = (400 \, \mathrm{kg})\left(1.47 \, \mathrm{kJ \, kg^{-1} \, °C^{-1}}\right)(6 - 90)°C$$

$$\therefore (Mh)_{G_{out}} = -4.94 \times 10^4 \, \mathrm{kJ}$$

The heat removed to the cooling water, Q, is equal to the enthalpy change of the cooling water between 2°C and 50°C:

$$Q = (2700 \, \mathrm{kg})(209.3 - 8.4) \, \mathrm{kJ \, kg^{-1}} = 5.42 \times 10^5 \, \mathrm{kJ}$$

These results can now be substituted into the energy balance equation:

$$\left(6.03 \times 10^5 \text{ kJ}\right) + (0 \text{ kJ}) - \left(4.03 \times 10^4 \text{ kJ}\right) - \left(-4.94 \times 10^4 \text{ kJ}\right) - Q_{loss} - 5.42 \times 10^5 \text{ kJ} = 0$$

$$\therefore Q_{loss} = 7.01 \times 10^4 \text{ kJ}$$

4. Finalize

The rate of heat loss to the surroundings is $7.0 \times 10^4 \text{ kJ h}^{-1}$.

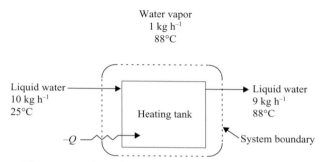

Figure 5.5 Flow sheet for continuous water heater.

absorbed during *endothermic reactions*, the enthalpy of the products is greater than the reactants, and ΔH_{rxn} is positive.

The specific heat of reaction Δh_{rxn} is a property of matter. The value of Δh_{rxn} depends on the reactants and products involved in the reaction and the temperature and pressure. Because any given molecule can participate in many reactions, it is not feasible to tabulate all possible Δh_{rxn} values. Instead, Δh_{rxn} can be calculated from the heats of combustion of the individual components involved in the reaction.

5.7.1 Heat of Combustion

Heat of combustion Δh_c is defined as the heat evolved during reaction of a substance with O_2 to yield certain oxidation products such as CO_2 gas, H_2O liquid, and N_2 gas. The *standard heat of combustion* Δh_c° is the specific enthalpy change associated with this reaction at standard conditions, usually 25°C and 1 atm pressure. By convention, Δh_c° is zero for the products of oxidation (i.e., CO_2 gas, H_2O liquid, N_2 gas, etc.); standard heats of combustion for other compounds are always negative. Table C.3 in Appendix C lists selected values; heats of combustion for other materials can be found in *Perry's Chemical Engineers' Handbook* [4] and *CRC Handbook of Chemistry and Physics* [2]. As an example, the standard heat of combustion for citric acid is given in Table C.3 as $-1962.0\,\text{kJ gmol}^{-1}$; this refers to the heat evolved at 25°C and 1 atm in the following reaction:

$$C_6H_8O_7(s) + 4.5O_2(g) \rightarrow 6CO_2(g) + 4H_2O(l)$$

Standard heats of combustion are used to calculate the *standard heat of reaction* ΔH_{rxn}° for reactions involving combustible reactants and combustion products:

$$\Delta H_{rxn}^{\circ} = \sum_{\text{reactants}} n\Delta h_c^{\circ} - \sum_{\text{products}} n\Delta h_c^{\circ} \qquad (5.19)$$

where n is the moles of reactant or product involved in the reaction, and Δh_c° is the standard heat of combustion per mole. The standard heat of reaction is the difference between the heats of combustion of reactants and products. Heats of combustion quantify the release of heat from complicated molecules as they are oxidized while heats of formation, often used in chemistry class calculations, account for the energy input to build complicated molecules from basic building blocks, hence the sign change between Eqs. 5.19 and 5.18.

5.7.2 Heat of Reaction at Nonstandard Conditions

Example 5.6 shows how to calculate the heat of reaction at standard conditions. However, most reactions do not occur at 25°C and the standard heat of reaction calculated using Eq. (5.20) may not be the same as the actual heat of reaction at the reaction temperature.

Consider the following reaction between compounds A, B, C, and D occurring at temperature T:

$$A + B \rightarrow C + D$$

The standard heat of reaction at 25°C is known from tabulated heat of combustion data. ΔH_{rxn} at temperature T can be calculated using the alternative reaction pathway outlined in Figure 5.6, in which reaction occurs at 25°C and the reactants and products are heated or cooled between 25°C and T before and after the reaction. Because the initial and final states

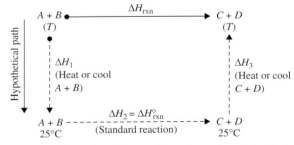

Figure 5.6 Flow sheet for cooling gluconic acid solution.

for the actual and hypothetical paths are the same, the total enthalpy change is also the same. Therefore:

$$\Delta H_{rxn}(at\,T) = \Delta H_1 + \Delta H_{rxn}^\circ + \Delta H_3 \qquad (5.20)$$

where ΔH_1 and ΔH_3 are changes in sensible heat and ΔH_{rxn}° is the standard heat of reaction at 25°C. ΔH_1 and ΔH_3 are evaluated using heat capacities and the methods described in Section 5.4.1.

Depending on the magnitude of ΔH_{rxn}° and the extent to which T deviates from 25°C, ΔH_{rxn} may not be much different from ΔH_{rxn}°. For example, consider the reaction for respiration of glucose:

$$C_6H_{12}O_6 + 6O_2 \rightarrow 6CO_2 + 6\,H_2O$$

ΔH_{rxn}° for this conversion is $-2805.0\,kJ$; if the reaction occurs at 37°C instead of 25°C, ΔH_{rxn} is $-2801.7\,kJ$. Contributions from sensible heat amount to only 3.3 kJ, which is insignificant compared with the total magnitude of ΔH_{rxn}° and can be ignored without much loss of

---

EXAMPLE 5.6 Calculation of Heat of Reaction From Heats of Combustion

Fumaric acid ($C_4H_4O_4$) is produced from malic acid ($C_4H_6O_5$) using the enzyme fumarase. Calculate the standard heat of reaction for the following enzyme transformation:

$$C_4H_6O_5 \rightarrow C_4H_4O_4 + H_2O$$

Solution

$\Delta h_c^\circ = 0$ for liquid water. From Eq. (5.20):

$$\Delta H_{rxn}^\circ = \left(n\Delta h_c^\circ\right)_{malic\ acid} - \left(n\Delta h_c^\circ\right)_{fumaric\ acid}$$

Appendix C lists the standard heats of combustion for these compounds:

$$\left(\Delta h_c^\circ\right)_{malic\ akid} = -1328.8\,kJ\,gmol^{-1}$$
$$\left(\Delta h_c^\circ\right)_{fumaric\ acid} = -1334.0\,kJ\,gmol^{-1}$$

Therefore, using a basis of 1 gmol of malic acid converted:

$$\Delta H_c^\circ = 1\,gmol(-1328.8\,kJ\,gmol^{-1}) - 1\,gmol(-1334.0\,kJ\,gmol^{-1})$$
$$\Delta H_c^\circ = 5.2\,kJ$$

As ΔH_{rxn}° is positive, the reaction is endothermic and heat is absorbed.

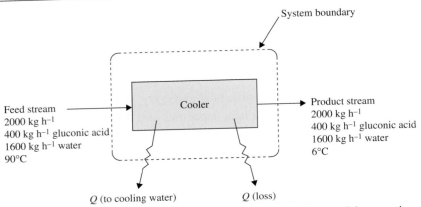

Figure 5.7 Hypothetical process path for calculating the heat of the reaction at nonstandard temperature.

accuracy. With reference to Figure 5.7, $\Delta H_1 = -4.8\,kJ$ for cooling 1 gmol glucose and 6 gmol O_2 from 37°C to 25°C; $\Delta H_3 = 8.1\,kJ$ for heating the products back to 37°C. Having opposite signs, ΔH_1 and ΔH_3 act to cancel each other. This situation is typical of most reactions in bioprocessing where the actual temperature of reaction is not sufficiently different from 25°C to warrant concern about sensible heat changes. When the heat of reaction is substantial compared with other types of enthalpy change, ΔH_{rxn} can be assumed equal to ΔH°_{rxn} irrespective of reaction temperature.

A major exception to this general rule is single-enzyme conversions. Because many single-enzyme reactions involve only small molecular rearrangements, heats of reaction are relatively small. For instance, per mole of substrate, the fumarase reaction of Example 5.6 involves a standard enthalpy change of only 5.2 kJ; other examples are 8.7 kJ $gmol^{-1}$ for the glucose isomerase reaction, $-26.2\,kJ\ gmol^{-1}$ for hydrolysis of sucrose, and $-29.4\,kJ$ per gmol glucose for hydrolysis of starch. For conversions such as these, sensible energy changes of 5 to 10 kJ are clearly significant and should not be ignored.

5.8 Heat of Reaction for Processes With Biomass Production

Biochemical reactions in cells do not occur in isolation but are linked in a complex array of metabolic transformations. Catabolic and anabolic reactions take place at the same time so that energy released in one reaction is used in other energy-requiring processes. Cells use chemical energy quite efficiently; however, some are inevitably released as heat. How can we estimate the heat of reaction associated with cell metabolism and growth?

5.8.1 Thermodynamics of Cell Growth

As described in Section 4.6.1, a macroscopic view of cell growth is represented by the equation:

$$C_w H_x O_y N_z + a O_2 + b H_g O_h N_i \rightarrow c CH_\alpha O_\beta N_\delta + d CO_2 + e H_2O \tag{4.4}$$

where a, b, c, d, and e are stoichiometric coefficients, $C_w H_x O_y N_z$ is the substrate, $H_g O_h N_i$ is the nitrogen source, and $CH_\alpha O_\beta N_\delta$ is dry biomass. Note the equation is normalized to 1 mole of substrate. Once the stoichiometric coefficients or yields are determined, Eq. (4.4) can be used as the reaction equation in energy balance calculations. We need, however, to determine the heat of reaction for this conversion.

Heats of reaction for cell growth can be estimated using stoichiometry and the concept of available electrons (Section 4.6.2). It has been found empirically that the energy content of organic compounds is related to their degree of reduction as follows:

$$\Delta h_c^\circ = -q\gamma x_C \tag{5.21}$$

where Δh_c° is the molar heat of combustion at standard conditions, q is the heat evolved per mole of available electrons transferred to O_2 during combustion, γ is the degree of reduction of the compound relative to N_2, and x_C is the number of carbon atoms in the molecular formula. The coefficient q relating Δh_c° and γ is relatively constant for many compounds. Patel and Erickson [5] assigned a value of 111 kJ released per gmol of electrons transferred to q; in another analysis, Roels [6] determined a value of 115 kJ released per gmol of electrons transferred. The correlation found by Roels is based on analysis of several chemical and biochemical compounds including biomass; the data are shown in Figure 5.8.

5.8.2 Heat of Reaction With O₂ as Electron Acceptor

The direct proportionality between heat of combustion and degree of reduction indicated in Eq. (5.21) and Figure 5.7 has important implications for determining the heat of reaction for aerobic cultures. The degree of reduction of a substance is related directly to the amount of O_2 required for its complete oxidation, and the heat produced by the reaction is directly proportional to the amount of O_2 consumed. O_2 accepts four electrons so if one mole of O_2 is consumed during respiration, four moles of electrons must be transferred. Using the parameter value of 115 kJ of energy released per gmol of electrons transferred, the amount of energy released from consumption of one gmol O_2 is (4×115) kJ, or 460 kJ. The overall result:

$$\Delta H_{rxn} \text{ for fully aerobic metabolism} \simeq -460 \text{ kJ gmol}^{-1} O_2 \text{ consumed} \tag{5.22}$$

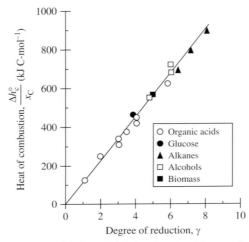

Figure 5.8 Relationship between degree of reduction and heat of combustion for various organic compounds. *From J.A. Roels, 1987, Thermodynamics of growth. In: J. Bu'Lock and B. Kristiansen, Eds., Basic Biotechnology, Academic Press, London.*

is verified by the experimental data of Cooney et al. [7] shown in Figure 5.9. Equation (5.22) is accurate for fully aerobic cultures under a wide range of conditions, including with product formation. Thus, once the amount of O_2 consumed during aerobic cell culture is known, the heat of reaction can be evaluated immediately.

5.8.3 Heat of Reaction With O_2 Not the Principal Electron Acceptor

If an anaerobic bioprocess uses electron acceptors other than O_2, the simple relationship for heat of reaction derived in Section 5.8.2 does not apply. Heats of combustion must be used to estimate the heat of reaction for anaerobic conversions. Consider the following reaction equation for anaerobic growth with product formation:

$$C_w H_x O_y N_z + b H_g O_h N_i$$
$$\rightarrow c\, CH_\alpha O_\beta N_\delta + d\, CO_2 + e\, H_2 O + f\, C_j H_k O_l N_m \tag{5.23}$$

where $C_j H_k O_l N_m$ is an extracellular product and f is its stoichiometric coefficient. With ammonia as nitrogen source and heats of combustion of H_2O and CO_2 zero, from Eq. (5.19) the equation for the standard heat of reaction is:

$$\Delta H_{rxn}^\circ = \left(n\Delta h_c^\circ \right)_{substrate} + \left(n\Delta h_c^\circ \right)_{NH_3} - \left(n\Delta h_c^\circ \right)_{biomass} - \left(n\Delta h_c^\circ \right)_{product} \tag{5.24}$$

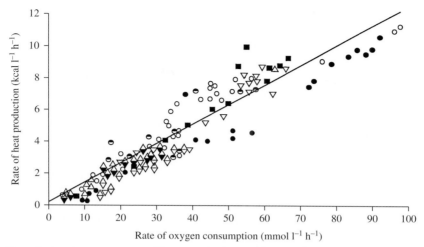

Figure 5.9 Correlation between rate of heat evolution and rate of O_2 consumption for a variety of microbial bioprocesses. Different symbols correspond with data from the following microorganisms and medium. *Data and symbol keys can be found in C.L. Cooney, D.I.C. Wang, and R.I. Mateles, Measurement of heat evolution and correlation with oxygen consumption during microbial growth. Biotechnol. Bioeng. 11, 269–281; Copyright © 1968. Reprinted by permission of John Wiley.*

where n is the number of moles and Δh_c° is the standard molar heat of combustion. Heats of combustion for substrate, NH_3, and product can be found in tables, but what is the heat of combustion of biomass?

The elemental composition of biomass does not vary a great deal, as shown in Table 4.11. If we assume an average biomass molecular formula of $CH_{1.8}O_{0.5}N_{0.2}$, the reaction equation for combustion of cells to CO_2, H_2O, and N_2 is:

$$CH_{1.8}O_{0.5}N_{0.2} + 1.2 O_2 \rightarrow CO_2 + 0.9 H_2O + 0.1 N_2$$

From Appendix C, the degree of reduction of biomass relative to N_2 is 4.80. Assuming an average of 5% ash associated with the biomass, the cell molecular weight is 25.9. The heat of combustion can be obtained by applying Eq. (5.21) with $q = 115 \, kJ \, gmol^{-1}$:

$$\left(\Delta h_c^\circ \right)_{\text{biomass}} = \left(-115 \text{ kJ gmol}^{-1} \right)(4.80)(1) \cdot \left| \frac{1 \text{ gmol}}{25.9 \text{ g}} \right|$$

$$\left(\Delta h_c^\circ \right)_{\text{biomass}} = -21.3 \text{ kJ g}^{-1} \tag{5.25}$$

Actual heats of combustion measured by Cordier et al. [8] for a range of microorganisms and culture conditions are listed in Table 5.1. The differences in Δh_c values for bacteria and yeast

Table 5.1: Heats of Combustion of Bacteria and Yeast

Organism	Substrate	Δh_c (kJ g^{-1})
Bacteria		
Escherichia coli	glucose	-23.04 ± 0.06
	glycerol	-22.83 ± 0.07
Enterobacter cloacae	glucose	-23.22 ± 0.14
	glycerol	-23.39 ± 0.12
Methylophilus methylotrophus	methanol	-23.82 ± 0.06
Bacillus thuringiensis	glucose	-22.08 ± 0.03
Yeast		
Candida lipolytica	glucose	-21.34 ± 0.16
Candida boidinii	glucose	-20.14 ± 0.18
	ethanol	-20.40 ± 0.14
	methanol	-21.52 ± 0.09
Kluyveromyces fragilis	lactose	-21.54 ± 0.07
	galactose	-21.78 ± 0.10
	glucose	-21.66 ± 0.19
	glucose*	-21.07 ± 0.07
		-21.30 ± 0.10
		-20.66 ± 0.26
		-21.22 ± 0.14

*Chemostat rather than batch culture: dilution rates were 0.036 h^{-1}, 0.061 h^{-1}, 0.158 h^{-1}, and 0.227 h^{-1}, respectively. From J.-L. Cordier, B.M. Butsch, B. Birou, and U. von Stockar, 1987, The relationship between elemental composition and heat of combustion of microbial biomass, *Appl. Microbiol. Biotechnol.* 25, 305–312.

reflect their slightly different elemental compositions. When the composition of an organism is unknown, the heat of combustion can be estimated using the following equations:

$$\Delta h_c \text{ for bacteria} \simeq -23.2 \text{ kJ g}^{-1} \tag{5.26}$$

$$\Delta h_c \text{ for yeast} \simeq -21.2 \text{ kJ g}^{-1} \tag{5.27}$$

These experimentally determined values compare well with that calculated in Eq. (5.25). Once the heat of combustion of the biomass is known, it can be used with the heats of combustion of the substrates and other products to estimate the heat of reaction for anaerobic cultures.

5.8.4 Magnitude of the Heat of Reaction in Different Cell Cultures

Equation (5.21) indicates that the degree of reduction of a substrate plays a large role in determining the heat of reaction. Substrates with a high degree of reduction γ_S (see Appendix C) contain large amounts of energy that cannot all be transferred to biomass and other products and are therefore dissipated as heat. This explains why the heat generated by cell growth on highly reduced substrates, such as methane, is much greater than in carbohydrate medium. For example, when methane ($\gamma_S = 8.0$) is used as the substrate for aerobic cell cultures, the amount of heat released per g of biomass produced is about four times more than less reduced substrates such as glucose ($\gamma_S = 4.0$). Similarly, the heat released using ethanol or methanol ($\gamma_S = 6.0$) as the substrate is two to three times that for glucose [6]. Another factor affecting heat release during culture is the intrinsic thermodynamic efficiency of the organism in transferring the energy contained in the substrate to the biomass and products. Different organisms growing on the same substrate can result in different heats of reaction.

Anaerobic processes do not oxidize substrates as completely as aerobic processes, therefore, heats of reaction for anaerobic cultures tend to be lower than for aerobic cultures. For example, aerobic growth of yeast on glucose generates about 2000 kJ of heat per gmol of glucose consumed; in contrast, growth of yeast on glucose under anaerobic conditions with production of ethanol generates only about 100 kJ gmol^{-1} glucose [9]. As a consequence, the cooling requirements for aerobic cultures are typically significantly greater than for anaerobic cultures.

5.9 Energy Balance Equation for Cell Culture

The heat of reaction typically dominates the energy balance in bioprocesses such that small enthalpy effects due to sensible heat changes and heats of mixing can generally be ignored. Here, we incorporate these observations into a simplified energy balance equation for cell processes.

Consider Eq. (5.9) applied to a continuous bioreactor. What are the major factors responsible for the enthalpy difference between the input and output streams in bioprocesses? Because cell culture media are usually dilute aqueous solutions with behaviors close to ideal solutions, even though the composition of the broth may vary as substrates are consumed and products formed, changes in the heats of mixing of these solutes are generally negligible. Similarly, even though there may be a temperature difference between the input and output streams, the overall change in enthalpy due to sensible heat is also small. Usually, the heat of reaction, latent heats of phase change, and shaft work are the only energy effects necessary to consider in bioprocess energy balances. Evaporation is the most likely phase change in bioreactor operation; if evaporation is controlled then latent heat effects can also be ignored.

Metabolic reactions typically generate 5 to 20 kJ of heat per second per cubic meter of bioreactor broth for growth on carbohydrate, and up to 60 kJ s^{-1} m^{-3} for growth on hydrocarbon substrates. By way of comparison, in aerobic cultures sparged with dry air, evaporation of the bioreactor broth removes only about 0.5 kJ s^{-1} m^{-3} as latent heat. The energy input due to shaft work varies between 0.5 and 5 kJ s^{-1} m^{-3} in large vessels and 10 to 20 kJ s^{-1} m^{-3} in small vessels; the heat effects of stirring are therefore more important in small-scale than in large-scale processes. Sensible heats and heats of mixing are generally several orders of magnitude smaller.

Therefore, for cell cultures, we can simplify energy balance calculations by substituting expressions for the heat of reaction and latent heat of vaporization for the first two terms of Eq. (5.9). From the definition of Eq. (5.17), ΔH_{rxn} is the difference between the product and reactant enthalpies. As the products are contained in the output flow and the reactants in the input, ΔH_{rxn} is approximately equal to the difference in enthalpy between the input and output streams. If evaporation is significant, the enthalpy of the vapor leaving the system will be $M_v \Delta h_v$ greater than the enthalpy of the liquid entering the system or formed by reaction, where M_v is the mass of liquid evaporated and Δh_v is the latent heat of vaporization.

Taking these factors into account, the modified steady-state energy balance equation for cell cultures is as follows:

$$-\Delta H_{rxn} - M_v \Delta h_v - Q + W_s = 0 \qquad (5.28)$$

ΔH_{rxn} has a negative sign in Eq. (5.28) because ΔH_{rxn} is equal to [enthalpy of products—enthalpy of reactants], whereas the energy balance equation refers to [enthalpy of input streams—enthalpy of output streams]. As sensible heat effects are considered negligible, the difference between ΔH_{rxn}° and ΔH_{rxn} at the reaction temperature can be ignored. Equation (5.28) applies even if some proportion of the reactants remains unconverted or if there are tie components in the system that do not react. At steady state, any material added to the system that does not participate in reactions must leave in the output stream. Therefore, ignoring enthalpy effects due to change in temperature or solution, the enthalpy of unreacted material in the output stream must be equal to its inlet enthalpy. Thus, unreacted material and tie components make no contribution to the energy balance.

It should be emphasized that Eq. (5.28) is greatly simplified and, as discussed in Section 5.7.2, may not be applicable to single-enzyme conversions. It is, however, a very useful equation for cell culture processes. Because the heat of reaction in aerobic systems is generally higher than for anaerobic systems (Section 5.8.5), the principal assumption of Eq. (5.28) that heats of mixing and sensible energy changes are negligible compared with ΔH_{rxn} has greater validity in aerobic cultures.

5.10 Cell Culture Energy Balance Worked Examples

For processes involving cell growth and metabolism, the enthalpy change accompanying reaction is relatively large. Energy balances for aerobic and anaerobic cultures can therefore be carried out using the modified energy balance equation (5.28). Because this equation contains no enthalpy terms, it is not necessary to define reference states. Application of Eq. (5.28) to anaerobic fermentation is illustrated in Example 5.7.

In Example 5.7, the water used as the solvent for components of the nutrient medium was ignored. This water was effectively a tie component, moving through the system unchanged and not contributing to the energy balance. In this problem, the cooling requirements could be determined directly from the heat of reaction.

For aerobic cultures, we can relate the heat of reaction to O_2 consumption, providing a short-cut method for determining ΔH_{rxn}. Heats of combustion are not required in these calculations. Also, as long as the amount of O_2 consumed is known, the mass balance for the problem need not be completed. The procedure for energy balance problems involving aerobic bioprocess is illustrated in Example 5.8.

5.11 Summary of Chapter 5

At the end of Chapter 5 you should:

- Know which forms of energy are common in bioprocesses
- Know the *general energy balance* in words and as a mathematical equation, and the simplifications that can be made for bioprocesses
- Be familiar with *heat capacity* tables and be able to calculate *sensible heat changes*
- Be able to calculate *latent heat changes*
- Understand *heats of mixing* for nonideal solutions
- Be able to determine *standard heats of reaction* from *heats of combustion*
- Know how to determine heats of reaction for aerobic and anaerobic cell cultures
- Be able to carry out energy balance calculations for biological systems with and without reaction

---

EXAMPLE 5.7 Continuous Ethanol Fermentation

Saccharomyces cerevisiae is grown anaerobically in continuous culture at 30°C. Glucose is used as the carbon source; ammonia is the nitrogen source. A mixture of glycerol and ethanol is produced. At steady state, the net mass flows to and from the reactor are as follows.

glucose in	36.0 kg h^{-1}
NH$_3$ in	0.40 kg h^{-1}
cells out	2.81 kg h^{-1}
glycerol out	7.94 kg h^{-1}
ethanol out	11.9 kg h^{-1}
CO$_2$ out	13.6 kg h^{-1}
H$_2$O out	0.15 kg h^{-1}

Estimate the cooling requirements.

Solution

1. Assemble
   (i) Units kg, kJ, h, °C
   (ii) Flow sheet
        The flow sheet for this process is shown in Figure 5.9
   (iii) System boundary
         The system boundary is shown in Figure 5.9

2. Analyze
   (i) Assumptions:
   - steady state
   - no leaks
   - system is homogeneous
   - heat of combustion for yeast is $-21.2 \, \text{kJ g}^{-1}$
   - ideal solutions
   - negligible sensible heat change
   - no shaft work
   - no evaporation
   (ii) Basis
        36.0 kg glucose, or 1 h
   (iii) Extra data
   - MW glucose = 180 g gmol^{-1}
   - MW NH$_3$ = 17 g gmol^{-1}
   - MW glycerol = 92 g gmol^{-1}
   - MW ethanol = 46 g gmol^{-1}

   Heats of combustion (Appendix C):

   $$\left(\Delta h_c^\circ\right)_{glucose} = -2805.0 \, \text{kJ gmol}^{-1}$$

   $$\left(\Delta h_c^\circ\right)_{NH_3} = -382.6 \, \text{kJ gmol}^{-1}$$

   $$\left(\Delta h_c^\circ\right)_{glycerol} = -1655.4 \, \text{kJ gmol}^{-1}$$

   $$\left(\Delta h_c^\circ\right)_{ethanol} = -1366.8 \, \text{kJ gmol}^{-1}$$

   (iv) Reaction

   $$\text{glucose} + \text{NH}_3 \rightarrow \text{biomass} + \text{glycerol} + \text{ethanol} + \text{CO}_2 + \text{H}_2\text{O}$$

   All components are involved in reaction.
   (v) Mass balance equation
       The mass balance is already complete: the total mass of components in equals the total mass of components out.
   (vi) Energy balance equation
        For cell metabolism, the modified steady-state energy balance equation is Eq. (5.28):

   $$-\Delta H_{rxn} - M_v \Delta h_v - Q + W_s = 0$$

3. Calculate

$W_s = 0\,kJ$; $M_v = 0\,gmol$. Therefore, the energy balance equation is reduced to:

$$-\Delta H_{rxn} - Q = 0$$

Evaluate the heat of reaction using Eq. (5.20). As the heat of combustion of H_2O and CO_2 is zero, the heat of reaction is:

$$\Delta H_{rxn} = \left(n\Delta h_c^\circ\right)_G + \left(n\Delta h_c^\circ\right)_A - \left(n\Delta h_c^\circ\right)_B - \left(n\Delta h_c^\circ\right)_{Gly} - \left(n\Delta h_c^\circ\right)_E$$

where G = glucose, A = ammonia, B = cells, Gly = glycerol, and E = ethanol. Because, in this problem, we are given the masses of reactants and products involved in the reaction, we can apply the equation for ΔH_{rxn} in mass terms:

$$\Delta H_{rxn} = \left(M\Delta h_c^\circ\right)_G + \left(M\Delta h_c^\circ\right)_A - \left(M\Delta h_c^\circ\right)_B - \left(M\Delta h_c^\circ\right)_{Gly} - \left(M\Delta h_c^\circ\right)_E$$

where Δh_c° is expressed per unit mass. Converting the Δh_c° data to $kJ\ kg^{-1}$:

$$\left(\Delta h_c^\circ\right)_G = -2805.0\,\frac{kJ}{gmol} \cdot \left|\frac{1\ gmol}{180\ g}\right| \cdot \left|\frac{1000\ g}{1\ kg}\right| = -1.558 \times 10^4\,kJ\ kg^{-1}$$

$$\left(\Delta h_c^\circ\right)_A = -382.6\,\frac{kJ}{gmol} \cdot \left|\frac{1\ gmol}{17\ g}\right| \cdot \left|\frac{1000\ g}{1\ kg}\right| = -2.251 \times 10^4\,kJ\ kg^{-1}$$

$$\left(\Delta h_c^\circ\right)_B = -21.2\,\frac{kJ}{g} \cdot \left|\frac{1000\ g}{1\ kg}\right| = -2.120 \times 10^4\,kJ\ kg$$

$$\left(\Delta h_c^\circ\right)_{Gly} = -1655.4\,\frac{kJ}{gmol} \cdot \left|\frac{1\ gmol}{92\ g}\right| \cdot \left|\frac{1000\ g}{1\ kg}\right| = -1.799 \times 10^4\,kJ\ kg^{-1}$$

$$\left(\Delta h_c^\circ\right)_E = -1366.8\,\frac{kJ}{gmol} \cdot \left|\frac{1\ gmol}{46\ g}\right| \cdot \left|\frac{1000\ g}{1\ kg}\right| = -2.971 \times 10^4\,kJ\ kg^{-1}$$

Therefore:

$$\Delta H_{rxn} = (36.0\ kg)\left(-1.558 \times 10^4\ kJ\ kg^{-1}\right) + (0.4\ kg)\left(-2.251 \times 10^4\ kJ\ kg^{-1}\right)$$
$$- (2.81\ kg)\left(-2.120 \times 10^4\ kJ\ kg^{-1}\right) - (7.94\ kg)\left(-1.799 \times 10^4\ kJ\ kg^{-1}\right)$$
$$- (11.9\ kg)\left(-2.971 \times 10^4\ kJ\ kg^{-1}\right)$$
$$\Delta H_{rxn} = -1.392 \times 10^4\ kJ$$

Substituting this result into the energy balance equation:

$$Q = 1.392 \times 10^4\ kJ$$

Q is positive, indicating that heat must be removed from the system.

4. Finalize

$1.4 \times 10^4\,kJ$ heat must be removed from the bioreactor per hour.

EXAMPLE 5.8 Citric Acid Production

Citric acid is manufactured using submerged culture of *Aspergillus niger* in a batch reactor operated at 30°C (Figure 5.10). Over a period of 2 days, 2500 kg glucose and 860 kg O_2 are consumed to produce 1500 kg citric acid, 500 kg biomass, and other products. Ammonia is used as the nitrogen source. Power input to the system by mechanical agitation of the broth is about 15 kW; approximately 100 kg of water is evaporated over the culture period. Estimate the cooling requirements.

Solution

1. Assemble
    (i) Units kg, kJ, h, °C
    (ii) Flow sheet
         The flow sheet is shown in Figure 5.11.
    (iii) System boundary
         The system boundary is shown in Figure 5.11.

2. Analyze
    (i) Assumptions
        • system is homogeneous
        • no leaks
        • ideal solutions
        • negligible sensible heat
        • heat of reaction at 30°C is −460 kJ $gmol^{-1}$ O_2 consumed
    (ii) Basis
         1500 kg citric acid produced, or 2 days
    (iii) Extra data
         Δh_v water at 30°C = 2430.7 kJ kg^{-1}
    (iv) Reaction

$$\text{glucose} + O_2 + NH_3 \rightarrow \text{ biomass} + CO_2 + H_2O + \text{ citric acid}$$

    All components are involved in reaction.
    (v) Mass balance
         The mass balance need not be completed, as the sensible energies associated with the inlet and outlet streams are assumed to be negligible.
    (vi) Energy balance
         The aim of the integral energy balance for batch culture is to calculate the amount of heat that must be removed to produce zero accumulation of energy in the system. Equation (5.29) is appropriate:

$$-\Delta H_{rxn} - M_v \Delta h_v - Q + W_s = 0$$

    where each term refers to the 2-day culture period.

3. Calculate

ΔH_{rxn} is related to the amount of O_2 consumed:

$$\Delta H_{rxn} = (-460 \text{ kJ gmol}^{-1})(860 \text{ kg}) \cdot \left|\frac{1000 \text{ g}}{1 \text{ kg}}\right| \cdot \left|\frac{1 \text{ gmol}}{32 \text{ g}}\right|$$

$$\Delta H_{rxn} = -1.24 \times 10^7 \text{ kJ}$$

Heat lost through evaporation is:

$$M_v \Delta h_v = (100 \text{ kg})(2430.7 \text{ kJ kg}^{-1}) = 2.43 \times 10^5 \text{ kJ}$$

Power input by mechanical agitation is 15 kW or 15 kJ s^{-1}. Over a period of 2 days:

$$W_s = (15 \text{ kJs}^{-1})(2 \text{ days}) \cdot \left|\frac{3600 \text{ s}}{1 \text{ h}}\right| \cdot \left|\frac{24 \text{ h}}{1 \text{ day}}\right| = 2.59 \times 10^6 \text{ kJ}$$

These results can now be substituted into the energy balance equation:

$$-(-1.24 \times 10^7 \text{ kJ}) - (2.43 \times 10^5 \text{ kJ}) - Q + (2.59 \times 10^6 \text{ kJ}) = 0$$
$$Q = 1.47 \times 10^7 \text{ kJ}$$

Q is positive, indicating that heat must be removed from the system. Note the relative magnitudes of the energy contributions from heat of reaction, shaft work, and evaporation; the effects of evaporation can often be ignored.

4. Finalize

1.5×10^7 kJ heat must be removed from the bioreactor per 1500 kg of citric acid produced.

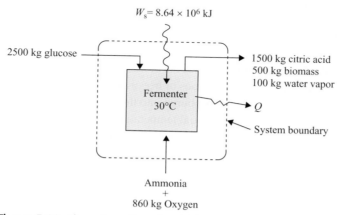

Figure 5.10 Flow sheet for microbial production of citric acid.

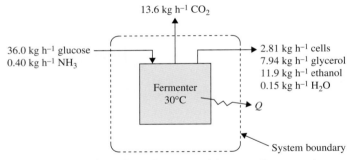

Figure 5.11 Flow sheet for anaerobic yeast fermentation.

Problems

5.1. Sensible energy change

Calculate the enthalpy change associated with the following processes:

a. *m*-Cresol is heated from 50°C to 100°C

b. Ethylene glycol is cooled from 40°C to 10°C

c. Succinic acid is heated from 10°C to 100°C

d. Air is cooled from 140°C to 50°C

5.2. Heat of vaporization

Nitrogen (N_2) is sometimes bubbled into bioreactors to maintain anaerobic conditions. It does not react and leaves in the bioreactor off-gas. However, it can strip water from the bioreactor so that water vapor also leaves in the off-gas. In a continuous bioreactor operated at 37°C, $30\,g\,h^{-1}$ water is evaporated in this way. How much heat must be put into the system to compensate for evaporative cooling and so maintain the temperature at a constant value?

5.3. Designer coffee mug

An award-winning manufacturer of designer homewares is developing a high-tech coffee mug based on thermodynamic principles. The aim is to keep the contents of the mug at a constant, optimum drinking temperature. When hot coffee is poured into the mug, the heat melts a reservoir of beeswax in an insulated jacket around the outside of the mug. Until all the beeswax is subsequently solidified, the temperature of the coffee will be maintained at 74°C, the melting point of beeswax. Other relevant properties of beeswax are: $\Delta h_f = 190\,J\,g^{-1}$; Cp solid = $1.6\,J\,g^{-1}\,°C^{-1}$. Freshly brewed coffee is poured into the mug at an average temperature of 92°C. If the mug holds 250 mL and is at an initial temperature of 25°C, what is the maximum amount of beeswax that can be used per mug if all the beeswax must be melted by the hot coffee? Assume that the other components of the coffee mug have negligible heat capacity and there are no other heat losses from the coffee.

5.4. Medium preparation

Nutrient medium in an insulated bioreactor is diluted and brought to sterilization temperature by injection with clean steam. Ten thousand kg of liquid medium at 130°C is required at the end of the process. Initially, the medium concentrate in the tank is at 40°C; the steam used is saturated at 220°C.

a. How much steam is required?

b. How much medium concentrate is required?

5.5. Enzyme conversion

An immobilized enzyme process is used in an ice-cream factory to hydrolyze lactose ($C_{12}H_{22}O_{11}$) to glucose ($C_6H_{12}O_6$) and galactose ($C_6H_{12}O_6$):

$$C_{12}H_{22}O_{11} + H_2O \rightarrow C_6H_{12}O_6 + C_6H_{12}O_6$$

Gel beads containing β-galactosidase are packed into a column reactor; 2500 kg of lactose enters the reactor per day as a 10% solution in water at 25°C. The reactor operates at steady state and 32°C; all of the lactose is converted. Because the heat of reaction for enzyme conversions is not as great as for cell culture, sensible heat changes and heats of mixing cannot be ignored.

	Δh_c (kJ gmol^{-1})	Cp (cal g^{-1} °C^{-1})	Δh_m (kcal gmol^{-1})
Lactose	−5652.5	0.30	3.7
Water	−	1.0	−
Glucose	−2805.0	0.30	5.6
Galactose	−2805.7	0.30	5.6

a. What is the standard heat of reaction for this enzyme conversion?

b. Estimate the heating or cooling requirements for this process. State explicitly whether heating or cooling is needed.

5.6. Production of glutamic acid

Immobilized cells of a genetically improved strain of *Brevibacterium lactofermentum* are used to convert glucose to glutamic acid for production of MSG (monosodium glutamate). The immobilized cells are unable to grow, but metabolize glucose according to the equation:

$$C_6H_{12}O_6 + NH_3 + 1.5O_2 \rightarrow C_5H_9O_4N + CO_2 + 3H_2O$$

A feed stream of 4% (w/w) glucose in water enters a 25,000-L reactor at 25°C at a flow rate of 2000 kg h^{-1}. A gaseous mixture of 12% NH_3 in air is sparged into the reactor at 1 atm and 15°C at a flow rate of 4 vvm (1 vvm means 1 vessel volume per minute). The product stream from the reactor contains residual sugar at a concentration of 0.5%.

a. Estimate the cooling requirements.

b. How important is cooling in this bioprocess? For example, assuming the reaction rate remains constant irrespective of temperature, if cooling were not provided and the reactor operated adiabatically, what would be the temperature? (In fact, the rate of conversion will decline rapidly at high temperatures due to cell death and enzyme deactivation.)

5.7. Bacterial production of alginate

Azotobacter vinelandii is investigated for production of alginate from sucrose. In a continuous bioreactor at 28°C with ammonia as nitrogen source, the yield of alginate was found to be

4 g alginate (g O_2)$^{-1}$. It is planned to produce alginate at a rate of 5 kg h^{-1}. Since the viscosity of alginate in aqueous solution is considerable, energy input due to mixing the broth cannot be neglected. The bioreactor is equipped with a flat-blade disc turbine; at a satisfactory mixing speed and air flow rate, the power requirements are estimated at 1.5 kW. Calculate the cooling requirements.

5.8. Acid fermentation

Propionibacterium species are tested for commercial-scale production of propionic acid. Propionic and other acids are synthesized in anaerobic culture using sucrose as the substrate and ammonia as the nitrogen source. Overall yields from sucrose are as follows:

propionic acid	40% (w/w)
acetic acid	20% (w/w)
butyric acid	5% (w/w)
lactic acid	3.4% (w/w)
biomass	12% (w/w)

Bacteria are inoculated into a vessel containing sucrose and ammonia; a total of 30 kg sucrose is consumed over a period of 10 days. What are the cooling requirements?

5.9. Ethanol fermentation

A crude bioreactor is set up in a shed in the backyard of a suburban house. Under anaerobic conditions with ammonia as the nitrogen source, about 0.45 g ethanol is formed per g glucose consumed. At steady state, the ethanol production rate averages 0.4 kg h^{-1}. The owner of this enterprise decides to reduce her electricity bill by using the heat released during the fermentation to warm water as an adjunct to the household hot water system. Cold water at 10°C is fed into a jacket surrounding the bioreactor at a rate of 2.5 L h^{-1}. To what temperature is the water heated? Heat losses from the system are negligible. Use a biomass composition of $CH_{1.75}O_{0.58}N_{0.18}$ plus 8% ash.

5.10. Production of bakers' yeast

Bakers' yeast is produced in a 50,000-L bioreactor under aerobic conditions. The carbon substrate is sucrose; ammonia is provided as the nitrogen source. The average biomass composition is $CH_{1.83}O_{0.55}N_{0.17}$ with 5% ash. Under conditions supporting efficient growth, biomass is the only major product and the biomass yield from sucrose is 0.5 g biomass (g sucrose)$^{-1}$. If the specific growth rate is 0.45 h^{-1}, estimate the rate of heat removal required to maintain constant temperature in the bioreactor when the yeast concentration is 10 g L^{-1}.

5.11. Culture kinetic parameters from thermal properties

A 250-L airlift bioreactor is used for continuous aerobic culture of *Rhizobium etli* bacteria. Succinic acid ($C_4H_6O_4$) is used as the carbon source; ammonium chloride is used as the nitrogen source. The products of the culture are biomass, carbon dioxide, and water only. The bioreactor is equipped with an external jacket through which water flows at a rate of 100 kg h^{-1}. The inlet water temperature is 20°C. The outside of the jacket is covered with insulation to prevent heat losses. At steady state, the cell concentration is 4.5 g L^{-1}, the rate of consumption of succinic acid is 395 g h^{-1}, and the outlet water temperature is 27.5°C.

a. What is the enthalpy change of the cooling water in the jacket? Use Cp water $= 4.2\,\text{kJ}\,\text{kg}^{-1}\,°\text{C}^{-1}$.

b. Assuming that all the heat generated by the culture is absorbed by the cooling water, estimate the rate of O_2 uptake by the cells. What other assumptions are involved in your answer?

c. Determine the biomass growth rate in units of $g\,h^{-1}$.

d. What is the specific growth rate of the cells in units of h^{-1}?

5.12. Production of snake antivenom

To satisfy strong market demand, a pharmaceutical company has developed a hybridoma cell line that synthesizes a monoclonal antibody capable of neutralizing the venom of the Australian death adder. In culture, the hybridoma cells exhibit mixed oxido-reductive energy metabolism.

a. The heat generated by the cells during growth is measured using a flow microcalorimeter attached to a small bioreactor. The relationship between the amount of heat generated and the amount of O_2 consumed is found to be $680\,\text{kJ gmol}^{-1}$, which is considerably higher than the $460\,\text{kJ gmol}^{-1}$ expected for fully oxidative metabolism. These data are used to estimate the heat of reaction associated with the anaerobic components of hybridoma cell activity. It is hypothesized that the level of production of lactic acid is a direct indicator of anaerobic metabolism, leading to the equation:

$$\Delta H_{rxn} \text{ (oxido-reductive)}$$
$$= \Delta H_{rxn} \text{ (fully oxidative)} + \Delta h_{rxn} \text{ (anaerobic)} \times \text{gmol lactic acid produced}$$

If the molar ratio of lactic acid production to O_2 consumption measured in the bioreactor is 5.5, estimate Δh_{rxn} for the anaerobic metabolic pathways.

b. For commercial antivenin production, the hybridoma culture is scaled up to a 500-L stirred bioreactor. Under these conditions, the specific rate of O_2 uptake is $0.3\,\text{mmol}\ (10^9\ \text{cells})^{-1}\,h^{-1}$, the specific rate of lactic acid production is $1.0\,\text{mmol}\ (10^9\ \text{cells})^{-1}\,h^{-1}$, and the maximum cell density is $7.5 \times 10^6\ \text{cells mL}^{-1}$.

  - Estimate the maximum rate of cooling required to maintain the culture temperature.
  - A new medium is developed along stoichiometric principles to bring the levels of nutrients provided to the cells closer to actual requirements. When the new medium is used, lactic acid production decreases significantly to only $0.05\,\text{mmol}\ (10^9\ \text{cells})^{-1}\,h^{-1}$. If the O_2 demand stays roughly the same, what will be the new cooling requirements?
  - The heat generated by the culture is absorbed by cooling water flowing in a coil inside the bioreactor. Water enters the coil at $20°C$ and leaves at $29°C$. What maximum flow rate of cooling water is required for cultures growing in the new medium? Use Cp water $= 4.19\,\text{kJ}\,\text{kg}^{-1}\,°\text{C}^{-1}$.

5.13. Ginseng production

Suspension cultures of *Panax ginseng* plant cells are used to produce ginseng biomass for the health tonic market. The cells are grown in batch culture in a 2500-L stirred bioreactor operated at $25°C$ and 1 atm pressure. Over a period of 12 days, $5.5 \times 10^6\,L$ of air at $25°C$ and 1 atm are pumped into the reactor; off-gas containing $5.1 \times 10^5\,L$ of O_2 leaves the vessel over the same period. About 135 kg of water is lost by evaporation during the culture. Estimate the cooling requirements.

5.14. Evaporative cooling

An engineer decides to design a special airlift reactor for growing *Torula utilis* cells. The reactor relies on evaporative cooling only. Water is evaporated from the broth by sparging with dry air; water vapor leaving the vessel is then condensed in an external condenser and returned to the culture. On average, 0.5 kg water is evaporated per day. If the rate of O_2 uptake by the cells is 140 mmol h^{-1} and the culture temperature must be maintained at 30°C, does evaporation provide adequate cooling for this system?

5.15. Penicillin process

Penicillium chrysogenum is used to produce penicillin in a 90,000-L bioreactor. The volumetric rate of O_2 uptake by the cells ranges from 0.45 to 0.85 mmol L^{-1} min^{-1} depending on time during the culture. Power input by stirring is 2.9 W L^{-1}. Estimate the cooling requirements.

5.16. Culture of methylotrophic yeast

a. Cells of *Pichia pastoris* yeast are cultured in medium containing glycerol ($C_3H_8O_3$) as the carbon source and ammonium hydroxide as the nitrogen source. Under aerobic conditions, the yield of biomass from substrate is 0.57 g biomass (g glycerol)$^{-1}$. Biomass, CO_2, and H_2O are the only major products of this culture. Estimate the cooling requirements per g of biomass produced.

b. When all the glycerol has been consumed, methanol (CH_4O) is added to the culture. There is enough ammonium hydroxide remaining in the medium to support growth of the biomass with methanol as substrate. The composition of the cells is essentially unchanged with a biomass yield from substrate of 0.44 g biomass (g methanol)$^{-1}$. In what way do the cooling demands change per g of biomass produced?

5.17. Algal culture for carotenoid synthesis

A newly discovered, carotenoid-producing microalga isolated from an artesian bore in the Simpson Desert is cultured under aerobic conditions in a 100-L bubble column reactor. The medium contains glucose as the primary carbon source and ammonia as the nitrogen source. The elemental formula for the biomass is $CH_{1.8}O_{0.6}N_{0.2}$; because carotenoids are present in the cells at levels of only a few μg per g biomass, they do not appreciably alter the biomass formula. The yield of biomass (including carotenoid) from substrate is 0.45 g biomass (g glucose)$^{-1}$; biomass, carbon dioxide, and water can be considered the only products formed. The reactor is operated continuously so that the rate of glucose consumption is 77 g h^{-1}. Estimate the cooling requirements. What assumptions are involved in your answer?

5.18. Checking the consistency of measured culture data

The data in the following table are measured during aerobic culture of *Corynebacterium glutamicum* on medium containing molasses, corn extract, and other nutrients for production of lysine.

Measured variable	Time period of cultivation (h)	
	0–12	12–36
Rate of uptake of reducing sugars (g L^{-1} h^{-1})	0.42	2.0
Rate of biomass production (g L^{-1} h^{-1})	0.29	0.21

Measured variable	Time period of cultivation (h)	
	0–12	12–36
Rate of lysine production (g L^{-1} h^{-1})	0.20	0.66
Rate of O_2 uptake (g L^{-1} h^{-1})	0.40	0.75
Rate of heat evolution (kJ L^{-1} h^{-1})	2.5	12.1

a. Use mass and energy balance principles to check whether the data for the 12–36-h culture period are consistent (e.g., to within ±10%).

b. It is suspected that amino acids present in the nutrient medium may initially provide an additional source of substrate in this culture. Are the data for the 0–12-h period consistent with this theory?

c. At the beginning of the culture, accurate measurement of O_2 uptake is difficult so that heat evolution is considered a more reliable indicator of O_2 consumption than direct O_2 measurements. If the amino acids provided in the nutrient medium can be considered to have the same average properties as glutamine, estimate the mass rate of amino acid uptake during the first 12 h of the culture.

5.19. Thermal mixing

Hot water at 90°C enters a mixing tank at a flow rate of 500 kg h^{-1}. At the same time, cool water at 18°C also flows into the tank. At steady state, the temperature of the discharge stream is 30°C and the mass of water in the tank is 1100 kg. Suddenly, the temperature of the cool water being added increases to 40°C. How long does it take the discharge stream to reach 40°C?

5.20. Laboratory heating

A one-room mobile laboratory used for field work is equipped with an electric furnace. The maximum power provided by the furnace is 2400 W. One winter morning when the furnace is switched off, the temperature in the laboratory falls to 5°C, which is equal to the outside air temperature. The furnace is turned on to its maximum setting, and after 30 min, the temperature in the laboratory reaches 20°C. The rate of heat loss from the room through the walls, ceiling, and windowpanes depends on the outside temperature according to the equation:

$$\hat{Q}_{loss} = k\left(T - T_o\right)$$

where $\hat{Q}_{loss}$ is the rate of heat loss, k is a constant, T is the inside temperature, and T_o is the outside temperature. If $k = 420$ kJ $°C^{-1}$ h^{-1}, estimate the "heat capacity" of the laboratory, that is, the energy required to raise the temperature of the room by 1°C.

5.21. Unsteady State Energy Balance: Boiling water

A beaker containing 2 L of water at 18°C is placed on a laboratory hot plate. The water begins to boil in 11 min.

a. Neglecting evaporation, write the energy balance for the process.

b. The hot plate delivers heat at a constant rate. Assuming that the heat capacity of water is constant, what is that rate?

References

[1] D.W. Green, R.H. Perry, Perry's Chemical Engineers' Handbook, Eighth Edition, McGraw-Hill, New York, Chicago, San Francisco, Lisbon, London, Madrid, Mexico City, Milan, New Delhi, San Juan, Seoul, Singapore, Sydney, Toronto, 2008, 1997, 1984, 1973, 1963, 1950, 1941, 1934. https://www.accessengineeringlibrary.com/content/book/9780071834087

[2] D.R. Lide, CRC Handbook of Chemistry and Physics, eighty-fourth ed., CRC Press, New York, 2003.

[3] National Research Council, International Critical Tables of Numerical Data, Physics, Chemistry and Technology, The National Academies Press, Washington, DC, 1930. https://doi.org/10.17226/20230

[4] B. Atkinson, F. Mavituna, Biochemical Engineering and Biotechnology Handbook, second ed., Macmillan, 1991.

[5] S.A. Patel, L.E. Erickson, Estimation of heats of combustion of biomass from elemental analysis using available electron concepts, Biotechnol. Bioeng. 23 (1981) 2051–2067.

[6] J.A. Roels, Energetics and Kinetics in Biotechnology, Elsevier Biomedical Press, 1983.

[7] C.L. Cooney, D.I.C. Wang, R.I. Mateles, Measurement of heat evolution and correlation with oxygen consumption during microbial growth, Biotechnol. Bioeng. 11 (1968) 269–281.

[8] J.-L. Cordier, B.M. Butsch, B. Birou, U. von Stockar, The relationship between elemental composition and heat of combustion of microbial biomass, Appl. Microbiol. Biotechnol. 25 (1987) 305–312.

[9] E.H. Battley, Energetics of Microbial Growth, John Wiley, 1987.

Suggestions for Further Reading

Process Energy Balances

R.M. Felder, R.W. Rousseau, Elementary Principles of Chemical Processes, 3rd ed., Wiley, 2005.

D.M. Himmelblau, J.B. Riggs, Basic Principles and Calculations in Chemical Engineering, 7th ed., Prentice Hall, 2004.

R.K. Sinnott, Chapter 3, Coulson and Richardson's Chemical Engineering, volume 6: Chemical Engineering Design, 4th ed., Elsevier, 2005.

Metabolic Thermodynamics

See also references 5–9.

Y. Guan, P.M. Evans, R.B. Kemp, Specific heat flow rate: an on-line monitor and potential control variable of specific metabolic rate in animal cell culture that combines microcalorimetry with dielectric spectroscopy, Biotechnol. Bioeng. 58 (1998) 464–477.

I. Marison, U. von Stockar, A calorimetric investigation of the aerobic cultivation of *Kluyveromyces fragilis* on various substrates, Enzyme Microbiol. Technol. 9 (1987) 33–43.

U. von Stockar, I.W. Marison, The use of calorimetry in biotechnology, Adv. Biochem. Eng./Biotechnol 40 (1989) 93–136.

Physical Processes

CHAPTER 6

Fluid Flow

Fluid mechanics is a core discipline area of engineering science concerned with the nature and properties of fluids in motion and at rest. Fluids play a central role in bioprocesses because most physical, chemical, and biological transformations take place in a fluid phase. As the behavior of fluids depends to a large extent on their physical characteristics, knowledge of fluid properties and techniques for their measurement is crucial. Fluids in bioprocessing often contain suspended solids, consist of more than one phase, and have non-Newtonian properties; all of these features complicate the analysis of flow behavior and present many challenges in bioprocess design. In bioreactors, fluid properties play a key role in determining the effectiveness of mixing, gas dispersion, mass transfer, and heat transfer. Together, these processes can exert a significant influence on system productivity and the success of process scale-up.

Fluid mechanics accounts for a substantial fraction of the chemical engineering literature; accordingly, complete treatment of the subject is beyond the scope of this book. Here, we content ourselves with study of those aspects of flow behavior particularly relevant to bioprocess fluids and the operation of bioprocessing equipment. Further information can be found in the references at the end of the chapter.

6.1 Classification of Fluids

A fluid is a substance that undergoes continuous deformation when subjected to a shearing force. Shearing forces act tangentially to the surfaces over which they are applied. For example, a simple shear force exerted on a stack of thin parallel plates will cause the plates to slide over each other, as in a pack of cards. Shear can also occur in other geometries; the effect of shear forces in planar and rotational systems is illustrated in Figure 6.1. Shearing in these examples causes *deformation*, which is a change in the relative positions of parts of a body. A shear force must be applied to produce fluid flow.

According to the preceding definition, fluids can be either gases or liquids. Two physical properties, viscosity and density, are used to classify fluids. If the density of a fluid changes with pressure, the fluid is *compressible*. Gases are generally classed as compressible fluids. The density of liquids is practically independent of pressure; liquids are *incompressible* fluids. Sometimes the distinction between compressible and incompressible fluids is not well

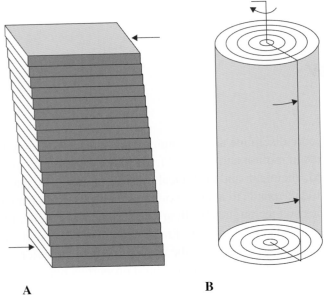

A   **B**

Figure 6.1 Laminar deformation due to (A) planar shear and (B) rotational shear. *From J.R. van Wazer, J.W. Lyons, K.Y. Kim, R.E. Colwell, Viscosity and Flow Measurement, John Wiley, 1963.*

defined; for example, a gas may be treated as incompressible if the variation in pressure and temperature is small.

Fluids are also classified based on viscosity. Viscosity is the property of fluids responsible for internal friction during flow. An *ideal* or *perfect* fluid is a hypothetical liquid or gas that is incompressible and has zero viscosity. The term *inviscid* applies to fluids with zero viscosity. All *real* fluids have finite viscosity and are therefore called *viscid* or *viscous* fluids. Fluids can be classified further as *Newtonian* or *non-Newtonian*. This distinction is explained in detail in later sections.

6.2 Fluids in Motion

Bioprocesses involve fluids in motion in vessels, pipes, and other equipment. Some general characteristics of fluid flow are described in the following sections.

6.2.1 Streamlines

When a fluid flows through a pipe or over a solid object, the velocity of the fluid varies depending on position. One way of representing variation in velocity is through *streamlines*, which follow the flow path. Constant velocity is shown by equidistant spacing of parallel streamlines as shown in Figure 6.2(a). The velocity profile for slow-moving fluid flowing over

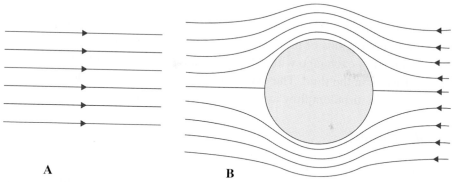

Figure 6.2 Streamlines for (A) constant fluid velocity and (B) steady flow over a submerged object.

a submerged object is shown in Figure 6.2(b); reduced spacing between the streamlines indicates that the velocity at the top and bottom of the object is greater than at the front and back.

Streamlines show only the net effect of fluid motion. Although streamlines suggest smooth continuous flow, the individual particles or parcels of fluid may actually be moving in an erratic fashion. The slower the flow, the more closely the streamlines represent actual motion. Slow fluid flow is therefore called *streamline* or *laminar* flow. In contrast, during fast motion, particles of fluid frequently cross and recross the streamlines. This is called *turbulent* flow, which is characterized by the formation of local regions of fluid rotation called *eddies*.

6.2.2 Shear Stress

Shear stresses develop in fluids when different parts of the fluid move relative to each other. Consider a flow stream with adjacent layers of fluid moving steadily at different velocities. Within the fluid there is continuous molecular movement so that an interchange of molecules occurs between the layers. This interchange influences the velocity of flow in the layers. For example, molecules moving from a slow layer will reduce the speed of an adjacent faster layer, while those coming from a fast layer will have an accelerating effect on a slower layer. This phenomenon is called *viscous drag* and gives rise to *viscous forces* in fluids as neighboring fluid layers transfer momentum and affect each other's velocity. Such forces are often referred to as *viscous shear stresses*, where shear stress is defined as force divided by the area of interaction between the fluid layers. Shear stresses in fluids are therefore induced by fluid flow and are a consequence of velocity differences within the fluid. The overall effect of viscous shear stress is to reduce the differences in velocity between adjacent fluid layers or streamlines. Shear stress can also be thought of as a source of resistance to changes in fluid motion.

In turbulent flow, the situation is more complex than that described for steady flow of adjacent fluid layers. Particles of fluid in turbulent flow undergo chaotic and irregular patterns of motion and are subject to large and abrupt changes in velocity. Slow-moving fluid can jump quickly into fast streams of motion; similarly, fast-moving fluid particles frequently arrive in relatively slow-moving regions of flow. The overall effect of this behavior is a huge increase in the rate of momentum exchange within the flow stream and, consequently, much greater effective shear stresses in the fluid. The extra shear stress contributions due to the strong velocity fluctuations in turbulent flow are known as *Reynolds stresses.*

6.2.3 Reynolds Number

The transition from laminar to turbulent flow depends not only on the velocity of the fluid but also on its viscosity and density and the geometry of the flow system. A dimensionless parameter used to characterize fluid flow is the *Reynolds number*. For full flow in pipes with circular cross-section, the Reynolds number *Re* is defined as:

$$Re = \frac{Du\rho}{\mu} \tag{6.1}$$

where D is the pipe diameter, u is the average linear velocity of the fluid, ρ is the fluid density, and μ is the fluid viscosity. For stirred vessels there is another definition of Reynolds number:

$$Re_i = \frac{N_i D_i^2 \rho}{\mu} \tag{6.2}$$

where Re_i is the *impeller Reynolds number*, N_i is the stirrer speed, D_i is the impeller diameter, ρ is the fluid density, and μ is the fluid viscosity. The Reynolds number is a dimensionless variable; the units and dimensions of the parameters on the right side of Eqs. (6.1) and (6.2) cancel.

The Reynolds number is named after Osborne Reynolds, who, in 1883, published a classical series of papers on the nature of flow in pipes. One of the most significant outcomes of Reynolds's experiments is that there is a *critical Reynolds number* marking the upper boundary for laminar flow in pipes. In smooth pipes, laminar flow is encountered at Reynolds numbers less than 2100. Under normal conditions, flow is turbulent at *Re* above about 4000. Between 2100 and 4000 is the *transition region* where flow changes between laminar and turbulent; flow characteristics in this region also depend on conditions at the entrance of the pipe and other variables. Flow in stirred tanks may be laminar or turbulent as a function of the impeller Reynolds number. The value of Re_i marking the transition depends on the geometry of the impeller and tank; however, for several commonly used mixing systems, flow is laminar at $Re_i < 10$ and turbulent at $Re_i > 10^4$. Thus, unlike for pipe flow, in many stirred tanks there is a relatively large flow-regime transition region of $10 \leq Re_i \leq 10^4$. Conceptually, *the Reynolds number represents the ratio of inertial forces to viscous forces* in the fluid and quantifies the relative importance of these two types of forces for given flow conditions. The Reynolds number is a dimensionless number used to categorize fluid systems in which the effect of viscosity is important in controlling the velocities or the flow pattern of a fluid and it is this ratio that determines whether flow remains laminar or becomes turbulent.

6.2.4 Hydrodynamic Boundary Layers

In most applications, fluid flow occurs in the presence of a stationary solid surface, such as the walls of a pipe or tank. That part of the fluid where flow is affected by the solid is called the *boundary layer*. As an example, consider flow of fluid parallel to the flat plate shown in Figure 6.3. Contact between the moving fluid and the plate causes the formation of a

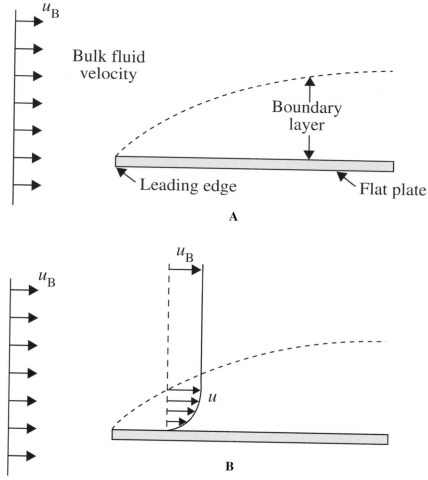

Figure 6.3 Fluid boundary layer for flow over a flat plate. (A) The boundary layer forms at the leading edge. (B) Compared with velocity u_B in the bulk fluid, the velocity u in the boundary layer is zero at the plate surface but increases with distance from the plate to reach u_B at the outer limit of the boundary layer.

boundary layer beginning at the leading edge and developing on both the top and bottom of the plate. Figure 6.3 shows only the upper stream; fluid motion below the plate will be a mirror image of that above.

When fluid flows over a stationary object, a thin film of fluid in contact with the surface adheres to it to prevent slippage over the surface. The fluid velocity at the surface of the plate in Figure 6.3 is therefore zero. When part of a flowing fluid has been brought to rest, the flow of adjacent fluid layers will be slowed by viscous drag. This phenomenon is illustrated in Figure 6.3(b). The velocity of fluid within the boundary layer, u, is represented by arrows; u is zero at the surface of the plate. Viscous drag forces are transmitted upward through the fluid from the stationary layer at the surface. The fluid layer just above the surface moves at a slow but finite velocity; layers further above move at increasing velocity as the drag forces associated with the stationary layer decrease. At the edge of the boundary layer, the fluid is unaffected by the presence of the plate and the velocity is close to that of the bulk flow, u_B. Formation of boundary layers is important not only in determining the characteristics of fluid flow but also for transfer of heat and mass between phases. These topics are discussed further in other chapters.

6.3 Viscosity

Viscosity is the most important property affecting the flow behavior of a fluid; it is related to the fluid's resistance to motion. Viscosity has a marked effect on pumping, mixing, mass transfer, heat transfer, and aeration of fluids; these in turn exert a major influence on bioprocess design and economics. The viscosity of bioprocess fluids is affected by the presence of cells, substrates, products, and gas.

Viscosity is an important aspect of *rheology*, the science of deformation and flow. Viscosity is the parameter used to relate the shear stress to the velocity gradient in fluids under laminar flow conditions. This relationship can be explained by considering the development of laminar flow between parallel plates, as shown in Figure 6.4. The plates are a relatively short distance apart, and initially, the fluid between them is stationary. The lower plate is then moved steadily to the right while the upper plate remains fixed. The shear (τ) is proportional to the gradient in velocity as a function of position y:

$$\tau \propto \frac{dv}{dy} \tag{6.3}$$

This proportionality can be converted into an equality with the use of a proportionality constant:

$$\tau = -\mu \frac{dv}{dy} \tag{6.4}$$

Stationary plate

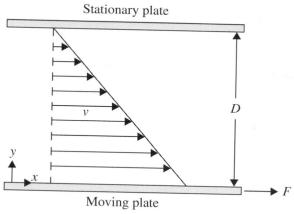

Moving plate

Figure 6.4 Velocity profile for Couette flow between parallel plates.

where μ is the proportionality constant. Equation (6.4) is called *Newton's law of viscosity*, and μ is the viscosity. Newton's law of viscosity is applicable only under laminar flow conditions. The minus sign is necessary in Eq. (6.4) because the velocity gradient is always negative if the direction τ is considered positive, as illustrated in Figure 6.4. $-dv/dy$ is called the *shear rate* and is often denoted by the symbol $\dot{\gamma}$. Equation (6.4) can therefore also be written as:

$$\tau = \mu \dot{\gamma} \tag{6.5}$$

Viscosity as defined in Eqs. (6.4) and (6.5) is sometimes called *dynamic viscosity*. Because τ has dimensions $L^{-1}MT^{-2}$ and $\dot{\gamma}$ has dimensions T^{-1}, μ must therefore have dimensions $L^{-1}MT^{-1}$. The SI unit of viscosity is the pascal second (Pa s), which is equal to $1\,N\,s\,m^{-2}$ or $1\,kg\,m^{-1}\,s^{-1}$. Other units include centipoise, cP. Of particular interest is the viscosity of water:

viscosity of water at $20°C \approx 1\,cP = 1\,mPa\,s = 10^{-3}\,Pa\,s = 10^{-3}\,kg\,m^{-1}\,s^{-1}$

A modified form of viscosity is the *kinematic viscosity*, which is usually given the Greek symbol ν:

$$\nu = \frac{\mu}{\rho} \tag{6.6}$$

where ρ is the fluid density.

Where dynamic viscosity provides information on the force needed to make a fluid flow at a certain rate, kinematic viscosity conveys how fast the fluid is moving. As kinematic viscosity incorporates fluid density as part of its measurement, it is a measure of velocity.

Fluids that obey Eq. (6.4) with constant μ are known as *Newtonian fluids*. The *flow curve* or *rheogram* for a Newtonian fluid is shown in Figure 6.5; a plot of τ versus $\dot{\gamma}$ gives a straight line with slope equal to μ. The viscosity of Newtonian fluids remains constant irrespective of

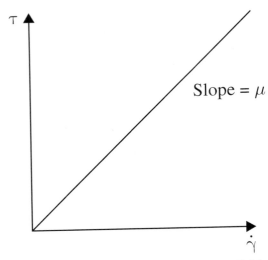

Figure 6.5 Flow curve for a Newtonian fluid.

changes in shear stress (force applied) or shear rate (velocity gradient). On the other hand, the ratio between shear stress and shear rate is not constant for *non-Newtonian fluids* but depends on the shear force exerted on the fluid. Accordingly, μ in Eq. (6.4) is not a constant for non-Newtonian fluids, and the velocity profile during Couette flow is not as simple as that shown in Figure 6.4.

6.4 Non-Newtonian Fluids

Most slurries, suspensions, and dispersions are non-Newtonian, as are homogeneous solutions of long-chain polymers and other large molecules. Many bioprocesses involve materials that exhibit non-Newtonian behavior, such as starches, extracellular polysaccharides, and culture broths containing suspended cells. Examples of non-Newtonian fluids are listed in Table 6.1.

Classification of non-Newtonian fluids depends on the relationship between shear stress and shear rate in the fluid. Types of non-Newtonian fluid commonly encountered in bioprocessing include *pseudoplastic*, *Bingham plastic*, and *Casson plastic*. The flow curves for these materials are shown in Figure 6.6. In each case, in contrast to Newtonian fluids, the flow curves are not straight lines because the ratio between the shear stress and shear rate is not constant. Nevertheless, an *apparent viscosity*, μ_a, can be defined for non-Newtonian fluids; μ_a is the ratio of shear stress and shear rate at a particular value of $\dot{\gamma}$:

$$\mu_a = \frac{\tau}{\dot{\gamma}} = K\dot{\gamma}^{n-1} \tag{6.7}$$

In Figure 6.6, μ_a is the slope of the dashed lines in the rheograms for pseudoplastic, Bingham plastic, and Casson plastic fluids; for Newtonian fluids, $\mu_a = \mu$. The apparent viscosity of a

Table 6.1: Examples of Newtonian and Non-Newtonian Fluids

Fluid type	Examples
Newtonian	All gases, water, dispersions of gas in water, low-molecular-weight liquids, aqueous solutions of low-molecular-weight compounds
Non-Newtonian	
Pseudoplastic	Rubber solutions, adhesives, polymer solutions, some greases, starch suspensions, cellulose acetate, mayonnaise, some soap and detergent slurries, some paper pulps, paints, wallpaper paste, biological fluids
Bingham plastic	Some plastic melts, margarine, cooking fats, some greases, toothpaste, some soap and detergent slurries, some paper pulps
Casson plastic	Blood, tomato sauce, orange juice, melted chocolate, printing ink

Adapted from B. Atkinson, F. Mavituna, Biochemical Engineering and Biotechnology Handbook, second ed., Macmillan, 1991.

non-Newtonian fluid is not a physical property in the same way that Newtonian viscosity is; apparent viscosity depends on the shear force exerted on the fluid. It is therefore meaningless to specify the apparent viscosity of a non-Newtonian fluid without also reporting the shear stress or shear rate at which it is measured.

Pseudoplastic fluids obey the *Ostwald–de Waele* or *power law*, relayed in the right-hand side term of Eq. (6.7), where τ is the shear stress, K is the *consistency index*, $\dot{\gamma}$ is the shear rate, and n is the *flow behavior index*. The parameters K and n characterize the rheology of power-law fluids. The flow behavior index n is dimensionless; the dimensions of K, $L^{-1}MTn^{-2}$, depend on n. As indicated in Figure 6.6, for pseudoplastic fluids $n < 1$. When $n > 1$, the fluid is dilatant rather than pseudoplastic. Dilatant is a recognized category of non-Newtonian fluid behavior; however, dilatant fluids do not occur often in bioprocessing. If $n = 1$ the fluid is Newtonian; in this case, Eqs. (6.5) and (6.7) are equivalent to $K = \mu$. As $n < 1$ for pseudo-plastic fluids, Eq. (6.7) indicates that the apparent viscosity decreases with increasing shear rate. Pseudoplastic fluids are therefore also known as *shear-thinning* fluids. Also included in Figure 6.6 are flow curves for plastic flow. Plastic fluids do not produce motion until some finite force or *yield stress* has been applied; this minimum stress is required to break down to some extent the internal "structure" of the fluid before any movement can occur.

6.4.1 Viscoelasticity

Viscoelastic fluids, such as some polymer solutions, exhibit an elastic response to changes in shear stress. When shear forces are removed from a moving viscoelastic fluid, the direction of flow may be reversed due to elastic forces developed during flow. Most viscoelastic fluids are also pseudoplastic and may exhibit other rheological characteristics such as yield stress. Mathematical analysis of viscoelasticity is therefore quite complex.

Fluid	Flow curve	Equation	Apparent viscosity μ_a
Newtonian	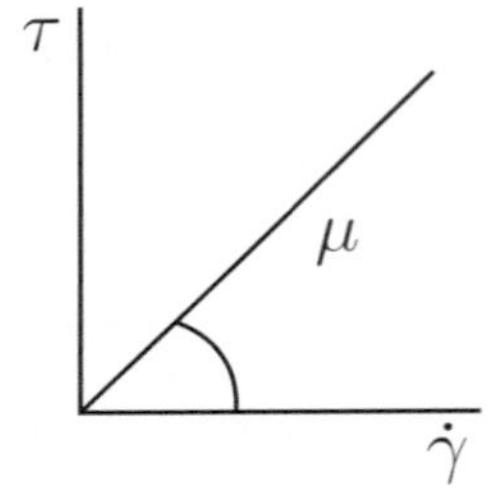	$\tau = \mu\dot{\gamma}$	Constant $\mu_a = \mu$
Pseudoplastic (power law)	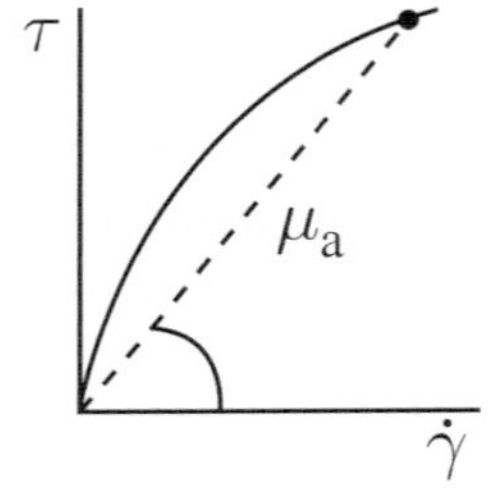	$\tau = K\dot{\gamma}^n$ $n < 1$	Decreases with increasing shear rate $\mu_a = K\dot{\gamma}^{n-1}$
Bingham plastic	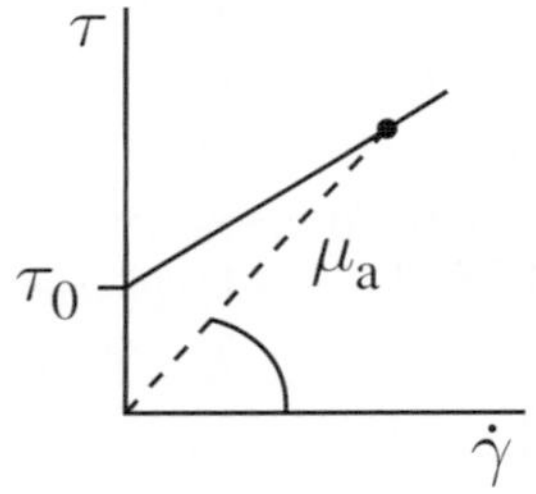	$\tau = \tau_0 + K_p\dot{\gamma}$	Decreases with increasing shear rate when yield stress τ_0 is exceeded $\mu_a = \dfrac{\tau_0}{\dot{\gamma}} + K_p$
Casson plastic	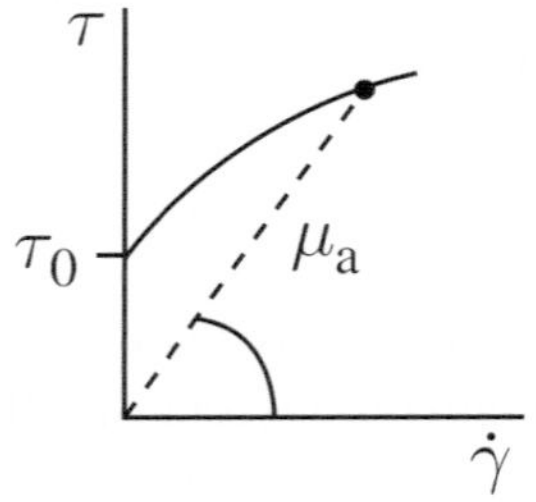	$\tau^{1/2} = \tau_0^{1/2} + K_p\dot{\gamma}^{1/2}$	Decreases with increasing shear rate when yield stress τ_0 is exceeded $\mu_a = \left[\left(\dfrac{\tau_0}{\dot{\gamma}}\right)^{1/2} + K_p\right]^2$

Figure 6.6 Classification of fluids according to their rheological behavior. *From B. Atkinson, F. Mavituna, Biochemical Engineering and Biotechnology Handbook, second ed., Macmillan, 1991.*

6.5 Rheological Properties of Cell Broths

Rheological data have been reported for a range of cell broths. This information has been obtained using various viscometers and measurement techniques; however, operating problems such as particle settling and broth centrifugation have been ignored in many cases. The rheology of dilute broths and cultures of yeast and non-chain-forming bacteria is usually Newtonian; animal cell suspensions with or without serum also remain Newtonian with viscosity close to that of water. On the other hand, most mycelial and plant cell suspensions are modeled as pseudoplastic fluids or, if there is a yield stress, as Bingham or Casson plastics. The rheological properties of selected microbial and plant cell suspensions are listed in Table 6.2. In most cases, the results are valid over only a limited range of shear conditions dictated largely by the choice of viscometer. If a bioprocess produces extracellular polymers such as in microbial production of pullulan and xanthan, the rheological characteristics of the broth depend strongly on the properties and concentration of these materials.

Table 6.2: Rheological Properties of Microbial and Plant Cell Suspensions

Culture	Shear rate (s^{-1})	Viscometer	Comments	Reference
Saccharomyces cerevisiae (pressed cake diluted with water)	2–100	rotating spindle	Newtonian below 10% solids ($\mu < 4$–$5\,cP$); pseudoplastic above 10% solids	[1]
Aspergillus niger (washed cells in buffer)	0–21.6	rotating spindle (guard removed)	pseudoplastic	[2]
Penicillium chrysogenum (whole broth)	1–15	turbine impeller	Casson plastic	[3]
P. chrysogenum (whole broth)	not given	coaxial cylinder	Bingham plastic	[4]
P. chrysogenum (whole broth)	not given	coaxial cylinder	pseudoplastic; K and n vary with CO_2 content of inlet gas	[5]
Endomyces sp. (whole broth)	not given	coaxial cylinder	pseudoplastic; K and n vary during batch culture	[6]
Streptomyces noursei (whole broth)	4–28	rotating spindle (guard removed)	Newtonian in batch culture; viscosity $40\,cP$ after $96\,h$	[7]
Streptomyces aureofaciens (whole broth)	2–58	rotating spindle/ coaxial cylinder	initially Bingham plastic due to high starch concentration in the medium; becomes Newtonian as starch is broken down; increasingly pseudoplastic as mycelium concentration increases	[8]

(Continued)

Table 6.2: Rheological Properties of Microbial and Plant Cell Suspensions—cont'd

Culture	Shear rate (s^{-1})	Viscometer	Comments	Reference
Aureobasidium pullulans (whole broth)	10.2–1020	coaxial cylinder	Newtonian at beginning of culture; increasingly pseudoplastic as concentration of product (exopolysaccharide) increases	[9]
Xanthomonas campestris	0.0035–100	cone-and-plate	pseudoplastic; K increases continually; n levels off when xanthan concentration reaches 0.5%; cell mass (max. 0.6%) has relatively little effect on viscosity	[10]
Cellulomonas uda (whole broth)	0.8–100	anchor impeller	shredded newspaper used as substrate; broth pseudoplastic with constant n until end of cellulose degradation; Newtonian thereafter	[11]
Nicotiana tabacum (whole broth)	not given	rotating spindle	pseudoplastic	[12]
Datura stramonium (whole broth)	0–1000	rotating spindle/ parallel plate	pseudoplastic and viscoelastic, with yield stress	[13]
Perilla frutescens (whole broth)	7.2–72	coaxial cylinder	Bingham plastic	[14]

6.6 Factors Affecting Broth Viscosity

The rheology of culture broths often changes throughout batch culture. Changes in the rheology of culture broths are caused by variation in one or more of the following properties:

- Cell concentration
- Cell morphology, including size, shape, mass, and vacuolation
- Flexibility and deformability of cells
- Osmotic pressure of the suspending fluid
- Concentration of polymeric substrate
- Concentration of polymeric product
- Rate of shear

Some of these parameters are considered in the following sections.

6.6.1 Cell Concentration

The viscosity of a suspension of spheres in Newtonian liquid can be predicted using the *Vand equation*:

$$\mu = \mu_L \left(1 + 2.5\psi + 7.25\psi^2\right) \qquad (6.8)$$

where μ_L is the viscosity of the suspending liquid and ψ is the volume fraction of solids. Equation (6.8) has been found to hold for yeast and spore suspensions at concentrations up to 14 vol% solids [15]. Many other cell suspensions do not obey Eq. (6.8); cell concentration can have a much stronger influence on rheological properties than is predicted by the Vand equation. As an example, Figure 6.7 shows how cell concentration affects the apparent viscosity of various pseudoplastic plant cell suspensions. In this case, a doubling in cell concentration causes the apparent viscosity to increase by a factor of up to 90. Similar results have been found for mold pellets in liquid culture [16]. When the viscosity is so strongly dependent on cell concentration, a steep drop in viscosity can be achieved by diluting the broth with water or medium. Periodic removal of part of the culture and refilling with fresh medium can thus be used to reduce the viscosity and improve fluid flow in viscous cell cultures.

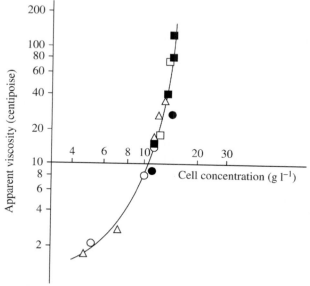

Figure 6.7 Relationship between the apparent viscosity and the cell concentration for plant cell suspensions forming aggregates of various sizes. (○) *Cudrania tricuspidata* 44 to 149 μm; (●) C. *tricuspidata* 149 to 297 μm; (□) *Vinca rosea* 44 to 149 μm; (■) V. *rosea* 149 to 297 μm; (△) *Nicotiana tabacum* 150 to 800 μm. From H. Tanaka, 1982, Oxygen transfer in broths of plant cells at high density. Biotechnol. Bioeng. 24, 425–442.

6.6.2 Cell Morphology

Small, monodispersed cells such as bacteria and yeast do not significantly affect the flow properties of culture broths. However, the morphological characteristics of other cell types, particularly filamentous microorganisms and plant cells, can exert a profound influence on broth rheology. Filamentous fungi and actinomycetes produce a variety of morphologies depending on the culture conditions. Individual cell filaments may be freely dispersed as shown in Figure 6.8(a), they may form loose clumps of branched and intertwined hyphae as in Figure 6.8(b), or they may develop highly compact cell aggregates and "hairy" pellets, such as those shown in Figure 6.8(c). These different morphologies have different rheological effects. Interactions between individual filaments and the formation of hyphal networks or loose clumps produce "structure" in the broth, resulting in high viscosity, pseudoplasticity, and yield stress behavior. In contrast, broths containing pelleted cells tend to be more Newtonian, depending on how readily the pellets are deformed during flow. Factors influencing the morphology of filamentous organisms and their tendency to clump include pH, growth rate, medium composition and ionic strength, dissolved oxygen tension, and agitation intensity.

Sample rheological data for pseudoplastic mycelial broths are shown in Figure 6.9. Suspensions of pelleted mycelia are more closely Newtonian in behavior than filamentous cells; as illustrated in Figure 6.9(a), the flow behavior index n for the pellets is closer to unity. As indicated in Figure 6.9(b), the consistency index, and therefore the apparent viscosity, can differ by several orders of magnitude depending on cell morphology.

6.6.3 Osmotic Pressure

The osmotic pressure of the culture medium affects cell turgor pressure. This in turn affects the hyphal flexibility of filamentous cells: increased osmotic pressure gives a lower turgor pressure, making the hyphae more flexible. Improved hyphal flexibility reduces broth viscosity and can also have a marked effect on yield stress.

6.6.4 Product and Substrate Concentrations

When the product of bioprocess is a polymer, continued excretion of the product in batch culture raises the broth viscosity. For example, during production of exopolysaccharide by *Aureobasidium pullulans*, the apparent viscosity measured at a shear rate of $1\,s^{-1}$ can reach as high as 24,000 cP [9]. Cell concentration usually has a negligible effect on the overall viscosity in these bioprocesses: the rheological properties of the fluid are dominated by the dissolved polymer. Other products having a similar effect on culture rheology include dextran, alginate, and xanthan gum.

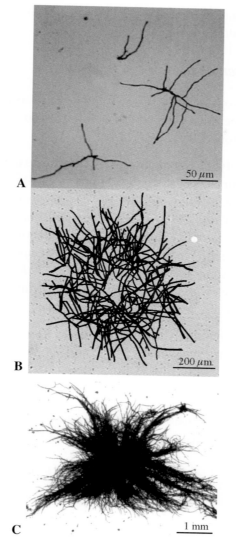

Figure 6.8 Different morphologies of filamentous microorganisms in submerged cultures. (A) Freely dispersed fungal filaments; (B) loose clump of filamentous cells; and (C) hairy pellet with compact core. The bars show approximate scales. *Images provided courtesy of C.R. Thomas and the Image Analysis Group, School of Chemical Engineering, University of Birmingham, UK.*

In contrast, when the culture medium contains a polymeric substrate such as starch, the apparent viscosity will decrease as growth progresses and the polymer is broken down. There could also be a progressive change from non-Newtonian to Newtonian behavior. In mycelial cultures this change is usually short-lived; as the cells grow and develop a structured filamentous network, the broth becomes increasingly pseudoplastic and viscous even though the polymeric substrate is being consumed.

6.7 Viscosity Measurement

Many different instruments or *viscometers* are available for measurement of rheological properties. Space does not permit a detailed discussion of viscosity measurement in this text; further information can be found elsewhere [17–22].

The objective of any viscosity measurement system is to create a controlled flow situation where easily measured parameters can be related to the shear stress τ and shear rate $\dot{\gamma}$. Usually, the fluid is set in rotational motion and the parameters measured are the torque M and the angular velocity Ω. These quantities are used to calculate τ and $\dot{\gamma}$ using approximate formulae that depend on the geometry of the apparatus; equations for particular viscometers can be found in other texts [2–6]. Once obtained, τ and $\dot{\gamma}$ are applied for evaluation of the viscosity of Newtonian fluids, or viscosity parameters such as K, n, and τ_0 for non-Newtonian fluids. Most modern viscometers use microprocessors to provide automatic readout of parameters such as shear stress, shear rate, and apparent viscosity.

Three types of viscometers commonly used in bioprocessing applications are cone-and-plate, coaxial cylinder, and impeller.

6.7.1 Cone-and-Plate Viscometer

The cone-and-plate viscometer consists of a flat horizontal plate and an inverted cone, the apex of which is in near contact with the plate as shown in Figure 6.10. The angle ϕ between

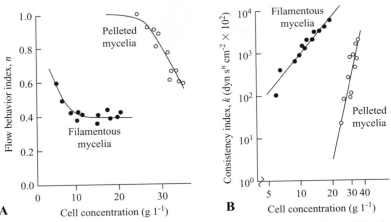

Figure 6.9 The effect of morphology on the rheology of mycelial broths. *From J.H. Kim, J.M. Lebeault, M. Reuss, Comparative study on rheological properties of mycelial broth in filamentous and pelleted forms, Eur. J. Appl. Microbiol. Biotechnol. 18, (1983), 11–16.*

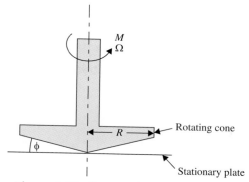

Figure 6.10 Cone-and-plate viscometer.

the plate and cone is very small, usually less than 3 degrees, and the fluid to be measured is located in this small gap. Large cone angles are not used in routine work for a variety of reasons, the most important being that analysis of the results for non-Newtonian fluids would be complex or impossible. The cone is rotated in the fluid and the angular velocity Ω and torque M are measured. It is assumed that the fluid undergoes steady laminar flow in concentric circles about the axis of rotation of the cone. This assumption is not always valid; however, for ϕ less than about 3 degrees, the error is small. Temperature can be controlled by circulating water from a constant temperature bath beneath the plate; this is effective provided the speed of rotation is not too high. Limitations of the cone-and-plate method for measurement of flow properties, including corrections for edge and temperature effects and turbulence, are discussed elsewhere [17–19].

6.7.2 Coaxial Cylinder Viscometer

The coaxial cylinder viscometer is a popular rotational device for measuring rheological properties. As shown in Figure 6.11, the instrument is designed to shear fluid located in the annulus between two concentric cylinders, one of which is held stationary while the other rotates. A cylindrical bob of radius R_i is suspended in sample fluid held in a stationary cylindrical cup of radius R_o. Liquid covers the bob to a height h from the bottom of the outer cup. As the inner cylinder rotates and the fluid undergoes steady laminar flow, the angular velocity Ω and torque M are measured. In some designs the outer cylinder rather than the inner bob rotates; in either case the motion is relative to angular velocity Ω.

Coaxial cylinder viscometers are used with Newtonian and non-Newtonian fluids. When the flow is non-Newtonian, the relationship between shear rate, rotational speed, and geometric factors is not simple and the calculations can be somewhat complicated. Limitations of the coaxial cylinder method, including corrections for end effects, slippage, temperature variation, and turbulence, are discussed elsewhere [17–20].

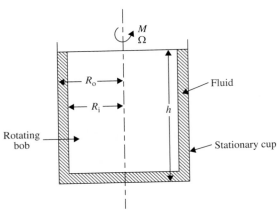

Figure 6.11 Coaxial cylinder viscometer.

6.7.3 Impeller Viscometer

Due to the difficulties associated with standard rotational viscometers, modified apparatus employing turbine and other impellers have been developed for the study of fermentation fluid rheology [3,22,23]. Instead of the rotating inner cylinder of Figure 6.11, a small impeller on a stirring shaft is used to shear the fluid sample. As the impeller rotates slowly in the fluid, accurate measurements of torque M and rotational speed N_i are made. The average shear rate in a stirred fluid is proportional to the stirrer speed:

$$\dot{\gamma} = kN_i \qquad (6.9)$$

where k is a constant that depends on the geometry of the impeller. For a turbine impeller under laminar flow conditions, the following relationship applies [3]:

$$\tau = \frac{2\pi Mk}{64D_i^3} \qquad (6.10)$$

where D_i is the impeller diameter. Equation (6.9) is experimentally derived; for turbine impellers k is approximately 10. Before carrying out viscosity measurements, the exact value of k for a particular apparatus can be evaluated using liquid with a known viscosity–shear rate relationship.

As Eq. (6.10) is valid only for laminar flow, viscosity measurements using the impeller method must be carried out under laminar flow conditions. Accordingly, if a turbine impeller is used, Re_i cannot be greater than about 10, as outlined in Section 6.2.3. From Eq. (6.2), Re_i is directly proportional to N_i, which, from Eq. (6.9), determines the value of $\dot{\gamma}$. Therefore, for Re_i restricted to low values because of the necessity for laminar flow, the range of shear rates that may be investigated is very limited. This range can be extended if anchor or helical agitators are used instead of conventional turbines (refer to later chapters for illustrations of these impellers) as laminar flow is maintained at higher Re_i. The value of k in Eq. (6.9) is also greater for anchor and helical agitators so that higher shear rates can be tested. As Eq. (6.10) is valid only for turbine impellers, the relationship between τ and M must be modified if alternative impellers are used. Application of anchor and helical impellers for viscosity measurement is described in more detail in the literature [11,23].

Because the flow patterns in stirred fluids are relatively complex, analysis of data from impeller viscometers is not absolutely rigorous from a theoretical point of view. However, the procedure is based on well-proven and widely accepted empirical correlations and is considered the most reliable technique for mycelial broths. As discussed below, the method eliminates many of the operating problems associated with conventional viscometers for the study of fermentation fluids.

6.7.4 Use of Viscometers With Cell Broths

Measurement of rheological properties is difficult when the fluid contains suspended solids such as cells. The viscosity of fermentation broths often appears time-dependent due to artifacts associated with the measuring device. With viscometers such as the cone-and-plate and coaxial cylinder, the following problems can arise:

1. The suspension is effectively centrifuged in the viscometer so that a region with lower cell density is formed near the rotating surface.
2. Solids settle out of suspension during measurement.
3. Cell clumps of about the same size as the gap in the coaxial cylinder viscometer, or about the same size as the cone angle in the cone-and-plate device, interfere with accurate measurement.
4. The measurement depends somewhat on the orientation of cells and cell clumps in the flow field.
5. Some types of cells begin to flocculate or deflocculate when the shear field is applied.
6. Cells can be destroyed during measurement.

The first problem is particularly troublesome because it is hard to detect and can give viscosity results that are too small by a factor of up to 100. For suspensions containing solids, the impeller method offers significant advantages compared with other measurement procedures. Stirring by the impeller prevents sedimentation, promotes uniform distribution of solids throughout the fluid, and reduces time-dependent changes in suspension composition. The impeller method has proved very useful for rheological measurements of microbial suspensions [22,24].

6.8 Summary of Chapter 6

This chapter covers a range of topics in rheology and fluid dynamics. At the end of Chapter 6 you should:

- Know the difference between *laminar* and *turbulent* flow
- Understand how shear stresses develop in laminar and turbulent flow
- Be familiar with the concept of the *Reynolds number*
- Be able to describe how fluid *boundary layers* develop in terms of *viscous drag*
- Be able to define *viscosity* in terms of *Newton's law*
- Know what *Newtonian* and *non-Newtonian* fluids are, and the difference between viscosity for Newtonian fluids and *apparent viscosity* for non-Newtonian fluids
- Know which factors affect broth viscosity

Problems

6.1. Conditions for turbulence

a. When the flow rate of water in a garden hose is measured using the bucket-and-stopwatch method, it takes 75 s to fill a 5-L bucket. If the inner diameter of the hose is 15 mm and the density and viscosity of the water are approximately $1000 \, kg \, m^{-3}$ and 1.0 cP, respectively, is flow in the hose laminar or turbulent?

b. A 5-cm-diameter wooden spoon is used to stir chicken consommé in a saucepan using a regular circular motion at a speed of 1.5 revolutions per second. At the temperature used to warm the broth, the liquid density is $970 \, kg \, m^{-3}$ and the viscosity is 0.45 cP. In this unbaffled system, turbulent flow occurs at impeller Reynolds numbers above about 10^5. Is flow in the saucepan likely to be turbulent?

6.2. Rheology of culture broth

The fungus *Aureobasidium pullulans* is used to produce an extracellular polysaccharide by catabolism of sucrose. After 120 hours of growth, measurements of shear stress and shear rate are made using a rotating cylinder viscometer.

Shear stress (dyn cm^{-2})	Shear rate (s^{-1})
44.1	10.2
235.3	170
357.1	340
457.1	510
636.8	1020

a. Plot the rheogram for this fluid.

b. Determine the appropriate non-Newtonian parameters.

6.3. Rheology of yeast suspensions

Apparent viscosities for pseudoplastic cell suspensions at varying cell concentrations are measured using a coaxial cylinder rotary viscometer. The results are shown in the following table.

Cell concentration (%)	Shear rate (s⁻¹)	Apparent viscosity (cP)
1.5	10	1.5
	100	1.5
3	10	2.0
	100	2.0
6	20	2.5
	45	2.4
10.5	10	4.7
	20	4.0
	50	4.1
	100	3.8
12	1.8	40
	4.0	30
	7.0	22
	20	15
	40	12
18	1.8	140
	7.0	85
	20	62
	40	55
21	1.8	710
	4.0	630
	7.0	480
	40	330
	70	290

Show on an appropriate plot how K and n vary with cell concentration.

6.4. Vand equation

A correlation between viscosity and cell concentration is required for suspension cultures of the hemp plant, *Cannabis sativa*. Morphologically, the cells are uniform and close to spherical in shape, and the culture exhibits Newtonian rheology at moderate cell densities. The viscosity of the whole broth including cells is measured using a rotating cylinder viscometer. Samples of broth are then centrifuged at 3000 rpm for 3 min in graduated centrifuge tubes for estimation of the cell volume fraction.

Broth viscosity (cP)	Total volume of broth in centrifuge tube (mL)	Volume of liquid decanted from centrifuge tube (mL)
1.0	10.0	9.6
1.0	11.6	11.0
1.1	11.2	10.4
1.2	9.2	8.3
1.4	10.4	9.2
1.6	10.9	9.3
2.3	11.0	8.8

For all the samples tested, the viscosity of the decanted liquid is 0.9 cP. Can the Vand equation be used to predict the viscosity of this plant cell suspension?

6.5. Viscosity and cell concentration

The filamentous bacterium *Streptomyces levoris* produces pseudoplastic culture broths. The rheological properties of the broth are measured at different cell concentrations using a turbine impeller viscometer.

Cell concentration (g L^{-1} dry weight)	Shear rate (s^{-1})	Shear stress (Pa)
6.0	5.0	1.0
	10	1.3
	20	1.8
	32	2.3
11.1	5.0	3.0
	10	4.3
	20	5.5
	30	6.7
15.5	5.0	9.2
	10	13.0
	20	17.4
	30	20.5
18.7	5.0	12.8
	10	16.9
	25	23.5

It is proposed to correlate the consistency index K and flow behavior index n with cell concentration using expressions of the form:

$$K \text{ or } n = Ax^B$$

where A and B are constants and x is the cell concentration.

a. Is this form of equation appropriate for K and n?

b. If not, why not? If so, evaluate A and B to determine the equations for K and n as a function of x.

c. Estimate the apparent viscosity of *S. levoris* broth if the cell concentration is 12.3 g L^{-1} and the shear rate is 8.5 s^{-1}.

References

[1] T.P. Labuza, D. Barrera Santos, R.N. Roop, Engineering factors in single-cell protein production. I. Fluid properties and concentration of yeast by evaporation, Biotechnol. Bioeng. 12 (1970) 123–134.

[2] T. Berkman-Dik, M. Özilgen, T.F. Bozoğlu, Salt, EDTA, and pH effects on rheological behavior of mold suspensions, Enzyme Microbiol. Technol. 14 (1992) 944–948.

[3] J.A. Roels, J. van den Berg, R.M. Voncken, The rheology of mycelial broths, Biotechnol. Bioeng. 16 (1974) 181–208.

[4] F.H. Deindoerfer, E.L. Gaden, Effects of liquid physical properties on oxygen transfer in penicillin fermentation, Appl. Microbiol. 3 (1955) 253–257.

[5] L.-K. Ju, C.S. Ho, J.F. Shanahan, Effects of carbon dioxide on the rheological behavior and oxygen transfer in submerged penicillin fermentations, Biotechnol. Bioeng. 38 (1991) 1223–1232.

[6] H. Taguchi, S. Miyamoto, Power requirement in non-Newtonian fermentation broth, Biotechnol. Bioeng. 8 (1966) 43–54.

[7]   F.H. Deindoerfer, J.M. West, Rheological examination of some fermentation broths, J. Biochem. Microbiol. Technol. Eng. 2 (1960) 165–175.

[8]   C.M. Tuffile, F. Pinho, Determination of oxygen-transfer coefficients in viscous streptomycete fermentations, Biotechnol. Bioeng. 12 (1970) 849–871.

[9]   A. LeDuy, A.A. Marsan, B. Coupal, A study of the rheological properties of a non-Newtonian fermentation broth, Biotechnol. Bioeng. 16 (1974) 61–76.

[10] M. Charles, Technical aspects of the rheological properties of microbial cultures, Adv. Biochem. Eng. 8 (1978) 1–62.

[11] P. Rapp, H. Reng, D.-C. Hempel, et al., Cellulose degradation and monitoring of viscosity decrease in cultures of *Cellulomonas uda* grown on printed newspaper, Biotechnol. Bioeng. 26 (1984) 1167–1175.

[12] A. Kato, S. Kawazoe, Y. Soh, Viscosity of the broth of tobacco cells in suspension culture, J. Ferment. Technol. 56 (1978) 224–228.

[13] R. Ballica, D.D.Y. Ryu, R.L. Powell, et al., Rheological properties of plant cell suspensions, Biotechnol. Prog. 8 (1992) 413–420.

[14] J.-J. Zhong, T. Seki, S.-I. Kinoshita, et al., Rheological characteristics of cell suspension and cell culture of *Perilla frutescens*, Biotechnol. Bioeng. 40 (1992) 1256–1262.

[15] F.H. Deindoerfer, J.M. West, Rheological properties of fermentation broths, Adv. Appl. Microbiol. 2 (1960) 265–273.

[16] J. Laine, R. Kuoppamäki, Development of the design of large-scale fermentors, Ind. Eng. Chem. Process Des. Dev. 18 (1979) 501–506.

[17] C.W. Macosko, Rheology: Principles, Measurements, and Applications, VCH, 1994.

[18] J. Ferguson, Z. Kembłowski, Applied Fluid Rheology, Elsevier Applied Science, 1991.

[19] R.W. Whorlow, Rheological Techniques, Ellis Horwood, 1980.

[20] J.R. van Wazer, J.W. Lyons, K.Y. Kim, et al., Viscosity and Flow Measurement, John Wiley, 1963.

[21] H.A. Barnes, J.F. Hutton, K. Walters, An Introduction to Rheology, Elsevier, 1989.

[22] B. Metz, N.W.F. Kossen, J.C. van Suijdam, The rheology of mould suspensions, Adv. Biochem. Eng 11 (1979) 103–156.

[23] J.H. Kim, J.M. Lebeault, M. Reuss, Comparative study on rheological properties of mycelial broth in filamentous and pelleted forms, Eur. J. Appl. Microbiol. Biotechnol. 18 (1983) 11–16.

[24] J.J.T.M. Bongenaar, N.W.F. Kossen, B. Metz, et al., A method for characterizing the rheological properties of viscous fermentation broths, Biotechnol. Bioeng. 15 (1973) 201–206.

Suggestions for Further Reading

Introductory Fluid Mechanics Textbooks

A. Vardy, Fluid Principles, McGraw-Hill, 1990.

F.M. White, Fluid Mechanics, 7th ed., McGraw-Hill, 2011.

Viscosity and Viscosity Measurement

D.G. Allen, C.W. Robinson, Measurement of rheological properties of filamentous fermentation broths, Chem. Eng. Sci. 45 (1990) 37–48.

B. Atkinson, F. Mavituna, 2nd ed., Biochemical Engineering and Biotechnology Handbook, Chapter 11, Macmillan, 1991.

E. Olsvik, B. Kristiansen, Rheology of filamentous fermentations, Biotech. Adv. 12 (1994) 1–39.

Turbulence

S. Kresta, Turbulence in stirred tanks: Anisotropic, approximate, and applied, Can. J. Chem. Eng. 76 (1998) 563–576.

M. Lesieur, Turbulence in Fluids, 3rd ed., Kluwer Academic, 1997.

J. Mathieu, J. Scott, Introduction to Turbulent Flow, Cambridge University Press, 2000.

S.B. Pope, Turbulent Flows, Cambridge University Press, 2000.

H. Wu, G.K. Patterson, Laser-Doppler measurements of turbulent-flow parameters in a stirred mixer, Chem. Eng. Sci. 44 (1989) 2207–2221.

CHAPTER 7

Mixing

The physical operation of mixing can determine the success of bioprocesses. Single- and multiple-phase mixing can occur in fluids with a range of rheologies. Mixing strongly influences cellular access to dissolved nutrients and O_2 and plays a critical role in controlling the bioreactor temperature. The equipment used for mixing has a significant effect on agitation efficiency, power requirements, and operating costs. A consequence of mixing operations is the development of hydrodynamic forces in the fluid. These forces are responsible for important processes in bioreactors such as bubble break-up and dispersion; however, cell damage can also occur and must be avoided. Problems with mixing are a major cause of productivity loss during scale-up of bioprocesses.

This chapter draws on material introduced in Chapter 6 about fluid properties and flow behavior. In turn, as mixing underpins effective heat and mass transfer in bioprocesses, this chapter provides the foundations for detailed treatment of these subjects in Chapters 8 and 9.

7.1 Functions of Mixing

Mixing is a physical operation that reduces nonuniformities in fluid by eliminating gradients of concentration, temperature, and other properties. Mixing is accomplished by interchanging material between different locations to produce a mingling of components. If a system is perfectly mixed, there is a homogeneous distribution of system properties. Mixing is used in bioprocesses to:

- Blend soluble components of liquid media such as sugars
- Disperse gases such as air through liquids in the form of small bubbles
- Maintain suspension of solid particles such as cells and cell aggregates
- Where necessary, disperse immiscible liquids to form an emulsion or suspension of fine droplets
- Promote heat transfer to or from liquids

Mixing is one of the most important operations in bioprocessing. To create an optimal environment for cell cultures, bioreactors must provide the cells with access to all substrates, including O_2 in aerobic cultures. It is not enough to just fill the bioreactor with nutrient-rich medium; unless the culture is mixed, zones of nutrient depletion will develop as the cells rapidly consume materials within their local environment. This problem is heightened if mixing does not maintain a uniform suspension of biomass; substrate concentrations can quickly drop to zero within layers of settled cells. We rely on good mixing to distribute any material added during the bioprocess, such as fresh medium to feed the cells or concentrated acid or alkali to control the culture pH. If these materials are not mixed rapidly throughout the reactor, their concentration can build up to toxic levels near the feed point with deleterious consequences for the cells in that region. Another important function of mixing is heat transfer. Bioreactors must be capable of transferring heat to or from the broth rapidly enough so that the desired temperature is maintained. Cooling water is used to take up excess heat from bioprocesses; the rate of heat transfer from the broth to the cooling water depends on mixing conditions.

Mixing can be achieved in many different ways. We will concentrate on the most common mixing technique in bioprocessing: mechanical agitation using an impeller.

7.2 Mixing Equipment

Mixing typically occurs in cylindrical stirred tanks, such as that shown in Figure 7.1. *Baffles*, which are vertical strips of metal mounted against the wall of the tank, are installed to reduce gross vortexing and swirling of the liquid. Mixing is achieved using an *impeller* mounted on a centrally located *stirrer shaft*. The stirrer shaft is driven rapidly by the *stirrer motor*; the effect of the rotating impeller is to pump the liquid and create a regular flow pattern. Liquid is forced away from the impeller, circulates through the vessel, and periodically returns to the impeller region. In gassed stirred tanks such as bioreactors used for aerobic culture, gas is introduced into the vessel by means of a *sparger* located beneath the impeller.

The mixing equipment exerts a significant influence on the outcome of the process. Aspects of this equipment are outlined in the following sections.

7.2.1 Vessel Geometry and Liquid Height

The shape of the stirred tanks effects the efficiency of mixing. Several base designs are shown in Figure 7.2. If possible, the base should be rounded at the edges rather than flat; this eliminates sharp corners and pockets where fluid currents may not penetrate, thus discouraging the formation of stagnant zones. The energy required to suspend solids in stirred tanks is sensitive to the shape of the vessel base: depending on the type of impeller and the flow pattern generated, the geometries shown in Figure 7.2(B) through (E) can be used to enhance particle suspension compared with the flat-base tank of Figure 7.2(A). In contrast, sloping sides or a conical base such as that shown in Figure 7.2(F) promotes settling of solids and should be avoided if solids suspension is required.

Other geometric specifications for stirred tanks are shown in Figure 7.3. Efficient mixing requires coordinating the tank height and diameter (D_T) with the impeller diameter (D_i). D_i is typically 1/4 to 1/2 of D_T while the height of liquid in the tank H_L should be no more than 1.0

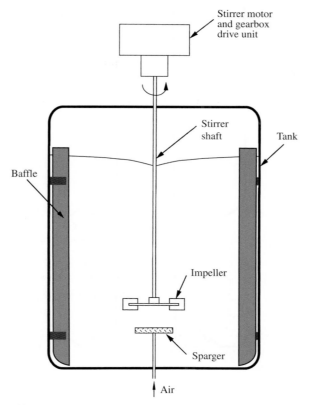

Figure 7.1 Typical configuration of a stirred tank.

to 1.25 D_T. The intensity of mixing decreases quickly as fluid moves away from the impeller zone, so large volumes of liquid in the upper portion of the vessel are difficult to mix and should be avoided.

Another aspect of tank geometry influencing mixing efficiency is the clearance C_i between the impeller and the lowest point of the tank floor (Figure 7.3). This clearance affects solids suspension, gas bubble dispersion, and hydrodynamic stability. C_i is typically 1/6 to 1/2 the tank diameter in most stirring operations.

7.2.2 Baffles

Baffles are standard equipment in stirred tanks. They assist mixing and create turbulence in the fluid by breaking up the circular flow generated by rotation of the impeller. Baffles are attached to the inside vertical walls of the tank by means of welded brackets. Four equally spaced baffles are usually sufficient to prevent liquid swirling and vortex formation. The optimum baffle width W_{BF} depends on the impeller design and fluid viscosity but

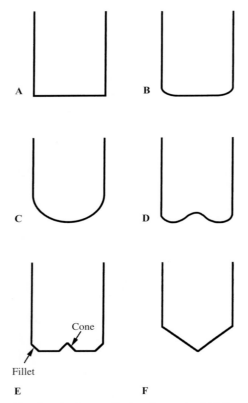

Figure 7.2 Different profiles for the base of stirred vessels: (A) flat; (B) dished; (C) round; (D) contoured; (E) cone-and-fillet; (F) conical.

is of the order 1/10 to 1/12 the tank diameter. For clean, low-viscosity liquids, baffles are attached perpendicular to the wall as illustrated in Figure 7.4(A). Alternatively, as shown in Figures 7.4(B) and (C), baffles may be mounted away from the wall with clearance $C_{BF} \approx$ 1/50 D_T or set at an angle. These arrangements prevent the development of stagnant zones and sedimentation along the inner edge of the baffle during mixing of viscous fluids or fluids containing suspended cells or particles.

7.2.3 Sparger

There are a large variety of sparger designs. These include simple open pipes, perforated tubes, porous diffusers, and complex two-phase injector devices. *Point spargers*, such as open pipe spargers, release bubbles at only one location in the vessel. Other sparger designs such as *ring spargers* have multiple gas outlets so that bubbles are released simultaneously from various locations. Bubbles leaving the sparger usually fall within a relatively narrow size range depending on the sparger type. However, as the bubbles rise from the sparger into the impeller zone, they are subjected to very high shear forces from operation of the stirrer that cause

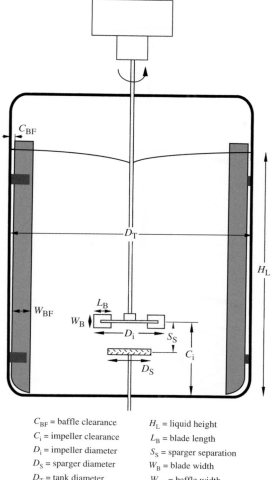

C_{BF} = baffle clearance H_L = liquid height
C_i = impeller clearance L_B = blade length
D_i = impeller diameter S_S = sparger separation
D_S = sparger diameter W_B = blade width
D_T = tank diameter W_{BF} = baffle width

Figure 7.3 Some geometric specifications for a stirred tank.

bubble break-up. The resulting small bubbles are flung into the bulk liquid for dispersion throughout the vessel. The diameter D_S of large ring spargers and the separation S_S between the sparger and impeller (Figure 7.3) can have an important influence on the efficiency of gas dispersion, though the type of sparger used has a relatively minor influence on the mixing process in most stirred tanks.

7.2.4 Stirrer Shaft

The primary function of the stirrer shaft is to transmit *torque* from the stirrer motor to the impeller. Torque is the tendency of a force to cause an object to rotate. In typical mixing operations, the impeller is attached to a vertical stirrer shaft that passes from the motor

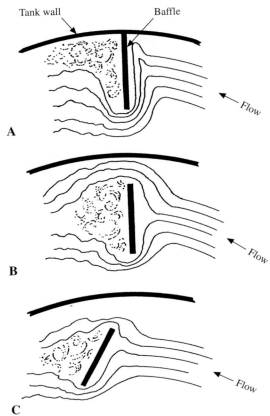

Figure 7.4 Baffle arrangements: (A) baffles attached to the wall for low-viscosity liquids; (B) baffles set away from the wall for moderate-viscosity liquids; (C) baffles set away from the wall and at an angle for high-viscosity liquids. *From F.A. Holland and F.S. Chapman, 1966, Liquid Mixing and Processing in Stirred Tanks, Reinhold, New York.*

through the top of the vessel. However, when headplate access is at a premium because of other devices and instruments, or if a shorter shaft is required to alleviate mechanical stresses (e.g., when mixing viscous fluids), the stirrer shaft may enter through the base of the vessel. The vessel configuration for a bottom-entering stirrer is shown in Figure 7.5. The main disadvantage of bottom-entering stirrers is the increased risk of fluid leaks due to failure or wear of the seals between the rotating stirrer shaft and the vessel floor.

7.3 Impellers

Many different types of impellers are available for mixing applications. A selection is illustrated in Figure 7.6. The choice of impeller depends on several factors, including the liquid viscosity, the need for turbulent shear flows (e.g., for bubble break-up and gas dispersion), and whether strong liquid currents are required. Figure 7.7 highlights impeller

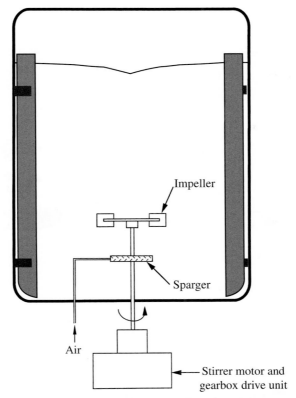

Figure 7.5 Stirred tank configuration for a bottom-entry stirrer.

recommendations specifically for a disk turbine, propeller, and pitched blade impeller. The recommended viscosity ranges for a number of common impellers are indicated in Figure 7.8. Impellers can also be classified broadly depending on whether they produce high levels of turbulence, or whether they have a strong pumping capacity for generation of large-scale flow currents. Both functions are required for good mixing, but they usually do not work together. The characteristics of several impellers in terms of their turbulence- and flow-generating properties are indicated in Figure 7.9. Several impellers used in bioprocesses are described in the following sections.

7.3.1 Rushton Turbine

The most frequently used impeller in the bioprocess industry is the six-flat-blade disc-mounted turbine shown in Figures 7.6 and 7.10. This impeller is also known as the *Rushton turbine*. The Rushton turbine has been the impeller of choice for bioprocesses since the 1950s, largely because it has been well studied and characterized, and because it is very effective for gas dispersion. Rushton turbines of diameter one-third the tank diameter have long been used as standard hardware for aerobic bioprocesses, but in recent years it has been

a   Disk turbine
b   Hollow blade stirrer
c,d Pitched blade impeller

e   Intermig impeller
f   Pitched paddle impeller
g   Propeller
h   Pitched blade propeller

i   Helical ribbon impeller
j-l Anchor impellers

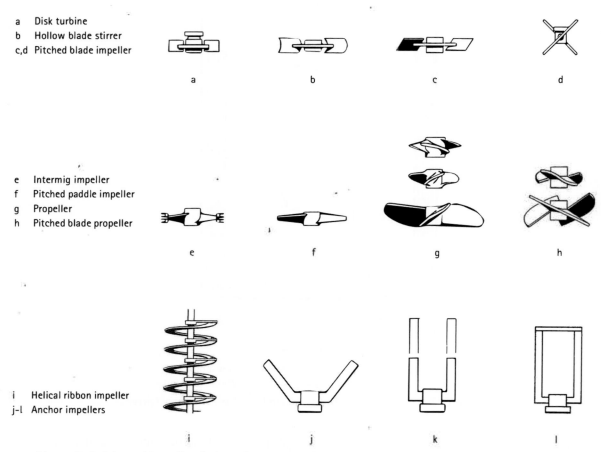

Figure 7.6 Selected impeller designs. From L. Hasler, R. Butz, and P. Grad, 2006, Transparency, *From L. Hasler, R. Butz, and P. Grad, 2006, Transparency, Form and Function: Fermenter Manufacturing-Art. DreiPunktVerlag. Germany. Courtesy of Bioengineering, Bioengineering AG, 8636 Wald, Switzerland.*

recognized that larger impellers of size up to one-half the tank diameter provide considerable benefits for improved mixing and gas distribution. Rushton turbines are effective for solids suspension, including in three-phase (solid–liquid–gas) systems. For three-phase mixing, an impeller clearance of 1/4 the tank diameter has been recommended for Rushton turbines, as this allows effective solids suspension, gas dispersion under the impeller, and adequate agitation in the upper parts of the vessel [1].

7.3.2 Propellers

A typical three-blade marine-type propeller is illustrated in Figure 7.11. The slope of the individual blades varies continuously from the outer tip to the inner hub. The *pitch* of a propeller is a measure of the angle of the propeller blades. It refers to the properties of the propeller as

The disk turbine with a radial primary flow direction is used predominantly where turbulent flow is needed. It is especially suitable for dispersing liquid / liquid substances and to achieve a high phase-related exchange surface which determines the rate of substance transport.

The propeller with an axial primary flow direction involves only very small shear forces. This leads to very gentle but nevertheless thorough mixing. For this reason, the propeller is recommended especially for the cultivation of animal cells. But it is also an excellent tool for homogenizing and suspending liquids a well as dispersing solids in liquids.

Pitched blade impellers are particularly suitable for homogenizing, suspending, and dispersing liquid / liquid and solid / liquid substances, as well as the gentle stirring of media.

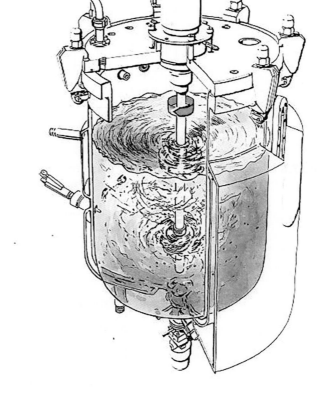

Figure 7.7 Illustration of a mixed bioreactor (right) and recommendations outlined for a disk turbine (top left), propeller (middle left), and pitched blade impeller (bottom left).
From L. Hasler, R. Butz, and P. Grad, 2006, Transparency, Form and Function: Fermenter Manufacturing-Art. DreiPunktVerlag. Germany. Courtesy of Bioengineering, Bioengineering AG, 8636 Wald, Switzerland.

a segment of a screw: pitch is the advance per revolution, or the distance that liquid is displaced along the impeller axis during one full turn. Propellers with square pitch, that is, pitch equal to the impeller diameter, are often used.

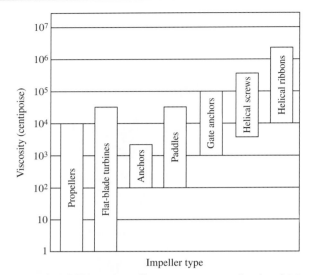

Figure 7.8 Viscosity ranges for different impellers. *From F.A. Holland and F.S. Chapman, 1966, Liquid Mixing and Processing in Stirred Tanks, Reinhold, New York.*

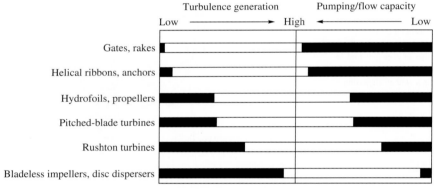

Figure 7.9 Characteristics of different impellers for generation of turbulence and liquid pumping.

Figure 7.10 Rushton turbine.

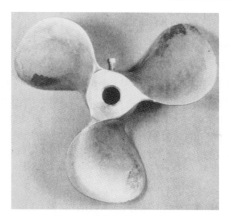

Figure 7.11 Propeller.

Propellers are axial-flow impellers. They may be operated for either downward or upward pumping of the fluid; downward pumping is more common. Propellers have high flow capacity and they are used with low-to-medium viscosity fluids and are usually installed with diameter around one-third the tank diameter. With gassing, propellers operated at high speed can generate flow and torque instabilities. Propellers are very effective for suspending solids, often outperforming Rushton turbines in that respect.

7.3.3 Pitched-Blade Turbines

Pitched-blade turbines have flat inclined blades, as shown in Figure 7.6. Commonly referred to as axial-flow impellers, pitched-blade turbines generate discharge streams with significant radial as well as axial velocity components. Pitched-blade turbines produce strong liquid flows and have a much higher pumping efficiency than Rushton turbines, making them very effective for blending applications. With gassing, ventilated cavities form behind the blades as a result of under-pressure in the trailing vortices in much the same way as for Rushton turbines. However, for pitched-blade turbines, there is only one vortex per blade. Pitched-blade turbines can be operated in either downward or upward pumping modes, downward being the more common.

7.3.4 Alternative Impeller Designs

So far, we have considered the characteristics of several traditional impellers that have been used in the chemical and bioprocessing industries for many decades. More recently, a variety of new agitator configurations has been developed commercially. These modern impellers have a range of technical features aimed at improving mixing in stirred tanks.

Curved-Blade Disc Turbines

Curved-blade disc turbines such as that shown in Figure 7.12 generate primarily radial flow, similar to the Rushton turbine. However, changing the shape of the blades has a significant effect on the impeller power requirements and gas-handling characteristics. Rotation with the concave side forward greatly discourages the development of trailing vortices behind the blades; therefore, with sparging, no large, ventilated cavities form on the convex surfaces. These impellers have the major advantage of being more difficult to flood and therefore can handle gas flow rates several times higher than Rushton turbines [2].

Hydrofoil Impellers

Two different hydrofoil impellers are shown in Figure 7.13. The blade angle and width are varied along the length of hydrofoil blades, and the leading edges are rounded like an airplane wing to reduce form drag and generate a positive lift. The shape of hydrofoil impellers allows for effective pumping and bulk mixing with strong axial velocities and low power consumption. Most hydrofoils are operated for downward pumping, but upward flow is also possible. A typical mean velocity vector plot for a downward-pumping hydrofoil impeller is shown in Figure 7.14. Liquid velocities in regions of the tank above the main circulation loops are considerably lower than below the impeller.

Downward-pumping hydrofoil impellers exhibit many of the hydrodynamic properties of downward-pumping pitched-blade turbines during aeration. They remain more prone to flooding than Rushton turbines. The ventilated cavities formed behind the blades of a downward-pumping hydrofoil impeller in low-viscosity fluid are illustrated in Figure 7.15.

The flow instabilities generated in gassed systems by downward-pumping hydrofoil impellers can be eliminated using wide-blade, upward-pumping hydrofoils. These impellers also have high gas-handling capacity and a limited tendency to flood at low stirrer speeds.

Figure 7.12 Scaba 6SRGT six-curved-blade disc turbine.
Photograph courtesy of Scaba AB, Sweden.

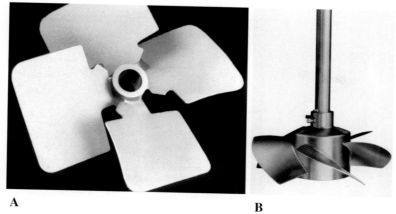

Figure 7.13 (A) Lightnin A315 hydrofoil impeller. (B) Prochem Maxflo T hydrofoil impeller. *Photograph courtesy of Lightnin Mixers, Australia. (A) Photograph courtesy of Chemineer Inc., Dayton, OH. (B)*

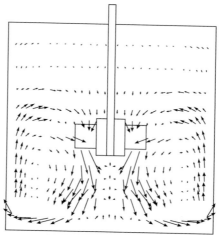

Figure 7.14 Mean velocity vector plot for a Prochem Maxflo T hydrofoil impeller. The velocities were measured using laser Doppler velocimetry. *Adapted from Z. Jaworski, A.W. Nienow, and K.N. Dyster, 1996, An LDA study of the turbulent flow field in a baffled vessel agitated by an axial, down-pumping hydrofoil impeller. Can. J. Chem. Eng. 74, 3–15.*

7.4 Stirrer Power Requirements

Electrical power is used to drive impellers in stirred vessels. The average power consumption per unit volume for industrial bioreactors ranges from $10\,kW\,m^{-3}$ for small vessels (approx. $0.1\,m^3$) to 1 to $2\,kW\,m^{-3}$ for large vessels (approx. $100\,m^3$). Friction in the stirrer motor gearbox and seals reduces the energy transmitted to the fluid; therefore, the electrical power consumed by stirrer motors is always greater than the mixing power. Energy costs for stirrer

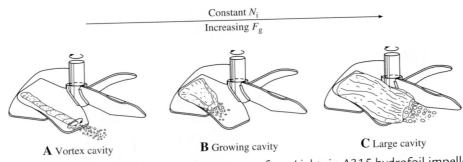

Constant N_i →
Increasing F_g

A Vortex cavity   **B** Growing cavity   **C** Large cavity

Figure 7.15 Changes in ventilated cavity structure for a Lightnin A315 hydrofoil impeller at constant stirrer speed N_i and increasing gas flow rate F_g. *Adapted from A. Bakker and H.E.A. van den Akker, 1994, Gas–liquid contacting with axial flow impellers. Trans. IChemE 72 A, 573–582.*

operation are a major financial commitment and an important design consideration in process economics.

The power required to achieve a given stirrer speed depends on the magnitude of the frictional forces and form drag that resist rotation of the impeller. Friction and form drag give rise to torque on the stirrer shaft; experimentally, the power input for stirring can be determined from measurements of the induced torque M:

$$P = 2\pi N_i M \tag{7.1}$$

where P is power and N_i is the stirrer speed.

General guidelines for estimating the power requirements in stirred tanks are outlined in the following sections.

7.4.1 Ungassed Newtonian Fluids

The power required to mix nonaerated fluids depends on the stirrer speed, the impeller shape and size, the tank geometry, and the density and viscosity of the fluid. The relationship between these variables is usually expressed in terms of dimensionless numbers such as the impeller Reynolds number Re_i:

$$R_{e_i} = \frac{N_i D_i^2 \rho}{\mu} \tag{6.2}$$

and the power number N_P:

$$N_P = \frac{P}{\rho N_i^3 D_i^5} \tag{7.2}$$

In Eqs. (6.2) and (7.2), N_i is stirrer speed, D_i is impeller diameter, ρ is fluid density, μ is fluid viscosity, and P is power. The power number N_P can be considered analogous to a drag coefficient for the stirrer system.

The relationship between Re_i and N_P has been determined experimentally for a range of impeller and tank configurations. The results for five impeller designs—Rushton turbine, downward-pumping pitched-blade turbine, marine propeller, anchor, and helical ribbon—are shown in Figures 7.16 and 7.17. Once the value of N_P is known, the power is calculated from Eq. (7.2) as:

$$P = N_P \rho N_i^3 D_i^5 \tag{7.3}$$

For a given impeller, the general relationship between power number and Reynolds number depends on the flow regime in the tank. The following three flow regimes can be identified in Figures 7.16 and 7.17.

1. *Laminar regime.* The laminar regime corresponds to $Re_i < 10$ for many impellers, including turbines and propellers. For stirrers with small wall-clearance such as anchor and helical ribbon mixers, laminar flow persists until $Re_i = 100$ or greater. In the laminar regime:

$$N_P \propto \frac{1}{Re_i} \qquad \text{or} \qquad P = k_1 \mu N_i^2 D_i^3 \tag{7.4}$$

where k_1 is a proportionality constant. Values of k_1 for the impellers represented in Figures 7.16 and 7.17 and are listed in Table 7.1 [3,4]. The power required for laminar flow is independent of the density of the fluid but directly proportional to the fluid viscosity.

2. *Turbulent regime.* The *power* number is independent of the Reynolds number in turbulent flow. Therefore:

$$P = N_P' \rho N_i^3 D_i^5 \tag{7.5}$$

where N_P' is the constant value of the power number in the turbulent regime. Approximate values of N_P' for the impellers in Figures 7.16 and 7.17 are listed in Table 7.1 [3,4]. Example 7.1 demonstrates the application of tank geometry in calculating power requirements.

7.4.2 Ungassed Non-Newtonian Fluids

Estimation of the power requirements for non-Newtonian fluids is more difficult. It is often impossible with highly viscous fluids to achieve fully developed turbulence; under these conditions, N_P is always dependent on Re_i and we cannot use the constant N_P' value in power calculations. In addition, because the viscosity of non-Newtonian liquids varies with shear conditions, the impeller Reynolds number used to correlate power requirements must be

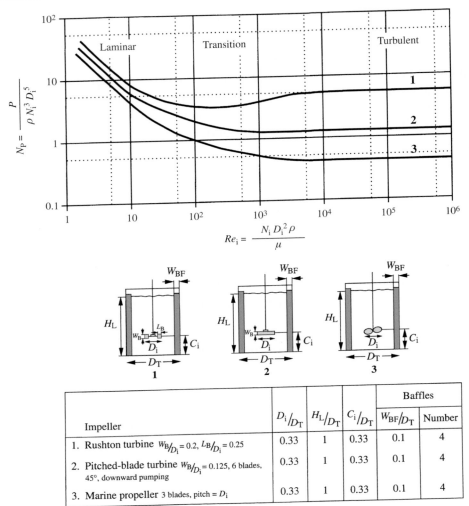

Figure 7.16 Correlations between the Reynolds number and power number for Rushton turbines, downward-pumping pitched-blade turbines, and marine propellers in fluids without gassing. *Data from J.H. Rushton, E.W. Costich, and H.J. Everett, 1950, Power characteristics of mixing impellers. Parts I and II. Chem. Eng. Prog. 46, 395–404, 467–476; and R.L. Bates, P.L. Fondy, and R.R. Corpstein, 1963, An examination of some geometric parameters of impeller power. Ind. Eng. Chem. Process Des. Dev. 2, 310–314.*

Impeller	D_i/D_T	H_L/D_T	C_i/D_T	Baffles W_{BF}/D_T	Number
1. Rushton turbine $W_B/D_i = 0.2$, $L_B/D_i = 0.25$	0.33	1	0.33	0.1	4
2. Pitched-blade turbine $W_B/D_i = 0.125$, 6 blades, 45°, downward pumping	0.33	1	0.33	0.1	4
3. Marine propeller 3 blades, pitch = D_i	0.33	1	0.33	0.1	4

redefined. Some power correlations have been developed using an impeller Reynolds number based on the apparent viscosity μ_a (Section 6.4). Therefore, from Eq. (6.7) for power-law fluids:

$$Re_i = \frac{N_i D_i^2 \rho}{K\dot{\gamma}^{n-1}} \qquad (7.6)$$

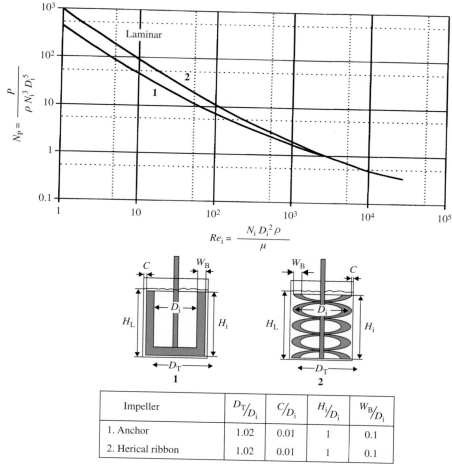

Figure 7.17 Correlations between the Reynolds number and power number for anchor and helical ribbon impellers in fluids without gassing. *From M. Zlokarnik and H. Judat, 1988, Stirring. In: W. Gerhartz, Ed., Ullmann's Encyclopedia of Industrial Chemistry, 5th ed., vol. B2, pp. 25-1–25-33, VCH, Weinheim, Germany.*

Impeller	D_T/D_i	C/D_i	H_i/D_i	W_B/D_i
1. Anchor	1.02	0.01	1	0.1
2. Herical ribbon	1.02	0.01	1	0.1

Table 7.1: Values of the Constants in Eqs. (7.4) and (7.5) for the Stirred Tank Geometries Defined in Figures 7.16 and 7.17.

Impeller type	k_1 $(Re_i = 1)$	N'_P $(Re_i = 10^5)$
Rushton turbine	70	5.0
Pitched-blade turbine	50	1.3
Marine propeller	40	0.35
Anchor	420	0.35
Helical ribbon	1000	0.35

where n is the flow behavior index and K is the consistency index. A problem with application of Eq. (7.6) is the evaluation of $\dot{\gamma}$. For stirred tanks, an approximate relationship is often used:

$$\dot{\gamma} = k N_i \tag{7.7}$$

where the value of the constant k depends on the geometry of the impeller. The relationship of Eq. (7.7) is discussed further in Section 7.13; however, for turbine impellers k is about 10. Substituting Eq. (7.7) into Eq. (7.6) gives an appropriate Reynolds number for pseudoplastic fluids:

$$Re_i = \frac{N_i^{2-n} D_i^2 \rho}{K k^{n-1}} \tag{7.8}$$

The relationship between the Reynolds number Re_i and the power number N_P for a Rushton turbine in a baffled tank containing pseudoplastic fluid is shown in Figure 7.18. The upper line was measured using Newtonian fluids for which Re_i is defined by Eq. (6.2); this line corresponds to part of the curve already shown in Figure 7.16. The lower line gives the Re_i–N_P relationship for pseudoplastic fluids with Re_i defined by Eq. (7.8). The laminar region extends to higher Reynolds numbers in pseudoplastic fluids than in Newtonian systems. At Re_i below 10 and above 200, the results for Newtonian and non-Newtonian fluids are essentially the same. In the intermediate range, pseudoplastic liquids require less power than Newtonian fluids to achieve the same Reynolds number.

There are several practical difficulties with application of Figure 7.18 to bioreactors. Flow patterns in pseudoplastic and Newtonian fluids differ significantly. Even if there is high turbulence near the impeller in pseudoplastic systems, the bulk liquid may be moving very slowly and consuming relatively little power.

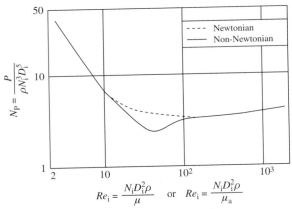

Figure 7.18 The correlation between the Reynolds number and power number for a Rushton turbine in ungassed non-Newtonian fluid in a baffled tank. *From A.B. Metzner, R.H. Feehs, H. Lopez Ramos, R.E. Otto, and J.D. Tuthill, 1961, Agitation of viscous Newtonian and non-Newtonian fluids. AIChE J. 7, 3–9.*

7.4.3 Gassed Fluids

Liquids into which gas is sparged have reduced power requirements for stirring. The presence of gas bubbles decreases the density of the fluid; however, the influence of density as expressed in Eq. (7.5) does not explain adequately all the power characteristics of gas–liquid systems. The gassing effects on power consumption are due to the bubbles having a profound impact on the hydrodynamic behavior of fluid around the impeller. As described for Rushton turbines, gas-filled cavities develop behind the stirrer blades in aerated liquids. These cavities decrease the drag forces generated at the impeller and significantly reduce the resistance to impeller rotation, thus causing a substantial drop in the power required to operate the stirrer compared with nonaerated conditions.

The relationship between the power drop with gassing and operating conditions such as the gas flow rate and stirrer speed is often represented using graphs such as those shown in Figure 7.19. P_g is the power required with gassing and P_0 is the power required without gassing. The operating conditions are represented by the dimensionless gas flow number Fl_g. Figure 7.19 shows reductions in power as a function of the gas flow number for three different impellers in low-viscosity fluid. Each curve represents experimental data for a given impeller operated at constant stirrer speed (i.e., for each curve, the value of Fl_g changes only because of changes in the gas flow rate).

When the curved-blade disc turbine shown in Figure 7.12 is rotated clockwise with the concave sides of the blades forward, the curvature of the blades ensures that no large, ventilated cavities can form on the convex surfaces. Consequently, in low-viscosity fluids the power consumption with gassing remains close to that without gassing until impeller flooding occurs. In non-Newtonian or viscous fluids, power losses may be greater at up to about 20% [2]. For the hydrofoil impellers shown in Figure 7.13 operated for downward pumping, depending on the stirrer speed, abrupt reductions in power can accompany the transition from indirect to direct loading as large cavities form behind the blades. However, this drop in power is usually less than with Rushton turbines under similar conditions [5]. In contrast, for upward-pumping hydrofoils, there is virtually no reduction in power draw with aeration over a wide range of gas flow rates [6].

Reduction in stirrer power consumption with gassing may seem a desirable feature because of the potential for energy and cost savings during operation of the impeller. However, when all the relevant factors are considered, stirrers with power requirements that are relatively insensitive to gassing are preferred. In the design of bioprocess equipment, the stirrer motor is usually sized to allow operation under nonaerated conditions. This is necessary to prevent motor burn-out if there is a failure of air supply during operation of the bioreactors. In addition, medium in bioreactors is often mixed without aeration during the heating and cooling cycles of *in situ* batch sterilization. Therefore, the decrease in impeller power consumption

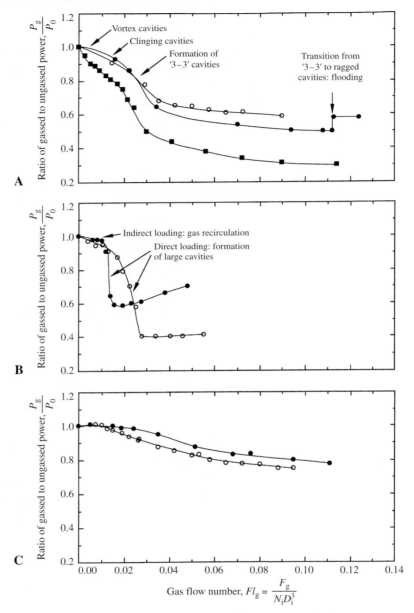

Figure 7.19 Variation in power consumption with gassing P_g relative to ungassed power consumption P_0 at constant stirrer speed. Fl_g is dimensionless gas flow number; F_g is volumetric gas flow rate; N_i is stirrer speed; D_i is impeller diameter.

with gassing represents an underutilization of the capacity of the stirrer motor. As outlined in Chapter 9, the rate of O_2 transfer from gas to liquid in aerated systems depends on the power input to the fluid; therefore, any reduction in power diminishes the effectiveness of mass transfer in the system with potential deleterious consequences for culture performance. Power losses may also reduce the ability of the stirrer to maintain complete suspension of solids. For example, the sudden reductions in power shown in Figure 7.20(a) and (b) can result in severe loss of suspension capacity, with the result that cells begin to settle out on the vessel floor. All these factors have promoted interest in the development of impellers such as the curved-blade disc turbine and upward-pumping hydrofoils, for which there is minimal reduction in power draw with gassing.

7.5 Power Input by Gassing

Gas sparging contributes to the total power input to bioreactors during operation under aerated conditions. The power input from sparging, P_v, can be calculated using the equation:

$$P_v = F_g \rho g H_L \tag{7.9}$$

where F_g is the volumetric flow rate of gas at the temperature and average pressure of the liquid in the tank, ρ is the liquid density, g is gravitational acceleration, and H_L is the liquid height. For aerated vessels stirred with an impeller, P_v is usually only a small fraction of the total power input and is often neglected. However, if high gas flow rates are used at low stirrer speeds, for example in reactors that rely mainly on gas sparging for mixing with the stirrer playing a relatively minor role, the contribution of P_v to the total power input can be more substantial.

7.6 Impeller Pumping Capacity

Fluid is pumped by the blades of rotating impellers. The volumetric flow rate of fluid leaving the blades varies with operating parameters such as the stirrer speed and size of the impeller but is also a characteristic of the impeller type or design. The effectiveness of impellers for pumping fluid is represented by a dimensionless number called the *flow number*:

$$Fl = \frac{Q}{N_i D_i^3} \tag{7.10}$$

where Fl is the flow number, Q is the volumetric flow rate of fluid leaving the impeller blades, N_i is the stirrer speed, and D_i is the impeller diameter. The flow number is a measure of the ability of the impeller to generate strong circulatory flows, such as those necessary for blending operations and solids suspension.

Typical fluid discharge velocity profiles for radial- and axial-flow turbines are shown in Figure 7.20. For radial-flow impellers, the mean discharge velocity is maximum at the center line of the blade and decays above and below the center line to form a bell-shaped curve, as illustrated in Figure 7.20(a). The discharge velocity profile for an axial-flow turbine is shown

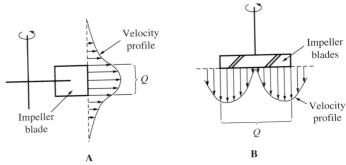

Figure 7.20 Graphic of discharge velocity profiles from the blades of: (A) a radial-flow impeller and (B) an axial-flow impeller.

Table 7.2: Values of the Turbulent Ungassed Flow Number *Fl* for a Selection of Different Impellers in Baffled Tanks Containing Low-Viscosity Fluid

Impeller	Fl	Reference
Rushton turbine	0.78	[7]
Propeller	0.73	[8]
3 blades		
Pitched-blade turbine	0.75	[9]
Downward pumping		
4 blades, 45°		
Pitched-blade turbine	0.81	[10]
Downward pumping		
6 blades, 45°		
Hydrofoil (Lightnin A315)	0.74	[10]
Downward pumping		
4 blades		
Hydrofoil (Prochem Maxflo T)	0.82	[4]
Downward pumping		
6 blades		

in Figure 7.20(b). In this case, Q is the volumetric flow rate of fluid leaving directly from the lower edges of the blades, excluding entrained flow from the surrounding region.

Values of the flow number Fl for several impellers operating in the turbulent regime in low-viscosity fluids without gassing are listed in Table 7.2. Fl is dependent on the vessel and blade geometry; however, this dependence and the variation of Fl between impeller types is not as great as for the turbulent power number N_P', as seen in Table 7.3.

The effectiveness of different impellers for generating flow can be compared relative to their power requirements. Combination of Eqs. (7.10) and (7.5) yields an expression for the impeller discharge flow rate per unit power consumed:

$$\frac{Q}{P} = \frac{Fl}{N_P' \rho N_i^2 D_i^2} \tag{7.11}$$

Therefore, if two different impellers of the same size are operated in the same fluid at the same stirrer speed, their pumping efficiencies can be compared using the ratio:

$$\frac{\left(\dfrac{Q}{P}\right)_{\text{impeller 1}}}{\left(\dfrac{Q}{P}\right)_{\text{impeller 2}}} = \frac{\left(\dfrac{Fl}{N_P'}\right)_{\text{impeller 1}}}{\left(\dfrac{Fl}{N_P'}\right)_{\text{impeller 2}}} \tag{7.12}$$

We can compare the pumping efficiencies of Rushton and pitched-blade turbines using Eq. (7.12) and values of Fl and N_P' from Tables 7.2 and 7.3. For a Rushton turbine with $D_i/D_T = 0.50$, $N_P' = 5.9$ and $Fl = 0.78$. For a six-blade downward-pumping pitched-blade turbine of the same size, $N_P' = 1.6$ and $Fl = 0.81$. Therefore:

$$\frac{\left(\dfrac{Q}{P}\right)_{\text{pitched-blade}}}{\left(\dfrac{Q}{P}\right)_{\text{Rushton}}} = \frac{\left(\dfrac{0.81}{1.6}\right)}{\left(\dfrac{0.78}{5.9}\right)} = 3.8 \tag{7.13}$$

This result indicates that pitched-blade turbines produce almost four times the flow for the same power input as Rushton turbines. The analysis provides an explanation for why Rushton turbines are considered to have relatively low pumping efficiency, while pitched-blade turbines are recognized for their high pumping capacity and effectiveness for blending operations. A comparison of Rushton turbines with hydrofoil impellers yields similar results. The analysis applies only for ungassed liquids; the effect of gassing on liquid pumping rates and power consumption varies considerably between different impellers.

7.7 Suspension of Solids

Bioreactors used for cell culture contain biomass solids. Microorganisms such as single-cell bacteria and yeast are small and finely divided; other cells such as mycelia and plant cells form macroscopic aggregates or clumps depending on the culture conditions. Cells contain a high percentage of water, therefore, the difference in density between the solid phase and the suspending liquid is generally very small. However, in some bioprocesses, cells

Table 7.3: Values of the Turbulent Ungassed Power Number N'_P for a Selection of Impellers and System Geometries

Impeller	System geometry	N'_P	Reference
Rushton turbine $D_i/D_T = 0.50$ $W_B/D_i = 0.20$ $L_B/D_i = 0.25$	Flat-bottom tank $H_U/D_T = 1.0$ Number of baffles = 4 $W_{BF}/D_T = 0.1$ $C_i/D_T = 0.25$	5.9	[6,7]
Pitched-blade turbine Downward pumping 6 blades, 45° $D_i/D_T = 0.40$	Flat-bottom tank $H_U/D_T = 1.0$ Number of baffles = 4 $W_{BF}/D_T = 0.1$ $C_i/D_T = 0.25$	1.8	[11]
Pitched-blade turbine Downward pumping 6 blades, 45° $D_i/D_T = 0.50$ $W_B/D_i = 0.20$	Flat-bottom tank $H_U/D_T = 1.0$ Number of baffles = 4 $W_{BF}/D_T = 0.1$ $C_i/D_T = 0.25$	1.6	[12]
Pitched-blade turbine Upward pumping 6 blades, 45° $D_i/D_T = 0.5$ $W_B/D_i = 0.20$	Flat-bottom tank $H_U/D_T = 1.0$ Number of baffles = 4 $W_{BF}/D_T = 0.1$ $C_i/D_T = 0.25$	1.6	[7,10]
Curved-blade disc turbine (Scaba 6SRGT) 6 blades $D_i/D_T = 0.33$ $W_B/D_i = 0.15$ $L_B/D_i = 0.28$	Flat-bottom tank $H_U/D_T = 1.0$ Number of baffles = 4 $W_{BF}/D_T = 0.1$ $C_i/D_T = 0.25$	1.5	[2]
Hydrofoil (Lightnin A315) Downward pumping 4 blades $D_i/D_T = 0.40$	Flat-bottom tank $H_U/D_T = 1.0$ Number of baffles = 4 $W_{BF}/D_T = 0.1$ $C_i/D_T = 0.25$	0.84	[3]
Hydrofoil (Prochem Maxflo T) Downward pumping 6 blades $D_i/D_T = 0.35$	Flat-bottom tank $H_U/D_T = 1.0$ Number of baffles = 4 $W_{BF}/D_T = 0.1$ $C_i/D_T = 0.45$	1.6	[4]

are immobilized on or in solid matrices of varying material density. These systems include anchorage-dependent animal cells cultured on microcarrier beads, and bacteria attached to sand grains for wastewater treatment.

One of the functions of mixing in bioreactors is to maintain the cells in suspension. Accumulation of biomass at the bottom of the vessel is highly undesirable, as cells within settled layers have poor access to nutrients and O_2. It is important, therefore, to know what operating conditions are required to completely suspend solids in stirred tanks.

7.7.1 Without Gassing

A common criterion used to define complete suspension of solids is that no particle should remain motionless on the bottom of the vessel for more than 1 to 2 seconds. Applying this criterion, the Zwietering equation [13]:

$$N_{JS} = \frac{S \nu_L^{0.1} D_p^{0.2} \left[g \left(\rho_p - \rho_L \right) / \rho_L \right]^{0.45} X^{0.13}}{D_i^{0.85}}$$

(7.14)

is generally accepted as the best correlation for N_{JS}, the stirrer speed required for just complete suspension of solids in the absence of gassing. In Eq. (7.14), S is a dimensionless parameter dependent on the impeller and tank geometry, ν_L is the liquid kinematic viscosity (Section 6.3), D_p is the diameter of the solid particles, g is gravitational acceleration, ρ_p is the particle density, ρ_L is the liquid density, X is the weight percentage of particles in the suspension, and D_i is the impeller diameter.

Zwietering's equation has been subjected to extensive testing over many years using a wide range of system properties. The exponents in Eq. (7.14) are independent of the tank size, impeller type, impeller-to-tank diameter ratio, and impeller off-bottom clearance; these geometric factors are reflected in the value of S. Table 7.4 lists some values of S for different impeller geometries; these data were obtained using flat-bottomed cylindrical vessels with four baffles of width 1/10 the tank diameter and liquid height equal to the tank diameter.

N_{JS} decreases significantly as the size of the impeller increases, not only because of the direct effect of D_i in Eq. (8.18) but also because the D_i/D_T ratio changes the value of S. For a fixed impeller off-bottom clearance, a general relationship is:

$$S \propto \left(\frac{D_T}{D_i} \right)^\alpha$$

(7.15)

where α is approximately 1.5 for Rushton turbines and 0.82 for propellers [16]. For many impellers, S is sensitive to the impeller off-bottom clearance ratio C_i/D_T. As shown in Table 7.4 for Rushton and pitched-blade turbines, S at constant D_i/D_T decreases as the impeller clearance is reduced, so that lower stirrer speeds are required for complete suspension. The shape of the base of the vessel (Figure 7.2) also influences the efficiency of solids

Table 7.4: Values of the Geometric Parameter S in Eq. (7.14) for Flat-Bottom Tanks

Impeller	D_i/D_T	C_i/D_T	S	Reference
Rushton	0.25	0.25	12	[14]
	0.33	0.17	5.8	[14]
	0.33	0.25	6.7	[14]
	0.33	0.50	8.0	[14]
	0.50	0.25	4.25	[14]
	0.50	0.17	3.9	[15]
Propeller	0.33	0.25	6.6	[15]
Pitched-blade turbine	0.33	0.20	5.7	[16]
Downward pumping, 4 blades, 45°	0.33	0.25	6.2	[16]
	0.33	0.33	6.8	[16]
	0.33	0.50	11.5	[16]
	0.50	0.25	5.8	[14]
Pitched-blade turbine	0.50	0.25	5.7	[15]
Downward pumping, 6 blades, 45°				
Pitched-blade turbine	0.50	0.25	6.9	[17]
Upward pumping, 6 blades, 45°				

EXAMPLE 7.1 Calculation of Power Requirements

A fermentation broth with viscosity 10^{-2} Pa s and density 1000 kg m^{-3} is agitated in a 50-m³ baffled tank using a marine propeller 1.3 m in diameter. The tank geometry is as specified in Figure 7.15. Calculate the power required for a stirrer speed of 4 s^{-1}.

Solution
From Eq. (6.2):

$$Re_i = \frac{4\,\text{s}^{-1}(1.3\,\text{m})^2\,1000\,\text{kg}\,\text{m}^{-3}}{10^{-2}\,\text{kg}\,\text{m}^{-1}\text{s}^{-1}} = 6.76 \times 10^5$$

From Figure 7.15, flow at this Re_i is fully turbulent. From Table 7.1, N_p' is 0.35. Therefore, from Eq. (7.5):

$$P = (0.35)1000\,\text{kg}\,\text{m}^{-3}\left(4\,\text{s}^{-1}\right)^3 (1.3\,\text{m})^5 = 8.3 \times 10^4\,\text{kg}\,\text{m}^2\,\text{s}^{-3}$$

Using 1 kg m² s^{-3} = 1 W. Therefore:

$$P = 83\text{kW}$$

suspension, with dished, contoured, and cone-and-fillet bases offering advantages in some cases for reducing S. The extent of this effect depends, however, on the type of flow pattern generated by the impeller [8].

Even if a stirrer is operated at speeds equal to or above N_{JS} to obtain complete particle suspension, this does not guarantee that the suspension is homogeneous throughout the tank. In general, speeds considerably higher than N_{JS} are required to achieve uniform particle concentration, for small particles such as dispersed cells with density similar to that of the suspending liquid, a reasonable degree of homogeneity can be expected at N_{JS} [1].

As illustrated in Example 7.2, complete suspension of cells and small cell clumps is generally achieved at low to moderate stirrer speeds.

7.7.2 With Gassing

The formation of ventilated cavities behind impeller blades reduces the power draw and liquid pumping capacity of the impeller. Therefore, higher stirrer speeds are often required for solids suspension in aerated systems.

Equations relating N_{JSg}, the stirrer speed required for just complete suspension of solids in the presence of gassing, to N_{JS} have been developed for various impellers [1,18]. For Rushton

---

EXAMPLE 7.2 Solids Suspension

Clump-forming fungal cells are cultured in a flat-bottomed 10-m³ bioreactor of diameter 2.4 m equipped with a Rushton turbine of diameter 1.2 m operated at 50 rpm. The impeller off-bottom clearance is 0.6 m. The density of the growth medium is 1000 kg m^{-3} and the viscosity is 0.055 Pa s. The density and diameter of the cell clumps are 1035 kg m^{-3} and 600 μm, respectively. The concentration of cells in the bioreactor reaches 40% w/w. Are the cells suspended under these conditions?

Solution

$D_i/D_T = 0.50$ and $C_i/D_T = 0.25$. Therefore, from Table 8.4, $S = 4.25$. $g = 9.81$ m s^{-2}. 1 Pa s = 1 kg m^{-1} s^{-1}. Therefore, using Eq. (7.9) to calculate the kinematic viscosity:

$$\nu_L = \frac{0.055 \text{ kg m}^{-1}\text{s}^{-1}}{1000 \text{ kg m}^{-3}} = 5.5 \times 10^{-5} \text{ m}^2 \text{ s}^{-1}$$

Substituting values into Eq. (8.18) gives:

$$N_{JS} = \frac{4.25\left(5.5 \times 10^{-5} \text{ m}^2 \text{ s}^{-1}\right)^{0.1}\left(600 \times 10^{-6} \text{ m}\right)^{0.2}\left[9.81 \text{ m s}^{-2}\dfrac{(1035 - 1000) \text{ kg m}^{-3}}{1000 \text{ kg m}^{-3}}\right]^{0.45} 40^{0.13}}{(1.2 \text{ m})^{0.55}}$$

$$N_{JS} = 0.31 \text{ s}^{-1} = 18.6 \text{ rpm}$$

The operating stirrer speed of 50 rpm is well above the stirrer speed required for solids suspension. The cells are therefore completely suspended.

EXAMPLE 7.3 Estimation of Mixing Time

A baffled bioreactor with tank diameter and liquid height equal to 1.2 m is stirred using a six-blade downward-pumping Prochem Maxflo T hydrofoil impeller. The impeller diameter is 0.42 m and the stirrer speed is 1.5 s^{-1}. The viscosity of the fermentation broth is 10^{-2} Pa s and the density is 1000 kg m^{-3}. Estimate the mixing time under nonaerated conditions.

Solution

From Eq. (6.2):

$$Re_i = \frac{1.5 \text{ s}^{-1}(0.42 \text{ m})^2 \, 1000 \text{ kg m}^{-3}}{10^{-2} \text{ kg m}^{-1} \text{ s}^{-1}} = 2.6 \times 10^4$$

Flow is turbulent for remote-clearance impellers at this Reynolds number; therefore Eq. (7.20) can be used to calculate t_m. From Table 7.3, N_p' for this hydrofoil impeller is equal to 1.6. Therefore:

$$t_m = \frac{5.4}{1.5 \text{ s}^{-1}} \left(\frac{1}{1.6}\right)^{1/3} \left(\frac{1.2 \text{ m}}{0.42 \text{ m}}\right)^2 = 25.1 \text{ s}$$

The mixing time is 25 s.

---

turbines of two different sizes with an impeller off-bottom clearance of one-quarter the tank diameter [1]:

$$N_{JSg} = N_{JS} + 2.4 \, F_{gv} \qquad \text{for} \qquad D_i/D_T = 0.33 \tag{7.16}$$

$$N_{JSg} = N_{JS} + 0.94 \, F_{gv} \qquad \text{for} \qquad D_i/D_T = 0.50 \tag{7.17}$$

where F_{gv} is the gas flow rate in units of vvm or volume of gas per volume of liquid per minute, and N_{JS} and N_{JSg} both have units of s^{-1} [8,18–20].

7.8 Mechanisms of Mixing

In this section, we consider the mechanisms controlling the rate of mixing in stirred tanks containing a single liquid phase. As illustrated schematically in Figures 7.21 and 7.22, large liquid circulation loops develop in stirred vessels. The velocity of fluid leaving the impeller must be sufficient to carry material into the most remote regions of the tank for effective mixing; fluid circulated by the impeller must also sweep the entire vessel in a reasonable time. In addition, turbulence must be developed in the fluid as mixing is certain to be poor unless flow is turbulent. All these factors are important in mixing, which can be described as a combination of three physical processes:

- Distribution
- Dispersion
- Diffusion

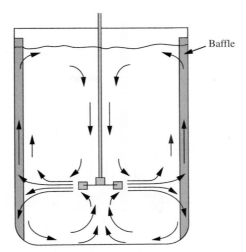

Figure 7.21 Flow pattern produced by a radial-flow impeller in a baffled tank.

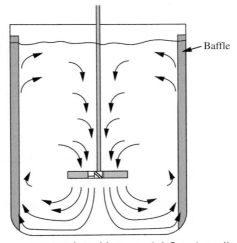

Figure 7.22 Flow pattern produced by an axial-flow impeller in a baffled tank.

Distribution is sometimes called *macromixing*; diffusion is also called *micromixing*. Dispersion can be classified as either micro- or macromixing depending on the scale of fluid motion.

The pattern of bulk fluid flow in a baffled vessel stirred by a centrally located radial-flow impeller is shown in detail in Figure 7.23. Near the impeller there is a zone of intense turbulence where fluid currents converge and exchange material. However, as fluid moves away from the impeller, flow becomes progressively slower and less turbulent. In large tanks, streamline or laminar flow may develop in these local regions. Under these conditions,

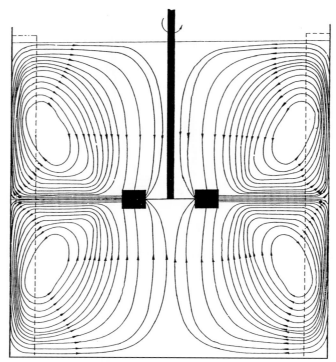

Figure 7.23 Flow pattern developed by a centrally located radial-flow impeller.
From R.M. Voncken, J.W. Rotte, and A.Th. ten Houten, 1965, Circulation model for continuous-flow, turbine-stirred, baffled tanks. In: Mixing—Theory Related to Practice, Proc. Symp. 10, AIChE–IChemE Joint Meeting, London.

because fluid elements move mostly parallel to each other in streamline flow (Chapter 6), mixing is not very effective away from the impeller zone.

Let us consider what happens when a small amount of liquid dye is dropped onto the top of the fluid in Figure 7.23. First, the dye is swept by circulating currents down to the impeller. At the impeller there is vigorous and turbulent motion of fluid; the dye is mechanically dispersed into smaller volumes and distributed between the large circulation loops. These smaller parcels of dye are then carried around the tank, dispersing all the while into those parts of the system not yet containing dye. Returning to the impeller, the dye aliquots are broken up into even smaller volumes for further distribution. After a time, dye is homogeneously distributed throughout the tank and achieves a uniform concentration.

The process whereby dye is transported to all regions of the vessel by bulk circulation currents is called *distribution*. Distribution is an important process in mixing but can be relatively slow. *Distribution is often the slowest step in the mixing process.* If the stirrer speed is sufficiently high, superimposed on the distribution process is turbulence. In turbulent flow, the fluid no longer travels along streamlines but moves erratically in the form of crosscurrents;

this enhances the mixing process at scales much smaller than the scale of bulk circulation. The kinetic energy of turbulent fluid is directed into regions of rotational flow called *eddies*; masses of eddies of various size coexist in turbulent flow. At steady state in a mixed tank, most of the energy from the stirrer is dissipated through the eddies as heat; energy lost in other processes (e.g., fluid collision with the tank walls) is generally negligible.

The process of breaking up the bulk flow into smaller and smaller eddies is called *dispersion*. Dispersion facilitates rapid transfer of material throughout the vessel. The degree of homogeneity possible as a result of dispersion is limited by the size of the smallest eddies that may be formed in a particular fluid. At steady state, the average rate of energy dissipation by turbulence over the entire tank is equal to the power input to the fluid by the impeller; this power input is the same as that estimated in Section 7.4. The greater the power input to the fluid, the smaller are the eddies. λ is also dependent on viscosity: at a given power input, smaller eddies are produced in low-viscosity fluids. For low-viscosity liquids such as water, λ is usually in the range 30 to 100 μm. For such fluids, this is the smallest scale of mixing achievable by dispersion.

Within eddies, flow of fluid is rotational and occurs in streamlines. Streamline flow does not facilitate mixing, therefore, to achieve mixing on a scale smaller than the Kolmogorov scale, we must rely on *diffusion*. Molecular diffusion is generally regarded as a slow process; however, over small distances it can be accomplished quite rapidly. Within eddies of diameter 30 to 100 μm, homogeneity is achieved in about 1 s for low-viscosity fluids. Consequently, if the power input to a stirred vessel produces eddies of this dimension, mixing on a molecular scale is accomplished virtually simultaneously.

7.9 Assessing Mixing Effectiveness

Achieving rapid mixing in a stirred tank requires the agitator provide good bulk circulation or macromixing. Micromixing at or near the molecular scale is also important but occurs relatively quickly compared with macromixing. Mixing effectiveness is therefore usually a reflection of the rate of bulk flow.

Mixing time is a useful parameter for assessing the overall speed of mixing in stirred vessels. The mixing time t_m is the time required to achieve a given degree of homogeneity starting from the completely segregated state. It can be measured by injecting a tracer into the vessel and following its concentration at a fixed point in the tank. Mixing time can also be determined by measuring the temperature response after addition of a small quantity of heated liquid.

Let us assume that a small pulse of tracer is added to fluid in a stirred tank already containing tracer material at concentration C_i. When flow in the system is circulatory, the tracer concentration measured at some fixed point in the tank can be expected to follow a pattern similar to

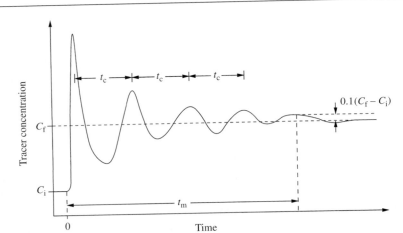

Figure 7.24 Concentration response after dye is injected into a stirred tank.

that shown in Figure 7.24. Before mixing is complete, a relatively high concentration will be detected every time the bulk flow brings tracer to the measurement point. The peaks in concentration will be separated by a period approximately equal to the average time required for fluid to traverse one bulk circulation loop. In stirred vessels this period is called the *circulation time*, t_c. After several circulations, the desired degree of homogeneity is reached.

Definition of the mixing time t_m depends on the degree of homogeneity required. Usually, mixing time is defined as the time after which the concentration of tracer differs from the final concentration C_f by less than 10% of the total concentration difference $(C_f - C_i)$. However, there is no single, universally applied definition of mixing time. Industrial-scale stirred vessels with working volumes between 1 and 100 m³ have mixing times between about 30 and 120 s, depending on conditions.

Intuitively, we can predict that the mixing time in stirred tanks will depend on variables such as the size of the tank and impeller, the fluid properties, and the stirrer speed. The relationship between mixing time and several of these variables has been determined experimentally for different impellers: results for a Rushton turbine in a baffled tank are shown in Figure 7.25. The dimensionless product $N_i t_m$, which is also known as the *homogenization number* or *dimensionless mixing time*, is plotted as a function of the impeller Reynolds number Re_i; t_m is the mixing time based on a 10% deviation from the total change in conditions, and N_i is the rotational speed of the stirrer. Conceptually, $N_i t_m$ represents the number of stirrer rotations required to homogenize the liquid after addition of a small pulse of tracer. At relatively low Reynolds numbers in the laminar–transition regime, $N_i t_m$ increases significantly with decreasing Re_i. However, as the Reynolds number is increased, $N_i t_m$ approaches a constant value that persists into the turbulent regime at Re_i above about 5×10^3. The relationship between $N_i t_m$ and Re_i for most other impellers is qualitatively similar to that shown in Figure 7.25 [21].

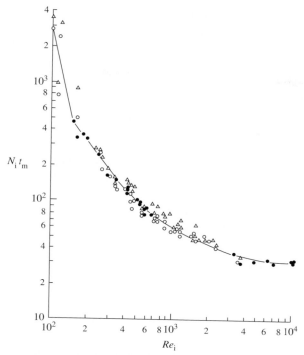

Figure 7.25 Variation of mixing time with Reynolds number for a Rushton turbine in a baffled tank. The impeller is located one-third the tank diameter off the floor of the vessel, the impeller diameter is one-third the tank diameter, the liquid height is equal to the tank diameter, and the tank has four baffles of width one-tenth the tank diameter. Several measurement techniques and tank sizes were used: (●) thermal method, vessel diameter 1.8 m (O) thermal method, vessel diameter 0.24 m; (△) decoloration method, vessel diameter 0.24 m. *Reprinted from C.J. Hoogendoorn and A.P. den Hartog, Model studies on mixers in the viscous flow region, Chem. Eng. Sci. 22, 1689–1699. Copyright 1967, with permission from Pergamon Press Ltd, Oxford.*

An equation has been developed for estimating the mixing time in stirred vessels under turbulent flow conditions. This expression for t_m can be applied irrespective of the type of impeller used [22]:

$$t_m = 5.9 D_T^{2/3} \left(\frac{\rho V_L}{P} \right)^{1/3} \left(\frac{D_T}{D_i} \right)^{1/3} \tag{7.18}$$

where t_m is the mixing time, D_T is the tank diameter, ρ is the liquid density, V_L is the liquid volume, P is the power input, and D_i is the impeller diameter. Equation (7.18) applies to baffled vessels stirred with a single impeller and with liquid height equal to the tank diameter. The relationship has been verified using a range of different impellers with $0.2 \leq D_i/D_T \leq 0.7$ in vessels of diameter up to 2.7 m. The equation is also valid under aerated conditions provided the impeller disperses the gas effectively (i.e., is not flooded) and P is the power drawn with gassing [23]. Equation (7.18) indicates that, for a tank of fixed diameter and liquid volume, mixing time is reduced if we use a large impeller and a high-power input.

For a cylindrical tank with liquid height equal to the tank diameter, the geometric formula for the volume of a cylinder is:

$$V_L = \frac{\pi}{4} D_T^3 \tag{7.19}$$

Also, as Eq. (7.18) applies under turbulent conditions, we can express P in terms of the turbulent power number N_P' using Eq. (7.5). Substituting Eqs. (7.5) and (7.19) into Eq. (7.18) gives:

$$t_m = \frac{5.4}{N_i} \left(\frac{1}{N_P'} \right)^{1/3} \left(\frac{D_T}{D_i} \right)^2 \tag{7.20}$$

Equation (7.20) indicates that mixing time reduces in direct proportion to stirrer speed. This is the same as saying that $N_i t_m$ for a given impeller and tank geometry is constant for turbulent flow, as discussed earlier with reference to Figure 7.25. At constant N_i, t_m is directly proportional to $(D_T/D_i)^2$, showing that mixing times can be reduced significantly using impellers with large D_i/D_T ratio. However, because of the strong influence of impeller diameter on power requirements, increasing D_i also raises the power consumption, so there will be a cost associated with using this strategy to improve mixing.

The parameters affecting mixing efficiency also affect stirrer power requirements, therefore, it is not always possible to achieve small mixing times without consuming enormous amounts of energy, especially in large vessels. Estimating mixing time in a bioreactor is examined in Example 7.3.

7.10 Scale-Up of Mixing Systems

Design of industrial-scale bioprocesses is usually based on the performance of small-scale prototypes. It is always better to know whether a particular process will work properly before it is constructed in full size: determining optimum operating conditions at production scale is expensive and time-consuming. Ideally, scale-up should be carried out so that conditions in the large vessel are as close as possible to those producing good results at the smaller scale. As mixing is of critical importance in bioreactors, it would seem desirable to keep the mixing time constant on scale-up. Unfortunately, as explained in this section, the relationship between mixing time and power consumption makes this rarely possible in practice.

Suppose a cylindrical 1-m³ pilot-scale stirred tank is scaled up to 100 m³. Let us consider the power required to maintain the same mixing time in the large and small vessels. From Eq. (7.18), equating t_m values gives:

$$5.9 D_{T1}{}^{2/3} \left(\frac{\rho_1 V_{L1}}{P_1} \right)^{1/3} \left(\frac{D_{T1}}{D_{i1}} \right)^{1/3} = 5.9 D_{T2}{}^{2/3} \left(\frac{\rho_2 V_{L2}}{P_2} \right)^{1/3} \left(\frac{D_{T2}}{D_{i2}} \right)^{1/3} \tag{7.21}$$

where subscript 1 refers to the small-scale system and subscript 2 refers to the large-scale system. If the tanks are geometrically similar, the ratio D_T/D_i is the same at both scales. Similarly, the fluid density will be the same before and after scale-up, and we can also cancel the constant multiplier from both sides. Therefore, Eq. (7.21) reduces to:

$$D_{T1}{}^{2/3} \left(\frac{V_{L1}}{P_1} \right)^{1/3} = D_{T2}{}^{2/3} \left(\frac{V_{L2}}{P_2} \right)^{1/3} \tag{7.22}$$

Cubing both sides of Eq. (7.22) and rearranging gives:

$$P_2 = P_1 \left(\frac{D_{T2}}{D_{T1}} \right)^2 \frac{V_{L2}}{V_{L1}} \tag{7.23}$$

The geometric relationship between V_L and D_T is given by Eq. (7.19) for cylindrical tanks with liquid height equal to the tank diameter. Solving Eq. (7.19) for D_T gives:

$$D_T = \left(\frac{4 V_L}{\pi} \right)^{1/3} \tag{7.24}$$

Substituting this expression into Eq. (7.23) for both scales gives:

$$P_2 = P_1 \left(\frac{V_{L2}}{V_{L1}} \right)^{5/3} \tag{7.25}$$

In our example of scale-up from 1 m³ to 100 m³, $V_{L2} = 100\, V_{L1}$. Therefore, the result from Eq. (7.25) is that $P_2 = {\sim}2000\, P_1$; that is, the power required to achieve equal mixing time in the 100-m³ tank is ~2000 times greater than in the 1-m³ vessel. This represents an extremely large increase in power, much greater than is economically or technically feasible with most equipment used for stirring. This example illustrates why the criterion of constant mixing time can hardly ever be applied for scale-up. It is inevitable that mixing times increase with scale because the implications for power consumption are impractical.

Reduced culture performance and productivity often accompany scale-up of bioreactors as a result of lower mixing efficiencies and consequent alteration of the physical environment.

7.11 Improving Mixing in Bioreactors

Longer mixing times are often unavoidable when stirred vessels are scaled up in size. In these circumstances, it is not possible to reduce mixing times sufficiently by simply raising the power input to the stirrer. In this section, we consider methods for improving mixing in stirred tanks that do not involve consumption of significantly greater amounts of energy.

7.11.1 Impeller and Vessel Geometry

Mixing can sometimes be improved by changing the system's physical configuration.

- Baffles should be installed; this is routine for stirred bioreactors and produces greater turbulence.
- For efficient mixing, the impeller should be mounted below the geometric center of the vessel. For example, mixing by radial impellers such as Rushton turbines is facilitated when the circulation currents below the impeller are smaller than those above (as shown in Figure 7.21), as this makes the upper and lower circulation loops asynchronous. Under these conditions, fluid particles leaving the impeller at the same instant but entering different circulation paths take different periods of time to return and exchange material. The rate of distribution throughout the vessel is increased when the same fluid particles from different circulation loops do not meet each other every time they return to the impeller region.
- For two- and three-phase mixing (i.e., in gas–liquid and gas–liquid–solid systems), good mixing includes achieving complete gas dispersion and solids suspension. These stirrer functions are sensitive to various aspects of tank geometry, including the impeller off-bottom clearance, type of sparger, clearance between the sparger and the impeller, and base profile of the tank. Optimization of these features of the system can yield considerable improvements in mixing effectiveness without necessarily requiring large amounts of extra power.
- The power required increases substantially when extra impellers are fitted. Furthermore, depending on the impeller design and the separation allowed between the impellers, mixing efficiency can actually be lower with multiple impellers than in single-impeller systems.

7.11.2 Feed Points

Severe mixing problems can occur in industrial-scale bioreactors when material is fed into the vessel during operation. Concentrated acid or alkali and antifoam agents are often pumped automatically into the broth for pH and foam control; bioreactors operated with continuous flow or in fed-batch mode also have fresh medium and nutrients added during the culture. If mixing and bulk distribution are slow, very high local concentrations of added material develop near the feed point. Such problems can be alleviated by installing multiple injection points to aid the distribution of substrate throughout the vessel. It is much less expensive to do this than to increase the stirrer speed and power input.

Location of the feed point (or feed points) is also important. In most commercial operations, material is fed into bioreactors using a single inlet delivering to the top surface of the liquid. However, mixing can be improved substantially by feeding directly into the impeller zone. This ensures rapid distribution and dispersion as convective currents and turbulence are strongest in this region.

7.12 Multiple Impellers

Typical bioreactors used for aerobic bioprocesses do not conform to the standard configuration for stirred tanks illustrated in Figure 7.1, where the liquid height is approximately equal to the tank diameter. Instead, aerobic cultures are carried out in tall vessels with liquid heights 2 to 5 times the tank diameter, an aspect ratio of 3:1 being common. The reasons for using this geometry are that relatively high hydrostatic pressures are produced in tall vessels filled with liquid, thus increasing the solubility of O_2, while rising air bubbles have longer contact time with the liquid, thus improving O_2 transfer from the gas phase.

Mixing in tall bioreactors is performed using more than one impeller mounted on the stirrer shaft. Each impeller generates its own circulation currents, but interaction between the fluid streams from different impellers can produce very complex flow patterns. An important parameter affecting the performance of multiple impellers is the spacing between them.

7.12.1 Multiple Rushton Turbines Without Gassing

The agitation systems in many bioprocess vessels consist of multiple Rushton turbines. In low-viscosity fluids, if Rushton turbines are spaced adequately apart, they each produce a radial discharge stream and generate independent large-scale circulation loops as illustrated in Figure 7.25. A vessel equipped with three Rushton turbines is mixed as if three separate stirred tanks were stacked one on top of the other. The power required by multiple impellers under conditions without gassing can be estimated from:

$$(P)_n = n(P)_1 \tag{7.26}$$

where $(P)_n$ is the power required by n impellers and $(P)_1$ is the power required by a single impeller. The minimum spacing between the impellers for the flow pattern in Figure 7.26, and for Eq. (7.26) to be valid, is not well defined, being reported variously as one to three impeller diameters or one tank diameter [23]. Equation (7.26) indicates that, at this spacing, each impeller draws the same power as if it were operating alone.

A drawback associated with operation of multiple Rushton turbines is compartmentation of the fluid. As indicated in Figure 7.26, the radial flow pattern generated by Rushton turbines creates separate circulation currents above and below each impeller, providing little opportunity for interaction between the fluid streams emanating from different impellers. Consequently, at a fixed stirrer speed, the overall rate of mixing is lower with multiple Rushton turbines than in a standard single-impeller system. Installation of two Rushton impellers in a vessel with liquid height twice the tank diameter does not achieve the same mixing time as one Rushton impeller in liquid with height equal to the tank diameter. In multiple impeller systems, *the rate of exchange flow between the fluid compartments generated by each impeller determines the rate of overall mixing in the vessel*. Therefore, improving the exchange between compartments has a high priority in mixing operations using multiple impellers.

When material is added to bioreactors with multiple Rushton turbines, the location of the feed point has a significant effect on mixing time. This is illustrated in Figure 7.27, in which mixing time is plotted as a function of the height of the injection point in a vessel stirred with dual Rushton turbines. The mixing time was lowest when the tracer was injected at the height where the circulation loops from the upper and lower impellers came together.

7.12.2 Other Impeller Combinations Without Gassing

Various combinations of impellers, including Rushton turbines, curved-blade disc turbines, propellers, and pitched-blade and hydrofoil impellers in both downward-pumping and upward-pumping modes, have been tested for multiple-impeller mixing. Combining

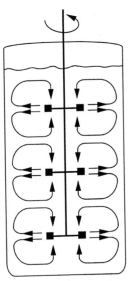

Figure 7.26 Independent circulation loops generated by multiple Rushton turbines in a tall bioreactor when the spacing between the impellers is relatively large.

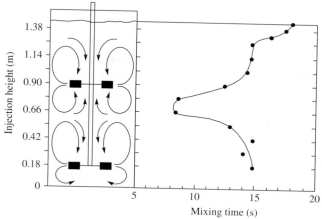

Figure 7.27 Variation in mixing time with height of the tracer injection point for dual Rushton turbines with clearance between the impellers equal to the tank diameter.

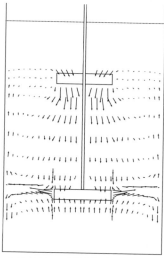

Figure 7.28 Mean velocity vector plot for a dual-impeller combination of a Rushton turbine in the lower position and a downward-pumping pitched-blade turbine above. The velocities were measured using laser Doppler velocimetry. *From V.P. Mishra and J.B. Joshi, 1994, Flow generated by a disc turbine. Part IV: Multiple impellers. Trans. IChemE 72A, 657–668.*

radial- and axial-flow impellers, or combining different axial-flow turbines, significantly reduces fluid compartmentation and enhances mixing compared with multiple Rushton turbines. However, the flow patterns generated can be complex and show strong sensitivity to impeller geometry, diameter, and clearance. A typical mean velocity vector plot for a dual-impeller system comprising a Rushton turbine in the lower position and a downward-pumping pitched-blade turbine above is shown in Figure 7.28.

7.12.3 Multiple-Phase Operation

When gassing is required in vessels with multiple impellers, Rushton or curved-blade disc turbines are often used in the lowest position closest to the sparger because of their superior ability to handle gas compared with axial-flow turbines. Combining hydrofoil or pitched-blade impellers above a Rushton turbine is very effective in aerated systems: the lower impeller breaks up the gas flow while the upper, high-flow impeller/s distribute the dispersed gas throughout the tank.

With gassing, the power relationship for multiple impellers is not as simple as that in Eq. (7.26). The main reason is that the power required by the individual impellers may be affected by the presence of gas, but each impeller is not affected to the same extent. At the lowest impeller, the formation of ventilated cavities at the impeller blades reduces the power drawn as described in Section 7.4.3.

Solids suspension in multiple impeller systems has been studied comparatively little. In general, irrespective of the liquid height, there appears to be no advantage associated with using more than one impeller to suspend solids, with respect to either the stirrer speed or power required. At the same time, however, the use of multiple impellers may improve the uniformity of particle concentration throughout the tank [19].

7.13 Effect of Rheological Properties on Mixing

Many bioprocess broths have high viscosities or exhibit non-Newtonian flow behaviors. These properties have a profound influence on mixing, making it more difficult to achieve small mixing times and homogeneous broth composition. The principal deleterious effects of high fluid viscosity and non-Newtonian rheology are reduced turbulence and the formation of stagnant zones in the vessel.

Turbulence is responsible for dispersing material at the scale of the smallest eddies. The existence of turbulence is indicated by the value of the impeller Reynolds number Re_i. Turbulence is damped at Re_i below about 10^4; as a consequence, mixing times increase significantly as shown in Figure 7.25. Re_i is inversely proportional to viscosity. Accordingly, nonturbulent flow and poor mixing are likely to occur during agitation of highly viscous fluids. Increasing the power input is an obvious solution; however, raising the power sufficiently to achieve turbulence is often impractical.

Most non-Newtonian fluids in bioprocessing are pseudoplastic. The apparent viscosity of these fluids depends on the shear rate, and as a result, their rheological behavior varies with the shear conditions. Metzner and Otto [24] proposed that the average shear rate $\dot{\gamma}_{av}$ in a stirred vessel is a linear function of the stirrer speed N_i:

$$\dot{\gamma}_{av} = k N_i \tag{7.27}$$

where k is a constant dependent on the type of impeller. Experimentally determined values of k are listed in Table 7.5. The validity of Eq. (7.27) was established in studies by Metzner et al. [24]. However, the shear rate in stirred vessels is far from uniform, being strongly dependent on distance from the impeller. Figure 7.29 shows estimated values of the shear rate in a pseudoplastic fluid as a function of radial distance from the tip of a Rushton turbine. The maximum shear rate close to the impeller is much higher than the average calculated using Eq. (7.27).

Table 7.5: Observed Values of k in Eq. (7.27)

Impeller type	k
Rushton turbine	10–13
Propeller	10
Anchor	20–25
Helical ribbon	30

From S. Nagata, 1975, *Mixing: Principles and Applications*, Kodansha, Tokyo.

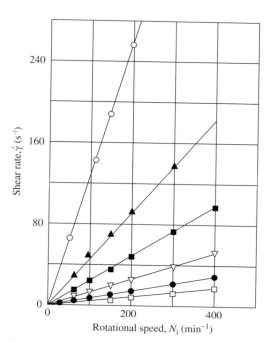

Figure 7.29 Shear rates in a pseudoplastic fluid as a function of the stirrer speed and radial distance from the impeller: (◯) impeller tip; (▲) 0.10 in.; (■) 0.20 in.; (▽) 0.34 in.; (●) 0.50 in.; (□) 1.00 in. *From A.B. Metzner and J.S. Taylor, 1960, Flow patterns in agitated vessels. AIChE J. 6, 109–114.*

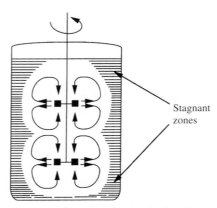

Figure 7.30 Mixing pattern for pseudoplastic fluid in a stirred bioreactor.

Pseudoplastic fluids are shear-thinning; that is, their apparent viscosity decreases with increasing shear. Accordingly, in stirred vessels, pseudoplastic fluids have relatively low apparent viscosity in the high-shear zone near the impeller, and relatively high apparent viscosity away from the impeller. As a result, flow patterns similar to that shown in Figure 7.30 can develop.

The effects of local fluid thinning in pseudoplastic fluids can be countered by modifying the geometry of the system and/or the impeller design. The use of multiple impellers, even when the liquid height is no greater than the tank diameter, improves the mixing of pseudoplastic fluids significantly. The combination of a Rushton turbine in the lower position and downward-pumping pitched-blade turbines above is particularly effective.

7.14 Role of Shear in Stirred Bioreactors

The development of shear stresses in fluids is related to the presence of velocity gradients. In turbulent flow, unsteady velocity components associated with eddies give rise to additional fluctuating velocity gradients that are superimposed on the mean flow. Consequently, turbulent shear stresses are much greater than those developed in laminar flow. Turbulent shear stress also varies considerably with time and position in the fluid.

Mixing in bioreactors used for aerobic cell culture must provide the shear conditions necessary to disperse gas bubbles and, if appropriate, break up liquid droplets or cell flocs. The dispersion of gas bubbles by impellers involves a balance between opposing forces. The interaction of bubbles with turbulent velocity fluctuations in the fluid causes the bubbles to stretch and deform. Bubble break-up occurs if the induced stresses exceed the stabilizing forces due to surface tension at the gas–liquid interface, which tend to restore the bubble to its spherical shape.

Table 7.6: Cell Shear Sensitivity

Cell type	Size	Shear sensitivity
Microbial cells	1 to 10 µm	Low
Microbial pellets/clumps	Up to 1 cm	Moderate
Plant cells	100 µm	Moderate/high
Plant cell aggregates	Up to 1 to 2 cm	High
Animal cells	20 µm	High
Animal cells on microcarriers	80 to 200 µm	Very high

Whereas bubble break-up is required in bioreactors to facilitate O_2 transfer, disruption of individual cells is undesirable. As indicated in Table 7.6, different cell types have different susceptibilities to damage in the bioreactor environment. This susceptibility is usually referred to as *shear sensitivity*, although the damage observed may not arise necessarily from the effects of fluid velocity gradients. In this context, the term "shear" is used imprecisely to mean any mechanism dependent on the hydrodynamic conditions in the vessel that results in cell damage.

Due to their relatively small size, bacteria and yeast are considered generally not to be shear sensitive. The main effect of agitation on filamentous fungi and actinomycetes is a change in morphology between pelleted and dispersed forms; the mean size of cell clumps also varies with hydrodynamic conditions. Insect, mammalian, and plant cells are particularly sensitive to hydrodynamic forces. Bioreactors used for culture of these cells must take this sensitivity into account while still providing adequate mixing and O_2 transfer. At the present time, the effects of hydrodynamic forces on cells are not understood completely. Cell disruption is an obvious outcome; however, more subtle *sublytic effects* such as slower growth or product synthesis, denaturation of extracellular proteins, and thickening of the cell walls may also occur. In some cases, cellular metabolism is stimulated by exposure to hydrodynamic forces.

7.15 Summary of Chapter 7

This chapter covers topics related to mixing and cell damage in bioreactors. At the end of Chapter 7 you should:

- Be familiar with the broad range of equipment used for mixing in stirred vessels, including different impeller designs, their operating characteristics, and their suitability for particular mixing applications
- Know how impeller size, stirrer speed, tank geometry, liquid properties, and gas sparging affect power consumption in stirred vessels
- Be able to determine the stirrer operating conditions for complete solids suspension
- Understand the mechanisms of mixing

- Know what is meant by *mixing time* and how it is measured
- Be able to describe the effects of scale-up on mixing, and options for improving mixing without the input of extra power
- Understand the factors affecting the performance of multiple-impeller systems
- Know the problems associated with mixing highly pseudoplastic or yield-stress fluids
- Understand how cells can be damaged in stirred and aerated bioreactors

Problems

7.1. Electrical power required for mixing

Laboratory-scale fermenters are usually mixed using small stirrers with electric motors rated between 100 and 500 W. One such motor is used to drive a 7-cm Rushton turbine in a small reactor containing fluid with the properties of water. The stirrer speed is 900 rpm. Estimate the power requirements for this process. How do you explain the difference between the amount of electrical power consumed by the motor and the power required by the stirrer?

7.2. Effect of viscosity on power requirements

A cylindrical bioreactor of diameter 3 m has four baffles. A Rushton turbine mounted in the reactor has a diameter of one-third the tank diameter. The liquid height is equal to the tank diameter and the density of the fluid is approximately $1\,g\,cm^{-3}$. The reactor is used to culture an anaerobic organism that does not require gas sparging. The broth can be assumed Newtonian. As the cells grow, the viscosity of the broth increases.

  a. The stirrer is operated at a constant speed of 90 rpm. Compare the power requirements when the viscosity is:

    (i) Approximately that of water

    (ii) 100 times greater than water

    (iii) 2×10^5 times greater than water

  b. The viscosity reaches a value 1000 times greater than water.

    (i) What stirrer speed is required to achieve turbulence?

    (ii) Estimate the power required to achieve turbulence.

    (iii) What power per unit volume is required for turbulence? Is it reasonable to expect to be able to provide this amount of power? Why or why not?

7.3. Power and scale-up

A pilot-scale fermenter of diameter and liquid height 0.5 m is fitted with four baffles of width one-tenth the tank diameter. Stirring is provided using a Scaba 6SRGT curved-blade disc turbine with diameter one-third the tank diameter. The density of the culture broth is $1000\,kg\,m^{-3}$ and the viscosity is 5 cP. Optimum culture conditions are provided in the pilot-scale fermenter when the stirrer speed is 185 rpm. Following completion of the pilot studies, a larger production-scale fermenter is constructed. The large fermenter has a capacity of 6 m³, is geometrically similar to the pilot-scale vessel, and is also equipped with a Scaba 6SRGT impeller of diameter one-third the tank diameter.

  a. What is the power consumption in the pilot-scale fermenter?

  b. If the production-scale fermenter is operated so that the power consumed per unit volume is the same as in the pilot-scale vessel, what is the power requirement after scale-up?

c. For the conditions in (b), what is the stirrer speed after scale-up?

d. If, instead of (b) and (c), the impeller tip speed $(=\pi N_i D_i)$ is kept the same in the pilot- and production-scale fermenters, what is the stirrer speed after scale-up?

e. For the conditions in (d), what power is required after scale-up?

7.4. Cell suspension and power requirements

A fermentation broth contains 40 wt% cells of average dimension 10 μm and density 1.04 g cm^{-3}. A marine propeller of diameter 30 cm is used for mixing. The density and viscosity of the medium are approximately the same as water. The fermentation is carried out without gas sparging in a vessel of diameter 75 cm.

a. Estimate the stirrer speed required to just completely suspend the cells.

b. What power is required for cell suspension?

You plan to improve this fermentation process by using a new cell strain immobilized in porous plastic beads of diameter 2 mm and density 1.75 g cm^{-3}. The particle concentration required for comparable rates of fermentation is 10% by weight.

c. How does changing over to the immobilized cell system affect the stirrer speed and power required for particle suspension?

7.5. Particle suspension and scale-up

Bacteria attached to particles of clinker are being tested for treatment of industrial waste. In the laboratory, the process is carried out under anaerobic conditions in a stirred bioreactor with liquid height equal to the tank diameter. The system is then scaled up to a geometrically similar vessel of volume 90 times that of the laboratory reactor. The suspending fluid and the particle size, density, and concentration remain the same as in the smaller vessel. The type of impeller is also unchanged, as is the impeller-to-tank diameter ratio.

a. How does the stirrer speed required for suspension of the particles change after scale-up?

b. Assuming operation in the turbulent regime for both vessels, what effect does scale-up have on:

(i) The power required for particle suspension?

(ii) The power per unit volume required for particle suspension?

7.6. Impeller diameter, mixing, and power requirements

Solids suspension and gas dispersion can both be achieved at lower stirrer speeds if the impeller diameter is increased. However, the power requirements for stirring increase with impeller size. In this exercise, the energy efficiencies of small and large impellers are compared for solids suspension and gas dispersion.

a. For ungassed Rushton turbines operating in the turbulent regime, how does the power required for complete solids suspension vary with impeller diameter if all other properties of the system remain unchanged?

b. Using the result from (a), compare the power requirements for complete solids suspension using Rushton turbines of diameters one-third and one-half the tank diameter.

c. All else being equal, how does the power required for complete gas dispersion by Rushton turbines vary with impeller diameter under turbulent flow conditions?

d. Compare the power requirements for complete gas dispersion using Rushton turbines of diameters one-third and one-half the tank diameter.

7.7. Efficiency of different impellers for solids suspension

Compared with a Rushton turbine of diameter one-half the tank diameter, what are the power requirements for solids suspension by a downward-pumping, six-blade pitched-blade turbine of the same size?

7.8. Power and mixing time with aeration

A cylindrical stirred bioreactor of diameter and liquid height 2 m is equipped with a Rushton turbine of diameter one-third the tank diameter. The bioreactor contains Newtonian culture broth with the same density as water and viscosity 4 cP.

  a. If the specific power consumption must not exceed 1.5 kW m^{-3}, determine the maximum allowable stirrer speed.
  b. What is the mixing time at the stirrer speed determined in (a)?
  c. The tank is now aerated. In the presence of gas bubbles, the approximate relationship between the ungassed turbulent power number $\left(N_P' \right)_0$ and the gassed turbulent power number $\left(N_P' \right)_g$ is: $\left(N_P' \right)_g = 0.5 \left(N_P' \right)_0$. What maximum stirrer speed is now possible in the sparged reactor?
  d. What is the mixing time with aeration at the stirrer speed determined in (c)?

7.9. Scale-up of mixing system

To ensure turbulent conditions during agitation with a turbine impeller, the Reynolds number must be at least 10^4.

  a. A laboratory fermenter of diameter and liquid height 15 cm is equipped with a 5-cm-diameter Rushton turbine operated at 800 rpm. If the density of the broth is close to that of water, what is the upper limit for the broth viscosity if turbulence is to be achieved?
  b. Estimate the mixing time in the laboratory fermenter.
  c. The fermenter is scaled up so the tank and impeller are 15 times the diameter of the laboratory equipment. The stirrer in the large vessel is operated at the same impeller tip speed ($= \pi N_i D_i$) as in the laboratory apparatus. How does scale-up affect the maximum viscosity allowable for maintenance of turbulent conditions?
  d. What effect does scale-up have on the mixing time?

7.10. Effect on mixing of scale-up at constant power per unit volume

A baffled pilot-scale bioreactor with a diameter of 75 cm has a liquid height equal to the vessel diameter. The bioreactor is scaled up to a geometrically similar production vessel. The working volume of the production bioreactor is 50 times greater than that at the pilot scale. If the agitation system is scaled up using the basis of constant power per unit volume, what effect will scale-up have on the mixing time?

7.11. Alternative impellers

Escherichia coli cells are cultured in an industrial-scale fermenter for production of supercoiled plasmid DNA used in gene therapy.

  a. A fermenter of diameter 2.3 m and working volume 10 m^3 is equipped with a Rushton turbine of diameter one-third the tank diameter. The impeller is operated at 60 rpm and the vessel is sparged with air. The density and viscosity of the fermentation fluid are close to those of water, that is, 1000 kg m^{-3} and 1 cP, respectively. The power with gassing is about 60% of the ungassed power.

    (i) Calculate the power draw.

    (ii) Estimate the mixing time.

b. To satisfy the burgeoning demand for gene therapy vectors and DNA vaccines, the fermentation factory is being expanded. Two new 10-m³ fermenters are being designed and constructed. It is decided to investigate the use of different impellers in an effort to reduce the power required but still achieve the same mixing time as that obtained with the Rushton turbine described in (a).

    (i) Compared with the Rushton turbine of diameter one-third the tank diameter, what power savings can be made using a Rushton turbine of diameter one-half the tank diameter?

    (ii) Compared with the Rushton turbine of diameter one-third the tank diameter, what power savings can be made using a Lightnin A315 hydrofoil impeller of diameter 0.4 times the tank diameter?

    (iii) The power with gassing is about 50% of the ungassed power for both the larger Rushton turbine and the A315 hydrofoil. What stirrer speeds are required with these impellers to achieve the same mixing time as that determined in (a) (ii)?

7.12. Turbulent shear damage

Microcarrier beads 120 μm in diameter are used to culture recombinant CHO cells for production of growth hormone. It is proposed to use a 20-cm Rushton turbine to mix the culture in a 200-L bioreactor. Oxygen and carbon dioxide are supplied by gas flow through the reactor headspace. The microcarrier suspension has a density of approximately $1010 \, \text{kg m}^{-3}$ and a viscosity of 10^{-3} Pa s.

a. Assuming that the power input by the stirrer is dissipated uniformly in the vessel, estimate the maximum allowable stirrer speed that avoids turbulent shear damage of the cells.

b. How is your estimate affected if the stirrer power is dissipated close to the impeller, within a volume equal to the impeller diameter cubed?

7.13. Avoiding cell damage

Suspended plant cell cultures derived from lemon trees are being used to produce citrus oil for the cosmetic industry. The cells are known to be sensitive to agitation conditions: cell damage occurs and oil production is detrimentally affected if the cumulative energy dissipation level exceeds $10^5 \, \text{J m}^{-3}$. The cells are grown in a bioreactor with continuous feeding of nutrient medium and withdrawal of culture broth. At the operating flow rate, the average residence time in the bioreactor is 2.9 days. The diameter of the vessel is 0.73 m, the liquid height is equal to the tank diameter, four 10% baffles are fitted, and stirring is carried out using a 25-cm curved-blade disc turbine with six blades. The vessel is aerated but the effect of gassing on the impeller power draw is negligible. At steady state, the cell concentration is 0.24 v/v, the broth density is 1 g cm⁻³, and the viscosity is 3.3 mPa s. What is the maximum stirrer speed that can be used without damaging the cells?

References

[1] C.M. Chapman, A.W. Nienow, M. Cooke, J.C. Middleton, Particle–gas–liquid mixing in stirred vessels. Part III: Three phase mixing, Chem. Eng. Res. Des. 61 (1983) 167–181.

[2] F. Saito, A.W. Nienow, S. Chatwin, I.P.T. Moore, Power, gas dispersion and homogenisation characteristics of Scaba SRGT and Rushton turbine impellers, J. Chem. Eng. Japan. 25 (1992) 281–287.

[3] C.M. McFarlane, X.-M. Zhao, A.W. Nienow, Studies of high solidity ratio hydrofoil impellers for aerated bioreactors. 2, Air–water studies. Biotechnol. Prog. 11 (1995) 608–618.

[4] Z. Jaworski, A.W. Nienow, K.N. Dyster, An LDA study of the turbulent flow field in a baffled vessel agitated by an axial, down-pumping hydrofoil impeller, Can. J. Chem. Eng. 74 (1996) 3–15.

[5] A.W. Nienow, Gas dispersion performance in bioreactor operation, Chem. Eng. Prog. 86 (2) (1990) 61–71.

[6] A.W. Nienow, Gas–liquid mixing studies: a comparison of Rushton turbines with some modern impellers, Trans. IChemE. 74A (1996) 417–423.

[7] K.N. Dyster, E. Koutsakos, Z. Jaworski, A.W. Nienow, An LDA study of the radial discharge velocities generated by a Rushton turbine: Newtonian fluids, $Re \geq 5$, Trans. IChemE. 71A (1993) 11–23.

[8] V.P. Mishra, K.N. Dyster, Z. Jaworski, A.W. Nienow, J. McKemmie, A study of an up- and a down-pumping wide blade hydrofoil impeller: Part I. LDA measurements, Can. J. Chem. Eng. 76 (1998) 577–588.

[9] S.M. Kresta, P.E. Wood, The mean flow field produced by a 45° pitched blade turbine: changes in the circulation pattern due to off bottom clearance, Can. J. Chem. Eng. 71 (1993) 42–53.

[10] A. Bakker, H.E.A. van den Akker, A computational model for the gas–liquid flow in stirred reactors, Trans. IChemE. 72A (1994) 594–606.

[11] J.J. Frijlink, A. Bakker, J.M. Smith, Suspension of solid particles with gassed impellers, Chem. Eng. Sci. 45 (1990) 1703–1718.

[12] W. Bujalski, A.W. Nienow, S. Chatwin, M. Cooke, The dependency on scale and material thickness of power numbers of different impeller types, W. Bujalski, A.W. Nienow, S. Chatwin, M. Cooke, The dependency on scale and material thickness of power numbers of different impeller types, in: Proc. Int. Conf. on Mechanical Agitation, 1 (1986).

[13] Th.N. Zwietering, Suspending of solid particles in liquid by agitators, Chem. Eng. Sci. 8 (1958) 244–253.

[14] C.M. Chapman, A.W. Nienow, M. Cooke, J.C. Middleton, Particle–gas–liquid mixing in stirred vessels. Part I: Particle–liquid mixing, Chem. Eng. Res. Des. 61 (1983) 71–81.

[15] A.W. Nienow, N. Harnby, M.F. Edwards, A.W. Nienow, The suspension of solid particles, Mixing in the Process Industries, in: N. Harnby, M.F. Edwards, A.W. Nienow, (Eds.), Mixing in the Process Industries, Butterworth-Heinemann, Oxford UK, 1992, pp. 364–393.

[16] C.W. Wong, J.P. Wang, S.T. Huang, Investigations of fluid dynamics in mechanically stirred aerated slurry reactors, Can. J. Chem. Eng. 65 (1987) 412–419.

[17] W. Bujalski, M. Konno, A.W. Nienow, Scale-up of 45° pitch blade agitators for gas dispersion and solid suspension, in: Mixing, Proc. 6th Eur. Conf. on Mixing, Pavia, Italy, 1988, pp. 389–398, BHRA The Fluid Engineering Centre, Cranfield, U.K.

[18] C.M. Chapman, A.W. Nienow, J.C. Middleton, Particle suspension in a gas sparged Rushton-turbine agitated vessel, Trans. IChemE. 59 (1981) 134–137.

[19] J.J. Frijlink, M. Kolijn, J.M. Smith, Suspension of solids with aerated pitched blade turbines, Fluid Mixing II, (1984) pp. 49–58.

[20] M.M.C.G. Warmoeskerken, M.C. van Houwelingen, J.J. Frijlink, J.M. Smith, Role of cavity formation in stirred gas–liquid–solid reactors, Chem. Eng. Res. Des. 62 (1984) 197–200.

[21] M. Zlokarnik, H. Judat, "Mixing", Ullmann's Encyclopedia of Industrial Chemistry, Vol. B2,25-1/33, VCH Verlagsgesellschaft mbH, Weinheim, (1988)

[22] A.W. Nienow, On impeller circulation and mixing effectiveness in the turbulent flow regime, Chem. Eng. Sci. 52 (1997) 2557–2565.

[23] V. Hudcova, V. Machon, A.W. Nienow, Gas–liquid dispersion with dual Rushton turbine impellers, Biotechnol. Bioeng 34 (1989) 617–628.

[24] A.B. Metzner, R.E. Otto, Agitation of non-Newtonian fluids, AIChE J 3 (1957) 3–10.

Suggestions for Further Reading

Mixing Equipment

J.Y. Oldshue, Fluid Mixing Technology, McGraw-Hill, New York, NY, 1983.

Power Requirements and Hydrodynamics in Stirred Vessels

See also references 7,10.

N. Harnby, M.F. Edwards, A.W. Nienow, Mixing in the Process Industries, in: N. Harnby, M.F. Edwards, A.W. Nienow, (Eds.), Mixing in the Process Industries, Butterworth-Heinemann, Oxford, UK. 1992.

C.M. McFarlane, A.W. Nienow, Studies of high solidity ratio hydrofoil impellers for aerated bioreactors: Parts 1–4, Biotechnol. Prog 11 (1995, 1996) 601–607.

A.W. Nienow, Hydrodynamics of stirred bioreactors, Appl. Mech. Rev 51 (1998) 3–32.

G.B. Tatterson, Fluid Mixing and Gas Dispersion in Agitated Tanks, McGraw-Hill, New York, NY, 1991.

G.B. Tatterson, Scaleup and Design of Industrial Mixing Processes, McGraw-Hill, New York, NY, 1994.

Multiple Impeller Systems

P.R. Gogate, A.A.C.M. Beenackers, A.B. Pandit, Multiple-impeller systems with a special emphasis on bioreactors: a critical review, Biochem. Eng. J 6 (2000) 109–144.

Cell Damage in Bioreactors

J. Chalmers, Animal cell culture, effects of agitation and aeration on cell adaptation, John Wiley, 200041–51

P.K. Namdev, E.H. Dunlop, Shear sensitivity of plant cells in suspensions, Appl. Biochem. Biotechnol. 54 (1995) 109–131.

E.T. Papoutsakis, Fluid-mechanical damage of animal cells in bioreactors, Trends in Biotechnol. 9 (1991) 427–437.

C.R. Thomas, Z. Zhang, E. Galindo, O.T. Ramírez, The effect of hydrodynamics on biological materialsAdvances in Bioprocess Engineering II, in: E. Galindo, O.T. Ramírez, (Eds.), Advances in Bioprocess Engineering II, Kluwer Academic Dordrecht: Springer, Netherlands, 1998, pp. 137–170.

J. Wu, Mechanisms of animal cell damage associated with gas bubbles and cell protection by medium additives, J. Biotechnol. 43 (1995) 81–94.

CHAPTER 8

Bioreactor Heat Transfer

In this chapter we are concerned with the process of heat flow between hot and cold systems. The rate at which heat is transferred depends directly on two variables: the temperature difference between the hot and cold bodies and the surface area available for heat exchange. The heat transfer rate is also influenced by many other factors, such as the geometry and physical properties of the system and, if fluid is present, the flow conditions. Fluids are often heated or cooled in bioprocessing. Typical examples are the removal of heat during bioreactor operation using cooling water and the heating of raw medium to sterilization temperature by steam.

Energy balances allow us to determine the heating and cooling requirements of bioreactors as shown in Chapter 5. Once the rate of heat transfer for a purpose is known, the surface area and other conditions needed to achieve this rate can be calculated using design equations. Estimating the heat transfer area is a central objective in design, as this parameter determines the size and cost of heat exchange equipment. In this chapter, the principles governing heat transfer are outlined for application in bioprocess design.

8.1 Heat Transfer Equipment

Heat exchange occurs most frequently between fluids in bioprocessing. Equipment is designed to allow the transfer of heat while preventing the fluids from physically contacting each other. In most heat exchangers, heat is transferred through a solid metal wall that separates the fluid streams. Sufficient surface area is provided so that the desired rate of heat transfer can be achieved. Heat transfer is facilitated by agitation and turbulent flow of the fluids.

8.1.1 Bioreactors

Two applications of heat transfer are common in bioreactor operation. The first is *in situ* batch sterilization of liquid medium. In this process, the reactor vessel containing medium is heated using steam and held at the sterilization temperature for a period of time. Cooling water is then used to bring the temperature back to normal operating conditions. Sterilization is discussed in more detail in Chapter 11. The other application of heat transfer is for temperature control during bioreactor operation. Most bioreactions take place within the range 25°C to 37°C and tight control of the temperature to within about 1°C is desirable. Metabolic activity of cells generates a substantial amount of heat which must be removed to avoid temperature increases.

The equipment used for heat exchange in bioreactors usually takes one of the forms illustrated in Figure 8.1. The bioreactor may have an external jacket or coil through which steam or cooling water is circulated. Alternatively, helical or baffle-type coils may be located internally. Another method is to pump liquid from the reactor through a separate heat exchange unit.

The surface area available for heat transfer is lower in the external jacket and coil designs than when internal coils are completely submerged in the reactor contents. External jackets provide sufficient heat transfer area for laboratory and other small-scale bioreactors; however, they are generally inadequate for large-scale bioprocesses. Internal coils are used frequently in production vessels; the coil can be operated with high cooling water velocities and the entire tube surface is exposed to the reactor contents providing a relatively large heat transfer area. The temperature of the cooling water rises as it flows through the tube and takes up heat from the media broth. The water temperature increases steadily from the inlet temperature T_{ci} to the

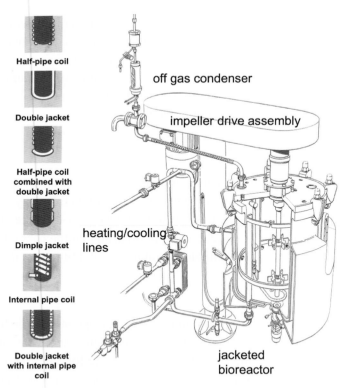

Figure 8.1 Heat transfer configurations for bioreactors (left). Jacketed bioreactor illustration detailing key design features (right). From L. Hasler, R. Butz, and P. Grad, 2006, Transparency, Form and Function: Fermenter Manufacturing-Art. DreiPunktVerlag. Germany. Courtesy of Bioengineering, Bioengineering AG, 8636 Wald, Switzerland.

outlet temperature T_{co}. If the contents of the bioreactor are well mixed, however, temperature gradients in the bulk fluid are negligible and the bioreactor temperature is uniform at T_F.

There are some disadvantages, however, with internal structures, as they interfere with mixing in the vessel and make cleaning of the reactor difficult. Another problem is the growth of cells as biofilms on the internal heat transfer surfaces. The coil must be able to withstand the thermal and mechanical stresses generated inside the bioreactor during sterilization and agitation; the possibility of nonsterile coolant leaking from fractured metal joints in the cooling coil significantly increases the risk of culture contamination. Due to these problems with internal coils, use of an external jacket is preferable for cell cultures with relatively low cooling requirements. However, bioprocesses with high heat loads may require an internal cooling coil together with an external jacket to achieve the necessary rate of cooling.

An external heat exchange unit is independent of the reactor, easy to scale up, and can provide greater heat transfer capacity than any of the other configurations.

8.1.2 General Equipment for Heat Transfer

Many types of heat exchanger equipment are used industrially. The simplest form of heat transfer equipment is the double-pipe heat exchanger (Figure 8.2). For larger capacities, more

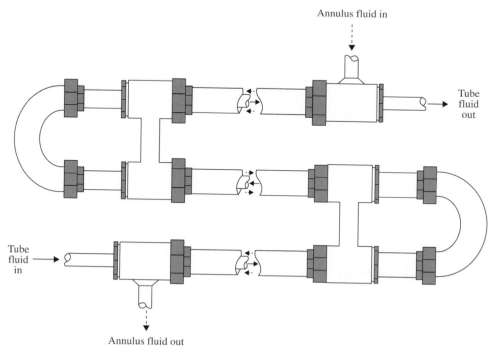

Figure 8.2 Double-pipe heat exchanger. From A.S. Foust, L.A. Wenzel, C.W. Clump, L. Maus, and L.B. Andersen, 1980, *Principles of Unit Operations*, 2nd ed., John Wiley, New York.

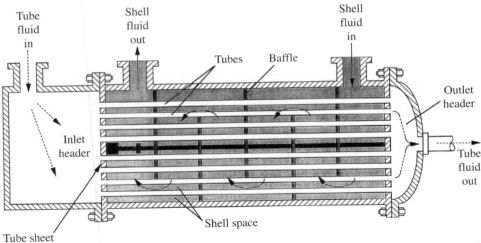

Figure 8.3 Single-pass shell-and-tube heat exchanger. From A.S. Foust, L.A. Wenzel, C.W. Clump, L. Maus, and L.B. Andersen, 1980, *Principles of Unit Operations*, 2nd ed., John Wiley, New York.

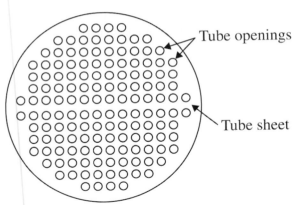

Figure 8.4 Tube sheet for a shell-and-tube heat exchanger.

elaborate shell-and-tube units (Figures 8.3 and 8.4) containing hundreds of square meters of heat exchange area are required.

8.2 Mechanisms of Heat Transfer

Heat transfer occurs by one or more of the following three mechanisms.

- *Conduction.* Heat conduction occurs by transfer of vibrational energy between molecules, or movement of free electrons. Conduction is particularly important in metals and occurs without observable movement of matter.

- *Convection. Convection* requires movement on a macroscopic scale; it is therefore confined to gases and liquids. *Natural convection* occurs when temperature gradients in the system generate localized density differences that result in flow currents. In *forced convection*, flow currents are set in motion by an external agent such as a stirrer or pump and are independent of density gradients. Higher rates of heat transfer are possible with forced convection compared with natural convection.
- *Radiation. Energy* is radiated is absorbed by matter, it appears as heat. Because radiation is important at much higher temperatures than those normally encountered in biological processing, it will not be considered further.

8.3 Conduction

In most heat transfer equipment, heat is exchanged between fluids separated by a solid wall. Heat transfer through the wall occurs by conduction. Here we consider equations describing the rate of conduction as a function of operating variables.

Conduction of heat through a homogeneous solid wall is depicted in Figure 8.5. The wall has thickness B; on one side of the wall the temperature is T_1 and on the other side the temperature is T_2. The area of wall exposed to each temperature is A. The rate of heat conduction through the wall is given by *Fourier's law*:

$$\hat{Q} = -kA\frac{dT}{dy} \tag{8.1}$$

where $\hat{Q}$ is the rate of heat transfer, k is the *thermal conductivity* of the wall, A is the surface area perpendicular to the direction of heat flow, T is temperature, and y is distance measured normal to A. dT/dy is the *temperature gradient* or change of temperature with distance through the wall. The negative sign in Eq. (8.1) indicates that heat always flows from hot to cold.

Fourier's law can also be expressed in terms of the *heat flux*, $\hat{q}$. Heat flux is defined as the rate of heat transfer per unit area normal to the direction of heat flow. Therefore, from Eq. (8.1):

$$\hat{q} = -k\frac{dT}{dy} \tag{8.2}$$

The rate of heat transfer $\hat{Q}$ has the same dimensions and units as power. The SI unit for $\hat{Q}$ is the watt (W); in imperial units $\hat{Q}$ is measured in Btu h^{-1}. Corresponding units of $\hat{q}$ are W m^{-2} and Btu h^{-1} ft^{-2}.

Thermal conductivity is a transport property of materials; values can be found in engineering and science handbooks. The unit for k includes W m^{-1} K^{-1} and Btu h^{-1} ft^{-1} °F^{-1}. The magnitude of k in Eqs. (8.1) and (8.2) reflects the ease with which heat is conducted; the higher the

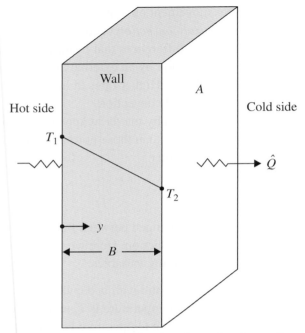

Figure 8.5 Heat conduction through a flat wall.

value of k, the faster the heat transfer. Table 8.1 lists thermal conductivities for some common materials; metals generally have higher thermal conductivities than other substances. Solids with low k values are used as *insulators* to minimize the rate of heat transfer, for example, from steam pipes or in buildings. Thermal conductivity varies somewhat with temperature; however, for small ranges of temperature, k can be considered constant.

8.3.1 Steady-State Conduction

Consider again the conduction of heat through the wall shown in Figure 8.5. At steady state, there can be neither accumulation nor depletion of heat within the wall; this means that the rate of heat flow $\hat{Q}$ must be the same at each point in the wall. If k is largely independent of temperature and A is also constant, the only variables in Eq. (8.1) are the temperature T and distance y. We can integrate Eq. (8.1) to obtain an expression for the rate of conduction as a function of the temperature difference across the wall.

Therefore, from the rules of integration:

$$\hat{Q}y = -kAT + K \qquad (8.3)$$

Table 8.1: Thermal Conductivities

Material	Temperature (°C)	(W m^{-1} °C^{-1})	(Btu h^{-1} ft^{-1} °F^{-1})
Solids: Metals			
Aluminum	300	230	133
Bronze	–	189	109
Copper	100	377	218
Iron (cast)	53	48	27.6
Iron (wrought)	18	61	35
Lead	100	33	19
Stainless steel	20	16	9.2
Steel (1% C)	18	45	26
Solids: Nonmetals			
Asbestos	0	0.16	0.09
	100	0.19	0.11
	200	0.21	0.12
Bricks (building)	20	0.69	0.40
Cork	30	0.043	0.025
Cotton wool	30	0.050	0.029
Glass	30	1.09	0.63
Glass wool	–	0.041	0.024
Rubber (hard)	0	0.15	0.087
Liquids			
Acetic acid (50%)	20	0.35	0.20
Ethanol (80%)	20	0.24	0.137
Glycerol (40%)	20	0.45	0.26
Water	30	0.62	0.356
	60	0.66	0.381
Gases			
Air	0	0.024	0.014
	100	0.031	0.018
Carbon dioxide	0	0.015	0.0085
Nitrogen	0	0.024	0.0138
O_2	0	0.024	0.0141
Water vapor	100	0.025	0.0145

From J.M. Coulson, J.F. Richardson, J.R. Backhurst, and J.H. Harker, 1999, *Coulson and Richardson's Chemical Engineering*, vol. 1, 6th ed., Butterworth-Heinemann, Oxford.
Note: To convert from W m^{-1} °C^{-1} to Btu h^{-1} ft^{-1} °F^{-1}, multiply by 0.578.

where K is the integration constant. K is evaluated by applying a single boundary condition; in this case we can use the boundary condition $T = T_1$ at $y = 0$, as indicated in Figure 8.5. Substituting this information into Eq. (9.3) gives:

$$K = k A T_1 \tag{8.4}$$

Using Eq. (9.4) to eliminate K from Eq. (9.3):

$$\hat{Q}\,y = -k\,A\left(T - T_1\right) \tag{8.5}$$

Equation (8.5) holds for all values of y including at $y = B$ where $T = T_2$ (refer to Figure 8.5) as $\hat{Q}$ at steady state is the same at all points in the wall. Substituting these values into Eq. (8.5) and rearranging gives the expression:

$$\hat{Q} = \frac{k\,A}{B}\left(T_1 - T_2\right) \tag{8.6}$$

Equation (8.6) allows us to calculate $\hat{Q}$ using the total temperature drop across the wall, ΔT, which is substituted for $T_1 - T_2$.

Equation (8.6) can then also be written in the form:

$$\hat{Q} = \frac{\Delta T}{R_w} \tag{8.7}$$

where R_w is the *thermal resistance* to heat transfer offered by the wall:

$$R_w = \frac{B}{kA} \tag{8.8}$$

The ΔT responsible for flow of heat is known as the *temperature-difference driving force*. Equation (8.7) is an example of the *general rate principle*, which equates the rate of a process to the ratio of the driving force and the resistance. Equation (8.8) can be interpreted as follows: the wall would pose more of a resistance to heat transfer if its thickness B were increased; on the other hand, the resistance is reduced if the surface area A is increased or if the material in the wall were replaced with a substance of higher thermal conductivity k.

8.3.2 Combining Thermal Resistances in Series

When a system contains several different heat transfer resistances in series, *the overall resistance is equal to the sum of the individual resistances*. For example, if the wall shown in Figure 8.5 were constructed of several layers of different materials, each layer would represent a separate resistance to heat transfer. Consider the three-layer system illustrated in Figure 8.6 with surface area A, layer thicknesses B_1, B_2, and B_3, thermal conductivities k_1, k_2, and k_3, and temperature drops across the layers of ΔT_1, ΔT_2, and ΔT_3. If the layers are in perfect thermal contact so there is no temperature drop across the interfaces, the temperature change across the entire structure is:

$$\Delta T = \Delta T_1 + \Delta T_2 + \Delta T_3 \tag{8.9}$$

The rate of heat conduction in this system is given by Eq. (8.7), with the overall resistance R_w equal to the sum of the individual resistances:

$$\hat{Q} = \frac{\Delta T}{R_w} = \frac{\Delta T}{\left(R_1 + R_2 + R_3\right)} \tag{8.10}$$

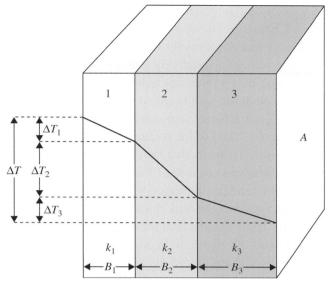

Figure 8.6 Heat conduction through three resistances in series.

where R_1, R_2, and R_3 are the thermal resistances of the individual layers:

$$R_1 = \frac{B_1}{k_1 A}, \quad R_2 = \frac{B_2}{k_2 A}, \quad R_3 = \frac{B_3}{k_3 A} \tag{8.11}$$

Equation (8.10) represents the important principle of *additivity of resistances*.

8.4 Heat Transfer Between Fluids

Convection and conduction both play important roles in heat transfer in fluids. In agitated, single-phase systems, convective heat transfer in the bulk fluid is linked directly to mixing and turbulence and is generally quite rapid. However, in the heat exchange equipment described in Section 8.1, additional resistances to heat transfer are encountered.

8.4.1 Thermal Boundary Layers

Figure 8.7 depicts the heat transfer situation at any point on the pipe wall of a heat exchanger. Figure 8.7(a) identifies a segment of pipe wall separating the hot and cold fluids; Figure 8.7(b) shows the magnified detail of fluid properties at the wall. Hot and cold fluids flow on either side of the wall; we will assume that both fluids are in turbulent flow. The bulk temperature of the hot fluid away from the wall is T_h; T_c is the bulk temperature of the cold fluid. T_{hw} and T_{cw} are the respective temperatures of the hot and cold fluids at the wall.

When fluid contacts a solid, a fluid boundary layer develops at the surface as a result of viscous drag as explained in Chapter 6. Therefore, the hot and cold fluids represented in Figure 8.7 consist of a turbulent core that accounts for the bulk of the fluid, and a thin sublayer or film near the wall where the velocity is relatively low. In the turbulent part of the fluid, rapidly moving eddies transfer heat quickly so that any temperature gradients in the bulk fluid can be neglected. The film of liquid at the wall is called the *thermal boundary layer* or *stagnant film*, although the fluid in it is not actually stationary. This viscous sublayer has an important effect on the rate of heat transfer. Most of the resistance to heat transfer to or from the fluid is contained in the film; the reason for this is that heat flow through the film must occur mainly by conduction rather than convection because of the reduced velocity of the fluid. The width of the film indicated by the broken lines in Figure 8.7(b) is the approximate distance from the wall at which the temperature reaches the bulk fluid temperature, either T_h or T_c. The thickness of the thermal boundary layer in most heat transfer situations is less than the hydrodynamic boundary layer described in Chapter 6. In other words, as we move away from the wall, the temperature normally reaches that of the bulk fluid before the velocity reaches that of the bulk flow stream.

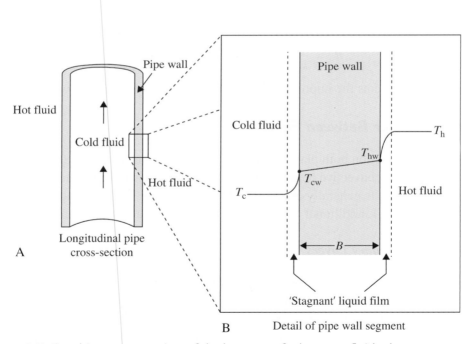

Figure 8.7 Graphic representation of the heat transfer between fluids that are separated by a pipe wall: (A) the longitudinal pipe cross-section identifying a segment of the pipe wall; (B) the magnified detail of the pipe wall segment showing boundary layers and temperature gradients at the wall.

8.4.2 Individual Heat Transfer Coefficients

Heat exchange between the fluids in Figure 8.7 encounters three major resistances in series: the hot-fluid film resistance at the wall, resistance due to the wall itself, and the cold-fluid film resistance. The rate of heat transfer through each thermal boundary layer in the fluid is given by an equation analogous to Eq. (8.6) for steady-state conduction:

$$\hat{Q} = hA\Delta T \tag{8.12}$$

where h is the *individual heat transfer coefficient*, A is the area for heat transfer normal to the direction of heat flow, and ΔT is the temperature difference between the wall and the bulk stream. $\Delta T = T_h - T_{hw}$ for the hot-fluid film; $\Delta T = T_{cw} - T_c$ for the cold-fluid film.

Unlike Eq. (8.6) for conduction, Eq. (8.12) does not contain a separate term for the thickness of the boundary layer; this thickness is difficult to measure and depends strongly on the prevailing flow conditions. Instead, the effect of film thickness is included in the value of h so that, unlike thermal conductivity, h is not a transport property of materials and its value cannot be found in handbooks. The heat transfer coefficient h is an empirical parameter incorporating the effects of system geometry, flow conditions, and fluid properties. As convective heat transfer involves fluid flow, it is a more complex process than conduction. Consequently, there is little theoretical basis for calculation of h; h must be determined experimentally or evaluated using published correlations derived from experimental data. Suitable correlations for heat transfer coefficients are presented in Section 8.5.1. The SI units for h are $W\ m^{-2}\ K^{-1}$; in the imperial system h is expressed as $Btu\ h^{-1}\ ft^{-2}\ °F^{-1}$. Magnitudes of h vary greatly; some typical values are listed in Table 8.2.

The rate of heat transfer $\hat{Q}$ in each fluid boundary layer can be written as the ratio of the temperature-difference driving force and the resistance, analogous to Eq. (8.7). Therefore, from Eq. (8.12), the two resistances to heat transfer on either side of the pipe wall are:

$$R_h = \frac{1}{h_h A} \tag{8.13}$$

and

$$R_c = \frac{1}{h_c A} \tag{8.14}$$

where R_h is the resistance to heat transfer in the hot fluid, R_c is the resistance to heat transfer in the cold fluid, h_h is the individual heat transfer coefficient for the hot fluid, h_c is the individual heat transfer coefficient for the cold fluid, and A is the surface area for heat transfer.

8.4.3 Overall Heat Transfer Coefficient

Application of Eq. (8.12) to calculate the rate of heat transfer in each boundary layer requires knowledge of ΔT for each fluid. This is usually difficult because we do not know T_{hw} and T_{cw};

Table 8.2: Individual Heat Transfer Coefficients

Process	Range of values of h	
	(W m^{-2} °C^{-1})	(Btu h^{-1} ft^{-2} °F^{-1})
Forced convection		
Heating or cooling air	10–500	2–100
Heating or cooling water	100–20,000	20–4000
Heating or cooling oil	60–2000	10–400
Boiling water flowing		
In a tube	5000–100,000	880–17,600
In a tank	2500–35,000	440–6200
Condensing steam, 1 atm		
On vertical surfaces	4000–11,300	700–2000
Outside horizontal tubes	9500–25,000	1700–4400
Condensing organic vapor	1100–2200	200–400
Superheating steam	30–110	5–20

Data from L.C. Thomas, 1992, *Heat Transfer*, Prentice Hall, Upper Saddle River, NJ.; J.P. Holman, 1997, *Heat Transfer*, 8th ed., McGraw-Hill, New York; and W.H. McAdams, 1954, *Heat Transmission*, 3rd ed., McGraw-Hill, New York.
Note: To convert from W m^{-2} °C^{-1} to Btu h^{-1} ft^{-2} °F^{-1}, multiply by 0.176.

it is much easier and more accurate to measure the bulk temperatures of fluids rather than wall temperatures. This problem is overcome by introducing the *overall heat transfer coefficient, U*, for the total heat transfer process through both fluid boundary layers and the wall. *U* is defined by the equation:

$$\hat{Q} = UA\Delta T \qquad (8.15)$$

where ΔT is the overall temperature difference between the bulk hot and cold fluids. The units of U are the same as for h (e.g., W m^{-2} K^{-1} or Btu h^{-1} ft^{-2} °F^{-1}). Equation (8.15) written in terms of the ratio of the driving force ΔT and the resistance yields an expression for the total resistance to heat transfer, R_T:

$$R_T = \frac{1}{UA} \qquad (8.16)$$

It was noted in Section 8.3.3 that when thermal resistances occur in series, the total resistance is the sum of the individual resistances. Applying this to the situation of heat exchange between fluids, R_T is equal to the sum of R_h, R_w, and R_c:

$$R_T = R_h + R_w + R_c \qquad (8.17)$$

Combining Eqs. (8.8), (8.13), (8.14), (8.16), and (8.17) gives:

$$\frac{1}{UA} = \frac{1}{h_h A} + \frac{B}{kA} + \frac{1}{h_c A} \qquad (8.18)$$

In Eq. (8.18), the surface area A appears in each term. When fluids are separated by a flat wall, the surface area for heat transfer through each boundary layer and the wall is the same, so that A can be cancelled from the equation. However, a minor complication arises for cylindrical geometry such as pipes. Let us assume that hot fluid is flowing inside a pipe while cold fluid flows outside, as shown in Figure 8.8. The inside diameter of the pipe is smaller than the outside diameter; therefore, the surface areas for heat transfer between the fluid and the pipe wall are different for the two fluids. The surface area of the wall of a cylinder is equal to the circumference multiplied by the length:

$$A = 2\pi RL \tag{8.19}$$

where R is the radius of the cylinder and L is its length. Therefore, the heat transfer area at the hot-fluid boundary layer inside the tube is $A_i = 2\pi R_i L$; the heat transfer area at the cold-fluid boundary layer outside the tube is $A_o = 2\pi R_o L$. The surface area available for conduction through the wall varies between A_i and A_o.

The variation of heat transfer area in cylindrical systems depends on the thickness of the pipe wall. For thin walls, the variation will be relatively small because R_i is similar to R_o. In engineering design, variations in surface area are incorporated into the equations for heat transfer. However, for the sake of simplicity, in our analysis we will ignore any differences in surface area; we will assume, in effect, that the pipes are thin-walled. Accordingly, for cylindrical as well as for flat geometry, we can cancel A from Eq. (8.18) and write a simplified equation for U:

$$\frac{1}{U} = \frac{1}{h_h} + \frac{B}{k} + \frac{1}{h_c} \tag{8.20}$$

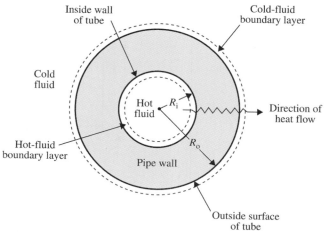

Figure 8.8 Effect of pipe wall thickness on the surface area for heat transfer.

The overall heat transfer coefficient characterizes the physical properties of the system and the operating conditions used for heat transfer. A small value of U for a particular process means that the system has only a limited ability to transfer heat. U can be increased by manipulating operating variables such as the fluid velocity in heat exchangers or the stirrer speed in bioreactors; these changes affect the value of h_c or h_h. The value of U is independent of A. To achieve a particular rate of heat transfer in an exchanger with small U, the heat transfer area must be relatively large; however, increasing A raises the cost of the equipment. If U is large, the heat exchanger is well designed and operating under conditions that enhance heat transfer.

8.4.4 Fouling Factors

Heat transfer equipment in service does not remain clean. Dirt and scale are deposited on one or both sides of the pipes, providing additional resistance to heat flow and reducing the overall heat transfer coefficient. The resistances to heat transfer when fouling affects both sides of the heat transfer surface are represented in Figure 8.9. Five resistances are present in series: the thermal boundary layer or liquid film on the hot-fluid side, a fouling layer on the hot-fluid side, the pipe wall, a fouling layer on the cold-fluid side, and the cold-fluid thermal boundary layer.

Each fouling layer has associated with it a heat transfer coefficient. For dirt and scale, the coefficient is called a *fouling factor*. Let h_{fh} be the fouling factor on the hot-fluid side and h_{fc}

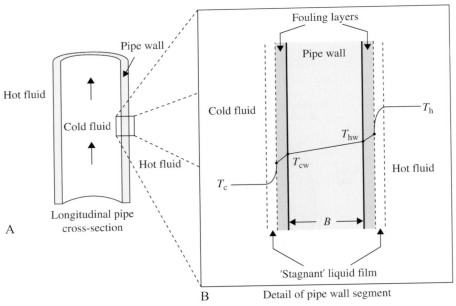

Figure 8.9 Heat transfer between fluids separated by a pipe wall with fouling deposits on both surfaces: (A) longitudinal pipe cross-section identifying a segment of the pipe wall; (B) magnified detail of the pipe wall segment showing boundary layers, fouling layers, and temperature gradients at the wall.

Table 8.3: Fouling Factors for Scale Deposits

Source of deposit	Fouling factor	
	(W m^{-2} °C^{-1})	(Btu h^{-1} ft^{-2} °F^{-1})
Water (temperatures up to 52°C, velocities over 1 m s^{-1})		
River water	2800	500
City or well water	5700	1000
Hard water	1900	330
Brackish water	5700	1000
Untreated cooling tower water	1900	330
Seawater	11,400	2000
Steam		
Good quality, oil free	11,400	2000
Liquids		
Industrial organic	5700	1000
Caustic solutions	2800	500
Vegetable oil	1900	330
Fuel oil	1100	200
Gases		
Compressed air	2800	500
Solvent vapor	5700	1000

Data from A.C. Mueller, 1985, Process heat exchangers. In: *Handbook of Heat Transfer Applications*, 2nd ed., W.M. Rohsenow, J.P. Hartnett, and E.N. Ganic (Eds.), pp. 4-78–4-173, McGraw-Hill, New York.
Note: To convert from W m^{-2} °C^{-1} to Btu h^{-1} ft^{-2} °F^{-1}, multiply by 0.176.

be the fouling factor on the cold-fluid side. When these additional resistances are present, they must be included in the expression for the overall heat transfer coefficient, U. Equation (8.20) becomes:

$$\frac{1}{U} = \frac{1}{h_{\mathrm{fh}}} + \frac{1}{h_{\mathrm{h}}} + \frac{B}{k} + \frac{1}{h_{\mathrm{c}}} + \frac{1}{h_{\mathrm{fc}}} \tag{8.21}$$

Adding fouling factors in Eq. (8.21) increases $1/U$, thus decreasing the value of U.

Accurate estimation of fouling factors is very difficult. The chemical nature of the deposit and its thermal conductivity depend on the fluid in the tube and the temperature; fouling thickness can also vary between cleanings. Typical values of fouling factors for various fluids are listed in Table 8.3.

8.5 Design Equations for Heat Transfer Systems

The basic equation for the design of heat exchangers is Eq. (8.15). If U, ΔT, and $\hat{Q}$ are known, this equation allows us to calculate A. Specification of A is a major objective of heat exchanger design; the surface area dictates the configuration and size of the equipment and its cost. In the following sections, we will consider procedures for determining U, ΔT, and $\hat{Q}$ for use in Eq. (8.15).

8.5.1 Calculation of Heat Transfer Coefficients

U can be determined as a combination of the individual film heat transfer coefficients, the properties of the separating wall, and, if applicable, any fouling factors, as described in Sections 8.4.3 and 8.4.4. The values of the individual heat transfer coefficients h_h and h_c depend on the thickness of the fluid boundary layers, which, in turn, depends on the flow velocity and fluid properties such as viscosity and thermal conductivity. Increasing the level of turbulence and decreasing the viscosity will reduce the thickness of the liquid film and hence increase the heat transfer coefficient.

Individual heat transfer coefficients for flow in pipes and stirred vessels can be evaluated using empirical correlations expressed in terms of dimensionless numbers. The general form of correlations for heat transfer coefficients is:

$$Nu = f\left(Re \text{ or } Re_i, Pr, Gr, \frac{D}{L}, \frac{\mu_b}{\mu_w} \right) \tag{8.22}$$

where f means "some function of," and:

$$Nu = \text{Nusselt number} = \frac{hD}{k_{fb}} \tag{8.23}$$

$$Re = \text{Reynolds number for pipe flow} = \frac{Du\rho}{\mu_b} \tag{8.24}$$

$$Re_i = \text{impeller Reynolds number} = \frac{N_i D_i^2 \rho}{\mu_b} \tag{8.25}$$

$$Pr = \text{Prandtl number} = \frac{C_p \mu_b}{k_{fb}} \tag{8.26}$$

And

$$Gr = \text{Grashof number for heat transfer} = \frac{D^3 g \rho^2 \beta \Delta T}{\mu_b^2} \tag{8.27}$$

The parameters in Eqs. (8.22) through (8.27) are as follows: h is the individual heat transfer coefficient, D is the pipe or tank diameter, k_{fb} is the thermal conductivity of the bulk fluid, u is the linear velocity of fluid in the pipe, ρ is the average density of the fluid, μ_b is the viscosity of the bulk fluid, N_i is the rotational speed of the impeller, D_i is the impeller diameter, Cp is the average heat capacity of the fluid, g is gravitational acceleration, β is the coefficient of thermal expansion of the fluid, ΔT is the variation of fluid temperature in the system, L is the pipe length, and μ_w is the viscosity of the fluid at the wall.

The Nusselt number, which contains the heat transfer coefficient h, represents the ratio of the rates of convective and conductive heat transfer. The Prandtl number represents the ratio of molecular momentum and thermal diffusivities; Pr contains physical constants which, for Newtonian fluids, are independent of flow conditions. A high Pr number (>5) indicates that heat transfer is favored to occur by fluid momentum rather than by fluid conduction. The Grashof number represents the ratio of buoyancy to viscous forces and appears in correlations only when the fluid is not well mixed. Under these conditions, the fluid density is no longer uniform and natural convection becomes an important heat transfer mechanism. In most bioprocessing applications, heat transfer occurs between fluids in turbulent flow in pipes and stirred vessels; therefore, forced convection dominates natural convection and the Grashof number is not important. The form of the correlation used to evaluate Nu and therefore h depends on the configuration of the heat transfer equipment, the flow conditions, and other factors.

A wide variety of heat transfer situations is met in practice and there are many correlations available to biochemical engineers designing heat exchange equipment. Different equations are needed to evaluate h_h and h_c depending on the flow geometry of the hot and cold fluids. Examples of correlations for heat transfer coefficients relevant to bioprocessing are given in the following sections. Other correlations can be found in the references listed at the end of this chapter.

8.5.2 Stirred Liquids

The heat transfer coefficient in stirred vessels depends on the degree of agitation and the properties of the fluid. When heat is transferred to or from a helical coil in the vessel, h for the tank side of the coil can be determined using the following equation [1,2]:

$$Nu = 0.9 Re_i^{0.62} Pr^{0.33} \left(\frac{\mu_b}{\mu_w} \right)^{0.14}$$

(8.28)

where Re_i is given by Eq. (8.25), and D in Nu refers to the inside diameter of the tank. For low-viscosity fluids such as water, the viscosity at the wall μ_w can be assumed equal to the bulk viscosity μ_b. For viscous media broths, because the broth temperature at the wall of the cooling coil is lower than in the bulk liquid, μ_w may be greater than μ_b so that Nu is reduced somewhat. Equation (8.28) can be applied with all types of impellers and has an accuracy of within about 20% in ungassed systems. A range of alternative correlations for helical coils in stirred vessels is available in the literature [3]

When heat is transferred to or from a jacket rather than a coil, the correlation is slightly modified [2]:

$$Nu = 0.36 Re_i^{0.67} Pr^{0.33} \left(\frac{\mu_b}{\mu_w} \right)^{0.14}$$

(8.29)

8.5.3 Logarithmic-Mean Temperature Difference

Application of the heat exchanger design equation, Eq. (8.15), requires knowledge of the temperature-difference driving force for heat transfer, ΔT. ΔT is the difference between the bulk temperatures of the hot and cold fluids. However, as we see in Figures 8.10, 8.11, 8.12, and 8.13, bulk fluid temperatures vary with position in heat exchangers, so that the temperature difference between the hot and cold fluids changes between one end of the equipment and the other. Thus, the driving force for heat transfer varies from point to point in the system. For application of Eq. (8.15), this difficulty is overcome using an average ΔT.

For single-pass heat exchangers in which flow is either cocurrent or countercurrent, the *logarithmic-mean temperature difference* is used. In this case, ΔT is given by the equation:

$$\Delta T = \frac{\Delta T_2 - \Delta T_1}{\ln\left(\Delta T_2 \,/\, \Delta T_1\right)} \tag{8.30}$$

where subscripts 1 and 2 denote the ends of the equipment. ΔT_1 and ΔT_2 are the temperature differences between the hot and cold fluids at the ends of the exchanger calculated using the values for T_{hi}, T_{ho}, T_{ci}, and T_{co}. For convenience and to eliminate negative numbers and their logarithms, subscripts 1 and 2 can refer to the ends of the exchanger. Equation (8.30) has been derived using the following assumptions:

- The overall heat transfer coefficient U is constant.
- The specific heat capacities of the hot and cold fluids are constant or, alternatively, any phase change in the fluids (e.g., boiling or condensation) occurs isothermally.
- Heat losses from the system are negligible.
- The system is at steady state in either countercurrent or cocurrent flow.

The most questionable of these assumptions is that of constant U since this coefficient varies with the temperature of the fluids and may be affected significantly by phase change.

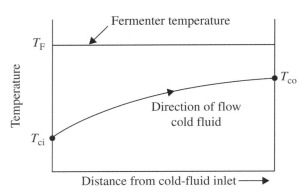

Figure 8.10 Cooling water temperature changes as a function of distance from inlet.

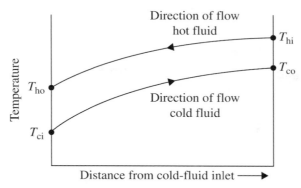

Figure 8.11 Temperature changes for countercurrent flow in a double-pipe heat exchanger.

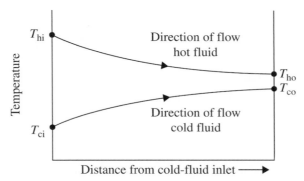

Figure 8.12 Temperature changes for cocurrent flow in a double-pipe heat exchanger.

However, if the change in U with temperature is gradual, or if temperature differences in the system are moderate, this assumption is not seriously in error. Other details of the derivation of Eq. (8.30) can be found in other texts [3–5].

The logarithmic-mean temperature difference as defined in Eq. (8.30) is also applicable when one fluid in the heat exchange system remains at a constant temperature. For the case of a bioreactor at constant temperature T_F cooled by water in a cooling coil (Figure 8.10), Eq. (8.30) can be simplified to:

$$\Delta T = \frac{T_{co} - T_{ci}}{\ln\left(\dfrac{T_F - T_{ci}}{T_F - T_{co}}\right)} \tag{8.31}$$

where T_{ci} is the inlet temperature of the cooling water and T_{co} is the outlet temperature of the cooling water.

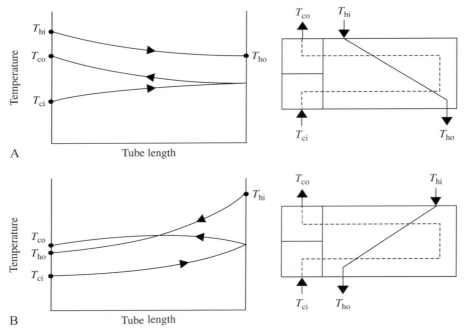

Figure 8.13 Temperature changes for a double tube-pass heat exchanger. (A) and (B) demonstrate complex temperature vs. position profiles in multipass heat exchangers which can combine cross-flow, cocurrent, and countercurrent regions.

In multipass shell-and-tube heat exchangers, for example, in the double tube-pass unit illustrated in Figure 8.14, flow is neither purely countercurrent nor purely cocurrent. Flow patterns in multiple-pass equipment can be complex, with cocurrent, countercurrent, and cross-flow all present. Under these conditions, the log-mean temperature difference does not accurately represent the average temperature difference in the system. For shell-and-tube heat exchangers with more than a single tube pass or a single shell pass, the log-mean temperature difference must be used with an appropriate correction factor to account for the geometry of the exchanger. Correction factors for a range of equipment configurations have been determined and are available in other references [3,4,6].

The log-mean temperature difference is not applicable to heat exchangers in which one or both fluids in the system change phase [3].

8.5.4 Energy Balance

Energy balances are applied during heat exchanger design to determine $\hat{Q}$ and all inlet and outlet temperatures required to specify ΔT. These energy balances are based on the general equations for flow systems derived in Chapter 5.

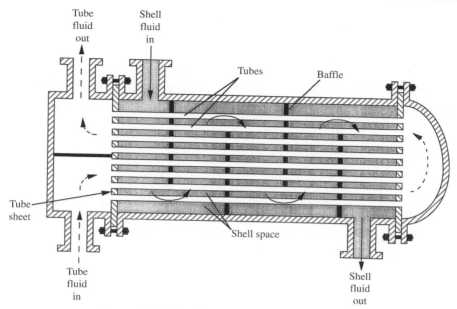

Figure 8.14 Double tube-pass heat exchanger.

Let us first consider the equations for double-pipe or shell-and-tube heat exchangers. From Eq. (5.8), under steady-state conditions ($dE/dt = 0$) and as no shaft work is performed during pipe flow $\left(W_s = 0 \right)$, the energy balance equation is:

$$\hat{M}_i h_i - \hat{M}_o h_o - \hat{Q} = 0 \tag{8.32}$$

where $\hat{M}_i$ is the mass flow rate in, $\hat{M}_o$ is the mass flow rate out, h_i is the specific enthalpy of the incoming stream, h_o is the specific enthalpy of the outgoing stream, and $\hat{Q}$ is the rate of heat transfer to or from the system. Unfortunately, the conventional symbols for individual heat transfer coefficient and specific enthalpy are the same: h. In this section, h in Eqs. (8.32) through (8.35) denotes specific enthalpy; otherwise in this chapter, h represents the individual heat transfer coefficient.

Equation (8.32) can be applied separately to each fluid in the heat exchanger. As the mass flow rate of each fluid does not vary between its inlet and outlet, for the hot fluid:

$$\hat{M}_h \left(h_{hi} - h_{ho} \right) - \hat{Q}_h = 0 \tag{8.33}$$

where subscript h denotes the hot fluid and $\hat{Q}_h$ is the rate of heat transfer from that fluid. Equations analogous to Eq. (8.33) can be derived for the cold fluid:

$$\hat{M}_c \left(h_{ci} - h_{co} \right) + \hat{Q}_c = 0 \tag{8.34}$$

where subscript c refers to the cold fluid. $\hat{Q}_c$ is the rate of heat flow into the cold fluid; therefore $\hat{Q}_c$ is added rather than subtracted in Eq. (8.34) according to the sign convention for energy flows explained in Chapter 5.

When there are no heat losses from the exchanger, all heat removed from the hot stream must be taken up by the cold stream. Thus, we can equate $\hat{Q}$ terms in Eqs. (8.33) and (8.34):

$$\hat{Q}_h = \hat{Q}_c = \hat{Q} \cdot$$

Therefore:

$$\hat{M}_h \left(h_{hi} - h_{ho} \right) = \hat{M}_c \left(h_{co} - h_{ci} \right) = \hat{Q} \tag{8.35}$$

The enthalpy differences in Eq. (8.35) can be expressed in terms of the heat capacity Cp and the temperature change for each fluid as long as there are no phase changes (*e.g.*, boiling or condensation) (Chapter 5). If we assume Cp is constant over the temperature range in the exchanger, Eq. (8.35) becomes:

$$\hat{M}_h C_{ph} \left(T_{hi} - T_{ho} \right) = \hat{M}_c C_{pc} \left(T_{co} - T_{ci} \right) = \hat{Q} \tag{8.36}$$

where C_{ph} is the heat capacity of the hot fluid, C_{pc} is the heat capacity of the cold fluid, T_{hi} is the inlet temperature of the hot fluid, T_{ho} is the outlet temperature of the hot fluid, T_{ci} is the inlet temperature of the cold fluid, and T_{co} is the outlet temperature of the cold fluid.

Eq. (8.36) can be used to determine $\hat{Q}$ using the inlet and outlet conditions of the fluid streams in the absence of phase change. This is illustrated in Example 8.1.

Equation (8.36) can also be applied to the situation of heat removal from a bioreactor for the purpose of temperature control. In this case, the temperature of the hot fluid (i.e., the media broth) is uniform throughout the system so that the left side of Eq. (8.36) is zero. If energy is absorbed by the cold fluid as sensible heat, the energy balance equation becomes:

$$\hat{M}_c C_{pc} \left(T_{co} - T_{ci} \right) = \hat{Q} \tag{8.37}$$

To use Eq. (8.37) for bioreactor design we must know $\hat{Q}$. $\hat{Q}$ is found by considering all significant heat sources and sinks in the system. An expression involving $\hat{Q}$ for bioprocess systems was presented in Chapter 5 and can be written as a function of time:

$$\frac{dE}{dt} = -\Delta \hat{H}_{rxn} - \hat{M}_v \Delta h_v - \hat{Q} + \hat{W}_s \tag{8.38}$$

where $\Delta \hat{H}_{rxn}$ is the rate of heat absorption or evolution due to metabolic reaction, $\hat{M}_v$ is the mass flow rate of evaporated liquid leaving the system, Δh_v is the latent heat of vaporization, and $\hat{W}_s$ is the rate of shaft work done on the system. For exothermic reactions $\Delta \hat{H}_{rxn}$ is negative; for endothermic reactions $\Delta \hat{H}_{rxn}$ is positive. In most bioprocess systems, the only source of shaft work is the stirrer; therefore $\hat{W}_s$ is the power dissipated by the impeller. Methods for

---

EXAMPLE 8.1 Heat Exchanger

Hot, freshly sterilized nutrient medium is cooled in a double-pipe heat exchanger before being used in a bioprocess. Medium leaving the sterilizer at 100°C enters the exchanger at a flow rate of 10 m³ h⁻¹; the desired outlet temperature is 30°C. Heat from the medium is used to raise the temperature of 25 m³ h⁻¹ of cooling water entering the exchanger at 15°C. The system operates at steady state. Assume that the nutrient medium has the properties of water.

a. What rate of heat transfer is required?
b. Calculate the final temperature of the cooling water as it leaves the heat exchanger.

Solution
The density of water and medium is 1000 kg m⁻³. Therefore:

$$\hat{M}_h = 10 \text{ m}^3 \text{ h}^{-1} \cdot \left| \frac{1 \text{ h}}{3600 \text{ s}} \right| \cdot (1000 \text{ kg m}^{-3}) = 2.78 \text{ kg s}^{-1}$$

$$\hat{M}_c = 25 \text{ m}^3 \text{ h}^{-1} \cdot \left| \frac{1 \text{ h}}{3600 \text{ s}} \right| \cdot (1000 \text{ kg m}^{-3}) = 6.94 \text{ kg s}^{-1}$$

From Table C.3 in Appendix C, the heat capacity of water can be taken as 75.4 J gmol⁻¹ °C⁻¹ for most of the temperature range of interest. Therefore:

$$C_{ph} = C_{pc} = 75.4 \text{ J gmol}^{-1} \text{ °C}^{-1} \cdot \left| \frac{1 \text{ gmol}}{18 \text{ g}} \right| \cdot \left| \frac{1000 \text{ g}}{1 \text{ kg}} \right| = 4.19 \times 10^3 \text{ J kg}^{-1} \text{ °C}^{-1}$$

a. From Eq. (8.36) for the hot fluid:

$$\hat{Q} = 2.78 \text{ kg s}^{-1} \left(4.19 \times 10^3 \text{ J kg}^{-1} \text{ °C}^{-1} \right) (100 - 30) \text{°C}$$

$$\hat{Q} = 8.15 \times 10^5 \text{ J s}^{-1} = 815 \text{ kW}$$

The rate of heat transfer required is 815 kJ s⁻¹.
b. For the cold fluid, from Eq. (8.36):

$$T_{co} = T_{ci} + \frac{\hat{Q}}{\hat{M}_c C_{pc}}$$

$$T_{co} = 15\text{°C} + \frac{8.15 \times 10^5 \text{ J s}^{-1}}{6.94 \text{ kg s}^{-1} \left(4.19 \times 10^3 \text{ J kg}^{-1} \text{ °C}^{-1} \right)} = 43.0\text{°C}$$

The exit cooling water temperature is 43°C.

---

estimating the power required for stirrer operation are described in Chapter 7. Equation (8.38) represents a considerable simplification of the energy balance. It is applicable to systems in which the heat of reaction dominates the energy balance so that contributions from sensible heat and heats of solution can be ignored. In large bioreactors, metabolic activity is the dominant source of heat; the energy input by stirring and heat losses from evaporation may also be worth considering in some cases. Other heat sources and sinks are relatively minor and can generally be neglected.

At steady state $dE/dt = 0$ Eq. (8.38) becomes Eq. 5.28 from Chapter 5, solving for $\hat{Q}$:

$$\hat{Q} = -\Delta\hat{H}_{rxn} - \hat{M}_v \Delta h_v + \hat{W}_s \tag{5.28}$$

Application of Eq. (5.28) to determine $\hat{Q}$ is illustrated in Examples 5.7 and 5.8. Once $\hat{Q}$ has been estimated, Eq. (8.37) is used to evaluate unknown operating conditions as shown in Example 8.2.

8.6 Application of the Design Equations

The equations in Sections 8.4 and 8.5 provide the essential elements for design of heat transfer systems. Figure 8.15 summarizes the relationships involved. Equation (8.15) is used as the design equation; $\hat{Q}$ and the inlet and outlet temperatures are available from energy balance calculations as described in Section 8.5.3. The overall heat transfer coefficient is evaluated from correlations such as those given in Section 8.5.1; additional terms are included if the heat transfer surfaces are fouled. The temperature-difference driving force is estimated from the bioprocess and cooling water temperatures. With these parameters at hand, the required heat transfer area can be determined.

8.6.1 Heat Exchanger Design for Bioprocess Systems

Calculating the equipment requirements for heating or cooling of bioreactors can sometimes be simplified by considering the relative importance of each heat transfer resistance.

- For large bioprocess vessels containing cooling coils, the fluid velocity in the vessel is generally much slower than in the coils; accordingly, the tube-side thermal boundary layer is relatively thin and most of the heat transfer resistance is located on the bioreactor side. Especially when there is no fouling in the tubes, the heat transfer coefficient for the cooling water can often be omitted when calculating U.
- Likewise, the pipe wall resistance can sometimes be ignored as conduction of heat through metal is generally very rapid. An exception is stainless steel, which is used widely in the bioprocess industry because it does not corrode in mild acid environments, withstands repeated exposure to clean steam, and does not have a toxic effect on cells. The low thermal conductivity of this material means that wall resistance may be important unless the pipe wall is very thin.

The correlations used to estimate bioreactor-side heat transfer coefficients, such as Eqs. (8.36) and (8.37), were not developed for bioprocess systems and must not be considered to give

EXAMPLE 8.2 Cooling Coil

A 150-m³ bioreactor is operated at 35°C to produce fungal biomass from glucose. The fungal biomass is to be sold as a high-protein, meat alternative. The rate of O_2 uptake by the culture is 1.5 kg m⁻³ h⁻¹; the agitator dissipates energy at a rate of 1 kW m⁻³. Cooling water available from a nearby river at 10°C is passed through an internal coil in the bioprocess tank at a rate of 60 m³ h⁻¹. If the system operates at steady state, what is the exit temperature of the cooling water?

Solution

The rate of heat generation by aerobic cultures can be calculated directly from the O_2 demand. As described in Chapter 5, approximately 460 kJ of heat is released for each gmol of O_2 consumed. Therefore, from Eq. (8.40) the metabolic heat load is:

$$\Delta \hat{H}_{rxn} = \frac{-460 \, kJ}{gmol} \cdot \left| \frac{1 \, gmol}{32 \, g} \right| \cdot \left| \frac{1000 \, g}{1 \, kg} \right| \cdot \left(1.5 \, kg \, m^{-3} \, h^{-1}\right) \cdot \left| \frac{1 \, h}{3600 \, s} \right| \cdot 150 \, m^3 = -898 \, kJ \, s^{-1}$$

$$\Delta \hat{H}_{rxn} = -898 \, kW$$

$\Delta \hat{H}_{rxn}$ is negative because the bioprocess is exothermic. The rate of heat dissipation by the agitator is:

$$1 \, kW \, m^{-3} \left(150 \, m^3\right) = 150 \, kW$$

We can now calculate $\hat{Q}$ from Eq. (5.28), assuming there is negligible evaporation from the bioreactor:

$$\hat{Q} = -\Delta \hat{H}_{rxn} - \hat{M}_v \Delta h_v + \hat{W}_s = (898 - 0 + 150) \, kW = 1048 \, kW$$

The density of the cooling water is 1000 kg m⁻³; therefore:

$$\hat{M}_c = 60 \, m^3 \, h^{-1} \cdot \left| \frac{1 \, h}{3600 \, s} \right| \cdot \left(1000 \, kg \, m^{-3}\right) = 16.7 \, kg \, s^{-1}$$

From Table C.3 in Appendix C, the heat capacity of water is 75.4 J gmol⁻¹ °C⁻¹. Therefore:

$$C_{pc} = 75.4 \, J \, gmol^{-1} \, °C^{-1} \cdot \left| \frac{1 \, gmol}{18 \, g} \right| \cdot \left| \frac{1000 \, g}{1 \, kg} \right| = 4.19 \times 10^3 \, J \, kg^{-1} \, °C^{-1}$$

We can now apply Eq. (8.36) after rearranging and solving for T_{co}:

$$T_{co} = T_{ci} + \frac{\hat{Q}}{\hat{M}_c C_{pc}}$$

$$T_{co} = 10°C + \frac{1048 \times 10^3 \, J \, s^{-1}}{16.7 \, kg \, s^{-1} \left(4.19 \times 10^3 \, J \, kg^{-1} \, °C^{-1}\right)} = 25.0°C$$

The outlet water temperature is 25°C.

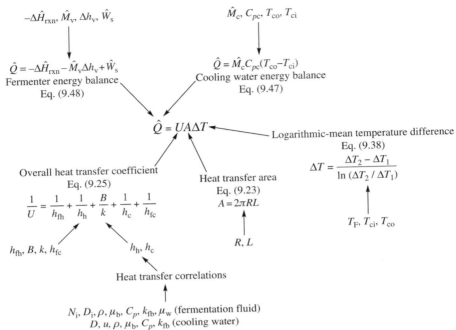

Figure 8.15 Summary of relationships and equations for design of a bioreactor cooling coil.

exact values. Most of the available correlations were developed using small-scale equipment; little information is available for industrial-size reactors.

Aerobic bioprocesses are typically carried out in tall vessels using multiple impellers; however, there have been relatively few studies of heat transfer using this vessel configuration. For helical cooling coils, geometric parameters such as the vertical separation between individual coils and the space between the coil and tank wall have been found to affect heat transfer, particularly for viscous fluids. Yet these factors are not included in most heat transfer correlations.

Equations (8.36) and (8.37) were developed for ungassed systems but are applied routinely for aerobic bioprocesses. Gassing alters the value of bioreactor-side heat transfer coefficients, but the magnitude of the effect and whether gassing causes an increase or decrease in heat transfer cannot yet be predicted. The rate of heat transfer in gas–liquid systems appears to depend on the distribution of bubbles in the vessel [5]; for example, accumulation of bubbles on or around the heat transfer surface is deleterious. Taking all of the above factors into account, correlations such as Eqs. (8.36) and (8.37) can provide only a starting point or rough estimate for evaluation of heat transfer coefficients in bioreactors.

When estimating heat transfer coefficients for non-Newtonian broths, the apparent viscosity can be substituted for μ_b in the correlation equations and dimensionless groups. However, this substitution is not straightforward when rheological parameters such as the flow behavior index n, the consistency index K, and the yield stress τ_0 change during the culture. The apparent viscosity also depends on the shear rate in the bioreactor, which varies greatly throughout the vessel. These factors make evaluation of heat transfer coefficients for non-Newtonian systems difficult.

Application of the heat exchanger design equations to specify a bioreactor cooling system is illustrated in Example 8.3.

Mixing and heat transfer are not independent functions in bioreactors. The impeller size and stirrer speed affect the value of the heat transfer coefficient in the bioprocess fluid; increased turbulence in the reactor decreases the thickness of the thermal boundary layer and facilitates rapid heat transfer. However, stirring also generates heat that must be removed from the reactor to maintain constant temperature. Although this energy contribution may be small in low-viscosity fluids because of the dominance of the heat of reaction, heat removal can be a severe problem in bioreactors containing highly viscous fluids. Turbulent flow and high heat transfer coefficients are difficult to achieve in viscous liquids without enormous power input, which itself generates an extra heat load. The effect of gas sparging on the heat load is usually neglected for stirred bioreactors. The energy associated with sparging can be evaluated as described in Section 7.5; however, the contribution to overall cooling requirements is minor.

EXAMPLE 8.3 Cooling-Coil Length in Bioreactor Design

A bioreactor used for antibiotic production must be kept at 35°C. After considering the O_2 demand of the organism and the heat dissipation from the stirrer, the maximum heat transfer rate required is estimated as 550 kW. Cooling water is available at 10°C; the exit temperature of the cooling water is calculated using an energy balance as 25°C. The heat transfer coefficient for the media broth is estimated from Eq. (8.28) as $2150\,W\,m^{-2}\,°C^{-1}$. The heat transfer coefficient for the cooling water is calculated as $14\,kW\,m^{-2}\,°C^{-1}$. It is proposed to install a helical cooling coil inside the bioreactor; the outer diameter of the pipe is 8 cm, the pipe wall thickness is 5 mm, and the thermal conductivity of the steel is $60\,W\,m^{-1}\,°C^{-1}$. An average internal fouling factor of $8500\,W\,m^{-2}\,°C^{-1}$ is expected; the bioreactor-side surface of the coil is kept relatively clean. What length of cooling coil is required?

Solution

$\hat{Q} = 550 \times 10^3\,W$. As the temperature in the bioreactor is constant, ΔT is calculated from Eq. (8.31):

$$\Delta T = \frac{(25 - 10)°C}{\ln\left(\dfrac{35 - 10}{35 - 25}\right)} = 16.4\,°C$$

(Continued)

U is calculated using Eq. (8.21) after omitting h_{fh}, as there is no fouling layer on the hot side of the coil:

$$\frac{1}{U} = \left(\frac{1}{2150\ \text{W m}^{-2}\,°\text{C}^{-1}} + \frac{5 \times 10^{-3}\ \text{m}}{60\ \text{W m}^{-1}\,°\text{C}^{-1}} + \frac{1}{14 \times 10^{3}\ \text{W m}^{-2}\,°\text{C}^{-1}} + \frac{1}{8500\ \text{W m}^{-2}\,°\text{C}^{-1}} \right)$$

$$= \left(4.65 \times 10^{-4} + 8.33 \times 10^{-5} + 7.14 \times 10^{-5} + 1.18 \times 10^{-4} \right) \text{m}^2\,°\text{C W}^{-1}$$

$$= 7.38 \times 10^{-4}\ \text{m}^2\,°\text{C W}^{-1}$$

$$U = 1355\ \text{W m}^{-2}\,°\text{C}^{-1}$$

Note the relative magnitudes of the four contributions to U: the cooling water film coefficient and the wall resistance make comparatively minor contributions and can often be neglected in design calculations.

We can now apply Eq. (8.15) to evaluate the required surface area A:

$$A = \frac{\hat{Q}}{U\,\Delta T} = \frac{550 \times 10^3\ \text{W}}{1355\ \text{W m}^{-2}\,°\text{C}^{-1}(16.4\,°\text{C})} = 24.75\ \text{m}^2$$

Equation (8.19) for the area of a cylinder can be used to evaluate the pipe length, L. As we have information for both the outer pipe diameter (8 cm) and the inner pipe diameter (outer pipe diameter − 2 × wall thickness = 8 cm − 2 × 5 mm = 7 cm), we can use an average pipe radius to determine L:

$$L = \frac{A}{2\pi R} = \frac{24.75\ \text{m}^2}{2\pi \left(\dfrac{0.5 \left(8 \times 10^{-2} + 7 \times 10^{-2} \right)}{2} \right) \text{m}} = 105.0\ \text{m}$$

The length of coil required is 105 m. The cost of such a length of pipe is a significant factor in the overall cost of the bioreactor.

Internal cooling coils typically add 15 to 25% to the cost of bioprocess vessels. The length of coil that can be used is limited by the size of the tank. Because the temperature of bioprocesses must be maintained within a very narrow range, if the culture generates a large heat load, providing sufficient cooling can be a challenge. Heat transfer can be the limiting factor in bioprocesses as discussed in Section 8.6.2. This is of particular concern for high-density cultures growing on carbon substrates with a high degree of reduction (Chapter 5), as large amounts of energy are released. Because biological reactions take place at near-ambient temperatures, the temperature of the cooling water is always close to that of the bioprocess; therefore, a relatively small driving force for heat transfer or ΔT value is inevitable. Improving the situation by refrigerating the cooling water involves a substantial increase in capital and operating costs and is generally not economically feasible. It may be possible to raise the cooling water flow rate, but this is also limited by equipment size and cost considerations. Although using an external heat exchanger is an option for increasing the heat transfer capacity of bioreactors, this introduces extra contamination risks and also adds significantly to the overall cost of the equipment. In some cases, the only feasible heat transfer solution for bioprocesses with high heat loads is to slow down the rate of metabolism, thus slowing the rate of heat generation. This might be achieved, for example, by reducing the cell density, decreasing the concentration of rate-limiting nutrients in the medium, or using a different carbon source.

8.6.2 Relationship between Heat Transfer and Cell Concentration

The design equation, Eq. (8.15), and the steady-state energy balance equation, Eq. (8.38), allow us to derive some important relationships for bioreactor operation. Because cell metabolism is usually the largest source of heat in bioreactors, the capacity of the system for heat removal can be linked directly to the maximum cell concentration in the reactor. Assuming that the heat dissipated from the stirrer and the cooling effects of evaporation are negligible compared with the heat of reaction, Eq. (8.38) becomes:

$$\hat{Q} = -\Delta\hat{H}_{rxn} \tag{8.39}$$

In aerobic bioprocesses, the heat of reaction is related to the rate of O_2 consumption by the cells. Approximately 460 kJ of heat is released for each gmol of O_2 consumed; therefore, if Q_{O_2} is the rate of O_2 uptake per unit volume in the bioreactor:

$$\Delta\hat{H}_{rxn} = \left(-460 \ kJ \ gmol^{-1}\right)Q_{O_2}V \tag{8.40}$$

where V is the reactor volume. Typical units for Q_{O_2} are gmol m^{-3} s^{-1}. $\Delta\hat{H}_{rxn}$ in Eq. (8.40) is negative because the reaction is exothermic. Substituting this equation into Eq. (8.39) gives:

$$\hat{Q} = \left(460 \ kJ \ gmol^{-1}\right)Q_{O_2}V \tag{8.41}$$

We can define q_{O_2} as the *specific O_2 uptake rate*, or the rate of O_2 consumption per cell, so $Q_{O_2} = q_{O_2}x$, where x is the cell concentration. Typical units for q_{O_2} are gmol (cdw g^{-1}) h^{-1}. Therefore:

$$\hat{Q} = \left(460 \ kJ \ gmol^{-1}\right)q_{O_2}xV \tag{8.42}$$

Substituting this into Eq. (8.15) gives:

$$\left(460 \ kJ \ gmol^{-1}\right)q_{O_2}xV = UA\Delta T \tag{8.43}$$

The fastest rate of heat transfer occurs when the temperature difference between the bioreactor contents and the cooling water is at its maximum. This occurs when $\Delta T = (T_F - T_{ci})$, where T_F is the bioprocess temperature and T_{ci} is the water inlet temperature. Therefore, assuming that the cooling water remains at temperature T_{ci}, we can derive from Eq. (8.43) an equation for the hypothetical maximum possible cell concentration supported by the heat transfer system:

$$x_{max} = \frac{UA\left(T_F - T_{ci}\right)}{\left(460 \ kJ \ gmol^{-1}\right)q_{O_2}V} \tag{8.44}$$

It is undesirable for the biomass concentration in bioreactors to be limited by the heat transfer capacity. Ideally, the extent of growth should be limited by other factors, such as the amount of substrate provided. Therefore, if the maximum cell concentration estimated using Eq. (8.44) is lower than that desired from the process, the heat transfer facilities must be improved. For example, the area A could be increased by installing a longer cooling coil, or the overall heat transfer coefficient could be improved by increasing the level of turbulence. Equation (8.44) was derived for bioreactors in which shaft work could be ignored; if the stirrer adds significantly to the total heat load, x_{max} will be smaller than that estimated using Eq. (8.44).

8.7 Hydrodynamic Considerations with Cooling Coils

Correlations, such as Eqs. (8.28) and (8.29), provide an estimate of the average heat transfer coefficient in bioreactors. However, the magnitude of heat transfer coefficients depends on the hydrodynamics of fluid flow near the heat transfer surface. As hydrodynamic conditions vary considerably in stirred vessels, there is a distribution of local coefficient values at different locations. As shown in Figure 8.16(A) for a vessel stirred with a single Rushton turbine, the heat transfer coefficient at the tank wall varies with height in the vessel and is maximum adjacent to the impeller. Heat transfer is most rapid in the region where the impeller discharge stream hits the wall; the heat transfer coefficient also increases with stirrer speed. The distribution of heat transfer coefficients in a vessel with dual Rushton turbines is shown in Figure 8.16(B). In this case, regions of high local values develop near each impeller, indicating elevated rates of heat transfer at two separate planes in the tank.

8.8 Summary of Chapter 8

After studying Chapter 8, you should:

- Be able to describe the equipment used for heat exchange in bioprocesses
- Understand the mechanisms of *conduction* and *convection* in heat transfer
- Know *Fourier's law* of conduction in terms of the *thermal conductivity* of materials
- Understand the importance of *thermal boundary layers* in convective heat transfer
- Know the *heat transfer design equation* and the meaning of the *overall heat transfer coefficient, U*
- Understand how the overall heat transfer coefficient can be expressed in terms of the individual resistances to heat transfer
- Know how *individual heat transfer coefficients* are estimated
- Know how to incorporate *fouling factors* into heat transfer analysis
- Be able to calculate basic design features of heat transfer systems
- Understand the interrelationships between heat transfer, hydrodynamics, and culture performance

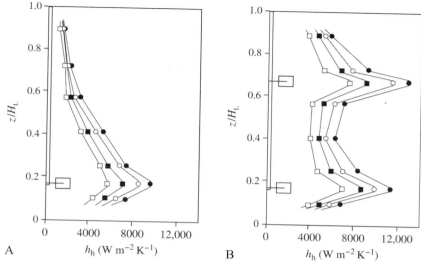

Figure 8.16 Profiles of the local heat transfer coefficient h_h at the wall of an ungassed stirred tank at different heights and stirrer speeds for: (A) a single Rushton turbine, and (B) dual Rushton turbines. The stirrer speeds were: ($\square$) 300 rpm; ($\blacksquare$) 400 rpm; (O) 500 rpm; and ($\bullet$) 600 rpm. z is the height from the vessel floor; H_L is the liquid height. The single impeller is located at height $z/H_L = 0.17$; the second impeller in (B) is located at $z/HL = 0.67$. Reprinted from J. Karcz, Studies of local heat transfer in a gas–liquid system agitated by double disc turbines in a slender vessel, *Chem. Eng. J.* 72, 217–227. © 1999, with permission from Elsevier Science.

Problems

8.1. Rate of conduction

   a. A furnace wall is constructed of firebrick 15 cm thick. The temperature inside the wall is 700°C; the temperature outside is 80°C. If the thermal conductivity of the brick under these conditions is $0.3\,W\,m^{-1}\,K^{-1}$, what is the rate of heat loss through $1.5\,m^2$ of wall surface?

   b. The $1.5\text{-}m^2$ area in part (a) is insulated with 4-cm-thick asbestos with thermal conductivity $0.1\,W\,m^{-1}\,K^{-1}$. What is the rate of heat loss now?

8.2. Overall heat transfer coefficient

Heat is transferred from one fluid to a second fluid across a metal wall. The film coefficients are 1.2 and $1.7\,kW\,m^{-2}\,K^{-1}$. The metal is 6 mm thick and has a thermal conductivity of $19\,W\,m^{-1}\,K^{-1}$. On one side of the wall there is a scale deposit with a fouling factor estimated at $830\,W\,m^{-2}\,K^{-1}$. What is the overall heat transfer coefficient?

8.3. Cocurrent versus countercurrent flow

Vegetable oil used in the manufacture of margarine needs to be cooled from 105°C to 55°C in a double-pipe heat exchanger. The inlet temperature of the cooling water is 20°C and the outlet temperature is 43°C. All else being equal, what heat transfer area is required if the heat exchanger is operated with cocurrent flow compared with that required for countercurrent flow?

8.4. Kitchen hot water

The hot water heater in your house is in a closet near the laundry. The kitchen is some distance away from the laundry on the opposite side of the house. When you turn on the hot water tap in the kitchen, it takes about 20 s for the hot water to arrive from the heater. In that time you are able to collect 2.5 L of cold water from the hot water tap. The diameter of the copper pipe delivering water from the heater to the kitchen is 10 mm.

a. What is the length of the hot water pipe to the kitchen?

b. Water leaves the heater at 75°C. If the ambient temperature is 12°C and the overall heat transfer coefficient for heat loss from the hot water pipe is 90 W m^{-2} K^{-1}, what is the temperature of the hot water arriving in your kitchen? Use Cp water $= 4.2$ kJ kg^{-1} °C^{-1}. (Iterative solution may be required.)

8.5. Double-pipe heat exchanger

In an amino acid bioprocess factory, molasses solution must be heated from 12°C to 30°C in a double-pipe heat exchanger at a rate of at least 19 kg min^{-1}. The heat is supplied by water at 63°C; an outlet water temperature of 33°C is desired. The overall heat transfer coefficient is 12 kW m^{-2} °C^{-1} and the exchanger is operated with countercurrent flow. The specific heat capacities of water and molasses solution are 4.2 kJ kg^{-1} °C^{-1} and 3.7 kJ kg^{-1} °C^{-1}, respectively.

a. Each stackable unit of a commercially available double-pipe assembly contains 0.02 m^2 of heat transfer surface. How many units need to be purchased and connected for this process?

b. For the number of units determined in (a), what flow rate of molasses can be treated?

c. For the number of units determined in (a), what flow rate of water is required?

8.6. Water heater on vacation

You are about to leave on a 4-day all-expenses-paid trip to Tahiti to attend a bioprocess engineering conference. It is midwinter in your hometown of Hobart, Tasmania, and you are trying to decide whether to turn off your storage hot water heater while you are away. The average midwinter temperature in July is 5°C, your hot water heater has a capacity of 110 L, and the temperature of the hot water is 68°C. You estimate that the overall heat transfer coefficient for loss of heat from the hot water tank to the atmosphere is 2 W m^{-2} K^{-1}. The surface area of the tank is 1.4 m^2. Which is more economical: turning off the heater to save the cost of maintaining the temperature at 68°C while you are away, or leaving the heater on to save the cost of heating up the entire contents of the tank from 5°C after you return home? Use Cp water $= 4.18$ kJ kg^{-1} °C^{-1}.

8.7. Fouling and pipe wall resistances

A copper pipe of diameter 3.5 cm and wall thickness 4 mm is used in an antibiotic factory to convey hot water 60 m at a flow rate of 20 L s^{-1}. The inlet temperature of the water is 90°C. Due to heat losses to the atmosphere, the outlet water temperature is 84°C. The ambient temperature is 25°C. The inside of the pipe is covered with a layer of fouling with a fouling factor of 7500 W m^{-2} °C^{-1}.

a. What is the rate of heat loss from the water?

b. What proportion of the total resistance to heat transfer is provided by the fouling layer?

c. If the copper pipe were replaced by a clean stainless steel pipe of the same thickness, what proportion of the total resistance to heat transfer would be provided by the pipe wall?

8.8. Effect of cooling-coil length on coolant requirements

A bioreactor is maintained at 35°C by water circulating at a rate of 0.5 kg s^{-1} in a cooling coil inside the vessel. The inlet and outlet temperatures of the water are 8°C and 15°C, respectively. The length of the cooling coil is increased by 50%. To maintain the same bioprocess temperature, the rate of heat removal must be kept the same. Determine the new cooling-water flow rate and outlet temperature by carrying out the following calculations. The heat capacity of the cooling water can be taken as 4.18 kJ kg^{-1} °C^{-1}.

a. From a steady-state energy balance on the cooling water, calculate the rate of cooling with the original coil.

b. Determine the mean temperature difference ΔT with the original coil.

c. Evaluate UA for the original coil.

d. If the length of the coil is increased by 50%, the area available for heat transfer, A', is also increased by 50% so that $A' = 1.5 A$. The value of the overall heat transfer coefficient is not expected to change very much. For the new coil, what is the value of UA'?

e. Estimate the new cooling-water outlet temperature. (Iterative solution may be required.)

f. By how much are the cooling water requirements reduced after the new coil is installed?

8.9. Bioreactor cooling coil

A bioprocess factory is set up in Belize to produce an anti-UV compound from a newly discovered bacterial strain. The anti-UV compound will be used in the Bahamas for manufacture of an improved sunscreen lotion. A cheap second-hand bioreactor is purchased from a cash-strapped pharmaceutical company in the Cayman Islands, but the reactor must be equipped with a new cooling coil. The bioprocess is to be carried out at 35°C. Cooling water is available from a nearby mountain stream at an average temperature of 15°C; however, local environmental laws in Belize prohibit the dumping of water back into the stream if it is above 25°C. Based on the heat of reaction for the bacterial culture, the cooling requirements are 15.5 kW. The overall heat transfer coefficient is estimated to be 340 W m^{-2} K^{-1}. If a company in Yucatan can supply stainless steel pipe with diameter of 4 cm for fabricating the cooling coil, what length is required?

8.10. Effect of fouling on heat transfer resistance

In current service, 20 kg s^{-1} of cooling water at 12°C must be circulated through a thin-walled coil inside a bioreactor to maintain the temperature at 37°C. The coil is 150 m long with pipe diameter of 12 cm; the exit water temperature is 28°C. After the inner and outer surfaces of the coil are cleaned, it is found that only 13 kg s^{-1} of cooling water is required to control the bioprocess temperature.

a. Calculate the overall heat transfer coefficient before cleaning.

b. What is the outlet water temperature after cleaning?

c. What fraction of the total resistance to heat transfer before cleaning was due to fouling deposits?

8.11. Suitability of an existing cooling coil

An enzyme manufacturer in the same industrial park as your antibiotic factory has a reconditioned 20-m^3 bioreactor for sale. You are in the market for a cheap 20-m^3 bioreactor; however, the vessel on offer is fitted with a 45-m helical cooling coil with pipe diameter of 7.5 cm. You propose to use the bioreactor for your newest production organism, which is known to have a maximum O_2 demand of $90\,\mathrm{mol\,m^{-3}\,h^{-1}}$ at its optimum culture temperature of $28°C$. You consider that the 3-m-diameter vessel should be stirred with a 1-m-diameter Rushton turbine operated at an average speed of 50 rpm. The media fluid can be assumed to have the properties of water. If $20\,\mathrm{m^3\,h^{-1}}$ cooling water is available at $12°C$, should you make an offer for the second-hand bioreactor and cooling coil?

8.12. Heat transfer and cooling water in bioreactor design

A 100-m³ bioreactor of diameter 5 m is stirred using a 1.7-m Rushton turbine operated at 80 rpm. The culture fluid has the following properties:

$$C_p = 4.2\ \mathrm{kJ\ kg^{-1}\ °C^{-1}}$$
$$k_{fb} = 0.6\ \mathrm{W\ m^{-1}\ °C^{-1}}$$
$$\rho = 10^3\ \mathrm{kg\ m^{-3}}$$
$$\mu_b = 10^{-3}\ \mathrm{N\ s\ m^{-2}}$$

Assume that the viscosity at the wall is equal to the bulk fluid viscosity. Heat is generated by the bioprocess at a rate of 2500 kW. This heat is removed to cooling water flowing in a helical stainless steel coil inside the vessel. The inner diameter of the coil pipe is 12 cm and the wall thickness is 6 mm. The thermal conductivity of the stainless steel is $20\,\mathrm{W\,m^{-1}\,°C^{-1}}$. There are no fouling layers present and the tube-side heat transfer coefficient can be neglected. The bioprocess temperature is $30°C$. Cooling water enters the coil at $10°C$ at a flow rate of $1.5 \times 10^5\,\mathrm{kg\,h^{-1}}$.

a. Calculate the bioreactor-side heat transfer coefficient.

b. Calculate the overall heat transfer coefficient, U.

c. What proportion of the total resistance to heat transfer is due to the pipe wall?

d. From the equations in Section 8.5, what is the contribution of shaft work to the cooling requirements for this bioreactor?

e. Calculate the outlet cooling-water temperature.

f. Estimate the length of cooling coil needed.

g. If the cooling water flow rate could be increased by 50%, what effect would this have on the length of cooling coil required?

8.13. Test for heat transfer limitation

An 8-m³ stirred bioreactor is used to culture *Gibberella fujikuroi* for production of gibberellic acid. The liquid medium contains $12\,\mathrm{g\,L^{-1}}$ glucose. Under optimal conditions, 1.0 g dry weight of cells is produced for every 2.2 g of glucose consumed. The culture-specific O_2 demand is 7.5 mmol per g dry weight per h. To achieve the maximum yield of gibberellic acid, the bioprocess temperature must be held constant at $32°C$. The bioreactor is equipped with a helical cooling coil of length 55 m and pipe diameter of 5 cm. Under usual bioreactor operating conditions, the overall heat transfer coefficient is $250\,\mathrm{W\,m^{-2}\,°C^{-1}}$. If the water pumped through the coil has an inlet temperature of $15°C$, does the heat transfer system support complete consumption of the substrate?

8.14. Optimum stirrer speed for removal of heat from viscous broth

The viscosity of a media broth containing exopolysaccharide is about 10,000 cP. The broth is stirred in an aerated 10-m³ bioreactor of diameter 2.3 m using a single 0.78-m-diameter Rushton turbine. Other properties of the broth are as follows:

$$C_p = 2 \text{ kJ kg}^{-1} \text{ }^\circ\text{C}^{-1}$$
$$k_{\text{fb}} = 2 \text{ W m}^{-1} \text{ }^\circ\text{C}^{-1}$$
$$\rho = 10^3 \text{ kg m}^{-3}$$

The bioreactor is equipped with an internal cooling coil providing a heat transfer area of 14 m²; the average temperature difference for heat transfer is 20°C. Neglect any variation of viscosity at the wall of the coil. Assume that the power dissipated in aerated broth is 40% lower than in ungassed liquid.

a. Using logarithmic coordinates, plot $\hat{Q}$ for several stirrer speeds between 0.5 and 10 s⁻¹.

b. From the equations presented in Section 8.5, calculate $\hat{W}_s$, the power dissipated by the stirrer, as a function of stirrer speed. Plot these values on the same graph as $\hat{Q}$.

c. If evaporation, heat losses, and other factors have a negligible effect on the heat load, the difference between $\hat{Q}$ and $\hat{W}_s$ is equal to the rate of removal of metabolic heat from the bioreactor. Plot the rate of metabolic heat removal as a function of stirrer speed.

d. At what stirrer speed is the removal of metabolic heat most rapid?

e. The specific rate of O_2 consumption is 6 mmol g⁻¹ h⁻¹. If the bioreactor is operated at the stirrer speed identified in (d), what is the maximum cell concentration?

f. How do you interpret the intersection of the curves for $\hat{Q}$ and $\hat{W}_s$ at high stirrer speed in terms of the capacity of the system to handle exothermic reactions?

References

[1] J.M. Coulson, J.F. Richardson, J.R. Backhurst, J.H. Harker, Coulson and Richardson's sixth ed., Chemical Engineering, vol. 1, Butterworth-Heinemann, 1999. (Chapter 9).

[2] T.H. Chilton, T.B. Drew, R.H. Jebens, Heat transfer coefficients in agitated vessels, Ind. Eng. Chem 36 (1944) 510–516.

[3] G.F. Hewitt, G.L. Shires, T.R. Bott, Process Heat Transfer, CRC Press, 1994.

[4] J.P. Holman, Heat Transfer, eighth ed., McGraw-Hill, 1997.

[5] G.J. Xu, Y.M. Li, Z.Z. Hou, L.F. Feng, K. Wang, Gas_liquid dispersion and mixing characteristics and heat transfer in a stirred vessel, Can. J. Chem. Eng. 75 (1997) 299–306.

[6] W.L. McCabe, J.C. Smith, P. Harriott, Unit Operations of Chemical Engineering, sixth ed., Section III, McGraw-Hill, 2001.

Suggestions for Further Reading

Heat Transfer Principles and Application

See also references 1–4.
Y.A. Çengel, Heat Transfer: A Practical Approach, 2nd ed., McGraw-Hill, 2003.
Perry's, *Perry's Chemical Engineers' Handbook,* 8th ed., Section 5, McGraw-Hill, 2008.
L.C. Thomas, Heat Transfer, Prentice Hall, 1992.

Heat Transfer Equipment

Perry's, *Perry's Chemical Engineers' Handbook,* 8th ed., Section 11, McGraw-Hill, 2008.

Fouling

T.R. Bott, Fouling of Heat Exchangers, Elsevier, 1995.

Heat Transfer in Bioprocessing

B. Atkinson, F. Mavituna, Chapter 14, and Biochemical Engineering and Biotechnology Handbook, 2nd ed., Macmillan, 1991.

T.J.S. Brain, K.L. Man, Heat transfer in stirred tank bioreactors, Chem. Eng. Prog. 85 (7) (1989) 76–80.

W.J. Kelly, A.E. Humphrey, Computational fluid dynamics model for predicting flow of viscous fluids in a large fermentor with hydrofoil flow impellers and internal cooling coils, Biotechnol. Prog 14. (1998) 248–258.

J.R. Swartz, Heat management in fermentation processes, Pergamon Press, 1985, and 299–303.

CHAPTER 9

Bioreactor Mass Transfer

Mass transfer occurs in mixtures when there are concentration differences. The different concentrations correlate with different potential energies and the universe tends toward dissipating potential energy driving forces. For example, when dye is dropped into a pail of water, mass transfer processes, driven by potential energy gradients, are responsible for the movement of dye molecules from regions of high concentration to low concentration until a uniform concentration is established.

The transport of substrates to cells and the transport of products away from cells are important mass transfer processes that can constrain system performance. An important example of mass transfer in bioprocessing is the supply of O_2 in bioreactors for aerobic growth. The concentration of O_2 at the surface of air bubbles is high compared with that in the bulk liquid; this concentration gradient promotes O_2 transfer from the bubbles into the medium where it is available to cells.

Mass transfer plays a vital role in many reaction systems. As the distance between the reactants and the site of reaction becomes greater, the rate of mass transfer is more likely to influence or control the reaction rate. Taking again the example of O_2 in aerobic cultures, if mass transfer of O_2 from the bubbles to the liquid phase is slow, the rate of cell metabolism can be limited by this O_2 transfer rate. As O_2 is a critical component of aerobic bioprocesses and is sparingly soluble in aqueous solutions, much of our interest in mass transfer lies with the transfer of O_2 across gas–liquid interfaces. However, liquid–solid mass transfer can also be important in systems containing clumps, pellets, flocs, or films of cells or enzymes. In these cases, nutrients in the liquid phase must be transported into the solid before they can be utilized in reactions. Unless mass transfer is rapid, the supply of nutrients will limit the rate of biological conversion.

Mass transfer occurs based on contributions from two distinct processes, molecular diffusion and convection. For example, in quiescent, unmixed fluids, mass transfer occurs because of molecular diffusion. However, many bioprocess systems contain moving fluid such that mass transfer by molecular motion is supplemented by convective transfer. The theory of mass transfer with applications relevant to bioprocessing will be considered in this chapter.

9.1 Molecular Diffusion

Molecular diffusion is the movement of component molecules in a mixture based on concentration differences. Diffusion of molecules occurs in the direction required to dissipate the concentration gradient, that is, from regions of high concentration to regions of low concentration. Steady state diffusion can occur if material is constantly added to the high-concentration region and constantly removed from the low-concentration region. This design is often exploited in mass transfer operations and reaction systems.

9.1.1 Diffusion Theory

In this text, we confine our discussion of diffusion to *binary mixtures*—that is, mixtures containing only two components. Consider a closed system containing molecular components A and B. Initially, the concentration of A in the system is not uniform; as indicated in Figure 9.1, concentration C_A varies from C_{A1} to C_{A2} as a function of distance y. In response to this concentration gradient, molecules of A will diffuse away from the region of high concentration (high potential energy) until eventually the whole system acquires uniform composition. If there is no large-scale fluid motion in the system, due to processes like stirring, mixing occurs solely by random molecular movement.

Assume that mass transfer of A occurs across area a perpendicular to the direction of diffusion. In single-phase systems, the rate of mass transfer due to molecular diffusion is given by *Fick's law of diffusion*, which states that the mass flux is proportional to the concentration gradient:

$$J_A = \frac{N_A}{a} = -\mathcal{D}_{AB}\frac{dC_A}{dy} \tag{9.1}$$

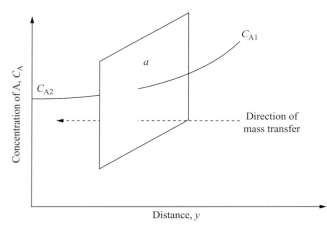

Figure 9.1 Concentration gradient of component A inducing mass transfer across area a.

where J_A is the *mass flux* of component A (mass A area^{-1} time^{-1}), N_A is the *rate of mass transfer* of component A (mass A time^{-1}), a is the area across which mass transfer occurs, $\mathcal{D}_{AB}$ is the *binary diffusion coefficient* or *diffusivity* of component A in a mixture of A and B, C_A is the concentration of component A, and y is distance. dC_A/dy is the *concentration gradient*, or potential energy driving force for A with respect to distance y. The mass flux is defined as the rate of mass transfer per unit area perpendicular to the direction of movement; typical units for J_A are gmol s^{-1} m^{-2}. Corresponding units for N_A are gmol s^{-1}, for C_A gmol m^{-3}, and for $\mathcal{D}_{AB}$ m^2 s^{-1}. Mass rather than mole units may be used for J_A, N_A, and C_A.

Equation (9.1) indicates that the rate of diffusion can be enhanced by increasing the area available for mass transfer, the concentration gradient, or the magnitude of the diffusion coefficient. The negative sign in Eq. (9.1) indicates that the direction of mass transfer is always from high concentration to low concentration, moving from regions of high potential energy to low potential energy.

The diffusion coefficient $\mathcal{D}_{AB}$ is a property of materials. $\mathcal{D}_{AB}$ is the proportionality constant linking the concentration driving force to mass flux and reflects the ease with which diffusion takes place. Diffusivities are several orders of magnitude smaller for diffusion in liquids than in gases. For example, $\mathcal{D}_{AB}$ for O_2 in air at 25°C and 1 atm is 2.1×10^{-5} m^2 s^{-1}, whereas $\mathcal{D}_{AB}$ for O_2 in water at 25°C and 1 atm is 2.5×10^{-9} m^2 s^{-1}. For dilute concentrations of glucose in water at 25°C, $\mathcal{D}_{AB}$ is 6.9×10^{-10} m^2 s^{-1}. The magnitude of $\mathcal{D}_{AB}$ depends on both the components of the mixture; for example, the diffusivity of carbon dioxide in water will be different from the diffusivity of carbon dioxide in another solvent such as ethanol. The value of $\mathcal{D}_{AB}$ is also dependent on temperature. $\mathcal{D}_{AB}$ values can be found in reference sources including Perry's Chemical Engineers' Handbook.

When diffusivity values are not available for the materials, temperatures, or pressures of interest, $\mathcal{D}_{AB}$ can be estimated using equations and correlations such as the Stokes-Einstein equation or the Wilke-Change correlation [1–3].

9.1.2 Analogy Between Mass, Heat, and Momentum Transfer

There is a mathematical and conceptual similarity between the processes of mass, heat, and momentum transfer occurring as a result of molecular motion. This is suggested by the form of the equations for mass, heat, and momentum fluxes:

$$J_A = -\mathcal{D}_{AB}\frac{dC_A}{dy} \tag{9.2}$$

$$\hat{q} = -k\frac{dT}{dy} \tag{8.2}$$

and

$$\tau = -\mu \frac{dv}{dy} \qquad (6.6)$$

The three processes represented above are different at the molecular level, but the basic equations have the same form. In each case, flux in the y-direction is directly proportional to the driving force (either dC_A/dy, dT/dy, or dv/dy), with the proportionality constant ($\mathscr{D}_{AB}$, k, or μ) quantifying a physical property of the material. The negative signs in Eqs. (9.2) and (6.6) indicate that transfer of mass, heat, or momentum is always down the driving force gradient. The analogy of Eqs. (9.2) and (6.6) is valid for transport of mass, heat, and momentum resulting from motion or vibration of molecules; however, there are limitations as noted elsewhere [3,4].

9.2 Role of Diffusion in Bioprocessing

Most industrial processes where mass transfer occurs utilize bulk fluid mixing. Bulk fluid motion results in more rapid, large-scale mixing than does molecular diffusion. Despite this, diffusive transport is still important in many areas of bioprocessing.

- *Scale of mixing*. Turbulence in fluids produces bulk mixing on a scale equal to the smallest eddy size, as discussed in Section 7.9. Within the smallest eddies, flow is largely streamlined so that further mixing must occur by diffusion of the fluid components. Mixing on a molecular scale therefore relies on diffusion as the final step in the mixing process.
- *Solid-phase reaction*. Reactions in biosystems are sometimes mediated by catalysts in solid form (e.g., clumps, flocs, and biofilms of cells) and by immobilized enzyme or cell particles. When cells or enzymes are incorporated into a solid particle, substrates must be transported into the solid before the reaction can take place. Mass transfer within solid particles is usually unassisted by bulk fluid convection; therefore, the only mechanism for intraparticle mass transfer is molecular diffusion. As the reaction proceeds, diffusion is also responsible for the removal of product molecules away from the site of reaction. When reaction is coupled with diffusion, the overall reaction rate can be constrained significantly by diffusion if it is slow, as discussed more fully in later Chapters.
- *Mass transfer across phase boundaries*. Mass transfer between phases occurs often in bioprocessing. O_2 transfer from gas bubbles to culture broth, penicillin recovery from aqueous to organic liquid, and glucose uptake from liquid medium into mold pellets are all typical examples. When different phases come into contact, the fluid velocity near the phase interface decreases significantly and diffusion becomes crucial for mass transfer.

9.3 Film Theory

The *two-film theory* is a useful model for mass transfer between phases. Mass transfer of solute from one phase to another involves transport from the bulk of one phase to the *phase boundary* or *interface*, then movement from the interface into the bulk of the second phase. Film theory is based on the concept that a fluid film or *mass transfer boundary layer* forms wherever there is contact between two phases.

Let us consider mass transfer of component A across the phase boundary represented in Figure 9.2. Assume that the two phases are immiscible liquids such as water and chloroform, and that A is initially at a higher concentration in the aqueous phase than in the organic phase. Each bulk phase is well mixed. The concentration of A in the bulk aqueous phase is C_{A1}; the concentration of A in the bulk organic phase is C_{A2}.

According to the film theory, the local fluid velocities approach zero at the interface even though the bulk phases are well mixed. This results in a thin film of stagnant fluid on either side of the interface; mass transfer through this film is based solely on molecular diffusion. The concentration of A changes near the interface as indicated in Figure 9.2; C_{A1i} is the interfacial concentration of A in the aqueous phase; C_{A2i} is the interfacial concentration of A in the organic phase. Most of the resistance to mass transfer resides in the liquid films rather than in the bulk liquid. It is generally assumed that there is negligible resistance to transport at the interface itself; this is equivalent to assuming that the phases are in equilibrium at the plane of contact. The difference between C_{A1i} and C_{A2i} at the interface accounts for the possibility that,

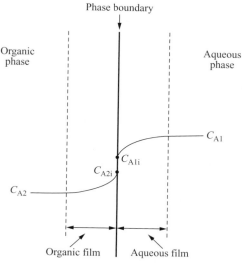

Figure 9.2 Film resistance to mass transfer between two immiscible liquids.

at equilibrium, A may be more soluble in one phase than in the other. For example, if A were acetic acid in contact at the interface with both water and chloroform, the equilibrium concentration in water would be greater than in chloroform by a factor of 5 to 10. Therefore, C_{A1i} would then be significantly higher than C_{A2i}.

Film theory is applied extensively in analysis of mass transfer, although the treatment here is greatly simplified. There are other conceptual models of mass transfer in fluids that lead to more accurate mathematical predictions than film theory [1,4]. Nevertheless, irrespective of how mass transfer is mathematically represented, diffusion is an important mechanism of mass transfer close to the interface between phases.

9.4 Convective Mass Transfer

The term *convective mass transfer* refers to mass transfer occurring in the presence of bulk fluid motion. Molecular diffusion will occur whenever there is a concentration gradient; however, if the bulk fluid is also moving, the overall rate of mass transfer will be higher due to the contribution of convective transport. Analysis of mass transfer is important in multiphase systems where interfacial boundary layers provide significant mass transfer resistance. Let us develop an expression for the rate of mass transfer that is applicable to mass transfer boundary layers.

The rate of mass transfer is directly proportional to the area available for transfer and the driving force for the transfer process. This can be expressed as:

$$\text{Transfer rate } \alpha \text{ transfer area} \times \text{driving force} \tag{9.3}$$

The proportionality coefficient in this equation is called the *mass transfer coefficient*, so that:

$$\text{Transfer rate} = \text{mass transfer coefficient} \times \text{transfer area} \times \text{driving force} \tag{9.4}$$

The driving force for mass transfer of component A through the phase boundary layer can be expressed in terms of the concentration difference of A across the fluid film. Therefore, the rate of mass transfer from the bulk fluid through the boundary layer to the interface is:

$$N_A = k a \Delta C_A = k a \left(C_{Ab} - C_{Ai} \right) \tag{9.5}$$

where N_A is the *volumetric rate of mass transfer* of component A (typical units: gmol m^{-3} s^{-1}), k is the mass transfer coefficient (typical units: m s^{-1}), a is the area available for mass transfer per volume (typical units: m^2 m^{-3} or m^{-1}), C_{Ab} is the bulk concentration of component A away from the phase boundary, and C_{Ai} is the concentration of A at the interface. Please note the definition of N_A and a, as written in Eq. 9.5, is on a volumetric basis, a change from Eq. 9.1.

Equation (9.5) indicates that the rate of convective mass transfer can be enhanced by increasing the area available for mass transfer, the concentration difference between the bulk fluid and the interface, and the magnitude of the mass transfer coefficient. Analogous with heat transfer processes (Eq. [8.11]), Eq. (9.5) can also be written in the form:

$$N_A = \frac{\Delta C_A}{R_m}$$

(9.6)

where ΔC_A is the driving force and R_m is the resistance to mass transfer:

$$R_m = \frac{1}{ka}$$

(9.7)

Mass transfer coupled with fluid flow is a more complicated process than diffusive mass transfer alone. The value of the mass transfer coefficient k reflects the contribution to mass transfer from all the processes in the system that influence the boundary layer. Like the heat transfer coefficient in Chapter 8, k depends on the combined effects of the flow velocity, the geometry of the mass transfer system, and fluid properties such as viscosity and diffusivity. Because the hydrodynamics of most practical systems are not easily characterized, k cannot be calculated reliably from first principles. Instead, it is measured experimentally or estimated using correlations available from the literature. In general, reducing the thickness of the boundary layer or increasing the diffusion coefficient in the film will enhance the value of k, thus improving the rate of mass transfer.

In bioprocessing, three mass transfer situations that involve multiple phases are *liquid–solid mass transfer, liquid–liquid mass transfer* between immiscible solutions, and *gas–liquid mass transfer*. Use of Eq. (9.5) to determine the rate of mass transfer in these systems is discussed in the following sections.

9.4.1 Liquid–Solid Mass Transfer

Mass transfer between a liquid and a solid is important in some bioprocessing applications. For example, the transport of substrates to immobilized, solid-phase cell or enzyme catalysts is a common process. Adsorption of molecules onto surfaces, such as in chromatography, requires transport from the liquid phase to a solid; liquid–solid mass transfer is also important in crystallization processes as molecules move from the liquid to the growing crystal. Conversely, the process of dissolving a solid in liquid requires liquid–solid mass transfer directed away from the solid surface.

Let us assume that component A is required for reaction at the surface of a solid. The situation at the interface between flowing liquid containing A and the solid is illustrated in Figure 9.3. Near the interface, the fluid velocity is reduced and a boundary layer develops. As A is consumed by reaction at the surface, the local concentration of A decreases and a concentration gradient is established through the film. The concentration difference between the bulk liquid and the phase interface drives mass transfer of A from the liquid to the solid, allowing the reaction to continue. Here for simplicity, we will assume the reaction occurs only at the

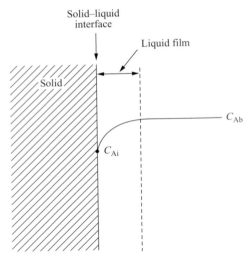

Figure 9.3 Concentration gradient for liquid–solid mass transfer.

surface of the solid. The concentration of A at the phase boundary is C_{Ai}; the concentration of A in the bulk liquid outside the film is C_{Ab}. If a is the liquid–solid interfacial area per unit volume, the volumetric rate of mass transfer can be determined from Eq. (9.5) as:

$$N_A = k_L a \left(C_{Ab} - C_{Ai} \right) \tag{9.8}$$

where k_L is the liquid-phase mass transfer coefficient.

Application of Eq. (9.8) requires knowledge of the mass transfer coefficient, the interfacial area between the phases, the bulk concentration of A, and the concentration of A at the interface. The value of the mass transfer coefficient is either measured experimentally or calculated using correlations. In most cases, the area a can be determined from the shape and size of the solid. Bulk concentrations are generally easy to measure; however, estimation of the interfacial concentration C_{Ai} is much more difficult, as measuring compositions at phase boundaries is not straightforward experimentally. To overcome this problem, we must consider the processes in the system that occur in conjunction with mass transfer. In the example of Figure 9.3, transport of A is linked to reaction at the surface of the solid, so C_{Ai} will depend on the rate of consumption of A at the interface. In practice, we can calculate the rate of mass transfer of A in this situation only if we have information about the rate of reaction at the solid surface. Simultaneous reaction and mass transfer occur in many bioprocesses as outlined in Chapter 11.

9.4.2 Liquid–Liquid Mass Transfer

Liquid–liquid mass transfer between immiscible solvents is most often encountered in the product recovery stages of bioprocessing. Organic solvents are used to isolate antibiotics, steroids, and alkaloids from culture broths, while two-phase aqueous systems are useful for protein purification. Liquid–liquid mass transfer is also important when hydrocarbons are used as substrates in cultivations, for example, to produce microbial biomass for single-cell protein. Water-immiscible organic solvents are of increasing interest for enzyme and whole-cell biocatalysis: two-phase reaction systems can be used to overcome problems with poor substrate solubility or product toxicity and can shift chemical equilibria for enhanced yields and selectivity in metabolic reactions.

The situation at the interface between two immiscible liquids is shown in Figure 9.2. Component A is present at bulk concentration C_{A1} in one phase; this concentration falls to C_{A1i} at the interface. In the other liquid, the concentration of A falls from C_{A2i} at the interface to C_{A2} in the bulk. The rate of mass transfer N_A in each liquid phase is expressed using Eq. (9.5):

$$N_{A1} = k_{L1}a\left(C_{A1} - C_{A1i}\right) \tag{9.9}$$

and

$$N_{A2} = k_{L2}a\left(C_{A2i} - C_{A2}\right) \tag{9.10}$$

where k_L is the liquid-phase mass transfer coefficient, and subscripts 1 and 2 refer to the different liquid phases. Application of Eqs. (9.9) and (9.10) is difficult because interfacial concentrations are not easy to measure experimentally. However, in this case, the difficult-to-measure C_{A1i} and C_{A2i} can be replaced with more easily measured concentrations by considering the physical situation at the interface and by algebraic manipulation of the equations.

First, let us recognize that at steady state, because there can be no accumulation of A at the interface or anywhere else in the system, any A transported through liquid 1 must also be transported through liquid 2. This means that N_{A1} in Eq. (9.9) must be equal to N_{A2} in Eq. (9.10) so that $N_{A1} = N_{A2} = N_A$. We can then rearrange Eqs. (9.9) and (9.10):

$$\frac{N_A}{k_{L1}a} = C_{A1} - C_{A1i} \tag{9.11}$$

and

$$\frac{N_A}{k_{L2}a} = C_{A2i} - C_{A2} \tag{9.12}$$

Often, it is assumed that there is negligible resistance to mass transfer at the interface—that is, within distances corresponding to molecular free paths on either side of the phase boundary. This is equivalent to assuming that the phases are in equilibrium *at the interface*; therefore, C_{A1i} and C_{A2i} are equilibrium concentrations.

The assumption of phase-boundary equilibrium has been subjected to many tests. As a result, it is known that there are special circumstances, such as when there is adsorption of material at the interface, for which the assumption is invalid. However, in ordinary situations, the

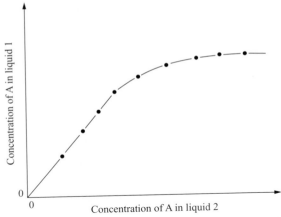

Figure 9.4 Equilibrium curve for solute A in two immiscible solvents 1 and 2.

evidence is that equilibrium does exist at the interface between phases. Note that we are not proposing to relate bulk concentrations C_{A1} and C_{A2} using equilibrium relationships, only C_{A1i} and C_{A2i}. If the bulk liquids were in equilibrium, there would be no driving force for mass transfer.

A typical equilibrium curve relating the concentrations of solute A in two immiscible liquid phases is shown in Figure 9.4. The equilibrium distribution of one solute between two phases can be described using the *distribution law*. At equilibrium, the ratio of the solute concentrations in the two phases is equal to the *distribution coefficient* or *partition coefficient, m*. When the concentration of A is low, the equilibrium curve is approximately a straight line so that m is constant. The distribution law is accurate only if both solvents are immiscible and there is no chemical reaction.

If C_{A1i} and C_{A2i} are equilibrium concentrations, they can be related using the partition coefficient m:

$$m = \frac{C_{A1i}}{C_{A2i}} \tag{9.13}$$

such that

$$C_{A1i} = m C_{A2i} \tag{9.14}$$

and

$$C_{A2i} = \frac{C_{A1i}}{m} \tag{9.15}$$

Equations (9.14) and (9.15) can now be used to eliminate the interfacial concentrations from Eqs. (9.11) and (9.12). First, we make a direct substitution:

$$\frac{N_A}{k_{L1}a} = C_{A1} - mC_{A2i} \tag{9.16}$$

and

$$\frac{N_A}{k_{L2}a} = \frac{C_{A1i}}{m} - C_{A2} \tag{9.17}$$

If we now multiply Eq. (9.12) by m:

$$\frac{mN_A}{k_{L2}a} = mC_{A2i} - mC_{A2} \tag{9.18}$$

and divide Eq. (9.11) by m:

$$\frac{N_A}{mk_{L1}a} = \frac{C_{A1}}{m} - \frac{C_{A1i}}{m} \tag{9.19}$$

and add Eq. (9.16) to Eq. (9.18), and Eq. (9.17) to Eq. (9.19), we eliminate the interfacial concentration terms completely:

$$N_A\left(\frac{1}{k_{L1}\,a} + \frac{m}{k_{L2}\,a}\right) = C_{A1} - mC_{A2} \tag{9.20}$$

$$N_A\left(\frac{1}{mk_{L1}\,a} + \frac{1}{k_{L2}a}\right) = \frac{C_{A1}}{m} - C_{A2} \tag{9.21}$$

Equations (9.20) and (9.21) combine the mass transfer resistances in the two liquid films and relate the rate of mass transfer N_A to the bulk fluid concentrations C_{A1} and C_{A2}. The bracketed expressions for the combined mass transfer coefficients are used to define the *overall liquid-phase mass transfer coefficient*, K_L. Depending on the terms used to represent the concentration difference, we can define two overall mass transfer coefficients:

$$\frac{1}{K_{L1}a} = \frac{1}{k_{L1}a} + \frac{m}{k_{L2}a} \tag{9.22}$$

and

$$\frac{1}{K_{L2}a} = \frac{1}{mk_{L1}a} + \frac{1}{k_{L2}a} \tag{9.23}$$

where K_{L1} is the overall mass transfer coefficient based on the bulk concentration in liquid 1, and K_{L2} is the overall mass transfer coefficient based on the bulk concentration in liquid 2.

We can now summarize the results to obtain two equations for the mass transfer rate at the interfacial boundary in liquid–liquid systems:

$$N_A = K_{L1}a\left(C_{A1} - mC_{A2}\right) \tag{9.24}$$

and

$$N_A = K_{L2} a \left(\frac{C_{A1}}{m} - C_{A2} \right)$$

(9.25)

where K_{L1} and K_{L2} are given by Eqs. (9.22) and (9.23). Use of either of these two equations requires knowledge of the concentrations of A in the bulk fluids, the partition coefficient m, the interfacial area a between the two liquid phases, and the value of either K_{L1} or K_{L2}. C_{A1} and C_{A2} are generally easy to measure. m can also be measured or found in handbooks of physical properties. The overall mass transfer coefficients can be measured experimentally or are estimated from correlations for k_{L1} and k_{L2} in the literature. The only remaining parameter is the interfacial area, a. In many applications of liquid–liquid mass transfer, it may be difficult to know how much interfacial area is available between the phases. For example, liquid–liquid extraction is often carried out in stirred tanks where an impeller is used to disperse and mix droplets of one phase through the other. The interfacial area in these circumstances will depend on the size, shape, and number of the droplets, which depend in turn on the intensity of agitation and properties of the fluid. Because these factors also affect the value of k_L, correlations for mass transfer coefficients in liquid–liquid systems are often given in terms of $k_L a$ as an aggregate parameter. The combined term $k_L a$ parameter is then referred to as the mass transfer coefficient for convenience.

Equations (9.24) and (9.25) indicate that the rate of mass transfer between two liquid phases is not dependent simply on the concentration difference: the equilibrium relationship is also an important factor. According to Eq. (9.24), the driving force for transfer of A from liquid 1 to liquid 2 is the difference between the bulk concentration C_{A1} and *the concentration of A in liquid 1 that would be in equilibrium with bulk concentration C_{A2} in liquid 2*. Similarly, the driving force for mass transfer according to Eq. (9.25) is the difference between C_{A2} and the concentration of A in liquid 2 that would be in equilibrium with C_{A1} in liquid 1.

9.4.3 Gas–Liquid Mass Transfer

Gas–liquid mass transfer is of great importance in bioprocessing because aerobic cell cultures require O_2. Transfer of a solute such as O_2 from gas to liquid phase is analyzed in a similar manner as previously presented for liquid–solid and liquid–liquid mass transfer.

Figure 9.5 shows the relevant factors at an interface between gas and liquid phases containing component A. Let us assume that A is transferred from the gas phase into the liquid phase. The concentration of A in the bulk liquid phase is C_{AL} and C_{ALi} at the interface. The concentration of A in the bulk gas phase is C_{AG} and C_{AGi} at the interface.

From Eq. (9.5), the rate of mass transfer of A through the gas boundary layer is:

$$N_{AG} = k_G a \left(C_{AG} - C_{AGi} \right)$$

(9.26)

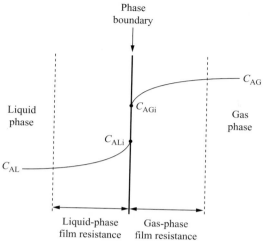

Figure 9.5 Concentration gradients for gas–liquid mass transfer.

and the rate of mass transfer of A through the liquid boundary layer is:

$$N_{AL} = k_L a \left(C_{ALi} - C_{AL} \right) \tag{9.27}$$

where k_G is the gas-phase mass transfer coefficient and k_L is the liquid-phase mass transfer coefficient. To eliminate C_{AGi} and C_{ALi}, we must manipulate the equations as discussed in Section 9.4.2.

If we assume that equilibrium exists at the interface, C_{AGi} and C_{ALi} can be related using the partition coefficient, which is related to the Henry's law coefficient. Therefore, we can write:

$$C_{AGi} = m C_{ALi} \tag{9.28}$$

or, alternatively:

$$C_{ALi} = \frac{C_{AGi}}{m} \tag{9.29}$$

where m is the partition coefficient. These equilibrium relationships can be incorporated into Eqs. (9.26) and (9.27) at steady state using procedures that parallel those applied previously for liquid–liquid mass transfer. The results are similar to Eqs. (9.20) and (9.21):

$$N_A \left(\frac{1}{k_G a} + \frac{m}{k_L a} \right) = C_{AG} - m C_{AL} \tag{9.30}$$

$$N_A \left(\frac{1}{m k_G a} + \frac{1}{k_L a} \right) = \frac{C_{AG}}{m} - C_{AL} \tag{9.31}$$

The combined mass transfer coefficients in Eqs. (9.30) and (9.31) can be used to define over-all mass transfer coefficients. The *overall gas-phase mass transfer coefficient* K_G is defined by the equation:

$$\frac{1}{K_G a} = \frac{1}{k_G a} + \frac{m}{k_L a} \tag{9.32}$$

and the overall liquid-phase mass transfer coefficient K_L is defined as:

$$\frac{1}{K_L a} = \frac{1}{m k_G a} + \frac{1}{k_L a} \tag{9.33}$$

The rate of mass transfer in gas–liquid systems can therefore be expressed using either of two equations:

$$N_A = K_G a \left(C_{AG} - m C_{AL} \right) \tag{9.34}$$

or

$$N_A = K_L a \left(\frac{C_{AG}}{m} - C_{AL} \right) \tag{9.35}$$

Equations (9.34) and (9.35) are usually expressed using equilibrium concentrations. $m C_{AL}$ is often written as C_{AG}^*, the gas-phase concentration of A in equilibrium with C_{AL}, and C_{AG}/m is often written as C_{AL}^*, the liquid-phase concentration of A in equilibrium with C_{AG}. Equations (9.34) and (9.35) become:

$$N_A = K_G a \left(C_{AG} - C_{AG}^* \right) \tag{9.36}$$

and

$$N_A = K_L a \left(C_{AL}^* - C_{AL} \right) \tag{9.37}$$

Equations (9.36) and (9.37) can be simplified for systems where most of the resistance to mass transfer lies in either the gas-phase interfacial film or the liquid-phase interfacial film. When solute A is very soluble in the liquid, for example, in transfer of ammonia to water, the liquid-side resistance is small compared to the resistance on the gas-side. From Eq. (9.7), if the liquid-side resistance is small, $k_L a$ must be relatively large. From Eq. (9.32), $K_G a$ is then approximately equal to $k_G a$. Using this result in Eq. (9.36) gives:

$$N_A = k_G a \left(C_{AG} - C_{AG}^* \right) \tag{9.38}$$

Conversely, if A is poorly soluble in the liquid (e.g., O_2 in aqueous solution), the liquid-phase mass transfer resistance dominates and $k_G a$ is much larger than $k_L a$. From Eq. (9.33), this means that $K_L a$ is approximately equal to $k_L a$ and Eq. (9.37) can be simplified to:

$$N_A = k_L a \left(C_{AL}^* - C_{AL} \right) \tag{9.39}$$

We will make further use of Eq. (9.39) in subsequent sections of this chapter because gas–liquid O_2 transfer plays a crucial role in many bioprocesses. Obtaining the experimental values of C_{AL} and C_{AL}^* for use in Eq. (9.39) is reasonably straightforward. However, as described in relation to liquid–liquid mass transfer in Section 9.4.2, it can be difficult to estimate the interfacial area a. The interfacial area when gas is sparged through a liquid will depend on the size and number of bubbles present, which in turn depend on many other factors such as medium composition, stirrer speed, and gas flow rate. Because k_L is also affected by these parameters, k_L and a are usually combined together and the combined term $k_L a$ referred to as the mass transfer coefficient.

9.5 O_2 Uptake in Cell Cultures

Cells in aerobic cultures take up O_2 from the liquid phase. The rate of O_2 transfer from gas to liquid is therefore of prime importance, especially in dense or fast-growing cell cultures where the demand for dissolved O_2 is high. An expression for the rate of O_2 transfer from gas to liquid is given by Eq. (9.39): N_A is the rate of O_2 transfer per unit volume of fluid (gmol m^{-3} s^{-1}), k_L is the liquid-phase mass transfer coefficient (m s^{-1}), a is the gas–liquid interfacial area per unit volume of fluid (m^2 m^{-3}), C_{AL} is the O_2 concentration in the culture broth (gmol m^{-3}), and C_{AL}^* is the O_2 concentration in the broth that would be in equilibrium with the gas phase (gmol m^{-3}). The equilibrium concentration C_{AL}^* is also known as the *saturation concentration* or *solubility* of O_2 in the broth. C_{AL}^* represents the maximum possible O_2 concentration that can occur in the liquid under the current gas-phase composition, temperature, and pressure. The difference $\left(C_{AL}^* - C_{AL} \right)$ between the maximum possible and actual O_2 concentrations in the liquid represents the *driving force* for mass transfer.

The solubility of O_2 in aqueous solutions at ambient temperature and pressure is less than 10 ppm (10 mg L^{-1}). This quantity of O_2 can be quickly consumed in aerobic cultures and must be replenished constantly by gas sparging. For perspective, typical cell broth uses an initial glucose concentration in the range of 10 to 100 g L^{-1} which is 3 to 4 orders of magnitude higher than the solubility of O_2 highlighting the challenge of O_2 and O_2 transfer. An actively respiring cell population can consume the entire O_2 content of a culture broth within a few seconds; therefore, the maximum amount of O_2 that can be dissolved in the medium must be transferred from the gas phase 10 to 15 times per minute. This is no easy task because the low solubility of O_2 in aqueous solutions guarantees that the concentration difference $\left(C_{AL}^* - C_{AL} \right)$ is always very small. Design of bioreactors for aerobic culture must take these factors into account and provide optimum mass transfer conditions.

9.5.1 Factors Affecting Cellular O_2 Demand

The rate O_2 is consumed by cells determines the rate at which O_2 must be transferred from the gas phase to the liquid phase. Many factors influence O_2 demand: the most important include cell species, culture growth phase, culture growth rate, and the nature of the carbon source provided in the medium. In batch culture, the rate of O_2 uptake varies with time. First, the concentration of cells increases with time during batch cultivation and the total rate of O_2 consumption is proportional to the number of cells present. In addition, the rate of O_2 consumption per cell, known as the *specific O_2 uptake rate*, can also vary with time, often reaching a maximum during the early stages of cell growth. If Q_{O2} is the O_2 uptake rate per volume of broth and q_{O2} is the specific O_2 uptake rate:

$$Q_{O2} = q_{O2}x \tag{9.40}$$

where x is cell concentration. Typical units for q_{O2} are gmol O_2 (g cdw)$^{-1}$ h^{-1}, and for Q_{O2}, gmol O_2 L^{-1} h^{-1}. Typical profiles of Q_{O2}, q_{O2}, and x during batch culture of microbial, plant, and animal cells are shown in Figure 9.6.

The inherent demand of an organism for O_2 (q_{O2}) depends on the physiological nature of the cell and its nutritional environment. However, when the level of dissolved O_2 in the medium falls below a certain point, the specific rate of O_2 uptake is also dependent on the O_2 concentration in the liquid, C_{AL}. The dependence of q_{O2} on C_{AL} is shown in Figure 9.7. If C_{AL} is above the *critical O_2 concentration* C_{crit}, q_{O2} is a constant maximum and independent of C_{AL}. If C_{AL} is below C_{crit}, q_{O2} is approximately linearly dependent on O_2 concentration. The nonlinear relationship can be interpreted in terms of Michaelis-Menten enzyme kinetics and the zeroth and first-order kinetics approximations at high and low substrates concentrations, relative to the half-saturation constant K_M, respectively.

O_2 limitations can be eliminated, allowing cell metabolism to function at its fastest, by keeping the dissolved O_2 concentration in the bioreactor above C_{crit}. The exact value of C_{crit} depends on the organism, but it usually falls between 5% and 10% of air saturation under typical operating conditions.

The substrates used in the growth medium can also affect the O_2 demand by affecting culture growth rates. For example, maximum O_2 consumption rates of 5.5, 6.1, and 12 mmol L^{-1} h^{-1} have been observed for *Penicillium* cultures growing on lactose, sucrose, and glucose, respectively [5]. O_2 requirements for cell growth also depend on the degree of reduction of the substrate as discussed in Section 5.8. The specific O_2 demand is greater for carbon substrates with higher degrees of reduction as more O_2 is required to balance the flux of electrons. Therefore, specific O_2 uptake rates tend to be higher in cultures growing on alcohol or alkane hydrocarbons as compared with carbohydrates.

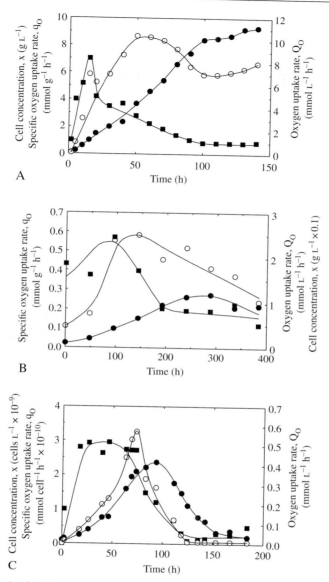

Figure 9.6 Variations in the O_2 uptake rate Q_O (O), the specific O_2 uptake rate q_O (■), and the cell concentration x (●) during batch culture of microbial, plant, and animal cells.

Typical maximum q_{O2} and Q_{O2} values observed during batch culture of various organisms are listed in Table 9.1. Plant and animal cells generally have lower rates of O_2 consumption than microbial cells because their specific growth rates tend to be much smaller.

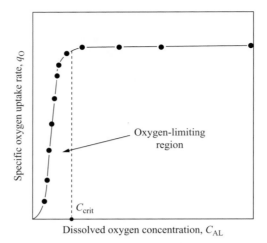

Figure 9.7 Relationship between the specific rate of O_2 consumption by cells and dissolved O_2 concentration.

Table 9.1: Typical O_2 Uptake Rates in Different Cell Cultures

Type of cell culture	Carbon source	Maximum specific O_2 uptake rate, qO_2 (mmol (g cdw)$^{-1}$ h^{-1})	Maximum specific O_2 uptake rate, qO_2 (mmol cell^{-1} h^{-1})	Maximum volumetric O_2 uptake rate, QO_2 (mmol L^{-1} h^{-1})	Reference
MICROBIAL					
Aerobacter aerogenes	Peptone	–	3.2×10^{-11}	7.4	[6]
Aspergillus niger	Glucose	1.6	–	8.8	[7]
Bacillus subtilis	Peptone	–	1.5×10^{-10}	–	[8]
Beneckea natriegens	*n*-Propanol	12	–	6.0	[9]
Escherichia coli	Peptone	–	3.2×10^{-11}	5.0	[6]
Penicillium chrysogenum	Lactose	1.2	–	30	[10]
Saccharomyces cerevisiae	Ethanol	10	–	40	[11]
Streptomyces aureofaciens	Corn starch	7.0	–	10	[12]
Streptomyces coelicolor	Glucose	7.4	–	5.5	[13]
Streptomyces griseus	Meat extract	4.1	–	16	[14]
Xanthomonas campestris	Glucose	4.5	–	11	[15]

(Continued)

Table 9.1: Typical O$_2$ Uptake Rates in Different Cell Cultures—cont'd

Type of cell culture	Carbon source	Maximum specific O$_2$ uptake rate, qO$_2$ (mmol (g cdw)$^{-1}$h^{-1})	Maximum specific O$_2$ uptake rate, qO$_2$ (mmol cell^{-1}h^{-1})	Maximum volumetric O$_2$ uptake rate, QO$_2$ (mmol L^{-1}h^{-1})	Reference
		PLANT			
Catharanthus roseus	Sucrose	0.45	–	2.7	[16]
Nicotiana tabacum	Sucrose	0.90	–	1.0	[17]
		ANIMAL			
Chinese hamster ovary (CHO)	Glucose/ glutamine	–	2.9×10^{-10}	0.60	[18]
Hybridoma	Glucose/ glutamine	–	2.9×10^{-10}	0.57	[19]

9.5.2 O$_2$ Transfer From Gas Bubble to Cell

O$_2$ molecules must overcome a series of transport resistances before being utilized by respiring cells. Eight mass transfer steps involved in transport of O$_2$ from the interior of gas bubbles to the site of intracellular reaction are represented in Figure 9.8. They are:

1. Transfer from the interior of the bubble to the gas–liquid interface
2. Movement across the gas–liquid interface
3. Diffusion through the liquid film (boundary layer) surrounding the bubble
4. Transport through the bulk liquid
5. Diffusion through the liquid film (boundary layer) surrounding the cells
6. Movement across the liquid–cell interface
7. If the cells are in a floc, clump, or solid particle, diffusion through the solid to the individual cell
8. Transport through the cytoplasm to the site of reaction

Note, transport through the gas boundary layer on the inside of the bubble has been neglected; this is based on the small resistance to mass transfer due to the large O$_2$ diffusion coefficient in the gas phase relative to the aqueous phase and the low solubility of O$_2$ in aqueous solutions. The liquid-film resistance dominates gas–liquid mass transfer (see Section 9.4). If the cells are individually suspended in liquid rather than in a clump, step (7) disappears.

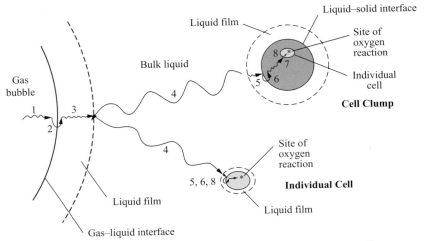

Figure 9.8 Steps for transfer of O_2 from gas bubble to cell.

The relative magnitudes of the various mass transfer resistances depend on the composition and rheological properties of the liquid, the mixing intensity, the size of the bubbles, the size of any cell clumps, interfacial adsorption characteristics, and other factors. However, for most bioreactors, the following analysis is valid.

1. Transfer through the bulk gas phase in the bubble is relatively fast.
2. The gas–liquid interface itself contributes negligible resistance.
3. The liquid film around the bubbles is a major resistance to O_2 transfer.
4. In a well-mixed bioreactor, concentration gradients in the bulk liquid are minimized and mass transfer resistance in this region is small. However, rapid mixing can be difficult to achieve in viscous culture broths; if this is the case, O_2 transfer resistance in the bulk liquid may be important.
5. Single cells are much smaller than gas bubbles; therefore, the liquid film surrounding each cell is much thinner than that around the gas bubbles. The mass transfer resistance from liquid films surrounding individual cell can generally be neglected. However, the liquid-film resistance around large clumps of cells can be significant.
6. Resistance at the liquid–cell interface is negligible.
7. When the cells are in clumps, intraparticle resistance is likely to be significant as O_2 must diffuse through the solid pellet to reach the cells in the interior. The magnitude of this resistance depends on the size and properties of the cell clumps.
8. Intracellular O_2 transfer resistance is negligible because of the small distances involved.

When cells are dispersed in the liquid and the bulk culture broth is well mixed, *the major resistance to O_2 transfer is the liquid film surrounding the gas bubbles*. Transport through this film becomes the rate-limiting step in the complete process and controls the overall mass transfer rate. Consequently, the rate of O_2 transfer from the bubbles to the cells is dominated by the rate of step (3). The mass transfer rate for this step is represented by Eq. (9.39).

At steady state, there can be no accumulation of O_2 at any location in the bioreactor; therefore, the rate of O_2 transfer from the bubbles must be equal to the rate of O_2 consumption by the cells. If we make N_A in Eq. (9.39) equal to Q_{O2} in Eq. (9.40), we obtain the following equation:

$$k_L a \left(C^*_{AL} - C_{AL} \right) = q_{O2} x \tag{9.41}$$

The mass transfer coefficient $k_L a$ is used to characterize the O_2 transfer capability of bioreactors. If $k_L a$ for a particular system is small, the ability of the reactor to deliver O_2 to the cells is limited. We can predict the response of a system to changes in mass transfer conditions using Eq. (9.41). For example, if the rate of cell metabolism remains unchanged but $k_L a$ is increased (e.g., by raising the stirrer speed to reduce the thickness of the boundary layer around the bubbles), the dissolved O_2 concentration C_{AL} must rise in order for the left side of Eq. (9.41) to remain equal to the right side. Similarly, if the rate of O_2 consumption by the cells accelerates while $k_L a$ is unaffected, C_{AL} must decrease.

We can use Eq. (9.41) to deduce some important relationships for bioreactors. First, let us estimate the maximum cell concentration that can be supported by the bioreactor's O_2 transfer system. For a given set of operating conditions, the maximum rate of O_2 transfer occurs when the concentration-difference driving force $\left(C^*_{AL} - C_{AL} \right)$ is highest—that is, when the concentration of dissolved O_2 C_{AL} is, for practical purposes, ~0 mg L^{-1}. Therefore from Eq. (9.41), the maximum cell concentration that can be supported by O_2 transfer in a batch bioreactor at quasi-steady state is:

$$x_{max} = \frac{k_L a \, C^*_{AL}}{q_{O2}} \tag{9.42}$$

It is generally undesirable for cell density to be limited by the rate of mass transfer. Therefore, if x_{max} estimated using Eq. (9.42) is lower than the cell concentration required in the cultivation process, $k_L a$ must be improved. Note that the cell concentration in Eq. (9.42) is a theoretical maximum corresponding to operation of the system at its maximum O_2 transfer rate.

Comparison of x_{max} values evaluated using Eqs. (8.54) and (9.42) can be used to gauge the relative effectiveness of heat and mass transfer processes in aerobic cultivations. For example, if x_{max} from Eq. (9.42) is small while x_{max} calculated from heat transfer considerations is large, we would know that mass transfer is more likely to limit biomass productivity than heat transfer. If both x_{max} values are greater than the target x for the process, heat and mass transfer can be considered adequate.

Equation (9.42) is a useful hypothetical relationship; however, as indicated in Figure 9.7, operation of culture systems at a dissolved O_2 concentration of ~ 0 mg L^{-1} is not advisable because the specific O_2 uptake rate depends on O_2 concentration in a manner analogous to

Michaelis-Menten kinetics. Accordingly, another important parameter is the minimum $k_L a$ required to maintain $C_{AL} > C_{crit}$ in the bioreactor. This can be determined from Eq. (9.41) as:

$$(k_L a)_{crit} = \frac{q_{O2} x}{\left(C_{AL}^* - C_{crit}\right)} \tag{9.43}$$

Example 9.1 illustrates the relationship between the maximum biomass concentration, the specific O_2 consumption rate, and the bioreactor O_2 mass transfer rate.

It is important to know the actual $k_L a$ realized by a bioreactor to assess the O_2 transfer capability for an application. Methods for measuring $k_L a$ in bioprocesses are outlined in Section 9.10. Application of Eqs. (9.42) and (9.43) also requires knowledge of the O_2 solubility C_{AL}^* and the specific O_2 uptake rate q_{O2}. Evaluation of these parameters is described in Sections 9.8 and 9.11.

EXAMPLE 9.1 Cell Concentration in Aerobic Culture

A strain of *Azotobacter vinelandii* is cultured in a 15-m³ stirred bioreactor for alginate production. Under current operating conditions, $k_L a$ is $0.17\,s^{-1}$. The solubility of O_2 in the broth is approximately $8 \times 10^{-3}\,kg\ m^{-3}$.

a. The specific rate of O_2 uptake is 12.5 mmol (g cdw)$^{-1}$ h^{-1}. What is the maximum cell concentration supported by O_2 transfer in the bioreactor?
b. The bacteria suffer growth inhibition after copper sulfate is accidentally added to the culture broth just after the start of the culture. This causes a reduction in the O_2 uptake rate to 3 mmol (g cdw)$^{-1}$ h^{-1}. What maximum cell concentration can now be supported by O_2 transfer in the bioreactor?

Solution

a. From Eq. (9.42):

$$x_{max} = \frac{0.17 s^{-1} \left(8 \times 10^{-3}\,kg\ m^{-3}\right)}{\dfrac{12.5\ mmol}{g\ h} \cdot \left|\dfrac{1\,h}{3600\,s}\right| \cdot \left|\dfrac{1\,gmol}{1000\,mmol}\right| \cdot \left|\dfrac{32\,g}{1\,gmol}\right| \cdot \left|\dfrac{1\,kg}{1000\,g}\right|}$$

$$x_{max} = 1.2 \times 10^4\,g\ m^{-3} = 12\ g\ L^{-1}$$

The maximum cell concentration supported by O_2 transfer in the bioreactor is $12\,g\,L^{-1}$.

b. Assume that addition of copper sulfate does not affect C_{AL}^* or $k_L a$.

$$x_{max} = \frac{0.17\ s^{-1} \left(8 \times 10^{-3}\,kg\ m^{-3}\right)}{\dfrac{3\ mmol}{g\ h} \cdot \left|\dfrac{1\,h}{3600\,s}\right| \cdot \left|\dfrac{1\,gmol}{1000\,mmol}\right| \cdot \left|\dfrac{32\,g}{1\,gmol}\right| \cdot \left|\dfrac{1\,kg}{1000\,g}\right|}$$

$$x_{max} = 5.0 \times 10^4\,g\ m^{-3} = 50\ g\ L^{-1}$$

The maximum cell concentration supported by O_2 transfer in the bioreactor after addition of copper sulfate is $50\,g\,L^{-1}$.

9.6 Factors Affecting O_2 Transfer in Bioreactors

The rate of O_2 transfer in culture broths is influenced by several physical and chemical factors that change the value of k_L, the value of a, or the driving force for mass transfer $\left(C_{AL}^* - C_{AL} \right)$. As a general guideline, k_L in culture liquids is about 1 to 4×10^{-4} m s^{-1}. If substantial improvement in mass transfer rates is required, it is usually most productive to focus on increasing the interfacial area a. Operating values of the combined coefficient $k_L a$ can span three orders of magnitude in bioreactors; this is due mainly to the large variation in a. Production-scale bioreactors typically have $k_L a$ values in the range 0.02 s^{-1} to 0.25 s^{-1}.

In this section, several aspects of bioreactor design and operation are discussed in terms of their effect on O_2 mass transfer.

9.6.1 Bubbles

The efficiency of gas–liquid mass transfer depends to a large extent on the characteristics of the bubbles dispersed in the liquid medium. Bubble behavior exerts a strong influence on the value of $k_L a$: some properties of bubbles affect mainly the magnitude of k_L, whereas others change the interfacial area a.

Large-scale, aerobic cultures are carried out most commonly in stirred bioreactors. In these vessels, O_2 is supplied to the medium by sparging air bubbles underneath the impeller. The action of the impeller then creates a dispersion of gas bubbles throughout the vessel. In small laboratory-scale bioreactors, all the liquid is close to the impeller; therefore, bubbles in these systems interact frequently with turbulent liquid currents. In contrast, bubbles in most industrial bioreactors spend a large proportion of their time floating relatively free and unimpeded through the liquid after initial dispersion at the impeller. Liquid in large bioreactors away from the impeller does not possess sufficient energy for continuous break-up of bubbles. This is a consequence of scale; most laboratory bioreactors operate with stirrer power between 10 and 20 kW m^{-3}, whereas large, agitated vessels operate at 0.5 to 5 kW m^{-3}. Consequently, virtually all large commercial-size, stirred tank reactors operate primarily in the free-bubble-rise regime [20].

The most important property of air bubbles in bioreactors is their size. More interfacial area a is provided, for a given volume of gas, if the gas is dispersed into many small bubbles rather than a few large ones. Therefore, a major goal in bioreactor design is a high level of gas dispersion. However, there are other important benefits associated with small bubbles. Small bubbles have correspondingly slow bubble-rise velocities; consequently they stay in the liquid longer, allowing more time for O_2 mass transfer. Small bubbles therefore create high *gas hold-up*, which is defined as the fraction of the working volume of the reactor occupied by entrained gas:

$$\varepsilon = \frac{V_G}{V_T} = \frac{V_G}{V_L + V_G}$$

(9.44)

where ε is the gas hold-up, V_T is the total fluid volume (gas + liquid), V_G is the volume of gas bubbles in the reactor, and V_L is the volume of liquid. High mass transfer rates are achieved at high gas hold-ups because the total interfacial area for O_2 transfer depends on the total volume of gas in the system as well as on the average bubble size [21]. Gas hold-up values are very difficult to predict and may be anything from very low (0.01) up to a maximum in commercial-scale stirred bioreactors of about 0.2.

To summarize the influence of bubble size on O_2 transfer, small bubbles are generally beneficial because they provide higher gas hold-ups and greater interfacial surface area compared with large bubbles. However, k_L for bubbles less than about 2 to 3 mm in diameter is reduced due to surface effects. Very small bubbles ≪1 mm should be avoided, especially in viscous broths.

9.6.2 Sparging, Stirring, and Medium Properties

In this section, we consider the physical processes in bioreactors and system properties that affect bubble size and the magnitude of $k_L a$.

Bubble Formation

Air bubbles are formed at the sparger in bioreactors. Several types of spargers can be used. *Porous spargers* of sintered metal, glass, or ceramic are applied mainly in small-scale systems; gas throughput is limited because the sparger poses a high resistance to flow. *Orifice spargers*, also known as perforated pipes, are constructed by making small holes in piping that is then fashioned into a ring or cross and placed in the reactor. *Point* or *nozzle spargers* are used in many agitated bioreactors from laboratory to production scale. These spargers consist of a single open pipe or partially closed pipe providing a point-source stream of air bubbles. Advantages compared with other sparger designs include low resistance to gas flow and small risk of blockage.

Gas Dispersion

The two-phase flow patterns in stirred vessels with gassing have been described in Chapter 7. The effectiveness of bubble break-up and dispersion depends on the relative rates of stirring and gas flow; the balance between these operating parameters determines whether *impeller flooding*, *impeller loading*, or *complete gas dispersion* occurs. Flooding should be avoided and occurs when the impeller is surrounded by gas and no longer contacts the liquid properly, resulting in poor mixing and gas dispersion.

Gas dispersion in stirred vessels takes place mainly in the immediate vicinity of the impeller. Because bubbles formed at the sparger are immediately drawn into the impeller zone, dispersion of gas in stirred vessels is largely independent of sparger design. Gas is increasingly recirculated around the tank with increased stirrer speed.

Bubble Coalescence

Coalescence of small bubbles into bigger bubbles generally reduces O_2 transfer because the total interfacial area and gas hold-up are reduced. The coalescence properties of liquids depend on the liquid composition. The presence of salts and ions suppresses coalescence as compared with pure water. This is an advantage for O_2 transfer. The addition of ions to water in sparged vessels markedly reduces the average bubble size and increases the gas hold-up, so much so that the interfacial area a in water containing salts may be up to 10 times greater than that obtained without salts [22].

Experimental results for the effect of solution composition on $k_L a$ are shown in Figure 9.9. The presence of solutes has a significant impact on the rate of O_2 transfer. The results for water and 5% Na_2SO_4 salt in water illustrate the effect of liquid coalescence properties: $k_L a$ is lower in water than in noncoalescing salt solution. The composition and therefore the coalescence properties of culture broths vary with time during cell cultures and $k_L a$ can also be expected to vary accordingly.

Viscosity

The rheology of fluids has a significant effect on bubble size, gas hold-up, and $k_L a$. The thickness of the fluid boundary layers surrounding the bubbles increases with increasing viscosity, thus increasing the resistance to O_2 transfer. High viscosity also dampens turbulence, changes the size and structure of the ventilated cavities at the impeller blades, and reduces the effectiveness of gas dispersion. $K_L a$ typically decreases with increasing liquid viscosity [22]. As an example, the $k_L a$ in a pseudoplastic carboxymethyl cellulose solution is significantly lower than in water or salt solution at the same bioreactor power input as shown in Figure 9.9.

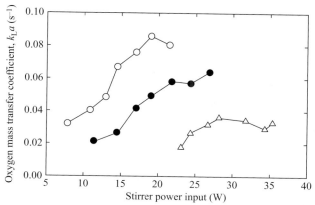

Figure 9.9 Effect of solution composition on $k_L a$ in a stirred tank at constant gas flow rate: (●) water (coalescing); (○) 5% Na_2SO_4 in water (noncoalescing); and (△) 0.7% w/w carboxymethyl cellulose in water (viscous, pseudoplastic). *Data from S.J. Arjunwadkar, K. Sarvanan, P.R. Kulkarni, and A.B. Pandit, 1998, Gas–liquid mass transfer in dual impeller bioreactor. Biochem. Eng. J. 1, 99–106.*

Stirrer Speed and Gas Flow Rate

Increasing the stirrer speed and gas flow rate improves the value of $k_L a$ under normal bioreactor operation. Typical data for $k_L a$ in Newtonian fluids of varying viscosity are shown in Figure 9.10. The strong dependence of $k_L a$ on stirrer speed is evident from Figure 9.10(a): in this system, doubling the stirrer speed N_i resulted in an average 4.6-fold increase in $k_L a$. Increasing the gas flow rate is a less effective strategy for improving $k_L a$: for example, in the system represented in Figure 9.10(b), doubling the gas flow rate increased $k_L a$ by only about 20%. Moreover, in most systems there is limited practical scope for increasing $k_L a$ by increasing the gas flow: impeller flooding occurs at high gas flow rates unless the impeller can disperse all the gas impinging on it. At very high gassing rates, the liquid contents can be blown out of the bioreactor.

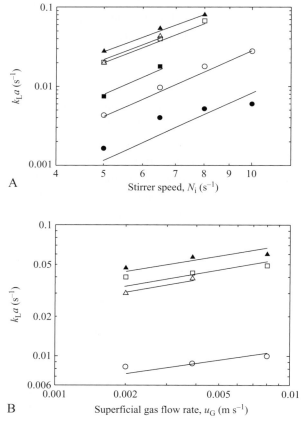

Figure 9.10 Dependence of $k_L a$ on operating conditions in a stirred tank. The data are plotted using logarithmic coordinates. (A) Effect of stirrer speed N_i at constant gas velocity. (B) Effect of gas flow rate u_G at constant stirrer speed. The symbols represent Newtonian fluids with viscosities: ($\triangle$) 0.91 mPa s; ($\blacktriangle$) 1.3 mPa s; ($\square$) 2.1 mPa s; ($\blacksquare$) 5.1 mPa s; ($\bigcirc$) 13.3 mPa s; and ($\bullet$) 70.2 mPa s. *From H. Yagi and F. Yoshida, 1975, Gas absorption by Newtonian and non-Newtonian fluids in sparged agitated vessels. Ind. Eng. Chem. Process Des. Dev. 14, 488–493.*

9.6.3 Antifoam Agents

Most cell cultures produce a variety of foam-producing and foam-stabilizing agents, such as proteins, polysaccharides, and fatty acids. Foaming is exacerbated by high gas flow rates and high stirrer speeds. Foaming causes a range of reactor operating problems; foam control is therefore an important consideration in bioreactor design. Foam overflowing from the bioreactor provides an entry route for contaminating organisms and can cause blockage of outlet gas lines. Additionally, fragile cells can be damaged by collapsing foam. A space of 20% to 30% of the tank volume must be left between the top of the liquid and the vessel headplate when setting up bioreactors to accommodate foam layers as well as the increase in fluid volume in aerated vessels due to gas hold-up.

Addition of special antifoam compounds to the medium is a common method of reducing foam build-up. However, antifoam agents affect the surface chemistry of bubbles and their tendency to coalesce and have a significant effect on $k_L a$. Most antifoam agents are strong surface-tension-lowering substances. Decrease in surface tension reduces the average bubble diameter, thus producing higher values of a. However, this is countered by a reduction in the mobility of the gas–liquid interface, which lowers the value of k_L. With most silicon-based antifoams, the decrease in k_L is generally larger than the increase in a so that, overall, $k_L a$ is reduced [24,25]. Typical data for surface tension and $k_L a$ as a function of antifoam concentration are shown in Figure 9.11. In this experiment, a reduction in $k_L a$ of almost 50% occurred after addition of only a small amount of antifoam.

Mechanical foam breakers avoid changing the properties of the liquid. Mechanical foam breakers, such as high-speed discs rotating at the top of the vessel and centrifugal foam destroyers, are suitable when foam development is moderate. However, some of these devices need large quantities of power to operate, especially in commercial-scale vessels; in addition, their limited foam-destroying capacity is a problem with highly foaming cultures.

9.6.4 Temperature

The temperature of aerobic bioprocess affects the solubility of O_2 C_{AL}^*, liquid viscosity, and the mass transfer coefficient k_L. Increasing the temperature causes C_{AL}^* to drop, so that the driving force for mass transfer $\left(C_{AL}^* - C_{AL} \right)$ is reduced. At the same time, the diffusivity of O_2 in the liquid film surrounding the bubbles is increased, resulting in an increase in k_L. The net effect of temperature on O_2 transfer depends on the range of temperature considered. For temperatures between 10°C and 40°C, which includes the operating range for most bioreactors, an increase in temperature is more likely to increase the rate of O_2 transfer. Above 40°C, the solubility of O_2 drops significantly, adversely affecting the driving force for mass transfer.

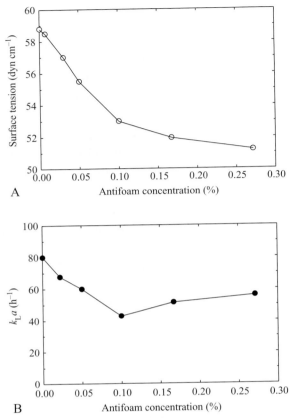

Figure 9.11 Effect of antifoam concentration on: (A) surface tension, and (B) $k_L a$, in a *Penicillium chrysogenum* culture broth. *Data from F.H. Deindoerfer and E.L. Gaden, 1955, Effects of liquid physical properties on oxygen transfer in penicillin fermentation. Appl. Microbiol. 3, 253–257; with unit conversion from R.K. Finn, 1954, Agitation–aeration in the laboratory and in industry. Bact. Rev. 18, 254–274.*

9.6.5 O_2 Partial Pressure

The O_2 partial pressure used to aerate bioreactors affects the value of C_{AL}^*. The equilibrium relationship between the O_2 partial pressure and C_{AL}^* O_2 is given by *Henry's law*:

$$p_{AG} = p_T\, y_{AG} = H\, C_{AL}^* \tag{9.45}$$

where p_{AG} is the *partial pressure* of component A in the gas phase, p_T is the total gas pressure, y_{AG} is the mole fraction of A in the gas phase, and C_{AL}^* is the solubility of component A in the liquid. H is *Henry's constant*, which is a function of temperature. If the partial pressure of O_2 in the gas increases at constant temperature, C_{AL}^* and therefore the mass transfer driving force $\left(C_{AL}^* - C_{AL} \right)$ increase.

O_2-enriched air or pure O_2 may be used instead of air to improve C^*_{AL} and O_2 transfer. The effect on O_2 solubility can be determined using Henry's law. At a fixed temperature, such that H remains constant, p_T and y_{AG} vary from condition 1 to condition 2 as:

$$\frac{C^*_{AL2}}{C^*_{AL1}} = \frac{p_{AG2}}{p_{AG1}} = \frac{p_{T2}\, y_{AG2}}{p_{T1}\, y_{AG1}}$$

(9.46)

According to the *International Critical Tables* [26], the mole fraction y_{AG1} of O_2 in air is 0.2099. If pure O_2 is used instead of air, y_{AG2} is 1. From Eq. (9.46), if the gases are applied at the same total pressure:

$$\frac{C^*_{AL2}}{C^*_{AL1}} = \frac{1}{0.2099} = 4.8$$

(9.47)

Therefore, sparging pure O_2 instead of air at the same total pressure and temperature increases the solubility of O_2 by a factor of 4.8. Alternatively, the solubility can be increased by sparging compressed air at higher total pressure p_T which also increases O_2 partial pressure. Both strategies increase the operating costs of the bioreactor.

9.6.6 Presence of Cells and Macromolecules

O_2 transfer is influenced by the presence of cells in culture broths. The effect depends on the morphology of the organism and the cell concentration. Cells with complex morphology, such as branched hyphae, generally lead to lower O_2 transfer rates by interfering with bubble break-up and promoting coalescence. Cells, proteins, and other molecules that adsorb at gas–liquid interfaces also cause *interfacial blanketing*, which reduces the effective contact area between gas and liquid. Macromolecules and very small particles accumulating at the bubble surface reduce the mobility of the interface, thus lowering k_L, but may also decrease coalescence, thereby increasing a. An example of change in $k_L a$ because of these combined factors is shown in Figure 9.12.

9.7 Measuring Dissolved O_2 Concentration

The concentration of dissolved O_2 in bioreactors is normally measured using a *dissolved O_2* (DO) *electrode*. There are two common types: *galvanic electrodes* and *polarographic electrodes*. Details of the construction and operating principles of O_2 probes can be found in the literature (e.g., [27], 28). Both designs have a membrane permeable to O_2 that separates the bioreactor broth from the electrode. O_2 diffuses through the membrane to the cathode, as illustrated in Figure 9.13, where it reacts to produce a current between the anode and cathode proportional to the O_2 partial pressure. An electrolyte solution in the electrode supplies ions that take part in the reaction and must be replenished at regular intervals. Repeated calibration of dissolved O_2 probes is necessary; fouling by cells attaching to the membrane surface,

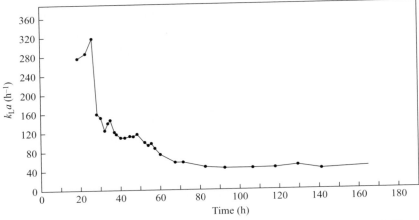

Figure 9.12 Variation in $k_L a$ during a 300-1 batch cultivation of a streptomycete. *From C.M. Tuffile and F. Pinho, 1970, Determination of oxygen transfer coefficients in viscous streptomycete fermentations. Biotechnol. Bioeng. 12, 849–871.*

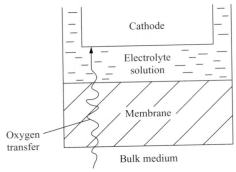

Figure 9.13 Diffusion of O_2 from the bulk liquid to the cathode of an O_2 electrode.

electronic noise due to air bubbles passing close to the membrane, and signal drift are the main operating challenges.

Movement of O_2 from the bulk medium to the cathode in O_2 probes is a mass transfer process. Operation of the probe relies on diffusion of O_2 across the membrane and electrolyte solution. This takes time, so the response of an electrode to sudden changes in dissolved O_2 level is subject to a time delay. This does not affect many applications of O_2 probes in bioreactors, as changes in dissolved O_2 tension during cell culture are normally relatively slow. However, electrode dynamics can have a significant influence on the measurement of O_2 uptake rates and $k_L a$ in culture broths.

Both galvanic and polarographic electrodes measure the O_2 *tension* or partial pressure of dissolved O_2 in the culture broth, not the dissolved O_2 concentration. To convert O_2 tension to dissolved O_2 concentration, it is necessary to know the solubility of O_2 in the liquid at the temperature and pressure of measurement.

9.8 Estimating O_2 Solubility

The concentration difference $\left(C_{AL}^* - C_{AL} \right)$ is the driving force for O_2 mass transfer. Air is used to provide O_2 in most industrial bioprocesses. Values for the solubility of O_2 in water at various temperatures and 1 atm air pressure are listed in Table 9.2. However, cultivations are not carried out using pure water, and the gas composition and pressure can be varied. The presence of dissolved material in the liquid and the O_2 partial pressure in the gas phase affect O_2 solubility.

9.8.1 Effect of O_2 Partial Pressure

O_2 solubility is directly proportional to the partial pressure of O_2 in the gas phase as indicated in Henry's law, Eq. (9.45). The solubility of O_2 in water as a function of these variables can be determined using Eq. (9.45) and the values for Henry's constant listed in Table 9.2.

9.8.2 Effect of Temperature

Variation of O_2 solubility with temperature is shown in Table 9.2 for water in the range 0°C to 40°C. O_2 solubility falls with increasing temperature. The solubility of O_2 from air in pure water between 0°C and 36°C has been correlated using the following equation [29]:

$$C_{AL}^* = 14.161 - 0.3943T + 0.007714T^2 - 0.0000646T^3 \tag{9.48}$$

where C_{AL}^* is O_2 solubility in units of mg L^{-1}, and T is temperature in °C.

Table 9.2: Solubility of O_2 in Water Under 1 atm Air Pressure

Temperature (°C)	O_2 solubility under 1 atm air pressure (kg m^{-3})	Henry's constant (m^3 atm gmol^{-1})
0	1.48×10^{-2}	0.454
10	1.15×10^{-2}	0.582
15	1.04×10^{-2}	0.646
20	9.45×10^{-3}	0.710
25	8.69×10^{-3}	0.774
26	8.55×10^{-3}	0.787
27	8.42×10^{-3}	0.797
28	8.29×10^{-3}	0.810
29	8.17×10^{-3}	0.822
30	8.05×10^{-3}	0.835
35	7.52×10^{-3}	0.893
40	7.07×10^{-3}	0.950

Calculated from data in *International Critical Tables*, 1928, vol. III, McGraw-Hill, New York, p. 257.

9.8.3 Effect of Solutes

The presence of solutes such as salts, acids, and sugars affects the solubility of O_2 in water as indicated in Tables 9.3 and 9.4. These data show that the solubility of O_2 is reduced by the addition of ions and sugars that are normally required in cultivation media. Quicker et al. [30] have developed an empirical correlation to correct values of O_2 solubility in water for the effects of cations, anions, and sugars:

$$\log_{10}\left(\frac{C_{AL0}^*}{C_{AL}^*}\right) = 0.5 \sum_i H_i z_i^2 C_{iL} + \sum_j K_j C_{jL} \frac{n!}{r!(n-r)!} \tag{9.49}$$

where

$C_{AL0}^* = O_2$ solubility at zero solute concentration (mol m^{-3})
$C_{AL}^* = O_2$ solubility in the presence of solutes (mol m^{-3})
H_i = constant for ionic component i (m^3 mol^{-1})
z_i = charge (valence) of ionic component i

Table 9.3: Solubility of Oxygen in Aqueous Solutions at 25°C under 1 atm Oxygen Pressure

Concentration (M)	Oxygen solubility at 25°C under 1 atm oxygen pressure (kg m^{-3})		
	HCl	½ H$_2$SO$_4$	NaCl
0	4.14×10^{-2}	4.14×10^{-2}	4.14×10^{-2}
0.5	3.87×10^{-2}	3.77×10^{-2}	3.43×10^{-2}
1.0	3.75×10^{-2}	3.60×10^{-2}	2.91×10^{-2}
2.0	3.50×10^{-2}	3.28×10^{-2}	2.07×10^{-2}

Calculated from data in *International Critical Tables*, 1928, vol. III, McGraw-Hill, New York, p. 271.

Table 9.4: Solubility of Oxygen in Aqueous Solutions of Sugars under 1 atm Oxygen Pressure

Sugar	Concentration (gmol per kg H$_2$O)	Temperature (°C)	Oxygen solubility under 1 atm oxygen pressure (kg m^{-3})
Glucose	0	20	4.50×10^{-2}
	0.7	20	3.81×10^{-2}
	1.5	20	3.18×10^{-2}
	3.0	20	2.54×10^{-2}
Sucrose	0	15	4.95×10^{-2}
	0.4	15	4.25×10^{-2}
	0.9	15	3.47×10^{-2}
	1.2	15	3.08×10^{-2}

Calculated from data in *International Critical Tables*, 1928, vol. III, McGraw-Hill, New York, p. 272.

Table 9.5: Values of H_i and K_j in Eq. (9.49) at 25°C

Cation	$H_i \times 10^3$ (m³ mol⁻¹)	Anion	$H_i \times 10^3$ (m³ mol⁻¹)	Sugar	$K_j \times 10^3$ (m³ mol⁻¹)
H⁺	−0.774	OH⁻	0.941	Glucose	0.119
K⁺	−0.596	Cl⁻	0.844	Lactose	0.197
Na⁺	−0.550	CO₃²⁻	0.485	Sucrose	0.149*
NH₄⁺	−0.720	SO₄²⁻	0.453		
Mg²⁺	−0.314	NO₃⁻	0.802		
Ca²⁺	−0.303	HCO₃⁻	1.058		
Mn²⁺	−0.311	H₂PO₄⁻	1.037		
		HPO₄²⁻	0.485		
		PO₄³⁻	0.320		

*Approximately valid for sucrose concentrations up to about 200 g L⁻¹.
From A. Schumpe, I. Adler, and W.-D. Deckwer, 1978, Solubility of oxygen in electrolyte solutions. *Biotechnol. Bioeng.* 20, 145–150; and G. Quicker, A. Schumpe, B. König, and W.-D. Deckwer, 1981, Comparison of measured and calculated oxygen solubilities in fermentation media. *Biotechnol. Bioeng.* 23, 635–650.

C_{iL} = concentration of ionic component i in the liquid (mol m⁻³)
K_j = constant for nonionic component j (m³ mol⁻¹)
C_{jL} = concentration of nonionic component j in the liquid (mol m⁻³)

Values of H_i and K_j for use in Eq. (9.49) are listed in Table 9.5. In a typical cultivation medium, the O_2 solubility is between 5% and 25% lower than in water as a result of solute effects.

9.9 Mass Transfer Correlations for O_2 Transfer

There are two approaches to evaluating mass transfer coefficients: calculation using empirical correlations and experimental measurement. It is more practical to evaluate $k_L a$ as a combined term as opposed to a separate determination of k_L and a. In this section, we consider calculation of $k_L a$ using published correlations. Experimental methods for measuring $k_L a$ are described in Section 9.10.

The value of $k_L a$ in bioreactors depends on the fluid properties and the prevailing hydrodynamic conditions. Relationships between $k_L a$ and parameters such as liquid density, viscosity, O_2 diffusivity, bubble diameter, and fluid velocity have been investigated extensively. Theoretically, these correlations allow prediction of mass transfer coefficients based on information gathered from previous experiments. In practice, however, the accuracy of published correlations for $k_L a$ applied to biological systems is generally poor. Most available correlations for O_2 transfer coefficients were determined using air in pure water; however, the presence of additives in water affects the value of $k_L a$ significantly and it is very difficult to make corrections for different liquid compositions.

Although published mass transfer correlations cannot typically be applied directly to bioprocess systems, there is a consensus in the literature about the form of the equations and the relationship between $k_L a$ and reactor operating conditions. The most successful correlations are dimensional equations of the form [22]:

$$k_L a = A \left(\frac{P_T}{V_L} \right)^\alpha u_G^\beta \tag{9.50}$$

where $k_L a$ is the O_2 transfer coefficient, P_T is the total power dissipated, and V_L is the liquid volume. u_G is the *superficial gas velocity*, which is defined as the volumetric gas flow rate divided by the cross-sectional area of the bioreactor. The hydrodynamic effects of flow and turbulence on bubble dispersion and the mass transfer boundary layer are represented by the power term, P_T. The total power dissipated is calculated as the sum of the power input by stirring under gassed conditions (Section 7.5.3) and, if it makes a significant contribution, the power associated with isothermal expansion of the sparged gas (Section 7.6). A, α, and β are constants. The values of α and β are largely insensitive to broth properties and usually fall within the range 0.2 to 1.0. In contrast, the dimensional parameter A varies significantly with liquid composition and is sensitive to the coalescing properties and cell content of culture broths. Because both exponents α and β are typically <1, increasing $k_L a$ by raising either the air flow rate or power input becomes progressively less efficient as the inputs increase.

For viscous and non-Newtonian fluids, a modified form of Eq. (9.50) can be used to incorporate explicitly the effect of viscosity on the mass transfer coefficient:

$$k_L a = B \left(\frac{P_T}{V_L} \right)^\alpha u_G^\beta \mu_a^{-\delta} \tag{9.51}$$

where B is a modified constant reflecting the properties of the liquid other than viscosity, μ_a is the apparent viscosity (Chapter 6), and δ is a constant typically in the range 0.5 to 1.3.

9.10 Measurement of $k_L a$

O_2 transfer coefficients are routinely determined experimentally because of the difficulties associated with using correlations to predict $k_L a$ in bioreactors (Section 9.9). This is not without its own problems, however, as discussed below. Whichever method is used to measure $k_L a$, the measurement conditions should match those applied in the bioreactor. Techniques for measuring $k_L a$ have been reviewed in the literature [22,31,32].

9.10.1 O_2 Balance Method

The steady-state O_2 balance method is the most reliable procedure for estimating $k_L a$ and allows determination from a single-point measurement. Importantly, this method can be applied to bioreactors during normal operation. It is strongly dependent, however, on accurate measurement of the inlet and outlet gas composition, flow rate, pressure, and temperature. Considerations for the design and operation of laboratory equipment to ensure accurate results are described by Brooks et al. [33]

To determine $k_L a$, the O_2 contents of the gas streams flowing to and from the bioreactor are measured. From a mass balance at steady state:

$$N_A = \frac{1}{V_L}\left[\left(F_g C_{AG}\right)_i - \left(F_g C_{AG}\right)_o\right]$$

(9.52)

where N_A is the volumetric rate of O_2 transfer, V_L is the volume of liquid in the bioreactor, F_g is the volumetric gas flow rate, C_{AG} is the gas-phase concentration of O_2, and subscripts i and o refer to the inlet and outlet gas streams, respectively. The first bracketed term on the right side represents the rate at which O_2 enters the bioreactor in the inlet gas stream; the second term is the rate at which O_2 leaves. The difference between them is the transfer rate of O_2 from the gas into the liquid phase. The units of O_2 concentration can be converted using the ideal gas law and incorporated into Eq. (9.52) to obtain an alternative expression:

$$N_A = \frac{1}{R V_L}\left[\left(\frac{F_g p_{AG}}{T}\right)_i - \left(\frac{F_g p_{AG}}{T}\right)_o\right]$$

(9.53)

where R is the ideal gas constant, p_{AG} is the O_2 partial pressure in the gas, and T is the absolute temperature. p_{AG} is usually measured using mass spectrometry or similar high-sensitivity technique because there is often not a great difference between the amounts of O_2 in the gas streams entering and leaving bioreactors. The temperature and flow rate of the gases must be measured carefully to determine an accurate value of N_A. C_{AL} can be measured in the culture broth using a dissolved O_2 electrode and C_{AL}^* is evaluated as described in Section 9.8. $k_L a$ can then be calculated using Eq. (9.39).

There are several assumptions inherent in the equations used in the O_2 balance method:

1. The liquid phase is well mixed.
2. The gas phase is well mixed.
3. The pressure is constant throughout the vessel.

Assumption (1) allows us to use a single C_{AL} value to represent the concentration of dissolved O_2 in the culture broth.

The O_2 balance method is not readily applicable to cultures with low cell growth and low O_2 uptake rates because the difference in O_2 content between the inlet and outlet gas streams can become diminishingly small, resulting in unacceptable levels of error. In these circumstances, other methods for measuring $k_L a$ must be considered.

Example 9.2 illustrates the calculation of the $k_L a$ value using the O_2 balance method.

EXAMPLE 9.2 Steady-State $k_L a$ Measurement

A 20-L stirred bioreactor containing *Bacillus thuringiensis* is used to produce a microbial insecticide. The O_2 balance method is applied to determine $k_L a$. The bioreactor operating pressure is 150 kPa and the culture temperature is 30°C. The O_2 tension in the broth is measured as 82% using a probe calibrated to 100% *in situ* using water and air at 30°C and 150 kPa. The solubility of O_2 in the culture fluid is the same as in water. Air is sparged into the vessel; the inlet gas flow rate measured outside the bioreactor at 1 atm pressure and 22°C is 0.23 L s^{-1}. The exit gas from the bioreactor contains 20.1% O_2 and has a flow rate of 8.9 L min^{-1}.

a. Calculate the volumetric rate of O_2 uptake by the culture.
b. What is the value of $k_L a$?

Solution

a. The bioreactor operating pressure is:

$$150 \times 10^3 \text{ Pa} \cdot \left| \frac{1 \text{ atm}}{1.013 \times 10^5 \text{ Pa}} \right| = 1.48 \text{ atm}$$

The O_2 partial pressure in the inlet air at 1 atm is 0.2099 atm. $R = 0.082057$ l atm K^{-1} gmol^{-1}. Using Eq. (9.53):

$$N_A = \frac{1}{0.082057 \text{ L atm K}^{-1} \text{ gmol}^{-1} (20 \text{ L})}$$

$$\left[\left(\frac{0.23 \text{ L s}^{-1} (0.2099 \text{ atm})}{(22 + 273.15) \text{ K}} \right) - \left(\frac{8.9 \text{ L min}^{-1} \cdot \left| \frac{1 \text{ min}}{60 \text{ s}} \right| \cdot (0.20 \text{ L})(1.48 \text{ atm})}{(30 + 273.15) \text{ K}} \right) \right]$$

$$N_A = \frac{1}{0.082057 \text{ L atm K}^{-1} \text{ gmol}^{-1} (20 \text{ L})} \left[(1.636 \times 10^{-4}) - (1.456 \times 10^{-4}) \right] 1 \text{ atm K}^1 \text{s}^{-1}$$

$$N_A = 1.1 \times 10^{-5} \text{ gmol L}^{-1} \text{s}^{-1}$$

Because, at steady state, the rate of O_2 transfer is equal to the rate of O_2 uptake by the cells, the volumetric rate of O_2 uptake by the culture is 1.1×10^{-5} gmol L^{-1} s^{-1}.

b. Assume that the gas phase is well mixed so that the O_2 concentration in the bubbles contacting the liquid is the same as in the outlet gas, that is, 20.1%. As the difference in O_2 concentration between the inlet and outlet gas streams is small, we can also consider the composition of the gas phase to be constant throughout the bioreactor. From Table 9.2, the solubility of O_2 in water at 30°C and 1 atm air pressure is 8.05×10^{-3} kg m^{-3} = 8.05×10^{-3} g L^{-1}. Using Eq. (9.46) to determine the solubility at the bioreactor operating pressure of 1.48 atm and gas-phase O_2 mole fraction of 0.201:

$$C^*_{AL2} = \frac{p_{T2} \, y_{AG2}}{p_{T1} \, y_{AG1}} C^*_{AL1} = \frac{(1.48 \text{ atm}) 0.20 \text{ L}}{(1 \text{ atm}) 0.2099} 8.05 \times 10^{-3} \text{ g L}^{-1} = 0.0114 \text{ g L}^{-1}$$

C_{AL} in the bioreactor is 82% of the O_2 solubility at 30°C and 1.48 atm air pressure. From Eq. (9.45), solubility is proportional to total pressure; therefore:

$$C_{AL} = 0.82 \frac{1.48 \text{ atm}}{1 \text{ atm}} 8.05 \times 10^{-3} \text{ g L}^{-1} = 9.77 \times 10^{-3} \text{ g L}^{-1}$$

Applying these results in Eq. (9.39):

$$k_L a = \frac{1.1 \times 10^{-5} \text{ gmol L}^{-1}\text{s}^{-1} \cdot \left| \dfrac{32 \text{ g}}{1 \text{ gmol}} \right|}{0.0114 \text{ g L}^{-1} - 9.77 \times 10^{-3} \text{g L}^{-1}}$$

$$k_L a = 0.22 \text{ s}^{-1}$$

The value of $k_L a$ is 0.22 s^{-1}.

9.10.2 Dynamic Method

The dynamic method for estimating $k_L a$ quantifies changes in dissolved O_2 tension using an O_2 electrode after a step change in aeration conditions in the bioreactor. The results are interpreted using unsteady-state, mass balance equations to obtain the value of $k_L a$. The main advantage of the dynamic method over the steady-state technique is the comparatively low cost of the analytical equipment. The measurement is also independent of the O_2 solubility and can be carried out even if C_{AL}^* is unknown. In practice, the dynamic method is best suited for measuring $k_L a$ in relatively small vessels.

While the dynamic method is simple and easy to perform experimentally, it requires knowledge about the response time of the dissolved O_2 electrode, the effect of liquid boundary layers at the probe surface, and gas-phase dynamics in the vessel.

Simple Dynamic Method

The simplest version of the dynamic method is described here. This method gives reasonable results for $k_L a$ only if the following assumptions are valid:

1. The liquid phase is well mixed.
2. The response time of the dissolved O_2 electrode is much smaller than $1/k_L a$.
3. The measurement is performed at sufficiently high stirrer speed to minimize liquid boundary layers at the surface of the O_2 probe.
4. Gas-phase dynamics can be ignored.

If any of the four assumptions do not hold, the simple dynamic method may not provide an accurate estimate of $k_L a$ and alternative procedures should be considered.

$k_L a$ estimation using the simple dynamic method requires the bioreactor that contains culture broth and is stirred and sparged at fixed rates so that the dissolved O_2 concentration C_{AL} is constant. At time t_0, the broth is deoxygenated, either by sparging N_2 into the vessel or, as indicated in Figure 9.14, by stopping the air sparge and allowing the culture to consume the available O_2. Air is then reintroduced into the broth at a constant flow rate and the increase in C_{AL} is measured using a dissolved O_2 probe as a function of time. It is important that the O_2 concentration remains above the critical level C_{crit} so that the rate of O_2 uptake by the cells remains independent of the dissolved O_2 tension. Assuming that reoxygenation of the broth is fast relative to cell growth, the dissolved O_2 level will soon reach a steady-state value $\overline{C}_{AL}$, which reflects a balance between O_2 supply and O_2 consumption in the system. C_{AL1} and C_{AL2} are two O_2 concentrations measured during reoxygenation at times t_1 and t_2, respectively. We can develop an equation for $k_L a$ in terms of these experimental data.

During the reoxygenation step, the system is not at steady state. The rate of change in dissolved O_2 concentration is equal to the rate of O_2 transfer from the gas to the liquid phase, minus the rate of O_2 consumption by the cells:

$$\frac{dC_{AL}}{dt} = k_L a \left(C_{AL}^* - C_{AL} \right) - q_{O2} x \tag{9.54}$$

where $q_{O2} x$ is the volumetric rate of O_2 consumption. We can determine an expression for $q_{O2} x$ by considering the final steady-state dissolved O_2 concentration, $\overline{C}_{AL}$. When $C_{AL} = \overline{C}_{AL}$, $dC_A/dt = 0$ because there is no change in C_{AL} with time. Therefore, from Eq. (9.54):

$$q_{O2} x = k_L a \left(C_{AL}^* - \overline{C}_{AL} \right) \tag{9.55}$$

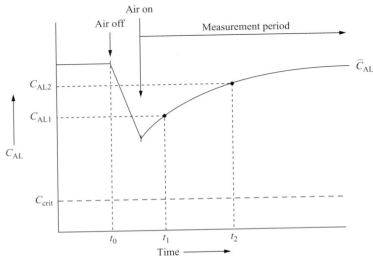

Figure 9.14 Variation of dissolved O_2 concentration for the dynamic measurement of $k_L a$.

Substituting this result into Eq. (9.54) and canceling the $k_L a C_{AL}^*$ terms gives:

$$\frac{dC_{AL}}{dt} = k_L a \left(\bar{C}_{AL} - C_{AL} \right)$$

(9.56)

Assuming $k_L a$ is constant with time, we can integrate Eq. (9.56) between t_1 and t_2. The resulting equation for $k_L a$ is:

$$k_L a = \frac{\ln \left(\dfrac{\bar{C}_{AL} - C_{AL1}}{\bar{C}_{AL} - C_{AL2}} \right)}{t_2 - t_1}$$

(9.57)

Using Eq. (9.57), $k_L a$ can be estimated using two points from Figure 9.14 or, more accurately, from several values of (C_{AL1}, t_1) and (C_{AL2}, t_2). When $\ln \left| \dfrac{\bar{C}_{AL} - C_{AL1}}{\bar{C}_{AL} - C_{AL2}} \right|$ is plotted against $(t_2 - t_1)$ as shown in Figure 9.15, the slope is $k_L a$. The value obtained for $k_L a$ reflects the operating stirrer speed, the air flow rate during the reoxygenation step, and the properties of the culture broth. Equation (9.57) can be applied to actively respiring cultures or to systems without O_2 uptake. In the latter case, $\bar{C}_{AL} = C_{AL}^*$ and N_2 sparging is required for the deoxygenation step of the procedure.

Electrode Response Time and Liquid Boundary Layers

The dynamic method relies on the measurement of changes in dissolved O_2 tension after a step change in bioreactor aeration conditions. Problems can arise with this approach if the O_2 electrode is slow to respond to the increase in liquid-phase O_2 levels as the culture broth is reoxygenated. If the electrode response is slower than the actual increase in O_2 concentration, the measured values of C_{AL} will reflect the response characteristics of the probe rather than the change in O_2 concentration in the bioreactor. Therefore, the *electrode response time* should always be measured as part of the dynamic method for determining $k_L a$. As well as the

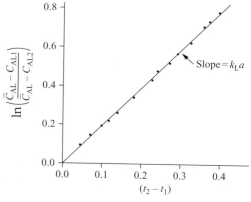

Figure 9.15 Evaluating $k_L a$ from data measured using the dynamic method.

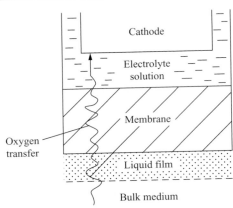

Figure 9.16 Development of a liquid film at the surface of an O_2 probe.

mass transfer resistances represented in Figure 9.13, development of a liquid boundary layer at the membrane-liquid interface, as illustrated in Figure 9.16, may also affect the response time of the electrode. Whether or not a boundary later is present depends on the flow conditions, liquid properties, and rate of oxygen consumption at the probe cathode.

The electrode response time and liquid boundary layer effects can be measured in test experiments. These experiments should be performed under conditions as close as possible to those applied for $k_L a$ measurement. The same gas flow rate and culture broth should be used. The bioreactor is prepared by sparging with air to give a constant dissolved O_2 tension, $\bar{C}_{AL}$. The O_2 electrode is equilibrated in a separate, vigorously agitated, N_2-sparged vessel providing a 0% O_2 environment. The probe is then transferred quickly from the N_2-sparged vessel to the bioreactor: this procedure exposes the probe to a step change in dissolved O_2 tension from 0% to $\bar{C}_{AL}$. The response of the probe is recorded. The procedure is repeated using a range of stirrer speeds in the bioreactor.

Typical results from the test experiments in low-viscosity fluid are shown in Figure 9.17. The electrode is transferred to the bioreactor at time zero. After being at 0% O_2, the probe takes some time to record a steady new signal corresponding to $\bar{C}_{AL}$. The response of the probe becomes faster as the stirrer speed N_i is increased, reflecting a progressive reduction in the thickness of the liquid boundary layer at the probe surface. At sufficiently high stirrer speed, no further change in electrode response is observed with additional increase in agitation rate, indicating that the boundary layer has been effectively minimized. Under these conditions, the response curve represents the dynamics of the electrode with minimal effects from the boundary layer.

The response of dissolved O_2 electrodes is usually assumed to follow first-order kinetics. Accordingly, the *electrode response time* τ_E is defined as the time taken for the probe to

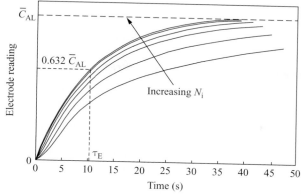

Figure 9.17 Typical electrode response curves after a step change in dissolved O_2 tension at different stirrer speeds, N_i. In this example, the electrode response time is just over 10 s.

indicate 63.2% of the total step change in dissolved O_2 level. As shown in Figure 9.17, the response time can be obtained from the response curves measured at high stirrer speeds in the practical absence of liquid boundary layer films. The response times are usually in the range of 10 to 100 s for commercially available, steam-sterilizable electrodes. However, faster non-autoclavable electrodes with response times of 2 to 3 s are available, as are microelectrodes that respond even more rapidly. Some O_2 electrodes have responses that deviate substantially from first-order (e.g., the electrode may have a tailing response so that it slows excessively as the new dissolved O_2 level is approached), or there may be a significant difference between the response times for upward and downward step changes. Such electrodes are not suitable for dynamic $k_L a$ measurements.

In viscous culture broths, it may be impossible to eliminate liquid boundary layers in the test experiments, even at high stirrer speeds. This makes estimation of the electrode response time difficult using culture fluid. Instead, τ_E may be evaluated using water, with the assumption that the electrode response does not depend on the measurement fluid.

The results from the test experiments are used to check the validity of two of the assumptions involved in the simple dynamic method. Assumption (2) is valid if the electrode response time is small compared with the rate of O_2 transfer; this is checked by comparing the value of τ_E with $1/k_L a$. The error in $k_L a$ has been estimated to be <6% for $\tau_E \leq 1/k_L a$ and <3% for $\tau_E \leq 0.2/k_L a$, so that commercial electrodes with response times between 2 and 3 s can be used to measure $k_L a$ values up to about 0.1 s^{-1} [22]. For assumption (3) to be valid, only stirrer speeds above that eliminating liquid boundary layers at the probe surface in culture broth can be used to determine $k_L a$. If either assumption (2) or (3) does not hold, the simple dynamic method cannot be used to measure $k_L a$. Factors involved in assumption (4) are outlined in the following section.

Gas-Phase Dynamics

The term *gas-phase dynamics* refers to changes with time in the properties of a gas disper-sion, including the number and size of the bubbles and the gas composition. Gas-phase dynamics can have a substantial influence on the results of $k_L a$ measurement. Because of the complexity and uncertainty associated with gas flow patterns, it is difficult to accurately account for these effects except in the simplest situations. Problems associated with gas-phase behavior can make the dynamic method an impractical technique for determining $k_L a$.

In the dynamic method, a change in aeration conditions is used as the basis for evaluating $k_L a$. While the inlet gas flow rate and composition may be altered quickly, this does not necessar-ily result in an immediate change in the gas hold-up and composition of the bubbles dispersed in the liquid. Depending on system variables such as the extent of gas recirculation, coales-cence properties of the liquid, and fluid viscosity, some time is required for a new gas hold-up and gas-phase composition to be established. Therefore, because the driving force for O_2 transfer depends on the gas-phase O_2 concentration, and as $k_L a$ varies with the volume of gas hold-up through its dependence on the interfacial area a, the O_2 transfer conditions and $k_L a$ itself are likely to be changing during the measurement period. The measured values of C_{AL} represent not only the kinetics of O_2 transfer but also the gas-phase dynamics in the bioreactor until a new steady state is established within the dispersed gas phase.

Let us consider the two methods commonly used to deoxygenate the culture broth for dynamic $k_L a$ measurement. Both these procedures affect the state of the gas dispersion at the start of the measurement period.

- N_2 *sparging*. N_2 is sparged into the broth at t_0 to achieve an initial reduction in dis-solved O_2 tension in a version of the simple dynamic method. Depending on the duration of N_2 sparging, we can assume that the gas hold-up contains more N_2 and less O_2 than in air. The bubbles will also contain carbon dioxide when active cells are present. If, at the commencement of the measurement period, the gas supply is sud-denly switched to air at the same flow rate, the gas hold-up volume, and therefore the gas–liquid interfacial area a, will remain roughly the same, but the composition of gas in the bubbles will start to change. The incoming air mixes with the preexisting N_2-rich hold-up until all the excess N_2 from the deoxygenation step is flushed out of the system. Until this process is complete, the measured C_{AL} values will be influenced by the changing gas-phase composition.
- *De-gassing*. An alternative procedure for deoxygenation of broth is to switch off the normal air supply to the bioreactor at t_0 (Figure 9.14), thus allowing O_2 consump-tion by active cells to reduce the dissolved O_2 tension. Depending on the extent of gas recirculation and the time required for bubbles to escape the liquid, the gas hold-up volume will be reduced during the deoxygenation step. When aeration is

recommenced, the gas hold-up and gas–liquid interfacial area a must be reestablished before k_La becomes constant. Until this occurs, the measured C_{AL} values will be affected by changes in the gas hold-up.

Transient gas-phase conditions are created using both strategies applied for deoxygenation during dynamic k_La measurement. Even if the bioreactor is relatively small, gas-phase transitions may continue to occur for a substantial proportion of the measurement period, influencing the accuracy of measured k_La values.

Example 9.3 uses the dynamic method to calculate the k_La using time-resolved O_2 measurements.

EXAMPLE 9.3 Estimating k_La Using the Simple Dynamic Method

A stirred bioreactor is used to culture hematopoietic cells isolated from umbilical cord blood. The liquid volume is 15 L. The simple dynamic method is used to determine k_La. The air flow is shut off for a few minutes and the dissolved O_2 level drops; the air supply is then reconnected at a flow rate of $0.25\,\text{L s}^{-1}$. The following results are obtained at a stirrer speed of 50 rpm.

Time (s)	5	20
O_2 tension (% air saturation)	50	66

When steady state is established, the dissolved O_2 tension is 78% air saturation. In separate test experiments, the electrode response to a step change in O_2 tension did not vary with stirrer speed above 40 rpm. The probe response time under these conditions was 2.8 s. When the k_La measurement was repeated using N_2 sparging to deoxygenate the culture, the results for O_2 tension as a function of time were similar to those listed. Estimate k_La.

Solution

$\bar{C}_{AL} = 78\%$ air saturation. Let us define $t_1 = 5\,\text{s}$, $C_{AL1} = 50\%$, $t_2 = 20\,\text{s}$, and $C_{AL2} = 66\%$. From Eq. (9.57):

$$k_La = \frac{\ln\left(\dfrac{78 - 50}{78 - 66}\right)}{(20 - 5)\ \text{s}} = 0.056\ \text{s}^{-1}$$

Before we can be confident about this value for k_La, we must consider the electrode response time, the presence of liquid films at the surface of the probe, and the influence of gas-phase dynamics. The results from the test experiments indicate that there are no liquid film effects at 50 rpm. For $\tau_E = 2.8\,\text{s}$ and $1/k_La = 17.9\,\text{s}$, $\tau_E = 0.16/k_La$. From Section 9.10.2, as $\tau_E < 0.2/k_La$, τ_E is small enough that the error associated with the electrode response can be neglected. Because the measured results for O_2 tension were similar using two different deoxygenation methods, the effect of gas-phase dynamics can also be neglected. Therefore, k_La is $0.056\,\text{s}^{-1}$.

9.11 Measurement of the Specific O_2 Uptake Rate, q_{O2}

Application of Eqs. (9.42) and (9.43) for analysis of the mass transfer performance of bioreactors requires knowledge of the culture's specific O_2 uptake rate, q_{O2}. This parameter reflects the requirement of an organism for O_2 to support growth and product synthesis. It can be measured using several experimental techniques.

The O_2 balance method for measuring k_La (Section 9.10.1) also allows us to evaluate q_O. At steady state, the volumetric rate of O_2 uptake by the cells is equal to the volumetric rate of gas–liquid O_2 transfer, N_A (Section 9.5.2). Once N_A is determined using Eq. (9.53), q_{O2} is found by dividing N_A by the cell concentration x. For example, for N_A with units of gmol L^{-1} s^{-1} and x with units of g cdw L^{-1}, q_{O2} is obtained with units of gmol (g cdw)$^{-1}$ s^{-1}. The accuracy of this technique depends strongly on accurate measurement of the inlet and outlet gas composition, flow rate, pressure, and temperature, and on the validity of the assumptions used to derive the mass transfer equations.

Dynamic methods can also be used to measure q_{O2}. A sample of culture broth containing a known cell concentration x is placed in a small chamber equipped with an O_2 electrode and stirrer as shown in Figure 9.18(A). The broth is sparged with air: at the commencement of the measurement period, the air flow is stopped and the vessel sealed to make it airtight. The decline in dissolved O_2 concentration due to O_2 uptake by the cells is recorded using the O_2 electrode as shown in Figure 9.18(B). The initial slope of the curve of C_{AL} versus time gives

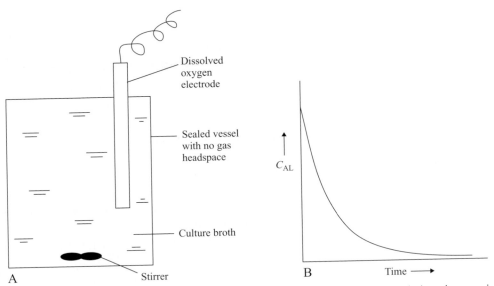

Figure 9.18 Dynamic method for measurement of q_O. (A) Sealed experimental chamber equipped with an O_2 electrode and stirrer and containing culture broth. (B) Measured data for dissolved O_2 concentration C_{AL} versus time.

the volumetric rate of O_2 uptake by the cells, Q_{O2}. Dividing Q_{O2} by the cell concentration x gives the specific rate of O_2 uptake, q_{O2}. All gas bubbles must be removed from the liquid before the measurements are started; the sealed chamber must also be airtight without any gas headspace so that additional O_2 cannot enter the liquid during data collection.

Factors similar to those outlined in Section 9.10.2 for dynamic measurement of $k_L a$ also affect the accuracy of q_{O2} obtained using this technique. The electrode response must be relatively fast and liquid boundary layers at the probe surface must be minimized by operating the stirrer at a sufficiently high speed. Fortunately, the concentration of cells in the chamber can be adjusted to make it easier to comply with these requirements. A relatively dilute cell suspension can be used to reduce the speed of O_2 uptake so that a relatively slow electrode response does not affect the results and any liquid boundary layers are more readily removed. The size of the vessel is also typically very small to minimize the effects of gas-phase dynamics and mixing characteristics.

9.12 Practical Aspects of O_2 Transfer in Large Bioreactors

Special difficulties are associated with measuring O_2 transfer rates and $k_L a$ in large bioreactors. These problems arise mainly because significant gradients of liquid- and gas-phase composition and other properties develop with increasing scale.

9.12.1 Liquid Mixing

In our discussion of O_2 transfer so far, we have assumed that the liquid phase is perfectly mixed and that $k_L a$ is constant throughout the entire reactor. This requires that turbulence and rates of turbulence kinetic energy dissipation are uniformly distributed. These conditions occur reasonably well in laboratory-scale stirred reactors, which are characterized by high turbulence throughout most of the vessel. In contrast, perfect mixing is difficult to achieve in commercial-scale reactors and turbulence is far from uniformly distributed. Most of the O_2 transfer in industrial-scale bioreactors takes place in the region near the impeller. Away from the impeller, the bubbles are in free rise and the liquid velocity is significantly reduced.

9.12.2 Gas Mixing

O_2 transfer measurements often assume the gas phase is well mixed—that is, the gas composition is uniform and equal to that in the outlet gas stream. In large bioreactors, these conditions may not be met due to substantial depletion of O_2 in the gas phase during passage of the bubbles from the bottom to the top of the tank. This creates an axial gradient of gas-phase concentration from the top to the bottom of the vessel. Modified mass transfer models that include the effects of plug flow or plug flow with axial dispersion in the gas phase have been used to better represent the gas mixing conditions in large-scale bioreactors [34,35].

9.12.3 Pressure Effects

Even when there is rapid mixing in large-scale bioreactors, variations in gas-phase pressure occur due to hydrostatic pressure changes from the top to the bottom of the vessel. The pressure at the bottom of tall vessels is higher than at the top due to the weight of the liquid. The hydrostatic pressure difference p_s is given by the equation:

$$p_s = \rho g H_L \tag{9.58}$$

where ρ is the liquid density, g is gravitational acceleration, and H_L is the liquid height. As the solubility of O_2 is sensitive to gas-phase pressure and O_2 partial pressure, significant variation in these conditions between the top and bottom of the vessel affects the value of C_{AL}^* used in mass transfer calculations. Allowance can be made for this in models of the mass transfer process [36]; alternatively, an average concentration-difference driving force $\left(C_{AL}^* - C_{AL} \right)$ across the system can be determined. A suitable average is the *logarithmic-mean concentration difference*, $\left(C_{AL}^* - C_{AL} \right)_{lm}$:

$$\left(C_{AL}^* - C_{AL} \right)_{lm} = \frac{\left(C_{AL}^* - C_{AL} \right)_o - \left(C_{AL}^* - C_{AL} \right)_i}{\ln\left[\dfrac{\left(C_{AL}^* - C_{AL} \right)_o}{\left(C_{AL}^* - C_{AL} \right)_i} \right]} \tag{9.59}$$

In Eq. (9.59), subscripts i and o represent conditions at the inlet and outlet ends of the vessel, respectively.

9.12.4 Interaction Between O_2 Transfer and Heat Transfer

Heat transfer is a critical function in bioreactors as discussed in Chapter 8. Culture broth must be cooled to remove the heat generated by metabolism and thus prevent the culture temperature rising to deleterious levels. Because the rate of metabolic heat generation in aerobic cultures is directly proportional to the rate of O_2 consumption by the cells, O_2 transfer and heat transfer are closely related. Rapid O_2 uptake can create major heat removal problems. Heat transfer can become the limiting factor affecting the maximum feasible rate of reaction especially for fast growing, aerobic cultures in large bioreactors.

The heat transfer requirements are usually largest toward the end of the culture cycle when the volumetric rate of O_2 uptake, Q_{O2}, is greatest. In some cases, it may be sensible to slow down the rate of O_2 consumption by the culture to avoid the necessity of installing expensive heat transfer equipment. Therefore, if strategies such as increasing the stirrer speed, gas flow rate, pressure, and O_2 partial pressure (Section 9.6) are undertaken to improve $k_L a$ and the O_2 transfer driving force, the consequent extra heat burden must be borne in mind. Heat and O_2 transfer are linked and should be considered together, especially in large-scale operations.

9.13 Alternative Methods for Oxygenation Without Sparging

In small-scale bioreactors or when shear-sensitive organisms such as animal cells are being cultured, alternative methods for providing O_2 are sometimes used. The large forces generated by bubbles bursting at the surface of sparged cell cultures can cause high rates of animal cell damage as outlined in Chapter 7. For that reason, aeration by other means may be required or preferred.

An alternative to gas sparging is *surface aeration* where gas containing O_2 is flushed through the headspace of the reactor above the liquid; O_2 is then transferred to the liquid through the upper surface of the culture broth. Surface aeration contributes to O_2 transfer even when the liquid is aerated by sparging; however, in vigorously agitated systems, its contribution is small compared with the high oxygenation rates achieved using entrained bubbles. The rate of O_2 transfer during surface aeration can be described using the equations in Section 9.4.3, with $k_L a$ in Eq. (9.39) representing the conditions in the liquid boundary layer at the liquid–headspace interface. The value of $k_L a$ and the rate of surface aeration increase with stirrer speed. The height of the impeller above the vessel floor may also be important because it affects the fluid velocity at the liquid surface.

Membrane tubing aeration is another bubble-free option for oxygenation of cultures. Aeration is achieved by gas exchange through silicone or microporous polypropylene or Teflon tubing immersed in the culture broth. Gas flowing in the tube diffuses through the tube walls and into the medium under a concentration-difference driving force; air or O_2-enriched air may be used in the tubing. For bubble-free aeration, the gas pressure inside the tubing must remain below the bubble point to avoid bubbles forming on the outside of the tube walls. The main resistances to O_2 transfer are the tube wall itself and the liquid film surrounding the outside of the tubing. The tubing assembly may be kept in motion as an effective stirrer to prevent the cells from settling and to promote mixing; this also enhances O_2 transfer through the liquid boundary layer at the tube surface.

9.14 O₂ Transfer in Shake Flasks

Shake flasks or *Erlenmeyer flasks* are employed commonly in laboratories for microbial cultures. A typical shake flask is shown in Figure 9.19. Shake flasks have flat bottoms and sloping sides and can be made of glass, plastic, or metal. The flask opening is of variable width and is fitted with a porous cap or plug closure. Gases are exchanged through the closure without the introduction of contaminating organisms. Flasks containing culture broth are placed on shaking tables or in incubator–shakers; the shaking movement is responsible for mixing and mass transfer in the flask.

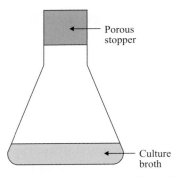

Figure 9.19 Typical shake flask for cell culture.

9.14.1 O_2 Transfer Through the Flask Closure

The first mass transfer resistance encountered in the delivery of O_2 to shake-flask cultures is the flask closure. A variety of porous materials is used to stopper shake flasks, including cotton, cotton wool bound with cheesecloth, and silicone sponge. The rate of O_2 transfer through the closure depends on the diffusion coefficient for O_2 in the material, the width of the neck opening, and the stopper depth. The diffusion coefficient varies with the porosity or bulk density of the stopper material. Wetting the stopper increases the resistance to gas transfer as well as the risk of culture contamination.

The flask closure affects the composition of gas in the headspace of the flask. In addition to impeding O_2 transfer into the culture, the closure impedes escape of carbon dioxide and other gases generated by the cells.

Although several types of flask stopper have been investigated for their O_2 transfer characteristics, it is unclear which type of closure gives the best results. In some cases, for example, with large flasks or if the liquid is shaken very vigorously, the flask closure becomes the dominant resistance for O_2 transfer to the cells.

9.14.2 O_2 Transfer Within the Flask

The culture broth in shake flasks is supplied with O_2 by surface aeration from the gas atmosphere in the flask. Within the flask, the main resistance to O_2 transfer is the liquid-phase boundary layer at the gas–liquid interface. The rate of O_2 transfer from the gas phase is represented by Eq. (9.39) and depends on the area available for O_2 transfer, the liquid velocity, the viscosity, diffusion coefficient, and surface mobility at the phase boundary, and the concentration-difference driving force $\left(C_{AL}^* - C_{AL} \right)$. Electrolytes and other components in the broth may affect the O_2 solubility and surface properties at the interface. O_2 transfer in shake flasks does not typically depend on the coalescing properties of the liquid because of the limited role of bubbles.

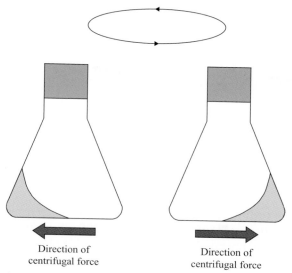

Figure 9.20 Distribution and movement of liquid in a shake flask on an orbital shaker.

The area available for O_2 transfer in shake flasks is not simply equal to the surface area of the resting fluid as represented in Figure 9.19. The liquid is distributed within the flask as shown in Figure 9.20 when a flask is shaken: liquid is thrown up onto the walls due to the centrifugal forces associated with flask rotation. As the liquid swirls around the flask, a thin film is deposited on the flask wall and is replaced with each rotation. The mass transfer area a at any given time includes this surface area of liquid film on the flask wall. The overall rate of O_2 transfer depends on the rate of generation of fresh liquid surface, or the frequency with which the liquid film is replenished.

Factors influencing the value of a include:

- Flask shape
- Flask size
- Surface properties of the flask walls (e.g., hydrophilic or hydrophobic)
- Shaking speed
- Flask displacement during shaking
- Liquid volume
- Liquid properties (e.g., viscosity)

The area of liquid film per unit volume of fluid is enhanced using large flasks with small liquid volumes on shakers with large displacement distance per rotation operated at high speed as indicated in Figure 9.21. The importance of limiting the liquid volume in shake flasks is evident in Figure 9.21; for example, using >100 mL of medium in a 250-mL flask reduces the mass transfer coefficient substantially compared with 50 mL of medium. It is also important

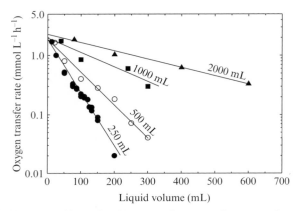

Figure 9.21 Effect of flask size and liquid volume on the rate of O_2 transfer. Shake flasks of nominal size—($\bullet$) 250 mL, ($\circ$) 500 mL, ($\blacksquare$) 1000 mL, and ($\blacktriangle$) 2000 mL—were tested on a reciprocating shaker with amplitude 3 in. operated at 96 rpm. The data are plotted using semi-logarithmic coordinates. *Data from M.A. Auro, H.M. Hodge, and N.G. Roth, 1957, Oxygen absorption rates in shaken flasks. Ind. Eng. Chem. 49, 1237–1238.*

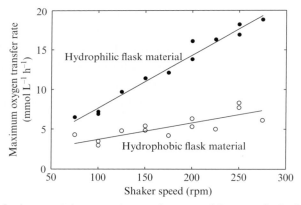

Figure 9.22 Effect of flask material properties on the rate of O_2 transfer in 250-mL shake flasks on an orbital shaker. *Data from U. Maier and J. Büchs, 2001, Characterisation of the gas–liquid mass transfer in shaking bioreactors. Biochem. Eng. J. 7, 99–106.*

that the flask material support the development of a liquid film on the walls; hydrophobic materials such as plastic are therefore not recommended when O_2 transfer is critical. The difference between O_2 transfer rates in hydrophilic and hydrophobic flasks is illustrated in Figure 9.22. When the culture has high O_2 requirements, gas–liquid mass transfer can be enhanced using shake flasks with baffles. These indentations break up the swirling motion of the liquid, increase the level of liquid splashing onto the flask walls, and thus improve aeration. A disadvantage of using baffles is the increased risk of wetting the flask closure.

Shake-flask culture and surface aeration are practical only at relatively small scales because surface-to-volume ratios decrease with increasing liquid volume. For some cultures, the maximum rate of O_2 transfer in shake flasks is not sufficient to meet the cellular O_2 demand, limiting culture performance.

9.15 Summary of Chapter 9

At the end of Chapter 9 you should:

- Be able to describe the *two-film theory* of mass transfer between phases
- Know *Fick's law* in terms of the *binary diffusion coefficient*, $\mathcal{D}_{AB}$
- Be able to describe in simple terms the mathematical analogy between mass, heat, and momentum transfer
- Know the equation for the rate of gas–liquid O_2 transfer in terms of the *mass transfer coefficient* $k_L a$ and the *concentration-difference driving force*
- Be able to identify the steps that are most likely to present major resistances to O_2 transfer from bubbles to cells
- Understand how O_2 transfer and $k_L a$ can limit the biomass density in bioreactors
- Know how $k_L a$ depends on bioreactor operating conditions such as the stirrer speed, power input, gas flow rate, and liquid properties such as viscosity
- Know how temperature, O_2 partial pressure, and the presence of dissolved and suspended material affect the rate of O_2 transfer and the solubility of O_2 in culture broths
- Be able to apply the *O_2 balance method* and the *simple dynamic method* for experimental determination of $k_L a$, with understanding of their advantages and limitations
- Know how the specific O_2 uptake rate q_{O2} can be measured in cell cultures
- Be familiar with techniques for culture aeration that do not involve gas sparging
- Understand the mechanisms of O_2 transfer in shake flasks

Problems

9.1. Rate-controlling processes in bioprocess

Serratia marcescens bacteria are used for the production of threonine. The maximum specific O_2 uptake rate of *S. marcescens* in batch culture is 5 mmol O_2 (g cdw)$^{-1}$ h^{-1}. It is planned to operate the bioreactor to achieve a maximum cell density of 40 g L^{-1}. At the cultivation temperature and pressure, the solubility of O_2 in the culture liquid is 8×10^{-3} kg m^{-3}. At a particular stirrer speed, $k_L a$ is 0.15 s^{-1}. Under these conditions, will the rate of cell metabolism be limited by mass transfer or depend solely on metabolic kinetics?

9.2. Test for O_2 limitation

An 8-m^3 stirred bioreactor is used to culture *Agrobacterium* sp. ATCC 31750 for production of curdlan. The liquid medium contains 80 g L^{-1} sucrose. Under optimal conditions, 1.0 g dry weight of cells is produced for every 4.2 g of sucrose consumed. The bioreactor is sparged with air at 1.5 atm pressure, and the specific O_2 demand is 7.5 mmol per g dry weight per h. To achieve the maximum yield of curdlan, the cultivation temperature is held constant at 32°C. The solubility of O_2 in the culture broth is 15% lower than in water due to solute effects. If the maximum $k_L a$ that can be achieved is $0.10 \, s^{-1}$, does the bioreactor's mass transfer capacity support complete consumption of substrate?

9.3. $k_L a$ required to maintain critical O_2 concentration

A genetically engineered strain of yeast is cultured in a bioreactor at 30°C for production of heterologous protein. The O_2 requirement is $80 \, \text{mmol L}^{-1} \, h^{-1}$; the critical O_2 concentration is $0.004 \, \text{mM}$. The solubility of O_2 in the culture broth is estimated to be 10% lower than in water due to solute effects.

    a. What is the minimum mass transfer coefficient necessary to sustain this culture with dissolved O_2 levels above critical if the reactor is sparged with air at approximately 1 atm pressure?

    b. What mass transfer coefficient is required if pure O_2 is used instead of air?

9.4. O_2 transfer with different impellers

A 10-m^3 stirred bioreactor with liquid height 2.3 m is used to culture *Trichoderma reesei* for production of cellulase. The density of the culture fluid is $1000 \, \text{kg m}^{-3}$. An equation for the O_2 transfer coefficient as a function of operating variables has been developed for *T. reesei* broth:

$$k_L a = 2.5 \times 10^{-3} \left(\frac{P_T}{V_L} \right)^{0.7} u_G^{0.3}$$

where $k_L a$ has units of s^{-1}, P_T is the total power input in W, V_L is the liquid volume in m^3, and u_G is the superficial gas velocity in $m \, s^{-1}$. The bioreactor is sparged using a gas flow rate of 0.6 vvm (vvm means volume of gas per volume of liquid per minute). The vessel is stirred with a single impeller but two alternative impeller designs, a Rushton turbine and a curved-blade disc turbine, are available. Both impellers are sized and operated so that their ungassed power draw is 9 kW.

    a. If the power loss with gassing is 50% for the Rushton and 5% for the curved-blade turbine, compare the kLa values achieved using each impeller.

    b. What is the percentage contribution to PT from gassing with the two different impellers?

    c. If the cell concentration is limited to $15 \, g \, L-1$ using the Rushton turbine because of mass transfer effects, estimate the maximum possible cell concentration with the curved-blade disc turbine.

It is decided to install the Rushton turbine, but to compensate for the effect on $k_L a$ of its loss of power with gassing by increasing the gas flow rate.

    • Estimate the gas flow rate required to obtain the same maximum cell concentration using the Rushton turbine as that achieved with the curved-blade disc turbine. Express your answer in vvm. What are your assumptions? (Iterative solution may be required.)

9.5. Foam control and O_2 transfer

Foaming is controlled routinely in bioreactors using a foam sensor and pump for automatic addition of antifoam agent. As shown in Figure 9P5.1, the foam sensor is located at the top of the vessel above the liquid surface. When a head of foam builds up so that foam contacts the lower tip of the sensor, an electrical signal is sent to the pump to add antifoam. The antifoam agent destroys the foam, the foam height is therefore reduced, contact with the foam sensor is broken,

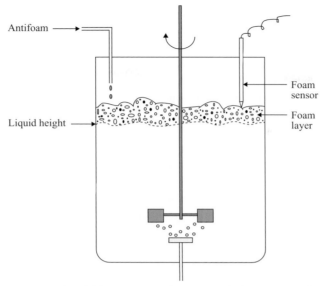

Figure 9P5.1 Stirred bioreactor with automatic foam control system.

and the pump supplying the antifoam agent is switched off. Further build-up of foam reactivates the control process. If the position of the foam sensor is fixed, when the gap between the liquid surface and sensor is reduced by raising the liquid height, a smaller foam build-up is tolerated before antifoam agent is added. Therefore, antifoam addition will be triggered more often if the working volume of the vessel is increased. Although a greater bioreactor working volume means that more cells and/or product are formed, addition of excessive antifoam agent could reduce $k_L a$ significantly, thereby increasing the likelihood of mass transfer limitations. A stirred bioreactor of diameter 1.5 m is used to culture *Bacillus licheniformis* for production of serine alkaline protease. The bioreactor is operated five times with automatic antifoam addition using five different liquid heights. The position of the foam sensor is the same in each run. The volume of antifoam added and the $k_L a$ at the end of the culture period are recorded.

Liquid height (m)	Antifoam added (L)	$k_L a$ (s^{-1})
1.10	0.16	0.016
1.29	0.28	0.013
1.37	1.2	0.012
1.52	1.8	0.012
1.64	2.4	0.0094

Under ideal conditions, the maximum specific O_2 uptake rate for *B. licheniformis* is 2.6 mmol g^{-1} h^{-1}. When glucose is used as the carbon source at an initial concentration of 20 g L^{-1}, a maximum of 0.32 g of cells are produced for each g of glucose consumed, and 0.055 g of protease is produced per g of biomass formed. The solubility of O_2 in the broth is estimated as 7.8 g m^{-3}.

a. Using the $k_L a$ values associated with each level of antifoam addition, estimate the maximum cell concentrations supported by O_2 transfer as a function of liquid height. Assume that antifoam exerts a much stronger influence on $k_L a$ than on other properties of the system such as O_2 solubility and specific O_2 uptake rate.

b. Calculate the maximum mass of cells and maximum mass of protease that can be produced based on the O_2 transfer capacity of the bioreactor as a function of liquid height.

c. Is protease production limited by O_2 transfer at any of the liquid heights tested?

d. What operating liquid height would you recommend for this bioprocess?

Explain your answer.

9.6. Improving the rate of O_2 transfer

Rifamycin is produced in a 17-m³ stirred bioreactor using a mycelial culture, *Nocardia mediterranei*. The bioreactor is sparged with air under slight pressure so the solubility of O_2 in the broth is $10.7\,\mathrm{g\,m^{-3}}$. Data obtained during operation of the bioreactor are shown in Figure 9P6.1. After about 147 h of culture, vegetable oil is added to the broth to disperse a thick build-up of foam. This has a severe effect on the O_2 transfer coefficient and reduces the dissolved O_2 tension.

a. Calculate the steady-state O_2 transfer rate before and after addition of the vegetable oil.

b. The relationship between $k_L a$ and the bioreactor operating conditions is:

$$k_L a \propto \left(\frac{P_T}{V_L} \right)^{0.5} u_G^{0.3}$$

Because increasing the gas flow rate would aggravate the problems with foaming, it is decided to restore the value of $k_L a$ by increasing the power input by stirring. If the power contribution from gas sparging is negligible, by how much does the stirrer power need to be increased to overcome the effects of the vegetable oil on $k_L a$?

c. To save the cost of increasing the power input, instead of (b), it is decided to improve O_2 transfer by sparging the bioreactor with O_2-enriched air. The total gas flow rate and pressure are unchanged. To restore the rate of O_2 transfer after vegetable oil addition to that before oil was added, what volume percentage of O_2 is required in the sparge gas if the desired dissolved O_2 concentration in the broth is $6.2 \times 10^{-3}\ \mathrm{kg\ m^{-3}}$?

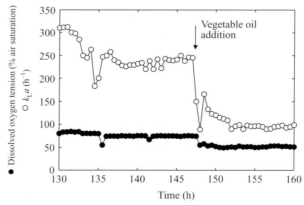

Figure 9P6.1 Online time-course data from a stirred bioreactor used for rifamycin production.

9.7. O₂ transfer for different cell types

The specific O_2 demands and critical O_2 concentrations for typical microbial, plant, and animal cell cultures are listed below.

Cell culture	q_{O2}	C_{crit} (mmol L⁻¹)
Escherichia coli	8.5 mmol (g dry weight)⁻¹ h⁻¹	0.0082
Vitis vinifera (grape)	0.60 mmol (g dry weight)⁻¹ h⁻¹	0.055
Chinese hamster ovary (CHO)	3.0×10^{-10} mmol cell⁻¹ h⁻¹	0.020

a. Estimate the $k_L a$ required to achieve cell concentrations of 25 g dry weight L⁻¹ for *E. coli* and *V. vinifera* and 3.0×10^9 cells L⁻¹ for CHO cells, while maintaining the dissolved O_2 concentration above critical. The O_2 solubility in the media used for the cultures is 7.2×10^{-3} kg m⁻³.

b. The relationship between $k_L a$ and the power input to a 1-m³ stirred bioreactor is:

$$k_L a \propto \left(\frac{P_T}{V_L}\right)^{0.5}$$

Compare the bioreactor power requirements for culture of the three different cell types under the conditions described in (a).

9.8. Single-point $k_L a$ determination using the O₂ balance method

A 200-L stirred bioreactor contains a batch culture of *Bacillus subtilis* bacteria at 28°C. Air at 20°C is pumped into the vessel at a rate of 1 vvm (vvm means volume of gas per volume of liquid per minute). The average pressure in the bioreactor is 1 atm. The volumetric flow rate of off-gas from the bioreactor is measured as 189 L min⁻¹. The exit gas stream is analyzed for O_2 and is found to contain 20.1% O_2. The dissolved O_2 concentration in the broth is measured using an O_2 electrode as 52% air saturation. The solubility of O_2 in the culture broth at 28°C and 1 atm air pressure is 7.8×10^{-3} kg m⁻³.

a. Calculate the O_2 transfer rate.

b. Determine the value of $k_L a$ for the system.

c. The O_2 analyzer used to measure the exit gas composition was incorrectly calibrated. If the O_2 content has been overestimated by 10%, what error is associated with the result for $k_L a$?

9.9. Steady-state $k_L a$ measurement

Escherichia coli bacteria are cultured at 35°C and 1 atm pressure in a 500-L bioreactor using the following medium:

Component	Concentration (g L⁻¹)
glucose	20
sucrose	8.5
$CaCO_3$	1.3
$(NH_4)_2SO_4$	1.3
Na_2HPO_4	0.09
KH_2PO_4	0.12

Air at 25°C and 1 atm is sparged into the vessel at a rate of 0.4 m³ min⁻¹. The dissolved O_2 tension measured using a polarographic electrode calibrated in situ in sterile culture medium is 45% air saturation. The gas flow rate leaving the bioreactor is measured using a rotary gas meter as 6.3 L s⁻¹. The O_2 concentration in the off-gas is 19.7%.

a. Estimate the solubility of O_2 in the culture broth. What are your assumptions?

b. What is the O_2 transfer rate?

c. Determine the value of $k_L a$.

d. Estimate the maximum cell concentration that can be supported by O_2 transfer in this bioreactor if the specific O_2 demand of the *E. coli* strain is 5.4 mmol (g cdw)⁻¹ h⁻¹.

e. If the biomass yield from the combined sugar substrates is 0.5 g cdw (g sugar)⁻¹, is growth in the culture limited by O_2 transfer or substrate availability?

9.10. O_2 transfer in a pressure vessel

A bioreactor of diameter 3.6 m and liquid height 6.1 m is used for production of ustilagic acid by *Ustilago zeae*. The pressure at the top of the bioreactor is 1.4 atma (absolute pressure). The vessel is stirred using dual Rushton turbines and the cultivation temperature is 29°C. The dissolved O_2 tension is measured using two electrodes: one electrode is located near the top of the tank, the other is located near the bottom. Both electrodes are calibrated *in situ* in sterile culture medium. The dissolved O_2 reading at the top of the bioreactor is 50% air saturation; the reading at the bottom is 65% air saturation. The bioreactor is sparged with air at 20°C at a flow rate of 30 m³ min⁻¹ measured at atmospheric pressure. Off-gas leaving the vessel at a rate of 20.5 m³ min⁻¹ contains 17.2% O_2. The solubility of O_2 in the culture broth is not significantly different from that in water. The density of the culture broth is 1000 kg m⁻³.

a. What is the O_2 transfer rate?

b. Estimate the pressure at the bottom of the tank.

c. The gas phase in large bioreactors is often assumed to exhibit plug flow. Under these conditions, no gas mixing occurs so that the gas composition at the bottom of the tank is equal to that in the inlet gas stream, while the gas composition at the top of the tank is equal to that in the outlet gas stream. For the gas phase in plug flow, estimate the O_2 solubility at the top and bottom of the tank.

d. What is the value of $k_L a$?

e. If the cell concentration is 16 g L⁻¹, what is the specific O_2 demand?

f. Industrial bioreactor vessels are rated for operation at elevated pressures so they can withstand steam sterilization. Accordingly, the bioreactor used for ustilagic acid production can be operated safely at a maximum pressure of 2.7 atma. Assuming that respiration by *U. zeae* and the value of $k_L a$ are relatively insensitive to pressure, what maximum cell concentration can be supported by O_2 transfer in the bioreactor after the pressure is raised?

9.11. Dynamic $k_L a$ measurement

The simple dynamic method is used to measure $k_L a$ in a bioreactor operated at 30°C and 1 atm pressure. Data for the dissolved O_2 concentration as a function of time during the reoxygenation step are as follows.

Time (s)	C_{AL} (% air saturation)
10	43.5
15	53.5
20	60.0
30	67.5
40	70.5
50	72.0
70	73.0
100	73.5
130	73.5

a. Calculate the value of $k_L a$.

b. What additional experiments are required to check the reliability of this $k_L a$ result?

9.12. $k_L a$ **measurement using the dynamic pressure method**

The dynamic pressure method is applied for measurement of $k_L a$ in a 3000-L stirred bioreactor containing a suspension culture of *Micrococcus glutamicus*. The stirrer is operated at 60 rpm and the gas flow rate is fixed at 800 L min^{-1}. The following dissolved O_2 concentrations are measured using a polarographic dissolved O_2 electrode after a step increase in bioreactor pressure.

Time (s)	C_{AL} (% air saturation)
6	50.0
10	56.1
25	63.0
40	64.7

The steady-state dissolved O_2 tension at the end of the dynamic response is 66% air saturation.

a. Estimate the value of $k_L a$.

b. An error is made determining the steady-state O_2 level, which is taken as 70% instead of 66% air saturation. What effect does this 6% error in $\bar{C}_{AL}$ have on the result for $k_L a$?

c. At the end of the $k_L a$ experiment, the electrode response time is measured by observing the output after a step change in dissolved O_2 tension from 0% to 100% air saturation. The stirrer speeds tested are 40, 50, and 60 rpm. Figure 9P12.1 at the bottom of page shows a chart recording of the results at 60 rpm; the results at 50 rpm are not significantly different. From this information, how much confidence do you have in the $k_L a$ measurements? Explain your answer.

9.13. **Surface versus bubble aeration**

Hematopoietic cells used in cancer treatment are cultured at 37°C in an 8.5-cm diameter bioreactor with working volume 500 mL. The culture fluid is mixed using a stirrer speed of 30 rpm. The reactor is operated at ambient pressure.

a. The dissolved O_2 tension is controlled at 50% air saturation using surface aeration only. A 50:20:30 mixture of air, O_2, and N_2 is passed at a fixed flow rate through the headspace. The specific O_2 uptake rate for hematopoietic cells is 7.7×10^{-12} g O_2 cell^{-1} h^{-1} and the cell concentration is 1.1×10^9 cells L^{-1}. Estimate the value of $k_L a$ for surface aeration.

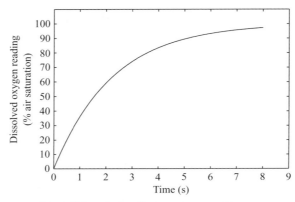

Figure 9P12.1: Chart recording of the electrode response at 60 rpm to a step change in dissolved O_2 tension.

b. Instead of surface aeration, the bioreactor is sparged gently with air. When the dissolved O_2 tension is maintained at the critical level of 8% air saturation, the cell concentration is 3.9×10^9 cells L^{-1}. What is the $k_L a$ for bubble aeration?

c. Surface aeration is preferred for this shear-sensitive culture, but the surface $k_L a$ needs improvement. For the gassing conditions applied in (a), estimate the vessel diameter required for surface aeration to achieve the same $k_L a$ obtained with sparging. What are your assumptions?

9.14. Shake-flask aeration

A mixed culture of heterotrophic microorganisms isolated from the Roman baths at Bath England is prepared for bioleaching of manganese ore. One hundred mL of molasses medium is used in 300-mL flasks with 4-cm-long silicone sponge stoppers. The width of the flask opening is 3.2 cm. The cultures are incubated at 30°C on an orbital shaker operated at 80 rpm.

a. With the flask closure removed, $k_L a$ for gas–liquid mass transfer is estimated using the dynamic method. During the reoxygenation step, the dissolved O_2 tension measured using a small, rapid-response electrode is 65% air saturation after 5 s and 75% after 30 s. The steady-state O_2 tension is 90% air saturation. What is the resistance to O_2 transfer in the flask? What are your assumptions?

b. An expression for the mass transfer coefficient K_c for the flask closure is:

$$K_c = \frac{\mathcal{D}_e A_c}{L_c V_G}$$

where $\mathcal{D}_e$ is the effective diffusion coefficient of O_2 in the closure material, A_c is the cross-sectional area of the closure, L_c is the closure length, and V_G is the volume of gas in the flask. If $\mathcal{D}_e$ for silicone sponge is 20.8 cm^2 s^{-1}, what resistance to O_2 transfer is provided by the flask closure?

a. What proportion of the total resistance to O_2 transfer does the flask closure represent?

b. It is decided to improve the rate of gas–liquid O_2 transfer so that the resistance to O_2 transfer within the flask is approximately equal to that of the flask closure. A study of the dependence of $k_L a$ on shake-flask operating parameters yields the relationship:

$$k_L a \propto N^{1.2} \left(\frac{V_F}{V_L} \right)^{0.85}$$

where N is the shaker speed in rpm, V_F is the flask size in mL, and V_L is the liquid volume in mL. If the shaker speed can be increased to a maximum of 150 rpm:

(i) What size flask is needed if the culture volume remains at 100 mL?

(ii) If 300-mL flasks are the only shake flasks available, what culture volume should be used?

References

[1] R.E. Treybal, Mass-Transfer Operations, third ed., McGraw-Hill, 1980.

[2] Perry's Chemical Engineers' Handbook, eighth ed., McGraw-Hill, 2008.

[3] T.K. Sherwood, R.L. Pigford, C.R. Wilke, Mass Transfer, McGraw-Hill, 1975.

[4] J.M. Coulson, J.F. Richardson, J.R. Backhurst, J.H. Harker, sixth ed. Coulson and Richardson's Chemical Engineering vol. 11999, (Chapters 10 and 12), Butterworth-Heinemann, 1999.

[5] M.J. Johnson, Metabolism of penicillin-producing molds, Ann. N.Y. Acad. Sci. 48 (1946) 57–66.

[6] C.E. Clifton, A comparison of the metabolic activities of *Aerobacter aerogenes, Eberthella typhi* and *Escherichia coli*, J. Bacteriol. 33 (1937) 145–162.

[7] G.C. Paul, M.A. Priede, C.R. Thomas, Relationship between morphology and citric acid production in submerged *Aspergillus niger* fermentations, Biochem. Eng. J 3 (1999) 121–129.

[8] O. Rahn, G.L. Richardson, Oxygen demand and oxygen supply, J. Bacteriol. 41 (1941) 225–249.

[9] P. Gikas, A.G. Livingston, Use of specific ATP concentration and specific oxygen uptake rate to determine parameters of a structured model of biomass growth, Enzyme Microb. Technol. 22 (1998) 500–510.

[10] E.B. Chain, G. Gualandi, G. Morisi, Aeration studies. IV. Aeration conditions in 3000-liter submerged fermentations with various microorganisms, Biotechnol. Bioeng. 8 (1966) 595–619.

[11] A.H.E. Bijkerk, R.J. Hall, A mechanistic model of the aerobic growth of *Saccharomyces cerevisiae*, Biotechnol. Bioeng. 19 (1977) 267–296.

[12] A.L. Jensen, J.S. Schultz, P. Shu, Scale-up of antibiotic fermentations by control of oxygen utilization, Biotechnol. Bioeng. 8 (1966) 525–537.

[13] K. Ozergin-Ulgen, F. Mavituna, Oxygen transfer and uptake in *Streptomyces coelicolor* A3(2) culture in a batch bioreactor, J. Chem. Technol. Biotechnol. 73 (1998) 243–250.

[14] W.H. Bartholomew, E.O. Karow, M.R. Sfat, R.H. Wilhelm, Oxygen transfer and agitation in submerged fermentations, Ind. Eng. Chem. 42 (1950) 1801–1809.

[15] A. Pinches, L.J. Pallent, Rate and yield relationships in the production of xanthan gum by batch fermentations using complex and chemically defined growth media, Biotechnol. Bioeng. 28 (1986) 1484–1496.

[16] P.A. Bond, M.W. Fowler, A.H. Scragg, Growth of *Catharanthus roseus* cell suspensions in bioreactors: online analysis of oxygen and carbon dioxide levels in inlet and outlet gas streams, Biotechnol. Lett. 10 (1988) 713–718.

[17] A. Kato, S. Nagai, Energetics of tobacco cells, *Nicotiana tabacum* L., growing on sucrose medium, Eur. J. Appl. Microbiol. Biotechnol. 7 (1979) 219–225.

[18] P. Ducommun, P.-A. Ruffieux, M.-P. Furter, I. Marison, U. von Stockar, A new method for on-line measurement of the volumetric oxygen uptake rate in membrane aerated animal cell cultures, J. Biotechnol. 78 (2000) 139–147.

[19] S. Tatiraju, M. Soroush, R. Mutharasan, Multi-rate nonlinear state and parameter estimation in a bioreactor, Biotechnol. Bioeng. 63 (1999) 22–32.

[20]   S.P.S. Andrew, Gas–liquid mass transfer in microbiological reactors, Trans. IChemE. 60 (1982) 3–13.

[21]   J.J. Heijnen, K. van't Riet, A.J. Wolthuis, Influence of very small bubbles on the dynamic k_LA measurement in viscous gas–liquid systems, Biotechnol. Bioeng. 22 (1980) 1945–1956.

[22]   K. van't Riet, Review of measuring methods and results in nonviscous gas–liquid mass transfer in stirred vessels, Ind. Eng. Chem. Process Des. Dev. 18 (1979) 357–364.

[23]   A. Ogut, R.T. Hatch, Oxygen transfer into Newtonian and non-Newtonian fluids in mechanically agitated vessels, Can. J. Chem. Eng. 66 (1988) 79–85.

[24]   Y. Kawase, M. Moo-Young, The effect of antifoam agents on mass transfer in bioreactors, Bioprocess Eng. 5 (1990) 169–173.

[25]   A. Prins, K. van't Riet, Proteins and surface effects in fermentation: foam, antifoam and mass transfer, Trends Biotechnol. 5 (1987) 296–301.

[26]   National Research Council. 1930. International Critical Tables of Numerical Data, Physics, Chemistry and Technology. Washington, DC: The National Academies Press. https://doi.org/10.17226/20230.

[27]   Y.H. Lee, G.T. Tsao, Dissolved oxygen electrodes, Adv. Biochem. Eng 13 (1979) 35–86.

[28]   V. Linek, J. Sinkule, V. Vacek, Dissolved oxygen probes, in : M. Moo-Youg (Ed.) Comprehensive Biotechnology, vol 4, Pergamon Press, (1985) 363–394.

[29]   G.A. Truesdale, A.L. Downing, G.F. Lowden, The solubility of oxygen in pure water and sea-water, J. Appl. Chem. 5 (1955) 53–62.

[30]   G. Quicker, A. Schumpe, B. König, W.-D. Deckwer, Comparison of measured and calculated oxygen solubilities in fermentation media, Biotechnol. Bioeng. 23 (1981) 635–650.

[31]   P.R. Gogate, A.B. Pandit, Survey of measurement techniques for gas–liquid mass transfer coefficient in bioreactors, Biochem. Eng. J. 4 (1999) 7–15.

[32]   M. Sobotka, A. Prokop, I.J. Dunn, A. Einsele, Review of methods for the measurement of oxygen transfer in microbial systems, Ann. Rep. Ferm. Proc. 5 (1982) 127–210.

[33]   J.D. Brooks, D.G. Maclennan, J.P. Barford, R.J. Hall, Design of laboratory continuous-culture equipment for accurate gaseous metabolism measurements, Biotechnol. Bioeng. 24 (1982) 847–856.

[34]   M. Nocentini, Mass transfer in gas–liquid, multiple-impeller stirred vessels, Trans. IChemE. 68 (1990) 287–294.

[35]   S. Shioya, I.J. Dunn, Model comparisons for dynamic kLa measurements with incompletely mixed phases, Chem. Eng. Commun. 3 (1979) 41–52.

[36]   K. Petera, P. Ditl, Effect of pressure profile on evaluation of volumetric mass transfer coefficient in kLa bioreactors, Biochem. Eng. J. 5 (2000) 23–27.

[37]   V. Linek, P. Beneš, V. Vacek, Dynamic pressure method for kLa measurement in large-scale bioreactors, Biotechnol. Bioeng. 33 (1989) 1406–1412.

[38]   V. Linek, T. Moucha, J. Doušová, J. Sinkule, Measurement of kLa by dynamic pressure method in pilot-plant fermentor, Biotechnol. Bioeng. 43 (1994) 477–482.

Suggestions for Further Reading

Mass Transfer Theory

See also references [1].

W.L. McCabe, J.C. Smith, P. Harriott, Chapter 17Unit Operations of Chemical Engineering, 7th ed., McGraw-Hill, 2005.

K. van't Riet, Mass transfer in fermentation, Trends Biotechnol. 1 (1983) 113–119.

Measurement of k_La and q_{O_2}

See also references 31–38.

V. Linek, J. Sinkule, P. Beneš, Critical assessment of gassing-in methods for measuring kLa in fermentors, Biotechnol. Bioeng. 38 (1991) 323–330.

K. Pouliot, J. Thibault, A. Garnier, G. Acuña Leiva, K_La evaluation during the course of fermentation using data reconciliation techniques, Bioprocess Eng. 23 (2000) 565–573.

P.-A. Ruffieux, U. von Stockar, I.W. Marison, Measurement of volumetric (OUR) and determination of specific (qO_2) oxygen uptake rates in animal cell cultures, J. Biotechnol. 63 (1998) 85–95.

M. Tobajas, E. García-Calvo, Comparison of experimental methods for determination of the volumetric mass transfer coefficient in fermentation processes, Heat Mass Transfer. 36 (2000) 201–207.

U. Maier, J. Büchs, Characterisation of the gas–liquid mass transfer in shaking bioreactors, Biochem. Eng. J. 7 (2001) 99–106.

C. Mrotzek, T. Anderlei, H.-J. Henzler, J. Büchs, Mass transfer resistance of sterile plugs in shaking bioreactors, Biochem. Eng. J. 7 (2001) 107–112.

J.S. Schultz, Cotton closure as an aeration barrier in shaken flask fermentations, Appl. Microbiol. 12 (1964) 305–310.

J.C. van Suijdam, N.W.F. Kossen, A.C. Joha, Model for oxygen transfer in a shake flask, Biotechnol. Bioeng. 20 (1978) 1695–1709.

PART 4

Reactions

CHAPTER 10

Homogeneous Reactions

A typical bioprocess transforms a relatively inexpensive set of substrates into value-added products using chemical and biochemical reactions. The *catalytic* reactions occur in vessels known as bioreactors which establish and control physicochemical environments to facilitate the reactions. By definition, a catalyst is a substance that affects the rate of reaction without altering the reaction equilibrium or being consumed in the reaction. Biocatalysts can be of viral, microbial, plant, or animal origin and can be individual enzymes, enzyme complexes, cellular organelles, or whole cells. Properties such as the reaction rate and yield of product from substrate characterize the performance of catalytic systems. Knowledge of these parameters is crucial for the design and operation of reactors. For bioprocesses producing commodity compounds like fuel ethanol, characteristics of the reaction rate and yield determine, to a large extent, the economic feasibility of the project.

Catalytic reactions can be categorized as *homogeneous* or *heterogeneous* reactions based on the physicochemical properties of the system. A reaction is homogeneous if it occurs in a single phase (e.g., liquid) and the temperature and all concentrations in the system are uniform. Most bioprocesses occurring in mixed vessels fall into this category; for classification purposes, suspended enzymes and cells are not considered a second phase. In contrast, heterogeneous reactions take place in the presence of multiple phases (e.g., solid and liquid) where gradients in concentration or temperature can occur. Analysis of heterogeneous reactions often requires application of transport principles in conjunction with reaction theory. Heterogeneous reactions are the subject of Chapter 11.

Here, we consider the basic aspects of reaction theory that allow us to quantify the extent and speed of homogeneous reactions and to identify the important factors controlling reaction rate.

10.1 Basic Reaction Theory

Reaction theory has two fundamental and distinct contributions: *reaction thermodynamics* and *reaction kinetics*. Reaction thermodynamics is concerned with energetics and how they affect *reaction spontaneity*, the *direction of chemical reactions*, as well as *how far* reactions can proceed before the energic driving force is dissipated. On the other hand, reaction kinetics is concerned with the *rate* at which reactions occur while being constrained by reaction thermodynamics.

10.1.1 Reaction Thermodynamics

Consider a reversible reaction represented by the following equation:

$$A + bB \rightleftharpoons yY + zZ \qquad (10.1)$$

A, B, Y, and Z are chemical species; b, y, and z are stoichiometric coefficients. If the components are placed in a closed system, the reaction proceeds, either to the right or left side of the equation, until *thermodynamic equilibrium* is achieved. At equilibrium, there is no net driving force for further change: the system has reached the limit of its capacity for chemical transformation. The equilibrium concentrations of reactants and products are related by the *equilibrium constant*, K_{eq}. For the reaction of Eq. (10.1):

$$K_{eq} = \frac{[Y]_e^y [Z]_e^z}{[A]_e [B]_e^b} \qquad (10.2)$$

where $[\ ]_e$ denotes the molar concentration (gmol L^{-1}) of A, B, Y, or Z at equilibrium. In aqueous systems, if water, H$^+$ ions, or solid substances are involved in the reaction, these components are accounted for by the equilibrium constant K_{eq} and are not explicitly listed on the right side of Eq. (10.2).

Reactions with large K_{eq} values favor large concentrations of products at equilibrium while reactions with small K_{eq} values favor large concentrations of reactants at equilibrium. K_{eq} varies with temperature as follows:

$$\ln K_{eq} = \frac{-\Delta G^\circ_{rxn}}{RT} \qquad (10.3)$$

where ΔG°_{rxn} is the *change in standard Gibbs energy* per mole of A reacted, R is the ideal gas constant, and T is absolute temperature. Values of R are listed in Appendix B. The superscript $^\circ$ in ΔG°_{rxn} denotes standard conditions. Usually, the standard condition for a substance is its most stable form at 1 atm pressure and 25°C; however, for biochemical reactions occurring in solution, other standard conditions may be used. [1] ΔG°_{rxn} is equal to the difference between the standard Gibbs energies of formation of the products and reactants:

$$\Delta G^\circ_{rxn} = yG^\circ_Y + zG^\circ_Z - G^\circ_A - bG^\circ_B \qquad (10.4)$$

where G°_i is the *standard Gibbs energy of formation* of chemical species i. Values of G°_i are available in chemistry and physics handbooks.

Gibbs energy G is related to enthalpy H, entropy S, and absolute temperature T as follows:

$$\Delta G = \Delta H - T\Delta S \qquad (10.5)$$

Therefore, from Eq. (10.3):

$$\ln K_{eq} = \frac{-\Delta H^{\circ}_{rxn}}{RT} + \frac{\Delta S^{\circ}_{rxn}}{R} \qquad (10.6)$$

Thus, for exothermic reactions which by definition have a negative ΔH°_{rxn}, K_{eq} decreases with increasing temperature. For endothermic reactions which by definition have a positive ΔH°_{rxn}, K_{eq} increases with increasing temperature.

A limited number of commercially important enzyme conversions, such as glucose isomerization and starch hydrolysis, are treated as reversible reactions (see Example 10.1). The reaction mixture at equilibrium typically contains significant amounts of both reactants and products. When ΔG°_{rxn} is negative and large in magnitude, the K_{eq} is also very large, and the reaction strongly favors the products relative to the reactants at equilibrium. These reactions are often approximated as *irreversible*.

Many industrial bioprocesses catalyzed by either enzymes or cells can be treated as irreversible. For example, the equilibrium constant for sucrose hydrolysis by the enzyme invertase is about 10^4 while the fermentation of glucose to ethanol and carbon dioxide, K_{eq} is about 10^{30}. The equilibrium ratio of products to reactants is so overwhelmingly large for these reactions that they are considered to proceed practically to completion (i.e., the reaction stops only when the concentration of reactants is nearly zero). Equilibrium thermodynamics provides insight into which reactions are favorable and what the equilibrium concentrations will be, however, many metabolic processes exist in a dynamic state, far from equilibrium.

10.1.2 Reaction Yield

The relationship between the amount of product formed per amount of reactant consumed is known here as the *yield*. There are different definitions of the term 'yield' in diffident technical disciplines. The definition of yield for this text will be defined here for convenience.

Consider the enzyme reaction:

$$\text{L-histidine} \rightarrow \text{urocanic acid} + NH_3 \qquad (10.7)$$

catalyzed by histidase. Based on the reaction stoichiometry, 1 gmol of urocanic acid is produced for each gmol of L-histidine consumed; the yield of urocanic acid from histidine is therefore 1 gmol urocanic acid (gmol L-histidine)$^{-1}$. However, let us assume that the histidase solution is contaminated with another enzyme, histidine decarboxylase. Histidine decarboxylase catalyzes the following reaction:

$$\text{L-histidine} \rightarrow \text{histamine} + CO_2 \qquad (10.8)$$

EXAMPLE 10.1 Effect of Temperature on Glucose Isomerization

Glucose isomerase is used extensively in the United States for production of high-fructose syrup from corn-derived sugar. The reaction is:

$$glucose \rightleftharpoons fructose$$

$\Delta H°_{rxn}$ for this reaction is 5.73 kJ gmol^{-1}; $\Delta S°_{rxn}$ is 0.0176 kJ gmol^{-1} K^{-1}.

a. Calculate the equilibrium constants at 50°C and 75°C.
b. A company aims to develop a sweeter mixture of sugars, that is, one with a higher concentration of fructose. Considering equilibrium only, would it be more desirable to operate the reaction at 50°C or 75°C?

Solution

a. Convert the temperatures from degrees Celsius to Kelvin:

$$T = 50°C = 323.15 \text{ K}$$

$$T = 75°C = 348.15 \text{ K}$$

$R = 8.3144$ J gmol^{-1} K^{-1} = 8.3144 × 10^{-3} kJ gmol^{-1} K^{-1}. Using Eq. (10.6):

$$\ln K_{eq}(50°C) = \frac{-5.73 \text{ kJ gmol}^{-1}}{(8.3144 \times 10^{-3} \text{ kJ gmol}^{-1} \text{ K}^{-1})323.15 \text{ K}} + \frac{0.0176 \text{ kJ gmol}^{-1} \text{ K}^{-1}}{8.3144 \times 10^{-3} \text{ kJ gmol}^{-1} \text{ K}^{-1}}$$

$$K_{eq}(50°C) = 0.98$$

Similarly for $T = 75°C$:

$$\ln K_{eq}(75°C) = \frac{-5.73 \text{ kJ gmol}^{-1}}{(8.3144 \times 10^{-3} \text{ kJ gmol}^{-1} \text{ K}^{-1})348.15 \text{K}} + \frac{0.0176 \text{ kJ gmol}^{-1} \text{K}^{-1}}{8.3144 \times 10^{-3} \text{ kJ gmol}^{-1} \text{K}^{-1}}$$

$$K_{eq}(75°C) = 1.15$$

b. As K_{eq} increases, the fraction of fructose in the equilibrium mixture increases. Therefore, from an equilibrium point of view, it is more desirable to operate the reactor at 75°C. However, other factors such as enzyme deactivation and energy requirements to maintain a relatively high temperature need to be considered.

If both enzymes are active, some L-histidine will react with histidase according to Eq. (10.7) and some will be decarboxylated according to Eq. (10.8). Analysis of the reaction mixture shows that 1 gmol of urocanic acid and 1 gmol of histamine are produced for every 2 gmol of histidine consumed. The *observed* or *apparent yield* of urocanic acid from L-histidine is 1 gmol urocanic acid (2 gmol L-histamine)$^{-1}$ = 0.5 gmol product (gmol substrate)$^{-1}$. The observed yield of 0.5 gmol product (gmol substrate)$^{-1}$ is different from the *stoichiometric, true*, or *theoretical yield* of 1 gmol product (gmol substrate)l^{-1} calculated from the reaction stoichiometry of Eq. (10.7) because the reactant was consumed in two separate reactions. An analogous situation arises if product rather than reactant is consumed in other reactions; in this case, the observed yield of product is lower than the theoretical yield. *When reactants or products are involved in additional reactions, the observed yield may be different from the theoretical yield.*

This analysis leads to two useful definitions of yield for reaction systems:

$$\begin{pmatrix} \text{true, stoichiometric, or} \\ \text{theoretical yield} \end{pmatrix} = \frac{\begin{pmatrix} \text{mass or moles of} \\ \text{product formed} \end{pmatrix}}{\begin{pmatrix} \text{mass or moles of reactant used} \\ \text{to form that particular product} \end{pmatrix}} \qquad (10.9)$$

and

$$(\text{observed or apparent yield}) = \frac{(\text{mass or moles of product present})}{(\text{mass or moles of reactant consumed})} \qquad (10.10)$$

There is a third type of yield applicable in certain situations. For reactions with incomplete conversion of reactant, it may be of interest to specify the amount of product formed per initial amount of reactant *provided to the reaction* rather than actually consumed. For example, consider the isomerization reaction catalyzed by glucose isomerase:

$$\text{glucose} \rightleftarrows \text{fructose} \qquad (10.11)$$

The reaction is carried out in a closed bioreactor with purified enzyme. At equilibrium, the sugar mixture contains 55 mol% glucose and 45 mol% fructose. The *theoretical yield* of fructose from glucose is 1 gmol fructose (gmol^{-1} glucose)$^{-1}$. The *observed yield* would also be 1 gmol fructose (gmol glucose)$^{-1}$. However, if the reaction is started with only glucose present, the equilibrium relationship of fructose to glucose is 0.45 gmol fructose (gmol glucose)$^{-1}$ because 0.55 gmol of glucose remains unreacted. This type of yield for incomplete reactions may be denoted as *gross yield* or in some cases, when appropriate stoichiometric units are considered, *conversion*:

$$\text{Gross yield} = \frac{(\text{Mass or moles of product present})}{(\text{Mass or moles of reactant provided to the reaction})} \qquad (10.12)$$

Example 10.2 illustrates constraints on reaction progression based on equilibrium considerations.

EXAMPLE 10.2 Incomplete Enzyme Reaction

An enzyme catalyzes the reaction:

$$A \rightleftarrows B$$

At equilibrium, the reaction mixture contains $6.3\,\mathrm{g\,L^{-1}}$ of A and $3.7\,\mathrm{g\,L^{-1}}$ of B.

a. What is the equilibrium constant?
b. If the reaction starts with A only, what is the yield of B from A at equilibrium?

Solution

a. From stoichiometry, the molecular weights of A and B must be equal. Therefore, the ratio of molar concentrations of A and B is equal to the ratio of mass concentrations. From Eq. (10.2):

$$K_{eq} = \frac{[B]_e}{[A]_e} = \frac{3.7\,\mathrm{g\,L^{-1}}}{6.3\,\mathrm{g\,L^{-1}}} = 0.59$$

b. Total mass is conserved in chemical reactions. Using a basis of 1 L, the total mass at equilibrium is $(6.3\,\mathrm{g} + 3.7\,\mathrm{g}) = 10\,\mathrm{g}$, if the reactions start with A only, 10 g of A must have been provided. 6.3 g of A remain at equilibrium, therefore 3.7 g of A must have been consumed during in reaction. Based on reaction stoichiometry, the true yield of B from A is 1 gmol B gmol^{-1} A. The observed yield is 3.7 g of B (3.7 g of A)$^{-1}$ =1 g B g^{-1} A or, in this case, 1 gmol B gmol^{-1} A. The gross yield is 3.7 g of B (10 g of A)$^{-1}$ = 0.37 g B g^{-1} A or 0.37 gmol B gmol^{-1} A.

10.1.3 Reaction Rate

Consider the general irreversible reaction:

$$a\,A + b\,B \rightarrow y\,Y + z\,Z \tag{10.13}$$

The rate of this reaction can be represented by the rate of consumption of compound A. Let us use the symbol R_A to denote the *total rate of reaction with respect to A*. The units of R_A are, for example, kg A s^{-1}.

How do we measure reaction rates? For a general reaction system that is open to the surroundings, the rate of reaction is related to the rate of change of mass in the system using the unsteady-state mass balance equation derived in Chapter 4:

$$\frac{dM}{dt} = \hat{M}_i - \hat{M}_o + R_G - R_C \tag{4.29}$$

where M is mass, t is time, $\hat{M}_i$ is the mass flow rate into the system, $\hat{M}_o$ is the mass flow rate out of the system, R_G is the mass rate of generation by reaction, and R_C is the mass rate of consumption by the reaction. Let us apply Eq. (4.29) to compound A, assuming that the reaction of Eq. (10.13) is the only reaction taking place. The rate of consumption R_C is equal to R_A and $R_G = 0$. The mass balance equation becomes:

$$\frac{dM_A}{dt} = \hat{M}_{Ai} - \hat{M}_{Ao} - R_A \tag{10.14}$$

Therefore, the rate of reaction R_A can be determined if we measure the rate of change in the mass of A in the system, dM_A/dt, and the rates of A entering and leaving the system, $\hat{M}_{Ai}$ and $\hat{M}_{Ao}$. In a *closed system* where $\hat{M}_{Ai} = \hat{M}_{Ao} = 0$, and Eq. (10.14) becomes:

$$R_A = \frac{-dM_A}{dt} \tag{10.15}$$

and the reaction rate can be measured simply by monitoring the rate of change in the mass of A with time. Most measurements of reaction rate are performed in closed systems and the data can be analyzed using Eq. (10.15). dM_A/dt is negative when A is consumed by the reaction; therefore, the minus sign in Eq. (10.15) is necessary to make R_A a positive quantity. Rate of reaction is sometimes called *reaction velocity*. Reaction velocity can also be measured in terms of components B, Y, or Z. In a closed system:

$$R_B = \frac{-dM_B}{dt} \quad R_Y = \frac{dM_Y}{dt} \quad R_Z = \frac{dM_Z}{dt} \tag{10.16}$$

where M_B, M_Y, and M_Z are masses of B, Y, and Z, respectively. When reporting reaction rate, the reactant being monitored should be specified. Because R_Y and R_Z are based on product accumulation, these reaction rates are called *production rates* or *productivity*.

Reaction rates can be expressed using different measurement bases each with their own set of units. There are three distinct reaction rates that are useful in different situations.

1. *Total rate.* Total reaction rate is defined in Eqs. (10.15) and (10.16) and has units of either mass or moles per unit time. Total rate is useful for specifying the output of a particular reactor or manufacturing plant. Production rates for factories are often expressed as total rates; for example: "The production rate is 100,000 tons per year." If additional reactors are built so that the reaction volume in the plant is increased, then the total reaction rate would increase. Similarly, if the amount of cells or enzyme used in each reactor were also increased, then the total production rate would increase even further.

2. *Volumetric rate.* It is often convenient to specify the reaction rate as a rate normalized to a volume basis. Units of volumetric rate are, for example, $kg\ m^{-3}\ s^{-1}$. If the reaction occurs in a closed system with volume V, r_A is the volumetric rate of reaction with respect to A and is calculated by dividing the total reaction rate for A by the volume:

$$r_A = \frac{R_A}{V} = \frac{-1}{V}\frac{dM_A}{dt} \tag{10.17}$$

When V is constant, Eq. (10.17) can be written:

$$r_A = \frac{-dC_A}{dt} \tag{10.18}$$

where C_A is the concentration of A in units of, for example, kg m^{-3}. Volumetric rates are particularly useful for comparing the performance of reactors of different size. A common objective in reactor design is to maximize the volumetric productivity so that the desired total production rate can be achieved using reactors of minimum size since reactors are very costly.

1. *Specific rate*. Biological reactions involve enzyme or cell catalysts. Because the total rate of conversion depends on the amount of catalyst present, it can be useful to specify the reaction rate as the rate per quantity of enzyme or cells. In a closed system, the specific reaction rate is the total rate R_A divided by the amount of enzyme or cells:

$$q_A = -\left(\frac{1}{E} \text{ or } \frac{1}{X} \right) \frac{dM_A}{dt} \tag{10.19}$$

where q_A is the specific rate of reaction with respect to A, E is the quantity of enzyme, X is the quantity of cells, and dM_A/dt is the total rate of change of the mass of A in the system. As the quantity of cells is often expressed as mass, the units of specific rate for a cell-catalyzed reaction would be, for example, kg A (kg cdw)$^{-1}$ s^{-1}. On the other hand, the mass of a particular enzyme added to a reaction is rarely known; most commercial enzyme preparations contain several components in unknown and variable proportions depending on the batch obtained from the manufacturer. To overcome these difficulties, enzyme quantity is often expressed as *units of activity* measured under specified conditions. One unit of enzyme is usually taken to be the amount that catalyzes the conversion of 1 µmole of substrate per minute at the optimal temperature, pH, and substrate concentration. Therefore, if E in Eq. (10.19) is expressed as units of enzyme activity, the specific rate of reaction could be reported as, for example, kg A (unit enzyme)$^{-1}$ s^{-1}. In a closed system where the volume of the reaction mixture remains constant, an alternative expression for the specific reaction rate is:

$$q_A = -\left(\frac{1}{e} \text{ or } \frac{1}{x} \right) \frac{dC_A}{dt} \tag{10.20}$$

where e is enzyme concentration and x is cell concentration.

Volumetric and total rates are not a direct reflection of catalyst performance; this is represented by the specific rate. Specific rates are employed when comparing different enzymes or cells. The specific rate is not dependent on the size of the system or the amount of catalyst present under most circumstances. Different strains of organism should be compared in terms of their specific reaction rates, not in terms of the volumetric or total rate.

In this book, the symbol R_A will be used to denote total reaction rate with respect to component A; r_A represents the volumetric rate and q_A represents the specific rate. These types of rates are interrelated and are central concepts in bioprocess design. The relationship between the three rates should be understood and an engineer should know how to convert from one type of rate to another ($R_A = r_A V = q_A x V$).

10.1.4 Reaction Kinetics

As reactions proceed in closed systems, the concentrations of the reactants decrease. In general, the rate of reaction depends on reactant concentration; therefore, the rate of reaction usually decreases with time. Reaction rates also vary with temperature; most reaction rates increase as the temperature rises. *Reaction kinetics* refers to the relationship between the rate of reaction and the conditions that affect it, such as reactant concentration and temperature. These relationships are conveniently described using *kinetic equations*.

Consider again the irreversible reaction of Eq. (10.13). The volumetric rate of this reaction can be expressed as a function of reactant concentrations using the following equation known as a *rate law*:

$$r_A = k C_A^\alpha C_B^\beta \tag{10.21}$$

where k is the *rate constant* or *rate coefficient* for the reaction, C_A is the concentration of reactant A, and C_B is the concentration of reactant B. The rate constant is independent of the concentrations of the reacting species, by definition, but is dependent on other system parameters, such as temperature. The reaction represented by Eq. (10.21) is said to be of *order* α with respect to reactant A and order β with respect to reactant B. The overall order of the reaction is ($\alpha + \beta$). If the orders in the rate law are consistent with the stoichiometric coefficients in the reaction equation, the reaction is said to *follow an elementary rate law*. However, this simplification is not always appropriate and the actual power law relationships in the rate law should be determined by experiment. The dimensions and units of k depend on the order of the reaction.

10.1.5 Effect of Temperature on Reaction Rate

Temperature has a significant effect on reactions. The change in the rate constant k with temperature is described by the *Arrhenius equation*:

$$k = A e^{-E/RT} \tag{10.22}$$

where k is the rate constant, A is the *Arrhenius constant* or *frequency factor*, E is the *activation energy* for the reaction, R is the ideal gas constant, and T is absolute temperature. The value of E is positive and large for many reactions so that, as T increases, k increases rapidly. Taking the natural logarithm of both sides of Eq. (10.22):

$$\ln k = \ln A - \frac{E}{RT} \tag{10.23}$$

Therefore, a plot of $\ln k$ versus $1/T$ gives a straight line with slope of $-E/R$ and an intercept of $\ln A$.

10.2 Calculation of Reaction Rates From Experimental Data

The volumetric rate of reaction in a closed system can be found by measuring the rate of change in the mass of reactant present as outlined in Section 10.1.3, provided the reactant is involved in only one reaction. Most kinetic studies of biological reactions are carried out in closed systems with a constant volume; therefore, Eq. (10.18) can be used to evaluate the volumetric reaction rate. The concentration of a particular reactant or product is measured as a function of time. For a reactant such as A in Eq. (10.13), the results will be like those shown in Figure 10.1(a): the concentration will decrease with time. The volumetric rate of reaction is equal to dC_A/dt, which can be evaluated as the slope of a curve drawn through the data. The slope of the curve in Figure 10.1(a) changes with time; the reaction rate is greater at the beginning of the experiment than at the end.

One way to determine reaction rate is to draw tangents to the curve of Figure 10.1(a) at various times and evaluate the slopes of the tangents; this is shown in Figure 10.1(b). Drawing tangents to curves is a highly subjective procedure prone to inaccuracy, even with special drawing tools designed for the purpose. More reliable techniques are available for *graphical differentiation* of rate data. Graphical differentiation is valid only if the data can be presumed to differentiate smoothly.

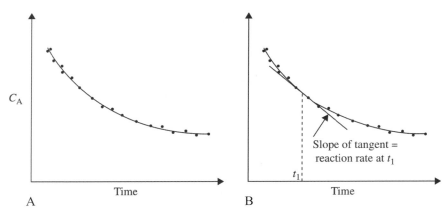

Figure 10.1 (A) Change in reactant concentration with time during reaction. (B) Graphical differentiation of concentration data by drawing a tangent.

10.2.1 *Average Rate–Equal Area Method*

This technique for determining rates is based on the *average rate–equal area construction* and will be demonstrated using data for O_2 uptake by cells in suspension culture. Results from the measurement of O_2 concentration in a closed system as a function of time are listed in the first two columns of Table 10.1. The average rate–equal area method involves the following steps.

- Tabulate values of ΔC_A and Δt for each time interval as shown in Table 10.1. ΔC_A values are negative because C_A decreases over each interval.
- Calculate the average O_2 uptake rate, $\Delta C_A / \Delta t$, for each time interval.
- Plot $\Delta C_A / \Delta t$ on linear graph paper. Over each time interval, draw a horizontal line to represent $\Delta C_A / \Delta t$ for that interval; this is shown in Figure 10.2.
- Draw a smooth curve to cut the horizontal lines in such a manner that the shaded areas above and below the curve are equal for each time interval. The curve thus developed gives values of dC_A / dt for all points in time. Results for dC_A / dt at the times of sampling can be read from the curve and are tabulated in Table 10.1.

The average rate–equal area method is not easily applied if the data has substantial scatter. If the concentration measurements were not very accurate, the horizontal lines representing $\Delta C_A / \Delta t$ might be located as shown in Figure 10.3. A curve equalizing the areas above and below each $\Delta C_A / \Delta t$ line would then show complex behavior as indicated by the dashed line in Figure 10.3. Experience suggests that this is not a realistic representation of reaction rate. The data of Figure 10.3 are better represented using a smooth curve to equalize as far as possible the areas above and below adjacent groups of horizontal lines.

Table 10.1: Graphical Differentiation Using the Average Rate–Equal Area Construction

Time (t, min)	O_2 concentration (C_A, ppm)	ΔC_A	Δt	$\Delta C_A / \Delta t$	dC_A / dt
0.0	8.00				−0.59
		−0.45	1.0	−0.45	
1.0	7.55				−0.38
		−0.33	1.0	−0.33	
2.0	7.22				−0.29
		−0.26	1.0	−0.26	
3.0	6.96				−0.23
		−0.20	1.0	−0.20	
4.0	6.76				−0.18
		−0.15	1.0	−0.15	
5.0	6.61				−0.14
		−0.12	1.0	−0.12	
6.0	6.49				−0.11
		−0.16	2.0	−0.08	
8.0	6.33				−0.06
		−0.08	2.0	−0.04	
10.0	6.25				−0.02

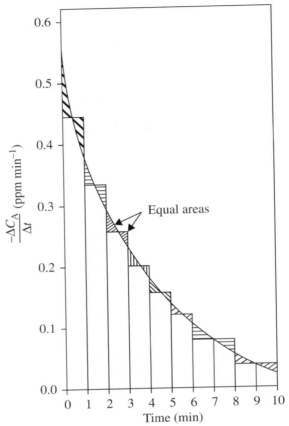

Figure 10.2 Graphical differentiation using the average rate–equal area construction.

10.2.2 Midpoint Slope Method

In this method, the raw data are smoothed and values tabulated at intervals. The midpoint slope method is illustrated using the same data as analyzed in Section 10.2.1.

- Plot the raw concentration data as a function of time and draw a smooth curve through the points. This is shown in Figure 10.4.
- Mark off the smoothed curve at time intervals of ε. ε should be chosen so that the number of intervals is less than the number of datum points measured; the less accurate the data, the fewer should be the intervals. In this example, ε is taken as 1.0 min until $t = 6$ min; thereafter $\varepsilon = 2.0$ min. The intervals are marked in Figure 10.4 as dashed lines. Values of ε are entered in Table 10.2.
- In the midpoint slope method, rates are calculated midway between two adjacent intervals of size ε. Therefore, the first rate determination is made for $t = 1$ min. Calculate the

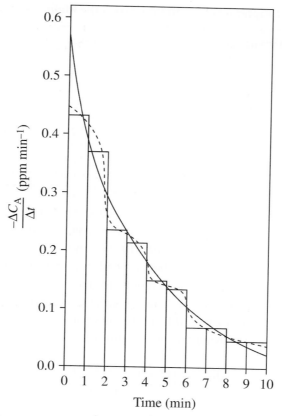

Figure 10.3 Average rate–equal area method for data with experimental error.

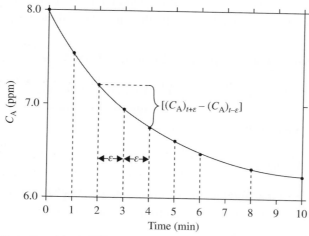

Figure 10.4 Graphical differentiation using the midpoint slope method.

Table 10.2: Graphical Differentiation Using the Midpoint Slope Method

Time (t, min)	O$_2$ concentration (CA, ppm)	ε	[(CA)t + ε – (CA)t – ε]	dCA/dt
0.0	8.00	1.0	–	–
1.0	7.55	1.0	–0.78	–0.39
2.0	7.22	1.0	–0.59	–0.30
3.0	6.96	1.0	–0.46	–0.23
4.0	6.76	1.0	–0.35	–0.18
5.0	6.61	1.0	–0.27	–0.14
6.0	6.49	1.0	–0.22	–0.11
8.0	6.33	2.0	–0.24	–0.06
10.0	6.25	2.0	–	–

differences $[(C_A)t + \varepsilon - (C_A)t - \varepsilon]$ from Figure 10.4, where $(C_A)t + \varepsilon$ denotes the concentration of A at time $t - \varepsilon$, and $(C_A)t - \varepsilon$ denotes the concentration at time $t - \varepsilon$. A difference calculation is illustrated in Figure 10.4 for $t = 3$ min. Note that the concentrations are not taken from the list of original data but are read from the smoothed curve through the points. When $t = 6$ min, $\varepsilon = 1.0$; concentrations for the difference calculation are read from the curve at $t - \varepsilon = 5$ min and $t + \varepsilon = 7$ min. For the last rate determination at $t = 8$ min, $\varepsilon = 2$ and the concentrations are read from the curve at $t - \varepsilon = 6$ min and $t + \varepsilon = 10$ min.

- Determine the slope or rate using the central-difference formula:

$$\frac{dC_A}{dt} = \frac{\left[\left(C_A \right)_{t+\varepsilon} - \left(C_A \right)_{t-\varepsilon} \right]}{2\varepsilon} \tag{10.24}$$

The results are listed in Table 10.2.

The values of dC_A/dt calculated using the two differentiation methods (Tables 10.1 and 10.2) compare favorably. Application of both methods using the same data allows checking of the results.

10.3 General Reaction Kinetics for Biological Systems

The kinetics of many biological reactions can be treated as zero-order, first-order, or a combination of these known as Michaelis–Menten or saturation kinetics. Mathematical expressions for biological kinetics are examined in this section.

10.3.1 Zero-Order Kinetics

If a reaction obeys zero-order kinetics, the reaction rate is independent of reactant concentration, in other words, the reaction rate is a function of reactant concentration raised to the zero power. The expression is written as:

$$r_A = k_0 s^0 = k_0 \tag{10.25}$$

where r_A is the volumetric rate of reaction with respect to component A and k_0 is the *zero-order rate constant*. k_0 as defined in Eq. (10.25) as a volumetric rate constant with units of, for example, mol m^{-3} s^{-1}. Because the volumetric rate of a catalytic reaction depends on the amount of catalyst present, when Eq. (10.25) is used to represent the rate of an enzyme or cell reaction, the value of k_0 includes the catalyst concentration as well as the specific rate of reaction. We could write:

$$k_0 = k_0'e \quad \text{or} \quad k_0 = k_0''x \tag{10.26}$$

where k_0' is the specific zero-order rate constant for enzyme reaction and e is the concentration of enzyme. Correspondingly, k_0'' is the specific zero-order rate constant for cell reaction and x is the cell concentration.

Let us assume that we have concentration data for reactant A as a function of time and wish to determine the appropriate kinetic constant for the reaction. The rate of reaction can be evaluated as the rate of change of C_A using the methods for graphical differentiation described in Section 10.2. Once r_A is found, if the reaction is zero-order, r_A will be constant and equal to k_0 at all times during the reaction. However, as an alternative approach, because the kinetic expression for zero-order reaction is relatively simple, rather than differentiate the concentration data, it is easier to integrate the rate equation to obtain an equation for C_A as a function of time. The experimental data can then be checked against the integrated equation. Combining Eqs. (10.18) and (10.25), the rate equation for a zero-order reaction in a closed, constant-volume system is:

$$\frac{-dC_A}{dt} = k_0 \tag{10.27}$$

Separating variables and integrating with initial condition $C_A = C_{A0}$ at $t = 0$ gives:

$$\int dC_A = \int -k_0 dt \tag{10.28}$$

or

$$C_A - C_{A0} = -k_0 t \tag{10.29}$$

Rearranging gives an equation for C_A:

$$C_A = C_{A0} - k_0 t \tag{10.30}$$

Therefore, when the reaction is zero-order, a plot of C_A versus time gives a straight line with slope $-k_0$. Application of Eq. (10.30) is illustrated in Example 10.3.

EXAMPLE 10.3 Kinetics of O$_2$ Uptake

Serratia marcescens is cultured in minimal medium in a small, stirred bioreactor. O$_2$ consumption is measured at a cell concentration of 22.7 g cultured in minimal me L^{-1}.

Time (min)	O$_2$ concentration (mmol L^{-1})
0	0.25
2	0.23
5	0.21
8	0.20
10	0.18
12	0.16
15	0.15

a. Determine the rate constant for O$_2$ uptake.
b. If the cell concentration is reduced to 12 g L^{-1}, what is the value of the rate constant?

Solution

a. As outlined in Section 10.5.1, microbial O$_2$ consumption is a zero-order reaction over a wide range of O$_2$ concentrations above C$_{crit}$. To test if the measured data are consistent with the zero-order kinetic model of Eq. (10.30), O$_2$ concentration is plotted as a function of time as shown in Figure 10.5. A zero-order model using a straight line to connect the points fits the data well. The slope is -6.7×10^{-3} mmol L^{-1} min^{-1}; therefore, $k_0 = 6.7 \times 10^{-3}$ mmol L^{-1} min^{-1}.
b. For the same cells cultured under the same conditions, from Eq. (10.26), k_0 is directly proportional to the number of cells present. Therefore, at a cell concentration of 12 g cdw L^{-1}:

$$k_0 = \frac{12 \text{ gL}^{-1}}{22.7 \text{ g L}^{-1}} (6.7 \times 10^{-3} \text{ mmol L}^{-1} \text{min}^{-1})$$

$$k_0 = 3.5 \times 10^{-3} \text{ mmol L}^{-1} \text{min}^{-1}$$

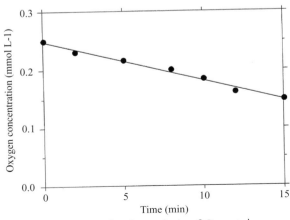

Figure 10.5 Kinetic analysis of O$_2$ uptake.

10.3.2 First-Order Kinetics

If a reaction obeys first-order kinetics, the relationship between reaction rate and reactant concentration scales with reactant concentration raised to the first order. The expression is as follows:

$$r_A = k_1 s^1 = k_1 s \tag{10.31}$$

where r_A is the volumetric rate of reaction with respect to A, k_1 is the *first-order rate constant*, and C_A is the concentration of A. k_1 has dimensions T^{-1} and units of, for example, s^{-1}. Like the zero-order rate constant in Section 10.3.1, the value of k_1 depends on the catalyst concentration.

If a first-order reaction takes place in a closed, constant-volume system, from Eqs. (10.18) and (10.31):

$$\frac{dC_A}{dt} = -k_1 C_A \tag{10.32}$$

Separating variables and integrating with initial condition $C_A = C_{A0}$ at $t = 0$ gives:

$$\int \frac{dC_A}{C_A} = \int -k_1 dt \tag{10.33}$$

or

$$\ln C_A = \ln C_{A0} - k_1 t \tag{10.34}$$

Rearranging gives an equation for C_A as a function of time:

$$C_A = C_{A0} e^{-k_1 t} \tag{10.35}$$

According to Eq. (10.34), if a reaction follows first-order kinetics, a plot of $ln C_A$ versus time gives a straight line with slope $-k_1$ and intercept $ln C_{A0}$. Analysis of first-order kinetics is illustrated in Example 10.4.

10.3.3 Michaelis–Menten Kinetics

The kinetics of many enzyme reactions can be reasonably represented by the *Michaelis–Menten equation* which is a form of *saturation* kinetics:

$$r_A = \frac{v_{max} C_A}{K_m + C_A} \tag{10.36}$$

EXAMPLE 10.4 Kinetics of Crude Oil Degradation

Soil contaminated with crude oil is treated using a mixed population of indigenous bacteria and fungi. The concentration of total petroleum hydrocarbons in a soil sample is measured as a function of time over a 6-week period.

Time (days)	Total petroleum hydrocarbon concentration (mg kg^{-1})
0	1375
7	802
14	695
21	588
28	417
35	356
42	275

a. Determine the rate constant for petroleum degradation.
b. Estimate the contaminant concentration after 16 days.

Solution

a. Test whether petroleum degradation can be modeled as a first-order reaction. A semi-log plot of total petroleum hydrocarbon concentration versus time is shown in Figure 10.6. The first-order model gives a straight line that fits the data well. The slope is equal to $-k_1$ and the intercept is C_{A0}. Therefore, $k_1 = 0.0355$ day^{-1} and $C_{A0} = 1192$ mg kg^{-1}.
b. Applying Eq. (10.35), the kinetic equation is:

$$C_A = 1192e^{-0.0355t}$$

where C_A has units of mg kg^{-1} and t has units of days. Therefore, after 16 days, $C_A = 675$ mg kg^{-1}.

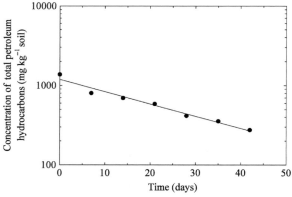

Figure 10.6 Graphic of the kinetic analysis of petroleum degradation.

where r_A is the volumetric rate of reaction with respect to reactant A, C_A is the concentration of A, v_{max} is the *maximum volumetric rate of reaction*, and K_m is the *Michaelis constant*. v_{max}

has the same dimensions as r_A; K_m has the same dimensions as C_A. Typical units for v_{max} are mol m^{-3} s^{-1}; typical units for K_m are mol m^{-3}. As defined in Eq. (10.36), v_{max} is a volumetric rate that is proportional to the amount of active enzyme present. The Michaelis constant K_m is also known as the *half saturation constant* and is equal to the reactant concentration at which $r_A = v_{max}/2$. K_m is independent of enzyme concentration but varies from one enzyme to another and with different substrates for the same enzyme. Values of K_m for some enzyme–substrate systems are listed in Table 10.3. K_m and other enzyme properties depend on the source of the enzyme.

If we adopt conventional symbols for biological reactions and call reactant A the *substrate*, Eq. (10.36) can be rewritten in the familiar form:

$$v = \frac{v_{max}\, s}{K_m + s} \tag{10.37}$$

where v is the volumetric rate of reaction (sometimes called *velocity*) and s is the substrate concentration. The semiempirical basis of the Michaelis–Menten equation will not be covered here; discussion of enzyme reaction models and the assumptions involved in derivation of Eq. (10.37) can be found elsewhere [2,3]. Suffice it to say that the simplest reaction sequence that accounts for the kinetic properties of many enzymes is:

$$E + S \underset{k_{-1}}{\overset{k_1}{\rightleftharpoons}} ES \overset{k_2}{\rightarrow} E + P \tag{10.38}$$

where E is enzyme, S is substrate, and P is product. ES is the *enzyme–substrate complex*. As expected in catalytic reactions, enzyme E is recovered at the end of the reaction. Binding of

Table 10.3: Michaelis Constants for Some Enzyme–Substrate Systems

Enzyme	Source	Substrate	K_m (mM)
Alcohol dehydrogenase	*Saccharomyces cerevisiae*	Ethanol	13.0
α-Amylase	*Bacillus stearothermophilus*	Starch	1.0
	Porcine pancreas	Starch	0.4
β-Amylase	Sweet potato	Amylose	0.07
Aspartase	*Bacillus cadaveris*	L-Aspartate	30.0
β-Galactosidase	*Escherichia coli*	Lactose	3.85
Glucose oxidase	*Aspergillus niger*	D-Glucose	33.0
	Penicillium notatum	D-Glucose	9.6
Histidase	*Pseudomonas fluorescens*	L-Histidine	8.9
Invertase	*Saccharomyces cerevisiae*	Sucrose	9.1
	Neurospora crassa	Sucrose	6.1
Lactate dehydrogenase	*Bacillus subtilis*	Lactate	30.0
Penicillinase	*Bacillus licheniformis*	Benzylpenicillin	0.049
Urease	Jack bean	Urea	10.5

From B. Atkinson and F. Mavituna, 1991, *Biochemical Engineering and Biotechnology Handbook*, 2nd ed., Macmillan, Basingstoke.

substrate to the enzyme in the first step is considered reversible with forward reaction constant k_1 and reverse reaction constant k_{-1}. Decomposition of the enzyme–substrate complex to give the product is an irreversible reaction with rate constant k_2; k_2 is known as the *turnover number* as it defines the number of substrate molecules converted to product per unit time by an enzyme saturated with substrate. The turnover number is sometimes referred to as the *catalytic constant* k_{cat}. The dimensions of k_{cat} are T^{-1} and the units are, for example, s^{-1}. Analysis of the reaction sequence yields the relationship:

$$v_{max} = k_{cat} e_a \tag{10.39}$$

where e_a is the concentration of active enzyme and v_{max} and e_a are expressed in Eq. (10.39) using molar units. Values of k_{cat} range widely for different enzymes from about $50\,min^{-1}$ to $10^7\,min^{-1}$ (brenda-enzymes.org).

The definition of K_m as the substrate concentration at which $v = v_{max}/2$ is equivalent to saying that K_m is the substrate concentration at which half of the enzyme's active sites are saturated with substrate. K_m is therefore considered a relative measure of the *substrate binding affinity* or the stability of the enzyme–substrate complex: lower K_m values imply higher enzyme affinity for the substrate. The *catalytic efficiency* of an enzyme is defined as the ratio k_{cat}/K_m with units of, for example, $mol^{-1}\,L\,s^{-1}$. Catalytic efficiency is often used to compare the utilization of different substrates by a particular enzyme and is a measure of the *substrate specificity* or relative suitability of a substrate for reaction with the enzyme.

Catalysts described by Michaelis–Menten kinetics become saturated at high substrate concentrations. Figure 10.7 shows the relationship between substrate concentration and reaction

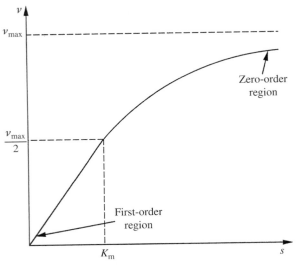

Figure 10.7 Michaelis–Menten plot.

rate defined by Eq. (10.37); the reaction rate v does not increase indefinitely with substrate concentration but approaches a limit, v_{max}. When $v = v_{max}$, all the enzyme is bound to substrate in the form of the enzyme–substrate complex. At high substrate concentrations $s \gg K_m$ and K_m in the denominator of Eq. (10.37) is negligibly small compared with s, so the equation can be simplified as:

$$v \approx \frac{v_{max} s}{s} \tag{10.40}$$

or

$$v \approx v_{max} \tag{10.41}$$

Therefore, at high substrate concentrations, the reaction rate approaches a constant value independent of substrate concentration; the reaction can be approximated as *zero order* with respect to substrate. On the other hand, at low substrate concentrations $s \ll K_m$, the value of s in the denominator of Eq. (10.37) is negligible compared with K_m, and Eq. (10.37) can be simplified to:

$$v \approx \frac{v_{max}}{K_m} s \tag{10.42}$$

The ratio of constants v_{max}/K_m is, in effect, a first-order rate coefficient (k_1) for the reaction. Therefore, at low substrate concentrations, there is an approximate linear dependence of reaction rate on s; in this concentration range, Michaelis–Menten reactions can be approximated as *first order* with respect to substrate.

The rate of enzyme reactions depends on the amount of enzyme present as indicated by Eq. (10.39). However, enzymes are not always available in pure form so that e_a may be unknown. In this case, the amount of enzyme can be expressed as *units of activity*; the specific activity of an enzyme–protein mixture could be reported, for example, as units of activity per mg of protein. The *international unit of enzyme activity*, which is abbreviated IU or U, is the amount of enzyme required to convert 1 μmol of substrate into products per minute under standard conditions. Alternatively, the SI unit for enzyme activity is the katal, which is defined as the amount of enzyme required to convert 1 mole of substrate per second. The abbreviation for katal is kat. Enzyme concentration can therefore be expressed using units of, for example, U mL^{-1} or kat L^{-1}.

The Michaelis–Menten equation is a satisfactory description of the kinetics of many enzymes, although there are exceptions such as glucose isomerase and amyloglucosidase. Procedures for checking whether a particular reaction follows Michaelis–Menten kinetics and for evaluating v_{max} and K_m from experimental data are described in Section 10.4. More complex kinetic equations must be applied if there are multiple substrates [2–4]. Modified kinetic expressions for enzymes subject to inhibition and other forms of regulation are described in Section 10.5.

10.3.4 Effect of System Parameters on Enzyme Reaction Rate

System conditions such as temperature and pH also influence the rate of an enzyme-catalyzed reaction. For an enzyme with a single rate-controlling step, the effect of temperature is described reasonably well using the Arrhenius expression of Eq. (10.22) with v_{max} substituted for k. An example showing the relationship between temperature and the rate of sucrose inversion by yeast invertase is given in Figure 10.8. Activation energies for enzyme reactions are of the order 40 to 80 kJ mol^{-1} [5]; as a rough guide, this means that a 10°C rise in temperature between 20°C and 30°C will increase the rate of reaction by a factor of 2 to 3.

The temperature range over which Arrhenius-type relationship between temperature and rate of reaction is observed for enzymes is quite limited. Many proteins start to denature around 50°C; if the temperature is raised higher than this, thermal deactivation occurs, and the reaction velocity drops quickly. Figure 10.9 illustrates how the Arrhenius relationship breaks down at high temperatures. In this experiment, the Arrhenius equation was obeyed between temperatures of about 0°C ($T = 273.15$ K; $1/T = 3.66 \times 10^{-3}$ K^{-1}) and about 53°C ($T = 326.15$ K; $1/T = 3.07 \times 10^{-3}$ K^{-1}). However, with further increase in temperature, the reaction rate declined rapidly due to thermal deactivation of the enzyme. Enzyme stability and the rate of

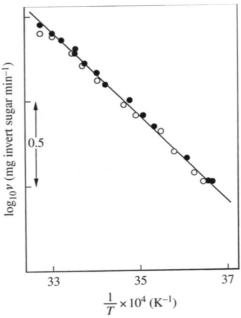

Figure 10.8 Arrhenius plot for inversion of sucrose by yeast invertase. *From I.W. Sizer, 1943, Effects of temperature on enzyme kinetics.* Adv. Enzymol. 3, 35–62.

enzyme deactivation are important factors affecting overall catalytic performance in enzyme reactors. This topic is discussed further in Section 10.5.

pH also has a pronounced effect on enzyme kinetics, as illustrated in Figure 10.10. Typically, the reaction rate is maximum at some optimal pH and declines sharply if the pH is shifted to either side of the optimum value. Ionic strength and water activity can also have considerable influence on the rate of enzyme reaction.

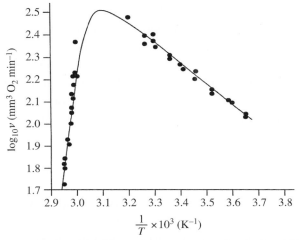

Figure 10.9 Arrhenius plot for catalase. The enzyme breaks down at high temperatures. *From I.W. Sizer, 1944, Temperature activation and inactivation of the crystalline catalase–hydrogen peroxide system.* J. Biol. Chem. *154, 461–473.*

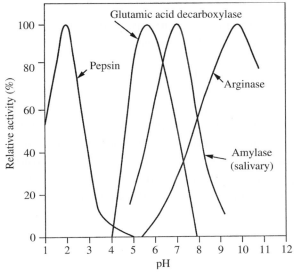

Figure 10.10 Effect of pH on enzyme activity. *From J.S. Fruton and S. Simmonds, 1958,* General Biochemistry, *2nd ed., John Wiley, New York.*

10.4 Determining Enzyme Kinetic Constants From Batch Data

Application of the Michaelis–Menten equation (see Section 10.3.3) requires two constants, v_{max} and K_m which must be evaluated. Estimating parameters for the Michaelis–Menten equation is not as straightforward as for zero- and first-order reactions. Several graphical methods have been developed over the years each with varying levels of simplicity and accuracy.

The first step in kinetic analysis of enzyme reactions is to obtain data for the rate of reaction v as a function of substrate concentration s. Rates of reaction can be determined from batch data as described in Section 10.2. Typically, only *initial rate data* are used. This means that several batch experiments are carried out with different initial substrate concentrations; from each set of data the reaction rate is evaluated at time zero. The initial rates and corresponding initial substrate concentrations are used as (v, s) pairs that can then be plotted in various ways to determine v_{max} and K_m. Initial rate data are preferred for analysis of enzyme reactions because the initial experimental conditions, such as the enzyme and substrate concentrations, are known.

10.4.1 Michaelis–Menten Plot

This simple procedure involves plotting (v, s) values directly as shown in Figure 10.7. v_{max} and K_m can be estimated roughly from this graph; v_{max} is the rate as $s \to \infty$ and K_m is the value of s at $v = v_{max}/2$. The accuracy of this method is usually poor because of the difficulty of extrapolating to v_{max}.

10.4.2 Lineweaver–Burk Plot

This method uses a linearization procedure to give a quasi-straight line from which v_{max} and K_m can be estimated. Inverting Eq. (10.37) gives:

$$\frac{1}{v} = \frac{K_m}{v_{max} s} + \frac{1}{v_{max}} \tag{10.43}$$

Therefore, a plot of $1/v$ versus $1/s$ gives an approximately straight line with slope K_m/v_{max} and intercept $1/v_{max}$. This double-reciprocal plot is known as the *Lineweaver–Burk plot* and is found frequently in the scientific literature. However, the linearization process used in this method distorts the experimental error in v so that these errors are amplified at low substrate concentrations. Consequently, the Lineweaver–Burk plot may give inaccurate parameter estimates [3].

10.4.3 Eadie–Hofstee Plot

If Eq. (10.43) is multiplied by $v\left(\dfrac{v_{max}}{K_m}\right)$ and rearranged, another linearized form of the Michaelis–Menten equation is obtained:

$$\frac{v}{s} = \frac{v_{max}}{K_m} - \frac{v}{K_m} \tag{10.44}$$

According to Eq. (10.44), a plot of v/s versus v gives a straight line with a slope of $-1/K_m$ and intercept v_{max}/K_m. This is called the *Eadie–Hofstee plot*. As with the Lineweaver–Burk plot, the Eadie–Hofstee linearization can distort errors in the data.

10.4.4 Langmuir Plot

Multiplying Eq. (10.43) by s produces the linearized form of the Michaelis–Menten equation according to Langmuir:

$$\frac{s}{v} = \frac{K_m}{v_{max}} + \frac{s}{v_{max}} \tag{10.45}$$

Therefore, a *Langmuir plot* of s/v versus s should give a straight line with slope $1/v_{max}$ and intercept K_m/v_{max}. Linearization of data for the Langmuir plot minimizes distortions in experimental error. Accordingly, its use for evaluation of v_{max} and K_m is recommended [6]. The Langmuir plot is also known as the *Hanes–Woolf plot*.

10.4.5 Direct Linear Plot

A different method for plotting enzyme kinetic data has been proposed by Eisenthal and Cornish-Bowden [7]. For each observation, the reaction rate v is plotted on the vertical axis against s on the negative horizontal axis. This is shown in Figure 10.11 for four pairs of (v, s) data. A straight line is then drawn to join corresponding $(-s, v)$ points. In the absence of experimental error, lines for each $(-s, v)$ pair intersect at a unique point, (K_m, v_{max}). When real data containing errors are plotted, a family of intersection points is obtained as shown in Figure 10.11. Each intersection gives one estimate of v_{max} and K_m; the median or middle v_{max} and K_m values are taken as the kinetic parameters for the reaction. This method is relatively insensitive to individual erroneous readings that may be far from the correct values. However, a disadvantage of the procedure is that deviations from Michaelis–Menten behavior are not easily detected. It is recommended, therefore, for enzymes that are known to follow Michaelis–Menten kinetics.

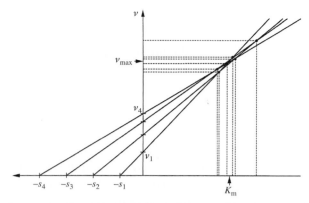

Figure 10.11 Direct linear plot for determination of enzyme kinetic parameters. *From R. Eisenthal and A. Cornish-Bowden, 1974, The direct linear plot: a new graphical procedure for estimating enzyme kinetic parameters. Biochem. J. 139, 715–720.*

10.5 Regulation of Enzyme Activity

Enzyme activity is often subject to diverse control mechanisms that regulate the flux of metabolites through metabolic pathways. Various mechanisms can control the activity of enzymes. Metabolites and other effector molecules interact with enzymes to change properties such as the availability of active sites for reaction, the affinity of the enzyme for substrate, and the ability of the enzyme to form and release product. These interactions can alter the activity of enzymes quickly *in situ*.

Any molecule that reduces the rate of an enzyme reaction is called an *inhibitor*. Reaction products, substrate analogues, synthetic molecules, metabolic intermediates, and heavy metals are examples of enzyme inhibitors; substrates can also act as inhibitors. Binding between the inhibitor and enzyme may be *reversible* or *irreversible*.

Enzyme inhibition and regulation are of greater relevance for the understanding of cellular metabolism. Several types of enzyme inhibition and activation have been identified, as outlined in the following sections. The many forms of enzyme regulation are treated in more detail elsewhere [2].

10.5.1 Reversible Inhibition

There are several mechanisms of reversible enzyme inhibition. The effects of this regulation on enzyme reaction rate are described using modified versions of the Michaelis–Menten equation.

Competitive Inhibition

Substances that cause competitive enzyme inhibition have some degree of molecular similarity with the substrate. These properties allow the inhibitor to compete with substrate for binding to the active reaction site on the enzyme, thus preventing some of the enzyme from forming the enzyme–substrate complex. Binding of inhibitor to enzyme is noncovalent and reversible so that an equilibrium is established between the free and bound forms of the inhibitor. Competitive inhibitors bind to free enzyme E but not to the enzyme–substrate complex ES. Interaction between the enzyme and inhibitor takes place at the same time as conversion of substrate S to product P according to the usual reaction sequence of Eq. (10.38):

$$E + S \rightleftarrows ES \rightarrow E + P \tag{10.46}$$
$$E + I \rightleftarrows EI$$

where I is the inhibitor and EI is the *enzyme–inhibitor complex*. The equation for the rate of enzyme reaction with competitive inhibition is:

$$v = \frac{v_{max} s}{K_m \left(1 + \dfrac{i}{K_i} \right) + s} \tag{10.47}$$

where i is the inhibitor concentration. K_i is the *inhibitor coefficient* or *dissociation constant* for the enzyme–inhibitor binding reaction:

$$K_i = \frac{[E]_e [I]_e}{[EI]_e} \tag{10.48}$$

where $[\ ]_e$ denotes molar concentration of E, I, or EI at equilibrium. Comparison of Eq. (10.47) with Eq. (10.37) for Michaelis–Menten kinetics without inhibition shows that competitive enzyme inhibition does not affect v_{max} but changes the apparent value of K_m:

$$K_{m,app} = K_m \left(1 + \frac{i}{K_i} \right) \tag{10.49}$$

where $K_{m,app}$ is the apparent value of K_m. The effect of competitive inhibition on the rate of enzyme reaction can be overcome by increasing the concentration of substrate.

Noncompetitive Inhibition

Noncompetitive inhibitors have an affinity for binding both the free enzyme and the ES complex, so that the inhibitor and substrate may bind simultaneously with the enzyme to form an inactive ternary complex EIS. The inhibitor binding site on the enzyme is located away from the active reaction site. Binding of the inhibitor is noncovalent and reversible and does not affect the affinity of the substrate for enzyme binding. The reactions involved in noncompetitive inhibition are:

$$
\begin{aligned}
E + S &\rightleftharpoons ES \rightarrow E + P \\
E + I &\rightleftharpoons EI \\
EI + S &\rightleftharpoons EIS \\
ES + I &\rightleftharpoons EIS
\end{aligned}
\tag{10.50}
$$

Noncompetitive inhibition reduces the effective maximum rate of enzyme reaction while K_m remains unchanged. The equation for the rate of enzyme reaction with noncompetitive inhibition is:

$$
v = \frac{v_{max}}{\left(1 + \dfrac{i}{K_i}\right)} \frac{s}{K_m + s}
\tag{10.51}
$$

where i is the inhibitor concentration and K_i is the inhibitor coefficient:

$$
K_i = \frac{[E]_e\,[I]_e}{[EI]_e} = \frac{[ES]_e\,[I]_e}{[EIS]_e}
\tag{10.52}
$$

From Eq. (10.51), the apparent maximum reaction velocity $v_{max,app}$ is:

$$
v_{max,app} = \frac{v_{max}}{\left(1 + \dfrac{i}{K_i}\right)}
\tag{10.53}
$$

The decline in v_{max} reflects the reduced concentration of ES in the reaction mixture as some ES is converted to the inactive complex, EIS. Inhibitor binding does not interfere with the formation of ES, therefore, noncompetitive inhibition cannot be overcome by increasing the substrate concentration.

Uncompetitive Inhibition

Uncompetitive inhibitors do not bind to free enzyme but affect enzyme reactions by binding to the enzyme–substrate complex at locations away from the active site. Inhibitor binding to ES is noncovalent and reversible and produces an inactive ternary complex EIS:

$$
\begin{aligned}
E + S &\rightleftharpoons ES \rightarrow E + P \\
ES + I &\rightleftharpoons EIS
\end{aligned}
\tag{10.54}
$$

Uncompetitive inhibition decreases both v_{max} and K_m. The equation for the rate of enzyme reaction with uncompetitive inhibition is:

$$v = \frac{v_{max}}{\left(1 + \dfrac{i}{K_i}\right)} \frac{s}{\left[\dfrac{K_m}{\left(1 + \dfrac{i}{K_i}\right)} + s\right]}$$

(10.55)

where

$$K_i = \frac{[ES]_e [I]_e}{[EIS]_e}$$

(10.56)

From Eq. (10.55):

$$v_{max,app} = \frac{v_{max}}{\left(1 + \dfrac{i}{K_i}\right)}$$

(10.57)

and

$$K_{m,app} = \frac{K_m}{\left(1 + \dfrac{i}{K_i}\right)}$$

(10.58)

As the inhibitor does not interfere with formation of the enzyme–substrate complex, uncompetitive inhibition cannot be overcome by increasing the substrate concentration.

Substrate inhibition is a special case of uncompetitive inhibition. High substrate concentrations can inhibit enzyme activity if more than one substrate molecule binds to an active site that is meant to accommodate only one. For example, different regions of multiple molecules of substrate may bind to different moieties within a single active site on the enzyme. If the resulting enzyme–substrate complex is unreactive, the rate of the enzyme reaction is reduced. The reaction sequence for substrate inhibition is:

$$E + S \rightleftarrows ES \rightarrow E + P$$
$$ES + S \rightleftarrows ESS$$

(10.59)

where ESS is an inactive substrate complex. The equation for the rate of enzyme reaction with substrate inhibition is:

$$v = \frac{v_{max}}{\left(1 + \dfrac{K_1}{s} + \dfrac{s}{K_2}\right)} \tag{10.60}$$

where K_1 is the dissociation constant for the reaction forming ES and K_2 is the dissociation constant for the reaction forming ESS:

$$K_1 = \frac{[E]_e[S]_e}{[ES]_e} \tag{10.61}$$

$$K_2 = \frac{[ES]_e[S]_e}{[ESS]_e} \tag{10.62}$$

Figure 10.12 shows the effect of substrate inhibition on the rate of enzyme reaction. The rate increases with substrate concentration at relatively low substrate levels, but passes through a maximum at substrate concentration s_{max} as inhibition causes a progressive decline in reaction rate with further increase in s. The value of s_{max} is related to K_1 and K_2 by the equation:

$$s_{max} = \sqrt{K_1 K_2} \tag{10.63}$$

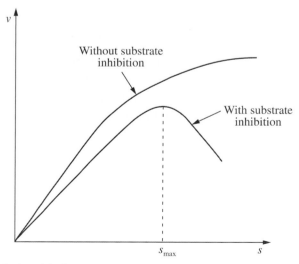

Figure 10.12 Relationship between enzyme reaction rate and substrate concentration with and without substrate inhibition.

Partial Inhibition

In the models for enzyme inhibition considered so far, binding between the inhibitor and enzyme was considered to block completely the formation of product P. However, it is possible that turnover may still occur after inhibitor binding even though the rate of product formation is reduced significantly, for example:

$$\text{EIS} \rightleftarrows \text{EI} + \text{P} \tag{10.64}$$

Partial inhibition does not occur often but is characterized by the maintenance of enzyme activity even at high inhibitor concentrations.

10.5.2 Irreversible Inhibition

Irreversible inhibition occurs when an inhibitor binds with an enzyme but does not dissociate from it under reaction conditions or within the reaction timeframe. In some cases, the inhibitor binds covalently so that product formation is prevented permanently. Michaelis–Menten kinetics do not apply to irreversible inhibition. The rate of inhibitor binding to the enzyme determines the effectiveness of the inhibition. Accordingly, the extent to which the enzyme reaction rate is slowed depends on the enzyme and inhibitor concentrations but is independent of substrate concentration.

10.5.3 Allosteric Regulation

Allosteric regulation of enzymes is crucial for the control of cellular metabolism. Allosteric regulation occurs when an activator or inhibitor molecule binds at a specific regulatory site on the enzyme and induces conformational or electrostatic changes that either enhance or reduce enzyme activity. Not all enzymes possess sites for allosteric binding; those that do are called *allosteric enzymes*. Ligands that bind to allosteric enzymes are known as *effectors*. *Homotropic* regulation occurs when a substrate also acts as an effector and influences the binding of further substrate molecules. *Heterotropic* regulation occurs when the effector and substrate are different entities.

Allosteric enzymes do not follow Michaelis–Menten kinetics. When enzyme binding sites display cooperativity so that effector binding influences substrate affinity, the *Hill equation* is the simplest model representing the relationship between enzyme activity and substrate concentration:

$$v = \frac{v_{max}\, s^n}{K_h + s^n} \tag{10.65}$$

where K_h is the *Hill constant* and n is the *Hill coefficient*. The Hill constant is related to the enzyme–substrate dissociation constant and reflects the affinity of the enzyme for a particular substrate. The units of K_h are those of (concentration) n. From Eq. (10.65), it follows that:

$$K_h = s^n \quad \text{when } v = \frac{v_{max}}{2} \tag{10.66}$$

The Hill coefficient n is an index of the cooperativity between binding sites on an enzyme. When $n = 1$, there is no cooperativity and Eq. (10.65) reduces to the Michaelis–Menten relationship of Eq. (10.37). For $n > 1$, the cooperativity is positive and the benefit from effector binding increases as the value of n increases. The maximum possible value of n is the number of substrate-binding sites on the enzyme. For $n < 1$, the cooperativity is negative and effector binding reduces substrate affinity and enzyme activity.

Graphical representation of Eq. (10.65) is shown in Figure 10.13. Values of $n > 1$ give the curve a sigmoidal shape; the greater the value of $n > 1$, the more pronounced is the sigmoidicity of the curve and the deviation from Michaelis–Menten kinetics ($n = 1$). The advantage of a sigmoidal saturation curve is that, in the intermediate range of substrate concentrations where the sigmoidal curve is steeper than for Michaelis–Menten kinetics, the sensitivity of the enzyme to substrate concentration is enhanced. In this region, a small change in substrate concentration produces a larger change in reaction velocity than for the same enzyme with no cooperativity between binding sites. Allosteric enzymes that exhibit negative cooperativity give reaction rate curves similar to that shown in Figure 10.13 for $n = 0.5$. For these enzymes, in the intermediate range of substrate concentration, enzyme activity is relatively insensitive to small changes in substrate concentration.

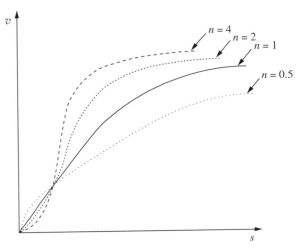

Figure 10.13 Reaction rate curves for allosteric enzymes exhibiting no ($n = 1$), positive ($n > 1$), and negative ($n < 1$) cooperativity.

A linearized form of the Hill equation is obtained by rearranging Eq. (10.65) and taking logarithms:

$$\ln\left(\frac{v}{v_{max} - v}\right) = n\ln s - \ln K_h \tag{10.67}$$

Therefore, the values of n and K_h for a particular allosteric enzyme can be determined from the slope and intercept of the straight line generated when $v/(v_{max} - v)$ is plotted versus s on log–log coordinates.

Allosteric enzymes are vitally important in metabolic control. They allow feedback inhibition of the enzyme cascades responsible for catabolism and biosynthesis. Typically, a metabolite produced at the end of a cascade will function as an inhibitor of an enzyme that is active earlier in the pathway. In this way, build-up of excess product through overactivity of the pathway can be moderated reducing the rate of product synthesis.

10.6 Kinetics of Enzyme Deactivation

Enzymes are protein molecules of complex configuration that can be destabilized by relatively weak forces. During enzyme-catalyzed reactions, enzyme deactivation occurs at a rate that is dependent on the structure of the enzyme and the reaction conditions. Environmental factors affecting enzyme stability include temperature, pH, ionic strength, mechanical forces, and the presence of denaturants such as solvents, detergents, and heavy metals. Because the amount of active enzyme can decline considerably during reaction, in many applications, the kinetics of enzyme deactivation are just as important as the kinetics of the reaction itself.

In the simplest model of enzyme deactivation, active enzyme E_a undergoes irreversible transformation to an inactive form E_i:

$$E_a \rightarrow E_i \tag{10.68}$$

The rate of deactivation is generally considered to be first order in active enzyme concentration:

$$r_d = k_d e_a \tag{10.69}$$

where r_d is the volumetric rate of deactivation, e_a is the active enzyme concentration, and k_d is the *deactivation rate constant*. In a closed system where enzyme deactivation is the only process affecting the concentration of active enzyme:

$$r_d = \frac{-de_a}{dt} = k_d e_a \tag{10.70}$$

Integration of Eq. (10.70) gives an expression for active enzyme concentration as a function of time:

$$e_a = e_{a0} e^{-k_d t} \tag{10.71}$$

where e_{a0} is the concentration of active enzyme at time zero. According to Eq. (10.71), the concentration of active enzyme decreases exponentially with time; the greatest rate of enzyme deactivation occurs when e_a is high.

As indicated in Eq. (10.39), the value of v_{max} for enzyme reaction depends on the concentration of active enzyme present. Therefore, as e_a declines due to deactivation, v_{max} is also reduced. We can estimate the variation of v_{max} with time by substituting into Eq. (10.39) the expression for e_a from Eq. (10.71):

$$v_{max} = k_{cat} e_{a0} e^{-k_d t} = v_{max0} e^{-k_d t} \tag{10.72}$$

where v_{max0} is the initial value of v_{max} before deactivation occurs.

The stability of enzymes is reported frequently in terms of *half-life*. Half-life is the time required for half the enzyme activity to be lost; after one half-life, the active enzyme concentration equals $e_{a0}/2$. Substituting $e_a = e_{a0}/2$ into Eq. (10.71), taking logarithms, and rearranging yields the following expression:

$$t_h = \frac{\ln 2}{k_d} \tag{10.73}$$

where t_h is the enzyme half-life.

The rate at which enzymes deactivate depends strongly on temperature. This dependency is generally well described using the Arrhenius equation (Section 10.1.5):

$$k_d = A e^{-E_d / RT} \tag{10.74}$$

where A is the Arrhenius constant or frequency factor, E_d is the *activation energy for enzyme deactivation*, R is the ideal gas constant, and T is absolute temperature. According to Eq. (10.74), as T increases, the rate of enzyme deactivation increases exponentially. Values of E_d are high, of the order 170 to 400 kJ gmol^{-1} for many enzymes [5]. Accordingly, a temperature increase of 10°C between 30°C and 40°C will increase the rate of enzyme deactivation by a factor of between 10 and 150. The stimulatory effect of increasing temperature on the rate of enzyme reaction has already been described in Section 10.3.4. However, as shown here, raising the temperature also reduces the amount of active enzyme present. Temperature has a critical effect on enzyme kinetics. Example 10.5 illustrates enzyme deactivation via the enzyme half-life parameter.

EXAMPLE 10.5 Enzyme Half-Life

Amyloglucosidase from *Endomycopsis bispora* is immobilized in very small polyacrylamide gel beads. The activities of immobilized and soluble enzyme are compared at 80°C. Initial rate data are measured at a fixed substrate concentration with the following results.

Time (min)	Enzyme activity (μmol mL^{-1} min^{-1})	
	Soluble enzyme	Immobilized enzyme
0	0.86	0.45
3	0.79	0.44
6	0.70	0.43
9	0.65	0.43
15	0.58	0.41
20	0.46	0.40
25	0.41	0.39
30	–	0.38
40	–	0.37

What is the half-life for each form of enzyme?

Solution

From Eq. (10.37), at any fixed substrate concentration, the rate of enzyme reaction v is directly proportional to v_{max}. Therefore, k_d can be determined from Eq. (10.72) using enzyme activity v instead of v_{max}. Making this change and expressing Eq. (10.72) using logarithms gives:

$$\ln v = \ln v_0 - k_d t$$

where v_0 is the initial enzyme activity before deactivation. So, if deactivation follows a first-order model, a semi-log plot of reaction rate versus time should give a straight line with slope $-k_d$. The experimental data are plotted in Figure 10.14.

From the slopes, k_d for soluble enzyme is 0.0296 min^{-1} and k_d for immobilized enzyme is 0.0051 min^{-1}. Applying Eq. (10.73) for half-life:

$$t_h(\text{soluble}) = \frac{\ln 2}{0.0296 \text{ min}^{-1}} = 23 \text{ min}$$

$$t_h(\text{immobilized}) = \frac{\ln 2}{0.0051 \text{ min}^{-1}} = 136 \text{ min}$$

Immobilization enhances the stability of the enzyme significantly.

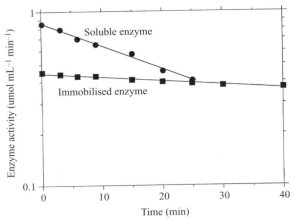

Figure 10.14 Kinetic analysis of enzyme deactivation.

10.7 Yields in Cell Culture

The basic concept of reaction yield was introduced in Section 10.1.2 for simple one-step reactions. However, cell growth lumps together hundreds or thousands of individual enzyme and chemical conversions. Despite this complexity, yield principles can be applied to cell metabolism to relate the consumption of substrate to the formation of biomass and other products. Yields that are frequently reported in this text are expressed using *yield coefficients*. Several yield coefficients, such as the yield of biomass from substrate, the yield of biomass from O_2, and the yield of product from substrate, are common. Yield coefficients allow us to quantify the nutrient requirements and production characteristics of organisms.

Some metabolic yield coefficients—the biomass yield $Y_{X/S}$, the product yield $Y_{P/S}$, and the respiratory quotient RQ—were introduced in Chapter 4. The definition of yield coefficients can be generalized as follows:

$$Y_{J/K} = \frac{-\Delta J}{\Delta K} \tag{10.75}$$

where $Y_{J/K}$ is the yield coefficient, J and K are substances involved in metabolism, ΔJ is the mass or moles of J produced, and ΔK is the mass or moles of K consumed. The negative sign is required in Eq. (10.75) because ΔK for a consumed substance is negative in value and yield is calculated as a positive quantity. A list of frequently used yield coefficients is given in Table 10.4. Note that in some cases, such as $Y_{P/X}$, both substances represented by the yield coefficient are products of metabolism. Although the term "yield" usually refers to the amount of product formed per amount of reactant, yields can also be used to relate other quantities.

Table 10.4: Some Metabolic Yield Coefficients

Symbol	Definition
$Y_{X/S}$	Mass or moles of biomass produced per unit mass or mole of substrate consumed; moles of biomass can be calculated from the "molecular formula" for biomass (see Section 4.6.1)
$Y_{P/S}$	Mass or moles of product formed per unit mass or mole of substrate consumed
$Y_{P/X}$	Mass or moles of product formed per unit mass or mole of biomass formed
$Y_{X/O2}$	Mass or moles of biomass formed per unit mass or mole of O_2 consumed
$Y_{C/S}$	Mass or moles of carbon dioxide formed per unit mass or mole of substrate consumed
RQ	Moles of carbon dioxide formed per mole of O_2 consumed; this yield is called the *respiratory quotient*
Y_{ATP}	Mass or moles of biomass formed per mole of ATP formed
Y_{kcal}	Mass or moles of biomass formed per kilocalorie of heat evolved during cultivation

Some yield coefficients are based on parameters such as the amount of ATP formed or heat evolved during metabolism.

10.7.1 Overall and Instantaneous Yields

A challenge with applying Eq. (10.75) is that values of ΔJ and ΔK change with time. In batch culture, ΔJ and ΔK can be calculated as the difference between initial and final states; this gives an *overall yield* representing an average value over the entire culture period. Alternatively, ΔJ and ΔK can be determined between two other points in time; this calculation might produce a different result for $Y_{J/K}$. Yields can vary during cultivation and it is sometimes necessary to evaluate the *instantaneous yield* at a particular point in time. If r_J and r_K are the volumetric rates of production and consumption of J and K, respectively, in a closed, constant-volume reactor, the instantaneous yield can be calculated as follows:

$$Y_{JK} = \lim_{\Delta K \to 0} \frac{-\Delta J}{\Delta K} = \frac{-dJ}{dK} = \frac{\dfrac{-dJ}{dt}}{\dfrac{dK}{dt}} = \frac{r_J}{r_K} \tag{10.76}$$

For example, $Y_{X/S}$ at a particular instant in time is defined as:

$$Y_{xs} = \frac{r_x}{r_s} = \frac{\text{growth rate}}{\text{substrate consumption rate}} \tag{10.77}$$

When yields for bioprocesses are reported, the time or time period to which they refer should also be stated.

Table 10.5: Observed Biomass Yields for Several Microorganisms and Substrates

Microorganism	Substrate	Observed biomass yield $Y'_{x/s}$ (g g^{-1})
Aerobacter cloacae	Glucose	0.44
Penicillium chrysogenum	Glucose	0.43
Candida utilis	Glucose	0.51
	Acetic acid	0.36
	Ethanol	0.68
Candida intermedia	n-Alkanes (C_{16}–C_{22})	0.81
Pseudomonas sp.	Methanol	0.41
Methylococcus sp.	Methane	1.01

From S.J. Pirt, 1975, *Principles of Microbe and Cell Cultivation*, Blackwell Scientific, Oxford.

10.7.2 Theoretical and Observed Yields

It is necessary to distinguish between theoretical and observed yields. This is particularly important for cell metabolism because there are many reactions occurring at the same time; theoretical and observed yields are therefore often different. Consider the example of biomass yield from substrate, $Y_{x/s}$. If the total mass of substrate consumed is S_T, some proportion of S_T equal to S_G will be used for growth while the remainder, S_R, is channeled into other products and metabolic activities not related to growth. Therefore, the observed biomass yield based on total substrate consumption is:

$$Y'_{x/s} = \frac{-\Delta X}{\Delta S_T} = \frac{-\Delta X}{\Delta S_G + \Delta S_R} \tag{10.78}$$

where ΔX is the amount of biomass produced and $Y'_{x/s}$ is the *observed biomass yield from substrate*. Values of observed biomass yields for several organisms and substrates are listed in Table 10.5. In comparison, the *true* or *theoretical biomass yield* from substrate is:

$$Y_{x/s} = \frac{-\Delta X}{\Delta S_G} \tag{10.79}$$

as ΔS_G is the mass of substrate directed into biomass production. Theoretical yields are sometimes referred to as *maximum possible yields* because they represent the yield in the absence of competing reactions. However, because of the complexity of metabolism, ΔS_G for cell growth is often unknown and the observed biomass yield may be the only biomass yield available. The concept of yields is illustrated in Example 10.6.

EXAMPLE 10.6 Yields in Acetic Acid Production

The reaction equation for aerobic production of acetic acid from ethanol is:

$$C_2H_5OH + O_2 \rightarrow CH_3CO_2H + H_2O$$
$$\underset{\text{(ethanol)}}{} \qquad \underset{\text{(acetic acid)}}{}$$

Acetobacter aceti bacteria are added to vigorously aerated medium containing $10\,g\,L^{-1}$ ethanol. After some time, the ethanol concentration is $2\,g\,L^{-1}$ and $7.5\,g\,L^{-1}$ of acetic acid is produced. How does the observed yield of acetic acid from ethanol compare with the theoretical yield?

Solution

Using a basis of 1 L, the observed yield over the entire culture period is obtained from application of Eq. (10.10):

$$Y'_{PS} = \frac{7.5g}{(10-2)g} = 0.94 \text{ g acetic acid (g ethanol)}^{-1}$$

The theoretical yield is based on the mass of ethanol actually used for synthesis of acetic acid. From the stoichiometric equation:

$$Y_{PS} = \frac{1 \text{ gmol acetic acid}}{1 \text{ gmol ethanol}} = \frac{60 \text{ g acetic acid}}{46 \text{ g ethanol}} = 1.30g \text{ acetic acid (g ethanol)}^{-1}$$

The observed yield is 72% of the theoretical yield.

10.8 Cell Growth Kinetics

The kinetics of cell growth are expressed using equations similar to equations used to describe enzyme kinetics. This is not surprising since cell metabolism depends on the integrated action of a multitude of enzymes.

10.8.1 Batch Growth

Batch cultivations are divided into a few phases of cell growth. A plot of cell concentration with time is presented in Figure 10.15. The different phases of growth are more readily distinguished when the logarithm of viable cell concentration is plotted against time; alternatively, a semi-log plot can be used where cell concentration is plotted using logarithm scaling and time is plotted with a standard linear scaling. The rate of cell growth varies with time depending on the growth phase. The first phase is lag phase which follows immediately after inoculation of the reactor; the rate of growth is essentially zero. Cells use the lag phase to adapt to their new environment: new enzymes or structural components may be synthesized but the concentration of cells does not increase. Following the lag period, growth starts in the acceleration growth phase. Exponential cell growth phase (sometimes called logarithm growth phase) appears as a straight line on a semi-log plot. At the end of the exponential growth

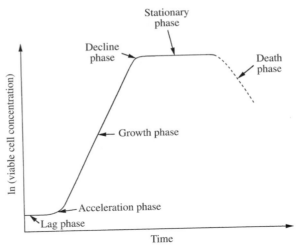

Figure 10.15 Typical batch cultivation growth curve.

Table 10.6: Summary of Batch Cell Growth Phases

Phase	Description	Specific growth rate
Lag	Cells adapt to the new environment; no or very little growth	$\mu \approx 0$
Acceleration	Growth starts	$\mu < \mu_{max}$
Growth	Growth achieves its maximum rate	$\mu \approx \mu_{max}$
Deceleration	Growth slows due to nutrient exhaustion or build-up of inhibitory products	$\mu < \mu_{max}$
Stationary	Growth ceases	$\mu = 0$
Death	Cells lose viability and lyse	$\mu < 0$

phase, as nutrients in the culture medium become depleted or inhibitory products accumulate, growth rate decreases, and the cells enter the deceleration growth phase. After this transition period, the stationary phase is reached, the cells cease growing, and the growth rate is zero. Some cultures exhibit a death phase as the cells lose viability or lyse. Table 10.6 provides a summary of growth and metabolic activity during the phases of batch culture.

During the growth and death phases, the rate of cell growth is described by the equation:

$$r_X = \mu x \qquad (10.80)$$

where r_X is the volumetric rate of biomass production with units of, for example, kg m^{-3} s^{-1}, x is the viable cell concentration with units of, for example, kg m^{-3}, and μ is the *specific growth rate*. The specific growth rate has dimensions T^{-1} and units of, for example, h^{-1}. Equation (10.80) has the same form as Eq. (10.31); cell growth is therefore classified as a *first-order autocatalytic reaction*.

In a closed system where growth is the only process affecting cell concentration, the rate of growth r_X is equal to the rate of change of cell concentration:

$$r_X = \frac{dx}{dt} \tag{10.81}$$

Combining Eqs. (10.80) and (10.81) gives:

$$\frac{dx}{dt} = \mu x \tag{10.82}$$

If μ is constant, we can integrate Eq. (10.82) directly. Separating variables give

$$\int \frac{dx}{x} = \int \mu dt \tag{10.83}$$

and integrating with initial condition $x = x_0$ at $t = 0$ gives:

$$\ln x = \ln x_0 + \mu t \tag{10.84}$$

where x_0 is the viable cell concentration at time zero. Rearranging gives an equation for cell concentration x as a function of time during batch growth:

$$x = x_0 e^{\mu t} \tag{10.85}$$

Equation (10.85) represents *exponential growth phase* only. According to Eq. (10.84), a plot of $\ln x$ versus time gives a straight line with slope μ. Because the relationship of Eq. (10.84) is strictly valid only if μ is unchanging, a plot of $\ln x$ versus t is often used to assess whether the specific growth rate is constant. As illustrated in Figure 10.15, μ is usually constant during the exponential growth phase. It is always advisable to prepare a semi-log plot of cell concentration vs. time before identifying phases of growth. If cell concentration is plotted on linear coordinates, growth often appears slow at the beginning of the culture because the concentration of cells is very small. We might be tempted to conclude there was a lag phase. However, when the same data are plotted using the logarithm of the cell concentration, it becomes clear that the culture did not experience a lag phase. Exponential growth always appears much slower at the beginning of the culture because the number of cells present is small relative to the final cell concentration.

Specific cell growth rates are often expressed in terms of the *doubling time* t_d. An expression for doubling time can be derived from Eq. (10.85). Starting with a cell concentration of x_0, the concentration at $t = t_d$ is $2x_0$. Substituting these values into Eq. (10.85):

$$2x_0 = x_0 e^{\mu t_d} \tag{10.86}$$

Cancelling x_0 gives:

$$2 = e^{\mu t_d} \tag{10.87}$$

Taking the natural logarithm of both sides:

$$\ln 2 = \mu t_d \tag{10.88}$$

or

$$t_d = \frac{\ln 2}{\mu} \tag{10.89}$$

Because the relationships used to derive Eq. (10.89) require the assumption of constant μ, doubling time is a valid representation of the growth rate only when μ is constant.

10.8.2 Balanced Growth

In an environment favorable for growth, cells regulate their metabolism and adjust the rates of internal reactions so that a condition of *balanced growth* occurs. Balanced growth means that the cell modulates the effects of external perturbations and keeps the biomass composition steady despite changes in environmental conditions.

The specific rate of production for each biomass component must be equal to the cell-specific growth rate μ for the biomass composition to remain constant:

$$r_Z = \mu z \tag{10.90}$$

where Z is a cellular constituent such as protein, RNA, DNA, or polysaccharide, r_Z is the volumetric rate of production of Z, and z is the concentration of Z in the reactor. Therefore, the doubling time for each cell component must be equal to t_d for balanced growth. Balanced growth cannot be achieved if environmental changes affect the rate of growth. In most cultures, balanced growth occurs during the exponential growth phase.

10.8.3 Effect of Substrate Concentration

The specific growth rate of the cells depends on the concentration of nutrients in the medium. Often, a single substrate exerts a dominant influence on the rate of growth; this component is known as the *growth-rate-limiting substrate* or, more simply, the *growth-limiting substrate*. The growth-limiting substrate is often the carbon or nitrogen source, although in some cases it is O_2 or another oxidant such as nitrate. During balanced growth, the specific growth rate is related to the concentration of the growth-limiting substrate by a homologue of the Michaelis–Menten equation, known as the *Monod equation*:

$$\mu = \frac{\mu_{max} s}{K_s + s} \tag{10.91}$$

In Eq. (10.91), s is the concentration of growth-limiting substrate, μ_{max} is the *maximum specific growth rate*, and K_S is the *substrate half saturation constant*. μ_{max} has dimensions T^{-1}; K_S has the same dimensions as substrate concentration. The form of Eq. (10.91) is shown in Figure 10.16. μ_{max} and K_S are intrinsic parameters of the cell–substrate system; values of K_S for several organisms are listed in Table 10.7.

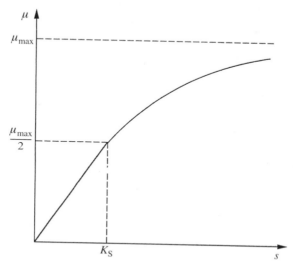

Figure 10.16 The relationship between the specific growth rate and the concentration of growth-limiting substrate in cell culture.

Table 10.7: K_S Values for Several Organisms

Microorganism (genus)	Limiting substrate	K_S (mg L^{-1})
Saccharomyces	Glucose	25
Escherichia	Glucose	4.0
	Lactose	20
	Phosphate	1.6
Aspergillus	Glucose	5.0
Candida	Glycerol	4.5
	O_2	0.042–0.45
Pseudomonas	Methanol	0.7
	Methane	0.4
Klebsiella	Carbon dioxide	0.4
	Magnesium	0.56
	Potassium	0.39
	Sulphate	2.7
Hansenula	Methanol	120.0
	Ribose	3.0
Cryptococcus	Thiamine	1.4×10^{-7}

From S.J. Pirt, 1975, *Principles of Microbe and Cell Cultivation*, Blackwell Scientific; and D.I.C. Wang, C.L. Cooney, A.L. Demain, P. Dunnill, A.E. Humphrey, and M.D. Lilly, 1979, *Fermentation and Enzyme Technology*, John Wiley, New York.

If μ is dependent on substrate concentration as indicated in Eq. (10.91), how can μ remain constant during cell growth as explained in Section 10.8.1? Afterall, substrate is being consumed and s is decreasing continuously throughout the growth period. The answer lies with the relative magnitudes of K_S and s in Eq. (10.91). Typical values of K_S are very small, on the order of mg per L for carbohydrate substrates and µg per L for compounds such as amino acids. The concentration of the growth-limiting substrate in culture media is normally much greater than K_S. As a result, $K_S \ll s$ in Eq. (10.91) and K_S can be neglected, the s terms in the numerator and denominator then cancel, and the specific growth rate is effectively independent of substrate concentration until s reaches values on order of $10 \cdot K_s$. Therefore, $\mu \approx \mu_{max}$ and biomass increases exponentially for most of the culturing period. This explains why μ remains constant and equal to μ_{max} until the medium is virtually exhausted of substrate. When s finally falls below $10 \cdot K_S$, the transition from growth to stationary phase can be very abrupt as the very small quantity of remaining substrate is consumed rapidly by the large number of cells.

The Monod equation is by far the most frequently used expression relating growth rate to substrate concentration. However, it is valid only for balanced growth and should not be applied when growth conditions are changing rapidly. There are also other restrictions; for example, the Monod equation has been found to have limited applicability at extremely low substrate levels. When growth is inhibited by high substrate or product concentrations, extra terms can

---

EXAMPLE 10.7 Microalgae Growth Kinetics

Microalga *Anabaena* is being grown in a photobioreactor for inexpensive protein production from CO_2 and sunlight. The medium initially has an excess of all nutrients and is sparged with air supplemented with 1% CO_2. Growth kinetics are based on light irradiance, however, *Anabaena* is subject to photoinhibition if the light irradiance is too high. A Monod-like growth expression can be modified to account for substrate inhibition, in this case, photoinhibition. This type of kinetic expression is often called the Haldane equation:

$$\mu(1) = \frac{\mu_{max}I}{k + I + \left(\dfrac{I}{k_{inh}}\right)^2}$$

Anabaena has the following parameters, $\mu_{max} = 1.5$ day^{-1}, I is the irradiance of photosynthetically active light in units of micro-Einsteins (µE, 10^{-6} mole photons m^{-2} s^{-1}), k is the equivalent of a Monod constant and has a value of 50 µE, and k_{inh} is the photoinhibition term and has a value of 70 µE. What is the optimal light irradiance to maximize the specific growth rate of *Anabaena*?

Solution
We will plot the specific growth rate, μ, as a function of irradiance, I, to determine the maximum I before photoinhibition lowers the growth rate. Figure 10.17 plots the specific growth rate as a function of irradiance. The maximum specific growth rate occurs at an $I = 60$ mE; if the irradiance is higher, it causes cellular stress which lowers the growth rate. Light is not the only "substrate" that causes inhibition at high abundances.

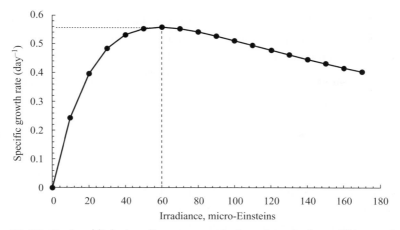

Figure 10.17 Optimal light irradiance to maximize microalgal specific growth rate.

be added to the Monod equation to account for these effects. Several other kinetic expressions have been developed for cell growth; these provide better correlations for experimental data in certain situations [8–11]. Example 10.7 shows a kinetic expression for the growth of microalgae as a function of light intensity; the equation includes the effects of too much light which inhibits growth.

10.9 Growth Kinetics With Plasmid Instability

A potential problem in culture of recombinant organisms is plasmid loss or inactivation. Plasmid instability occurs in individual cells which, by reproducing, can generate a large plasmid-free population in the reactor and reduce the overall rate of synthesis of plasmid-encoded products. Plasmid instability occurs because of DNA mutation or defective plasmid segregation. For segregational stability, the total number of plasmids present in the culture must double once per generation and the plasmid copies must be distributed equally between mother and daughter cells.

A simple model has been developed for batch culture to describe changes in the fraction of plasmid-bearing cells as a function of time [12]. The important parameters in this model are the probability of plasmid loss per generation of cells, and the difference in growth rate between plasmid-bearing and plasmid-free cells. Exponential growth of the host cells is assumed. If x^+ is the concentration of plasmid-carrying cells and x^- is the concentration of plasmid-free cells, the rates at which the two cell populations grow are:

$$r_{X^+} = (1 - p)\mu^+ x^+ \tag{10.92}$$

and

$$r_{X^-} = p\mu^+ x^+ + \mu^- x^- \tag{10.93}$$

where r_{X^+} is the rate of growth of the plasmid-bearing population, r_{X^-} is the rate of growth of the plasmid-free population, p is the probability of plasmid loss per cell division ($p \le 1$), μ^+ is the specific growth rate of plasmid-carrying cells, and μ^- is the specific growth rate of plasmid-free cells.

The model assumes that all plasmid-containing cells are identical in growth rate and probability of plasmid loss; this is the same as assuming that all plasmid-containing cells have the same copy number. By comparing Eq. (10.92) with Eq. (10.80), we can see that the rate of growth of the plasmid-bearing population is reduced by ($p\mu^+ x^+$). This is because some of the progeny of plasmid-bearing cells do not contain plasmid and do not join the plasmid-bearing population. On the other hand, growth of the plasmid-free population has two contributions as indicated in Eq. (10.93). Existing plasmid-free cells grow with specific growth rate μ^-; in addition, this population is supplemented by the generation of plasmid-free cells due to defective plasmid segregation by plasmid-carrying cells.

At any time, the fraction of cells in the culture with plasmid is:

$$F = \frac{x^+}{x^+ + x^-} \tag{10.94}$$

In batch culture where rates of growth can be determined by monitoring cell concentration, $r_{X^+} = dx^+/dt$ and $r_{X^-} = dx^-/dt$. Therefore, Eqs. (10.92) and (10.93) can be integrated simultaneously with initial conditions $x^+ = x_0^+$ and $x^- = x_0^-$ at $t = 0$. After n generations of plasmid-containing cells:

$$F = \frac{1 - \alpha - p}{1 - \alpha - 2^{n(\alpha + p - 1)} p} \tag{10.95}$$

where

$$\alpha = \frac{\mu^-}{\mu^+} \tag{10.96}$$

and

$$n = \frac{\mu^+ t}{\ln 2} \tag{10.97}$$

The value of F depends on α, the ratio of the specific growth rates of plasmid-free and plasmid-carrying cells. In the absence of selection pressure, the presence of plasmid usually reduces the growth rate of organisms due to the additional metabolic requirements imposed by the plasmid DNA. Therefore, α is usually >1.

Some values of α from the literature are listed in Table 10.8; typically, $1.0 < \alpha < 2.0$. Under selection pressure α may equal zero; if the plasmid encodes biosynthetic enzymes for essential nutrients, the loss of plasmid may result in $\mu^- = 0$. When this is the case, F remains close

Table 10.8: Relative Growth Rates of Plasmid-Free and Plasmid-Carrying Cells

Organism	Plasmid	$\alpha = \dfrac{\mu^-}{\mu^+}$
Escherichia coli C600	F' lac	0.99–1.10
E. coli K12 EC1055	R1*drd*-19	1.03–1.12
E. coli K12 IR713	TP120 (various)	1.50–2.31
E. coli JC7623	Col E1	1.29
	Col E1 derivative TnA insertion (various)	1.15–1.54
	Col E1 deletion mutant (various)	1.06–1.65
Pseudomonas aeruginosa PA01	TOL	2.00

Source: From T. Imanaka and S. Aiba, 1981, A perspective on the application of genetic engineering: stability of recombinant plasmid. *Ann. N.Y. Acad. Sci.* 369, 1–14.

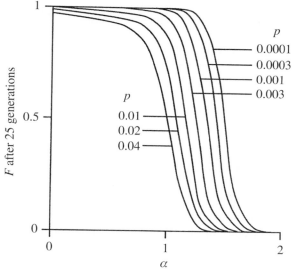

Figure 10.18 Fraction of plasmid-carrying cells in batch culture after 25 generations. *From T. Imanaka and S. Aiba, 1981, A perspective on the application of genetic engineering: stability of recombinant plasmid. Ann. N.Y. Acad. Sci. 369, 1–14.*

to 1 as the plasmid-free population cannot reproduce. F also depends on p, the probability of plasmid loss per generation, which can be as high as 0.1 if segregation occurs. When mutation or random insertions or deletions are the only cause of plasmid instability, p is usually much lower at about 10^{-6}.

Batch culture of microorganisms can sometimes require 25 cell generations or more. Results for F after 25 generations have been calculated from Eq. (10.95) and are shown in Figure 10.18 as a function of p and α. F deteriorates substantially as α increases from 1.0 to 2.0. Further application of Eq. (10.95) is illustrated in Example 10.8.

EXAMPLE 10.8 Plasmid Instability in Batch Culture

A plasmid-containing strain of *E. coli* is used to produce recombinant protein in a 250-L bioreactor. The probability of plasmid loss per generation is 0.005. The specific growth rate of plasmid-free cells is $1.4\,h^{-1}$; the specific growth rate of plasmid-bearing cells is $1.2\,h^{-1}$. Estimate the fraction of plasmid-bearing cells after 18 h of growth if the inoculum contains only cells with plasmid.

Solution

The number of generations of plasmid-carrying cells after 18 h is calculated from Eq. (10.97):

$$n = \frac{(1.2\,h^{-1})18h}{\ln 2} = 31$$

Substituting this into Eq. (10.95) with $p = 0.005$ and $\alpha = 1.4\,h^{-1}/1.2\,h^{-1} = 1.17$:

$$F = \frac{1 - 1.17 - 0.005}{1 - 1.17 - 2^{31\{1.17+0.005-1\}}0.005} = 0.45$$

Therefore, after 18 h only 45% of the cells contain plasmid.

More complex models that recognize the segregated nature of plasmid populations are available [13–16].

10.10 Production Kinetics in Cell Culture

The goal of many bioprocesses is to produce low-molecular-weight compounds, such as ethanol, amino acids, antibiotics, or vitamins. The kinetics of these processes are essential for economic viability. Bioprocess products can be classified according to the relationship between product synthesis and energy generation in the cell as indicated in Table 10.9 [10,17]. Compounds in the first category are linked directly to cellular energy metabolism; these materials are synthesized using pathways that produce ATP. The second class of product is partly linked to energy generation but requires additional energy and substrate for synthesis. Formation of other products such as antibiotics involves secondary metabolism and reactions far removed from energy metabolism.

Irrespective of the class of product, the rate of product formation in cell culture can be expressed as a function of biomass concentration:

$$r_P = q_P x \tag{10.98}$$

where r_P is the volumetric rate of product formation with units of, for example, kg product $m^{-3}\,s^{-1}$, x is biomass concentration, and q_P is the *specific rate of product formation* with units like kg product (kg biomass)$^{-1}\,s^{-1}$. q_P can be evaluated at any time during the bioprocess as

Table 10.9: Classification of Low-Molecular-Weight Fermentation Products

Class of metabolite	Examples
Products directly associated with generation of energy in the cell	Ethanol, acetic acid, gluconic acid, acetone, butanol, lactic acid, other products of anaerobic fermentation
Products indirectly associated with energy generation	Amino acids and their products, citric acid, nucleotides
Products for which there is no clear direct or indirect coupling to energy generation	Penicillin, streptomycin, vitamins

the ratio of production rate and biomass concentration; q_P is not necessarily constant during batch cultivation. Depending on whether the product is linked to energy metabolism or not, we can develop equations for q_P as a function of growth rate and other metabolic parameters.

10.10.1 *Product Formation Directly Coupled With Energy Metabolism*

For products directly linked to ATP synthesis, the rate of production is related to the cellular energy demand. Growth is usually the major energy-requiring function of cells; therefore, if production is coupled to energy metabolism, the product will be formed whenever there is growth. However, ATP is also required for other activities called *maintenance*. Examples of maintenance functions include cell motility, turnover of cellular components, and adjustment of membrane potential and internal pH. Maintenance activities are carried out by living cells even in the absence of growth. Products synthesized in energy pathways will be produced whenever maintenance functions are active. Kinetic expressions for the rate of product formation must account for growth-associated and maintenance-associated production, as in the following equation:

$$r_P = Y_{P/X} r_X + m_P x \qquad (10.99)$$

In Eq. (10.99), r_X is the volumetric rate of biomass formation, $Y_{P/X}$ is the theoretical or true yield of product from biomass, m_P is the *specific rate of product formation due to maintenance*, and x is biomass concentration. m_P has typical units of kg product (kg biomass)$^{-1}$ s^{-1}. Equation (10.99) states that the rate of product formation depends partly on the rate of growth and partly also on cell concentration. Applying Eq. (10.80) to Eq. (10.99):

$$r_P = \left(Y_{P/X} \mu + m_P \right) x \qquad (10.100)$$

Comparison of Eqs. (10.98) and (10.100) shows that, for products coupled to energy metabolism, q_P is equal to a combination of growth-associated and nongrowth-associated terms:

$$q_P = Y_{P/X} \mu + m_P \qquad (10.101)$$

10.10.2 Product Formation Not Coupled With Energy Metabolism

Product synthesis not directly linked to energy metabolism is difficult to relate to growth. However, the rate of nongrowth-associated, product synthesis is often directly proportional to the biomass concentration, so that the production rate defined in Eq. (10.98) can be applied with a constant q_P. An example is penicillin synthesis; equations for the rate of penicillin production as a function of biomass concentration and specific growth rate have been derived by Heijnen *et al* [18].

10.11 Kinetics of Substrate Uptake in Cell Culture

Cells consume substrate from the external environment and channel it into different metabolic pathways. Some substrate may be directed into growth and product synthesis; another fraction may generate cellular energy for maintenance activities. Substrate requirements for maintenance vary considerably depending on the organism and culture conditions; a complete accounting of substrate uptake should include maintenance. The specific rate of substrate uptake for maintenance activities is known as the *maintenance coefficient*, m_S. The typical units of m_S are kg substrate (kg biomass)$^{-1}$ s^{-1}. Some examples of maintenance coefficients for various microorganisms are listed in Table 10.10. The ionic strength of the culture medium exerts a considerable influence on the value of m_S; significant amounts of energy are needed to maintain concentration gradients across cell membranes. The physiological significance of m_S has been the subject of much debate; there are indications that m_S for a particular organism may not be cons van Verseveld, Stoichiometry of microbial growth, tant at all growth rates.

**Table 10.10: Maintenance Coefficients for Several Microorganisms
With Glucose as Energy Source**

Microorganism	Growth conditions	m_S (kg substrate [kg biomass]$^{-1}$h^{-1})
Saccharomyces cerevisiae	Anaerobic	0.036
	Anaerobic, 1.0 M NaCl	0.360
Azotobacter vinelandii	Nitrogen fixing, 0.2 atm dissolved O_2 tension	1.5
	Nitrogen fixing, 0.02 atm dissolved O_2 tension	0.15
Klebsiella aerogenes	Anaerobic, tryptophan-limited, 2 g L^{-1} NH$_4$Cl	2.88
	Anaerobic, tryptophan-limited, 4 g L^1 NH$_4$Cl	3.69
Lactobacillus casei		0.135
Aerobacter cloacae	Aerobic, glucose-limited	0.094
Penicillium chrysogenum	Aerobic	0.022

From S.J. Pirt, 1975, *Principles of Microbe and Cell Cultivation*, Blackwell Scientific, Oxford.

The uptake rate of substrate can be expressed as a function of biomass concentration using an equation analogous to Eq. (10.98):

$$r_S = q_S x \tag{10.102}$$

where r_S is the volumetric rate of substrate consumption with units of, for example, kg substrate m^{-3} s^{-1}, q_S is the *specific rate of substrate uptake* with units such as kg substrate (kg biomass)$^{-1}$ s^{-1}, and x is the biomass concentration. In this section, we will develop equations for q_S as a function of growth rate and other relevant metabolic parameters.

10.11.1 Substrate Uptake in the Absence of Extracellular Product Formation

In some cell cultures, there is little or no extracellular product formation; for example, biomass itself is the product like with the manufacture of bakers' yeast or single-cell protein. In the absence of extracellular product synthesis, we assume that all substrate entering the cell is used for growth and maintenance functions. The rates of these cellular activities are related as follows:

$$r_s = \frac{r_X}{Y_{X/S}} + m_s x \tag{10.103}$$

where r_X is the volumetric rate of biomass production, $Y_{X/S}$ is the true yield of biomass from substrate, m_s is the maintenance coefficient, and x is the biomass concentration. Equation (10.103) states that the rate of substrate uptake depends partly on the rate of growth but also with cell concentration. When r_X is expressed using Eq. (10.80), Eq. (10.103) becomes:

$$r_s = \left(\frac{\mu}{Y_{X/S}} + m_s \right) x \tag{10.104}$$

If we now express μ as a function of substrate concentration using Eq. (10.91), Eq. (10.104) becomes:

$$r_s = \left[\frac{\mu_{max} s}{(K_s + s) Y_{X/S}} + m_s \right] x \tag{10.105}$$

When s is zero, Eq. (10.105) predicts that substrate consumption will proceed at a rate equal to $m_s x$. Substrate uptake in the absence of substrate is impossible; this feature of Eq. (10.105) is therefore unrealistic. The problem arises because of implicit assumptions we have made about the nature of maintenance activities. It can be shown, however, that Eq. (10.105) is a realistic description of substrate uptake while there is external substrate available. When the substrate is exhausted, maintenance energy is generally supplied by endogenous metabolism.

10.11.2 Substrate Uptake With Extracellular Product Formation

Patterns of substrate flux in cells synthesizing products depend on whether product formation is coupled to energy metabolism. When products are formed in energy-generating pathways, for example, in anaerobic culture, product synthesis is an unavoidable consequence of cell growth and maintenance. As illustrated in Figure 10.19(a), there is no separate flow of substrate into the cell for product synthesis; product is formed from the substrate consumed to support growth and maintenance. Substrate consumed for maintenance does not contribute directly to growth; it therefore constitutes a separate substrate flux into the cell. In contrast, when production is not linked to energy metabolism, all or some of the substrate required for product synthesis is in addition to and separate from that substrate needed for growth and maintenance, as illustrated in Figure 10.19(b).

Substrate consumption equations do not include a separate term for product synthesis when products are directly linked to energy generation. The substrate requirements for product formation are already included in the terms representing growth- and maintenance-associated substrate uptake. The rate of substrate uptake is related to growth and maintenance requirements by Eqs. (10.104) and (10.105).

In cultures where product synthesis is indirectly coupled to cellular energy metabolism, the rate of substrate consumption is a function of three factors: the growth rate, the rate of

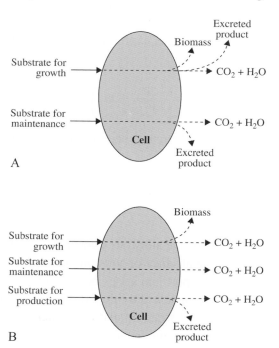

Figure 10.19 Substrate uptake with product formation: (A) product formation directly coupled to energy metabolism; (B) product formation not directly coupled to energy metabolism.

product formation, and the rate of substrate uptake for maintenance. These different cellular functions can be related using yield and maintenance coefficients:

$$r_S = \frac{r_X}{Y_{X/S}} + \frac{r_P}{Y_{P/S}} + m_S x \tag{10.106}$$

where r_S is the volumetric rate of substrate consumption, r_X is the volumetric rate of biomass production, r_P is the volumetric rate of product formation, $Y_{X/S}$ is the true yield of biomass from substrate, $Y_{P/S}$ is the true yield of product from substrate, m_S is the maintenance coefficient, and x is the biomass concentration. If we express r_X and r_P using Eqs. (10.80) and (10.98), respectively:

$$r_S = \left(\frac{\mu}{Y_{XS}} + \frac{q_P}{Y_{PS}} + m_S \right) x \tag{10.107}$$

10.12 Effect of Culture Conditions on Cell Kinetics

Temperature has a large effect on metabolic rate. Temperature influences reaction rates directly as described by the Arrhenius equation (Section 10.1.5); it can also change the configuration of cell constituents, especially proteins and membrane components. In general, the effect of temperature on growth is similar to that described in Section 10.3.4 for enzymes. An approximate twofold increase in the specific growth rate of cells occurs for every 10°C rise in temperature, until structural breakdown of cell proteins and lipids starts to occur. Like other rate constants, the maintenance coefficient m_S has an Arrhenius-type temperature dependence [19]; this can have a significant kinetic effect on cultures where turnover of macromolecules is an important contribution to maintenance requirements. In contrast, temperature has only a minor effect on the biomass yield coefficient, $Y_{X/S}$ [19]. Other cellular responses to temperature are described elsewhere [1,20,21].

The growth rate of cells depends on medium pH in much the same way as enzyme activity (Section 10.3.4). Maximum growth rate is usually maintained over 1 to 2 pH units but declines with further variation. pH can also affect the profile of product synthesis and can change maintenance energy requirements [1,20,21].

10.13 Determining Cell Kinetic Parameters From Batch Data

To apply the equations presented in Sections 10.8 through 10.11 to real bioprocesses, we must know the kinetic and yield parameters for the system and have information about the rates of growth, substrate uptake, and product formation. Batch culture is the most frequently applied method for investigating kinetic behavior, but it is not always the best.

10.13.1 Rates of Growth, Product Formation, and Substrate Uptake

Determining growth rates in cell culture requires measurement of cell concentration. Many different experimental procedures are applied for biomass estimation [20,22]. Direct measurement can be made of cell number, dry or wet cell mass, packed cell volume, or culture turbidity; alternatively, indirect estimates are obtained from measurements of product formation, heat evolution, or cell composition. Cell viability is usually evaluated using agar medium plating or staining techniques. Each method for biomass estimation can give slightly different results. For example, the rate of growth determined using cell dry weight may differ from that based on cell number because dry weight in the culture can increase without a corresponding increase in the number of cells.

EXAMPLE 10.9 Hybridoma Doubling Time

A mouse–mouse hybridoma cell line is used to produce monoclonal antibody. Growth in batch culture is monitored with the following results.

Time (days)	Cell concentration (cells mL^{-1} × 10^{-6})
0.0	0.45
0.2	0.52
0.5	0.65
1.0	0.81
1.5	1.22
2.0	1.77
2.5	2.13
3.0	3.55
3.5	4.02
4.0	3.77
4.5	2.20

a. Determine the specific growth rate during the growth phase.
b. What is the culture doubling time?

Solution

a. The data are plotted as a semi-log graph in Figure 10.20. No lag phase is evident. As Eq. (10.84) applies only when μ is constant, that is, during exponential growth, we must determine which datum points belong to the exponential growth phase. In Figure 10.20, the final three points appear to belong to the decline and death phases of the culture. Fitting a straight line to the remaining data up to and including $t = 3.0$ days gives a slope of 0.67 day^{-1}. Therefore, $\mu = 0.67$ day^{-1}.

b. From Eq. (10.89):

$$t_d = \frac{\ln 2}{0.67 \text{ day}^{-1}} = 1.0 \text{ day}$$

This doubling time applies only during the exponential growth phase of the culture.

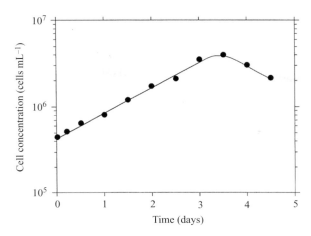

Figure 10.20 Calculation of the specific growth rate for hybridoma cells.

A common method can be applied to calculate μ during exponential growth phase. Assuming that growth is represented by the first-order model of Eq. (10.80), the integrated relationship of Eqs. (10.84) or (10.85) allows us to obtain μ directly. During the exponential growth phase when μ is essentially constant, a plot of ln x versus time gives a straight line with slope μ. This is illustrated in Example 10.9.

The volumetric rates of product formation and substrate uptake, r_P and r_S, can be evaluated by graphical differentiation of product and substrate concentration data, respectively. The specific product formation rate q_P and specific substrate uptake rate q_S are obtained by dividing the respective volumetric rates by the cell concentration.

10.13.2 μ_{max} and K_S

The Monod equation for the specific growth rate, Eq. (10.91), is analogous mathematically to the Michaelis–Menten expression for enzyme kinetics. In principle, the techniques described in Section 10.4 for determining v_{max} and K_m for enzyme reactions can be applied for the evaluation of μ_{max} and K_S. However, because values of K_S in cell culture are usually very small, accurate determination of this parameter from batch data is difficult. Better estimation of K_S can be made using continuous culture of cells as discussed in Chapter 12. The measurement of μ_{max} from batch data is relatively straightforward. If all nutrients are present in excess, the specific growth rate during exponential growth is equal to the maximum specific growth rate. Therefore, the specific growth rate calculated using the procedure of Example 10.9 can be considered equal to μ_{max}.

10.14 Effect of Maintenance on Yields

True yields such as $Y_{X/S}$, $Y_{P/X}$, and $Y_{P/S}$ are often difficult to evaluate. True yields are essentially stoichiometric coefficients; however, the stoichiometry of biomass production and product formation is known only for relatively simple bioprocesses. If the metabolic pathways are complex, stoichiometric calculations become very complicated. However, theoretical yields can be related to observed yields such as $Y'_{X/S}$, $Y'_{P/X}$, and $Y'_{P/S}$ which are more easily determined.

10.14.1 Observed Yields

Expressions for observed yield coefficients are obtained by applying Eq. (10.76):

$$Y'_{X/S} = \frac{-dX}{dS} = \frac{r_X}{r_s} \tag{10.108}$$

$$Y'_{P/X} = \frac{dP}{dX} = \frac{r_p}{r_X} \tag{10.109}$$

and

$$Y'_{P/S} = \frac{-dP}{dS} = \frac{r_p}{r_s} \tag{10.110}$$

where X, S, and P are masses of cells, substrate, and product, respectively, and r_X, r_S, and r_P are observed rates evaluated from experimental data. Therefore, observed yield coefficients can be determined by plotting ΔX, ΔS, or ΔP against each other and evaluating the slope as illustrated in Figure 10.21. Alternatively, observed yield coefficients at a particular instant in time can be calculated as the ratio of rates evaluated at that instant. Observed yields are not necessarily constant throughout batch culture; in some cases, they exhibit significant dependence on growth rate and environmental parameters such as substrate concentration. Nevertheless, for many cultures, the observed biomass yield $Y'_{X/S}$ is approximately constant during batch growth. Because of the errors in experimental data, uncertainty is common with measured yield coefficients.

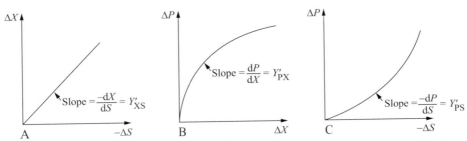

Figure 10.21 Evaluation of observed yields in batch culture from cell, substrate, and product concentrations. (A) biomass per substrate yield, (B) product per biomass yield, (C) product per substrate yield.

10.14.2 Biomass Yield From Substrate

Equations for the true biomass yield can be determined for systems without extracellular product formation or when product synthesis is directly coupled to energy metabolism. Substituting expressions for r_X and r_S from Eqs. (10.80) and (10.104) into Eq. (10.108) gives:

$$Y'_{X/S} = \frac{\mu}{\left(\dfrac{\mu}{Y_{X/S}} + m_S\right)} \tag{10.111}$$

Inverting Eq. (10.111) produces the expression:

$$\frac{1}{Y'_{X/S}} = \frac{1}{Y_{X/S}} + \frac{m_s}{\mu} \tag{10.112}$$

Therefore, if $Y_{X/S}$ and m_S are relatively constant, a plot of $1/Y'_{X/S}$ versus $1/\mu$ gives a straight line with slope m_S and intercept $1/Y_{X/S}$. Equation (10.112) is not generally applied to batch growth data, μ does not vary from μ_{max} for much of the culture period so it is difficult to plot $Y'_{X/S}$ as a function of specific growth rate. As a general guideline, the true biomass yield for microbial growth on glucose under aerobic conditions is around 0.45 g biomass (g glucose)$^{-1}$ while anaerobic conditions have values around 0.1 g biomass (g glucose)$^{-1}$.

In processes such as the production of bakers' yeast and single-cell protein where the desired product is biomass, it is essential to maximize the observed yield of cells from substrate. The true yield $Y_{X/S}$ is limited by stoichiometric considerations. However, from Eq. (10.111), $Y'_{X/S}$ can be improved by decreasing the maintenance coefficient. This is achieved by increasing the specific growth rate, lowering the temperature of the bioprocess, using a medium of lower ionic strength, or applying a different organism or strain with lower maintenance energy requirements.

10.14.3 Product Yield From Biomass

The observed yield of product from biomass $Y'_{P/X}$ is defined in Eq. (10.109). When product synthesis is directly coupled to energy metabolism, r_P is given by Eq. (10.100). Substituting this and Eq. (10.80) into Eq. (10.109) gives:

$$Y'_{PX} = Y_{PX} + \frac{m_p}{\mu} \tag{10.113}$$

The extent of deviation of $Y'_{P/X}$ from $Y_{P/X}$ depends on the relative magnitudes of m_P and μ. To increase the observed yield of product for a particular process, m_P should be increased and μ decreased. Equation (10.113) does not apply to products not directly coupled to energy metabolism; we do not have a general expression for r_P in terms of the true yield coefficient for this class of product.

10.14.4 Product Yield From Substrate

The observed product yield from substrate $Y'_{P/S}$ is defined in Eq. (10.110). For products coupled to energy generation, expressions for r_P and r_S are available from Eqs. (10.100) and (10.104). Therefore:

$$Y'_{P/S} = \frac{Y_{P/X}\mu + m_P}{\left(\dfrac{\mu}{Y_{X/S}} + m_s\right)} \tag{10.114}$$

In many anaerobic bioprocesses, such as ethanol production, the yield of product from substrate is a critical factor affecting process economics. At high $Y'_{P/S}$, more ethanol is produced per mass of carbohydrate consumed and the overall cost of production is reduced. Growth rate has a strong effect on $Y'_{P/S}$ for ethanol. Because $Y'_{P/S}$ is low when $\mu=\mu_{max}$, it is desirable to reduce the specific growth rate of the cells. Low growth rates can be achieved by depriving the cells of some essential nutrient, for example, a nitrogen source, or by immobilizing the cells to prevent growth. Increasing the rate of maintenance activity relative to growth will also enhance the product yield. This can be done by using a culture medium with high ionic strength, raising the temperature, or selecting a mutant or different organism with high maintenance requirements.

The effect of growth rate and maintenance on $Y'_{P/S}$ is difficult to determine for products not directly coupled with energy metabolism unless information is available about the effect of these parameters on q_P.

10.15 Kinetics of Cell Death

The kinetics of cell death are an important consideration in the design of sterilization processes and for analysis of cell cultures where substantial loss of viability occurs. In a lethal environment, cells in a population do not die all at once; deactivation of the culture occurs over a finite period of time depending on the initial number of viable cells and the severity of the conditions. Loss of cell viability can be described mathematically in much the same way as enzyme deactivation (Section 10.6). Cell death is a first-order process:

$$r_d = k_d N \tag{10.115}$$

where r_d is the rate of cell death, N is the number of viable cells, and k_d is the *specific death constant*. Alternatively, the rate of cell death can be expressed in terms of cell concentration rather than cell number:

$$r_d = k_d x \tag{10.116}$$

where k_d is the specific death constant based on cell concentration and x is the concentration of viable cells.

In a closed system where cell death is the only process affecting the concentration of viable cells, the rate of cell death is equal to the rate of decrease in the number of cells. Therefore, using Eq. (10.115):

$$r_d = \frac{-dN}{dt} = k_d N \qquad (10.117)$$

If k_d is constant, we can integrate Eq. (10.117) to derive an expression for N as a function of time:

$$N = N_0 \, e^{-k_d t} \qquad (10.118)$$

where N_0 is the number of viable cells at time zero. Taking natural logarithms of both sides of Eq. (10.118) gives:

$$\ln N = \ln N_0 - k_d t \qquad (10.119)$$

According to Eq. (10.119), if first-order death kinetics apply, a plot of $\ln N$ versus t gives a straight line with slope $-k_d$. Experimental measurements have confirmed the relationship of Eq. (10.119) for many organisms in a vegetative state; as an example, results for the thermal death of *Escherichia coli* at various temperatures are shown in Figure 10.22.

Like other kinetic constants, the value of the specific death constant k_d depends on temperature. This effect can be described using the Arrhenius relationship of Eq. (10.74). Typical

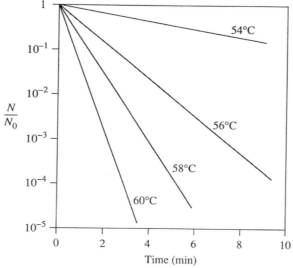

Figure 10.22 Relationship between temperature and the rate of thermal death for vegetative *Escherichia coli* cells. *From S. Aiba, A.E. Humphrey, and N.F. Millis, 1965,* Biochemical Engineering, *Academic Press, New York.*

E_d values for the thermal destruction of microorganisms are high, of the order 250 to 290 kJ gmol^{-1} [23]. Therefore, small increases in temperature have a significant effect on k_d and the rate of cell death. The concept of thermal death kinetics is illustrated in Example 10.10.

EXAMPLE 10.10 Thermal Death Kinetics

The number of viable spores of a new strain of *Bacillus subtilis* is measured as a function of time at various temperatures.

Time (min)	Number of spores at:			
	$T = 85°C$	$T = 90°C$	$T = 110°C$	$T = 120°C$
0.0	2.40×10^9	2.40×10^9	2.40×10^9	2.40×10^9
0.5	2.39×10^9	2.38×10^9	1.08×10^9	2.05×10^7
1.0	2.37×10^9	2.30×10^9	4.80×10^8	1.75×10^5
1.5	–	2.29×10^9	2.20×10^8	1.30×10^3
2.0	2.33×10^9	2.21×10^9	9.85×10^7	–
3.0	2.32×10^9	2.17×10^9	2.01×10^7	–
4.0	2.28×10^9	2.12×10^9	4.41×10^6	–
6.0	2.20×10^9	1.95×10^9	1.62×10^5	–
8.0	2.19×10^9	1.87×10^9	6.88×10^3	–
9.0	2.16×10^9	1.79×10^9	–	–

a. Determine the activation energy for thermal death of *B. subtilis* spores.
b. What is the specific death constant at 100°C?
c. Estimate the time required to kill 99% of spores in a sample at 100°C.

Solution

a. A semi-log plot of the number of viable spores versus time is shown in Figure 10.23 for each of the four temperatures. From Eq. (10.119), the slopes of the lines in Figure 10.23 are equal to $-k_d$ at the various temperatures. Fitting straight lines to the data gives the following results:
$k_d (85°C) = 0.012 \text{ min}^{-1}$
$k_d (90°C) = 0.032 \text{ min}^{-1}$
$k_d (110°C) = 1.60 \text{ min}^{-1}$
$k_d (120°C) = 9.61 \text{ min}^{-1}$
The relationship between k_d and absolute temperature is given by Eq. (10.74). Therefore, a semi-log plot of k_d versus $1/T$ should yield a straight line with slope $= -E_d/R$ where T is absolute temperature. Temperature is converted from degrees Celsius to Kelvin; the results for k_d are plotted against $1/T$ in units of K^{-1} in Figure 10.24. The slope is $-27,030$ K. $R = 8.3144$ J K^{-1} gmol^{-1}. Therefore:

$$E_d = 27,030 \text{ K} (8.3144 \text{ J K}^{-1} \text{gmol}^{-1}) = 2.25 \times 10^5 \text{ J gmol}^{-1}$$
$$E_d = 225 \text{ kJ gmol}^{-1}$$

b. The equation for the line in Figure 10.24 is:

$$k_d = 6.52 \times 10^{30}\, e^{-27.030/T}$$

where k_d has units of min^{-1} and T has units of K. Therefore, at $T = 100°C = 373.15\,K$, $k_d = 0.23\,min^{-1}$.

c. From Eq. (10.119):

$$t = \frac{-(\ln N - \ln N_0)}{k_d}$$

or

$$t = \frac{-\ln\left(\dfrac{N}{N_0}\right)}{k_d}$$

For N equal to 1% of N_0, $N/N_0 = 0.01$. At $100°C$, $k_d = 0.23\,min^{-1}$ and the time required to kill 99% of the spores is:

$$t = \frac{-\ln(0.01)}{0.23\ min^{-1}} = 20\ min$$

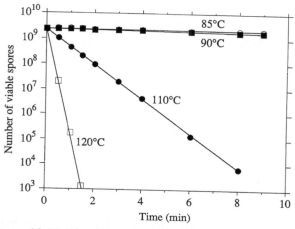

Figure 10.23 The thermal death of *Bacillus subtilis* spores.

When contaminating organisms in culture media are being killed by heat sterilization, nutrients in the medium may also be destroyed. The sensitivity of nutrient molecules to temperature is also described by the Arrhenius equation of Eq. (10.74). Values of the activation energy E_d for thermal destruction of vitamins and amino acids are 84 to $92\,kJ\ gmol^{-1}$; for proteins, E_d is about $165\,kJ\ gmol^{-1}$ [23]. Because these values are somewhat lower than typical E_d values for the thermal death of microorganisms, raising the temperature has a greater effect on

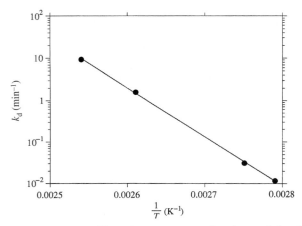

Figure 10.24 Calculation of kinetic parameters for thermal death of spores.

cell death than nutrient destruction. This means that sterilization at higher temperatures for shorter periods of time has the advantage of killing cells with limited destruction of medium components.

10.16 Summary of Chapter 10

At the end of Chapter 10 you should:

- Understand the difference between *reversible* and *irreversible reactions*, and the limitations of equilibrium thermodynamics in representing industrial cell culture and enzyme reactions
- Know the meaning of *total rate* (R_A), *volumetric rate*), (r_A), and *specific rate* (q_A) when describing the rate of reactions
- Be able to calculate reaction rates from batch concentration data using *graphical differentiation*
- Be familiar with kinetic relationships for *zero-order, first-order*, and *Michaelis–Menten* reactions
- Be able to quantify the effect of temperature on biological reaction rates
- Know how to determine the enzyme kinetic parameters v_{max} and K_m from batch concentration data
- Know how the kinetics of *enzyme deactivation* affect the rate of enzyme reactions
- Be able to calculate *yield coefficients* for cell cultures
- Understand the mathematical relationships that describe *cell growth kinetics*
- Be able to evaluate the specific rates of growth, product formation, and substrate uptake in batch cell cultures
- Know the effect of *maintenance activities* on growth, product synthesis, and substrate utilization in cells
- Understand the kinetics of *cell death* and the thermal destruction of cells

Problems

10.1 Reaction equilibrium

Calculate equilibrium constants for the following reactions under standard conditions:

a. glutamine $+ H_2O \rightarrow$ glutamate $+ NH_4^+$ $\quad \Delta G^\circ_{rxn} = -14.1 \, kJ \, mol^{-1}$

b. malate $\rightarrow$ fumarate $+ H_2O$ $\quad \Delta G^\circ_{rxn} = 3.2 \, kJ \, mol^{-1}$

Can either of these reactions be considered irreversible?

10.2 Equilibrium yield

The following reaction catalyzed by phosphoglucomutase occurs during breakdown of glycogen:

$$\text{glucose 1-phosphate} \rightleftarrows \text{glucose 6-phosphate}$$

A reaction is started by adding phosphoglucomutase to 0.04 gmol of glucose 1-phosphate in 1 L of solution at 25°C. The reaction proceeds to equilibrium, at which the concentration of glucose 1-phosphate is 0.002 M and the concentration of glucose 6-phosphate is 0.038 M.

a. Calculate the equilibrium constant.

b. What is the theoretical yield?

c. What is the yield based on the amount of reactant supplied?

10.3 Reaction rate

a. The volume of a bioreactor is doubled while keeping the cell concentration and other cultivation conditions the same.

   (i) How is the volumetric productivity affected?

   (ii) How is the specific productivity affected?

   (iii) How is the total productivity affected?

b. If, instead of (a), the cell concentration was doubled, what effect would this have on the volumetric, specific, and total productivities?

c. A bioreactor produces 100 kg of lysine per day.

   (i) If the volumetric productivity is $0.8 \, g \, L^{-1} \, h^{-1}$, what is the volume of the bioreactor?

   (ii) The cell concentration is $20 \, g \, L^{-1}$ dry weight. Calculate the specific productivity.

10.4 Enzyme kinetics

Lactase, also known as β-galactosidase, catalyzes the hydrolysis of lactose to produce glucose and galactose from milk and whey. Experiments are carried out to determine the kinetic parameters for the enzyme. The initial rate data are as follows.

Lactose concentration (mol $L^{-1} \times 10^2$)	Initial reaction velocity (mol $L^{-1} \, min^{-1} \times 10^3$)
2.50	1.94
2.27	1.91
1.84	1.85
1.35	1.80
1.25	1.78
0.730	1.46
0.460	1.17
0.204	0.779

Evaluate v_{max} and K_m.

10.5   Enzyme kinetics after site-specific mutagenesis

Cyclophosphamide (CPA) is an anticancer prodrug that requires activation in the liver by cytochrome P450 2B enzymes for production of cytotoxic metabolites. Site-specific mutagenesis is used to alter the amino acid sequence of P450 2B1 in an attempt to improve the kinetics of CPA activation. The rate of reaction of CPA is studied using rat P450 2B1 and a site-specific variant of P450 2B1 produced using *Escherichia coli*. The results are as follows.

	Initial reaction velocity (mol min^{-1} mol^{-1} P450)	
Initial CPA concentration (mM)	Rat P450 2B1	Variant P450 2B1
0.3	5.82	17.5
0.5	9.03	24.5
0.8	12.7	24.0
1.5	17.1	23.9
3	20.2	27.3
5	27.8	33.1
7	31.5	27.7

a. Determine the kinetic constants for the rat and site-specific variant enzymes.

b. When CPA is administered to cancer patients, the peak plasma concentration of the drug is relatively low at around 100 to 200 µM. Are the kinetic properties of the variant P450 2B1 better than those of rat P450 2B1 for CPA activation in this situation? Explain your answer.

c. Did manipulation of the enzyme using site-specific mutagenesis improve the catalytic efficiency for CPA activation?

10.6   Kinetic properties of pheromone-degrading enzyme

An enzyme that degrades pheromone is isolated from the sensory hairs of the silk moth, *Antheraea polyphemus*. The kinetics of the reaction are studied at pH 7.2 using a fixed enzyme concentration and temperatures of 10°C to 40°C. The following table lists results for reaction velocity as a function of substrate concentration.

	Initial reaction velocity (µmol L^{-1} s^{-1})			
Pheromone concentration (µmol L^{-1})	T = 10°C	T = 20°C	T = 30°C	T = 40°C
0.5	3.0×10^{-6}	5.5×10^{-6}	4.2×10^{-6}	7.7×10^{-6}
1.0	5.1×10^{-6}	9.2×10^{-6}	9.5×10^{-6}	1.2×10^{-5}
1.0	4.2×10^{-6}	9.7×10^{-6}	8.9×10^{-6}	1.6×10^{-5}
1.5	6.1×10^{-6}	1.1×10^{-5}	1.3×10^{-5}	1.8×10^{-5}
2.2	7.1×10^{-6}	1.6×10^{-5}	9.8×10^{-6}	2.1×10^{-5}
5.5	1.1×10^{-5}	1.5×10^{-5}	1.9×10^{-5}	3.8×10^{-5}
5.5	9.8×10^{-6}	1.9×10^{-5}	2.6×10^{-5}	3.0×10^{-5}
11	1.2×10^{-5}	2.2×10^{-5}	2.9×10^{-5}	3.9×10^{-5}
11	9.5×10^{-6}	2.1×10^{-5}	2.5×10^{-5}	3.6×10^{-5}

a. Determine v_{max} and K_m at the four reaction temperatures.

b. Determine the activation energy for this enzyme reaction.

10.7 Enzyme substrate specificity

Xylose isomerase extracted from the thermophile *Thermoanaerobacterium thermosulfurigenes* uses glucose and xylose as substrates. Applying each substrate separately, the rates of reaction are measured at 60°C using an enzyme concentration of 0.06 mg mL^{-1}.

Glucose concentration (mM)	Initial reaction velocity (μmol mL^{-1} min^{-1})	Xylose concentration (mM)	Initial reaction velocity (μmol mL^{-1} min^{-1})
80	0.151	20	0.320
100	0.194	50	0.521
300	0.385	100	0.699
500	0.355	300	1.11
750	0.389	500	1.19
1000	0.433	700	1.33
1400	0.445	900	1.13

a. What is the relative catalytic efficiency of the enzyme using xylose as substrate compared with glucose?

b. What conclusions can you draw about the substrate specificity of this enzyme?

10.8 Effect of temperature on the hydrolysis of starch

α-Amylase from malt is used to hydrolyze starch. The dependence of the initial reaction rate on temperature is determined experimentally. Results measured at fixed starch and enzyme concentrations are listed in the following table.

Temperature (°C)	Rate of glucose production (mmol m^{-3} s^{-1})
20	0.31
30	0.66
40	1.20
60	6.33

a. Determine the activation energy for this reaction.

b. α-Amylase is used to break down starch in baby food. It is proposed to conduct the reaction at a relatively high temperature so that the viscosity is reduced. What is the reaction rate at 55°C compared with 25°C?

c. Thermal deactivation of this enzyme is described by the equation:

$$k_d = 2.25 \times 10^{27} e^{-41,630/RT}$$

where k_d is the deactivation rate constant in h^{-1}, R is the ideal gas constant in Cal gmol^{-1} K^{-1}, and T is temperature in K. What is the half-life of the enzyme at 55°C compared with 25°C? Which of these two operating temperatures is more practical for processing baby food? Explain your answer.

10.9 Optimum temperature for enzymatic hydrolysis of cellulose

The production of glucose from lignocellulosic materials is a major hurdle in the production of biofuels from renewable resources. β-Glucosidase from *Aspergillus niger* is investigated as a potential component in an enzyme cocktail for cellulose hydrolysis at elevated temperatures. The activity of $5.1\,mg\,L^{-1}$ of β-glucosidase is measured at temperatures from 30°C to 70°C using cellobiose as substrate. The results are as follows.

Temperature (°C)	v_{max} (mmol L^{-1} min^{-1})
30	456
45	1250
50	1590
60	2900
70	567

   a. Estimate the optimum temperature for cellobiose hydrolysis.
   b. Evaluate the Arrhenius parameters for thermal activation of β-glucosidase.
   c. What is the maximum rate of cellobiose hydrolysis at 55°C?

10.10 Enzyme reaction and deactivation

Lipase is being investigated as an additive to laundry detergent for removal of stains from fabric. The general reaction is:

$$\text{fats} \rightarrow \text{fatty acids} + \text{glycerol}$$

The Michaelis constant for pancreatic lipase is 5 mM. At 60°C, lipase is subject to deactivation with a half-life of 8 min. Fat hydrolysis is carried out in a well-mixed batch reactor that simulates a top-loading washing machine. The initial fat concentration is 45 gmol m^{-3}. At the beginning of the reaction, the rate of hydrolysis is 0.07 mmol L^{-1} s^{-1}. How long does it take for the enzyme to hydrolyze 80% of the fat present?

10.11 Effect of amino acid composition and metal binding on enzyme stability

Genomic analysis has revealed that enzymes from hyperthermophilic microorganisms contain lower amounts of the thermo-labile amino acids, glutamine (Gln) and asparagine (Asn), than enzymes from mesophiles. It is thought that this adaptation may contribute to the thermostability of enzymes in high-temperature environments above 50°C. A variant form of xylose isomerase containing a low (Gln+Asn) content is developed using site-directed mutagenesis. Experiments are conducted to measure the activity of native and variant xylose isomerase at 68°C using an enzyme concentration of 0.5 mg mL^{-1}. Because binding with metal ions also has a potential stabilizing effect on this enzyme, the activity of native xylose isomerase is measured in the absence and presence of 0.015 mM Mn^{2+}. The reactions are performed with excess substrate and give the following results.

Native enzyme		Variant enzyme		Native enzyme+Mn^{2+}	
Time (min)	Ratio of activity to initial activity	Time (min)	Ratio of activity to initial activity	Time (min)	Ratio of activity to initial activity
0	1	0	1	0	1
10	0.924	5	0.976	10	0.994
30	0.795	20	0.831	20	0.834
45	0.622	50	0.608	50	0.712

Native enzyme		Variant enzyme		Native enzyme+Mn2+	
Time (min)	Ratio of activity to initial activity	Time (min)	Ratio of activity to initial activity	Time (min)	Ratio of activity to initial activity
60	0.513	140	0.236	180	0.275
120	0.305	200	0.149	240	0.197
150	0.225	–	–	–	–

a. Does deactivation of this enzyme at 68°C follow first-order kinetics under each of the conditions tested?

b. Compare the half-life of the variant enzyme with those of the native enzyme with and without Mn^{2+}.

c. What conclusions can you draw about the relative effectiveness of the strategies used to enhance the stability of this enzyme?

10.12 Growth parameters for recombinant

Escherichia coli is used for production of recombinant porcine growth hormone. The bacteria are grown aerobically in batch culture with glucose as the growth-limiting substrate. Cell and substrate concentrations are measured as a function of culture time with the following results.

Time (h)	Cell concentration, x (kg m^{-3})	Substrate concentration, s (kg m^{-3})
0.0	0.20	25.0
0.33	0.21	24.8
0.5	0.22	24.8
0.75	0.32	24.6
1.0	0.47	24.3
1.5	1.00	23.3
2.0	2.10	20.7
2.5	4.42	15.7
2.8	6.9	10.2
3.0	9.4	5.2
3.1	10.9	1.65
3.2	11.6	0.2
3.5	11.7	0.0
3.7	11.6	0.0

a. Plot μ as a function of time.

b. What is the value of μ_{max}?

c. What is the observed biomass yield from substrate? Is $Y'_{X/S}$ constant during the culture?

10.13 Growth parameters for hairy roots

Hairy roots are produced by genetic transformation of plants using *Agrobacterium rhizogenes*. The following biomass and sugar concentrations are obtained during batch culture of *Atropa belladonna* hairy roots in a bubble column bioreactor.

Time (days)	Biomass concentration (g L^{-1} dry weight)	Sugar concentration (g L^{-1})
0	0.64	30.0
5	1.95	27.4
10	4.21	23.6
15	5.54	21.0
20	6.98	18.4
25	9.50	14.8
30	10.3	13.3
35	12.0	9.7
40	12.7	8.0
45	13.1	6.8
50	13.5	5.7
55	13.7	5.1

   a. Plot μ as a function of culture time. When does the maximum specific growth rate occur?

   b. Plot the specific rate of sugar uptake as a function of time.

   c. What is the observed biomass yield from substrate? Is $Y'_{X/S}$ constant during the culture?

10.14 Kinetics of diatom growth and silicate uptake

Growth and nutrient uptake in batch cultures of the freshwater diatom, *Cyclotella meneghiniana*, are studied under silicate-limiting conditions. Unbuffered freshwater medium containing 25 µM silicate is inoculated with cells. Samples are taken over a period of 4 days for measurement of cell and silicate concentrations. The results are as follows.

Time (days)	Cell concentration (cells L^{-1})	Silicate concentration (µM)
0	4.41×10^5	8.00
0.5	5.53×10^5	7.97
1.0	1.31×10^6	7.72
1.5	3.00×10^6	7.59
2.0	4.82×10^6	6.96
2.5	1.12×10^7	5.33
3.0	1.67×10^7	4.63
4.0	2.57×10^7	1.99

   a. Does this culture exhibit exponential growth?

   b. What is the value of μ_{max}?

   c. Is there a lag phase?

   d. What is the observed biomass yield from substrate?

   e. Is the observed biomass yield from substrate constant during the culture?

10.15 Algal batch growth kinetics

Chlorella sp. algae are used to produce lipids for the manufacture of biodiesel. The kinetics of algal growth are determined in batch culture under phosphate-limiting conditions at an illumination intensity of approximately 60 µmol m^{-2} s^{-1} and a photoperiod of 14 h light:10 h dark. The following results are obtained from five different cultures at temperatures from 19°C to 28.5°C, including data from a replicate culture at 25°C.

| | $T = 19°C$ | | $T = 20°C$ | | $T = 25°C$ | | $T = 25°C$ | | $T = 28.5°C$ |
Time (days)	Cell concentration (cells L^{-1})	Time (days)	Cell concentration (cells L^{-1})	Time (days)	Cell concentration (cells L^{-1})	Time (days)	Cell concentration (cells L^{-1})	Time (days)	Cell concentration (cells L^{-1})
0	1.30×10^5	0	1.30×10^5	0	1.35×10^5	0	1.35×10^5	0	1.30×10^5
0.2	1.87×10^5	0.5	3.99×10^5	0.5	2.58×10^5	0.5	2.95×10^5	1.0	1.87×10^5
0.5	2.44×10^5	1.0	6.97×10^5	1.0	9.49×10^5	1.0	1.02×10^6	1.5	2.44×10^5
1.0	6.12×10^5	1.5	1.14×10^6	1.5	2.91×10^6	1.5	1.34×10^6	2.6	6.12×10^5
1.4	1.34×10^6	2.1	4.88×10^6	2.1	7.66×10^6	2.1	9.03×10^6	–	–
1.8	1.72×10^6	2.1	3.98×10^6	2.6	1.77×10^7	2.6	6.26×10^7	–	–
2.2	3.92×10^6	3.5	4.12×10^7	3.8	2.03×10^8	3.8	2.96×10^8	–	–
2.5	7.01×10^6	–	–	–	–	–	–	–	–
3.0	1.75×10^7	–	–	–	–	–	–	–	–
3.8	4.70×10^7	–	–	–	–	–	–	–	–

a. Do the cultures exhibit exponential growth at each of the temperatures tested?

b. Evaluate the maximum specific growth rates and doubling times for each of the five algal cultures.

c. Use the data to determine an expression for μ_{max} as a function of temperature during thermal activation of growth.

d. Estimate the maximum specific growth rate at 22°C.

e. Evaluate the number of algal cells produced if a 1.6-m³ bioreactor is inoculated at a cell density of 2×10^5 cells L^{-1} and the culture grows exponentially at 22°C for a period of 2 days and 6 h.

10.16 Kinetics of batch cell culture with nisin production

Lactococcus lactis subsp. *lactis* produces nisin, a biological food preservative with bactericidal properties. Medium containing $10 \, g \, L^{-1}$ sucrose is inoculated with bacteria and the culture is monitored for 24 h. The pH is controlled at 6.80 using 10 M NaOH. The following results are recorded.

Time (h)	Cell concentration (g L⁻¹ dry weight)	Sugar concentration (g L⁻¹)	Nisin concentration (IU mL⁻¹)
0.0	0.02	9.87	0
1.0	0.03	9.77	278
2.0	0.03	9.38	357
3.0	0.042	9.45	564
4.0	0.21	9.02	662
4.5	0.33	8.55	695
5.0	0.35	9.12	1213
5.5	0.38	8.09	1341
6.0	1.13	7.58	1546
6.5	1.95	6.24	1574
7.0	3.66	1.30	1693
7.5	4.09	1.14	1678
8.0	4.23	0.05	1793
8.5	4.07	0.02	1733
9.0	3.85	0.02	1567
9.5	2.66	0.015	1430
10.0	2.42	0.018	1390
11.0	2.03	0	–
12.0	1.54	0	1220
14.0	1.89	0	995
16.0	1.45	0	–
24.0	1.67	0	617

a. Plot all the data on a single graph, using multiple vertical axes to show the cell, sugar, and nisin concentrations as a function of time.

(i) What is the relationship between nisin production and growth?

(ii) *L. lactis* produces lactic acid as a product of energy metabolism. Release of lactic acid into the medium can lower the pH significantly and cause premature cessation of growth. Do the measured data show any evidence of this? Explain your answer.

    (iii) Is 24 h an appropriate batch culture duration for nisin production? Explain your answer.

    (iv) What is the observed overall yield of biomass from substrate? Is this a meaningful parameter for characterizing the extent of sugar conversion to biomass in this culture? Explain your answer.

b. Plot the data for cell concentration as a function of time on semi-logarithmic coordinates.

    (i) Does the culture undergo a lag phase? If so, what is the duration of the lag phase?

    (ii) Does exponential growth occur during this culture? If so, over what time period?

    (iii) Develop an equation for cell concentration as a function of time during the growth phase.

    (iv) What is the maximum specific growth rate?

    (v) What is the culture doubling time? When does this doubling time apply?

    (vi) Does the culture reach stationary phase? Explain your answer.

    (vii) Over what period does the culture undergo a decline phase?

    (viii) Determine the specific cell death constant for the culture.

    (ix) At the end of the growth phase, what is the observed yield of biomass from substrate based on the initial concentrations of biomass and substrate? Compare this with the answer to (a) (iv) above and comment on any difference.

    (x) At the end of the growth phase, what is the observed yield of nisin from substrate? Use a conversion factor of 40×10^6 IU per g of nisin.

    (xi) At the end of the growth phase, what is the observed yield of nisin from biomass in units of mg g^{-1} dry weight?

c. Analyze the data for sugar concentration to determine the rate of substrate consumption as a function of culture time. Plot the results.
- What is the maximum rate of substrate consumption?
- When does the maximum rate of substrate consumption occur?

d. Plot the specific rate of substrate consumption as a function of time.
- What is the maximum specific rate of substrate consumption? When does it occur? How much confidence do you have in your answer?
- What is the specific rate of substrate consumption during the growth phase?

e. Analyze the data for nisin concentration to determine the rate of nisin production as a function of culture time. Plot the results.
- Estimate the maximum nisin productivity. Express your answer in units of mg L^{-1}h^{-1}.
- When does the maximum nisin productivity occur?

f. Plot the specific rate of nisin production as a function of time.
- What is the maximum specific rate of nisin production? Express your answer in units of mg h^{-1}g^{-1} dry weight.
- When does this maximum occur?

g. Plot the rate of biomass production versus the rate of substrate consumption for the culture.
- During growth when the rate of biomass production is positive rather than negative, can the observed biomass yield from substrate be considered constant?
- How does the value for $Y'_{X/S}$ determined graphically for the entire growth phase compare with the results in (a) and (b) above? Explain any differences.

h. Plot the rate of nisin production versus the rate of substrate consumption, and the rate of nisin production versus the rate of biomass production, using two separate graphs. During the culture period when the rate of nisin production is positive rather than negative, can the observed yield coefficients $Y'_{P/S}$ and $Y'_{P/X}$ be considered constant during batch culture?

10.17 Yeast culture and astaxanthin production

The yeast *Phaffia rhodozyma* produces the carotenoid pigment astaxanthin in the absence of light. The cells are grown in batch culture using medium containing $40\,g\,L^{-1}$ glucose. Cell, substrate, and product concentrations are measured as a function of culture time. The results are as follows.

Time (h)	Cell concentration (g L⁻¹ dry weight)	Sugar concentration (g L⁻¹)	Astaxanthin concentration (mg L⁻¹)
0	0.01	40	0
5	0.019	39.96	0.19
10	0.029	39.90	0.32
12	0.029	39.91	0.45
20	0.074	39.65	0.60
24	0.124	39.14	0.72
30	0.157	38.95	0.81
35	0.356	38.22	0.87
45	0.906	35.24	1.01
50	2.28	32.40	1.29
60	3.67	17.05	1.35
70	7.17	0	1.33
78	6.59	0	1.48

a. Does the culture exhibit a lag phase?

b. Does growth of *Phaffia rhodozyma* follow first-order kinetics?

c. Evaluate μ_{max} and the culture doubling time. Over what period are these parameters valid representations of growth?

d. From a plot of astaxanthin concentration versus time, estimate the initial rate of astaxanthin synthesis. Express your answer in units of mg $L^{-1}\,h^{-1}$.

e. Using the result from (d), estimate the initial specific rate of astaxanthin production. What confidence do you have in this result?

f. Plot Δx versus Δs to determine the observed biomass yield from glucose. Use the initial cell and substrate concentrations to calculate Δx and Δs. During the growth phase, how does the observed biomass yield from glucose vary with culture time?

g. What is the overall product yield from glucose?

h. What is the overall product yield from biomass?

10.18 Ethanol fermentation by yeast and bacteria

Ethanol is produced by anaerobic fermentation of glucose by *Saccharomyces cerevisiae*. For the strain of *S. cerevisiae* employed, the maintenance coefficient is 0.18 kg glucose (kg biomass)$^{-1}\,h^{-1}$, $Y_{X/S}$ is 0.11 kg biomass (kg glucose)$^{-1}$, $Y_{P/X}$ is 3.9 kg ethanol (kg biomass)$^{-1}$, and μ_{max} is $0.4\,h^{-1}$. It is decided to investigate the possibility of using *Zymomonas mobilis* bacteria instead of yeast for making ethanol. *Z. mobilis* is known to produce ethanol under anaerobic conditions using a different metabolic pathway to that employed by yeast. Typical values of $Y_{X/S}$ are lower than for yeast at about 0.06 kg biomass (kg glucose)$^{-1}$; on the other hand, the maintenance coefficient is higher at 2.2 kg glucose (kg biomass)$^{-1}$ h^{-1}. $Y_{P/X}$ for *Z. mobilis* is 7.7 kg ethanol (kg biomass)$^{-1}$; μ_{max} is 0.3 h^{-1}.

a. From stoichiometry, what is the maximum theoretical yield of ethanol from glucose?

b. $Y'_{P/S}$ is maximum and equal to the theoretical yield when there is no growth and all substrate entering the cell is used for maintenance activities. If ethanol is the sole extracellular product of energy-yielding metabolism, calculate m_P for each organism.

c. *S. cerevisiae* and *Z. mobilis* are cultured in batch bioreactors. Predict the observed product yield from substrate for the two cultures.

d. What is the efficiency of ethanol production by the two organisms? Efficiency is defined as the observed product yield from substrate divided by the maximum or theoretical product yield.

e. How does the specific rate of ethanol production by *Z. mobilis* compare with that by *S. cerevisiae*?

f. Using Eq. (10.101), compare the proportions of growth-associated and nongrowth-associated ethanol production by *Z. mobilis* and *S. cerevisiae*. For which organism is nongrowth-associated production more substantial?

g. To achieve the same volumetric ethanol productivity from the two cultures, what relative concentrations of yeast and bacteria are required?

h. In the absence of growth, the efficiency of ethanol production is the same in both cultures. Under these conditions, if yeast and bacteria are employed at the same concentration, what size bioreactor is required for the yeast culture compared with that required for the bacterial culture to achieve the same total productivity?

i. Predict the observed biomass yield from substrate for the two organisms. For which organism is biomass disposal less of a problem?

j. Make a recommendation about which organism is better suited for industrial ethanol production. Explain your answer.

10.19 Plasmid loss during culture maintenance

A stock culture of plasmid-containing *Streptococcus cremoris* cells is maintained with regular subculturing for a period of 28 days. After this time, the fraction of plasmid-carrying cells is found to be 0.66. The specific growth rate of plasmid-free cells at the storage temperature is 0.033 h^{-1}; the specific growth rate of plasmid-containing cells is 0.025 h^{-1}. If all the cells initially contained plasmid, estimate the frequency of plasmid loss per generation.

10.20 Medium sterilization

A steam sterilizer is used to sterilize liquid medium for a bioprocess. The initial concentration of contaminating organisms is 10^8 per L. For design purposes, the final acceptable level of contamination is usually taken to be 10^{-3} cells; this corresponds to a risk that one batch in a thousand will remain contaminated even after sterilization is complete. For how long should 1 m^3 of medium be treated if the sterilization temperature is:

a. 80°C?

b. 121°C?

c. 140°C?

To be safe, assume that the contaminants present are spores of *Bacillus stearothermophilus*, one of the most heat-resistant microorganisms known. For these spores, the activation energy for thermal death is $283\,kJ\,gmol^{-1}$ and the Arrhenius constant is $10^{36.2}\,s^{-1}$ [24].

10.21 Effect of medium osmolarity on growth and death of hybridoma cells

Medium osmolarity is an important variable in the design of serum-free media for mammalian cell culture. A murine hybridoma cell line synthesizing IgG_1 antibody is grown in culture media adjusted to three different osmolarities by addition of NaCl or sucrose. Growth and viability of the cells in batch culture are monitored over a period of 300 h. Results for viable cell concentration as a function of culture time are as follows.

	Viable cell concentration (cells mL^{-1})		
Time (h)	Osmolarity = 290 mOsm L^{-1}	Osmolarity = 380 mOsm L^{-1}	Osmolarity = 435 mOsm L^{-1}
0	4.2×10^4	4.2×10^4	4.2×10^4
25	9.8×10^4	6.3×10^4	4.8×10^4
50	2.2×10^5	1.2×10^5	7.0×10^4
75	6.5×10^5	2.2×10^5	9.7×10^4
100	1.1×10^6	4.1×10^5	1.6×10^5
125	1.0×10^6	5.1×10^5	2.5×10^5
150	9.7×10^5	7.9×10^5	4.1×10^5
175	7.8×10^5	7.7×10^5	5.9×10^5
200	7.0×10^5	6.0×10^5	5.4×10^5
225	7.0×10^5	5.5×10^5	4.4×10^5
250	6.5×10^5	3.5×10^5	2.6×10^5
275	4.7×10^5	1.7×10^5	1.2×10^5
300	4.2×10^5	1.3×10^5	8.8×10^4

a. Does medium osmolarity affect whether the hybridoma cells undergo a lag phase?
b. How does medium osmolarity affect the maximum specific growth rate of the cells?
c. How does medium osmolarity affect the maximum cell concentration achieved and the time taken to reach maximum cell density?
d. What effect does medium osmolarity have on the specific death rate of the cells during the culture death phase?

References

[1] B. Atkinson, F. Mavituna, Biochemical Engineering and Biotechnology Handbook, second ed., Macmillan, 1991.
[2] R.L. Stein, Kinetics of Enzyme Action. John Wiley, 2011.
[3] A. Cornish-Bowden, C.W. Wharton, Enzyme Kinetics. IRL Press, 1988.
[4] M. Dixon, E.C. Webb, Enzymes, second ed., Longmans, 1964.
[5] I.W. Sizer, Effects of temperature on enzyme kinetics. Adv. Enzymol 3 (1943) 35–62.
[6] A. Moser, Rate equations for enzyme kinetics. VCH, 1985.199226
[7] R. Eisenthal, A. Cornish-Bowden, The direct linear plot: a new graphical procedure for estimating enzyme kinetic parameters. Biochem. J. 139 (1974) 715–720.
[8] A. Moser, Kinetics of batch fermentations. VCH, 1985.243283

[9]   J.E. Bailey, D.F. Ollis, Biochemical Engineering Fundamentals, second ed., (Chapter 7), McGraw-Hill, 1986.

[10] J.A. Roels, N.W.F. Kossen, On the modelling of microbial metabolism. Prog. Ind. Microbiol. 14 (1978) 95–203.

[11] M.L. Shuler, F. Kargi, Bioprocess Engineering: Basic Concepts, second ed., (Chapter 6), Prentice Hall, 2002.

[12] T. Imanaka, S. Aiba, A perspective on the application of genetic engineering: stability of recombinant plasmid, Ann. N.Y. Acad. Sci., 369 (1981) 1–14.

[13] N.S. Cooper, M.E. Brown, C.A. Caulcott, A mathematical model for analysing plasmid stability in microorganisms. J. Gen. Microbiol. 133 (1987) 1871–1880.

[14] D.F. Ollis, H.-T. Chang, Batch fermentation kinetics with (unstable) recombinant cultures, Biotechnol. Bioeng. 24 (1982) 2583–2586.

[15] J.E. Bailey, M. Hjortso, S.B. Lee, F. Srienc, Kinetics of product formation and plasmid segregation in recombinant microbial populations, Ann. N.Y. Acad. Sci 413 (1983) 71–87.

[16] K.D. Wittrup, J.E. Bailey, A segregated model of recombinant multicopy plasmid propagation, Biotechnol. Bioeng. 31 (1988) 304–310.

[17] A.H. Stouthamer, H.W. van Verseveld, Stoichiometry of microbial growth, In: M. Moo-Young (ed.), Comprehensive Biotechnology, vol. 1, Pergamon, (1985) 215–238.

[18] J.J. Heijnen, J.A. Roels, A.H. Stouthamer, Application of balancing methods in modeling the penicillin fermentation, Biotechnol. Bioeng. 21 (1979) 2175–2201.

[19] J.J. Heijnen, J.A. Roels, A macroscopic model describing yield and maintenance relationships in aerobic fermentation processes, Biotechnol. Bioeng. 23 (1981) 739–763.

[20] S.J. Pirt, Principles of Microbe and Cell Cultivation, Blackwell Scientific, 1975.

[21] R.G. Forage, D.E.F. Harrison, D.E. Pitt, Effect of environment on microbial activity, In: M. Moo-Young (ed.), Comprehensive Biotechnology, vol. 1, Pergamon (1985) 251–280.

[22] D.I.C. Wang, C.L. Cooney, A.L. Demain, P. Dunnill, A.E. Humphrey, M.D. Lilly, Fermentation and Enzyme Technology, John Wiley, 1979.

[23] C.L. Cooney, Media sterilization. In: M. Moo-Young (ed.), Comprehensive Biotechnology, vol. 2, Pergamon (1985) 287–298.

[24] F.H. Deindoerfer, A.E. Humphrey, Analytical method for calculating heat sterilization times, Appl. Microbiol. 7 (1959) 256–264.

Suggestions for Further Reading

A.L. Lehninger, Bioenergetics. W.A. Benjamin, 1965.

G.F. Froment, K.B. Bischoff, J. De Wilde, Chemical Reactor Analysis and Design, 3rd ed., Chapter 1, John Wiley, 2010.

C.D. Holland, R.G. Anthony, Fundamentals of Chemical Reaction Engineering, 2nd ed., Chapter 1, Prentice Hall, 1989.

O. Levenspiel, Chemical Reaction Engineering, 3rd ed., Chapters 2 and 3, John Wiley, 1999.

S.W. Churchill, The Interpretation and Use of Rate Data: The Rate Concept, McGraw-Hill, 1974.

O.A. Hougen, K.M. Watson, R.A. Ragatz, Chemical Process Principles, Part I, 2nd ed., Chapter 1, John Wiley, 1962.

Enzyme Kinetics and Deactivation

See also references [2–6].

H. Bisswanger, Enzyme Kinetics: Principles and Methods, 2nd ed., Wiley-VCH, 2008.

R.A. Copeland, Enzymes, 2nd ed., Wiley-VCH, 2000.

R.W. Lencki, J. Arul, R.J. Neufeld, Effect of subunit dissociation, denaturation, aggregation, coagulation, and decomposition on enzyme inactivation kinetics. Parts I and II, Biotechnol. Bioeng 40 (1992) 1421–1434.

Cell Kinetics and Yield

See also references [1], and [17–22].

J.A. Roels, *Energetics and Kinetics in Biotechnology*, Elsevier Biomedical, 1983.

A.H. Stouthamer, O.K. Sebek, A.I. Laskin, Energy production, growth, and product formation by microorganisms. Genetics of Industrial Microorganisms, in: O.K. Sebek, A.I. Laskin, (Eds.), Genetics of Industrial Microorganisms, American Society for Microbiology (1979),

K. van't Riet, J. Tramper, Basic Bioreactor Design, Marcel Dekker, 1991.

Growth Kinetics With Plasmid Instability

See also references [12–16].

M.A. Hjortso, J.E. Bailey, Plasmic stability in budding yeast populations: steady-state growth with selection pressure, Biotechnol. Bioeng 26 (1984) 528–536.

C.A. Sardonini, D. DiBiasio, A model for growth of *Saccharomyces cerevisiae* containing a recombinant plasmid in selective media, Biotechnol. Bioeng 29 (1987) 469–475.

F. Srienc, J.L. Campbell, J.E. Bailey, Analysis of unstable recombinant *Saccharomyces cerevisiae* population growth in selective medium, Biotechnol. Bioeng 18 (1986) 996–1006.

Cell Death Kinetics

See also references [23] and [23].

S. Aiba, A.E. Humphrey, N.F. Millis, Biochemical Engineering, Academic Press, 1965.

J.W. Richards, Introduction to Industrial Sterilization, Academic Press, 1968.

CHAPTER 11

Heterogeneous Reactions

The reaction systems analyzed so far have been assumed to be homogeneous: local variations in metabolite concentration or the rate of reaction were not considered. Yet, many bioprocesses have concentrations of substrates and products that differ from point to point. Concentration gradients can arise in single-phase systems when mixing is poor. Additionally, when different phases are present, concentration gradients can be expected within boundary layers around gas bubbles and solids. More severe gradients are found inside solid biocatalysts such as cell flocs, biofilms, and in immobilized cell and enzyme particles.

Reactions occurring in the presence of multiple phases, often with significant concentration or temperature gradients, are called *heterogeneous reactions*. We confine our attention in this chapter to concentration effects because biological reactions are not typically associated with large temperature gradients. When heterogeneous reactions occur within solid catalysts, not all reactive molecules are available for immediate conversion. Reaction takes place only after the reactants are transported to the site of reaction. Thus, mass transfer processes can have a considerable influence on the overall reaction rate.

Reaction kinetic analyses become more complex when there are spatial variations in concentration because many reaction rates depend on reactant concentrations. The principles of homogeneous reaction and the equations outlined in Chapter 10 remain valid for heterogeneous systems; however, the concentrations used in the equations must be those present at the site of reaction. For solid biocatalysts, the concentration of substrate at each point inside the solid must be known for evaluation of local reaction rates. In most cases, these concentrations cannot be measured directly; fortunately, they can be estimated using diffusion–reaction theory.

In this chapter, methods are presented for analyzing reactions affected by mass transfer. The mathematics required is more advanced than that applied elsewhere in this book. The practical outcome of this chapter is simple criteria for assessing mass transfer limitations in heterogeneous reaction systems. These criteria can be used directly in experimental design and data analysis.

11.1 Heterogeneous Reactions in Bioprocessing

Reactions involving solid-phase catalysts are important in bioprocessing. Macroscopic flocs, clumps, and biofilms are produced naturally by many bacteria and fungi, e.g., mycelial pellets are common in antibiotic-producing bioprocesses. Some cells grow as biofilms on reactor walls; others form flocs such as in waste treatment processes. Plant cell suspensions invariably contain aggregates and microorganisms in soil particles play a crucial role in geochemical cycling and environmental bioremediation. In tissue engineering, animal cells are cultured in three-dimensional scaffolds for applications like surgical transplantation and organ repair. In all of these systems, the rate of reaction depends on the rate of reactant mass transfer outside and within the solid catalyst.

Cells or enzymes can also be incorporated with a solid phase using *immobilization* techniques. Many procedures are available for artificial immobilization of cells and enzymes; the results of two commonly used methods are illustrated in Figure 11.1. Cells and enzymes can be immobilized by entrapment within gels such as alginate, agarose, or carrageenan as shown in Figure 11.1(A). Cells or enzymes are mixed with liquified gel before it is hardened or cross-linked and formed into small particles. The gel polymer must be porous to allow diffusion of reactants and products to and from the interior of the particle. An alternative to gel immobilization is entrapment within porous solids such as ceramics, porous glass, and resin beads as shown in Figure 11.1(B). Enzymes or cells can diffuse or migrate into the pores of these particles and attach to the internal surfaces; substrate must then diffuse through the pores for reaction to occur. In both immobilization systems, the sites of reaction are distributed throughout the particle. Thus, a catalyst particle of higher activity can be formed by increasing the loading of cells or enzyme per volume of matrix.

Immobilized biocatalysts have advantages in large-scale processing. One of the most important is continuous operation using the same catalytic material. For enzymes, an additional advantage is that immobilization can enhance stability, increases the enzyme half-life, and improve reaction specificity (Section 10.6). Additional examples of immobilization methods and the rationale can be found in the references at the end of this chapter.

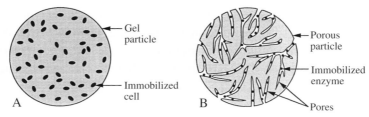

Figure 11.1 Immobilized biocatalysts: (A) cells entrapped in soft gel; (B) enzymes attached to the internal surfaces of a porous solid.

Engineering analysis of heterogeneous reactions applies equally well to naturally occurring solid catalysts and artificially immobilized cells and enzymes.

11.2 Concentration Gradients and Reaction Rates in Solid Catalysts

Consider a spherical catalyst of radius R immersed in well-mixed liquid containing substrate A. The substrate concentration is uniform and equal to C_{Ab} in the bulk liquid away from the particle. If the particle were inactive, after some time the concentration of substrate inside the solid would reach a value in equilibrium with C_{Ab}. However, when substrate A is consumed by reaction, C_A decreases within the particle as shown in Figure 11.2. The concentration profile will be symmetrical with a minimum at the center if immobilized cells or enzymes are distributed uniformly within the catalyst. Mass transfer of substrate to reaction sites in the particle is driven by the concentration difference between the bulk solution and particle interior.

In the bulk liquid, substrate is transported rapidly by convective currents. However, substrate molecules must be transported from the bulk liquid across the relatively stagnant boundary layer to the solid surface; this process is called *external mass transfer*. A concentration gradient develops across the boundary layer from C_{Ab} in the bulk liquid to C_{As} at the solid–liquid interface. If the particle were nonporous and all the enzymes or cells were confined to its outer surface, external mass transfer would be the only transport process required. More often, reaction takes place inside the particle so *internal mass transfer* is also critical.

The concentration gradient shown in Figure 11.2 is typical for spherical systems although, other variations are possible. If mass transfer is slower than reaction, all the substrate entering

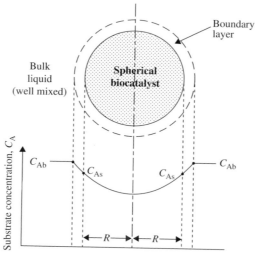

Figure 11.2 Typical substrate concentration profile for a spherical biocatalyst.

the particle may be consumed before reaching the center. In this case, the concentration falls to zero within the solid as illustrated in Figure 11.3(A). Cells or enzymes near the center are starved of substrate and the core of the particle becomes inactive. The *partition coefficient* for the substrate is not equal to unity in the examples of Figures 11.3(B) and 11.3(C). This means that, at equilibrium and in the absence of reaction, the concentration of substrate in the solid is higher or lower than in the liquid. The discontinuity of substrate concentration at the solid–liquid interface shows that substrate distributes preferentially to the solid phase in Figure 11.3(B). Conversely, Figure 11.3(C) shows the concentrations when substrate distributes preferentially to the liquid phase. The effect of mass transfer on intraparticle concentration can be magnified or diminished by substrate partitioning. However, partition effects can often be neglected because most materials used for cell and enzyme immobilization are very porous and contain a high percentage of water. In our treatment of heterogeneous reaction, we will assume that partitioning is not significant.

11.2.1 True and Observed Reaction Rates

Reaction rates can vary with position in solid catalysts because substrate concentrations vary with position. Even for zero-order reactions, the reaction rate changes with position if substrate is exhausted. Each cell or enzyme molecule within the solid responds to the local

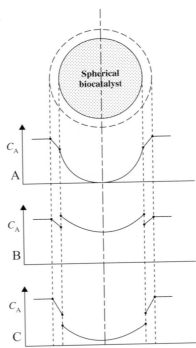

Figure 11.3 Variations in substrate concentration profile in spherical biocatalysts.

substrate concentration with a reaction rate determined by its kinetic properties. This local rate of reaction is known as the *true rate* or *intrinsic rate*. Like any reaction rate, intrinsic rates can be expressed as total, volumetric, or specific rates as described in Section 10.1. The relationship between true reaction rate and local substrate concentration follows the principles outlined in Chapter 10 for homogeneous reactions.

True local rates of reaction are difficult to measure in solid catalysts without altering the reaction conditions. It is relatively easy, however, to measure the overall reaction rate for the entire catalyst. In a closed system, the rate of disappearance of substrate from the bulk liquid must equal the overall rate of conversion by reaction; in heterogeneous systems, this is called the *observed rate*. It is important to remember that the observed rate is not usually equal to the true activity of any cell or enzyme in the particle. Because substrate levels are reduced inside solid catalysts compared with those in the external medium, we expect the observed rate to be lower than the rate that would occur if the entire particle were exposed to the bulk liquid. The relationship between observed reaction rate and bulk substrate concentration is not as simple as in homogeneous reactions. Equations for the observed rate of heterogeneous reactions also involve mass transfer considerations.

True reaction rates depend on the kinetic parameters of the cells or enzyme. For example, the rate of reaction of an immobilized enzyme demonstrating Michaelis–Menten kinetics (Section 10.3.3) depends on the values of v_{max} and K_m for the enzyme in its immobilized state. The kinetic parameters can be altered during immobilization due to cell or enzyme damage, conformational change, or steric hindrance, therefore, values measured before immobilization may not apply. These parameters are sometimes called *true kinetic parameters* or *intrinsic kinetic parameters*. True kinetic parameters for immobilized biocatalysts can be difficult to determine because any measured reaction rates incorporate mass transfer effects. The problem of evaluating true kinetic parameters is discussed further in Section 11.9.

11.2.2 Interaction between Mass Transfer and Reaction

Rates of reaction and substrate mass transfer are not independent in heterogeneous systems. The rate of mass transfer depends on the concentration gradient in the system; this, in turn, depends on the rate of substrate depletion by reaction. On the other hand, the rate of reaction depends on the availability of substrate, which depends on the rate of mass transfer.

An objective of analyzing heterogeneous reactions is to determine the relative influences of mass transfer and reaction on observed reaction rates. One can conceive, that if a reaction proceeds slowly even in the presence of adequate substrate, it is likely that mass transfer is rapid enough to meet the demands of the reaction. In this case, the observed rate is dominated by the reaction process rather than the mass transfer process. Conversely, if the reaction tends to be rapid, it is likely that mass transfer is too slow and will limit the observed rate.

Reactions that are significantly affected are called *mass transfer-limited* or *diffusion-limited* reactions. It is also possible to distinguish the relative influence of internal and external mass transfer on the observed rate of reaction. Improving mass transfer and eliminating mass transfer restrictions are desired objectives in heterogeneous catalysis. Once the effect and location of major mass transfer resistances are identified, we can then devise strategies for their alleviation.

11.3 Internal Mass Transfer and Reaction

Let us now concentrate on the processes of mass transfer and reaction occurring within a solid biocatalyst; external mass transfer will be examined later. The equations used in this analysis depend on the geometry of the system and the reaction kinetics. First, let us consider the case of cells or enzymes immobilized in spherical particles.

11.3.1 Steady-State Shell Mass Balance

Mathematical analysis of heterogeneous reactions involves a technique called the *shell mass balance*. Here, we will perform a shell mass balance on a spherical catalyst particle of radius R. Imagine a control volume that is a thin spherical shell of thickness Δr located at radius r from the center, as shown in Figure 11.4. It may be helpful to think of this shell as the thin wall of a ping-pong ball encased inside and concentric with a larger baseball of radius R. Substrate diffusing into the sphere must cross the shell-shaped control volume to reach the center.

A mass balance of substrate is performed around the shell by considering the processes of mass transfer and reaction occurring at radius r. The control volume for the mass balance

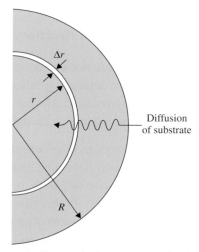

Figure 11.4 Shell mass balance on a spherical particle.

is the shell only; the remainder of the sphere is ignored for the moment. Substrate diffuses into the shell at radius $(r + \Delta r)$ and leaves at radius r; within the shell, immobilized cells or enzyme consume substrate by reaction. The shell control volume is analyzed using the general mass balance equation derived in Chapter 4:

$$\left\{\begin{array}{c} \text{mass in} \\ \text{through} \\ \text{system} \\ \text{boundaries} \end{array}\right\} - \left\{\begin{array}{c} \text{mass out} \\ \text{through} \\ \text{system} \\ \text{boundaries} \end{array}\right\} + \left\{\begin{array}{c} \text{mass} \\ \text{generated} \\ \text{within} \\ \text{system} \end{array}\right\} - \left\{\begin{array}{c} \text{mass} \\ \text{consumed} \\ \text{within} \\ \text{system} \end{array}\right\} = \left\{\begin{array}{c} \text{mass} \\ \text{accumulated} \\ \text{within} \\ \text{system} \end{array}\right\} \quad (4.1)$$

Certain assumptions are made so that each term in the equation can be expressed mathematically [1].

1. *The particle is isothermal.* The kinetic parameters for enzyme and cell reactions vary considerably with temperature. If temperature in the particle is nonuniform, different values of the kinetic parameters must be applied. However, as temperature gradients generated by immobilized cells and enzymes are generally negligible, assuming constant temperature throughout the particle is reasonable and greatly simplifies the mathematical analysis.
2. *Mass transfer occurs by diffusion only.* We will assume that the particle is impermeable to flow, so that convection within the pores is negligible. This assumption is valid for many solid-phase biocatalysts. Depending on pore size, pressure gradients can induce convection of liquid through the particle and this enhances significantly the supply of nutrients. The analysis of mass transfer and reaction presented in this chapter must be modified when convective transport occurs [2–6].
3. *Diffusion can be described using Fick's law with constant effective diffusivity.* We will assume that diffusive transport through the particle is governed by Fick's law (Section 9.1.1). Interaction of the substrate with other concentration gradients, and phenomena affecting the transport of charged species, are not considered. Fick's law will be applied using the *effective diffusivity* of substrate in the solid, $\mathcal{D}_{Ae}$. The value of $\mathcal{D}_{Ae}$ is a function of the molecular diffusion characteristics of the substrate, the tortuousness of the diffusion path within the solid, and the fraction of the particle volume available for diffusion. We will assume that $\mathcal{D}_{Ae}$ is constant and independent of substrate concentration in the particle; this means that $\mathcal{D}_{Ae}$ does not change with position.
4. *The particle is homogeneous.* Immobilized enzymes or cells are assumed to be distributed uniformly within the particle. Properties of the immobilization matrix are also considered to be uniform.
5. *The substrate partition coefficient is unity.* This assumption is valid for most substrates and particles and ensures that there is no discontinuity of concentration at the solid–liquid interface.

6. *The particle is at steady state.* This assumption is usually valid if there is no change in activity of the catalyst due to, for example, enzyme deactivation, cell growth, or cell differentiation.
7. *Substrate concentration varies with a single spatial variable.* We will assume that the substrate concentration varies only in the radial direction for the sphere of Figure 11.4, and that substrate diffuses radially through the particle from the external surface toward the center.

Equation (4.1) can be applied to our shell mass balance according to these assumptions. Substrate is transported into and out of the shell by diffusion; therefore, the first and second terms are expressed using Fick's law with constant effective diffusivity. The third term is zero as no substrate is generated. Substrate is consumed by reaction inside the shell at a rate equal to the volumetric rate of reaction r_A multiplied by the volume of the shell. According to assumption (6), the system is at steady state. Therefore, its mass and composition must be unchanging, substrate cannot accumulate in the shell, and the right side of Eq. (4.1) is zero. As outlined below, after substituting the appropriate expressions and applying calculus to reduce the dimension of the shell to an infinitesimal thickness, the result of the shell mass balance is a second-order differential equation for substrate concentration as a function of radius in the particle.

For a shell mass balance on substrate A, the terms of Eq. (4.1) are as follows:

Rate of input by diffusion: $\left. \left(\mathscr{D}_{Ae} \dfrac{dC_A}{dr} 4\pi r^2 \right) \right|_{r+\Delta r}$

Rate of output by diffusion: $\left. \left(\mathscr{D}_{Ae} \dfrac{dC_A}{dr} 4\pi r^2 \right) \right|_{r}$

Rate of generation: 0

Rate of consumption by reaction: $r_A \, 4\pi r^2 \Delta r$

Rate of accumulation at steady state: 0

$\mathscr{D}_{Ae}$ is the effective diffusion coefficient of substrate A in the particle, C_A is the concentration of A in the particle, r is the distance measured radially from the center, Δr is the thickness of the shell, and r_A is the rate of reaction *per unit volume of particle*. Each term in the mass balance equation has dimensions MT^{-1} or NT^{-1} and units of, for example, kg h^{-1} or gmol s^{-1}. The first two terms are derived from Fick's law of Eq. (9.1); the surface area of the spherical shell available for diffusion is $4\pi r^2$. The term

$$\left. \left(\mathscr{D}_{Ae} \dfrac{dC_A}{dr} 4\pi r^2 \right) \right|_{r+\Delta r}$$

means $\left(\mathcal{D}_{Ae} \dfrac{dC_A}{dr} 4\pi r^2 \right)$ evaluated at radius $(r + \Delta r)$, and

$$\left(\mathcal{D}_{Ae} \dfrac{dC_A}{dr} 4\pi r^2 \right)\Bigg|_{r}$$

means $\left(\mathcal{D}_{Ae} \dfrac{dC_A}{dr} 4\pi r^2 \right)$ evaluated at r. The volume of the shell is estimated as $4\pi r^2 \Delta r$ which

assumes the shell can be approximated as a slab due to the incredibly small height Δr.

After substituting these terms into Eq. (4.1), we obtain the following steady-state mass balance equation:

$$\left(\mathcal{D}_{Ae} \dfrac{dC_A}{dr} 4\pi r^2 \right)\Bigg|_{r+\Delta r} - \left(\mathcal{D}_{Ae} \dfrac{dC_A}{dr} 4\pi r^2 \right)\Bigg|_{r} - r_A 4\pi r^2 \Delta r = 0 \tag{11.1}$$

Dividing each term by $4\pi\Delta r$ gives:

$$\dfrac{\left(\mathcal{D}_{Ae} \dfrac{dC_A}{dr} r^2 \right)\Big|_{r+\Delta r} - \left(\mathcal{D}_{Ae} \dfrac{dC_A}{dr} r^2 \right)\Big|_{r}}{\Delta r} - r_A r^2 = 0 \tag{11.2}$$

Eq. (11.2) can be written in the form:

$$\dfrac{\Delta\left(\mathcal{D}_{Ae} \dfrac{dC_A}{dr} r^2 \right)}{\Delta r} - r_A r^2 = 0 \tag{11.3}$$

where $\Delta\left(\mathcal{D}_{Ae} \dfrac{dC_A}{dr} r^2 \right)$ means the change in $\left(\mathcal{D}_{Ae} \dfrac{dC_A}{dr} r^2 \right)$ across Δr.

Equation (11.3) is valid for a spherical shell of thickness Δr. To develop an equation that applies to any *point* in the sphere, we must shrink Δr to zero. As Δr appears only in the first term of Eq. (11.3), taking the limit of Eq. (11.3) as $\Delta r \to 0$ gives:

$$\lim_{\Delta r \to 0} \dfrac{\Delta\left(\mathcal{D}_{Ae} \dfrac{dC_A}{dr} r^2 \right)}{\Delta r} - r_A r^2 = 0 \tag{11.4}$$

Invoking the definition of the derivative, Eq. (11.4) is identical to the second-order differential equation:

$$\frac{d}{dr}\left(\mathcal{D}_{Ae} \frac{dC_A}{dr} r^2 \right) - r_A r^2 = 0 \tag{11.5}$$

$\mathcal{D}_{Ae}$ is independent of r, based on assumption (3), and can be moved outside of the differential:

$$\mathcal{D}_{Ae} \frac{d}{dr}\left(\frac{dC_A}{dr} r^2 \right) - r_A r^2 = 0 \tag{11.6}$$

Equation (11.6) is a differential equation representing diffusion and reaction in a spherical biocatalyst. Equation (11.6) is a second-order differential equation; applying the product rule expands the equation to the form:

$$\mathcal{D}_{Ae}\left(\frac{d^2 C_A}{dr^2} r^2 + 2r \frac{dC_A}{dr} \right) - r_A r^2 = 0 \tag{11.7}$$

In principle, Eq. (11.7) can be solved by integration to yield an expression for the concentration profile in the particle, C_A as a function of r. However, we cannot integrate Eq. (11.7) as it stands because the reaction rate r_A is, in many cases, a function of C_A. Let us consider solutions of Eq. (11.7) with r_A for first-order, zero-order, and Michaelis–Menten kinetics.

11.3.2 Concentration Profile: First-Order Kinetics and Spherical Geometry

For first-order kinetics, Eq. (11.7) becomes:

$$\mathcal{D}_{Ae}\left(\frac{d^2 C_A}{dr^2} r^2 + 2r \frac{dC_A}{dr} \right) - k_1 C_A r^2 = 0 \tag{11.8}$$

where k_1 is the intrinsic, first-order rate constant with units of s^{-1}, for example. k_1 depends on the mass of cells or enzymes in the particle. According to assumptions (1), (3), and (4) in Section 11.3.1, k_1 and $\mathcal{D}_{Ae}$ for a given particle are assumed constant. Accordingly, the only variables in Eq. (11.8) are C_A and r and the equation is ready for integration. Equation (11.8) is a second-order differential equation so, we need two boundary conditions to solve it uniquely. These are:

$$C_A = C_{As} \quad \text{at} \quad r = R \tag{11.9}$$

$$\frac{dC_A}{dr} = 0 \quad \text{at} \quad r = 0 \tag{11.10}$$

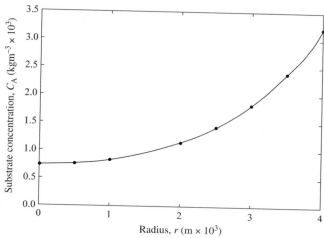

Figure 11.5 Substrate concentration profile in an immobilized enzyme bead.

where C_{As} is the concentration of substrate at the outer surface of the particle. We will assume for now that C_{As} is known or can be measured. Equation (11.10) is called the *symmetry condition*. As indicated in Figures 11.1 and 11.2, for a particle with uniform properties, the substrate concentration profile is symmetrical about the center of the sphere, with a minimum at the center so that $dC_A/dr = 0$ at $r = 0$. Integration of Eq. (11.8) with the boundary conditions Eqs. (11.9) and (11.10) results in the following equation for substrate concentration as a function of radius [7]:

$$C_A = C_{As} \frac{R}{r} \frac{\sinh\left(r\sqrt{k_1 / \mathcal{D}_{Ae}} \right)}{\sinh\left(R\sqrt{k_1 / \mathcal{D}_{Ae}} \right)} \tag{11.11}$$

Sinh is the abbreviation for *hyperbolic sine*; $\sinh x$ is defined as:

$$\sinh x = \frac{e^x - e^{-x}}{2} \tag{11.12}$$

Equation (11.11) may appear complex but contains simple exponential terms relating C_A and r, with $\mathcal{D}_{Ae}$ representing the rate of mass transfer and k_1 representing the rate of reaction. Example 11.1 illustrates the calculation of concentration of substrate as a function of radius in an immobilized system.

11.3.3 Concentration Profile: Zero-Order Kinetics and Spherical Geometry

For zero-order kinetics, Eq. (11.7) becomes:

$$\mathcal{D}_{Ae}\left(\frac{d^2C_A}{dr^2}r^2 + 2r\frac{dC_A}{dr} \right) - k_0 r^2 = 0 \tag{11.13}$$

EXAMPLE 11.1 Concentration Profile for Immobilized Enzyme

Enzyme is immobilized in agarose beads of diameter 8 mm. The concentration of enzyme in the beads is 0.018 kg of protein per m^3 of gel. Ten beads are immersed in a well-mixed solution containing substrate at a concentration of 3.2×10^{-3} kg m^{-3}. The effective diffusivity of substrate in the agarose gel is 2.1×10^{-9} m^2 s^{-1}. The kinetics of the enzyme reaction can be approximated as first-order with specific rate constant of 3.11×10^5 s^{-1} per kg of protein. Mass transfer effects outside the particles are negligible. Plot the steady-state substrate concentration profile inside the beads as a function of particle radius.

Solution

$R = 4 \times 10^{-3}$ m; $\mathcal{D}_{Ae} = 2.1 \times 10^{-9}$ m^2 s^{-1}. In the absence of external mass transfer effects, $C_{As} = 3.2 \times 10^{-3}$ kg m^{-3}. To determine the substrate concentration profile, we consider mass transfer and reaction in a single bead.

$$\text{Volume per bead} = \frac{4}{3}\pi R^3 = \frac{4}{3}\pi \left(4 \times 10^{-3} \text{ m}\right)^3 = 2.68 \times 10^{-7} \text{ m}^3$$

$$\text{Amount of enzyme per bead} = 2.68 \times 10^{-7} \text{ m}^3 \left(0.018 \text{ kg m}^{-3}\right) = 4.82 \times 10^{-9} \text{ kg}$$

Therefore:

$$k_1 = 3.11 \times 10^5 \text{ s}^{-1} \text{ kg}^{-1} \left(4.82 \times 10^{-9} \text{ kg}\right) = 0.0015 \text{ s}^{-1}$$

Intraparticle substrate concentrations are calculated as a function of radius using Eq. (11.11). Terms in Eq. (11.11) include:

$$R\sqrt{\frac{k_1}{\mathcal{D}_{Ae}}} = \left(4 \times 10^{-3} \text{ m}\right)\sqrt{\frac{0.0015 \text{ s}^{-1}}{2.1 \times 10^{-9} \text{ m}^2 \text{ s}^{-1}}} = 3.381$$

and

$$\sinh\left(R\sqrt{k_1 / \mathcal{D}_{Ae}}\right) = \frac{e^{3.381} - e^{-3.381}}{2} = 14.68$$

Results for C_A as a function of r are as follows.

r (m)	C_A (kg m^{-3})
0.005×10^{-3}	7.37×10^{-4}
0.5×10^{-3}	7.59×10^{-4}
1.0×10^{-3}	8.28×10^{-4}
2.0×10^{-3}	1.14×10^{-3}
2.5×10^{-3}	1.42×10^{-3}
3.0×10^{-3}	1.82×10^{-3}
3.5×10^{-3}	2.39×10^{-3}
4.0×10^{-3}	3.20×10^{-3}

The results are plotted in Figure 11.5. The substrate concentration is reduced inside the particle to reach a minimum of 7.4×10^{-4} kg m^{-3} at the center.

where k_0 is the intrinsic, zero-order rate constant with units of, for example, gmol s^{-1} m^{-3} of particle. Like k_1 for first-order reactions, k_0 varies with cell or enzyme loading in the particle.

Zero-order reactions are remarkable because the reaction rate is independent of substrate concentration, provided substrate is present. We must account for the possibility that the substrate becomes depleted within the particle when solving Eq. (11.13). If we assume the substrate is depleted at some radius R_0, the rate of reaction for $0 < r \leq R_0$ is zero. Everywhere else inside the particle (i.e., $r > R_0$), the volumetric reaction rate is constant and equal to k_0 irrespective of substrate concentration. For this scenario, the boundary conditions are modified to:

$$C_A = C_{As} \quad \text{at} \quad r = R \tag{11.9}$$

$$\frac{dC_A}{dr} = 0 \quad \text{at} \quad r = R_0 \tag{11.14}$$

Solution of Eq. (11.13) with these boundary conditions gives the following expression for C_A as a function of r [7]:

$$C_A = C_{As} + \frac{k_0 R^2}{6 \mathcal{D}_{Ae}} \left(\frac{r^2}{R^2} - 1 + \frac{2R_0^3}{rR^2} - \frac{2R_0^3}{R^3} \right) \tag{11.15}$$

Equation (11.15) is difficult to apply in practice because R_0 is generally not known. However, the equation can be simplified if C_A remains >0 everywhere so that R_0 no longer exists (Figure 11.6). Substituting $R_0 = 0$ into Eq. (11.15) gives:

$$C_A = C_{As} + \frac{k_0}{6 \mathcal{D}_{Ae}} \left(r^2 - R^2 \right) \tag{11.16}$$

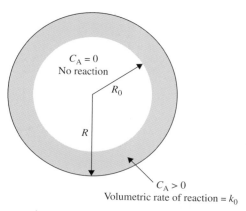

Figure 11.6 Concentration and reaction zones in a spherical particle with zero-order reaction. Substrate is depleted at radius R_0.

In bioprocessing applications, it is important that the core of catalyst particles does not become starved of substrate. The likelihood of this happening increases with the size of the particle and with increases in cell or enzyme loading. We can calculate the maximum particle radius for zero-order reactions for which C_A remains nonzero everywhere except at the very center. Therefore, calculating R from Eq. (11.16) with $C_A = 0$ at $r = 0$:

$$R_{max} = \sqrt{\frac{6 \mathcal{D}_{Ae} C_{As}}{k_0}} \tag{11.17}$$

where R_{max} is the maximum particle radius which avoids a dead zone. Example 11.2 illustrates the application of this criterion for enzyme particle design.

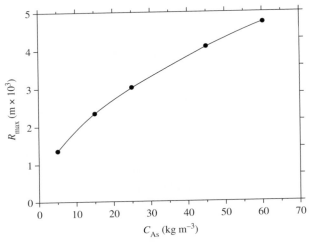

Figure 11.7 Maximum particle radius as a function of external substrate concentration.

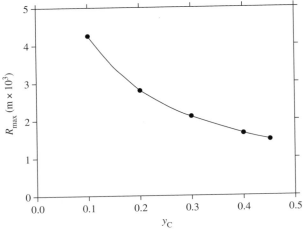

Figure 11.8 Maximum particle radius as a function of cell density.

EXAMPLE 11.2 Maximum Particle Size for Zero-Order Reaction

Nonviable yeast cells are immobilized in alginate beads. The beads are stirred in glucose medium under anaerobic conditions. The effective diffusivity of glucose in the beads depends on cell density according to the relationship:

$$\mathcal{D}_{Ae} = 6.33 - 7.17 y_C$$

where $\mathcal{D}_{Ae}$ is the effective diffusivity $\times 10^{10}$ m^2 s^{-1} and y_C is the weight fraction of yeast in the gel. The rate of glucose consumption can be assumed to be zero-order; the reaction rate constant at a yeast density in the beads of 15 wt% is 0.5 g min^{-1} L^{-1}. For the catalyst to be utilized effectively, the concentration of glucose inside the particles should remain above zero.

a. Plot the maximum allowable particle size as a function of the bulk glucose concentration between 5 g L^{-1} and 60 g L^{-1}.
b. For medium containing 30 g L^{-1} glucose, plot R_{max} as a function of cell loading between 10 and 45 wt%.

Solution

a. Using the equation provided, at $y_C = 0.15$, $\mathcal{D}_{Ae} = 5.25 \times 10^{-10}$ m^2 s^{-1}. Converting k_0 to units of kg, m, s:

$$k_0 = 0.5 \text{ g min}^{-1} L^{-1} \cdot \left| \frac{1 \text{ kg}}{1000 \text{ g}} \right| \cdot \left| \frac{1 \text{ min}}{60 \text{ s}} \right| \cdot \left| \frac{1000 \, L}{1 \text{ m}^3} \right|$$

$$k_0 = 8.33 \times 10^{-3} \text{ kg s}^{-1} \text{m}^{-3}$$

Assume that C_{As} is equal to the bulk glucose concentration. R_{max} is calculated from Eq. (11.17).

C_{As} (kg m^{-3})	R_{max} (m)
5	1.38×10^{-3}
15	2.38×10^{-3}
25	3.07×10^{-3}
45	4.13×10^{-3}
60	4.76×10^{-3}

The results are plotted in Figure 11.7. At low external glucose concentrations, the particles are restricted to small radii. As C_{As} increases, the driving force for diffusion increases so that larger particles may be used.

b. $C_{As} = 30$ g L^{-1} = 30 kg m^{-3}. As y_C varies, the values of $\mathcal{D}_{Ae}$ and k_0 are affected. Changes in $\mathcal{D}_{Ae}$ can be calculated from the equation provided. We assume that k_0 is directly proportional to the cell density as described in Eq. (10.26), that is, there is no steric hindrance or interaction between the cells as y_C increases. R_{max} is calculated as a function of y_C using Eq. (11.17) and the corresponding values of $\mathcal{D}_{Ae}$ and k_0.

yC	$\mathcal{D}_{Ae}\ (\mathrm{m^2 s^{-1}})$	$k_0\ (\mathrm{kg\ m^{-3}\ s^{-1}})$	$R_{max}\ (\mathrm{m})$
0.1	5.61×10^{-10}	5.56×10^{-3}	4.26×10^{-3}
0.2	4.90×10^{-10}	1.11×10^{-2}	2.82×10^{-3}
0.3	4.18×10^{-10}	1.67×10^{-2}	2.12×10^{-3}
0.4	3.46×10^{-10}	2.22×10^{-2}	1.67×10^{-3}
0.45	3.10×10^{-10}	2.50×10^{-2}	1.49×10^{-3}

The results are plotted in Figure 11.8. As y_C increases, $\mathcal{D}_{Ae}$ declines and k_0 increases. Reducing $\mathcal{D}_{Ae}$ lowers the rate of diffusion into the particles; raising k_0 increases the demand for substrate. Therefore, increasing the cell density exacerbates the limiting effect of mass transfer on the reaction rate. To ensure adequate supply of substrate under these conditions, the particle size must be reduced.

11.3.4 Concentration Profile: Michaelis–Menten Kinetics and Spherical Geometry

If reaction in the particle follows Michaelis–Menten kinetics, r_A takes the form of Eq. (10.36) and Eq. (11.7) becomes:

$$\mathcal{D}_{Ae} \left(\frac{d^2 C_A}{dr^2} r^2 + 2r \frac{dC_A}{dr} \right) - \frac{v_{max} C_A}{K_m + C_A} r^2 = 0 \tag{11.18}$$

where v_{max} and K_m are the intrinsic kinetic parameters for the reaction. In Eq. (11.18), v_{max} has units of, for example, kg s^{-1} m^{-3} of particle; the value of v_{max} depends on the cell or enzyme density.

Owing to the nonlinearity of the Michaelis–Menten expression, simple analytical integration of Eq. (11.18) is not possible. However, results for C_A as a function of r can be obtained using numerical methods. Because Michaelis–Menten kinetics lie between zero- and first-order kinetics (Section 10.3.3), the explicit solutions found in Sections 11.3.2 and 11.3.3 for first- and zero-order reactions can be used to estimate the upper and lower limits of the concentration profile for Michaelis–Menten reactions.

Concentration profiles calculated from the equations presented in this section have been verified experimentally in several studies. Using special microelectrodes with tip diameters of the order of 1 to 10 μm, it is possible to measure the concentrations of O_2 and ions inside soft solids and cellular biofilms. As an example, the O_2 concentrations measured in agarose beads containing immobilized enzyme are shown in Figure 11.9. The experimental data are very close to the calculated concentration profiles. Similar results have been found in other systems [8–10].

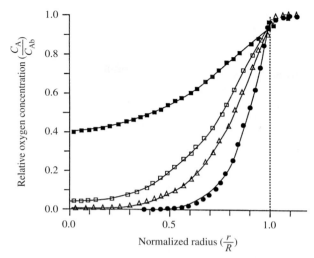

Figure 11.9 Measured and calculated O_2 concentrations in a spherical agarose bead containing immobilized enzyme. Particle diameter = 4 mm; C_{Ab} = 0.2 mol m^{-3}. The enzyme loadings are: 0.0025 kg m^{-3} of gel (■); 0.005 kg m^{-3} of gel (□); 0.0125 kg m^{-3} of gel (△); and 0.025 kg m^{-3} of gel (●). Measured concentrations are shown using symbols; calculated profiles are shown as lines. *From C.M. Hooijmans, S.G.M. Geraats, and K.Ch.A.M. Luyben, 1990, Use of an oxygen microsensor for the determination of intrinsic kinetic parameters of an immobilized oxygen reducing enzyme. Biotechnol. Bioeng. 35, 1078–1087.*

11.3.5 Concentration Profiles for Other Geometries

Our attention has so far focused on spherical catalysts. However, equations similar to Eq. (11.7) can be obtained from shell mass balances on other geometries. The other geometry most relevant to bioprocessing is the flat plate. A typical substrate concentration profile for this geometry, without external boundary layer effects, is illustrated in Figure 11.10. Equations for flat-plate geometry are used to analyze reactions in cell biofilms attached to inert solids; in this case, the biofilm constitutes the flat plate. Even if the surface supporting the biofilm is curved rather than flat, if the biofilm thickness b is very small compared with the radius of curvature, equations for flat-plate geometry are applicable.

The flat plate is assumed to have infinite length to simplify the mathematical treatment and to keep the problem one-dimensional (as required by assumption (7) of Section 11.3.1). In practice, this assumption is reasonable if its length is much greater than its thickness. If not, it must be assumed that the ends of the plate are sealed to eliminate axial concentration

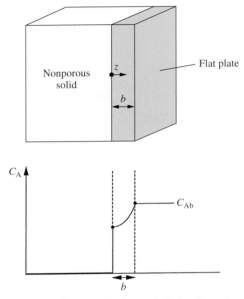

Figure 11.10 Substrate concentration profile in an infinite flat plate without boundary-layer effects.

gradients. The boundary conditions for integrating the differential equation for diffusion and reaction in a flat plate are analogous to Eqs. (11.9) and (11.10):

$$C_A = C_{As} \quad \text{at} \quad z = b \tag{11.19}$$

$$\frac{dC_A}{dz} = 0 \quad \text{at} \quad z = 0 \tag{11.20}$$

where C_{As} is the concentration of A at the solid–liquid interface, z is distance measured from the inner surface of the plate (Figure 11.10), and b is the plate thickness. Equations for the steady-state concentration profiles for first- and zero-order kinetics and spherical and flat-plate geometries are summarized in Table 11.1.

11.3.6 Prediction of the Observed Reaction Rate

Equations for intracatalyst substrate concentrations, such as those in Table 11.1, allow us to predict the overall or observed rate of reaction in the catalyst. Let us consider the situation for spherical particles and first-order, zero-order, and Michaelis–Menten kinetics. Analogous equations can be derived for other geometries.

1. *First-order kinetics.* The rate of reaction at any position in a spherical catalyst depends on the first-order kinetic constant k_1 and the concentration of substrate at that position. The overall rate for the entire particle is equal to the sum of all the rates at every location in the

Table 11.1: Steady-State Concentration Profiles

First-order reaction: $r_A = k_1 C_A$

Sphere[a]

$$C_A = C_{As} \frac{R}{r} \frac{\sinh\left(r\sqrt{k_1 / \mathcal{D}_{Ae}}\right)}{\sinh\left(R\sqrt{k_1 / \mathcal{D}_{Ae}}\right)}$$

Flat plate[b]

$$C_A = C_{As} \frac{\cosh\left(z\sqrt{k_1 / \mathcal{D}_{Ae}}\right)}{\cosh\left(b\sqrt{k_1 / \mathcal{D}_{Ae}}\right)}$$

Zero-order reaction: $r_A = k_0$

Sphere[c]

$$C_A = C_{As} + \frac{k_0}{6\mathcal{D}_{Ae}}\left(r^2 - R^2\right)$$

Flat plate[c]

$$C_A = C_{As} + \frac{k_0}{2\mathcal{D}_{Ae}}\left(z^2 - b^2\right)$$

[a]Sinh is the abbreviation for hyperbolic sine. Sinh x is defined as: $\sinh x = \dfrac{e^x - e^{-x}}{2}$

[b]Cosh is the abbreviation for hyperbolic cosine. Cosh x is defined as: $\cosh x = \dfrac{e^x + e^{-x}}{2}$
[c]For $CA > 0$ everywhere within the catalyst.

solid. This sum is equivalent mathematically to integrating the expression $k_1 C_A$ over the entire particle volume, taking into account the variation of C_A with radius expressed in Eq. (11.11). The result is an equation for the observed reaction rate $r_{A,obs}$ in a single particle:

$$r_{A,obs} = 4\pi R \mathcal{D}_{Ae} C_{As} \left[R\sqrt{k_1 / \mathcal{D}_{Ae}} \coth\left(R\sqrt{k_1 / \mathcal{D}_{Ae}} \right) - 1 \right] \tag{11.21}$$

where coth is the abbreviation for *hyperbolic cotangent* defined as:

$$\coth x = \frac{e^x + e^{-x}}{e^x - e^{-x}} \tag{11.22}$$

Note that, because the observed reaction rate was found by integrating the local reaction rate over the volume of a single particle, $r_{A,obs}$ in Eq. (11.21) is the observed reaction rate *per particle* with units of, for example, kg s^{-1}.

2. *Zero-order kinetics.* As long as substrate is present, zero-order reactions occur at a fixed rate that is independent of substrate concentration. Therefore, if $C_A > 0$ for r > 0, the observed rate of reaction is equal to the zero-order rate constant k_0 multiplied by the particle volume:

$$r_{A,obs} = \frac{4}{3}\pi R^3 k_0 \tag{11.23}$$

However, if C_A falls to zero at some nonzero radius R_0 in the particle, the inner volume $(4/3)\pi R_0^3$ is inactive. In this case, the rate of reaction is equal to k_0 multiplied by the active particle volume:

$$r_{A,obs} = \left(\frac{4}{3}\pi R^3 - \frac{4}{3}\pi R_0^3 \right)k_0 = \frac{4}{3}\pi \left(R^3 - R_0^3 \right)k_0 \tag{11.24}$$

$r_{A,obs}$ calculated using Eqs. (11.23) and (11.24) is the observed reaction rate per particle with units of, for example, kg s^{-1}.

3. *Michaelis–Menten kinetics.* Observed rates for Michaelis–Menten reactions cannot be expressed explicitly because we do not have an equation for C_A as a function of radius. $r_{A,obs}$ can be evaluated, however, using numerical methods.

11.4 The Thiele Modulus and Effectiveness Factor

Charts based on the equations of the previous section allow us to determine $r_{A,obs}$ relative to r_{As}^*, the reaction rate that would occur if all cells or enzymes were exposed to the external substrate concentration. Differences between $r_{A,obs}$ and r_{As}^* show the extent to which a reaction is affected by internal mass transfer. The theoretical basis for comparing $r_{A,obs}$ and r_{As}^* is described in the following sections.

11.4.1 First-Order Kinetics

If a catalyst particle is unaffected by mass transfer, the concentration of substrate inside the particle is constant and equal to the surface concentration, C_{As}. Thus, the rate of reaction per particle for first-order kinetics without internal mass transfer effects is $k_1 C_{As}$ multiplied by the particle volume:

$$r_{As}^* = \frac{4}{3}\pi R^3 k_1 C_{As} \tag{11.25}$$

The extent to which $r_{A,obs}$ is different from r_{As}^* is expressed using the dimensionless number, *internal effectiveness factor* (η_i):

$$\eta_i = \frac{r_{A,obs}}{r_{As}^*} = \frac{\left(\text{observed rate} \right)}{\left(\begin{array}{c} \text{rate that would occur if } C_A = C_{As} \\ \text{everywhere in the particle} \end{array} \right)} \tag{11.26}$$

In the absence of mass transfer limitations, $r_{A,obs} = r_{As}^*$ and $\eta_i = 1$. $\eta_i < 1$ when mass transfer reduces $r_{A,obs}$. H_i is a dimensionless number so the same units should be used for $r_{A,obs}$ and r_{As}^*, for example, kg s^{-1} per particle, gmol s^{-1} m^{-3}, and so on. We can substitute expressions for

$r_{A,obs}$ and r_{As}^* from Eqs. (11.21) and (11.25) into Eq. (11.26) to derive an expression for η_{i1}, the internal effectiveness factor for first-order reaction:

$$\eta_{i1} = \frac{3\mathcal{D}_{Ae}}{R^2 k_1}\left[R\sqrt{k_1/\mathcal{D}_{Ae}}\,\coth\left(R\sqrt{k_1/\mathcal{D}_{Ae}}\,\right) - 1\right] \tag{11.27}$$

Thus, the internal effectiveness factor for first-order reaction depends on R, k_1, and $\mathcal{D}_{Ae}$. These parameters are usually grouped together to form a dimensionless number called the *Thiele modulus*. There are several definitions of the Thiele modulus in the literature. Application of the original modulus was cumbersome because a separate definition was required for different reaction kinetics and catalyst geometries [11]. Generalized moduli that apply to any catalyst shape and reaction kinetics have since been proposed [12–14].

The generalized Thiele modulus φ is defined as:

$$\phi = \frac{V_p}{S_x}\frac{r_A\,|C_{As}}{\sqrt{2}}\left(\int_{C_{A,eq}}^{C_{As}} \mathcal{D}_{Ae}\,r_A dC_A\right)^{-1/2} \tag{11.28}$$

where V_p is the catalyst volume, S_x is the catalyst external surface area, C_{As} is the substrate concentration at the surface of the catalyst, r_A is the reaction rate, $r_A\,|C_{As}$ is the reaction rate when $C_A = C_{As}$, $\mathcal{D}_{Ae}$ is the effective diffusivity of substrate, and $C_{A,eq}$ is the equilibrium substrate concentration. Many net, bioprocess transformations can be treated as essentially irreversible, so that $C_{A,eq}$ is zero for many biological applications. From geometry, $V_p/S_x = R/3$ for spheres and b for flat plates. Expressions determined from Eq. (11.28) for first-order, zero-order, and Michaelis–Menten kinetics are listed in Table 11.2 as φ_1, φ_0, and φ_m, respectively. φ represents the important parameters affecting mass transfer and reaction in heterogeneous systems: the catalyst size (R or b), the effective diffusivity ($\mathcal{D}_{Ae}$), the surface substrate concentration (C_{As}), and the intrinsic rate parameters (k_0, k_1, or v_{max} and K_m). Only the Thiele modulus for first-order reaction does not depend on the substrate concentration.

The parameters R, k_1, and $\mathcal{D}_{Ae}$ in Eq. (11.27) can be grouped together as φ_1, the result is:

$$\eta_{i1} = \frac{1}{3\phi_1^2}\left(3\phi_1\coth 3\phi_1 - 1\right) \tag{11.29}$$

where coth is defined in Eq. (11.22). Equation (11.29) applies to spherical geometry and first-order reaction; the analogous equation for flat plates is listed in Table 11.3. Plots of η_{i1} versus φ_1 for sphere, cylinder, and flat-plate catalysts are shown in Figure 11.11. The curves coincide exactly for $\varphi_1 \rightarrow 0$ and $\varphi_1 \rightarrow \infty$, and fall within 10% to 15% for the remainder of the range. Figure 11.11 can be used to evaluate η_{i1} for any catalyst shape, provided that φ_1 is calculated using Eq. (11.28). Because of the errors involved in estimating the parameters defining φ_1, it

Table 11.2: Generalized Thiele Moduli

First-order reaction: $r_A = k_1 C_A$

$$\phi_1 = \frac{V_p}{S_x}\sqrt{\frac{k_1}{\mathcal{D}_{Ae}}}$$

Sphere

$$\phi_1 = \frac{R}{3}\sqrt{\frac{k_1}{\mathcal{D}_{Ae}}}$$

Flat plate

$$\phi_1 = b\sqrt{\frac{k_1}{\mathcal{D}_{Ae}}}$$

Zero-order reaction: $r_A = k_0$

$$\phi_0 = \frac{1}{\sqrt{2}}\frac{V_p}{S_x}\sqrt{\frac{k_0}{\mathcal{D}_{Ae}C_{As}}}$$

Sphere

$$\phi_0 = \frac{R}{3\sqrt{2}}\sqrt{\frac{k_0}{\mathcal{D}_{Ae}C_{As}}}$$

Flat plate

$$\phi_0 = \frac{b}{\sqrt{2}}\sqrt{\frac{k_0}{\mathcal{D}_{Ae}C_{As}}}$$

Michaelis–Menten reaction: $rA = \dfrac{V_{max}C_A}{K_m + C_A}$

$$\phi_m = \frac{1}{\sqrt{2}}\frac{V_p}{S_x}\sqrt{\frac{V_{max}}{\mathcal{D}_{Ae}C_{As}}\left(\frac{1}{1+\beta}\right)}\left[1+\beta\ln\left(\frac{\beta}{1+\beta}\right)\right]^{-1/2}$$

$$\beta = \frac{K_m}{C_{As}}$$

Sphere

$$\phi_m = \frac{R}{3\sqrt{2}}\sqrt{\frac{V_{max}}{\mathcal{D}_{Ae}C_{As}}\left(\frac{1}{1+\beta}\right)}\left[1+\beta\ln\left(\frac{\beta}{1+\beta}\right)\right]^{-1/2}$$

Flat plate

$$\phi_m = \frac{b}{\sqrt{2}}\sqrt{\frac{V_{max}}{\mathcal{D}_{Ae}C_{As}}\left(\frac{1}{1+\beta}\right)}\left[1+\beta\ln\left(\frac{\beta}{1+\beta}\right)\right]^{-1/2}$$

---

has been suggested that effectiveness factor curves such as those in Figure 11.11 be viewed as diffuse bands rather than precise functions [15].

If the first-order Thiele modulus is known, we can use Figure 11.11 to find the internal effectiveness factor. Equations (11.25) and (11.26) can then be applied to predict the observed reaction rate for the catalyst. At low values of $\varphi_1 < 0.3$, $\eta_{i1} \approx 1$ and the rate of reaction is not adversely affected by internal mass transfer. However, as φ_1 increases above 0.3, η_{i1} falls as mass transfer limitations become substantial. Therefore, the value of the Thiele modulus indicates immediately whether the rate of reaction is diminished due to diffusional effects,

Table 11.3: Effectiveness Factors (φ for Each Geometry and Kinetic Order Is Defined in Table 11.2)

First-order reaction: $r_A = k_1 C_A$	
Sphere[a]	$\eta_{i1} = \dfrac{1}{3\phi_1^2}\left(3\phi_1 \coth 3\phi_1 - 1\right)$
Flat plate[b]	$\eta_{i1} = \dfrac{\tanh\phi_1}{\phi_1}$

Zero-order reaction: $r_A = k_0$		
Sphere[c] $\eta_{i0} = 1 - \left[\dfrac{1}{2} + \cos\left(\dfrac{\Psi + 4\pi}{3}\right)\right]^3$ where $\Psi = \cos^{-1}\left(\dfrac{2}{3\phi_0^2} - 1\right)$	$\eta_{i0} = 1$ for $\varphi_0 > 0.577$	for $0 < \varphi_0 \leq 0.577$
Flat plate	$\eta_{i0} = 1$ $\eta_{i0} = \dfrac{1}{\phi_0}$	for $0 < \varphi_0 \leq 1$ for $\varphi_0 > 1$

[a]Coth is the abbreviation for hyperbolic cotangent. Coth x is defined as: $\coth x = \dfrac{e^x + e^{-x}}{e^x - e^{-x}}$

[b]Tanh is the abbreviation for hyperbolic tangent. Tanh x is defined as: $\tanh x = \dfrac{e^x - e^{-x}}{e^x + e^{-x}}$

[c]Cos is the abbreviation for cosine. The notation $\cos^{-1}x$ (or arccos x) denotes any angle the cosine of which is x. Angles used to determine cos and $\cos^{-1}$ are in radians.

or whether the catalyst is performing at its maximum rate at the prevailing surface substrate concentration. For strong diffusional limitations at $\varphi_1 > 10$, η_{i1} for all geometries can be estimated as:

$$\eta_{i1} \approx \frac{1}{\phi_1} \tag{11.30}$$

11.4.2 Zero-Order Kinetics

If substrate is present throughout the catalyst, evaluation of the internal effectiveness factor η_{i0} for zero-order reactions is straightforward. The reaction rate is independent of substrate concentration and the reaction proceeds at the same rate that would occur if $C_A = C_{As}$ throughout the particle. Therefore, from Eq. (11.26), $\eta_{i0} = 1$ and:

$$r_{A,obs} = r_{As}^* = \frac{4}{3}\pi R^3 k_0 \tag{11.31}$$

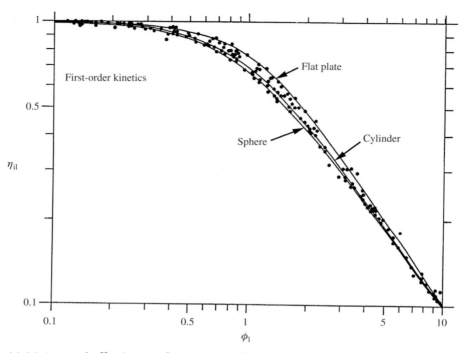

Figure 11.11 Internal effectiveness factor η_{i1} as a function of the generalized Thiele modulus φ_1 for first-order kinetics and spherical, cylindrical, and flat-plate geometries. The dots represent calculations on finite or hollow cylinders and parallelepipeds. *From R. Aris, 1975, Mathematical Theory of Diffusion and Reaction in Permeable Catalysts, vol. 1, Oxford University Press, London.*

If C_A falls to zero within the particle, the effectiveness factor must be evaluated differently. In this case, $r_{A,obs}$ is given by Eq. (11.24) and the internal effectiveness factor is:

$$\eta_{i0} = \frac{\frac{4}{3}\pi\left(R^3 - R_0^3\right)k_0}{\frac{4}{3}\pi R^3 k_0} = 1 - \left(\frac{R_0}{R}\right)^3 \tag{11.32}$$

To evaluate η_{i0} for zero-order kinetics, first we must know whether substrate is depleted in the catalyst and, if it is, the value of R_0. This information is not usually available because we cannot easily measure intraparticle concentrations. Fortunately, further mathematical analysis [7] overcomes this problem by representing the system in terms of measurable properties such as R, $\mathcal{D}_{Ae}$, C_{As}, and k_0 rather than R_0. These parameters define the Thiele modulus for zero-order reaction, φ_0. The results are summarized in Table 11.3 and Figure 11.12. C_A remains >0 and $\eta_{i0} = 1$ for $0 < \varphi_0 \leq 0.577$. For $\varphi_0 > 0.577$, η_{i0} declines as more and more of the particle becomes inactive. Effectiveness factors for flat-plate systems are also shown in Table 11.3 and

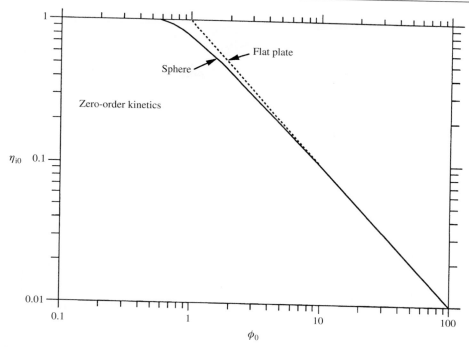

Figure 11.12 Internal effectiveness factor η_{i0} as a function of the generalized Thiele modulus φ_0 for zero-order kinetics and spherical and flat-plate geometries.

Figure 11.12. In flat biofilms, $\varphi_0 = 1$ represents the threshold condition for substrate depletion. The η_{i0} curves for spherical and flat-plate geometries coincide exactly at small and large values of φ_0 as shown in Figure 11.12.

11.4.3 Michaelis–Menten Kinetics

For a spherical catalyst, the rate of Michaelis–Menten reaction in the absence of internal mass transfer effects is:

$$r_{As}^* = \frac{4}{3}\pi R^3 \left(\frac{v_{max} C_{As}}{K_m + C_{As}} \right) \tag{11.33}$$

Our analysis cannot proceed further, however, because we do not have an equation for $r_{A,obs}$. Accordingly, we cannot develop an analytical expression for the effectiveness factor for Michaelis–Menten kinetics, η_{im}, as a function of φ_m. Diffusion–reaction equations for Michaelis–Menten kinetics are generally solved by numerical computation. As an example, the results for flat-plate geometry are shown in Figure 11.13 as a function of the parameter β, which is equal to the ratio K_m/C_{As}.

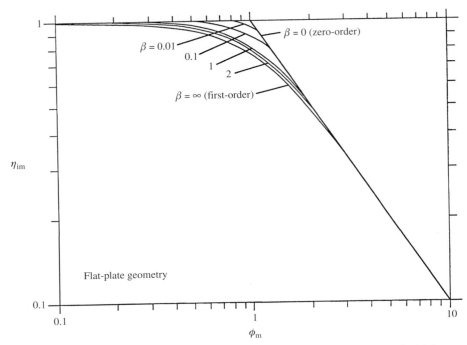

Figure 11.13 Internal effectiveness factor η_{im} as a function of the generalized Thiele modulus φ_m and parameter β for Michaelis–Menten kinetics and flat-plate geometry. $\beta = K_m/C_{As}$. *From R. Aris, 1975, Mathematical Theory of Diffusion and Reaction in Permeable Catalysts, vol. 1, Oxford University Press, London.*

We can obtain approximate values for η_{im} by considering the zero- and first-order asymptotes of the Michaelis–Menten equation. The curves for η_{im} fall between the lines for zero- and first-order reactions depending on the value of β as indicated in Figure 11.13. As $\beta \to \infty$, Michaelis–Menten kinetics can be approximated as first-order and the internal effectiveness factor can be evaluated from Figure 11.11 with $k_1 = v_{max}/K_m$. When $\beta \to 0$, zero-order kinetics and Figure 11.12 apply with $k_0 = v_{max}$. Effectiveness factors for β between zero and infinity must be evaluated using numerical methods.

Use of the generalized Thiele modulus for Michaelis–Menten reactions (Table 11.2) eliminates almost all variation in the internal effectiveness factor with changing β, except in the vicinity of $\varphi_m = 1$. The generalized modulus also brings together effectiveness-factor curves for all shapes of catalyst at the two asymptotes $\varphi_m \to 0$ and $\varphi_m \to \infty$. Therefore, Figure 11.13 is valid for spherical catalysts if φ_m is much less or much greater than 1. It should be noted, however, that variation between geometries in the intermediate region around $\varphi_m = 1$ can be significant [16].

If the values of φ_m and β are such that Michaelis–Menten kinetics cannot be approximated as either zero- or first-order, η_{im} may be estimated using an equation proposed by Moo-Young and Kobayashi [17]:

$$\eta_{im} = \frac{\eta_{i0} + \beta\eta_{i1}}{1 + \beta} \tag{11.34}$$

where $\beta = K_m/C_{As}$. η_{i0} is the zero-order internal effectiveness factor obtained using values of φ_0 evaluated with $k_0 = v_{max}$, and η_{i1} is the first-order effectiveness factor obtained using φ_1 calculated with $k_1 = v_{max}/K_m$. For flat-plate geometry, the largest deviation of Eq. (11.34) from exact values of η_{im} is 0.089; this occurs at $\varphi_m = 1$ and $\beta = 0.2$. For spherical geometry, the greatest deviations occur around $\varphi_m = 1.7$ and $\beta = 0.3$; the maximum error in this region is 0.09. Further details can be found in the original paper [17]. Example 11.3 compares reaction rates for free and immobilized enzymes.

EXAMPLE 11.3 Reaction Rates for Free and Immobilized Enzyme

Invertase is immobilized in ion-exchange resin of average diameter of 1 mm. The amount of enzyme in the beads is measured by protein assay as $0.05\,\mathrm{kg\,m^{-3}}$ of resin. A small column reactor is packed with 20 cm^3 of beads. Seventy-five mL of a 16-mM sucrose solution is pumped rapidly through the bed. In another reactor, an identical quantity of free enzyme is mixed into the same volume of sucrose solution. The kinetic parameters for the immobilized enzyme can be assumed to be equal to those for the free enzyme: K_m is 8.8 mM and the turnover number is 2.4×10^{-3} gmol of sucrose (g enzyme)$^{-1}$ s^{-1}. The effective diffusivity of sucrose in the ion-exchange resin is 2×10^{-6} cm^2 s^{-1}.

a. What is the rate of reaction using free enzyme?
b. What is the rate of reaction using immobilized enzyme?

Solution

The invertase reaction is:

$$C_{12}H_{22}O_{11} + H_2O \rightarrow C_6H_{12}O_6 + C_6H_{12}O_6$$
$$\underset{\text{sucrose}}{} \qquad \underset{\text{glucose}}{} \quad \underset{\text{fructose}}{}$$

Convert the data provided to units of gmol, m, s:

$$K_m = \frac{8.8 \times 10^{-3}\,\mathrm{gmol}}{L} \cdot \left|\frac{1000\,L}{1\,m^3}\right| = 8.8\,\mathrm{gmol\,m^{-3}}$$

$$\mathcal{D}_{Ae} = 2 \times 10^{-6}\,\mathrm{cm^2\,s^{-1}} \cdot \left|\frac{1\,m}{100\,cm}\right|^2 = 2 \times 10^{-10}\,\mathrm{m^2 s^{-1}}$$

$$R = \frac{1\,mm}{2} \cdot \left|\frac{1\,m}{10^3\,mm}\right| = 5 \times 10^{-4}\,\mathrm{m}$$

If flow through the reactor is rapid, we can assume that C_{As} is equal to the bulk sucrose concentration C_{Ab}:

$$C_{As} = C_{Ab} = 16\,\text{mM} = \frac{16 \times 10^{-3}\,\text{gmol}}{\text{L}} \cdot \left|\frac{1000\,\text{L}}{1\,\text{m}^3}\right| = 16\,\text{gmolm}^{-3}$$

Also:

$$\text{Mass of enzyme} = \frac{0.05\,\text{kg}}{\text{m}^3}(20\,\text{cm}^3) \cdot \left|\frac{1\,\text{m}}{100\,\text{cm}}\right|^3 = 10^{-6}\,\text{kg}$$

c. In the free enzyme reactor:

$$\text{Enzyme concentration} = \frac{10^{-6}\,\text{kg}}{75\,\text{cm}^3} \cdot \left|\frac{100\,\text{cm}}{1\,\text{m}}\right|^3 = 1.33 \times 10^{-2}\,\text{kgm}^{-3}$$

From Eq. (10.39), v_{max} is obtained by multiplying the turnover number by the concentration of active enzyme. Assuming that all of the enzyme present is active:

$$v_{max} = \frac{2.4 \times 10^{-3}\,\text{gmol}}{\text{gs}}(1.33 \times 10^{-2}\,\text{kgm}^{-3}) \cdot \left|\frac{1000\,\text{g}}{1\,\text{kg}}\right| = 3.19 \times 10^{-2}\,\text{gmol}\,\text{s}^{-1}\,\text{m}^{-3}$$

The volumetric rate of reaction is given by the Michaelis–Menten equation. As the free enzyme reaction takes place at uniform sucrose concentration C_{Ab}:

$$v = \frac{v_{max}C_{Ab}}{K_m + C_{Ab}} = \frac{3.19 \times 10^{-2}\,\text{gmol}\,\text{s}^{-1}\,\text{m}^{-3}\,(16\,\text{gmol}\,\text{m}^{-3})}{(8.8\,\text{gmol}\,\text{m}^{-3} + 16\,\text{gmol}\,\text{m}^{-3})} = 2.06 \times 10^{-2}\,\text{gmol}\,\text{s}^{-1}\,\text{m}^{-3}$$

Multiplying by the liquid volume gives the total rate of reaction:

$$v = 2.06 \times 10^{-2}\,\text{gmol}\,\text{s}^{-1}\text{m}^{-3}(75\,\text{cm}^3) \cdot \left|\frac{1\,\text{m}}{100\,\text{cm}}\right|^3 = 1.55 \times 10^{-6}\,\text{gmol}\,\text{s}^{-1}$$

The total rate of reaction using free enzyme is 1.55×10^{-6} gmol s^{-1}.

d. In the equations for heterogeneous reaction, v_{max} is expressed on a catalyst volume basis. Therefore:

$$v_{max} = \frac{2.4 \times 10^{-3}\,\text{gmol}}{\text{gs}}(0.05\,\text{kgm}^{-3}) \cdot \left|\frac{1000\,\text{g}}{1\,\text{kg}}\right| = 0.12\,\text{gmol}\,\text{s}^{-1}\text{m}^{-3}\,\text{of resin}$$

To determine the effect of mass transfer, we must calculate η_{im}. The method used depends on the values of β and φ_m:

$$\beta = \frac{K_m}{C_{As}} = \frac{8.8\,\text{gmol}\,\text{m}^{-3}}{16\,\text{gmol}\,\text{m}^{-3}} = 0.55$$

From Table 11.2 for Michaelis–Menten kinetics and spherical geometry:

$$\phi_m = \frac{R}{3\sqrt{2}}\sqrt{\frac{v_{max}}{\mathcal{D}_{Ae}C_{As}}\left(\frac{1}{1+\beta}\right)}\left[1+\beta\ln\left(\frac{\beta}{1+\beta}\right)\right]^{-1/2}$$

$$\phi_m = \frac{5\times10^{-4}\,\text{m}}{3\sqrt{2}}\sqrt{\frac{0.12\,\text{gmol s}^{-1}\text{m}^{-3}}{\left(2\times10^{-10}\,\text{m}^2\,\text{s}^{-1}\right)\left(16\,\text{gmol m}^{-3}\right)}\left(\frac{1}{1+0.55}\right)}\left[1+0.55\ln\left(\frac{0.55}{1+0.55}\right)\right]^{-1/2}$$

$$\phi_m = 0.71$$

Because both β and φ_m have intermediate values, Figure 11.13 cannot be applied for spherical geometry. Instead, we can use Eq. (11.34). From Table 11.2 for zero-order kinetics and spherical geometry:

$$\phi_0 = \frac{R}{3\sqrt{2}}\sqrt{\frac{k_0}{\mathcal{D}_{Ae}C_{As}}}$$

Using $k_0 = v_{max}$:

$$\phi_0 = \frac{R}{3\sqrt{2}}\sqrt{\frac{v_{max}}{\mathcal{D}_{Ae}C_{As}}}$$

$$\phi_0 = \frac{5\times10^{-4}\,\text{m}}{3\sqrt{2}}\sqrt{\frac{0.12\,\text{gmol s}^{-1}\,\text{m}^{-3}}{\left(2\times10^{-10}\,\text{m}^2\,\text{s}^{-1}\right)\left(16\,\text{gmol m}^{-3}\right)}}$$

$$\phi_0 = 0.72$$

From Figure 11.12 or Table 11.3, at this value of φ_0, $\eta_{i0} = 0.93$. Similarly, for first-order kinetics and spherical geometry with $k_1 = v_{max}/K_m$:

$$\phi_1 = \frac{R}{3}\sqrt{\frac{v_{max}}{K_m\mathcal{D}_{Ae}}}$$

$$\phi_1 = \frac{5\times10^{-4}\,\text{m}}{3}\sqrt{\frac{0.12\,\text{gmol m}^{-3}\,\text{s}^{-1}}{\left(8.8\,\text{gmol m}^{-3}\right)\left(2\times10^{-10}\,\text{m}^2\,\text{s}^{-1}\right)}}$$

$$\phi_1 = 1.4$$

From Figure 11.11 or Table 11.3, $\eta_{i1} = 0.54$. Substituting these results into Eq. (11.34) gives:

$$\eta_{im} = \frac{0.93 + 0.55(0.54)}{1 + 0.55} = 0.79$$

The total rate of reaction for immobilized enzyme without diffusional limitations is evaluated using the Michaelis–Menten expression with substrate concentration C_{As}:

$$r_{As}^* = \frac{v_{max}C_{As}}{K_m + C_{As}} = \frac{0.12\,\text{gmol s}^{-1}\,\text{m}^{-3}(20\,\text{cm}^3)\cdot\left|\frac{1\,\text{m}}{100\,\text{cm}}\right|^3(16\,\text{gmol m}^{-3})}{\left(8.8\,\text{gmol m}^{-3} + 16\,\text{gmol m}^{-3}\right)}$$

$$= 1.55\times10^{-6}\,\text{gmol s}^{-1}$$

Not surprisingly, as the kinetic parameters were assumed to be equal, this is the same rate as that found using the same quantity of free enzyme. The effectiveness factor indicates that the rate of reaction for the immobilized enzyme is 79% of that occurring in the absence of mass transfer effects. Applying Eq. (11.26):

$$r_{A,obs} = \eta_{im} r_{As}^* = 0.79(1.55 \times 10^{-6} \text{ gmol s}^{-1}) = 1.22 \times 10^{-6} \text{ gmol s}^{-1}$$

The total rate of reaction using immobilized enzyme is 1.22×10^{-6} gmol s^{-1}.

11.4.4 The Observable Thiele Modulus

Diffusion–reaction theory as presented in the previous sections allows us to quantify the effect of mass transfer on the rate of enzyme and cell reactions. However, a shortcoming of the methods outlined so far is that they are useful only if we know the true kinetic parameters for the reaction: k_0, k_1, or v_{max} and K_m. In many cases, these values are not known and can be difficult to evaluate. One way to circumvent this problem is to apply the *observable Thiele modulus* Φ, sometimes called *Weisz's modulus* [18], which is defined as:

$$\Phi = \left(\frac{V_p}{S_x}\right)^2 \frac{r_{A,obs}}{\mathcal{D}_{Ae} C_{As}} \tag{11.35}$$

where V_p is the catalyst volume, S_x is the catalyst external surface area, $r_{A,obs}$ is the observed reaction rate *per unit volume of catalyst*, $\mathcal{D}_{Ae}$ is the effective diffusivity of substrate, and C_{As} is the substrate concentration at the catalyst surface. Expressions for Φ for spheres and flat plates are listed in Table 11.4. Evaluation of the observable Thiele modulus does not rely on prior knowledge of the kinetic parameters; Φ is defined in terms of the measured or observed reaction rate, $r_{A,obs}$.

For the observable Thiele modulus to be useful, we need to relate Φ to the internal effectiveness factor η_i. Some mathematical consideration of the equations already presented in this

Table 11.4: Observable Thiele Moduli

Sphere	$\Phi = \left(\dfrac{R}{3}\right)^2 \dfrac{r_{A,obs}}{\mathcal{D}_{Ae} C_{As}}$
Flat plate	$\Phi = b^2 \dfrac{r_{A,obs}}{\mathcal{D}_{Ae} C_{As}}$

chapter yields the following relationships for first-order, zero-order, and Michaelis–Menten kinetics:

$$\text{First order:} \quad \Phi = \phi_1^2 \eta_{i1} \tag{11.36}$$

$$\text{Zero order:} \quad \Phi = 2\phi_0^2 \eta_{i0} \tag{11.37}$$

$$\text{Michaelis-Menten:} \quad \Phi = 2\phi_m^2 \eta_{im}(1 + \beta)\left[1 + \beta \ln\left(\frac{\beta}{1 + \beta}\right)\right] \tag{11.38}$$

Equations (11.36) through (11.38) apply to all catalyst geometries and allow us to develop plots of Φ versus η_i using the relationships between φ and η_i developed in the previous sections and represented in Figures 11.11 through 11.13. The results for spherical catalysts and first-order, zero-order, and Michaelis–Menten kinetics are given in Figure 11.14; similar results for flat-plate geometry are shown in Figure 11.15. All curves for β between zero and infinity are bracketed by the first- and zero-order lines. For each value of β, curves for all geometries coincide in the asymptotic regions $\Phi \rightarrow 0$ and $\Phi \rightarrow \infty$. At intermediate values

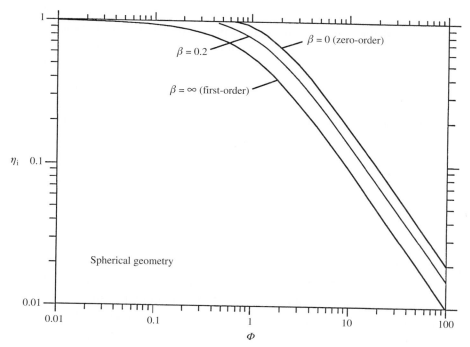

Figure 11.14 Internal effectiveness factor η_i as a function of the observable Thiele modulus Φ for spherical geometry and first-order, zero-order, and Michaelis–Menten kinetics. $\beta = K_m/C_{As}$. From W.H. Pitcher, 1975, *Design and operation of immobilized enzyme reactors*. In: R.A. Messing (Ed.), *Immobilized Enzymes for Industrial Reactors*, pp. 151–199, Academic Press, New York.

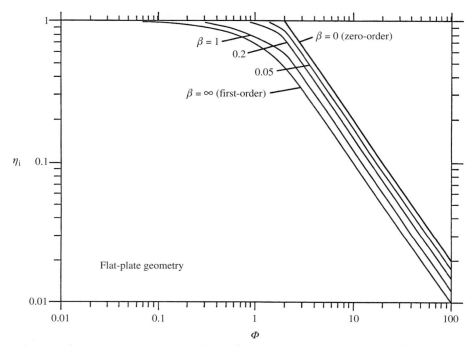

Figure 11.15 Internal effectiveness factor η_i as a function of the observable Thiele modulus Φ for flat-plate geometry and first-order, zero-order, and Michaelis–Menten kinetics. $\beta = K_m/C_{As}$. *From W.H. Pitcher, 1975, Design and operation of immobilized enzyme reactors. In: R.A. Messing (Ed.), Immobilized Enzymes for Industrial Reactors, pp. 151–199, Academic Press, New York.*

of Φ, the variation between effectiveness factors for different geometries can be significant. For $\Phi > 10$:

$$\text{First-order kinetics:} \quad \eta_{i1} \approx \frac{1}{\Phi} \tag{11.39}$$

$$\text{Zero-order kinetics:} \quad \eta_{i0} \approx \frac{2}{\Phi} \tag{11.40}$$

Although Φ is an observable modulus and is independent of kinetic parameters, use of Figures 11.14 and 11.15 for Michaelis–Menten reactions requires knowledge of K_m for evaluation of β. This makes application of the observable Thiele modulus difficult for Michaelis–Menten kinetics. However, we know that effectiveness factors for Michaelis–Menten reactions lie between the first- and zero-order curves of Figures 11.14 and 11.15; therefore, we can always estimate the upper and lower bounds of η_{im}.

11.4.5 Weisz's Criteria

The following general observations can be made from Figures 11.14 and 11.15.

1. If $\Phi < 0.3$, $\eta_i \approx 1$ and internal mass transfer limitations are insignificant.
2. If $\Phi > 3$, η_i is substantially <1 and internal mass transfer limitations are significant.

These statements are known as *Weisz's criteria*. They are valid for all geometries and reaction kinetics. For Φ in the intermediate range $0.3 \leq \Phi \leq 3$, closer analysis is required to determine the influence of mass transfer on reaction rate. Example 11.4 illustrates application of the Weisz's criteria for O_2 transfer to immobilized cells.

EXAMPLE 11.4 Internal O_2 Transfer to Immobilized Cells

Baby hamster kidney cells are immobilized in alginate beads. The average particle diameter is 5 mm. The bulk O_2 concentration in the medium is 8×10^{-3} kg m^{-3}, the rate of O_2 consumption by the cells is 8.4×10^{-5} kg s^{-1} m^{-3} of catalyst, and the effective diffusivity of O_2 in the beads is 1.88×10^{-9} m^2 s^{-1}. Assume that the O_2 concentration at the surface of the catalyst is equal to the bulk concentration, and that O_2 uptake follows zero-order kinetics.

a. Are internal mass transfer effects significant?
b. What reaction rate would be observed if diffusional resistances were eliminated?

Solution

c. To assess the effect of internal mass transfer on the reaction rate, we need to calculate the observable Thiele modulus. From Table 11.4 for spherical geometry:

$$\Phi = \left(\frac{R}{3}\right)^2 \frac{r_{A,obs}}{\mathcal{D}_{Ae}C_{As}}$$

For:

$$R = \frac{5 \times 10^{-3} \text{ m}}{2} = 2.5 \times 10^{-3} \text{ m}$$

and $C_{As} = C_{Ab}$:

$$\Phi = \left(\frac{2.5 \times 10^{-3} \text{ m}}{3}\right)^2 \frac{8.4 \times 10^{-5} \text{ kg s}^{-1} \text{ m}^{-3}}{\left(1.88 \times 10^{-9} \text{ m}^2 \text{ s}^{-1}\right)\left(8 \times 10^{-3} \text{ kg m}^{-3}\right)} = 3.9$$

From Weisz's criteria, internal mass transfer effects are significant.

d. For spherical catalysts and zero-order reaction, from Figure 11.14, at $\Phi = 3.9$, $\eta_{i0} = 0.40$. From Eq. (11.26), the reaction rate without diffusional restrictions is:

$$r_{As}^* = \frac{r_{A,obs}}{\eta_i} = \frac{8.4 \times 10^{-5} \text{ kg s}^{-1} \text{ m}^{-3}}{0.40} = 2.1 \times 10^{-4} \text{ kg s}^{-1} \text{ m}^{-3} \text{ of catalyst}$$

Table 11.5: Minimum Intracatalyst Substrate Concentration for Zero-Order Kinetics
(Φ for each Geometry Is Defined in Table 11.4)

Sphere	
$C_{A,min} = 0$	for $\Phi \geq 0.667$
$C_{A,min} = C_{As}\left(1 - \dfrac{3}{2}\Phi\right)$	for $\Phi < 0.667$
	Flat plate
$C_{A,min} = 0$	for $\Phi \geq 2$
$C_{A,min} = C_{As}\left(1 - \dfrac{1}{2}\Phi\right)$	for $\Phi < 2$

11.4.6 Minimum Intracatalyst Substrate Concentration

It is sometimes of interest to know the minimum substrate concentration, $C_{A,min}$, inside solid catalysts. We can use this information to check, for example, that the concentration does not fall below some critical value for cell metabolism. $C_{A,min}$ is estimated easily for zero-order reactions. If Φ is such that $\eta_{i0} < 1$, we know immediately that $C_{A,min}$ is zero because $\eta_{i0} = 1$ if $C_A > 0$ throughout the particle. For $\eta_{i0} = 1$, simple manipulation of the equations already presented in this chapter allows us to estimate $C_{A,min}$ for zero-order reactions. The results are summarized in Table 11.5.

11.5 External Mass Transfer

Many of the equations in Sections 11.3 and 11.4 contain the term C_{As}, the concentration of substrate A at the external surface of the catalyst. This term made its way into the analysis in the boundary conditions used for solution of the shell mass balance. So far, we have assumed that C_{As} is a known quantity. However, because surface concentrations are very difficult to measure experimentally, we must find ways to estimate C_{As} using theoretical principles.

During reaction, a reduction in external substrate concentration may occur as a liquid boundary layer develops at the surface of the solid catalyst (Figure 11.2). In the absence of a boundary layer, the substrate concentration at the solid surface C_{As} is equal to the substrate concentration in the bulk liquid C_{Ab}, which is easily measured. When the boundary layer is present, C_{As} takes some value less than C_{Ab}. The rate of mass transfer across the boundary layer is given by the following equation derived from Eq. (9.8) in Chapter 9:

$$N_A = k_S\, a\left(C_{Ab} - C_{As}\right) \tag{11.41}$$

where N_A is the rate of mass transfer, k_S is the *liquid-phase mass transfer coefficient* with dimensions LT^{-1}, and a is the external surface area of the catalyst. If N_A is expressed per volume of catalyst with units of, for example, kgmol s^{-1} m^{-3}, to be consistent, a must also be expressed on a catalyst-volume basis with units of, for example, m^2 m^{-3} or m^{-1}. Therefore, using the previous notation of this chapter, a in Eq. (11.41) is equal to S_x/V_p for the catalyst. At steady state, the rate of substrate transfer across the boundary layer must be equal to the rate of substrate consumption by the catalyst, $r_{A,obs}$. Therefore:

$$r_{A,obs} = k_S \frac{S_x}{V_p} \left(C_{Ab} - C_{As} \right)$$
(11.42)

where $r_{A,obs}$ is the observed reaction rate per volume of catalyst. Rearranging gives:

$$\frac{C_{As}}{C_{Ab}} = 1 - \frac{V_p}{S_x} \frac{r_{A,obs}}{k_S C_{Ab}}$$
(11.43)

Equation (11.43) can be used to evaluate C_{As} before applying the equations in the previous sections to calculate internal substrate concentrations and effectiveness factors. The magnitude of external mass transfer effects can be gauged from Eq. (11.43). $C_{As}/C_{Ab} \approx 1$ indicates no or negligible external mass transfer limitations, as the substrate concentration at the surface is approximately equal to that in the bulk. On the other hand, $C_{As}/C_{Ab} \ll 1$ indicates a very steep concentration gradient in the boundary layer and severe external mass transfer effects. We can define from Eq. (11.43) an *observable modulus for external mass transfer*, Ω:

$$\Omega = \frac{V_p}{S_x} \frac{r_{A,obs}}{k_S C_{Ab}}$$
(11.44)

Expressions for Ω for spherical and flat-plate geometries are listed in Table 11.6. The following criteria are used to assess the extent of external mass transfer effects.

1. If $\Omega \ll 1$, $C_{As} \approx C_{Ab}$ and external mass transfer effects are insignificant.
2. Otherwise, $C_{As} < C_{Ab}$ and external mass transfer effects are significant.

Table 11.6: Observable Moduli for External Mass Transfer

Sphere	$\Omega = \dfrac{R}{3} \dfrac{r_{A,obs}}{k_S C_{Ab}}$
Flat plate	$\Omega = b \dfrac{r_{A,obs}}{k_S C_{Ab}}$

For reactions affected by both internal and external mass transfer restrictions, we can define a *total effectiveness factor* η_T:

$$\eta_T = \frac{r_{A,obs}}{r_{Ab}^*} = \frac{(\text{observed rate})}{\left(\begin{array}{c} \text{rate that would occur if } C_A = C_{Ab} \\ \text{everywhere in the particle} \end{array}\right)} \quad (11.45)$$

η_T can be related to the internal effectiveness factor η_i by rewriting Eq. (11.45) as:

$$\eta_T = \left(\frac{r_{A,obs}}{r_{As}^*}\right)\left(\frac{r_{As}^*}{r_{Ab}^*}\right) = \eta_i \, \eta_e \quad (11.46)$$

where η_e is the *external effectiveness factor* and η_i is defined in Eq. (11.26). Therefore, η_e has the following meaning:

$$\eta_e = \frac{r_{As}^*}{r_{Ab}^*} = \frac{\left(\begin{array}{c} \text{rate that would occur if } C_A = C_{As} \\ \text{everywhere in the particle} \end{array}\right)}{\left(\begin{array}{c} \text{rate that would occur if } C_A = C_{Ab} \\ \text{everywhere in the particle} \end{array}\right)} \quad (11.47)$$

Expressions for η_e for first-order, zero-order, and Michaelis–Menten kinetics are listed in Table 11.7. For zero-order reactions, $\eta_{e0} = 1$ as long as $C_{As} > 0$ and $C_{Ab} > 0$. However, $\eta_{e0}=1$ does not imply that an external boundary layer does not exist. Because r_{As}^* and r_{Ab}^* are independent of C_A for zero-order kinetics, $\eta_{e0} = 1$ even when there is a reduction in concentration across the boundary layer. Furthermore, $\eta_{e0} = 1$ does not imply that eliminating the external boundary layer could not improve the observed reaction rate. Removing the boundary layer would increase the value of C_{As}, thus establishing a greater driving force for internal mass transfer and reducing the likelihood of C_A falling to zero inside the particle.

Example 11.5 demonstrates the influence of external mass transfer resistance on reaction rates.

Table 11.7: External Effectiveness Factors

First-order reaction: $r_A = k_1 C_A$	$\eta_{e1} = \dfrac{C_{As}}{C_{Ab}}$
Zero-order reaction: $r_A = k_0$	$\eta_{e0} = 1$
Michaelis–Menten reaction: $r_A = \dfrac{v_{max} C_A}{K_m + C_A}$	$\eta_{em} = \dfrac{C_{As}(K_m + C_{Ab})}{C_{Ab}(K_m + C_{As})}$

EXAMPLE 11.5 Effect of Mass Transfer on Bacterial Denitrification

Denitrifying bacteria are immobilized in gel beads and used in a stirred reactor for removal of nitrate from groundwater. At a nitrate concentration of $3\,g\,m^{-3}$, the conversion rate is $0.011\,g\,s^{-1}\,m^{-3}$ of catalyst. The effective diffusivity of nitrate in the gel is $1.5 \times 10^{-9}\,m^2\,s^{-1}$, the beads are 6 mm in diameter, and the liquid–solid mass transfer coefficient is $10^{-5}\,m\,s^{-1}$. K_m for the immobilized bacteria is approximately $25\,g\,m^{-3}$.

a. Does external mass transfer influence the reaction rate?
b. Are internal mass transfer effects significant?
c. By how much would the reaction rate be improved if both the internal and external mass transfer resistances were eliminated?

Solution

d.

$$R = \frac{6 \times 10^{-3}\,m}{2} = 3 \times 10^{-3}\,m$$

The effect of external mass transfer is found by calculating Ω. From Table 11.6 for spherical geometry:

$$\Omega = \frac{R\ r_{A,obs}}{3\ k_s C_{Ab}} = \frac{3 \times 10^{-3}\ m}{3}\ \frac{0.011\,g\,s^{-1}\,m^{-3}}{\left(10^{-5}\,m\,s^{-1}\right)\left(3\,g\,m^{-3}\right)} = 0.37$$

As this value of Ω is relatively large, external mass transfer effects are significant.

e. From Eq. (11.43):

$$\frac{C_{As}}{C_{Ab}} = 1 - \Omega = 0.63$$

$$C_{As} = 0.63\,C_{Ab} = 0.63\left(3\,g\,m^{-3}\right) = 1.9\,g\,m^{-3}$$

The observable Thiele modulus is calculated with $C_{As} = 1.9\,g\,m^{-3}$ using the equation for spheres from Table 11.4:

$$\Phi = \left(\frac{R}{3}\right)^2 \frac{r_{A,obs}}{\mathcal{D}_{Ae} C_{As}} = \left(\frac{3 \times 10^{-3}\ m}{3}\right)^2 \frac{0.011\,g\,s^{-1}\,m^{-3}}{\left(1.5 \times 10^{-9}\,m^2\,s^{-1}\right)\left(1.9\,g\,m^{-3}\right)} = 3.9$$

From Weisz's criteria (Section 11.4.5), as $\Phi > 3$, internal mass transfer effects are significant.

f. The reaction rate in the absence of mass transfer effects is $0\,r_{Ab}^*$, which is related to $r_{A,obs}$ by Eq. (11.45). Therefore, we can calculate r_{Ab}^* if we know η_T. Because the nitrate concentration is much smaller than K_m, we can assume that the reaction operates in the first-order regime. From Table 11.7 for first-order kinetics:

$$\eta_{e1} = \frac{C_{As}}{C_{Ab}} = 0.63$$

From Figure 11.14 for spherical geometry, at $\Phi = 3.9$, $\eta_{i1} = 0.21$. Therefore, using Eq. (11.46):

$$\eta_{T1} = \eta_{i1}\eta_{e1} = 0.21(0.63) = 0.13$$

From Eq. (11.45):

$$r_{Ab}^* = \frac{r_{A,obs}}{\eta_{T1}} = \frac{0.011\,\text{g s}^{-1}\,\text{m}^{-3}}{0.13} = 0.085\,\text{g s}^{-1}\,\text{m}^{-3}$$

If the internal and external mass transfer resistances were eliminated, the reaction rate would be increased by a factor of 0.085/0.011, or about 7.7.

11.6 Liquid–Solid Mass Transfer Correlations

The mass transfer coefficient k_S must be known before we can account for external mass transfer effects. k_S depends on liquid properties such as viscosity, density, and diffusivity as well as the reactor hydrodynamics. It is difficult to determine k_S accurately, especially for particles that are neutrally buoyant. However, values can be estimated using correlations from the literature; these are usually accurate under the conditions specified to be within 10% to 20%. Selected correlations can be found in engineering texts and resources [19–23].

11.7 Effective Diffusivity

The value of the effective diffusivity $\mathcal{D}_{Ae}$ reflects the ease with which compound A is transported within the catalyst matrix and is strongly dependent on the pore structure of the solid. Effective diffusivities are normally lower than corresponding molecular diffusivities in water because porous solids offer more resistance to diffusion. Some experimental $\mathcal{D}_{Ae}$ values for selected biocatalyst systems are listed in Table 11.8.

Techniques for measuring effective diffusivity are described in several papers [27–31,34]; however, accurate measurement of $\mathcal{D}_{Ae}$ is difficult in most systems. During the measurement, external mass transfer limitations must be overcome using high liquid flow rates around the catalyst. If experimental values of $\mathcal{D}_{Ae}$ are higher than the molecular diffusion coefficient in water, this may indicate the presence of convective fluid currents within the catalyst [3–6]. As cells can pose a significant barrier to diffusion, $\mathcal{D}_{Ae}$ for immobilized cell preparations must be determined with the biomass present.

Table 11.8: Effective Diffusivity Values

Substance	Solid	Temperature	$\mathcal{D}_{Ae}\ (m^2s^{-1})$	$\dfrac{\mathcal{D}_{Ae}}{\mathcal{D}_{Aw}}$ [a]	Reference
O_2	Agar (2% w/v) containing *Candida lipolytica* cells	30°C	1.94×10^{-9}	0.70	[24]
	Microbial aggregates from domestic waste treatment plant	20°C	1.37×10^{-9}	0.62	[25]
	Trickling-filter slime	25°C	0.82×10^{-9}	–	[26]
Glucose	Microbial aggregates from domestic waste treatment plant	20°C	0.25×10^{-9}	0.37	[25]
	Glass fiber discs containing *Saccharomyces uvarum* cells	30°C	0.30×10^{-9}	0.43	[3]
	Calcium alginate (3 wt%)	30°C	0.62×10^{-9}	0.87	[27]
	Calcium alginate (2.4–2.8 wt%) containing 50 wt% bakers' yeast	30°C	0.26×10^{-9}	0.37	[27]
Sucrose	Calcium alginate (2% w/v)	25°C	0.48×10^{-9}	0.86	[28]
	Calcium alginate (2% w/v) containing 12.5% (v/v) *Catharanthus roseus* cells	25°C	0.14×10^{-9}	0.25	[28]
L-Tryptophan	Calcium alginate (2%)	30°C	0.67×10^{-9}	1.0	[29]
	κ-Carrageenan (4%)	30°C	0.58×10^{-9}	0.88	[30]
Lactate	Agar (1%) containing 1% Ehrlich ascites tumor cells	37°C	1.4×10^{-9}	0.97	[31]
	Agar (1%) containing 6% Ehrlich ascites tumor cells	37°C	0.7×10^{-9}	0.48	[31]
Ethanol	κ-Carrageenan (4%)	30°C	1.01×10^{-9}	0.92	[30]
	Calcium alginate (1.4–3.8 wt%)	30°C	1.25×10^{-9}	0.92	[27]
	Calcium alginate (2.4–2.8 wt%) containing 50 wt% bakers' yeast	30°C	0.45×10^{-9}	0.33	[27]
Nitrate	Compressed film of nitrifying organisms	–	1.4×10^{-9}	0.90	[25]
Ammonia	Compressed film of nitrifying organisms	–	1.3×10^{-9}	0.80	[25]
Bovine serum albumin	Agarose (5.5% v/v)	20°C	0.023×10^{-9}	0.38	[32]
	Agarose (7.2% v/v)	20°C	0.016×10^{-9}	0.28	[32]
Lysozyme	Cation exchange resin, pH 7	–	0.05×10^{-9}	0.45	[33]

[a] $\mathcal{D}_{Aw}$ is the molecular diffusivity in water at the temperature of measurement.

11.8 Minimizing Mass Transfer Effects

To improve overall rates of reaction in bioprocesses, mass transfer restrictions must be minimized or eliminated. In this section, we consider practical ways of achieving this objective based on the equations presented in Tables 11.4 and 11.6.

11.8.1 Minimizing Internal Mass Transfer Effects

Internal mass transfer effects are eliminated when the internal effectiveness factor η_i is equal to 1. η_i approaches unity as the observable Thiele modulus Φ decreases. From Table 11.4, Φ is decreased by:

1. Reducing the observed reaction rate $r_{A,obs}$
2. Reducing the size of the catalyst
3. Increasing the effective diffusivity $\mathcal{D}_{Ae}$
4. Increasing the surface substrate concentration C_{As}

All of these changes impact directly on the effectiveness of substrate mass transfer within the particle.

Paradoxically, reducing the reaction rate $r_{A,obs}$ improves the effectiveness of mass transfer aimed at increasing the reaction rate. When the catalyst is very active with a high demand for substrate, mass transfer is likely to be slow relative to reaction, so that steep concentration gradients are produced. However, limiting the reaction rate by operating at suboptimum conditions or using an organism or enzyme with low intrinsic activity does not achieve the overall goal of higher conversion rates. Because $r_{A,obs}$ is the reaction rate per unit volume of catalyst, another way of reducing $r_{A,obs}$ is to reduce the cell or enzyme loading in the solid. This reduces the demand for substrate per particle so that mass transfer has a better chance of supplying substrate at a sufficient rate. Therefore, if the same mass of cells or enzyme is distributed between more particles, the rate of conversion will increase. The trade-off, however, is that using more particles may mean that a larger reactor is required.

As Φ is proportional to the catalyst size squared (R^2 for spheres or b^2 for flat plates), reducing the catalyst size has a more dramatic effect on Φ than changes in any other variable. It is therefore a good way to improve the reaction rate. In principle, mass transfer limitations can be overcome completely if the particle size is decreased sufficiently. However, it is often extremely difficult in practice to reduce particle dimensions to this extent [35]. Even if mass transfer effects are eliminated, operation of large-scale reactors with tiny, highly compressible gel particles raises new problems. Some degree of internal mass transfer restriction must usually be tolerated.

11.8.2 Minimizing External Mass Transfer Effects

External mass transfer effects decrease as the observable modulus Ω is reduced. From the equations in Table 11.6, this is achieved by:

1. Reducing the observed reaction rate $r_{A,obs}$
2. Reducing the size of the catalyst
3. Increasing the mass transfer coefficient k_S
4. Increasing the bulk substrate concentration C_{Ab}

Decreasing the catalyst size and increasing the mass transfer coefficient reduce the thickness of the boundary layer and facilitate external mass transfer. k_S is increased most readily by raising the liquid velocity outside of the catalyst. However, as k_S is a function of other system properties such as the liquid viscosity, density, and substrate diffusivity, changes in these variables can also reduce boundary layer effects to some extent. External mass transfer is more rapid at high bulk substrate concentrations; the higher the concentration, the greater is the driving force for mass transfer across the boundary layer. Decreasing $r_{A,obs}$ as described in the previous section also reduces external mass transfer limitations by reducing the demand for substrate.

In large-scale reactors, problems with external mass transfer may be unavoidable if sufficiently high liquid velocities cannot be achieved. However, it is advisable to eliminate fluid boundary layers in the laboratory to simplify analysis of the data. Several laboratory reactor configurations allow almost complete elimination of interparticle and interphase concentration gradients [36]. Recycle reactors such as that shown in Figure 11.16 have been employed

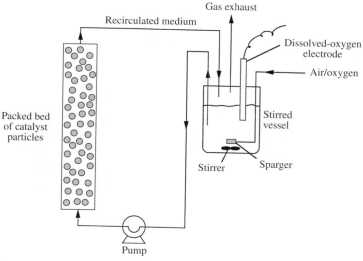

Figure 11.16 Batch recirculation reactor for measuring the rate of O_2 uptake by immobilized cells or enzymes.

extensively for study of immobilized cell and enzyme reactions; operation with high liquid velocity through the bed reduces boundary layer effects. Another suitable configuration is the *spinning basket reactor* [37,38] illustrated in Figure 11.17. By rotating the baskets at high speed, high relative velocities are achieved between the particles and the surrounding fluid. The slip velocities obtained in this apparatus are significantly greater than those found using freely suspended particles. Another laboratory design aimed at increasing the slip velocity is the stirred vessel shown in Figure 11.18. In this system, catalyst particles are held relatively stationary in a wire mesh cage while liquid is agitated at high speed.

The elimination of external mass transfer effects can be verified by calculating the observable parameter Ω as described in Section 11.5. However, if the mass transfer coefficient is not known accurately, an experimental test may be used instead [39]. Consider again the apparatus of Figure 11.16. If boundary layers around the particles affect the rate of reaction, increasing the liquid velocity through the bed will improve conversion rates by reducing the boundary layer thickness and bringing C_{As} closer to C_{Ab}. At sufficiently high liquid velocity, external mass transfer effects may be removed. When this occurs, further increases in pump speed do not change the overall reaction rate, as illustrated in Figure 11.19. Therefore, if we can identify a liquid velocity u_L^* at which the reaction rate becomes independent of liquid velocity, operation at $u_L > u_L^*$ will ensure that $\eta_e = 1$ and external mass transfer effects are eliminated. For stirred reactors, a similar relationship holds between $r_{A,obs}$ and agitation speed.

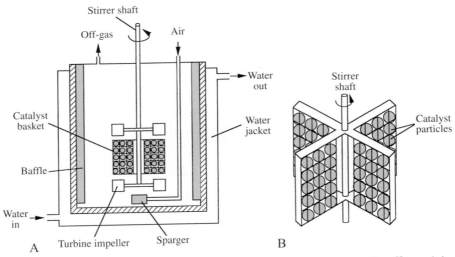

Figure 11.17 Spinning basket reactor for minimizing external mass transfer effects: (a) reactor configuration; (b) detail of the spinning basket. *From D.G. Tajbl, J.B. Simons, and J.J. Carberry, 1966, Heterogeneous catalysis in a continuous stirred tank reactor. Ind. Eng. Chem. Fund. 5, 171–175.*

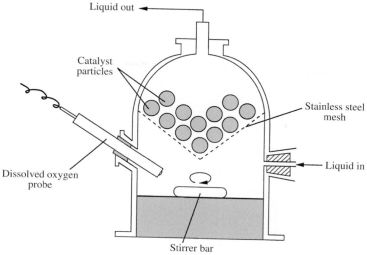

Figure 11.18 Stirred laboratory reactor for minimizing external mass transfer effects. *From K. Sato and K. Toda, 1983, Oxygen uptake rate of immobilized growing Candida lipolytica. J. Ferment. Technol. 61, 239–245.*

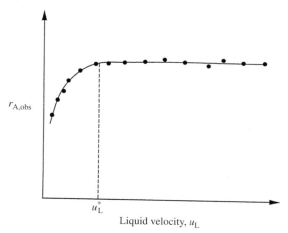

Figure 11.19 Relationship between observed reaction rate and external liquid velocity for reduction of external mass transfer effects.

11.9 Evaluating True Kinetic Parameters

The intrinsic kinetics of zero-order, first-order, and Michaelis–Menten reactions are represented by the parameters k_0, k_1, and v_{max} and K_m. In general, it cannot be assumed that the values of these parameters will be the same before and after cell or enzyme immobilization: significant changes can occur during the immobilization process. As an example, Figure 11.20

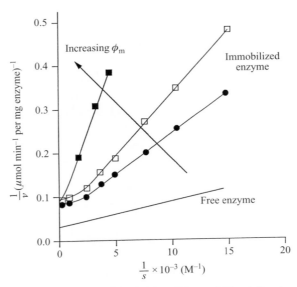

Figure 11.20 Lineweaver–Burk plots for free and immobilized β-galactosidase. Enzyme concentrations within the gel are: 0.10 mg mL^{-1} (●); 0.17 mg mL^{-1} (□); and 0.50 mg mL^{-1} (■). Data from P.S. Bunting and K.J. Laidler, 1972, Kinetic studies on solid-supported β-galactosidase. Biochemistry 11, 4477–4483.

shows Lineweaver–Burk plots (Section 10.4.2) for free and immobilized β-galactosidase enzyme. According to Eq. (10.43), the slopes and intercepts of the lines in Figure 11.20 indicate the values of K_m/v_{max} and $1/v_{max}$, respectively. Compared with the results shown for free enzyme, the steeper slopes and higher intercepts obtained for the immobilized enzyme indicate that immobilization reduces v_{max}; this is a commonly observed result. The value of K_m can also be affected [9].

Kinetic parameters for homogeneous reactions can be determined directly from experimental rate data as described in Sections 10.3 and 10.4. However, evaluating the true kinetic parameters of immobilized cells and enzymes is somewhat more difficult. The observed rate of reaction is not the true rate at all points in the catalyst; mass transfer processes effectively "mask" the true kinetic behavior. Accordingly, v_{max} and K_m for immobilized catalysts cannot be estimated using the classical plots described in Section 10.4. Under the influence of mass transfer, these plots no longer give straight lines over the entire range of substrate concentration [40,41].

Lineweaver–Burk plots for immobilized enzymes can be nonlinear as illustrated in Figure 11.20. However, the deviation from linearity can be obscured by the scatter in real experimental data and the distortion of errors due to the Lineweaver–Burk linearization (Section 10.4.2). Studies have shown that the effects of diffusion are more pronounced in Eadie–Hofstee plots than in Lineweaver–Burk or Langmuir plots; however, all three for immobilized enzymes can be approximated by straight lines over certain intervals.

We should not conclude that the immobilized enzyme fails to obey Michaelis–Menten kinetics even though the Lineweaver–Burk plots for immobilized β-galactosidase are nonlinear in Figure 11.20. The kinetic form of reactions is generally maintained after immobilization of cells and enzymes [9]. The nonlinearity of the Lineweaver–Burk plots is due instead to the effect of mass transfer on the measured reaction rate.

Several methods have been proposed for determining v_{max} and K_m in heterogeneous catalysts [40–43]. The most straightforward approach is experimental: it involves reducing the particle size and catalyst loading and increasing the external liquid velocity to eliminate all mass transfer resistances. The measured rate data can then be analyzed for kinetic parameters as if the reaction were homogeneous.

11.10 General Comments on Heterogeneous Reactions in Bioprocessing

Before concluding this chapter, some general observations and guidelines for heterogeneous reactions are outlined.

- *Importance of O_2 mass transfer limitations.* In aerobic reactions, mass transfer of O_2 is more likely to limit the rate of reaction than mass transfer of most other substrates. The reason is the poor solubility of O_2 in aqueous solutions. Whereas the sugar concentration in a typical culture broth is around 10 to $100\,kg\,m^{-3}$, the dissolved O_2 concentration under 1 atm of air pressure is limited to about $8 \times 10^{-3}\,kg\,m^{-3}$. Therefore, because C_{As} is so low for O_2, the observable Thiele modulus Φ (refer to Table 11.4) can be several orders of magnitude greater than for other substrates. In anaerobic systems, the substrate most likely to be affected by mass transfer limitations is more difficult to identify.
- *Relationship between the internal effectiveness factor and the substrate concentration gradient.* Depending on the reaction kinetics, the severity of intraparticle concentration gradients can be inferred from the value of the internal effectiveness factor. For first-order kinetics, $\eta_{i1} = 1$ implies that concentration gradients do not exist in the catalyst because the reaction rate is proportional to the substrate concentration. Conversely, if $\eta_{i1} < 1$, we can conclude that concentration gradients are present. For zero-order reactions, $\eta_{i0} = 1$ does not imply that gradients are absent because, as long as $C_A > 0$, the reaction rate is unaffected by any reduction in substrate concentration. Concentration gradients can be so steep that C_A is reduced to almost zero within the catalyst, but η_{i0} will remain equal to 1. On the other hand, $\eta_{i0} < 1$ implies that the concentration gradient is severe and that some fraction of the particle volume is starved of substrate.
- *Relative importance of internal and external mass transfer limitations.* For porous catalysts, it has been demonstrated with realistic values of mass transfer and diffusion

parameters that external mass transfer limitations do not exist unless internal limitations are also present [44]. Concentration differences between the bulk liquid and external catalyst surface are never observed without larger internal gradients developing within the particle. On the other hand, if internal limitations are known to be present, external limitations may or may not be important depending on conditions. Significant external mass transfer effects can occur when reaction does not take place inside the catalyst, for example, if cells or enzymes are attached only to the exterior surface.

- *Operation of catalytic reactors.* Certain solid-phase properties are desirable for operation of immobilized cell and enzyme reactors. For example, in packed bed reactors, large, rigid, and uniformly shaped particles promote well-distributed and stable liquid flow. Solids in packed columns should also have sufficient mechanical strength to withstand their own weight. These requirements are in direct conflict with those needed for rapid intraparticle mass transfer, as diffusion is facilitated in particles that are small, soft, and porous. Because blockages and large pressure drops through the bed must be avoided, mass transfer rates are usually compromised. In stirred reactors, soft, porous gels are readily destroyed at the agitation speeds needed to eliminate external boundary layer effects.

- *Product effects.* Products formed by reaction inside catalysts must diffuse away based on the concentration gradient between the interior of the catalyst and the bulk liquid. The concentration profile for product is the reverse of that for substrate: the concentration is highest at the center of the catalyst and lowest in the bulk liquid. If activity of the cells or enzyme is affected by product inhibition, high intraparticle product concentrations may inhibit progress of the reaction. Immobilized enzymes that produce or consume H^+ ions are often affected in this way. Because enzyme reactions are very sensitive to pH (Section 10.3.4), small local variations in intraparticle pH due to slow diffusion of ions can have a significant influence on the reaction rate [45].

11.11 Summary of Chapter 11

At the end of Chapter 11, you should:

- Know what heterogeneous reactions are and where they occur in bioprocessing
- Understand the difference between *observed* and *true reaction rates*
- Know how concentration gradients arise in solid-phase catalysts
- Understand the concept of the *effectiveness factor*
- Be able to apply the *generalized Thiele modulus* and *observable Thiele modulus* to determine the effect of internal mass transfer on reaction rate
- Know how to minimize internal and external mass transfer restrictions
- Understand that it is generally difficult to determine true kinetic parameters for heterogeneous reactions

Problems

11.1. Diffusion and reaction in a waste treatment lagoon

Industrial wastewater is often treated in large shallow lagoons. Consider such a lagoon as shown in Figure 11P1.1 covering land of area A. Microorganisms form a sludge layer of thickness L at the bottom of the lagoon; this sludge remains essentially undisturbed by movement of the liquid. At steady state, wastewater is fed into the lagoon so that the bulk concentration of digestible substrate in the liquid remains constant at s_b. Cells consume substrate diffusing into the sludge layer, thereby establishing a concentration gradient across thickness L. The microorganisms can be assumed to be distributed uniformly in the sludge. As indicated in Figure 11P1.1, distance from the bottom of the lagoon is measured by coordinate z.

a. Set up a shell mass balance on substrate by considering a thin slice of sludge of thickness Δz perpendicular to the direction of diffusion. The rate of microbial reaction per unit volume of sludge is:

$$r_S = k_1 s$$

where s is the concentration of substrate in the sludge layer (gmol cm^{-3}) and k_1 is the first-order rate constant (s^{-1}). The effective diffusivity of substrate in the sludge is $\mathcal{D}_{Se}$. Obtain a differential equation relating s and z. *Hint*: Area A is constant for flat-plate geometry and can be canceled from all terms of the mass balance equation.

b. External mass transfer effects at the liquid–sludge interface are negligible. What are the boundary conditions for this problem?

c. The differential equation obtained in (a) is solved by making the substitution:

$$s = N e^{pz}$$

where N and p are constants.

    (i) Substitute this expression for s into the differential equation derived in (a) to obtain an equation for p. Remember that $\sqrt{p^2} = p$ or $-p$.

    (ii) Because there are two possible values of p, let:

$$s = N e^{pz} + M e^{-pz}$$

    (iii) Apply the boundary condition at $z = 0$ to this expression to obtain a relationship between N and M.

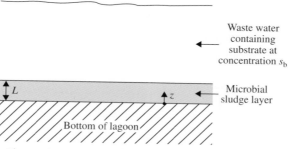

Figure 11P1.1 Lagoon for wastewater treatment.

(iv) Use the boundary condition at $z = L$ to find N and M explicitly. Obtain an expression for s as a function of z.

(v) Use the definition of $\cosh x$:

$$\cosh x = \frac{e^x + e^{-x}}{2}$$

to prove that:

$$\frac{s}{s_b} = \frac{\cosh\left(z\sqrt{k_1 / \mathcal{D}_{se}}\right)}{\cosh\left(L\sqrt{k_1 / \mathcal{D}_{se}}\right)}$$

d. At steady state, the rate of substrate consumption must be equal to the rate at which substrate enters the sludge. As substrate enters the sludge by diffusion, the overall rate of reaction can be evaluated using Fick's law:

$$r_{A,obs} = \mathcal{D}_{Se}A\frac{ds}{dz}\bigg|_{z=L}$$

where $dz\big|_{z=L}$ means ds/dz evaluated at $z = L$. Use the equation for s from (c) to derive an equation for $r_{A,obs}$. *Hint:* The derivative of $\cosh(ax) = a\sinh(ax)$ where a is a constant and:

$$\sinh x = \frac{e^x - e^{-x}}{2}$$

e. Show from the result of (d) that the internal effectiveness factor is given by the expression:

$$\eta_{il} = \frac{\tanh\phi_1}{\phi_1}$$

where

$$\phi_1 = L\sqrt{\frac{k_1}{\mathcal{D}_{Se}}}$$

and

$$\tanh x = \frac{\sinh x}{\cosh x}$$

f. Plot the substrate concentration profiles through a sludge layer of thickness 2 cm for the following sets of conditions:

	Condition		
	1	2	3
k_1 (s^{-1})	4.7×10^{-8}	2.0×10^{-7}	1.5×10^{-4}
$\mathcal{D}_{Se}$ (cm^2 s^{-1})	7.5×10^{-7}	2.0×10^{-7}	6.0×10^{-6}

Take s_b to be 10^{-5} gmol cm^{-3}. Label the profiles with corresponding values of φ_1 and η_{i1}. Comment on the general relationship between φ_1, the shape of the concentration profile, and the value of η_{i1}.

11.2. O$_2$ concentration profile in an immobilized enzyme catalyst

L-Lactate 2-monooxygenase from *Mycobacterium smegmatis* is immobilized in spherical agarose beads. The enzyme catalyzes the reaction:

$$C_3H_6O_3 + O_2 \rightarrow C_2H_4O_2 + CO_2 + H_2O$$
$$\underset{\text{lactic acid}}{} \qquad \underset{\text{acetic acid}}{}$$

Agarose beads 4 mm in diameter are immersed in a well-mixed solution containing 0.5 mM O$_2$. A high lactic acid concentration is provided so that O$_2$ is the rate-limiting substrate. The effective diffusivity of O$_2$ in agarose is 2.1×10^{-9} m^2 s^{-1}. K_m for the immobilized enzyme is 0.015 mM and v_{max} is 0.12 mol s^{-1} per kg of enzyme. The beads contain 0.012 kg of enzyme m^{-3} of gel. External mass transfer effects are negligible.

a. Plot the O$_2$ concentration profile inside the beads.

b. What fraction of the catalyst volume is active?

c. Determine the largest bead size that allows the maximum overall rate of conversion.

11.3. Effect of O$_2$ transfer on recombinant cells

Recombinant *E. coli* cells contain a plasmid derived from pBR322 incorporating genes for the enzymes β-lactamase and catechol 2,3-dioxygenase from *Pseudomonas putida*. To produce the desired enzymes, the organism requires aerobic conditions. The cells are immobilized in spherical beads of carrageenan gel. The effective diffusivity of O$_2$ is 1.4×10^{-9} m^2 s^{-1}. O$_2$ uptake is zero-order with intrinsic rate constant 10^{-3} mol s^{-1} m^{-3} of particle. The concentration of O$_2$ at the surface of the catalyst is 8×10^{-3} kg m^{-3}. Cell growth is negligible.

a. What is the maximum particle diameter for aerobic conditions throughout the catalyst particles?

b. For particles half the diameter calculated in (a), what is the minimum O$_2$ concentration in the beads?

c. The density of cells in the gel is reduced by a factor of five. If the specific activity of the cells is independent of cell loading, what is the maximum particle size for aerobic conditions?

11.4. Ammonia oxidation by immobilized cells

Thiosphaera pantotropha is being investigated for aerobic oxidation of ammonia to nitrite for wastewater treatment. The organism is immobilized in spherical agarose particles of diameter 3 mm. The effective diffusivity of O$_2$ in the particles is 1.9×10^{-9} m^2 s^{-1}. The immobilized cells are placed in a flow chamber for measurement of O$_2$ uptake rate. Using published correlations, the liquid–solid mass transfer coefficient for O$_2$ is calculated as 6×10^{-5} m s^{-1}. When the bulk O$_2$ concentration is 6×10^{-3} kg m^{-3}, the observed rate of O$_2$ consumption is 2.2×10^{-5} kg s^{-1} m^{-3} of catalyst.

a. What effect does external mass transfer have on the respiration rate?

b. What is the effectiveness factor?

c. For optimal activity of *T. pantotropha*, O$_2$ levels must be kept above the critical level, 1.2×10^{-3} kg m^{-3}. Is this condition satisfied?

11.5. Microcarrier culture and external mass transfer

Mammalian cells form a monolayer on the surface of microcarrier beads of diameter 120 μm and density 1.2×10^3 kg m^{-3}. The culture is maintained in spinner flasks in serum-free medium of viscosity 10^{-3} N s m^{-2} and density 10^3 kg m^{-3}. The diffusivity of O_2 in the medium is 2.3×10^{-9} m^2 s^{-1}. The observed rate of O_2 uptake is 0.015 mol s^{-1} m^{-3} at a bulk O_2 concentration of 0.2 mol m^{-3}. What effect does external mass transfer have on the reaction rate?

11.6. Immobilized enzyme reaction kinetics

Invertase catalyzes the reaction:

$$\underset{\text{sucrose}}{C_{12}H_{22}O_{11}} + H_2O \rightarrow \underset{\text{glucose}}{C_6H_{12}O_6} + \underset{\text{fructose}}{C_6H_{12}O_6}$$

Invertase from *Aspergillus oryzae* is immobilized in porous resin particles of diameter 1.6 mm at a density of 0.1 μmol of enzyme g^{-1}. The effective diffusivity of sucrose in the resin is 1.3×10^{-11} m^2 s^{-1}. The resin is placed in a spinning basket reactor operated so that external mass transfer effects are eliminated. At a sucrose concentration of 0.85 kg m^{-3}, the observed rate of conversion is 1.25×10^{-3} kg s^{-1} m^{-3} of resin. K_m for the immobilized enzyme is 3.5 kg m^{-3}.

a. Calculate the effectiveness factor.
b. Determine the true first-order reaction constant for immobilized invertase.
c. Assume that the specific enzyme activity is not affected by steric hindrance or conformational changes as the enzyme loading increases. This means that k_1 should be directly proportional to the enzyme concentration in the resin. Plot the changes in effectiveness factor and observed reaction rate as a function of enzyme loading from 0.01 μmol g^{-1} to 2.0 μmol g^{-1}. Comment on the relative benefit of increasing the concentration of enzyme in the resin.

11.7. Mass transfer effects in plant cell culture

Suspended *Catharanthus roseus* cells form spherical aggregates approximately 1.5 mm in diameter. O_2 uptake is measured using the apparatus of Figure 11.16; medium is recirculated with a superficial liquid velocity of 0.83 cm s^{-1}. At a bulk concentration of 8 mg L^{-1}, O_2 is consumed at a rate of 0.28 mg per g wet weight of cells per hour. Assume that the density and viscosity of the medium are similar to water and the specific gravity of wet cells is 1. The effective diffusivity of O_2 in the aggregates is 9×10^{-6} cm^2 s^{-1}, or half that in the medium. O_2 uptake follows zero-order kinetics.

a. Does external mass transfer affect the O_2 uptake rate?
b. To what extent does internal mass transfer affect O_2 uptake?
c. Roughly, what would you expect the profile of O_2 concentration to be within the aggregates?

11.8. Respiration in mycelial pellets

Aspergillus niger cells are observed to form self-immobilized aggregates of average diameter 5 mm. The effective diffusivity of O_2 in the aggregates is 1.75×10^{-9} m^2 s^{-1}. In a fixed-bed reactor, the O_2 consumption rate at a bulk O_2 concentration of 8×10^{-3} kg m^{-3} is 8.7×10^{-5} kg s^{-1} m^{-3} of biomass. The liquid–solid mass transfer coefficient is 3.8×10^{-5} m s^{-1}.

a. Is O_2 uptake affected by external mass transfer?
b. What is the external effectiveness factor?

    c. What reaction rate would be observed if both internal and external mass transfer resistances were eliminated?

    d. If only external mass transfer effects were removed, what would be the reaction rate?

11.9. **Effect of mass transfer on glyphosate removal**

Cultured soybean (*Glycine max*) cells are immobilized in spherical agarose beads of average radius of 3.5 mm. In phytoremediation experiments, the beads are placed in a column reactor and medium containing 60 mg L^{-1} glyphosate is recirculated through the bed using a peristaltic pump. Consumption of glyphosate is considered a first-order reaction. The rate of glyphosate uptake is measured as a function of liquid recirculation rate.

Liquid flow rate (cm³ s⁻¹)	Glyphosate uptake rate per unit volume of catalyst (kg s⁻¹ m⁻³)
10	8.3×10^{-6}
15	1.9×10^{-5}
20	3.6×10^{-5}
25	6.1×10^{-5}
30	8.3×10^{-5}
35	8.4×10^{-5}
45	8.4×10^{-5}
55	8.4×10^{-5}

    a. What is the minimum liquid flow rate required to eliminate external mass transfer effects? Explain your answer.

    b. The immobilized cell reactor is operated using a recirculation flow rate of 40 cm³ s⁻¹. If the effective diffusivity of glyphosate in the beads is 0.9×10^{-9} m² s⁻¹, estimate the maximum volumetric rate of glyphosate uptake for this system in the absence of mass transfer limitations.

11.10. **Mass transfer effects in tissue-engineered cartilage**

A layer of cartilage 3 mm thick is produced using human chondrocytes seeded at high density into a polymer mesh scaffold. The scaffold is attached to a nonporous polymer base simulating the mechanical properties of bone. O_2 is provided to the cells by sparging nutrient medium with air in a separate vessel and bathing the surface of the cartilage tissue with the medium in a perfusion bioreactor. The concentration of dissolved O_2 in the medium is maintained at 7×10^{-3} kg m⁻³ and the effective diffusivity of O_2 in cartilage is 1.0×10^{-9} m² s⁻¹. The rate of O_2 consumption by the chondrocytes is measured as 6.6×10^{-6} kg s⁻¹ m⁻³ of cartilage. The mass transfer coefficient for transport of O_2 through the liquid boundary layer at the cartilage surface is 2×10^{-5} m s⁻¹. O_2 uptake follows zero-order kinetics.

    a. Is the cartilage culture affected by O_2 mass transfer limitations?

    b. Estimate the O_2 uptake rate in the absence of internal and external mass transfer effects.

    c. What is the lowest O_2 concentration within the cartilage tissue?

    d. The flow rate of medium over the tissue is increased so that any boundary layers at the surface of the cartilage are eliminated. The thickness of the cartilage layer is also reduced to 2 mm while keeping the chondrocyte density constant. Estimate the culture O_2 demand under these conditions. (Iterative solution may be required.)

    e. When external boundary layers are removed, what maximum thickness of cartilage can be used while avoiding O_2 depletion in the tissue?

11.11. O$_2$ uptake by immobilized bacteria

Bacteria are immobilized in spherical agarose beads of diameter 3 mm. The effective diffusivity of O$_2$ in the beads is 1.9×10^{-9} m^2 s^{-1}. The immobilized cells are placed in a stirred bioreactor and the rate of O$_2$ uptake is measured as 2.2×10^{-5} kg s^{-1} m^{-3} of catalyst at a bulk O$_2$ concentration of 6×10^{-3} kg m^{-3}. At the operating stirrer speed in the bioreactor, the liquid–solid mass transfer coefficient is estimated as 6×10^{-5} m s^{-1}. O$_2$ uptake can be considered a zero-order reaction.

a. What effect does external mass transfer have on the rate of O$_2$ uptake?

b. What are the values of the internal, external, and total effectiveness factors?

c. The O$_2$ concentration within the particles must be kept above 4×10^{-4} kg m^{-3} for optimal culture performance. Is this condition satisfied?

11.12. Three-dimensional culture of mesenchymal stem cells

Human mesenchymal stem cells isolated from adipose tissue are immobilized in spherical alginate beads of diameter 3.5 mm. The beads are placed in a column reactor similar to that shown in Figure 11.16. The pump is set to deliver a medium recirculation flow rate of 0.033 1 s^{-1}. At this liquid flow rate, the mass transfer coefficient for external O$_2$ transfer is calculated using literature correlations for spherical particles in a packed bed as 3.6×10^{-5} m s^{-1}. The effective diffusivity of O$_2$ in the alginate beads containing cells is 9×10^{-6} cm^2 s^{-1}, and the bulk dissolved O$_2$ concentration is 8×10^{-3} kg m^{-3}. The rate of O$_2$ consumption per unit volume of catalyst is measured using the dissolved O$_2$ electrode as 7.6×10^{-5} kg s^{-1} m^{-3}. Consumption of O$_2$ by the cells can be considered a zero-order reaction.

a. Is the rate of O$_2$ uptake affected by external mass transfer effects?

b. Is the rate of O$_2$ uptake affected by internal mass transfer effects?

c. Are cells at the center of the beads supplied with O$_2$?

d. What is the maximum O$_2$ consumption rate that could be achieved using this immobilized cell system if the dissolved O$_2$ concentration in the bulk liquid, 8×10^{-3} kg m^{-3}, were present at all points within all of the beads?

e. In the absence of an external boundary layer, what is the maximum bead diameter that supports a finite (>0) concentration of O$_2$ at all points in the catalyst?

11.13. Diffusion and reaction of glucose and O$_2$

Microbial cells are immobilized in spherical gel beads and cultured in a batch reactor under aerobic conditions. The medium used contains 20 g L^{-1} of glucose. The effective diffusivity of glucose in the beads is 0.42×10^{-9} m^2 s^{-1}; the effective diffusivity of O$_2$ is 1.8×10^{-9} m^2 s^{-1}. O$_2$ uptake follows zero-order kinetics; as the culture progresses and the concentration of glucose in the medium is reduced to low levels, glucose uptake can be considered a first-order reaction. The gas–liquid mass transfer capacity of the reactor is such that the dissolved O$_2$ concentration in the medium is maintained at the saturation value of 8×10^{-3} kg m^{-3} throughout the culture period. According to the stoichiometry of growth for this culture, for every mole of glucose consumed, 2.7 moles of O$_2$ are required. O$_2$ uptake is strongly limited by internal mass transfer effects; however, vigorous mixing in the reactor is effective in eliminating external boundary layers. Estimate the medium glucose concentration that must be reached before mass transfer of glucose exerts a limiting effect on the reaction rate similar to that prevailing for O$_2$. What are the implications of your answer for operation of immobilized cell reactors?

11.14. Uptake of growth factor by neural stem cell spheroids

Neural stem cells cultured in vitro form compact spheroids of diameter 500 μm and density 1.15 g cm^{-3}. The spheroids are suspended and gently agitated in nutrient medium of viscosity 1 cP and density 1 g cm^{-3}. The stem cells consume complex growth factor proteins that are provided in the medium at a concentration of 10 ng mL^{-1}. The diffusivity of growth factor in aqueous solution is 3.3×10^{-6} cm^2 s^{-1}; the effective diffusivity of growth factor in the spheroids is 5×10^{-11} m^2 s^{-1}. When 10 mg of spheroids are cultured in 1 mL of nutrient medium, the medium is exhausted of growth factor after 15 h. Cell growth during this period is negligible. Utilization of growth factor by the cells can be considered to follow first-order kinetics.

a. Does external mass transfer affect the rate of growth factor uptake?

b. Does internal mass transfer affect the rate of growth factor uptake?

c. By what factor would the rate of growth factor consumption be increased in the absence of all mass transfer resistances?

11.15. Maximum reaction rate for immobilized enzyme

Penicillin-G amidase is immobilized in commercial macroporous carrier beads of diameter 100 μm. The immobilized enzyme is used for large-scale penicillin hydrolysis in a well-mixed enzyme reactor that eliminates external boundary layer effects. Penicillin-G substrate is provided as a 2.68-mM solution. The molecular diffusivity of penicillin-G in water is 4.0×10^{-6} cm^2 s^{-1}; the effective diffusivity in the carrier beads is estimated to be 45% of that value. The observed rate of penicillin-G conversion is 125 U cm^{-3} of catalyst, where 1 U is defined as 1 μmol min^{-1}.

a. By what factor could the rate of penicillin-G conversion be increased if internal mass transfer effects were eliminated?

b. The Michaelis constant K_m for freely suspended penicillin-G amidase is 13 μM. In the absence of information about the kinetic parameters after immobilization, this value is assumed to apply to the immobilized enzyme. Estimate the maximum rate of conversion of penicillin-G that could be achieved in the absence of mass transfer effects.

References

[1] S.F. Karel, S.B. Libicki, C.R. Robertson, The immobilization of whole cells: engineering principles, Chem. Eng. Sci. 40 (1985) 1321–1354.

[2] R. Wittler, H. Baumgartl, D.W. Lübbers, K. Schügerl, Investigations of oxygen transfer into *Penicillium chrysogenum* pellets by microprobe measurements, Biotechnol. Bioeng. 28 (1986) 1024–1036.

[3] V. Bringi, B.E. Dale, Experimental and theoretical evidence for convective nutrient transport in an immobilized cell support, Biotechnol. Prog. 6 (1990) 205–209.

[4] A. Nir, L.M. Pismen, Simultaneous intraparticle forced convection, diffusion and reaction in a porous catalyst, Chem. Eng. Sci. 32 (1977) 35–41.

[5] A.E. Rodrigues, J.M. Orfao, A. Zoulalian, Intraparticle convection, diffusion and zero order reaction in porous catalysts, Chem. Eng. Commun. 27 (1984) 327–337.

[6] G. Stephanopoulos, K. Tsiveriotis, The effect of intraparticle convection on nutrient transport in porous biological pellets, Chem. Eng. Sci. 44 (1989) 2031–2039.

[7] K. van't Riet, J. Tramper, Basic Bioreactor Design, Marcel Dekker, 1991.

[8] C.M. Hooijmans, S.G.M. Geraats, E.W.J. van Neil, L.A. Robertson, J.J. Heijnen, K. Ch, A.M. Luyben, Determination of growth and coupled nitrification/denitrification by immobilized *Thiosphaera pantotropha* using measurement and modeling of oxygen profiles, Biotechnol. Bioeng. 36 (1990) 931–939.

[9] C.M. Hooijmans, S.G.M. Geraats, K. Ch, A.M. Luyben, Use of an oxygen microsensor for the determination of intrinsic kinetic parameters of an immobilized oxygen reducing enzyme, Biotechnol. Bioeng. 35 (1990) 1078–1087.

[10] D. de Beer, J.C. van den Heuvel, Gradients in immobilized biological systems, Anal. Chim. Acta. 213 (1988) 259–265.

[11] E.W. Thiele, Relation between catalytic activity and size of particle, Ind. Eng. Chem. 31 (1939) 916–920.

[12] R. Aris, A normalization for the Thiele modulus, Ind. Eng. Chem. Fund. 4 (1965) 227–229.

[13] K.B. Bischoff, Effectiveness factors for general reaction rate forms, AIChE J. 11 (1965) 351–355.

[14] G.F. Froment, K.B. Bischoff, J. De Wilde, Chapter 3: Chemical Reactor Analysis and Design, third ed., John Wiley, 2010.

[15] R. Aris, Introduction to the Analysis of Chemical Reactors, Prentice Hall, 1965.

[16] R. Aris, Mathematical Theory of Diffusion and Reaction in Permeable Catalysts, 1, Oxford University Press, 1975.

[17] M. Moo-Young, T. Kobayashi, Effectiveness factors for immobilized-enzyme reactions, Can. J. Chem. Eng. 50 (1972) 162–167.

[18] P.B. Weisz, Diffusion and chemical transformation: an interdisciplinary excursion, Science. 179 (1973) 433–440.

[19] T.K. Sherwood, R.L. Pigford, C.R. Wilke, Mass Transfer, McGraw-Hill, 1975.

[20] W.L. McCabe, J.C. Smith, P. Harriott, (Section IV)Unit Operations of Chemical Engineering., seventh ed., McGraw-Hill, 2005.

[21] P.L.T. Brian, H.B. Hales, Effects of transpiration and changing diameter on heat and mass transfer to spheres, AIChE J. 15 (1969) 419–425.

[22] W.E. Ranz, W.R. Marshall, Evaporation from drops. Parts I and II, Chem. Eng. Prog. 48 (1952) 141–146.

[23] M. Moo-Young, H.W. Blanch, Design of biochemical reactors: mass transfer criteria for simple and complex systems, Adv. Biochem. Eng. 19 (1981) 1–69.

[24] K. Sato, K. Toda, Oxygen uptake rate of immobilized growing *Candida lipolytica*, J. Ferment. Technol. 61 (1983) 239–245.

[25] J.V. Matson, W.G. Characklis, Diffusion into microbial aggregates, Water Res. 10 (1976) 877–885.

[26] Y.S. Chen, H.R. Bungay, Microelectrode studies of oxygen transfer in trickling filter slimes, Biotechnol. Bioeng. 23 (1981) 781–792.

[27] A. Axelsson, B. Persson, Determination of effective diffusion coefficients in calcium alginate gel plates with varying yeast cell content, Appl. Biochem. Biotechnol. 18 (1988) 231–250.

[28] H.T. Pu, R.Y.K. Yang, Diffusion of sucrose and yohimbine in calcium alginate gel beads with or without entrapped plant cells, Biotechnol. Bioeng. 32 (1988) 891–896.

[29] H. Tanaka, M. Matsumura, I.A. Veliky, Diffusion characteristics of substrates in Ca-alginate gel beads, Biotechnol. Bioeng. 26 (1984) 53–58.

[30] C.D. Scott, C.A. Woodward, J.E. Thompson, Solute diffusion in biocatalyst gel beads containing biocatalysis and other additives, Enzyme Microb. Technol. 11 (1989) 258–263.

[31] T.J. Chresand, B.E. Dale, S.L. Hanson, R.J. Gillies, A stirred bath technique for diffusivity measurements in cell matrices, Biotechnol. Bioeng. 32 (1988) 1029–1036.

[32] E.M. Johnson, D.A. Berk, R.K. Jain, W.M. Deen, Hindered diffusion in agarose gels: test of effective medium model, Biophys. J. 70 (1996) 1017–1026..

[33] S.R. Dziennik, E.B. Belcher, G.A. Barker, M.J. DeBergalis, S.E. Fernandez, A.M. Lenhoff, Nondiffusive mechanisms enhance protein uptake rates in ion exchange particles, Proc. Natl. Acad. Sci. USA. 100 (2003) 420–425.

[34] S.H. Omar, Oxygen diffusion through gels employed for immobilization: Parts 1 and 2, Appl. Microbiol. Biotechnol. 40 (1993) 1–6.

[35] B.J. Rovito, J.R. Kittrell, Film and pore diffusion studies with immobilized glucose oxidase, Biotechnol. Bioeng. 15 (1973) 143–161.

[36] Y.T. Shah, Gas–Liquid–Solid Reactor Design, McGraw-Hill, 1979.

[37] J.J. Carberry, Designing laboratory catalytic reactors, Ind. Eng. Chem. 56 (1964) 39–46.

[38] D.G. Tajbl, J.B. Simons, J.J. Carberry, Heterogeneous catalysis in a continuous stirred tank reactor, Ind. Eng. Chem. Fund. 5 (1966) 171–175.

[39] J.R. Ford, A.H. Lambert, W. Cohen, R.P. Chambers, Recirculation reactor system for kinetic studies of immobilized enzymes, Biotechnol. Bioeng. Symp. 3 (1972) 267–284.

[40] B.K. Hamilton, C.R. Gardner, C.K. Colton, Effect of diffusional limitations on Lineweaver–Burk plots for immobilized enzymes, AIChE J. 20 (1974) 503–510.

[41] J.-M. Engasser, C. Horvath, Effect of internal diffusion in heterogeneous enzyme systems: evaluation of true kinetic parameters and substrate diffusivity, J. Theor. Biol. 42 (1973) 137–155.

[42] D.S. Clark, J.E. Bailey, Structure–function relationships in immobilized chymotrypsin catalysis, Biotechnol. Bioeng. 25 (1983) 1027–1047.

[43] G.K. Lee, R.A. Lesch, P.J. Reilly, Estimation of intrinsic kinetic constants for pore diffusion-limited immobilized enzyme reactions, Biotechnol. Bioeng. 23 (1981) 487–497.

[44] E.E. Petersen, Chemical Reaction Analysis, Prentice Hall, 1965.

[45] P.S. Stewart, C.R. Robertson, Product inhibition of immobilized *Escherichia coli* arising from mass transfer limitation. App, Environ. Microbiol. 54 (1988) 2464–2471.

Uncited references

[22]; [23]

Suggestions for Further Reading

Immobilized Cells and Enzymes

K. Buchholz, V. Kasche, U.T. Bornscheuer, Biocatalysts and Enzyme Technology, Wiley-VCH, 2005.

J.A.M. de Bont, J. Visser, B. Mattiasson, J. Tramper, Physiology of Immobilized Cells, Elsevier, 1990.

J.M. Guisan, Immobilization of Enzymes and Cells, Humana Press, 2010.

E. Katchalski-Katzir, Immobilized enzymes—learning from past successes and failures, Trends Biotechnol. 11 (1993) 471–478.

V. Nedović, R. Willaert, Fundamentals of Cell Immobilisation Biotechnology, Kluwer Academic, 2010.

W. Tischer, V. Kasche, Immobilized enzymes: crystals or carriers? Trends Biotechnol. 17 (1999) 326–335.

C. Webb, G.A. Dervakos, Studies in Viable Cell Immobilization, Academic Press, 1996.

Engineering Analysis of Mass Transfer and Reaction

See also references [1].

J.-M. Engasser, C. Horvath, Diffusion and kinetics with immobilized enzymes, Appl. Biochem. Biotechnol. 1 (1976) 127–220.

C.N. Satterfield, Mass Transfer in Heterogeneous Catalysis, MIT Press, 1970.

PART 5

Bioreactors and Downstream Processes

Reactor Engineering

The bioreactor is the heart of most cell or enzyme bioconversion processes. Designing bioreactors is a complex task, relying on scientific and engineering principles as well as experience-based guidelines. Decisions made in reactor design can have a significant impact on overall process performance, yet there are no simple or standard design procedures available for specifying all aspects of the vessel and its operation. Knowledge of reaction kinetics is essential for understanding how biological reactors work. Application of other areas of bioprocess engineering, such as mass and energy balances, mixing, mass transfer, and heat transfer, are also essential.

The performance of bioreactors depends on several key aspects of their design and operation.

- *Reactor configuration.* Should the reactor be a stirred tank or an air pressure-driven vessel without mechanical agitation?
- *Reactor size.* What size reactor is required to achieve the desired production goals?
- *Mode of operation.* Will the reactor be operated batch-wise or as a continuous process? Should substrate be fed intermittently? Should the reactor be operated alone or in series?
- *Reactor operating parameters.* What reaction conditions, such as temperature, pH, and dissolved O_2 tension, should be maintained in the vessel? How will these parameters be monitored and controlled to optimize process performance? How will contamination be avoided?

This chapter brings together principles covered in previous chapters. Here, we will consider a range of bioreactor configurations and their applications. Mass balance techniques and design equations will be used to predict the outcome of enzyme and cellular reactions for different modes of bioreactor operation. The design of sterilization systems is also included as a critical element of industrial bioprocessing.

12.1 Bioreactor Design Perspective

Defining system objectives is necessary first step for rational bioreactor design. Simple goals like "Produce 1 g of monoclonal antibody per day," or "Produce 10,000 tons of amino acid per year," provide a starting point for the design process. Other objectives are also very relevant: the product should be made at the lowest possible cost to maximize the company's commercial advantage. However, in some cases, economic objectives are overridden by safety concerns, the requirement for high product purity, regulatory agency constraints, or the need to minimize environmental impacts. The final reactor design will reflect all these process requirements and, in most cases, will represent a compromise or tradeoff between conflicting demands.

We will briefly consider bioprocessing considerations for different types of products and examine the importance of reactor engineering in improving process performance and economics. As shown in Figure 12.1, the value of bioprocess products covers a range spanning many orders of magnitude. Identifying which aspects of a process are cost-determining is a first step to manage and reduce the cost of a bioprocess. The breakdown of production costs varies with products; however, a general scheme is shown in Figure 12.2. The following considerations are important:

1. Research and development
2. Upstream processing, including production in bioreactors
3. Downstream processing, including waste treatment or disposal
4. Administration and marketing

Cost-determining factors strongly influence the economic viability of a process (Figure 12.3). Different categories products have different distributions of major costs. Consequently, different strategies are used to improve process economics. New, high-value biotechnology products such as therapeutic recombinant proteins and antibodies have enormous investment

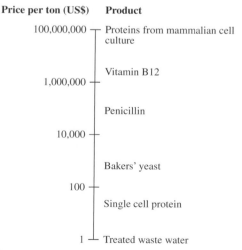

Figure 12.1 Range of value of bioprocess products. *From P.N. Royce, 1993, A discussion of recent developments in fermentation monitoring and control from a practical perspective.* Crit. Rev. Biotechnol. *13, 117–149.*

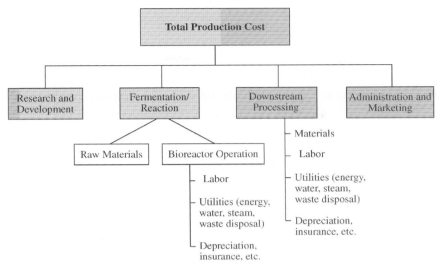

Figure 12.2 Contributions to total production cost in bioprocessing.

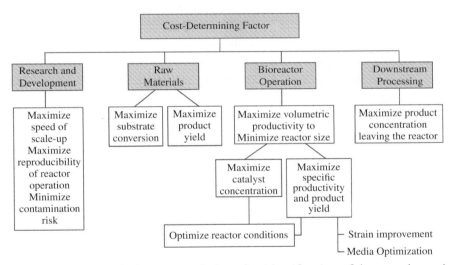

Figure 12.3 Strategies for bioreactor design after identification of the cost-determining factors in the process.

requirements for research and development (1), including approval by government regulatory agencies. Getting the product into the marketplace quickly and safely is the most important cost-saving measure for these products; any savings made by improving the efficiency of the reactor are generally negligible in comparison. If the cost of research and development dominates, design of the reactor is directed toward achieving rapid scale-up; this is more important than maximizing conversion or minimizing operating costs. For new biotechnology products intended for therapeutic use, regulatory guidelines require that the entire production scheme be validated, and process control guaranteed for consistent quality and safety; the reproducibility of reactor operation is therefore critical.

The cost of commodity products is typically dominated by bioreactor costs (2). Examples of commodity products include biomass such as bakers' yeast and single-cell protein, catabolic metabolites such as ethanol and lactic acid, and bioconversion products such as high-fructose corn syrup. The cost of raw materials, the efficiency of the bioconversion, and energy costs for operating the bioreactors as well as the geographical location of the production facility with its associated labor costs can have large impacts on process economics. For relatively low-value, metabolites such as ethanol, citric acid, biomass, and lactic acid, raw material costs range from 40% of the cost of reaction for citric acid to about 70% for ethanol produced from molasses [1,2]. When the cost of raw materials is significant, maximizing the conversion of substrate and yield of product in the reactor has high priority. When bioreactor operation dominates the cost structure, the reactor should be as small as possible as this reduces both the capital cost of the equipment and the cost of day-to-day operations.

Intracellular products and specialty products intended for human medical applications such as antibiotics, vitamins, amino acids, therapeutic proteins, and monoclonal antibodies have high downstream processing costs (3) relative to upstream costs. The purity of the products is essential to their safety and ultimately for their marketability. If downstream processing is expensive, the reactor is designed and operated to maximize the product concentration leaving the vessel; this cuts the cost of recovering the product from dilute solutions. While not trivial, upstream considerations like substrate costs are not as important because of the high final selling price.

The cost of administration and marketing (4) is often relatively small relative to the other costs for most bioproducts. Although, there are exceptions. Markets for food products are often very competitive and marketing costs for bioproducts like yogurt or cheese can be substantial. Nevertheless, for most bioprocess products outside of the new, high-value medical category, production costs represent a substantial contribution to the final price. Even if bioreactor operation itself is not cost-determining, aspects of reactor design may still be important [1].

12.2 Bioreactor Configurations

The cylindrical tank, either stirred or unstirred, is the most common reactor in bioprocessing. Yet, many different bioreactor configurations are used in industry (Figure 12.4). Much of the challenge in reactor design lies in the provision of adequate mixing and aeration for O_2 requiring bioprocesses. Figure 12.5 depicts different strategies for sparging and aeration of bioreactors. In contrast, reactors for anaerobic culture are often relatively simple in construction without sparging or agitation. In the following discussion of bioreactor configurations, we will assume that aerobic operation is a requirement.

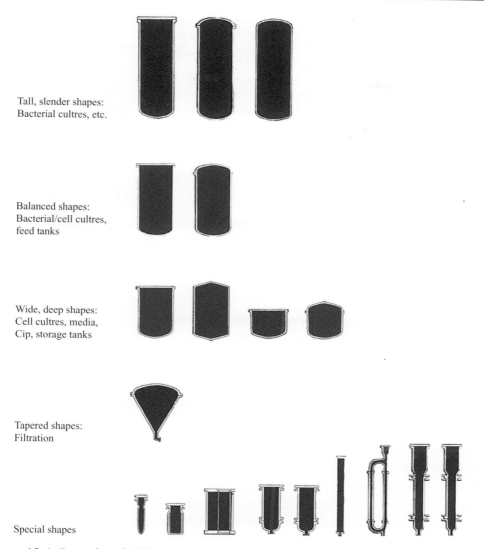

Tall, slender shapes:
Bacterial cultres, etc.

Balanced shapes:
Bacterial/cell cultres,
feed tanks

Wide, deep shapes:
Cell cultres, media,
Cip, storage tanks

Tapered shapes:
Filtration

Special shapes

Figure 12.4 Examples of different bioreactor configurations. *From L. Hasler, R. Butz, and P. Grad, 2006, Transparency, Form and Function: Fermenter Manufacturing-Art. DreiPunktVerlag. Germany. Courtesy of Bioengineering, Bioengineering AG, 8636 Wald, Switzerland.*

12.2.1 Stirred Tank Reactor

A conventional stirred, aerated bioreactor is shown schematically in Figure 12.6. Mixing and bubble dispersion are achieved by mechanical agitation; this requires a relatively high input of energy per unit volume. Baffles are used in stirred reactors to reduce vortexing. A wide variety of impeller sizes and shapes is available to produce different flow patterns inside the

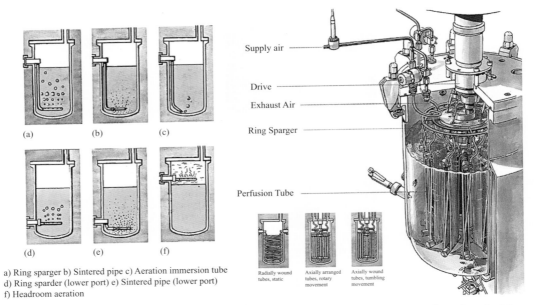

a) Ring sparger b) Sintered pipe c) Aeration immersion tube
d) Ring sparder (lower port) e) Sintered pipe (lower port)
f) Headroom aeration

Supply air

Drive

Exhaust Air

Ring Sparger

Perfusion Tube

Radially wound tubes, static

Axially arranged tubes, rotary movement

Axially wound tubes, tumbling movement

Figure 12.5 Various sparger and aeration designs for bioreactors (left panel) and an inside look at a ring sparger with perfusion tubes (right). *From L. Hasler, R. Butz, and P. Grad, 2006, Transparency, Form and Function: Fermenter Manufacturing-Art. DreiPunktVerlag. Germany. Courtesy of Bioengineering, Bioengineering AG, 8636 Wald, Switzerland.*

vessel. The mixing and mass transfer functions of stirred reactors are described in detail in Chapters 7 and 9.

The aspect ratio of stirred vessels (i.e., the ratio of height to diameter) can be varied over a wide range. The least expensive shape to build has an aspect ratio of about 1; this shape has the smallest surface area and therefore requires the least material to construct for a given volume. However, when aeration is required, the aspect ratio is usually increased. This provides for longer contact times between the rising bubbles and the liquid and produces a greater hydrostatic pressure at the bottom of the vessel for increased O_2 solubility.

Typically, only 70% to 80% of the volume of stirred reactors is filled with liquid. This allows adequate headspace for disengagement of liquid droplets from the exhaust gas and to accommodate any foam that may develop. Foam can occur in bioprocesses due to the introduction of gases into the culture medium and is further stabilized by proteins produced by organisms in the culture itself. If foaming is a problem, a supplementary impeller called a *foam breaker* may be installed as shown in Figure 12.4 [3]. Chemical antifoam agents are also used; however, because antifoams reduce the rate of O_2 transfer they are used sparingly (Section 9.6.3).

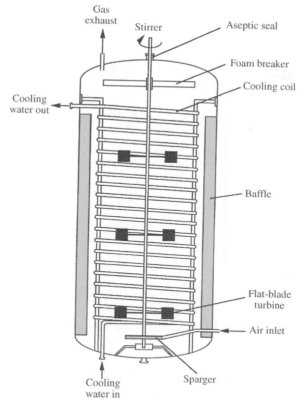

Figure 12.6 Typical stirred tank bioreactor for aerobic culture.

Temperature control and heat transfer in stirred vessels can be accomplished using internal cooling coils as shown in Figure 12.6 or other strategies. Important mass transfer considerations associated with different mixing strategies can be found in Chapter 7.

12.2.2 Bubble Column Reactor

Alternatives to the stirred reactor include vessels with no mechanical agitation. Bubble column reactors use gas sparging and viscous drag to achieve aeration and mixing: this strategy requires considerably less energy than mechanical stirring. Bubble column reactors are used industrially for production of bakers' yeast, beer, and vinegar, and for treatment of wastewater.

Bubble column reactors are structurally simple. They are generally cylindrical vessels with height greater than twice the diameter as shown in Figure 12.7. Other than a sparger for entry of compressed air, bubble columns typically have no internal structures. A height-to-diameter

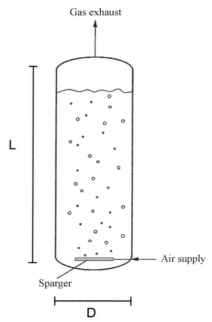

Figure 12.7 Bubble column bioreactor.

ratio of about 3:1 is common in bakers' yeast production; for other applications, towers with height-to-diameter ratios of 6:1 have been used. Perforated horizontal plates are sometimes installed in tall bubble columns to break up and redistribute coalesced bubbles. The advantages of bubble columns include low capital cost, lack of moving parts, and satisfactory heat and mass transfer performance. As in stirred vessels, foaming can be a problem requiring mechanical dispersal or addition of antifoam to the medium.

Bubble column hydrodynamics and mass transfer characteristics depend entirely on the behavior of the bubbles released from the sparger. Different flow regimes occur depending on the gas flow rate, sparger design, column diameter, and medium properties such as viscosity. *Homogeneous flow* occurs only at low gas flow rates and when bubbles leaving the sparger are evenly distributed across the column cross-section. In homogeneous flow, all bubbles rise with the same upward velocity and there is little or no back mixing of the gas phase. Liquid mixing in this flow regime is limited, as it arises solely from entrainment in the wakes of the bubbles. In contrast, under normal operating conditions at higher gas velocities, large chaotic circulatory flow cells develop and *heterogeneous flow* occurs as illustrated in Figure 12.8. In this regime, bubbles and liquid tend to rise up the center of the column while a corresponding downflow of liquid occurs near the walls. This liquid circulation entrains bubbles so that some back mixing of gas occurs.

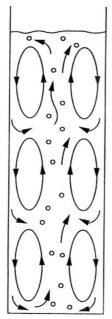

Figure 12.8 Heterogeneous flow in a bubble column.

The liquid mixing time in bubble columns depends on the flow regime. For heterogeneous flow, the following equation has been proposed [4] for the upward liquid velocity at the center of the column for $0.1 < D < 7.5\,\text{m}$ and $0 < u_G < 0.4\,\text{m s}^{-1}$:

$$u_L = 0.9\left(g\,D\,u_G\right)^{0.33} \tag{12.1}$$

where u_L is the linear liquid velocity, g is gravitational acceleration, D is the column diameter, and u_G is the gas superficial velocity. u_G is equal to the volumetric gas flow rate at atmospheric pressure divided by the reactor cross-sectional area. From this equation for u_L, an expression for the mixing time t_m (Chapter 7) in bubble columns can be obtained [5]:

$$t_m = 11\frac{H}{D}\left(g\,u_G\,D^{-2}\right)^{-0.33} \tag{12.2}$$

where H is the height of the bubble column.

The gas–liquid mass transfer coefficient in bioreactors depends largely on the diameter of the bubbles and the gas hold-up, as discussed in Chapter 9. In bubble columns containing inviscid liquids, these variables depend solely on the gas flow rate. The following correlation has been proposed for inviscid media in heterogeneous flow [4,5]:

$$k_L a \approx 0.32\,u_G^{0.7} \tag{12.3}$$

where $k_L a$ is the combined volumetric mass transfer coefficient and u_G is the gas superficial velocity. Equation (12.3) is valid for bubbles with mean diameter of about 6 mm [4].

12.2.3 Airlift Reactor

Airlift reactors achieve mixing without mechanical agitation, similar to bubble column reactors. A distinguishing feature between airlift and bubble column reactors is the more defined liquid flow patterns in airlift reactors. The upflowing and downflowing streams are physically separated as shown in Figure 12.9. The gas is sparged into a vessel feature called the *riser*. The resulting gas hold-up and decreased fluid density cause liquid in the riser to move upward. As gas bubbles disengage from the liquid at the top of the riser, heavier bubble-free liquid recirculates through the *downcomer*. Thus, liquid circulation in airlift reactors results from the density difference between the riser and downcomer.

Airlift reactors have been applied for production of single-cell protein from methanol, in municipal and industrial waste treatment, and for plant and animal cell culture. Large airlift reactors with capacities ranging up to thousands of cubic meters have been constructed. Tall internal loop airlifts built underground are known as *deep shaft reactors*; the very high

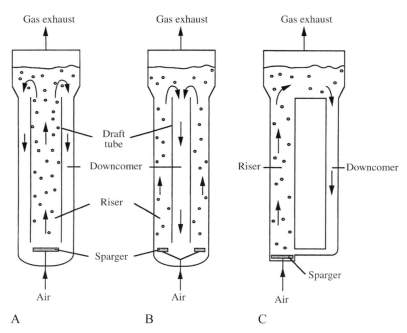

Figure 12.9 Airlift reactor configurations: (A) and (B) internal loop vessels; (C) external loop airlift.

hydrostatic pressure at the bottom of these vessels improves gas–liquid mass transfer considerably. The height of airlift reactors is typically about 10 times the diameter; for deep shaft systems, the height-to-diameter ratio may be increased up to 100.

The most common airlift configurations are illustrated in Figure 12.9. *Internal loop vessels* have the riser and downcomer and are separated by an internal baffle or *draft tube*; air may be sparged into either the draft tube or the annulus, as shown in Figure 12.9(A) and (B). The *external loop* or *outer loop* airlift configuration is shown in Figure 12.9(C); separate vertical tubes are connected by short horizontal sections at the top and bottom. Because the riser and downcomer are further apart in external loop vessels, gas disengagement is more effective than in internal loop devices. Fewer bubbles are carried into the downcomer, the density difference between fluids in the riser and downcomer is greater, and circulation of liquid in the vessel is faster.

Airlift reactors generally provide better mixing than bubble columns, except at low liquid velocities. Compared with bubble columns, the airlift configuration confers a degree of stability to the liquid flow patterns produced; therefore, higher gas flow rates can be used without incurring operating problems such as slug flow or spray formation. Several empirical correlations have been developed for calculation of the liquid velocity, circulation time, and mixing time in airlift reactors [6–11]. A following relationship for the mass transfer coefficient for external loop airlifts has been proposed:

$$k_L a < 0.32 u_G^{0.7} \tag{12.4}$$

Several other empirical mass transfer correlations have been developed for Newtonian and non-Newtonian fluids in airlift reactors [6].

12.2.4 Stirred and Air-Driven Reactors: Comparison of Operating Characteristics

Adequate mixing and mass transfer can be achieved in stirred tanks, bubble columns, and airlift vessels for low-viscosity fluids. When a large bioreactor (50–500 m³) is required for low-viscosity culture, a bubble column is an attractive choice because it is relatively simple and inexpensive to install and operate. Mechanically agitated reactors are impractical at volumes greater than about 500 m³ as the power required to achieve adequate mixing becomes prohibitive.

Stirred vessels are more suitable for viscous liquids because greater power can be input by mechanical agitation. Nevertheless, mass transfer rates decline rapidly in stirred vessels at viscosities over 50 to 100 cP [5]. Air-driven reactors are not effective at generating reasonable mass transfer rates with high-viscosity fluids.

Heat transfer can be another important consideration in the choice between air-driven and stirred reactors. Mechanical agitation generates more heat than sparging of compressed gas. When the heat of reaction is high, such as for production of single-cell protein from methanol, the removal of additional heat from stirrer friction can be problematic, so air-driven reactors may be preferred.

Stirred tank and air-driven vessels account for most bioreactors used with aerobic cultures. However, other reactor configurations are used in different applications.

12.2.5 Packed Bed

Packed bed reactors are used with immobilized or particulate biocatalysts. The reactor consists of a tube, usually vertical, packed with catalyst particles. Medium can be fed at either the top or bottom of the column and forms a continuous liquid phase between the particles. Packed bed reactors have been applied commercially with immobilized cells and enzymes for production of aspartate and fumarate, conversion of penicillin to 6-aminopenicillanic acid, and resolution of amino acid isomers.

Mass transfer between the liquid medium and solid catalyst is facilitated by high liquid flow rates through the bed. Packed beds are often operated with liquid recycle as shown in Figure 12.10. The catalyst is prevented from leaving the column by screens at the liquid outlet. The particles used are relatively incompressible to withstand their own weight in the column without deforming and occluding the liquid flow. Aeration is generally accomplished in a separate vessel; if air is sparged directly into the bed, bubble coalescence produces gas

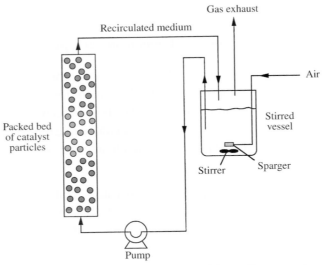

Figure 12.10 Packed bed reactor with medium recycle.

pockets between the particles and flow *channeling* or maldistribution occurs. Packed beds are unsuitable for processes that produce large quantities of carbon dioxide or other gases that can become trapped in the packing.

12.2.6 Fluidized Bed

Packed beds operated in upflow mode with catalyst beads of appropriate size and density can expand at high liquid flow rates due to the upward motion of the particles. These are known fluidized bed reactors and illustrated in Figure 12.11. The particles in fluidized beds are in constant motion, so that air can be introduced directly into the column without clogging the bed or causing flow channeling. Fluidized bed reactors are used in waste treatment with sand or similar material supporting mixed microbial populations. They are also used with flocculating organisms in the brewing industry and to produce vinegar.

12.2.7 Trickle Bed

The trickle bed reactor is another variation of the packed bed. The liquid is sprayed onto the top of the stationary packing and trickles down through the bed in small rivulets as illustrated in Figure 12.12. Air can be introduced at the base of the column because the liquid phase is not continuous. Gases move with relative ease around the packing. Trickle bed reactors are used widely for aerobic wastewater treatment.

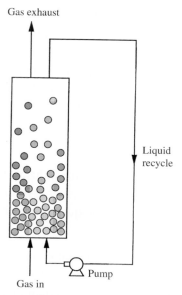

Figure 12.11 Fluidized bed reactor.

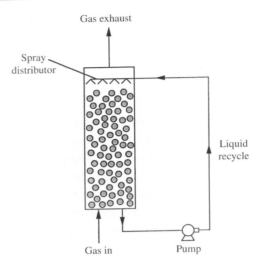

Figure 12.12 Trickle bed reactor.

12.3 Practical Considerations for Bioreactor Construction and Operation

Industrial bioreactors for sterile operation are usually designed as steel pressure vessels capable of withstanding full vacuum and up to about 3 atm of positive pressure at 150°C to 180°C. A hole is provided at the top of large vessels to allow workers entry into the tank for cleaning and maintenance; on smaller vessels, the top is removable (Figure 12.13). Flat headplates are commonly used with laboratory-scale bioreactors; for larger vessels, a domed construction is less expensive. Large bioreactors are equipped with a lighted vertical sight-glass for inspecting the contents of the reactor. Nozzles for medium, antifoam, and acid and alkali addition, air exhaust pipes, a pressure gauge, and a rupture disc for emergency pressure release are normally located on the headplate. Ports for pH, temperature, and dissolved O_2 sensors are a minimum requirement; a steam-sterilizable sample outlet should also be provided. The vessel must be fully draining via a harvest nozzle located at the lowest point of the reactor.

12.3.1 Aseptic Operation

Most bioprocesses outside of the food and beverage industry are carried out using pure cultures and aseptic conditions. Keeping the bioreactor free of unwanted organisms is especially important for slow-growing cultures that can be quickly overrun by contamination or for producing compounds with medical uses.

Typically, 3% to 5% of bioprocesses in an industrial plant are lost due to failure of sterilization procedures [12]. Higher contamination rates occur in processes using complex, nutrient-rich media and relatively slow-growing cells such as mammalian cells; a contamination rate

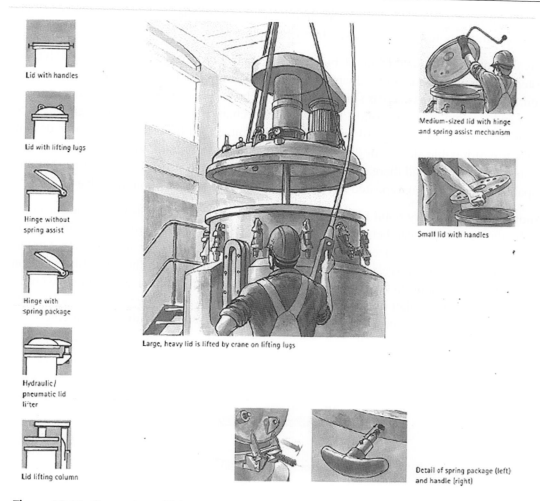

Lid with handles

Lid with lifting lugs

Hinge without spring assist

Hinge with spring package

Hydraulic / pneumatic lid lifter

Lid lifting column

Medium-sized lid with hinge and spring assist mechanism

Small lid with handles

Large, heavy lid is lifted by crane on lifting lugs

Detail of spring package (left) and handle (right)

Figure 12.13 Illustration of lid access configurations for bioreactors. *From L. Hasler, R. Butz, and P. Grad, 2006, Transparency, Form and Function: Fermenter Manufacturing-Art. DreiPunktVerlag. Germany. Courtesy of Bioengineering, Bioengineering AG, 8636 Wald, Switzerland.*

of 17% has been reported for industrial-scale production of β-interferon from human fibroblasts [13].

Industrial bioreactors are designed for *in situ* steam sterilization under pressure. The vessel should have a minimum number of internal structures such as ports or nozzles to ensure that steam reaches all parts of the equipment. The reactor should be free of crevices and stagnant areas where liquid or solids can accumulate; for this reason, polished welded joints are used in preference to other coupling methods. After sterilization, all nutrient medium and air entering the bioreactor must be sterile. Filters preventing the passage of microorganisms are fitted to exhaust gas lines; this serves to contain the culture inside the bioreactor and insures against contamination should there be a drop in operating pressure.

The flow of liquids to and from the bioreactor is controlled using valves. Because valves are a potential entry point for contaminants, their construction must be suitable for aseptic operation. Common designs such as simple gate and globe valves tend to leak around the valve stem and accumulate broth solids in the closing mechanism. Pinch and diaphragm valves such as those shown in Figures 12.14 and 12.15 are recommended for bioreactors. These designs make use of flexible sleeves or diaphragms so that the closing mechanism is isolated from the contents of the pipe and there are no dead spaces in the valve structure. Rubber or neoprene capable of withstanding repeated sterilization cycles is used to fashion the valve closure.

Another potential entry point for contamination with stirred reactors is where the stirrer shaft enters the vessel. The gap between the rotating stirrer shaft and the bioreactor body must be sealed. Several types of stirrer seals have been developed to prevent contamination. On large bioreactors, mechanical seals are commonly used where one part of the assembly is stationary while the other rotates on the shaft [14]. The precision-machined surfaces of the two components are pressed together by springs or expanding bellows and cooled and lubricated with water. Stirrer seals are especially critical if the reactor is designed with a bottom-entering

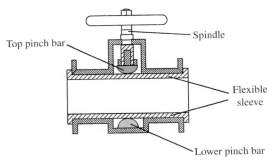

Figure 12.14 Pinch valve.

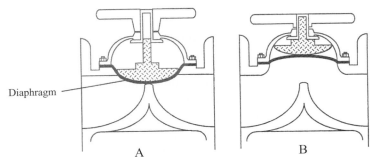

Figure 12.15 Weir-type diaphragm valve in (A) closed and (B) open positions.

stirrer; in this case, double mechanical seals may be installed to prevent fluid leakage. On smaller vessels, magnetic drives can be used to couple the stirrer shaft with the motor; with these devices, the stirrer shaft does not pierce the bioreactor body. A magnet in a housing on the outside of the bioreactor is driven by the stirrer motor; inside, another magnet is attached to the end of the stirrer shaft and held in place by bearings. Sufficient power can be transmitted using magnetic drives to agitate vessels up to 800 L in size [15].

12.3.2 Bioreactor Inoculation and Sampling

Consideration must be given in the design of bioreactor to aseptic inoculation and sample removal. Inocula for large-scale bioprocesses are transferred from smaller reactors: to prevent contamination during this operation, both vessels are maintained under positive air pressure. The simplest aseptic transfer method is to pressurize the inoculum vessel using sterile air: culture is then effectively pushed into the larger bioreactor. An example of the pipe and valve connections required for this type of transfer is shown in Figure 12.16.

Sampling ports are fitted to bioreactor to allow removal of broth for analysis. An arrangement for sampling that preserves aseptic operation is shown in Figure 12.17. Figure 12.18 demonstrates the use of a pierceable septum for inoculation or sterile medium addition to a bioreactor.

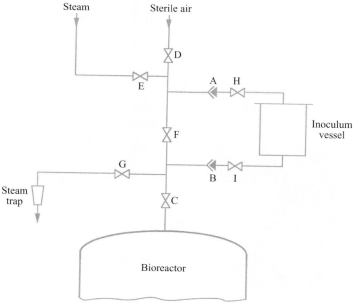

Figure 12.16 Pipe and valve connections for aseptic transfer of inoculum to a large-scale bioreactor. *From A. Parker, 1958, Sterilization of equipment, air and media. In: R. Steel (Ed.), Biochemical Engineering, pp. 97–121, Heywood, London.*

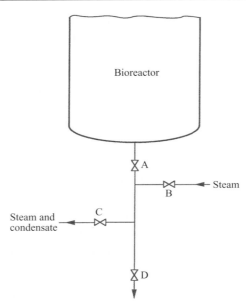

Figure 12.17 Pipe and valve connections for a simple bioreactor sampling port.

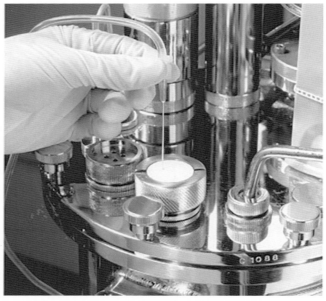

Figure 12.18 Sterile inoculation of a bioreactor using a septum system. Photo from QualiTru Sampling Systems (formerly known as QMI), Used with permission.

12.3.3 Materials of Construction

Bioreactors are constructed from materials that can withstand repeated steam sterilization, pressurization, and cleaning cycles. Materials contacting the culture media should also be nonreactive and nonabsorptive. Glass is used to construct small bioreactors with volumes $\leq 30\,L$. The advantages of glass are that it is smooth, nontoxic, corrosion-proof, and transparent for easy inspection of the vessel contents. Entry ports for adding medium, inoculum, and air, as well as instruments such as pH and temperature sensors, are usually designed into stainless steel headplates for the glass vessels.

Pilot- and large-scale bioreactors are often made of corrosion-resistant stainless steel. Interior steel surfaces are polished to a bright "mirror" finish to facilitate cleaning and sterilization of the reactor; welds on the interior of the vessel are ground flush before polishing.

12.3.4 Impeller, Baffle, and Sparger Design

The impeller and baffle designs determine the effectiveness of mixing and gas mass transfer in stirred bioreactors. Their design features are described in Chapter 7. In air-driven bioreactors without mechanical stirring, the sparger plays a direct role in achieving mixing and O_2 mass transfer. Several sparger configurations used in bioreactors are described in Chapters 7 and 9, and previously in Section 12.2.

12.3.5 Evaporation Control

Aerobic cultures are sparged continuously with air. Most components of air are inert and leave directly through the exhaust gas line. However, when dry air is used to sparge a bioreactor, water is stripped from the medium and leaves as vapor in the off-gas and evaporative water losses can be significant. The higher the air flow rate the larger the problem.

Air sparged into bioreactors can be prehumidified by bubbling through columns of water outside the bioreactor to minimize evaporative loss. Humid air has less capacity for removing water from the bioreactor than dry air. Bioreactors are also equipped with water-cooled condensers that condense vapors entrained in the off-gas stream and return them to the vessel [16].

12.4 Monitoring and Control of Bioreactors

The environment inside bioreactors should allow optimal catalytic activity. Parameters such as temperature, pH, dissolved O_2 concentration, medium flow rate, stirrer speed, and sparging rate have a significant effect on the outcome of cultivations. Bioreactor properties are often monitored and control action taken to rectify any deviations from the desired values.

12.4.1 Bioreactor Monitoring

Controlling bioprocesses requires knowledge of system variables that affect the process. These parameters can be grouped into three categories: physical, chemical, and biological; examples of typical process variables are given in Table 12.1. Many of the physical measurements listed are well established in the bioprocess industry; others are the focus of research efforts.

The time required to complete a measurement should be consistent with the rate of change of the variable for effective control of cell cultures being monitored. For example, in a typical bioreactor, the time scale for changes in pH and dissolved O_2 tension is several minutes. Ideally, measurements should be made *in situ* and online so that the result is available for timely control action. Examples of measurements that can be made online in industrial bioprocesses are temperature, pressure, pH, dissolved O_2 tension, flow rate, stirrer speed, power consumption, foam level, broth weight, and gas composition. Figure 12.19 illustrates the online interaction of a particular biological molecule with a biosensor that generates a measurable electrical signal detected by the transducer. Figure 12.20 shows typical results from online measurement of dilution rate and carbon dioxide evolution in a production-scale bioreactor. In many cases, *signal conditioning* or *smoothing* must be carried out to reduce the noise in data before they can be applied for process control or modeling. Instruments for taking online measurements are relatively commonplace and detailed descriptions can be found elsewhere [17–19].

Many important variables, such as biomass concentration and broth composition, are generally not measured online because of the lack of appropriate instruments. Instead, samples

Table 12.1: Parameters Measured or Controlled in Bioreactors

Physical	Chemical	Biological
Temperature	pH	Biomass concentration
Pressure	Dissolved O_2 tension	Enzyme concentration
Reactor weight	Dissolved CO_2 concentration	Biomass composition (DNA, RNA, protein, lipid, carbohydrate, ATP/ADP/AMP, NAD/NADH levels, etc.)
Liquid level	Redox potential	
Foam level	Exit gas composition	
Agitator speed	Conductivity	
Power consumption	Broth composition (substrate, product, ion concentrations, etc.)	
Gas flow rate	Viability	
Medium flow rate	Morphology	
Culture viscosity		
Gas hold-up		

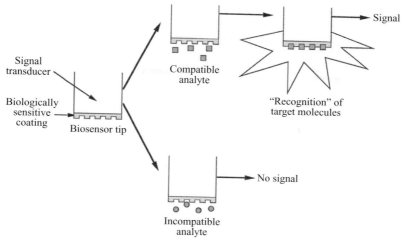

Figure 12.19 Operating principle for biosensors. *From E.A.H. Hall, 1991, Biosensors, Prentice Hall, Englewood Cliffs, NJ.*

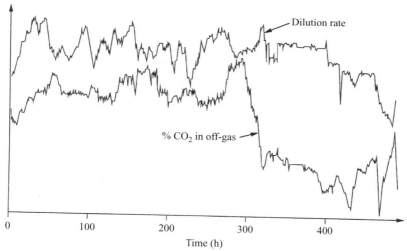

Figure 12.20 Online measurements of dilution rate and off-gas carbon dioxide during an industrial mycelial bioprocess. *From G.A. Montague, A.J. Morris, and A.C. Ward, 1989, Fermentation monitoring and control: a perspective.* Biotechnol. Genet. Eng. Rev. *7, 147–188.*

are removed from the reactor for offline analysis. Control actions based on time-consuming laboratory measurements are not as effective as when online results are used.

Another approach to online chemical and biological measurements involves the use of automatic sampling devices linked to analytical equipment for high-performance liquid

Table 12.2: Reliability of Bioreactor Equipment

Equipment	Reliability	Mean time between failures (weeks)
Temperature probe	–	150–200
pH probe	98%	9–48
Dissolved O_2 probe	50%–80%	9–20
Mass spectrometer	–	10–50
Paramagnetic O_2 analyzer	–	24
Infrared CO_2 analyzer	–	52

From S.W. Carleysmith, 1987, Monitoring of bioprocessing, In: J.R. Leigh (Ed.), *Modelling and Control of Fermentation Processes*, pp. 97–117, Peter Peregrinus, London; and P.N. Royce, 1993, A discussion of recent developments in fermentation monitoring and control from a practical perspective, *Crit. Rev. Biotechnol.* 13, 117–149.

chromatography (HPLC), image analysis, nuclear magnetic resonance (NMR) spectroscopy, flow cytometry, or fluorometry [19].

The reliability and frequency of failure of some bioreactor instruments are listed in Table 12.2. Further details about online measurement techniques can be found elsewhere [20–23].

12.4.2 Process Modeling

Modern approaches to bioreactor control require a reasonably accurate mathematical model of the reaction and reactor environment. Models are mathematical relationships between variables. Traditionally, models are based on a combination of "theoretical" relationships that provide the structure of the model, and experimental observations that provide the numerical values of coefficients. As an example, a frequently used mathematical model for batch cultivation consists of the Monod equation for growth and an expression for the rate of substrate consumption as a function of biomass concentration:

$$\frac{dx}{dt} = \mu x = \frac{\mu_{max} s \, x}{K_S + s} \tag{12.5}$$

$$\frac{-ds}{dt} = \frac{\mu x}{Y_{xs}} + m_s x \tag{12.6}$$

This model represents a combination of Eqs. (10.80), (10.91), and (10.103). The form of the equations was determined from experimental observation of many different culture systems. In principle, once values of the parameters μ_{max}, K_S, $Y_{X/S}$, and m_S are determined, we can use the model to estimate the cell concentration x and substrate concentration s as a function of time. A common problem with bioreactor models is that the model parameters can be difficult to measure, or they tend to change with time during the reaction process. As a simple example, online measurements of carbon dioxide in bioreactor off-gas can be applied with an appropriate process model to estimate the biomass concentration during penicillin production

[24–27]. As shown in Figure 12.21, the results were satisfactory for 100-m³ fed-batch cultivation over a period of more than 8 days. However, if major fluctuations occur in operating conditions, or if cell properties and model parameters change with time, the model may become inadequate and the accuracy of the estimation will decline [28].

Development of a comprehensive model covering all key aspects of a particular bioprocess and capable of predicting the effects of a wide range of culture variables is a demanding exercise. Accurate models applicable to a range of process conditions are rare.

12.4.3 Feedback Control

Let us assume that we wish to maintain the pH in a bioreactor at a constant value against a variety of disturbances, for example, different metabolic activities. One of the simplest control schemes is a conventional *feedback control loop*, the basic elements of which are shown in Figure 12.22. A measurement device senses the value of the pH and sends the signal to a

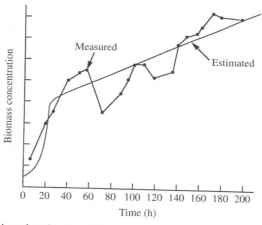

Figure 12.21 Measured and estimated biomass concentrations during a large-scale fed-batch penicillin cultivation. *From G.A. Montague, A.J. Morris, and J.R. Bush, 1988, Considerations in control scheme development for fermentation process control. IEEE Contr. Sys. Mag. 8, April, 44–48.*

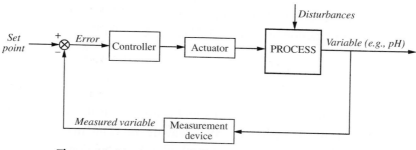

Figure 12.22 Components of a feedback control loop.

controller. The measured value is compared with the desired value known as the *set point*. The difference between the measured and desired values is the *error*, which is used by the controller to determine what action must be taken to correct the process. The controller is an automatic electronic, pneumatic, or computer device. The controller produces a signal that is transmitted to the actuator, which executes the control action.

In a typical system for pH control, an electrode serves as the measurement device and a pump connected to a reservoir of acid or alkali serves as the actuator. Simple *on–off control* is generally sufficient for pH; if the measured pH falls below the set point, the controller switches on the pump that adds alkali to the bioreactor. When enough alkali is added and the pH returned to the set value, the pump is switched off. Small deviations from the set point are usually tolerated in on–off control to avoid rapid switching and problems due to measurement delay.

If the actuator can function over a continuous range, such as a variable-speed pump for supply of cooling water, it is common to use *proportional-integral-derivative* or *PID control*. The control action is determined in proportion to the error, the integral of the error, and the derivative of the error with respect to time. The relative weightings given to these functions determine the response of the controller and the overall "strength" of the control action. Proper tuning of PID controllers usually provides excellent regulation of measured variables. Considerations for adjusting PID controllers are covered in texts on process control, for example [29–36].

12.5 Ideal Reactor Operation

This chapter has so far considered different bioreactor types, aspects of their construction, and control considerations. Another important aspect of reactor performance is their mode of operation. There are three principal modes of bioreactor operation: batch, fed-batch, and continuous. The choice of operating mode has a significant effect on substrate conversion, product concentration, susceptibility to contamination, and process reliability.

Properties such as the final biomass, substrate, and product concentrations and the time required for conversion can be predicted for different reactor operating schemes using mass balances. The rate of change of a component mass in a reactor can be related to other system rates including the rate of reaction using Eq. (4.29) from Chapter 4:

$$\frac{dM}{dt} = \hat{M}_i - \hat{M}_o + R_G - R_C \tag{4.29}$$

where M is the mass of component A in the vessel, t is time, $\hat{M}_i$ is the mass flow rate of A entering the reactor, $\hat{M}_o$ is the mass flow rate of A leaving, R_G is the mass rate of generation of A by reaction, and R_C is the mass rate of consumption of A by reaction.

12.5.1 Batch Operation of a Mixed Bioreactor

Batch processes operate in closed systems; substrates are added at the beginning of the process and products are removed only at the end. Aerobic reactions are not batch operations in the strictest sense of the definition; the low solubility of O_2 in aqueous media means that it must be supplied continuously while carbon dioxide and other off-gases are removed continuously. However, bioreactors with neither input nor output of liquid or solid material are classified as batch reactors. The liquid volume in batch reactors can be considered constant if there are no leaks or evaporation from the vessel.

Most commercial bioreactors are mixed vessels operated in batch. The classic mixed reactor is the stirred tank; however, mixed reactors can also be of bubble column, airlift, or other configurations if the concentrations of substrate, product, and catalyst inside the vessel are uniform. The cost of running a batch reactor depends on the time taken to achieve the desired product concentration or level of substrate conversion; operating costs are reduced if the reaction is completed quickly. It is therefore useful to be able to design processes with competitive properties like the short batch times.

Enzyme Reaction

Let us apply Eq. (4.29) to the limiting substrate in a batch enzyme reactor such as that shown in Figure 12.23. $\hat{M}_i = \hat{M}_0 = 0$ because there is no substrate flow into or out of the vessel. The mass of substrate in the reactor, M, is equal to the substrate concentration s multiplied by the liquid volume V. As substrate is not generated in the reaction, $R_G = 0$. The rate of substrate consumption R_C is equal to the volumetric rate of enzyme reaction v multiplied by V;

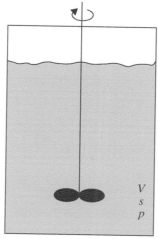

Figure 12.23 Flow sheet for a stirred batch enzyme reactor.

for Michaelis–Menten kinetics, v is given by Eq. (10.37). Therefore, from Eq. (4.29), the mass balance is:

$$\frac{d(sV)}{dt} = \frac{-v_{max}s}{K_m + s} V \tag{12.7}$$

where v_{max} is the maximum rate of enzyme reaction and K_m is the Michaelis constant (Section 10.3.3). Because V is constant in batch reactors, we can take V outside of the differential operator and cancel it from both sides of the equation to give:

$$\frac{ds}{dt} = \frac{-v_{max}s}{K_m + s} \tag{12.8}$$

Integration of this differential equation provides an expression for the batch reaction time. Assuming that v_{max} and K_m are constant during the reaction, separating variables gives:

$$-\int dt = \int \frac{K_m + s}{v_{max}s} ds \tag{12.9}$$

Integrating with initial condition $s = s_0$ at $t = 0$ gives:

$$t_b = \frac{K_m}{v_{max}} \ln \frac{s_0}{s_f} + \frac{s_0 - s_f}{v_{max}} \tag{12.10}$$

where t_b is the batch reaction time required to reduce the substrate concentration from s_0 to s_f. The batch reaction time required to produce a certain concentration of product can be determined from Eq. (12.10) and stoichiometric relationships. as seen in Figure 12.24. Application of Eq. (12.10) is demonstrated in Example 12.1.

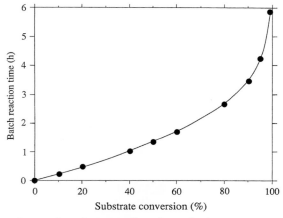

Figure 12.24 Batch reaction time as a function of substrate conversion for enzyme reaction in a mixed vessel.

EXAMPLE 12.1 Time Course for Batch Enzyme Conversion

An enzyme is used to produce a compound used in the manufacture of sunscreen lotion. v_{max} for the enzyme is 2.5 mmol m^{-3} s^{-1}; K_m is 8.9 mM. The initial concentration of substrate is 12 mM. Plot the time required for batch reaction as a function of substrate conversion.

Solution

$s_0 = 12$ mM. Converting the units of v_{max} to mM h^{-1}:

$$v_{max} = 2.5 \text{ mmol m}^{-3}\text{s}^{-1} \cdot \left| \frac{3600 \text{ s}}{1 \text{ h}} \right| \cdot \left| \frac{1 \text{ m}^3}{1000 \text{ L}} \right| = 9 \text{ mmolL}^{-1}\text{h}^{-1} = 9 \text{ mM h}^{-1}$$

The results from application of Eq. (12.10) are tabulated on the next page and plotted in Figure 12.20.

Substrate conversion (%)	s_f (mM)	t_b (h)
0	12.0	0.00
10	10.8	0.24
20	9.6	0.49
40	7.2	1.04
50	6.0	1.35
60	4.8	1.71
80	2.4	2.66
90	1.2	3.48
95	0.60	4.23
99	0.12	5.87

At high substrate conversions, the time required to achieve an incremental increase in conversion is greater than at low conversions. Accordingly, the benefit gained from conversions above 80 to 90% must be weighed against the significantly greater reaction time, and therefore reactor operating costs, involved.

Enzymes are subject to deactivation as discussed in Section 10.6. Accordingly, the concentration of active enzyme in the reactor, and therefore the value of v_{max}, may change with time. When deactivation is significant, the decrease in v_{max} with time can be expressed using Eq. (10.72) such that Eq. (12.8) becomes:

$$\frac{ds}{dt} = \frac{-v_{max0}\, e^{-K_d t} s}{K_m + s} \tag{12.11}$$

where v_{max0} is the value of v_{max} before deactivation occurs and k_d is the first-order deactivation rate constant. Separating variables gives:

$$-\int e^{-K_d t}\, dt = \int \frac{K_m + s}{v_{max0}\, s}\, ds \tag{12.12}$$

Integrating Eq. (12.12) with initial condition $s = s_0$ at $t = 0$ gives:

$$t_b = \frac{-1}{k_d} \ln\left[1 - k_d\left(\frac{K_m}{v_{max0}} \ln\frac{s_0}{s_f} + \frac{s_0 - s_f}{v_{max0}}\right)\right] \tag{12.13}$$

where t_b is the batch reaction time and s_f is the final substrate concentration. The batch reaction time calculation with enzyme deactivation is demonstrated in Example 12.2.

For reactions with immobilized enzyme, Eq. (12.8) must be modified to account for the effect of mass transfer limitations on the rate of enzyme reaction:

$$\frac{ds}{dt} = -\eta_T \frac{v_{max} s}{K_m + s} \tag{12.14}$$

where η_T is the total effectiveness factor incorporating internal and external mass transfer restrictions (Section 11.5), s is the bulk substrate concentration, and v_{max} and K_m are the intrinsic kinetic parameters. Integration of Eq. (12.14) allows evaluation of t_b; however, because η_T is a function of s, integration is not straightforward.

Cell Culture

A similar analysis can be applied to cell cultures to evaluate batch times. A mass balance on cells in a well-mixed batch reactor can be written using Eq. (4.29). A typical flow sheet for this system is shown in Figure 12.25. $\hat{M}_i = \hat{M}_o = 0$ because cells do not enter or leave the vessel. The mass of cells in the reactor, M, is equal to the cell concentration x multiplied by

---

EXAMPLE 12.2 Batch Reaction Time With Enzyme Deactivation

The enzyme of Example 12.1 deactivates with half-life of 4.4 h. Compare with Figure 12.20 the batch reaction time required to achieve 90% substrate conversion.

Solution

$s_0 = 12\,mM$; $v_{max0} = 9\,mM\,h^{-1}$; $K_m = 8.9\,mM$; $s_f = (0.1\,s_0) = 1.2\,mM$. The deactivation rate constant is calculated from the half-life t_h using Eq. (10.73):

$$k_d = \frac{\ln 2}{t_h} = \frac{\ln 2}{4.4\ h} = 0.158\ h^{-1}$$

Substituting values into Eq. (12.13) gives:

$$t_b = \frac{-1}{0.158\ h^{-1}} \ln\left[1 - 0.158\ h^{-1}\left(\frac{8.9\ mM}{9\ mM\ h^{-1}} \ln\frac{12\ mM}{1.2\ mM} + \frac{(12 - 1.2)\ mM}{9\ mM\ h^{-1}}\right)\right] = 5.0\ h$$

The time required to achieve 90% conversion is 5.0 h. Therefore, enzyme deactivation increases the batch reaction time from 3.5 h (Example 12.1) to 5.0 h.

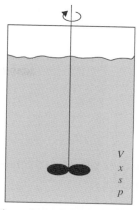

Figure 12.25 Flow sheet for a stirred batch bioreactor.

the liquid volume V. The total rate of cell growth R_G is equal to $r_x V$ where r_x is the volumetric rate of biomass production. From Eq. (10.80), $r_x = \mu x$ where μ is the specific growth rate. If cell death takes place in the reactor alongside growth, $R_C = r_d V$ where r_d is the volumetric rate of cell death. From Eq. (10.116), r_d can be expressed using the first-order equation $r_d = k_d x$, where k_d is the specific death constant. Therefore, Eq. (4.29) for cells in a batch reactor is:

$$\frac{d(xV)}{dt} = \mu x V - k_d x V \tag{12.15}$$

For V constant, Eq. (12.15) becomes:

$$\frac{dx}{dt} = \left(\mu - k_d \right) x \tag{12.16}$$

Because μ in batch culture remains approximately constant and equal to μ_{max} for most of the growth period (Section 10.8.3), if k_d likewise remains constant, we can integrate Eq. (12.16) directly to find the relationship between batch culture time and cell concentration. Using the initial condition $x = x_0$ at $t = 0$:

$$x = x_0 \, e^{\left(\mu_{max} - k_d \right)t} \tag{12.17}$$

If x_f is the final biomass concentration after batch growth time t_b, rearrangement of Eq. (12.17) gives:

$$t_b = \frac{1}{\mu_{max} - k_d} \ln \frac{x_f}{x_0} \tag{12.18}$$

If the rate of cell death is negligible compared with growth, $k_d \ll \mu_{max}$, Eqs. (12.17) and (12.18) reduce to:

$$x = x_0 \, e^{\mu_{max} t} \tag{12.19}$$

and

$$t_b = \frac{1}{\mu_{max}} \ln \frac{x_f}{x_0} \tag{12.20}$$

Equations (12.18) and (12.20) allow us to calculate the batch culture time required to achieve cell density x_f starting from cell density x_0.

The time required for batch culture can also be related to substrate conversion and product concentration using expressions for the rates of substrate uptake and product formation derived in Chapter 10. First, let us apply Eq. (4.29) to the growth-limiting substrate in a batch culture. $\hat{M}_i = \hat{M}_o = 0$ because substrate does not flow into or out of the reactor; the mass of substrate in the reactor, M, is equal to sV where s is substrate concentration and V is liquid volume. Substrate is not generated; therefore $R_G = 0$. R_C is equal to $r_s V$ where r_s is the volumetric rate of substrate uptake. The expression for r_s depends on whether extracellular product is formed by the culture and the relationship between product synthesis and energy generation in the cell as discussed in Section 10.11. If product is formed but is not directly coupled with energy metabolism, r_s is given by Eq. (10.107) and:

$$R_C = r_s V = \left(\frac{\mu}{Y_{X/S}} + \frac{q_P}{Y_{P/S}} + m_S \right) x V \tag{12.21}$$

where μ is the specific growth rate, $Y_{X/S}$ is the true biomass yield from substrate, q_P is the specific rate of product formation not directly linked with energy metabolism, $Y_{P/S}$ is the true product yield from substrate, and m_s is the maintenance coefficient. Therefore, from Eq. (4.29), the mass balance equation for substrate is:

$$\frac{d(sV)}{dt} = -\left(\frac{\mu}{Y_{X/S}} + \frac{q_P}{Y_{P/S}} + m_s \right) x V \tag{12.22}$$

For μ equal to μ_{max} and V constant, we can write Eq. (12.22) as:

$$\frac{ds}{dt} = -\left(\frac{\mu_{max}}{Y_{X/S}} + \frac{q_P}{Y_{P/S}} + m_s \right) x \tag{12.23}$$

Because x is a function of time, we must substitute an expression for x into Eq. (12.23) before it can be integrated. When $\mu = \mu_{max}$ and assuming that cell death is negligible, x is given by Eq. (12.19). Therefore, Eq. (12.23) becomes:

$$\frac{ds}{dt} = -\left(\frac{\mu}{Y_{X/S}} + \frac{q_P}{Y_{P/S}} + m_s \right) x_0 e^{\mu_{max} t} \tag{12.24}$$

If the bracketed terms are constant during growth, Eq. (12.24) can be integrated directly with initial condition $s = s_0$ at $t = 0$ to obtain the following equation:

$$t_b = \frac{1}{\mu_{max}} \ln \left[1 + \frac{s_0 - s_f}{\left(\dfrac{1}{Y_{X/S}} + \dfrac{q_P}{\mu_{max} Y_{P/S}} + \dfrac{m_S}{\mu_{max}} \right) x_0} \right] \tag{12.25}$$

where t_b is the batch culture time and s_f is the final substrate concentration. Evaluating q_P for products indirectly coupled or not related at all to energy metabolism requires further analysis (Sections 10.10.2 and 10.10.3). However, if no product is formed or if production is directly linked with energy metabolism, the expression for the rate of substrate consumption r_s does not contain a separate term for product synthesis (Sections 10.11.1 and 10.11.2) and Eq. (12.25) can be simplified to:

$$t_b = \frac{1}{\mu_{max}} \ln \left[1 + \frac{s_0 - s_f}{\left(\dfrac{1}{Y_{X/S}} + \dfrac{m_S}{\mu_{max}} \right) x_0} \right] \tag{12.26}$$

If maintenance requirements can be neglected, the equation for t_b becomes:

$$t_b = \frac{1}{\mu_{max}} \ln \left[1 + \frac{Y_{X/S}}{x_0} \left(s_0 - s_f \right) \right] \tag{12.27}$$

To obtain an expression for batch culture time as a function of product concentration, we must apply Eq. (4.29) to the product. Again, $\hat{M}_i = \hat{M}_o = 0$. The mass of product in the reactor, M, is equal to pV where p is product concentration and V is liquid volume. Assuming that product is not consumed, $R_C = 0$. R_G is equal to $r_P V$ where r_P is the volumetric rate of product formation. According to Eq. (10.98), for all types of product $r_P = q_P x$ where q_P is the specific rate of product formation. Therefore:

$$R_G = r_P V = q_P x V \tag{12.28}$$

From Eq. (4.29), the mass balance equation for product is:

$$\frac{d(pV)}{dt} = q_p x V \tag{12.29}$$

If cell death is negligible and μ is equal to μ_{max}, x is given by Eq. (12.19). Therefore, for V constant, we can write Eq. (12.29) as:

$$\frac{dp}{dt} = q_p x_0 e^{\mu_{max} t} \tag{12.30}$$

If q_p is constant, Eq. (12.30) can be integrated directly with initial condition $p = p_0$ at $t = 0$ to obtain the following equation for batch culture time as a function of the final product concentration p_f:

$$t_b = \frac{1}{\mu_{max}} \ln\left[1 + \frac{\mu_{max}}{x_0 q_p}\left(p_f - p_0\right)\right] \tag{12.31}$$

To summarize the equations for batch culture time, the time required to achieve a certain biomass density can be evaluated using Eq. (12.18) or (12.20). If cell death is negligible, the time needed for a particular level of substrate conversion is calculated using Eq. (12.25), (12.26), or (12.27) depending on the type of product formed and the importance of maintenance metabolism. For negligible cell death and constant q_p, the batch time required to achieve a particular product concentration can be found from Eq. (12.31). Application of these equations is illustrated in Example 12.3.

EXAMPLE 12.3 Batch Culture Time

Zymomonas mobilis is used to convert glucose to ethanol in a batch bioreactor under anaerobic conditions. The yield of biomass from substrate is 0.06 g biomass (g glucose)$^{-1}$; $Y_{P/X}$ is 7.7 g product (g biomass)$^{-1}$. The maintenance coefficient is 2.2 g glucose (g biomass^{-1}) h^{-1}; the specific rate of product formation due to maintenance is 1.1 g ethanol (g biomass)$^{-1}$ h^{-1}. The maximum specific growth rate of *Z. mobilis* is 0.3 h^{-1}. Five grams of bacteria are inoculated into 50 L of medium containing 12 g L^{-1} glucose. Determine the batch culture times required to:

a. Produce 10 g of biomass
b. Achieve 90% substrate conversion
c. Produce 100 g of ethanol

Solution
$Y_{X/S} = 0.06$ g biomass (g glucose)$^{-1}$; $Y_{P/X} = 7.7$ g ethanol (g biomass)$^{-1}$; $\mu_{max} = 0.3$ h^{-1}; $m_s = 2.2$ g glucose (g biomass^{-1}) h^{-1}; $m_p = 1.1$ g ethanol (g biomass)$^{-1}$ h^{-1}; $s_0 = 12$ g glucose L^{-1}.

$$x_0 = \frac{5 \text{ g}}{50 \text{ L}} = 0.1 \text{ g L}^{-1}$$

a. If 10 g of biomass are produced, the final amount of biomass present is $(10 + 5)\,g = 15\,g$. Therefore:

$$x_f = \frac{15\,g}{50\,L} = 0.3\,g\,L^{-1}$$

From Eq. (12.20):

$$t_b = \frac{1}{0.3\,h^{-1}}\ln\frac{0.3\,g\,L^{-1}}{0.1\,g\,L^{-1}} = 3.7\,h$$

The batch culture time required to produce 10 g of biomass is 3.7 h.

b. If 90% of the substrate is converted, $s_f = (0.1\,s_0) = 1.2\,g\,L^{-1}$. Ethanol synthesis is directly coupled to energy metabolism in the cell; therefore, Eq. (12.26) applies:

$$t_b = \frac{1}{0.3\,h^{-1}}\,\ln\left[1 + \frac{(12 - 1.2)\,g\,L^{-1}}{\left(\dfrac{1}{0.06\,g\,g^{-1}} + \dfrac{2.2\,g\,g^{-1}h^{-1}}{0.3\,h^{-1}}\right)0.1\,g\,L^{-1}}\right] = 5.7\,h$$

The batch culture time required to achieve 90% substrate conversion is 5.7 h.

c. q_P is calculated using Eq. (10.101). For batch culture with $\mu = \mu_{max}$:
$q_P = 7.7\,g$ ethanol (g biomass)$^{-1}$ $(0.3\,h^{-1}) + 1.1\,g$ ethanol (g biomass)$^{-1}$ $h^{-1} = 3.4\,g$ ethanol (g biomass)$^{-1}$ h^{-1}
As no product is present initially, $p_0 = 0$. For production of 100 g ethanol:

$$p_f = \frac{100\,g}{50\,L} = 2.0\,g\,L^{-1}$$

From Eq. (12.31):

$$t_b = \frac{1}{0.3\,h^{-1}}\,\ln\left[1 + \frac{0.3\,h^{-1}}{(0.1\,g\,L^{-1})(3.4\,h^{-1})}(2\,g\,L^{-1})\right] = 3.4\,h$$

The batch culture time required to produce 100 g of ethanol is 3.4 h.

12.5.2 Total Time for Batch Reactor Cycle

The parameter t_b represents the time required for batch enzyme or cell conversion. In practice, batch operations involve lengthy unproductive periods in addition to t_b. Following the enzyme reaction or cell cultivation, time t_{hv} is taken to harvest the contents of the reactor and time t_p is needed to clean, sterilize, or otherwise prepare the reactor for the next batch. For cell culture,

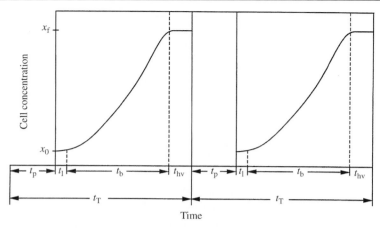

Figure 12.26 Preparation, lag, reaction, and harvest times in operation of a batch bioreactor.

a lag time of duration t_l often occurs after inoculation of the cells (Section 10.8.1), during which no growth or product formation occurs. These time periods are illustrated for bioprocesses in Figure 12.26. The total *downtime* t_{dn} associated with batch reactor operation is:

$$t_{dn} = t_{hv} + t_p + t_l \tag{12.32}$$

and the total batch reaction time t_T is:

$$t_T = t_b + t_{dn} \tag{12.33}$$

12.5.3 Fed-Batch Operation of a Mixed Reactor

Fed-batch operation of reactors involve intermittent or continuous feeding of nutrients to supplement the reactor contents providing some control over the substrate concentration and growth rates. By starting with a relatively dilute solution of substrate and adding more nutrients as the conversion proceeds, growth rates can be controlled at moderate values. This is important, for example, in cultures where high O_2 demand during fast growth exceeds the capacity of mass transfer, or when high substrate concentrations are inhibitory or switch on undesirable metabolic pathways. The flow rate and timing of the feed phase of operation are often determined by monitoring parameters such as the dissolved O_2 tension or exhaust gas composition. Fed-batch culture is used extensively in the production of bakers' yeast to overcome catabolite repression and control O_2 demand; it is also used routinely for penicillin production. As enzyme reactions are rarely carried out as fed-batch operations, we will only consider fed-batch reactors for cell culture.

Fed-batch reactors have two distinct culturing phases, the batch phase and the feeding phase. The initial batch culturing phase can be described using the design equations in Section 12.5.1.

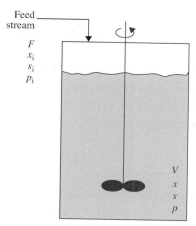

Figure 12.27 Flow sheet for a stirred fed-batch bioreactor.

The batch phase is operated using only a fraction of the total bioreactor volume so there is space available for the additional medium. The equations describing this phase are identical to the previously discussed batch growth. The equations here will focus on the feed phase of operation. A flow sheet for a well-mixed, fed-batch bioreactor is shown in Figure 12.27. The volumetric flow rate of entering feed is F; the concentrations of biomass, growth-limiting substrate, and product in this stream are x_i, s_i, and p_i, respectively. We will assume that F is constant. Owing to input of the feed, the liquid volume V is not constant. Equations for fed-batch culture are derived by carrying out unsteady-state mass balances.

The unsteady-state mass balance equation for total mass in a flow reactor can be developed from Eq. (4.29):

$$\frac{dM}{dt} = \hat{M}_i - \hat{M}_o + R_G - R_C \tag{4.29}$$

where total mass is conserved so R_G and R_C are equal to 0. Total mass can be expressed as the product of volume and density. Similarly, mass flow rate can be expressed as the product of volumetric flow rate and density.

$$\text{Total mass in the tank:} M = \rho V; \quad \text{therefore} \frac{dM}{dt} = \frac{d(\rho V)}{dt}$$
$$\text{Mass flow rate in:} \ \hat{M}_i = F_i \rho_i$$
$$\text{Mass flow rate out:} \ \hat{M}_o = F_o \rho_o$$

Substituting these terms into Eq. (6.29):

$$\frac{d(\rho V)}{dt} = F_i \rho_i - F_o \rho_o$$

which is a differential equation representing an unsteady-state mass balance on total mass. ρ is the density of the reactor contents, V is the liquid volume in the reactor, F_i and F_o are the input and output mass flow rates, and ρ_i and ρ_o are the densities of the input and output streams, respectively. For the fed-batch reactor of Figure 12.23, $F_o = 0$ and $F_i = F$. With dilute solutions such as those used in bioprocessing, we can assume that ρ is constant and that $\rho_i = \rho$; density can then be taken outside of the differential and canceled through the equation. Therefore, Eq. (4.29) for fed-batch bioreactors is:

$$\frac{dV}{dt} = F \tag{12.34}$$

A similar mass balance can be performed for cells based on Eq. (4.29). In fed-batch operations, $\hat{M}_o = 0$. $\hat{M}_i$ is equal to the feed flow rate F multiplied by the cell concentration x_i in the feed. The mass of cells in the reactor, M, is equal to xV where x is the cell concentration and V is the liquid volume; the rate of biomass generation R_G is equal to $\mu x V$ where μ is the specific growth rate; and the rate of cell death R_C is equal to $k_d x V$ where k_d is the specific death constant. Applying these terms in Eq. (4.29) gives:

$$\frac{d(xV)}{dt} = Fx_i + \mu x V - k_d x V \tag{12.35}$$

As the V in fed-batch culture is not constant, it cannot be taken outside of the differential in Eq. (12.35). We can expand the differential using the product rule. After grouping terms this gives:

$$x\frac{dV}{dt} + V\frac{dx}{dt} = Fx_i + (\mu - k_d)xV \tag{12.36}$$

Applying Eq. (12.34) to Eq. (12.36):

$$xF + V\frac{dx}{dt} = Fx_i + (\mu - k_d)xV \tag{12.37}$$

Dividing through by V and rearranging gives:

$$\frac{dx}{dt} = \frac{F}{V}x_i + x(\mu - k_d - \frac{F}{V}) \tag{12.38}$$

Let us define the *dilution rate D* with dimensions T^{-1}:

$$D = \frac{F}{V} \tag{12.39}$$

In fed-batch systems, V increases with time; therefore, if F is constant, D decreases as the reaction proceeds. Applying Eq. (12.39) to Eq. (12.38):

$$\frac{dx}{dt} = Dx_i + x(\mu - k_d - D) \tag{12.40}$$

Equation (12.40) can be simplified for most applications. Usually, the feed material is sterile so that $x_i = 0$. If, in addition, the rate of cell death is negligible compared with growth so that $k_d \ll \mu$, Eq. (12.40) becomes:

$$\frac{dx}{dt} = x(\mu - D) \tag{12.41}$$

Let us now apply Eq. (4.29) to the limiting substrate in our fed-batch reactor. In this case, $\hat{M}_o$ and R_G are zero and the mass flow rate of substrate entering the reactor, $\hat{M}_i$, is equal to Fs_i. The mass of substrate in the reactor, M, is equal to the substrate concentration s multiplied by the volume V. For cell cultivations producing product not directly coupled with energy metabolism, R_C is given by Eq. (12.21). Substituting these terms into Eq. (4.29) gives:

$$\frac{d(sV)}{dt} = Fs_i - \left(\frac{\mu}{Y_{X/S}} + \frac{q_P}{Y_{P/S}} + m_s \right) xV \tag{12.42}$$

where μ is the specific growth rate, $Y_{X/S}$ is the true biomass yield from substrate, q_P is the specific rate of product formation, $Y_{P/S}$ is the true product yield from substrate, and m_s is the maintenance coefficient. Expanding the differential and applying Eqs. (12.34) and (12.39) gives:

$$\frac{ds}{dt} = D(s_i - s) - \left(\frac{\mu}{Y_{X/S}} + \frac{q_P}{Y_{P/S}} + m_s \right) x \tag{12.43}$$

Equations (12.41) and (12.43) are differential equations for the rates of change of the cell and substrate concentrations in fed-batch reactors. As D is a function of time, integration of these equations is more complicated than for batch reactors. However, we can derive analytical expressions for fed-batch culture if we simplify Eqs. (12.41) and (12.43). Here, we examine the situation where the reactor is operated first in batch until a high cell density is achieved and the substrate is virtually exhausted. When this condition is reached, fed-batch operation is started with medium flow rate F. As a result, the cell concentration x is maintained high and approximately constant so that $dx/dt \approx 0$. From Eq. (12.41), if $dx/dt \approx 0$, $\mu \approx D$. Therefore, substituting $\mu \approx D$ into the Monod expression of Eq. (10.91):

$$D \approx \frac{\mu_{max} s}{K_s + s} \tag{12.44}$$

Rearrangement of Eq. (12.44) gives an expression for the substrate concentration as a function of dilution rate:

$$s \approx \frac{D K_S}{\mu_{max} - D} \tag{12.45}$$

Let us assume that the culture does not produce product or, if there is product formation, that it is directly linked with energy generation. If maintenance requirements can also be neglected, Eq. (12.43) can be simplified to:

$$\frac{ds}{dt} = D(s_i - s) - \frac{\mu x}{Y_{X/s}} \tag{12.46}$$

When the cell density in the reactor is high, virtually all substrate entering the vessel is consumed immediately; therefore, $s \ll s_i$ and $ds/dt \approx 0$. Applying these relationships with $\mu \approx D$ to Eq. (12.46), we obtain:

$$x \approx Y_{X/S} S_i \tag{12.47}$$

For product synthesis directly coupled with energy metabolism, Eq. (12.47) allows us to derive an approximate expression for the product concentration in fed-batch reactors. Assuming that the feed does not contain product:

$$p \approx Y_{PS} S_i \tag{12.48}$$

Even though the cell concentration remains virtually unchanged with $dx/dt \approx 0$, because the liquid volume increases with time in fed-batch reactors, the total mass of cells also increases. Consider the rate of increase of total biomass in the reactor, dX/dt, where X is equal to xV. Using the results of Eqs. (12.34) and (12.47) with $dx/dt \approx 0$:

$$\frac{dX}{dt} = \frac{d(xV)}{dt} = x\frac{dV}{dt} + V\frac{dx}{dt} = Y_{X/S}\, s_i\, F \tag{12.49}$$

Equation (12.49) can now be integrated with initial condition $X = X_0$ at the start of liquid flow to give:

$$X = X_0 + (Y_{X/S}\, s_i\, F)t_{fb} \tag{12.50}$$

where t_{fb} is the fed-batch time after commencement of feeding. Equation (12.50) indicates that, for $Y_{X/S}$, s_i, and F constant, the total biomass in fed-batch bioreactors increases as a linear function of time.

Under conditions of high biomass density and almost complete depletion of substrate, a *quasi-steady-state* condition prevails in fed-batch reactors where $dx/dt \approx 0$, $ds/dt \approx 0$, and $dp/dt \approx 0$. Eqs. (12.47) and (12.45) can then be used to calculate the biomass and substrate concentrations in reactors where cell death and maintenance requirements are negligible and product is either absent or directly coupled with energy metabolism. Equation (12.48) allows calculation of the product concentration for metabolites directly coupled with energy metabolism. At quasi-steady state, the specific growth rate μ and the dilution rate F/V are approximately equal; therefore as V increases, the growth rate decreases. When fed-batch operation is used for production of biomass such as bakers' yeast, it is useful to be able to predict the total mass of cells in the reactor as a function of time. An expression for total biomass is given by Eq. (12.50). Note that under quasi-steady-state conditions, x, s, and p are almost constant, but μ, V, D, and X are changing. Further details of fed-batch operation are given by Pirt [37].

12.5.4 Continuous Operation of a Mixed Reactor

Bioreactors are operated continuously in some bioprocess industries such as brewing, production of bakers' yeast, and waste treatment. Enzyme conversions can also be carried out using continuous systems. The flow sheet for a continuous mixed reactor is shown in Figure 12.28. If the vessel is well mixed, the product stream has the same composition as the liquid in the reactor. Therefore, when continuous reactors are used with freely suspended cells or enzymes, the catalyst is continuously withdrawn from the vessel in the product stream. The continuous removal of cells is balanced by the continuous growth of cells under steady state conditions. However, the continuous removal of enzyme is a serious shortcoming as additional enzyme is not produced in the reaction so their concentration continuously decreases. Continuous reactors are used with free enzymes only if the enzyme is inexpensive and can be added continuously to maintain the catalyst concentration. Continuous operation may be feasible if the enzyme is immobilized and retained inside the vessel.

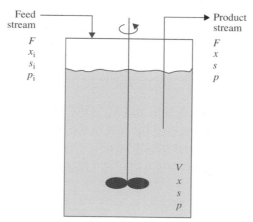

Figure 12.28 Flow sheet for a continuous stirred tank bioreactor.

Different steady-state operating strategies are available for continuous bioreactors. In a *chemostat*, the liquid volume is kept constant by setting the inlet and outlet flow rates equal. The dilution rate is therefore constant and steady state is achieved as concentrations in the chemostat adjust to the feed rate. In a *turbidostat*, the liquid volume is kept constant by setting the outlet flow rate equal to the inlet flow rate; however, the inlet flow rate is adjusted to keep the biomass concentration constant. Thus, in a turbidostat, the dilution rate adjusts to the steady-state value required to achieve the desired biomass concentration. Turbidostats require more complex monitoring and control systems than chemostats and are not used at large scales. Accordingly, we will concentrate here on chemostat operation. Chemostats are often referred to using the abbreviation CSTR, meaning *continuous stirred tank reactor*.

Characteristic operating parameters for continuous reactors are the dilution rate D defined in Eq. (12.39) and the average *residence time* τ. These parameters are related as follows:

$$\tau = \frac{1}{D} = \frac{V}{F} \tag{12.51}$$

In continuous reactor operation, the amount of material that can be processed over a given period of time is represented by the flow rate F. Therefore, for a given throughput, the reactor size V and associated capital and operating costs are minimized when τ is made as small as possible. Continuous reactor theory allows us to determine relationships between τ (or D) and the steady-state cell, substrate, and product concentrations in the reactor.

Enzyme Reaction

Let us apply Eq. (4.29) to the limiting substrate in a continuous enzyme reactor operated at steady state. The mass flow rate of substrate entering the reactor, $\hat{M}_i$, is equal to Fs_i; $\hat{M}_o$, the mass flow rate of substrate leaving, is Fs. As substrate is not generated in the reactor, $R_G = 0$. The rate of substrate consumption R_C is equal to the volumetric rate of enzyme reaction v multiplied by V; for Michaelis–Menten kinetics, v is given by Eq. (10.37). The left side of Eq. (4.29) is zero when the system is operated at steady state. Therefore, the steady-state substrate mass balance equation for continuous enzyme reaction is:

$$Fs_i - Fs - \frac{v_{max}\,s}{K_m + s}V = 0 \tag{12.52}$$

where v_{max} is the maximum rate of reaction and K_m is the Michaelis constant. For reactions with free enzyme, we assume that the enzyme lost in the product stream is replaced continuously so that v_{max} remains constant and steady state is achieved. Dividing through by V and applying the definition of the dilution rate from Eq. (12.39) gives:

$$D(s_i - s) = \frac{v_{max}\,s}{K_m + s} \tag{12.53}$$

If v_{max}, K_m, and s_i are known, Eq. (12.53) can be used to calculate the dilution rate required to achieve a particular level of substrate conversion. The steady-state product concentration can then be evaluated from stoichiometry.

For reactions with immobilized enzymes, Eq. (12.53) must be modified to account for mass transfer effects on the rate of enzyme reaction:

$$D(s_i - s) = \frac{\eta_T v_{max} s}{K_m + s} \tag{12.54}$$

where η_T is the total effectiveness factor (Section 11.5), s is the bulk substrate concentration, and v_{max} and K_m are the intrinsic kinetic parameters. η_T can be calculated for constant s using the theory for heterogeneous reactions outlined in Chapter 11. Example 12.4 demonstrates the use of immobilized enzymes in a CSTR.

EXAMPLE 12.4 Immobilized Enzyme Reaction in a CSTR

Mushroom tyrosinase is immobilized in spherical beads of diameter 2 mm for conversion of tyrosine to dihydroxyphenylalanine (DOPA) in a continuous, well-mixed bubble column. The Michaelis constant for the immobilized enzyme is 2 gmol tyrosine m^{-3}. A solution containing 15 gmol tyrosine m^{-3} is fed into the reactor; the desired level of substrate conversion is 99%. The reactor is loaded with beads at a density of 0.25 m^3 m^{-3}; all enzyme is retained within the vessel. The intrinsic v_{max} for the immobilized enzyme is 1.5×10^{-2} gmol tyrosine s^{-1} per m^3 of catalyst. The effective diffusivity of tyrosine in the beads is 7×10^{-10} m^2 s^{-1}. External mass transfer effects are negligible. Immobilization stabilizes the enzyme so that deactivation is minimal over the operating period. Determine the reactor volume needed to treat 18 m^3 of tyrosine solution per day.

Solution

$K_m = 2$ gmol tyrosine m^{-3}; $v_{max} = 1.5 \times 10^{-2}$ gmol tyrosine s^{-1} m^{-3}; $R = 10^{-3}$ m; $\mathcal{D}_{Ae} = 7 \times 10^{-10}$ $m^2 s^{-1}$; $s_i = 15$ gmol tyrosine m^{-3}. Converting the feed flow rate to units of m^3 s^{-1}:

$$F = 18 \ m^3 \ day^{-1} \cdot \left| \frac{1 \ day}{24 \ h} \right| \cdot \left| \frac{1 \ h}{3600 \ s} \right| = 2.08 \times 10^{-4} \ m^3 \ s^{-1}$$

For 99% conversion of tyrosine, the outlet and therefore the internal substrate concentration $s = (0.01 \ s_i) = 0.15$ gmol tyrosine m^{-3}. As $s \ll K_m$, we can assume that the reaction follows first-order kinetics (Section 10.3.3) with $k_1 = v_{max}/K_m = 7.5 \times 10^{-3}$ s^{-1}. The first-order Thiele modulus for spherical catalysts is calculated from Table 11.2:

$$\phi_1 = \frac{R}{3} \sqrt{\frac{k_1}{\mathcal{D}_{Ae}}} = \frac{10^{-3} m}{3} \sqrt{\frac{7.5 \times 10^{-3} \ s^{-1}}{7 \times 10^{-10} \ m^2 \ s^{-1}}} = 1.09$$

(Continued)

From Table 11.3 or Figure 11.11, $\eta_{i1} = 0.64$. As there is negligible external mass transfer resistance, $\eta_{e1} = 1$ and, from Eq. (11.46), $\eta_T = 0.64$. Substituting values into Eq. (12.54) gives:

$$D(15 - 0.15)\, \text{gmol m}^{-3} = \frac{0.64(1.5 \times 10^{-2}\ \text{gmol s}^{-1}\text{m}^{-3})(0.15\ \text{gmol m}^{-3})}{2\ \text{gmol m}^{-3} + 0.15\ \text{gmol m}^{-3}}$$

$$D = 4.51 \times 10^{-5}\, \text{s}^{-1}$$

From Eq. (12.51):

$$V = \frac{F}{D} = \frac{2.08 \times 10^{-4}\ \text{m}^3\,\text{s}^{-1}}{4.51 \times 10^{-5}\ \text{s}^{-1}} = 4.6\ \text{m}^3$$

The reactor volume needed to treat 18 m^3 of tyrosine solution per day is 4.6 m^3.

Cell Culture

Let us consider the reactor of Figure 12.28 operated as a continuous bioreactor and apply Eq. (4.29) for steady-state mass balances on biomass, substrate, and product.

For biomass, $\hat{M}_i$ in Eq. (4.29) is the mass flow rate of cells entering the reactor: $\hat{M}_i = Fx_i$. $\hat{M}_o$ is the mass flow rate of cells leaving: $\hat{M}_o = Fx$. The other terms in Eq. (4.29) are the same as in Section 12.5.1: $R_G = \mu x V$, where μ is the specific growth rate and V is the liquid volume, and $R_C = k_d x V$, where k_d is the specific death constant. At steady state, the left side of Eq. (4.29) is zero. Therefore, the steady-state mass balance equation for biomass is:

$$Fx_i - Fx + \mu x V - k_d x V = 0 \tag{12.55}$$

Usually, the feed stream in continuous culture is sterile so that $x_i = 0$. If, in addition, the rate of cell death is negligible compared with growth, $k_d \ll \mu$ and Eq. (12.55) becomes:

$$\mu x V = Fx \tag{12.56}$$

Canceling x from both sides, dividing by V, and applying the definition of dilution rate from Eq. (12.39) gives:

$$\mu = D \tag{12.57}$$

Applying Eq. (12.57) to the Monod expression of Eq. (10.91) gives a design equation for controlling the steady-state concentration of limiting substrate in the reactor:

$$s = \frac{D K_S}{\mu_{max} - D} \tag{12.58}$$

Let us now apply Eq. (4.29) at steady state to the limiting substrate. In this case, $\hat{M}_i = Fs_i$, $\hat{M}_o = Fs$, and $R_G = 0$. If product is formed but is not directly coupled with energy metabolism, R_C is given by Eq. (12.21). Therefore, the steady-state mass balance equation for substrate is:

$$Fs_i - Fs - \left(\frac{\mu}{Y_{X/S}} + \frac{q_P}{Y_{P/S}} + m_s \right) xV = 0 \tag{12.59}$$

where μ is the specific growth rate, $Y_{X/S}$ is the true biomass yield from substrate, q_P is the specific rate of product formation not directly linked with energy metabolism, $Y_{P/S}$ is the true product yield from substrate, and m_S is the maintenance coefficient. In Eq. (12.59), we can divide through by V, substitute the definition of dilution rate from Eq. (12.39), and replace μ with D according to Eq. (12.57). Rearrangement then gives the following design equation for the steady-state cell concentration x:

$$x = \frac{D(s_i - s)}{\dfrac{D}{Y_{XS}} + \dfrac{q_P}{Y_{PS}} + m_S} \tag{12.60}$$

Equation (12.60) can be simplified if there is no product synthesis or if production is directly linked with energy metabolism:

$$x = \frac{D(s_i - s)}{\dfrac{D}{Y_{XS}} + m_S} \tag{12.61}$$

If, in addition, maintenance effects can be ignored, Eq. (12.61) becomes:

$$x = (s_i - s)Y_{X/S} \tag{12.62}$$

Substituting for s from Eq. (12.58), we obtain an expression for the steady-state cell concentration in a CSTR:

$$x = \left(s_i - \frac{D K_S}{\mu_{max} - D} \right) Y_{X/S} \tag{12.63}$$

Equation (12.63) is valid at steady state in the absence of maintenance requirements and when product synthesis is either absent or directly linked with energy metabolism.

We can also apply Eq. (4.29) for a steady-state mass balance on cultivation product. In this case, $M_i = Fp_i$ and $M_o = Fp$. R_G is given by Eq. (12.28) and $R_C = 0$. Therefore, Eq. (4.29) becomes:

$$Fp_i - Fp + q_P xV = 0 \tag{12.64}$$

where q_P is the specific rate of formation for all classes of product. Dividing through by V, substituting the definition of the dilution rate from Eq. (12.39), and rearranging gives an expression for the steady-state product concentration as a function of biomass concentration x:

$$p = p_i + \frac{q_P x}{D} \tag{12.65}$$

x in Eq. (12.65) can be evaluated from Eq. (12.60), (12.61), or (12.62). Evaluation of q_P depends on the type of product formed (Section 10.10).

Equation (12.58) is an explicit expression for the steady-state substrate concentration in a chemostat. The steady-state biomass concentration x can be evaluated from Eq. (12.60), (12.61), or (12.62); the choice of expression for biomass depends on the relative significance of maintenance metabolism and the type of product, if any, produced. If q_P is known, the product concentration in the reactor can be evaluated from Eq. (12.65). In the simplest case when products are either absent or directly linked to energy generation and maintenance effects can be neglected, the chemostat is represented by Eqs. (12.58) and (12.63). The form of these equations is shown in Figure 12.29. At low feed rates (i.e., $D \rightarrow 0$) nearly all the

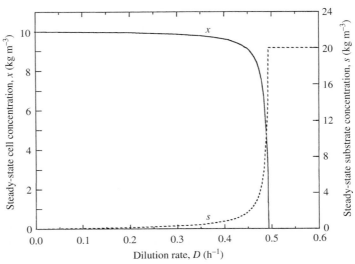

Figure 12.29 Steady-state cell and substrate concentrations as a function of dilution rate in a chemostat. The curves were calculated using the following parameter values: $\mu_{max} = 0.5\,h^{-1}$, $K_S = 0.2\,kg$ substrate m^{-3}, $Y_{X/S} = 0.5\,kg$ biomass (kg substrate)$^{-1}$, and $s_i = 20\,kg$ substrate m^{-3}.

substrate is consumed at steady state so that, from Eq. (12.62), $x \approx s_i \cdot Y_{X/S}$. As D increases, s increases slowly at first and then more rapidly as D approaches μ_{max}; correspondingly, x decreases so that $x \to 0$ as $D \to \mu_{max}$. The condition at high dilution rate whereby x reduces to zero is known as *washout*; washout of cells occurs when the rate of cell removal in the reactor outlet stream is greater than the rate of generation by growth. For systems with negligible maintenance requirements and either zero or energy-associated product formation, the critical dilution rate D_{crit} at which the steady-state biomass concentration just becomes zero can be estimated by substituting $x = 0$ into Eq. (12.63) and solving for D:

$$D_{crit} = \frac{\mu_{max} s_i}{K_S + s_i} \tag{12.66}$$

For most cell cultures, $K_S \ll s_i$; therefore $D_{crit} \approx \mu_{max}$. The operating dilution rate must always be less than D_{crit} to avoid washout of cells from the chemostat. Near washout, the system is very sensitive to small changes in D that cause relatively large shifts in x and s.

The rate of biomass production in a chemostat is equal to the rate at which cells leave the reactor, Fx. The volumetric productivity Q_X is therefore equal to Fx divided by V. Applying the definition of the dilution rate from Eq. (12.39):

$$Q_X = \frac{Fx}{V} = Dx \tag{12.67}$$

where Q_X is the volumetric rate of biomass production. Similarly, the volumetric rate of product formation Q_P is:

$$Q_P = \frac{Fp}{V} = Dp \tag{12.68}$$

When maintenance requirements are negligible and product formation is either absent or energy-associated, we can substitute into Eq. (12.67) the expression for x from Eq. (12.63):

$$Q_X = D\left(s_i - \frac{D K_S}{\mu_{max} - D} \right) Y_{XS} \tag{12.69}$$

This relationship between Q_X and D is shown in Figure 12.30. The rate of biomass production reaches a maximum at the optimum dilution rate for biomass productivity D_{opt}; therefore, at D_{opt} the slope $dQ_X/dD = 0$. Differentiating Eq. (12.69) with respect to D and equating to zero provides an expression for D_{opt}:

$$D_{opt} = \mu_{max}\left(1 - \sqrt{\frac{K_S}{K_S + s_i}} \right) \tag{12.70}$$

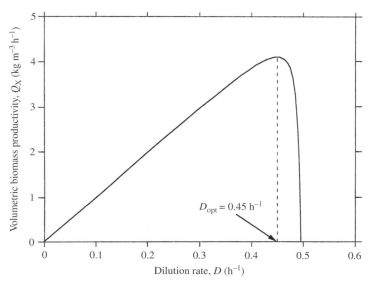

Figure 12.30 Steady-state volumetric biomass productivity as a function of dilution rate in a chemostat. The curve was calculated using the following parameter values: $\mu_{max} = 0.5\,h^{-1}$, $K_S = 0.2\,kg$ substrate m^{-3}, $Y_{X/S} = 0.5\,kg$ biomass (kg substrate)$^{-1}$, and $s_i = 20\,kg$ substrate m^{-3}.

Operation of a chemostat at D_{opt} gives the maximum rate of biomass production from the reactor. However, because D_{opt} is usually very close to D_{crit}, it may not be practical to operate at D_{opt}. Small variations of dilution rate due to perturbations in pump performance in this region can cause large fluctuations in x and s and, unless the dilution rate is controlled very precisely, washout may occur.

Excellent agreement between chemostat theory and experimental results has been found for many culture systems. When deviations occur, they are due primarily to imperfect operation of the reactor. For example, if the vessel is not well mixed, some liquid will have higher residence time in the reactor than the rest and the concentrations will not be uniform. Under these conditions, the equations derived in this section do not hold. Similarly, if cells adhere to glass or metal surfaces in the reactor and produce wall growth, biomass will be retained in the vessel and washout will not occur even at high dilution rates. Other deviations occur if inadequate time is allowed for the system to reach steady state. Examples 12.5 and 12.6 calculate steady state values for biomass, substrate, and product, as well as demonstrate calculations for productivity.

12.5.5 Chemostat With Immobilized Cells

Consider the continuous stirred tank immobilized cell bioreactor shown in Figure 12.31. Spherical particles containing cells are kept suspended and well mixed by the stirrer. The concentration of immobilized cells per unit volume of liquid in the reactor is x_{im}. Let us assume

---

EXAMPLE 12.5 Steady-State Concentrations in a Chemostat

The *Zymomonas mobilis* cells of Example 12.3 are used for chemostat culture in a 60 m³ bioreactor. The feed contains $12 \, g \, L^{-1}$ of glucose; K_S for the organism is 0.2 g glucose L^{-1}.

a. What flow rate is required for a steady-state substrate concentration of 1.5 g glucose L^{-1}?
b. At the flow rate of (a), what is the cell density?
c. At the flow rate of (a), what concentration of ethanol is produced?

Solution

$Y_{X/S} = 0.06 \, g$ biomass (g glucose)$^{-1}$; $Y_{P/X} = 7.7 \, g$ ethanol (g biomass)$^{-1}$; $\mu_{max} = 0.3 \, h^{-1}$; $K_S = 0.2 \, g$ glucose L^{-1}; $m_S = 2.2 \, g$ glucose (g biomass)$^{-1} \, h^{-1}$; $s_i = 12 \, g$ glucose L^{-1}; $V = 60 \, m^3$. From Example 12.3, $q_P = 3.4 \, g$ ethanol (g biomass)$^{-1} \, h^{-1}$. From the general definition of yield in Section 10.7.1, we can deduce that $Y_{P/S} = Y_{P/X} Y_{X/S} = 0.46 \, g$ ethanol (g glucose)$^{-1}$.

a. $s = 1.5 \, g$ glucose L^{-1}. Combining the Monod expression of Eq. (10.91) with Eq. (12.57) gives an expression for D:

$$D = \frac{\mu_{max} s}{K_S + s} = \frac{(0.3 \, h^{-1})(1.5 \, g \, L^{-1})}{0.2 \, g \, L^{-1} + 1.5 \, g \, L^{-1}} = 0.26 \, h^{-1}$$

From the definition of dilution rate in Eq. (12.39):

$$F = DV = (0.26 \, h^{-1})(60 \, m^3) = 15.6 \, m^3 \, h^{-1}$$

The flow rate giving a steady-state substrate concentration of 1.5 g glucose L^{-1} is 15.6 m³ h⁻¹.

b. When product synthesis is coupled with energy metabolism, as is the case for ethanol, x is evaluated using Eq. (12.61). Therefore:

$$x = \frac{(0.26 \, h^{-1})(12 - 1.5) \, g \, L^{-1}}{\left(\dfrac{0.26 \, h^{-1}}{0.06 \, g \, g^{-1}} + 2.2 \, g \, g^{-1} h^{-1} \right)} = 0.42 \, g \, L^{-1}$$

The steady-state cell density at a flow rate of 15.6 m³ h⁻¹ is 0.42 g biomass L^{-1}.

c. Assuming that ethanol is not present in the feed, $p_i = 0$. The steady-state product concentration is given by Eq. (12.65):

$$p = \frac{(3.4 \, h^{-1})(0.42 \, g \, L^{-1})}{0.26 \, h^{-1}} = 5.5 \, g \, L^{-1}$$

The steady-state ethanol concentration at a flow rate of 15.6 m³ h⁻¹ is 5.5 g ethanol L^{-1}.

---

that x_{im} is constant; this is achieved if all particles are retained in the vessel and any cells produced by immobilized cell growth are released into the medium. The concentration of suspended cells is x_s. We will assume that the intrinsic specific growth rates of suspended and immobilized cells are the same and equal to μ. Suspended cells are removed from the reactor in the product stream; immobilized cells are retained inside the vessel. For simplicity, let us

---

EXAMPLE 12.6 Substrate Conversion and Biomass Productivity in a Chemostat

A 5-m^3 bioreactor is operated continuously using a feed substrate concentration of 20 kg m^{-3}. The microorganism cultivated in the reactor has the following characteristics: $\mu_{max} = 0.45\,h^{-1}$; $K_S = 0.8\,kg$ substrate m^{-3}; $Y_{X/S} = 0.55\,kg$ biomass (kg substrate)$^{-1}$.

a. What feed flow rate is required to achieve 90% substrate conversion?
b. How does the biomass productivity at 90% substrate conversion compare with the maximum possible?

Solution

a. For 90% substrate conversion, $s = (0.1\,s_i) = 2\,kg\,m^{-3}$. Combining the Monod expression of Eq. (10.91) with Eq. (12.57) gives an expression for D:

$$D = \frac{\mu_{max}s}{K_s + s} = \frac{(0.45\ h^{-1})(2\ kg\ m^{-3})}{0.8\ kg\ m^{-3} + 2\ kg\ m^{-3}} = 0.32\ h^{-1}$$

From Eq. (12.39):

$$F = DV = (0.32\ h^{-1})(5\ m^3) = 1.6\ m^3\ h^{-1}$$

The feed flow rate required for 90% substrate conversion is 1.6 m^3 h^{-1}.

b. Assuming that maintenance requirements and product formation are negligible, from Eq. (12.69):

$$Q_X = 0.32\ h^{-1}\left[20\ kg\ m^{-3} - \frac{(0.32\ h^{-1})(0.8\ kg\ m^{-3})}{(0.45\ h^{-1} - 0.32\ h^{-1})}\right]0.55\ kg\ biomass\ (kg\ substrate)^{-1} = 3.17\ kg\ m^{-3}\ h^{-1}$$

The maximum biomass productivity occurs at D_{opt}, which can be evaluated using Eq. (12.70):

$$D_{opt} = 0.45\ h^{-1}\left(1 - \sqrt{\frac{0.8\ kg\ m^{-3}}{0.8\ kg\ m^{-3} + 20\ kg\ m^{-3}}}\right) = 0.36\ h^{-1}$$

The maximum biomass productivity is determined from Eq. (12.69) with $D = D_{opt}$:

$$Q_{X.max} = 0.36\ h^{-1}\left[20\ kg\ m^{-3} - \frac{(0.36\ h^{-1})(0.8\ kg\ m^{-3})}{(0.45\ h^{-1} - 0.36\ h^{-1})}\right]0.55\ kg\ biomass\ (kg\ substrate)^{-1} = 3.33\ kg\ m^{-3}\ h^{-1}$$

Therefore, the biomass productivity at 90% substrate conversion is 3.17/3.33×100% = 95% of the theoretical maximum.

---

assume that cell death and maintenance requirements are negligible, the reactor feed is sterile, and any product synthesis is directly coupled with energy metabolism.

The system shown in Figure 12.31 reaches steady state. The relationships between the operating variables and the concentrations inside the reactor can be determined using mass balances. Let us consider a mass balance on suspended cells. At steady state, the mass balance equation is similar to Eq. (12.55) except that x_i is zero for sterile feed and cell death is

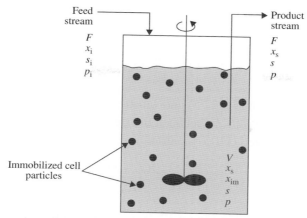

Figure 12.31 Flow sheet for a continuous stirred tank bioreactor with immobilized cells.

assumed to be negligible. In addition, as suspended cells are produced by growth of both the suspended and immobilized cell populations, the equation for the immobilized cell bioreactor must contain two cell generation terms instead of one:

$$-Fx_s + \mu x_s V + \mu x_{im} V = 0 \tag{12.71}$$

If diffusional limitations affect the growth rate of the immobilized cells, μx_{im} must be replaced by $\eta_T \mu x_{im}$ where η_T is the total effectiveness factor defined in Section 11.5. Dividing through by V and applying the definition of the dilution rate from Eq. (12.39) gives:

$$Dx_s = \mu(x_s + \eta_T x_{im}) \tag{12.72}$$

or

$$D = \mu \left(1 + \frac{\eta_T x_{im}}{x_s}\right) \tag{12.73}$$

For $x_{im} = 0$, Eq. (12.73) reduces to Eq. (12.57) for a chemostat containing suspended cells only: $\mu = D$.

The steady-state mass balance equation for the limiting substrate can be derived from Eq. (4.29) with $\hat{M}_i = Fs_i$, $\hat{M}_o = Fs$, and $R_G = 0$. Both cell populations consume substrate: in the absence of product- and maintenance-associated substrate requirements, the rate of substrate consumption R_C can be related directly to the growth rates of the immobilized and suspended cells using the biomass yield coefficient $Y_{X/S}$. If we assume that the value of $Y_{X/S}$ is the same for all cells, by analogy with Eq. (12.59), the mass balance equation for substrate is:

$$Fs_i - Fs - \frac{\mu x_s}{Y_{X/S}} V - \frac{\eta_T \mu x_{im}}{Y_{X/S}} V = 0 \tag{12.74}$$

Dividing through by V, applying the definition of the dilution rate from Eq. (12.39), and rearranging gives:

$$D(s_i - s) = \frac{\mu}{Y_{X/S}}(x_s + \eta_T x_{im})$$ (12.75)

By manipulating Eqs. (12.73) and (12.75) and substituting the Monod expression for μ from Eq. (10.91), the following relationship between the steady-state substrate concentration s, dilution rate D, and immobilized cell concentration x_{im} is obtained:

$$\frac{\mu_{max}s}{K_S + s} = \frac{D(s_i - s)Y_{X/S}}{(s_i - s)Y_{X/S} + \eta_T x_{im}}$$ (12.76)

The form of Eq. (12.76) is shown in Figure 12.28 as a graph of the percentage substrate conversion versus dilution rate. For a chemostat with suspended cells only (i.e., $x_{im} = 0$), at steady state $D = \mu$ and the maximum operating dilution rate D_{crit} is limited by the maximum specific growth rate of the cells. From Eq. (12.73), for any $x_{im} > 0$, D at steady state in the immobilized cell reactor is greater than μ. Accordingly, the dilution rate is no longer limited by the maximum specific growth rate of the cells and, as shown in Figure 12.28, immobilized cell chemostats can be operated at D considerably greater than D_{crit} without washout. At a given dilution rate, the presence of immobilized cells also improves substrate conversion and reduces the amount of substrate lost in the product stream. However, reaction rates with immobilized cells can be reduced significantly by the effects of mass transfer in and around the particles. As illustrated in Figure 12.32, at the same concentration of immobilized cells, the substrate conversion at $\eta_T = 1$ is greater than at lower values of η_T when mass transfer limitations are significant.

12.5.6 Chemostat Cascade

The coupling of two or more chemostats in series produces a multistage process in which conditions such as pH, temperature, and medium composition can be varied in each reactor. This is advantageous if the reactor conditions required for growth are different from those required for product synthesis, for example, in the production of recombinant proteins and many metabolites not directly linked with energy metabolism. One way of operating a two-stage chemostat cascade is shown in Figure 12.33. In this process, the product stream from the first reactor feeds directly into the second reactor. Substrate leaving the first reactor at concentration s_1 is converted in the second tank so that $s_2 < s_1$ and $p_2 > p_1$. In some applications, the second chemostat is supplemented with fresh medium containing nutrients, inducers, or inhibitors for optimal product formation.

Design equations for cell and enzyme chemostat cascades can be derived as a simple extension of the theory developed in Section 12.5.4; the same mass balance principles are applied,

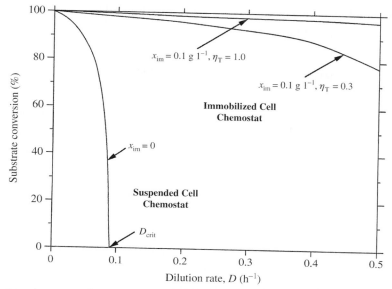

Figure 12.32 Steady-state substrate conversion in a chemostat as a function of dilution rate with and without immobilized cells. The curves were calculated using the following parameter values: $\mu_{max} = 0.1\,h^{-1}$, $K_S = 10^{-3}$ g substrate L^{-1}, $Y_{X/S} = 0.5$ g biomass (g substrate)$^{-1}$, and $s_i = 8 \times 10^{-3}$ g substrate L^{-1}.

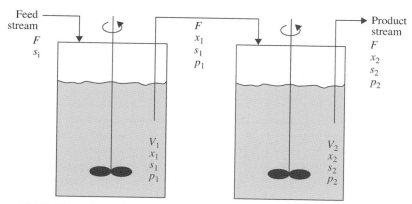

Figure 12.33 Flow sheet for a cascade of two continuous stirred tank bioreactors.

and steady state is assumed for each reactor. Details can be found in other references [5,37,38]. In bioreactor cascades, cells entering the second and subsequent vessels may go through periods of unbalanced growth as they adapt to the new environmental conditions in each reactor. Therefore, the use of simple unstructured metabolic models such as those outlined in this text does not always give accurate results. Nevertheless, it can be shown that the total reactor residence time required to achieve a given degree of substrate conversion is significantly lower using two chemostats in series than if only one chemostat were used. In other words, the total reactor volume required is reduced using two smaller tanks in series compared with a single large tank. Usually, only two to four reactors in series are justified as the benefits associated with adding successive stages diminish significantly [39].

12.5.7 Chemostat With Cell Recycle

The cell concentration in a single chemostat can be increased by recycling the biomass in the product stream back to the reactor. With more catalyst present in the vessel, higher volumetric rates of substrate utilization and product formation can be achieved. The critical dilution rate for washout is also increased, thus allowing greater operating flexibility.

There are several ways by which cells can be recycled in bioprocesses. *External biomass feedback* is illustrated in Figure 12.34. In this scheme, a cell separator such as a centrifuge or settling tank is used to concentrate the biomass leaving the reactor. A portion of the concentrate is recycled back to the CSTR with flow rate F_r and cell concentration x_r. Such systems can be operated under steady-state conditions and are used extensively in biological waste treatment. Another way of achieving biomass feedback is *perfusion culture* or *internal biomass feedback*. This operating scheme is used often for mammalian cell culture and is illustrated in Figure 12.35. Depletion of nutrients and accumulation of inhibitory products may limit batch cell densities for mammalian cell lines to about 10^6 cells mL^{-1}. The cell density and therefore the volumetric productivity of these cultures can be increased by retaining the biomass in the reactor while fresh medium is added and spent broth removed. As indicated in Figure 12.35, cells in a perfusion reactor are physically retained in the vessel by a mechanical device such as a filter. Liquid throughput is thus achieved without continuous removal or

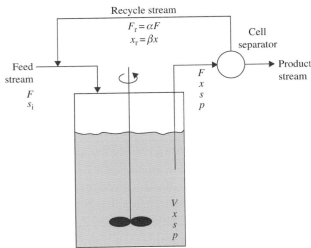

Figure 12.34 Flow sheet for a continuous stirred tank bioreactor with external biomass feedback.

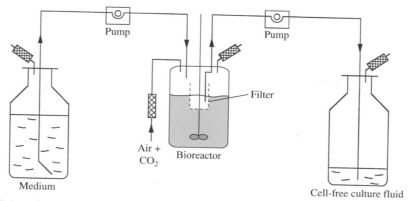

Figure 12.35 Flow sheet for a continuous perfusion reactor with internal biomass feedback.

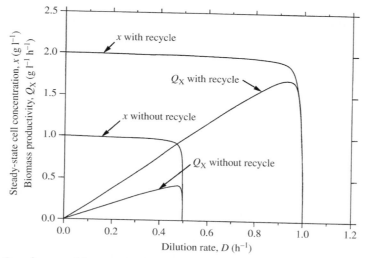

Figure 12.36 Steady-state biomass concentration and volumetric biomass productivity for a chemostat with and without cell recycle. The curves were calculated using the following parameter values: $\mu_{max} = 0.5\,h^{-1}$, $K_S = 0.01\,g$ substrate L^{-1}, $Y_{X/S} = 0.5\,g$ biomass (g substrate)$^{-1}$, $s_i = 2\,g$ substrate L^{-1}, $\alpha = 0.5$, and $\beta = 2.0$.

dilution of the cells so that cell concentrations greater than 10^7 cells mL^{-1} can be obtained. However, a common problem associated with perfusion systems is blocking of the filter.

With growing cells, if all the biomass is recycled or retained in the reactor, the cell concentration will increase with time and steady state will not be achieved. Therefore, for steady-state operation, some proportion of the biomass must be removed from the system. Chemostat reactors with cell recycle are analyzed using the same mass balance techniques applied in Section 12.5.4 [37,38]. Typical results for biomass concentration and biomass productivity with and without cell recycle are shown in Figure 12.36. With cell recycle, because the recycled cells are an additional source of biomass in the reactor, washout occurs at dilution rates greater than the maximum specific growth rate. If α is the recycle ratio:

$$\alpha = \frac{F_r}{F} \tag{12.77}$$

and β is the biomass concentration factor:

$$\beta = \frac{x_r}{x} \tag{12.78}$$

the critical dilution rate with cell recycle is increased by a factor of $1/(1+\alpha-\alpha\beta)$ relative to that in a simple chemostat without cell recycle. Figure 12.36 also shows that, at the same dilution rate, the biomass productivity is also greater in recycle systems by the same factor.

12.5.8 *Evaluation of Kinetic and Yield Parameters in Chemostat Culture*

In a steady-state chemostat with sterile feed and negligible cell death, as indicated in Eq. (12.57), the specific growth rate μ is equal to the dilution rate D. This relationship is useful for determining the kinetic and yield parameters that characterize cellular reactions. Combining the Monod expression of Eq. (10.91) with Eq. (12.57) gives an equation for D in chemostat cultures:

$$D = \frac{\mu_{max} s}{K_s + s} \tag{12.79}$$

where μ_{max} is the maximum specific growth rate, K_s is the substrate constant, and s is the steady-state substrate concentration in the reactor.

Equation (12.79) is analogous mathematically to the Michaelis–Menten expression for enzyme kinetics, Eq. (10.37). If s is measured at various dilution rates, the same techniques described in Section 10.4 for determining v_{max} and K_m can be applied for evaluation of μ_{max} and K_S. Rearrangement of Eq. (12.79) gives the following linearized equations that can be used for Lineweaver–Burk, Eadie–Hofstee, and Langmuir plots, respectively:

$$\frac{1}{D} = \frac{K_S}{\mu_{max} s} + \frac{1}{\mu_{max}} \tag{12.80}$$

$$\frac{D}{s} = \frac{\mu_{max}}{K_S} - \frac{D}{K_S} \tag{12.81}$$

and

$$\frac{s}{D} = \frac{K_S}{\mu_{max}} + \frac{s}{\mu_{max}} \tag{12.82}$$

For example, according to Eq. (12.80), μ_{max} and K_S can be determined from the slope and intercept of a plot of $1/D$ versus $1/s$. The comments made in Sections 10.4.2 through 10.4.4 about the distortion of experimental error with linearization of measured data apply also to Eqs. (12.80) through (12.82).

Chemostat operation is also convenient for determining true yields and maintenance coefficients for cell cultures. An expression relating these parameters to the specific growth rate is given by Eq. (10.112). In chemostat culture with $\mu = D$, Eq. (10.112) becomes:

$$\frac{1}{Y'_{X/S}} = \frac{1}{Y_{X/S}} + \frac{m_S}{D}$$

(12.83)

where $Y'_{X/S}$ is the observed biomass yield from substrate, $Y_{X/S}$ is the true biomass yield from substrate, and m_S is the maintenance coefficient. Therefore, as shown in Figure 12.37, a plot

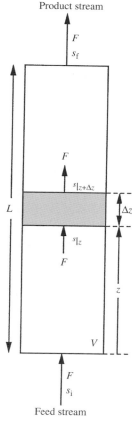

Figure 12.37 Flow sheet for a continuous plug flow tubular reactor.

of $1/Y'_{X/S}$ versus $1/D$ gives a straight line with slope m_S and intercept $1/Y_{X/S}$. In a chemostat with sterile feed, the observed biomass yield from substrate $Y'_{X/S}$ is obtained as follows:

$$Y'_{X/S} = \frac{x}{s_i - s} \tag{12.84}$$

where x and s are measured steady-state cell and substrate concentrations, respectively, and s_i is the inlet substrate concentration.

12.5.9 Continuous Operation of a Plug Flow Reactor

Plug flow operation is an alternative to mixed operation for continuous reactors. No mixing occurs in an ideal plug flow reactor; liquid entering the reactor passes through as a discrete "plug" and does not interact with neighboring fluid elements. This is achieved at high flow rates that minimize back mixing and variations in liquid velocity. Plug flow occurs most readily in column or tubular reactors such as that shown in Figure 12.38. Plug flow reactors can be operated in up flow or downflow mode or, in some cases, horizontally. Plug flow tubular reactors are known by the abbreviation PFTR.

Liquid in a PFTR flows at constant velocity; therefore, all parts of the liquid have identical residence time in the reactor. Concentration gradients of substrate and product develop in the direction of flow in the vessel. At the feed end of the PFTR, the substrate concentration will be high and the product concentration low. At the outlet end of the tube, the substrate concentration will be low and the product concentration high.

Consider operation of the PFTR shown in Figure 12.38. The volumetric liquid flow rate through the vessel is F and the feed stream contains substrate at concentration s_i. The

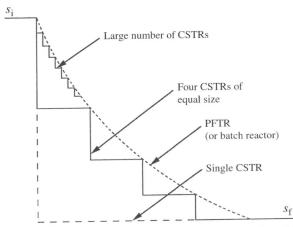

Figure 12.38 Concentration changes in PFTR, single CSTR, and multiple CSTR vessels.

substrate concentration at the reactor outlet is s_f. This exit concentration can be related to the inlet conditions and reactor residence time using mass balance techniques. Let us first consider plug flow operation for enzyme reaction.

Enzyme Reaction

To develop equations for a plug flow enzyme reactor, we will consider a small section of the reactor of length Δz as indicated in Figure 12.38. This section is located at distance z from the feed point. Let us perform a steady-state mass balance on substrate around this control volume using Eq. (4.29). $\hat{M}_i$ is the mass flow rate of substrate entering the control volume; therefore $\hat{M}_i = Fs|_z$ where F is the volumetric flow rate through the reactor and $s|_z$ is the substrate concentration at axial position z. Similarly, $\hat{M}_o$, the mass flow rate of substrate leaving the control volume, is $FS|_{z+\Delta z}$. Substrate is not generated in the reaction, so $R_G = 0$. The rate of substrate consumption R_C is equal to the volumetric rate of enzyme reaction v multiplied by the volume of the section; for Michaelis–Menten kinetics, v is given by Eq. (10.37). The section volume is equal to $A\Delta z$ where A is the cross-sectional area of the reactor. At steady state, the left side of Eq. (4.29) is zero. Therefore, the mass balance equation for substrate is:

$$Fs|_z - Fs|_{z+\Delta z} - \frac{v_{max}S}{K_m + s} A\Delta z = 0 \tag{12.85}$$

where v_{max} is the maximum rate of enzyme reaction and K_m is the Michaelis constant. Dividing through by $A\Delta z$ and rearranging gives:

$$\frac{F\left(s|_{z+\Delta z} - s|_z\right)}{A\Delta z} = \frac{-v_{max}s}{K_m + s} \tag{12.86}$$

The volumetric flow rate F divided by the reactor cross-sectional area A is equal to the superficial velocity through the column, u. Therefore:

$$\frac{u\left(s|_{z+\Delta z} - s|_z\right)}{\Delta z} = \frac{-v_{max}s}{K_m + s} \tag{12.87}$$

F and A are constant so u is also constant. Equation (12.87) applies to any control volume in the reactor of thickness Δz. To obtain an equation that is valid at any point in the reactor, we must take the limit as $\Delta z \to 0$:

$$u\left(\lim_{\Delta z \to 0} \frac{s|_{z+\Delta z} - s|_z}{\Delta z}\right) = \frac{-v_{max}s}{K_m + s} \tag{12.88}$$

Applying the definition of a derivative:

$$u\frac{ds}{dz} = \frac{-v_{max}s}{K_m + s} \tag{12.89}$$

Equation (12.89) is a differential equation for the substrate concentration gradient through the length of a plug flow reactor. Assuming that u and the kinetic parameters are constant, Eq. (12.89) is ready for integration. Separating variables and integrating with the boundary condition $s = s_i$ at $z = 0$ gives an expression for the reactor length L required to achieve an outlet substrate concentration of s_f:

$$L = u \left[\frac{K_m}{v_{max}} \ln \frac{s_i}{s_f} + \frac{s_i - s_f}{v_{max}} \right] \tag{12.90}$$

The residence time τ for continuous reactors is defined in Eq. (12.51). If we divide V and F in Eq. (12.51) by A, we can express the residence time for plug flow reactors in terms of parameters L and u:

$$\tau = \frac{V}{F} = \frac{\left(\dfrac{V}{A}\right)}{\left(\dfrac{F}{A}\right)} = \frac{L}{u} \tag{12.91}$$

Therefore, Eq. (12.90) can be written as:

$$\tau = \frac{K_m}{v_{max}} \ln \frac{s_i}{S_f} + \frac{s_i - s_f}{v_{max}} \tag{12.92}$$

Equations (12.90) and (12.92) allow us to calculate the reactor length and residence time required to achieve conversion of substrate from concentration s_i to s_f at flow rate u. Note that the form of Eq. (12.92) is identical to that of Eq. (12.10) for a batch enzyme reactor. As in batch reactors where the concentrations of substrate and product vary continuously with elapsed time, the concentrations in plug flow reactors change continuously with axial position as material moves from the inlet to the outlet of the vessel. Thus, plug flow operation can interpreted as a way of simulating batch culture properties in a continuous flow system.

Plug flow operation is generally impractical for enzyme conversions unless the enzyme is immobilized and retained inside the vessel. For immobilized enzyme reactions affected by diffusion, Eq. (12.89) must be modified to account for mass transfer effects:

$$u \frac{ds}{dz} = -\eta_T \frac{v_{max} s}{K_m + s} \tag{12.93}$$

where η_T is the total effectiveness factor representing internal and external mass transfer limitations (Section 11.5), s is the bulk substrate concentration, and v_{max} and K_m are intrinsic enzyme kinetic parameters. Because η_T is a function of s, Eq. (12.93) cannot be integrated directly.

Plug flow operation with immobilized enzymes is most likely to be approached in packed bed reactors such as that shown in Figure 12.10. However, the presence of packing in the column can cause substantial back mixing and axial dispersion of the liquid and thus interfere with ideal plug flow. Nevertheless, application of the equations developed in this section can give satisfactory results for the design of fixed bed immobilized enzyme reactors. Example 12.7 calculates product synthesis in an immobilized enzyme PFTR.

EXAMPLE 12.7 Plug Flow Reactor for Immobilized Enzyme

Immobilized lactase is used to hydrolyze lactose in dairy waste to glucose and galactose. The enzyme is immobilized in resin particles and packed into a 0.5-m^3 column. The total effectiveness factor for the system is close to unity, K_m for the immobilized enzyme is 1.32 kg lactose m^{-3}, and v_{max} is 45 kg lactose m^{-3} h^{-1}. The lactose concentration in the feed stream is 9.5 kg m^{-3}. A substrate conversion of 98% is required. The column is operated under plug flow conditions for a total of 310 days per year.

a. At what flow rate should the reactor be operated?
b. How many tons of glucose are produced per year?

Solution

a. For 98% substrate conversion, $s_f = (0.02\ s_i) = 0.19$ kg lactose m^{-3}. Substituting parameter values into Eq. (12.86) gives:

$$\tau = \frac{1.32 \text{ kg m}^{-3}}{45 \text{ kg m}^{-3} \text{ h}^{-1}} \ln\left(\frac{9.5 \text{ kg m}^{-3}}{0.19 \text{ kg m}^{-3}}\right) + \frac{(9.5 - 0.19) \text{ kg m}^{-3}}{45 \text{ kg m}^{-3} \text{ h}^{-1}} = 0.32 \text{ h}$$

From Eq. (12.51):

$$F = \frac{V}{\tau} = \frac{0.5 \text{ m}^3}{0.32 \text{ h}} = 1.56 \text{ m}^3 \text{ h}^{-1}$$

The flow rate required is 1.6 m^3 h^{-1}.

b. The rate of lactose conversion is equal to the difference between the inlet and outlet mass flow rates of lactose:

$$F(s_i - s_f) = 1.56 \text{ m}^3 \text{ h}^{-1}(9.5 - 0.19) \text{ kg m}^{-3} = 14.5 \text{ kg h}^{-1}$$

Converting this to an annual rate based on 310 days of operation per year and a molecular weight for lactose of 342:

$$\text{Rate of lactose conversion} = 14.5 \text{ kg h}^{-1} \cdot \left|\frac{24 \text{ h}}{1 \text{ day}}\right| \cdot \left|\frac{310 \text{ days}}{1 \text{ year}}\right| \cdot \left|\frac{1 \text{ kgmol}}{342 \text{ kg}}\right| = 315 \text{ kgmol year}^{-1}$$

(Continued)

The enzyme reaction for lactase is:

$$lactose + H_2O \rightarrow glucose + galactose$$

Therefore, from reaction stoichiometry, 315 kgmol of glucose are produced per year. The molecular weight of glucose is 180; therefore:

$$Rate\ of\ glucose\ production = 315\ kgmol\ year^{-1} \cdot \left|\frac{180\ kg}{1\ kgmol}\right| \cdot \left|\frac{1\ ton}{1000\ kg}\right| = 56.7\ ton\ year^{-1}$$

The amount of glucose produced per year is 57 tons.

Microbial Cell Culture

Analysis of plug flow reactors hosting cell culture follows the same procedure as for hosting an enzyme reaction. If the cell-specific growth rate is constant and equal to μ_{max} throughout the reactor and cell death can be neglected, the equations for the reactor residence time are analogous to those derived in Section 12.5.1 for batch cultivation:

$$\tau = \frac{1}{\mu_{max}} \ln \frac{x_f}{x_i} \tag{12.94}$$

where τ is the reactor residence time defined in Eq. (12.51), x_i is the biomass concentration at the inlet, and x_f is the biomass concentration at the outlet. The form of Eq. (12.94) is identical to that of Eq. (12.20) for batch culture time.

Plug flow operation is not suitable for cultivation of suspended cells unless the biomass is recycled or there is continuous inoculation of the vessel. Plug flow operation with cell recycle is used in large-scale wastewater treatment; however, these applications are limited. Plug flow reactors may be suitable for immobilized cell reactions when the catalyst is packed into a fixed bed as shown in Figure 12.8. Even so, operating problems such as those mentioned in Section 12.2.5 mean that PFTRs are rarely employed for industrial bioprocesses.

12.5.10 Comparison Between Major Modes of Reactor Operation

The relative performance of batch, chemostat, and PFTR reactors can be considered from a theoretical point of view in terms of the substrate conversions and product concentrations achieved using vessels of the same size. As the total reactor volume is not fully utilized at all times during fed-batch operation, it is difficult to include this mode of operation in a general comparison.

The kinetic characteristics of PFTRs are the same as batch reactors (Section 12.5.8): the residence time required for conversion in a plug flow reactor is therefore the same as in a mixed vessel operated in batch. It can also be shown theoretically that as the number of stages in a chemostat cascade increases, the conversion characteristics of the entire system approach those of an ideal plug flow or mixed batch reactor. This is shown diagrammatically in Figure 12.35. The smooth dashed curve represents the progressive decrease in substrate concentration with time spent in a PFTR or batch reactor; the concentration is reduced from s_i at the inlet to s_f at the outlet. In a single well-mixed chemostat operated with the same inlet and outlet concentrations because conditions in the vessel are uniform, there is a step change in substrate concentration as soon as the feed enters the reactor. In a cascade of chemostats, the concentration is uniform in each reactor but there is a stepwise drop in concentration between each stage. As illustrated in Figure 12.39, the greater the number of units in a chemostat cascade, the closer the concentration profile approaches plug flow or batch behavior.

The benefits associated with different reactor designs or modes of operation depend on the reaction kinetics. For zero-order reactions, there is no difference between single batch, chemostat, and PFTR reactors in terms of the overall conversion rate. However, for most reactions including first-order and Michaelis–Menten conversions, the rate of reaction decreases as the concentration of substrate decreases. The reaction rate is therefore high at the start of batch reaction or at the entrance to a plug flow reactor because the substrate concentration there is at its greatest. Subsequently, the reaction velocity falls gradually as substrate is consumed. In contrast, substrate entering a chemostat is immediately diluted to the final or outlet steady-state concentration so that the rate of reaction is comparatively low for the entire reactor.

Accordingly, for first-order and Michaelis–Menten reactions, chemostats achieve lower substrate conversions and lower product concentrations than batch reactors or PFTRs of the same volume. In practice, batch processing is usually preferred to PFTR systems for enzyme reactions because of the operating problems mentioned in Section 12.2.5. However, as discussed in Section 12.5.2, the total time for batch operation depends on the duration of the downtime between batches as well as on the actual conversion time. The length of downtime varies considerably from system to system, therefore, we cannot account for it here in a general way.

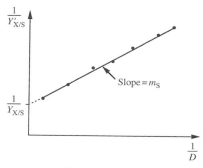

Figure 12.39 Graphical determination of the maintenance coefficient m_S and true biomass yield $Y_{X/S}$ using data from chemostat culture.

The downtime between batches should be minimized as much as possible to maintain high overall production rates.

The comparison between reactors yields a different result if the reaction is autocatalytic, such as in cell cultures. At the beginning of batch culture, the rate of substrate conversion is low because relatively few cells are present: as catalyst is produced during cultivation, the volumetric reaction rate increases as the conversion proceeds. However, in chemostat operations, substrate entering the vessel is exposed immediately to a relatively high biomass concentration so that the reaction rate is also high. This rate advantage may disappear if the steady-state substrate concentration in the chemostat is so small that, despite the higher biomass levels present, the rate is lower than the average in a batch or PFTR device.

Despite the productivity benefits associated with chemostats, an overwhelming majority of commercial cell cultivations are conducted in batch processes. The reasons lie with the practical advantages associated with batch culture. The risk of contamination is lower in batch systems than in continuous flow reactors; equipment and control failures during long-term continuous operation are also potential problems. Continuous cultivation is feasible only when the cells are genetically stable; if developed strains revert to more rapidly growing mutants, the culture can become dominated over time by the revertant cells. As freshly produced inocula are used in batch processes, closer control over the genetic characteristics of the culture is achieved. Continuous culture is not suitable to produce metabolites normally formed near stationary phase when the culture growth rate is low; as mentioned above, the productivity in a batch reactor is likely to be greater than in a chemostat under these conditions. Production can be much more flexible using batch processing; for example, different products each with small market volumes can be made in different batches. In contrast, continuous cultivations must be operated for lengthy periods to reap the full benefits of their high productivity.

12.6 Sterilization

Commercial cell cultivations typically require thousands of liters of liquid medium and millions of liters of air. These raw materials must be provided free of contaminating organisms for processes operated with pure cultures. There are many methods available for process sterilization, including chemical treatment, exposure to ultraviolet, gamma, and X-ray radiation, sonication, filtration, and heating, however, only the last two are used in large-scale operations. Aspects of bioreactor design and construction for aseptic operation were described in Sections 12.3.1 and 12.3.2. Here, we consider the design of sterilization systems for liquids and gases.

12.6.1 Batch Heat Sterilization of Liquids

Liquid medium is commonly sterilized in batch in the vessel where it will be used. The liquid is heated to the sterilization temperature by introducing steam into the coils or jacket of the vessel (Figure 8.1); alternatively, steam is bubbled directly into the medium, or the vessel is heated electrically. If direct steam injection is used, allowance must be made for dilution of the medium by condensate, which typically adds 10% to 20% to the liquid volume. The quality of the steam used for direct injection must also be sufficiently high to avoid contaminating the medium with metal ions or organics. A typical temperature–time profile for batch sterilization is shown in (Figure 12.40A). Depending on the rate of heat transfer from the steam or electrical element, raising the temperature of the medium in large bioreactors can take a considerable period of time. Once the holding or sterilization temperature is reached, the temperature is held constant for time t_{hd}. Cooling water in the coils or jacket of the bioreactor is then used to reduce the medium temperature to the required level.

For operation of batch sterilization systems, we must be able to estimate the holding time required to achieve the desired level of cell destruction. As well as destroying contaminant organisms, heat sterilization is also capable of destroying nutrients in the medium.

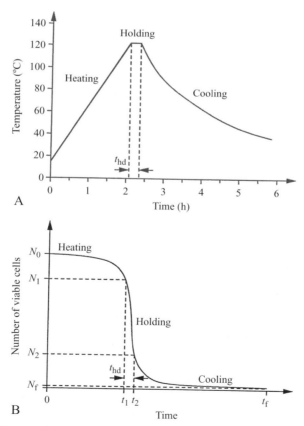

Figure 12.40 (A) Variation of temperature with time for batch sterilization of liquid medium. (B) Reduction in the number of viable cells with time during batch sterilization.

Let us denote the number of contaminants present in the raw medium N_0. As indicated in Figure 12.40(B), during the heating period this number is reduced to N_1. At the end of the holding period, the cell number is N_2. The final cell number after cooling is N_f. Ideally, N_f is zero: at the end of the sterilization cycle as we want no contaminants to be present. However, because absolute sterility would require an infinitely long sterilization time, it is theoretically impossible to achieve. Normally, the target level of contamination is expressed as a fraction of a cell, which is related to the *probability* of contamination. For example, we could aim for $N_f = 10^{-3}$; this means that we accept the risk that one batch in 1000 will not be sterile at the end of the process. Once N_0 and N_f are known, we can determine the holding time required to reduce the number of cells from N_1 to N_2 by considering the kinetics of cell death.

The rate of heat sterilization is governed by the equations for thermal death outlined in Section 10.15. In a batch vessel where cell death is the only process affecting the number of viable cells, from Eq. (10.117):

$$\frac{dN}{dt} = -k_d N \tag{12.95}$$

where N is the number of viable cells, t is time, and k_d is the specific death constant. Equation (12.95) applies to each stage of the batch sterilization cycle: heating, holding, and cooling. However, because k_d is a strong function of temperature, direct integration of Eq. (12.95) is valid only when the temperature is constant, that is, during the holding period. The result is:

$$\ln \frac{N_1}{N_2} = k_d\, t_{hd} \tag{12.96}$$

or

$$t_{hd} = \frac{\ln \dfrac{N_1}{N_2}}{k_d} \tag{12.97}$$

where t_{hd} is the holding time, N_1 is the number of viable cells at the start of holding, and N_2 is the number of viable cells at the end of holding. k_d is evaluated as a function of temperature using the Arrhenius equation from Chapter 10:

$$k_d = A\, e^{-E_d/RT} \tag{10.74}$$

where A is the Arrhenius constant or frequency factor, E_d is the activation energy for thermal cell death, R is the ideal gas constant, and T is absolute temperature.

To use Eq. (12.97), we must know N_1 and N_2. These numbers are determined by considering the effect of cell death during the heating and cooling periods when the temperature is not constant. Combining Eqs. (12.95) and (10.74) gives:

$$\frac{dN}{dt} = -Ae^{-E_d/RT}N \tag{12.98}$$

Integration of Eq. (12.98) gives for the heating period:

$$\ln\frac{N_0}{N_1} = \int_0^{t_1} Ae^{-E_d/RT}dt \tag{12.99}$$

and for the cooling period:

$$\ln\frac{N_2}{N_f} = \int_{t_2}^{t_f} Ae^{-E_d/RT}dt \tag{12.100}$$

where t_1 is the time at the end of heating, t_2 is the time at the end of holding, and t_f is the time at the end of cooling. We cannot complete integration of these equations until we know how the temperature varies with time during the heating and cooling periods.

Unsteady-state temperatures during heating and cooling can be determined from the heat transfer properties of the system. Depending on how heating and cooling are achieved, the general form of equations for temperature as a function of time is either linear, exponential, or hyperbolic, as shown in Figure 12.41 and Table 12.3. Applying an appropriate expression for

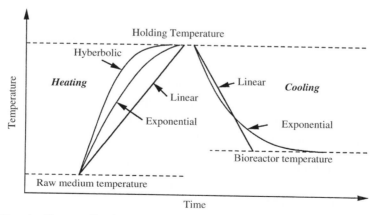

Figure 12.41 Graph of generalized temperature–time profiles for the heating and cooling stages of a batch sterilization cycle. *From F.H. Deindoerfer and A.E. Humphrey, 1959, Analytical method for calculating heat sterilization times.* Appl. Microbiol. *7, 256–264.*

Table 12.3: General Equations for Temperature as a Function of Time During the Heating and Cooling Periods of Batch Sterilization

Heat transfer method	Temperature–time profile
Heating	
Direct sparging with steam	$T = T_0\left(1 + \dfrac{\dfrac{h\hat{M}_s t}{M_m C_p T_0}}{1 + \dfrac{\hat{M}_s}{M_m}t}\right)$ (hyperbolic)
Electrical heating	$T = T_0\left(1 + \dfrac{\hat{Q}t}{M_m C_p T_0}\right)$ (linear)
Heat transfer from isothermal steam	$T = T_s\left(1 + \dfrac{T_0 - T_s}{T_s}e^{\left(\frac{-UAt}{\hat{M}_m C_p}\right)}\right)$ (exponential)
Cooling	
Heat transfer to nonisothermal cooling water	$T = T_{ci}\left\{1 + \dfrac{T_0 - T_{ci}}{T_{ci}}e^{\left(\frac{-\hat{M}_w C_{pw} t}{M_m C_p}\right)\left[1 - e^{\left\lvert\frac{-UA}{\hat{M}_w C_{pw}}\right\rvert}\right]}\right\}$ (exponential)

A, surface area for heat transfer; C_p, specific heat capacity of medium; C_{pw}, specific heat capacity of cooling water; h, specific enthalpy difference between the steam and raw medium; M_m, initial mass of medium; $\hat{M}_s$, mass flow rate of steam; $\hat{M}_w$, mass flow rate of cooling water; $\hat{Q}$, rate of heat transfer; T, temperature; T_0, initial medium temperature; T_{ci}, inlet temperature of cooling water; T_s, steam temperature; t, time; U, overall heat transfer coefficient.

From F.H. Deindoerfer and A.E. Humphrey, 1959, Analytical method for calculating heat sterilization times. *Appl. Microbiol.* 7, 256–264.

T in Eq. (12.99) from Table 12.3 allows the cell number N_1 at the start of the holding period to be determined. Similarly, applying an appropriate expression for T in Eq. (12.100) for cooling allows evaluation of N_2 at the end of the holding period. Use of these results for N_1 and N_2 in Eq. (12.97) completes the holding time calculation.

Normally, cell death at temperatures below about 100°C is minimal. However, when heating and cooling are relatively slow, temperatures can remain elevated and close to the holding temperature for considerable periods of time. As a result, the reduction in cell numbers outside of the holding period is significant. Usually, the holding time is of the order of minutes, whereas heating and cooling of large volumes of liquid can take hours. Further information and sample calculations for batch sterilization are given in [40] and [41].

12.6.2 *Continuous Heat Sterilization of Liquids*

Continuous sterilization, particularly a high-temperature, short-exposure-time process, can reduce thermal damage to the medium significantly compared with batch sterilization, while achieving high levels of cell deactivation. Other advantages include improved steam economy and more reliable scale-up. The amount of steam needed for continuous sterilization is 20% to 25% of that used in batch processes; the time required is also significantly reduced because heating and cooling are virtually instantaneous. Typical equipment configurations for continuous sterilization are shown in Figure 12.42. The rates of heating and cooling in continuous sterilizers are typically more rapid than in batch systems (Figure 12.43). Accordingly, in the design of continuous sterilizers, contributions to cell death outside of the holding period are generally ignored.

12.6.3 *Filter Sterilization of Liquids*

Sometimes, culture media or selected medium ingredients are sterilized by filtration rather than by heat. Medium containing heat-labile components such as serum, differentiation

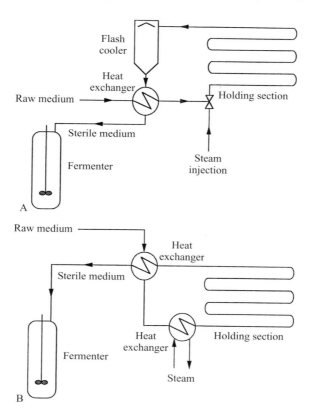

Figure 12.42 Continuous sterilizing equipment: (A) continuous steam injection with flash cooling; (B) heat transfer using heat exchangers.

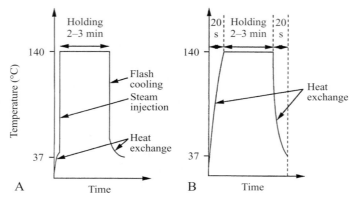

Figure 12.43 Variation of temperature with time in the continuous sterilizers of Figure 12.38: (A) continuous steam injection with flash cooling; (B) heat transfer using heat exchangers.

factors, enzymes, or other proteins is easily destroyed by heat and must be sterilized by other means. Typically, the membranes used for filter sterilization of liquids are made of cellulose esters or other polymers and have pore diameters between 0.2 and 0.45 µm. As medium passes through the filter, bacteria and other particles with dimensions greater than the pore size are screened out and collect on the surface of the membrane. The small pore sizes used in liquid filtration mean that the membranes readily become blocked unless the medium is prefiltered to remove any large particles.

Filtration using pore sizes of 0.2 to 0.45 µm may not be as effective or reliable as heat sterilization because viruses and mycoplasma are able to pass through the membranes. However, the availability of 0.1-µm sterilizing-grade membrane filters for mycoplasma removal has seen the widespread adoption of filter sterilization in many mammalian cell culture processes where complex, proteinaceous media are used routinely.

12.6.4 Sterilization of Air

The number of microbial cells in air is of the order 10^3 to $10^4 \, \mathrm{m}^{-3}$ [40]. Filtration is the most common method used to sterilize air in large-scale bioprocesses; heat sterilization of gases is economically impractical. *Depth filters* consisting of compacted beds or pads of fibrous material such as glass wool have been used widely in the bioprocess industry. Distances between the fibers in depth filters are typically 2 to 10 µm, or up to 10 times greater than the dimensions of the bacteria and spores to be removed. Airborne particles penetrate the bed to various depths before their passage through the filter is arrested; the depth of the filter required to produce air of sufficient quality depends on the operating flow rate and the incoming level of contamination.

Increasingly, depth filters are being replaced in industrial applications by membrane cartridge filters. These filters use steam-sterilizable or disposable polymeric membranes that act as surface filters trapping contaminants as on a sieve. Membrane filter cartridges typically contain a pleated, hydrophobic filter with small and uniformly sized pores of diameter 0.45 μm or less. The hydrophobic nature of the surface minimizes problems with filter wetting while the pleated configuration allows a high filtration area to be packed into a small cartridge volume. Prefilters built into the cartridge or located upstream reduce fouling of the membrane by removing large particles, oil, water droplets, and foam from the incoming gas.

Filters are also used to sterilize off-gases leaving bioreactors. In this case, the objective is to prevent release into the atmosphere of any organisms entrained in aerosols in the headspace of the reactor. The concentration of cells in unfiltered bioreactor off-gas is several times greater than in air.

12.7 Summary of Chapter 12

Chapter 12 contains a variety of qualitative and quantitative information about the design and operation of bioreactors. After studying this chapter, you should:

* Be able to assess in general terms the influence of reactor engineering on total production costs in bioprocessing
* Be familiar with a range of bioreactor configurations in addition to the standard *stirred tank*, including *bubble column*, *airlift*, *packed bed*, *fluidized bed*, and *trickle bed* designs
* Understand the practical aspects of bioreactor construction, particularly those aimed at maintaining aseptic conditions
* Know the types of measurement required for *bioreactor monitoring* and the problems associated with the lack of online analytical methods for important bioreactor parameters
* Be familiar with established and alternative approaches to *bioreactor control*
* Be able to estimate *batch reaction times* for enzyme and cell reactions
* Be able to predict the performance of *fed-batch reactors* operated under *quasi-steady state conditions*
* Be able to analyze the operation of *continuous stirred tank reactors* and *continuous plug flow reactors*
* Understand the theoretical advantages and disadvantages of batch and continuous reactors for enzyme and cell reactions
* Know how to use steady-state chemostat data to determine cell kinetic and yield parameters
* Know the design procedures for batch and continuous heat sterilization of liquid medium
* Be able to describe methods for filter sterilization of liquid medium and bioreactor gases

Problems

12.1. Economics of batch enzyme conversion

An enzyme is used to convert substrate to a commercial product in a 1600-L batch reactor. v_{max} for the enzyme is 0.9 g substrate L^{-1} h^{-1}; K_m is 1.5 g substrate L^{-1}. The substrate concentration at the start of the reaction is 3 g L^{-1}. According to the reaction stoichiometry, conversion of 1 g of substrate produces 1.2 g of product. The cost of operating the reactor including labor, maintenance, energy, and other utilities is estimated at $4800 per day. The cost of recovering the product depends on the substrate conversion achieved and the resulting concentration of product in the final reaction mixture. For conversions between 70% and 100%, the cost of downstream processing can be approximated using the equation:

$$C = 155 - 0.33X$$

where C is the cost in $ per kg of product treated and X is the percentage substrate conversion. Product losses during processing are negligible. The market price for the product is $750 kg^{-1}. Currently, the enzyme reactor is operated with 75% substrate conversion; however, it is proposed to increase this to 90%. Estimate the effect that this will have on the economics of the process.

12.2. Batch production of aspartic acid using cell-bound enzyme

Aspartase enzyme is used industrially for the manufacture of aspartic acid, a component of low-calorie sweetener. Fumaric acid ($C_4H_4O_4$) and ammonia are converted to aspartic acid ($C_4H_7O_4N$) according to the equation:

$$C_4H_4O_4 + NH_3 \rightleftarrows C_4H_7O_4N$$

Under investigation is a process using aspartase in intact *Bacillus cadaveris* cells. In the substrate range of interest, the conversion can be described using Michaelis–Menten kinetics with $K_m = 4.0$ g L^{-1}. The substrate solution contains 15% (w/v) ammonium and fumarate; enzyme is added in the form of lyophilized cells and the reaction is stopped when 85% of the substrate is converted. At 32°C, v_{max} for the enzyme is 5.9 g aspartic acid L^{-1} h^{-1} and its half-life is 10.5 days. At 37°C, v_{max} increases to 8.5 g aspartic acid L^{-1} h^{-1} but the half-life is reduced to 2.3 days.

a. Which operating temperature would you recommend?

b. The average downtime between batch reactions is 28 h. At the temperature chosen in (a), calculate the reactor volume required to produce 5000 tons of aspartic acid per year.

12.3. Prediction of batch culture time

A strain of *Escherichia coli* has been genetically engineered to produce a therapeutic human protein. A batch culture is started by inoculating 12 g of cells into a 100-L bubble column bioreactor containing 10 g L^{-1} glucose. The culture does not exhibit a lag phase. The maximum specific growth rate of the cells is 0.9 h^{-1}; the biomass yield from glucose is 0.575 g biomass (g glucose)$^{-1}$.

a. Estimate the time required to reach stationary phase (why would the cells enter stationary phase?).

b. What would be the final cell concentration if the cultivation is stopped after only 70% of the substrate is consumed?

12.4. Fed-batch scheduling

Nicotiana tabacum cells are cultured to high density for production of polysaccharide gum. The reactor used is a stirred tank that initially contains 100 L of medium. The maximum specific growth rate of the culture is 0.18 day^{-1} and the yield of biomass from substrate is 0.5 g biomass (g substrate)$^{-1}$. The concentration of growth-limiting substrate in the medium is 3% (w/v). The reactor is inoculated with 1.5 g L^{-1} of cells and operated in batch until the substrate is exhausted; medium flow is then started at a rate of 4 L day^{-1}. Fed-batch operation occurs for 40 days under quasi-steady-state conditions.

a. Estimate the batch culture time and the final biomass concentration achieved after the batch culture period.

b. What is the final mass of cells in the reactor?

c. The bioreactor is available 275 days per year with a downtime between runs of 24 h. How much plant cell biomass is produced annually?

12.5. Fed-batch production of cheese starter culture

Lactobacillus casei is propagated under anaerobic conditions to provide a starter culture for manufacture of Swiss cheese. The culture produces lactic acid as a by-product of energy metabolism. The system has the following characteristics:

$Y_{X/S} = 0.23$ kg biomass (kg lactose)$^{-1}$

$K_S = 0.15$ kg lactose m^{-3}

$\mu_{max} = 0.35$ h^{-1}

$m_S = 0.135$ kg lactose (kg biomass)$^{-1}$ h^{-1}

A stirred bioreactor is operated in fed-batch mode under quasi-steady-state conditions with a feed flow rate of 4 m^3 h^{-1} and feed substrate concentration of 80 kg lactose m^{-3}. After 6 h, the liquid volume is 40 m^3.

a. What was the initial culture volume?

b. What is the concentration of substrate at quasi-steady state?

c. What is the concentration of cells at quasi-steady state?

d. What mass of cells is produced after 6 h of fed-batch operation?

12.6. Continuous enzyme conversion in a fixed bed reactor

A system is being developed to remove urea from the blood of patients with renal failure. A prototype fixed bed reactor is set up using urease immobilized in 2-mm-diameter gel beads. Buffered urea solution is recycled rapidly through the bed so that the system is well mixed and external mass transfer effects are negligible. The urease reaction is:

$$(NH_2)_2CO + 3 H_2O \rightarrow 2 NH_4^+ + HCO_3^- + OH^-$$

K_m for the immobilized urease is 0.54 g urea L^{-1}. The volume of beads in the reactor is 250 cm^3, the total amount of urease is 10^{-4} g, and the turnover number is 11,000 g NH$_4^+$ (g enzyme)$^{-1}$ s^{-1}. The effective diffusivity of urea in the gel is 7 × 10^{-6} cm^2 s^{-1}. The reactor is operated continuously with a liquid volume of 1 L. The feed stream contains 0.42 g L^{-1} of urea and the desired urea concentration after enzyme treatment is 0.02 g L^{-1}. Ignoring enzyme deactivation, what volume of urea solution can be treated in 30 min?

12.7. Batch and continuous biomass production

Pseudomonas methylotrophus is used to produce single-cell protein from methanol in a 1000-m³ airlift bioreactor. The biomass yield from substrate is 0.41 g biomass (g methanol)$^{-1}$, K_S is 0.7 mg methanol L^{-1}, and the maximum specific growth rate is 0.44 h^{-1}. The medium contains 4% (w/v) methanol. A substrate conversion of 98% is desirable. The reactor may be operated in either batch or continuous mode. If operated in batch, an inoculum of 0.01% (w/v) is used and the downtime between batches is 20 h. If continuous operations are used at steady state, a downtime of 25 days is expected per year. Neglecting maintenance requirements, compare the annual biomass production achieved using batch and continuous reactors.

12.8. Bioreactor design for immobilized enzymes

6-Aminopenicillanic acid used to produce semi-synthetic penicillin is prepared by enzymatic hydrolysis of bioreactor-derived penicillin-G. Penicillin-G acylase immobilized in alginate is being considered for the process. The immobilized enzyme particles are sufficiently small so that mass transfer does not affect the reaction rate. The starting concentration of penicillin-G is 10% (w/v); because of the high cost of the substrate, 99% conversion is required. Under these conditions, enzymatic conversion of penicillin-G can be considered a first-order reaction. It has not been decided whether a batch, CSTR, or plug flow reactor would be most suitable. The downtime between batch reactions is expected to be 20 h. For the batch and CSTR reactors, the reaction rate constant is 0.8×10^{-4} s^{-1}; in the PFTR, the packing density of enzyme beads can be up to four times greater than in the other reactors. Determine the smallest reactor required to treat 400 tons of penicillin-G per year.

12.9. Chemostat culture with protozoa

Tetrahymena thermophila protozoa have a doubling time of 6.5 h when grown using bacteria as the growth-limiting substrate. The yield of protozoal biomass is 0.33 g biomass (g bacteria)$^{-1}$ and the effective $K_s = 12$ mg L^{-1}. The protozoa are cultured at steady state in a chemostat using a feed stream containing 10 g L^{-1} of nonviable bacteria.

a. What is the maximum dilution rate for operation of the chemostat?

b. What is the concentration of *T. thermophila* when the operating dilution rate is one-half of the maximum?

c. What is the concentration of bacteria when the dilution rate is three-quarters of the maximum?

d. What is the biomass productivity when the dilution rate is one-third of the maximum?

12.10. Two-stage chemostat for fungal secondary metabolite production

A two-stage chemostat system is used for production of a secondary metabolite. The volume of each reactor is 0.5 m³; the flow rate of feed is 50 L h^{-1}. Mycelial growth occurs in the first reactor; the second reactor is used for product synthesis. The concentration of substrate in the feed is 10 g L^{-1}. Kinetic and yield parameters for the organism are:

$Y_{X/S} = 0.5$ kg biomass (kg substrate)$^{-1}$

$K_S = 1.0$ kg substrate m^{-3}

$\mu_{max} = 0.12$ h^{-1}

$m_S = 0.025$ kg substrate (kg biomass)$^{-1}$ h^{-1}

$q_P = 0.16$ kg product (kg biomass)$^{-1}$ h^{-1}

$Y_{P/S} = 0.85$ kg product (kg substrate)$^{-1}$

Assume that product synthesis is negligible in the first reactor and growth is negligible in the second reactor.

a. Determine the cell and substrate concentrations entering the second reactor.

b. What is the overall substrate conversion?

c. What is the final concentration of product?

12.11. Growth of cyanobacteria in a continuous bioreactor with cell recycle

Ammonium-limited growth of the cyanobacteria *Oscillatoria agardhii* is studied using a 1-L chemostat. The temperature of the culture is controlled at 25°C and continuous illumination is applied at an intensity of 37 μE m^{-2} s^{-1}. Sterile ammonia solution containing 0.1 mg L^{-1} N is fed to the cells. The maximum specific growth rate of the cyanobacteria is 0.5 day^{-1} and the K_s for N is 0.5 μg L^{-1}. The yield of biomass on nitrogen is 18 mg of dry weight per mg of N. The reactor is operated to give a residence time of 60 h. No extracellular products are formed.

a. What is the overall substrate conversion?

b. What is the biomass productivity?

Intensification of the culture process is investigated using cell recycle. While keeping the feed rate of fresh medium the same, a recycle stream containing three times the concentration of cells in the reactor outflow is fed back to the vessel at a volumetric flow rate that is one-quarter the flow rate of the fresh feed. The substrate concentration s in a chemostat with cell recycle is given by the equation:

$$s = \frac{D K_S (1 + \alpha - \alpha\beta)}{\mu_{max} - D(1 + \alpha - \alpha\beta)}$$

where D is the dilution rate based on the fresh feed flow rate, μ_{max} and K_S are the maximum specific growth rate and substrate constant for the organism, respectively, α is the volumetric recycle ratio, and β is the biomass concentration factor for the recycle stream.

c. How much does using a cell recycle reduce the size of the bioreactor required to achieve the same level of substrate conversion as that obtained without cell recycle?

12.12. Kinetic analysis of bioremediating bacteria using a chemostat

A strain of *Ancylobacter* bacteria capable of growing on 1,2-dichloroethane is isolated from sediment in the river Rhine. The bacteria are to be used for onsite bioremediation of soil contaminated with chlorinated halogens. Kinetic parameters for the organism are determined using data obtained from chemostat culture. A 1-L bioreactor is used with a feed stream containing 100 μM of 1,2-dichloroethane. Steady-state substrate concentrations are measured as a function of chemostat flow rate.

Flow rate (mL h^{-1})	Substrate concentration (μM)
10	17.4
15	25.1
20	39.8
25	46.8
30	69.4
35	80.1
50	100

  a. Determine μ_{max} and K_S for this organism.

  b. Determine the maximum practical operating flow rate.

12.13. Kinetic and yield parameters of an auxotrophic mutant

An *Enterobacter aerogenes* auxotroph capable of overproducing threonine has been isolated. The kinetic and yield parameters for this organism are investigated using a 2-L chemostat fed with medium containing $10\,g\,L^{-1}$ glucose. Steady-state cell and substrate concentrations are measured at a range of reactor flow rates.

Flow rate (L h⁻¹)	Cell concentration (g L⁻¹)	Substrate concentration (g L⁻¹)
1.0	3.15	0.010
1.4	3.22	0.038
1.6	3.27	0.071
1.7	3.26	0.066
1.8	3.21	0.095
1.9	3.10	0.477

Determine the maximum specific growth rate, the substrate constant, the maintenance coefficient, and the true biomass yield from glucose for this culture.

12.14. Chemostat culture for metabolic engineering

A chemostat of working volume 400 mL is used to obtain steady-state data for metabolic flux analysis. *Lactobacillus rhamnosus* is cultured at pH 6.0 using sterile medium containing $12\,g\,L^{-1}$ glucose and no lactic acid. Lactic acid production is coupled with energy metabolism in this organism. Concentrations of biomass, glucose, and lactic acid are measured at five different operating flow rates.

Flow rate (mL h⁻¹)	Steady-state concentration (g L⁻¹)		
	Biomass	Glucose	Lactic acid
48	1.26	0.078	16.4
112	1.33	0.285	13.1
140	1.33	0.466	12.8
164	1.30	0.706	11.4
216	0.94	4.09	8.07

  a. Determine μ_{max} and K_S.

  b. Determine m_S and $Y_{X/S}$.

  c. Determine m_P and $Y_{P/X}$.

  d. Determine $Y_{P/S}$.

  e. Determine the maximum practical operating flow rate for this system.

  f. Determine the operating flow rate for maximum biomass productivity.

  g. At the flow rate determined in (f), what are the steady-state rates of biomass and lactic acid production and substrate consumption?

Uncited references

[20]; [21]; [24–26]; [30–36]

References

[1]   P.N. Royce, A discussion of recent developments in fermentation monitoring and control from a practical perspective, Crit. Rev. Biotechnol. 13 (1993) 117–149.

[2]   B.L. Maiorella, H.W. Blanch, C.R. Wilke, Economic evaluation of alternative ethanol fermentation processes, Biotechnol. Bioeng. 26 (1984) 1003–1025.

[3]   L.A. Boon, F.W.J.M.M. Hoeks, R.G.J.M. van der Lans, W. Bujalski, M.O. Wolff, A.W. Nienow, Comparing a range of impellers for "stirring as foam disruption", Biochem. Eng. J. 10 (2002) 183–195.

[4]   J.J. Heijnen, K. van't Riet, Mass transfer, mixing and heat transfer phenomena in low viscosity bubble column reactors, Chem. Eng. J. 28 (1984) B21–B42.

[5]   K. van't Riet, J. Tramper, Basic Bioreactor Design, Marcel Dekker, 1991.

[6]   M.Y. Chisti, Airlift Bioreactors, Elsevier Applied Science, 1989.

[7]   P. Verlaan, J. Tramper, K. van't Riet, K.Ch.A.M. Luyben, A hydrodynamic model for an airlift-loop bioreactor with external loop, Chem. Eng. J 33 (1986) B43–B53.

[8]   U. Onken, P. Weiland, Hydrodynamics and mass transfer in an airlift loop fermentor, Eur. J. Appl. Microbiol. Biotechnol. 10 (1980) 31–40.

[9]   P. Verlaan, J.-C. Vos, K. van't Riet, Hydrodynamics of the flow transition from a bubble column to an airlift-loop reactor, J. Chem. Tech. Biotechnol 45 (1989) 109–121.

[10] M.H. Siegel, J.C. Merchuk, K. Schügerl, Airlift reactor analysis: interrelationships between riser, downcomer, and gas–liquid separator behavior, including gas recirculation effects, AIChE J 32 (1986) 1585–1596.

[11] A.B. Russell, C.R. Thomas, M.D. Lilly, The influence of vessel height and top-section size on the hydrodynamic characteristics of airlift fermentors. Biotechnol. Bioeng. 43 (1994) 69–76.

[12] B. Atkinson, F. Mavituna, Biochemical Engineering and Biotechnology Handbook, second ed., Macmillan, 1991.

[13] M. Morandi, A. Valeri, Industrial scale production of β-interferon. Adv. Biochem. Eng./Biotechnol 37 (1988) 57–72.

[14] B. Cameron, Mechanical seals for bioreactors. Chem. Engr 442 (November) (1987) 41–42.

[15] Y. Chisti, Assure bioreactor sterility, Chem. Eng. Prog. 88 (September) (1992) 80–85.

[16] T. Akiba, T. Fukimbara, Fermentation of volatile substrate in a tower-type fermenter with a gas entrainment process, J. Ferment. Technol. 51 (1973) 134–141.

[17] P.F. Stanbury, A. Whitaker, S.J. Hall, Principles of Fermentation Technology, second ed., (Chapter 8), Butterworth-Heinemann, 1995.

[18] B. Kristiansen, Instrumentation, in: J. Bu'Lock, B. Kristiansen, (Eds.), Basic Biotechnology, Academic Press (1987), pp. 253–281.

[19] S.W. Carleysmith, Monitoring of bioprocessing, In: J.R. Leigh (Ed.), Modelling and Control of Fermentation Processes, Peter Peregrinus (1987), pp. 97–117.

[20] J. Cooper, T. Cass. Biosensors, Oxford University Press (2003).

[21] B.R. Eggins, Chemical Sensors and Biosensors, John Wiley, 2002.

[22] A. Hayward, On-line, *in-situ* measurements within fermenters, in: B. McNeil, L.M. Harvey, (Eds.), Practical Fermentation Technology, John Wiley (2008), pp. 271–288.

[23] D. Dochain (Ed.), Automatic Control of Bioprocesses, John Wiley (2008).

[24] P.N. Royce, N.F. Thornhill, Analysis of noise and bias in fermentation oxygen uptake rate data, Biotechnol. Bioeng. 40 (1992) 634–637.

[25] M. Meiners, W. Rapmundt, Some practical aspects of computer applications in a fermentor hall, Biotechnol. Bioeng. 25 (1983) 809–844.

[26] R.T.J.M. van der Heijden, C. Hellinga, K. Ch, A.M. Luyben, G. Honderd, State estimators (observers) for the on-line estimation of non-measurable process variables, Trends Biotechnol 7 (1989) 205–209.

[27] G.A. Montague, A.J. Morris, J.R. Bush, Considerations in control scheme development for fermentation process control, IEEE Contr. Sys. Mag 8 (April) (1988) 44–48.

[28] G.A. Montague, A.J. Morris, M.T. Tham, Enhancing bioprocess operability with generic software sensors, J. Biotechnol 25 (1992) 183–201.

[29] D.E. Seborg, T.F. Edgar, D.A. Mellichamp, F.J. Doyle, Process Dynamics and Control, third ed., John Wiley, 2011.

[30] J.-I. Horiuchi, Fuzzy modeling and control of biological processes, J. Biosci. Bioeng 94 (2002) 574–578.

[31] K.B. Konstantinov, T. Yoshida, Knowledge-based control of fermentation processes, Biotechnol. Bioeng. 39 (1992) 479–486.

[32] J. Glassey, G. Montague, P. Mohan, Issues in the development of an industrial bioprocess advisory system, Trends Biotechnol 18 (2000) 136–141.

[33] R. Babuška, M.R. Damen, C. Hellinga, H. Maarleveld, Intelligent adaptive control of bioreactors, J. Intellig. Manuf 14 (2003) 255–265.

[34] Z.K. Nagy, Model based control of a yeast fermentation bioreactor using optimally designed artificial neural networks, Chem. Eng. J. 127 (2007) 95–109.

[35] L.Z. Chen, S.K. Nguang, X.D. Chen, Modelling and Optimization of Biotechnological Processes: Artificial Intelligence Approaches, Springer-Verlag, 2010.

[36] T. Becker, T. Enders, A. Delgado, Dynamic neural networks as a tool for the online optimization of industrial fermentation, Bioprocess Biosyst. Eng. 24 (2002) 347–354.

[37] S.J. Pirt, Principles of Microbe and Cell Cultivation, Blackwell Scientific, 1975.

[38] M.L. Shuler, F. Kargi, Bioprocess Engineering: Basic Concepts, second ed., (Chapter 9), Prentice Hall, 2002.

[39] F. Reusser, Theoretical design of continuous antibiotic fermentation units, Appl. Microbiol. 9 (1961) 361–366.

[40] S. Aiba, A.E. Humphrey, N.F. Millis, Biochemical Engineering, Academic Press, 1965.

[41] J.W. Richards, Introduction to Industrial Sterilisation, Academic Press, 1968.

Suggestions for Further Reading

Bioreactor Configurations and Operating Characteristics

See also references [5] through [5].

M.Y. Chisti, M. Moo-Young, Airlift reactors: characteristics, applications and design considerations, Chem. Eng. Comm 60 (1987) 195–242.

C.L. Cooney, Bioreactors: design and operation, Science. 219 (1983) 728–733.

W.-D. Deckwer, Bubble column reactors, VCH, (1985) 445–464

Practical Considerations for Reactor Design

See also references [14] and [14].

Y. Chisti, Build better industrial bioreactors, Chem. Eng. Prog 88 (January) (1992) 55–58.

Bioreactor Monitoring and Control

See also references [17] through [17].

S. Albert, R.D. Kinley, Multivariate statistical monitoring of batch processes: an industrial case study of fermentation supervision, Trends Biotechnol 19 (2001) 53–62.

J.S. Alford, Bioprocess control: advances and challenges, Comp. Chem. Eng. 30 (2006) 1464–1475.

P. Harms, Y. Kostov, G. Rao, Bioprocess monitoring. Curr, Opinion Biotechnol 13 (2002) 124–127.

B. Lennox, G.A. Montague, H.G. Hiden, G. Kornfeld, P.R. Goulding, Process monitoring of an industrial fed-batch fermentation, Biotechnol. Bioeng. 74 (2001) 125–135.

C.-F. Mandenius, Recent developments in the monitoring, modeling and control of biological production systems, Bioprocess Biosyst. Eng. 26 (2004) 347–351.

J.R. Richards, Principles of control system design, in: J.R. Leigh, (Ed.), Modelling and Control of Fermentation Processes, Peter Peregrinus (1987), 189–214.

K. Schügerl, Bioreaction Engineering, Bioprocess Monitoring, John Wiley, 1997.

K. Shimizu, An overview on the control system design of bioreactors, Adv. Biochem. Eng./Biotechnol 50 (1993) 65–84.

B. Sonnleitner, Measurement, monitoring, modelling and control, in: C. Ratledge, B. Kristiansen, (Eds.), Basic Biotechnology, Cambridge University Press (2006), 251–270.

Sterilization

See also references [40] and [40].

F.G. Bader, Sterilization: prevention of contamination. In: A.L. Demain, N.A. Solomon, (Eds.), Manual of Industrial Microbiology and Biotechnology, American Society of Microbiology (1986) 345–362.

R.S. Conway, Selection criteria for fermentation air filters, Comprehensive Biotechnology, 2 (1985) 279–286.

C.L. Cooney (1985) Media sterilization. In: Cooney, CL, Humphrey AE (eds) Pergamon Press, London. pp. 209-211.

F.H. Deindoerfer, A.E. Humphrey, Analytical method for calculating heat sterilization times, Appl. Microbiol 7 (1959) 256–264.

CHAPTER 13

Unit Operations for Downstream Processing

Bioprocesses convert raw materials into useful products. Individual steps in the process that alter the properties of materials are called unit operations. Although the specific objectives of different bioprocesses vary, each processing scheme can be viewed as a series of component operations that appear again and again in different systems. For example, most bioprocesses involve one or more of the following unit operations: adsorption, centrifugation, chromatography, crystallization, dialysis, distillation, drying, evaporation, filtration, flocculation, flotation, homogenization, humidification, microfiltration, milling, precipitation, sedimentation, solvent extraction, and ultrafiltration.

As an illustration, the sequence of unit operations used for the manufacture of some enzymes is shown in the flow sheet of Figure 13.1. Although the same operations are involved in other processes, the order in which they are carried out, the conditions used, and the materials handled account for the differences in results. The engineering principles for design of unit operations are independent of specific industries or applications.

In a typical bioprocess process, raw materials are altered most significantly by the reactions occurring in the bioreactor. However, physical changes before and after bioprocess are also important to prepare the substrates for reaction and to extract and purify the desired product from the culture broth. The term "unit operation" usually refers to processes that cause physical modifications to materials, such as a change of phase or component concentration. Chemical or biochemical transformations are the subject of reaction engineering.

13.1 Overview of Downstream Processing

In bioprocess broths, the desired product is present within a complex mixture of components. Any treatment of the spent culture broth to concentrate and to purify the product is known as *downstream processing*. In most cases, downstream processing requires only physical modification of the broth material rather than further chemical or biochemical transformation. Nevertheless, there are several reasons why downstream processing is often technically challenging.

- *Bioprocess products are formed in dilute solution.* Water is the main component of cell culture media and, therefore, of harvested bioprocess broth. Many other components are

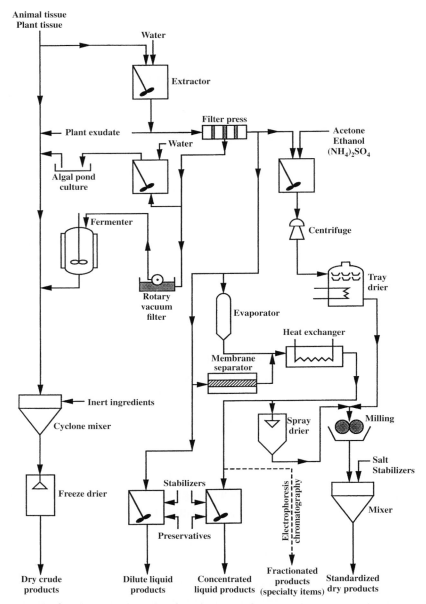

Figure 13.1 Typical unit operations used in the manufacture of enzymes. *From B. Atkinson, and F. Mavituna, 1991,* Biochemical Engineering and Biotechnology Handbook, *2nd ed., Macmillan, Basingstoke; and W.T. Faith, C.E. Neubeck, and E.T. Reese, 1971, Production and applications of enzymes.* Adv. Biochem. Eng. *1, 77–111.*

also present in the broth mixture, providing a wide range of contaminating substances that must be removed to isolate the desired product. In general, purification from dilute solutions involves more recovery steps and higher costs than when the product is available in a concentrated form with fewer impurities.

*   *Biological products are labile.* Products produced in bioprocesses are sensitive to temperature and can be degraded by exposure to solvents, strong acids and bases, and high salt concentrations. This restricts the range of downstream processing operations that can be applied for product recovery.
*   *Harvested culture broths are susceptible to contamination.* Typically, once the broth is removed from the controlled environment of the bioreactor, aseptic conditions are no longer maintained, and the material is subject to degradation by the activity of contaminating organisms. Unless downstream processing occurs rapidly and without delay, product quality can deteriorate significantly.

Although each recovery scheme will be different, the sequence of steps in downstream processing can be generalized depending on whether the biomass itself is the desired product (e.g., bakers' yeast), whether the product is contained within the cells (e.g., enzymes and recombinant proteins), or whether the product accumulates outside the cells in the bioprocess liquor (e.g., ethanol, antibiotics, and monoclonal antibodies). General schemes for these three types of downstream processing operation are represented in Figure 13.2 and involve the following major steps.

1. *Cell removal.* The removal of cells from the bioprocess liquor is a common first step in product recovery. If the cells are the product, little or no further downstream processing is required. If the product is contained within the biomass, harvesting the cells from the large volume of bioprocess liquid removes many of the impurities present and concentrates the product substantially. Removal of the cells can also assist the recovery of products from the liquid phase. Filtration, microfiltration, and centrifugation are typical unit operations for cell removal.
2. *Cell disruption and cell debris removal.* These downstream processing steps are required when the product is located inside the cells. Unit operations such as high-pressure homogenization are used to break open the cells and release their contents for subsequent purification. The cell debris generated during cell disruption is separated from the product by filtration, microfiltration, or centrifugation.
3. *Primary isolation.* Many techniques are available for primary isolation of bioprocess products from cell homogenate or cell-free broth. The methods used depend on the physical and chemical properties of the product and surrounding material. The aim of primary isolation is to remove components with properties that are substantially different from those of the product. Typically, processes for primary isolation are relatively nonselective; however, significant increases in product quality and concentration can be accomplished.

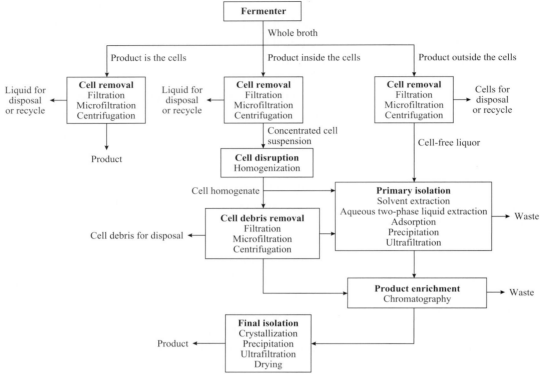

Figure 13.2 Generalized downstream processing schemes for cells as product, products located inside the cells, and products located outside the cells in the fermentation liquor.

Unit operations such as solvent extraction, aqueous two-phase liquid extraction, adsorption, precipitation, and ultrafiltration are used for primary isolation. There are special challenges associated with the design and operation of primary isolation processes in large-scale production systems. When the product is extracellular, large volumes of culture liquid must be treated at this stage. As intermediate storage of this liquid is impractical and disposal expensive, the processes and equipment used for primary isolation must be robust and reliable to minimize broth spoilage and product deterioration in the event of equipment breakdown or process malfunction. It is essential that the operations used for primary isolation be able to treat the bioprocess liquor at the rate it is generated. A desirable feature of primary isolation processes is that a significant reduction in liquid volume is achieved. This reduces the equipment size and operating costs associated with subsequent recovery steps.

4. *Product enrichment*. Processes for product enrichment are highly selective and are designed to separate the product from impurities with properties close to those of the product. Chromatography is a typical unit operation used at this stage of product resolution.

5. *Final isolation*. The form of the product and final purity required vary considerably depending on the product application. Ultrafiltration for liquid products, and crystallization followed by centrifugation or filtration and drying for solid products, are typical operations used for final processing of high-quality materials such as pharmaceuticals.

A typical profile of product concentration and quality through the various stages of downstream processing is given in Table 13.1.

The performance of downstream processing operations can be characterized quantitatively using two parameters, the *concentration factor* δ and the *separation factor* α. These parameters are defined as:

$$\delta = \frac{\text{concentration of product after treatment}}{\text{concentration of product before treatment}} \tag{13.1}$$

and

$$\alpha = \frac{\left(\dfrac{\text{concentration of product}}{\text{concentration of contaminant}}\right)\text{after treatment}}{\left(\dfrac{\text{concentration of product}}{\text{concentration of contaminant}}\right)\text{before treatment}} \tag{13.2}$$

A concentration factor of >1 indicates that the product is enriched during the treatment process. The separation factor differs from the concentration factor by representing the change in product concentration relative to that of some key contaminating compound. Individual downstream processing operations may achieve high concentration factors even though separation of the desired product from a particular contaminant remains relatively poor. On the other hand, highly selective recovery methods give high values of α, but this may be accompanied by only a modest increase in product concentration.

Downstream processing can account for a substantial proportion of the total production cost of a bioprocess product. For example, the ratio of production cost to product recovery cost is approximately 60:40 for antibiotics such as penicillin. For newer antibiotics this ratio is

Table 13.1: Typical Profile of Product Quality During Downstream Processing

Step	Typical unit operation	Product concentration (g L^{-1})	Product quality (%)
Harvest broth	–	0.1–5	0.1–1
Cell removal	Filtration	1–5	0.2–2
Primary isolation	Extraction	5–50	1–10
Product enrichment	Chromatography	50–200	50–80
Final isolation	Crystallization	50–200	90–100

Adapted from P.A. Belter, E.L. Cussler, and W.-S. Hu, 1988, *Bioseparations: Downstream Processing For Biotechnology*, John Wiley, New York.

reversed; product recovery is more costly than production. Products such as recombinant proteins and monoclonal antibodies require expensive downstream processing that can account for 80 to 90% of the total process cost. Starting product levels before recovery have a strong influence on downstream costs; purification is more expensive when the initial concentration of product in the biomass or bioprocess broth is low. As illustrated in Figure 13.3 for several products of bioprocessing, the higher the starting concentration, the cheaper is the final material.

As each downstream processing step involves some loss of product, total losses can be substantial for multistep procedures. For example, if 80% of the product is retained at each purification step, after a six-step process only $(0.8)^6$ or about 26% of the initial product remains. If the starting concentration is very low, more recovery stages are required with higher attendant losses and costs. This situation can be improved either by enhancing the biosynthesis of product or by developing better downstream processing techniques that minimize product loss.

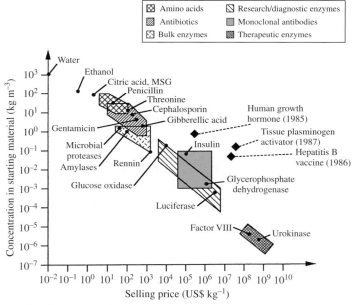

Figure 13.3 Relationship between selling price and concentration before downstream recovery for several products of bioprocessing. *From J.L. Dwyer, 1984, Scaling up bio-product separation with high-performance liquid chromatography.* Bio/Technology 2, 957–964; and J. van Brunt, 1988, How big is big enough? Bio/Technology 6, 479–485.

13.2 Overview of Cell Removal Operations

One of the first steps in downstream processing is the removal of cells from the culture liquid. This is the case if the cells themselves are the product, or if the product is an intra- or extra-cellular metabolite. Although whole broth processing without cell removal is possible, it is not commonly pursued.

The major process options for cell removal are *filtration*, *microfiltration*, and *centrifugation*. Broadly, separations using filtration and microfiltration are based on particle size, whereas centrifugation relies on particle density. Cell removal from microbial bioprocess broths can be technically very challenging because of the small size, low density, and gelatinous nature of many microorganisms. In contrast, due mainly to the low numbers of cells generated, cell removal after mammalian cell culture is comparatively straightforward.

Prior to filtration or centrifugation, it may be necessary to pretreat or *precondition* the bioprocess broth to improve the efficiency of cell separation. Heating is used to denature proteins and enhance the filterability of mycelial broths, such as in penicillin production. Pretreatments that reduce the viscosity of the broth are beneficial for both filtration and centrifugation. Alternatively, electrolytes or polymeric flocculants may be added to promote aggregation of cells and colloids into larger, denser particles that are easier to filter or centrifuge. Filter aids, which are solid particles used to increase the speed of filtration, are often applied to bioprocess broths prior to filtering.

The general features of unit operations used for cell removal are described here.

- *Filtration.* In conventional filtration, cell solids are retained on a filter cloth to form a porous cake while liquid filtrate passes through the cloth. The process generates a relatively dry cake of packed cells; however, the liquid filtrate usually contains a small proportion of solids that escape through the filter cloth. Large-scale filtration is difficult to perform under sterile conditions. Filtration is not a practical option for cell removal if filter aid is required to achieve an acceptable filtration rate and the product is intracellular or the cells themselves are the bioprocess product. This is because contamination of the filtered cells with foreign particles is inappropriate unless the cells are waste by-products; the presence of filter aid can also cause equipment problems if the cells must be disrupted mechanically to release intracellular material. Microfiltration and centrifugation are better options for cell removal under these circumstances.
- *Microfiltration.* Microfiltration uses microporous membranes and cross-flow filtration methods to recover the cells as a fluid concentrate. The maximum cell concentrations generated range from about 10% w/v for gelatinous solids up to 60 to 70% w/v for more rigid particles. Microfiltration is an attractive option for harvesting cells containing intracellular products or cells as product because filter aids are also not required. Because cell recovery using microfiltration is typically close to 100%, the liquid filtrate

or permeate produced by microfiltration is of much greater clarity than that generated by conventional filtration using filter aids. This is an advantage when the product is extracellular, as the filtrate contains minimal contaminating components. A potential disadvantage is that unacceptably high amounts of extracellular product may be entrained in the cell concentrate stream. Microfiltration can be carried out under sterile conditions. *Centrifugation*. This can be an effective strategy for cell recovery when the cells are too small and difficult to filter using conventional filtration. In principle, centrifugation is suitable for recovery of cells as product and for preprocessing of both intracellular and extracellular components. Centrifugation of bioprocess broths produces a thick, concentrated cell sludge or cream that contains more extracellular liquid than is produced in conventional filtration. The use of flocculants and other broth-conditioning agents improves the performance of centrifuges; however, the presence of these additives in either the cell concentrate, or liquid discharge may be undesirable for particular products. Steam-sterilizable centrifuges are available for separations that must be carried out under aseptic conditions. Aerosol generation associated with the operation of high-speed centrifuges can create health and safety problems depending on the organism and products being separated. The equipment costs for centrifugation tend to be greater than for filtration and microfiltration.

- *Other operations*. Although filtration, microfiltration, and centrifugation account for the vast majority of cell separations in bioprocessing, other methods are also available. *Gravity settling* or *sedimentation* is suitable for cells that form aggregates or flocs of sufficient size and density to settle quickly under gravity. Polyvalent agents and polymers may be added to increase particle coagulation and improve the rate of sedimentation. Gravity settling is used typically in large-scale waste treatment processes for liquid clarification, and in beer brewing processes with flocculent strains of yeast. Another cell recovery process is *foam flotation*, which relies on the selective adsorption or attachment of cells to gas bubbles rising through liquid. Surfactants may be used to increase the number of cells associated with the bubbles. The cells are recovered by skimming the foam layers that collect on top of the liquid. The effectiveness of foam flotation has been reported for a range of different microorganisms [1].

13.3 Filtration

Conventional filtration separates solid particles from a fluid–solid mixture by forcing the fluid through a *filter medium* or *filter cloth* that retains the particles. Solids are deposited on the filter and, as the deposit or *filter cake* increases in depth, pose a resistance to further filtration. Filtration can be performed using either vacuum or positive-pressure equipment. The pressure difference exerted across the filter to separate fluid from the solids is called the filtration *pressure drop*.

The ease of filtration depends on the properties of the solid and fluid. Filtration of crystalline, incompressible solids in low-viscosity liquids is relatively straightforward. In contrast, culture broths can be difficult to filter because of the small size and gelatinous nature of the cells and the viscous non-Newtonian behavior of the broth. Most microbial filter cakes are *compressible*, that is, the porosity of the cake declines as the pressure drop across the filter increases. This can be a major problem causing reduced filtration rates and product loss. Filtration of bioprocess broths is carried out typically under non-aseptic conditions; therefore, the process must be efficient and reliable to avoid contamination and the degradation of labile products. Membrane filtration will be specifically discussed in Section 13.9.

13.3.1 Filter Aids

Filter aids such as diatomaceous earth have found widespread use in the bioprocess industry to improve the efficiency of filtration. Diatomaceous earth, also known as kieselguhr, is the fused exoskeletal remains of diatoms. Packed beds of granulated kieselguhr have very high porosity; as little as 15% of the total volume of packed kieselguhr is solid, the rest is empty space. Such high porosity facilitates liquid flow around the particles and improves the rate of filtration. Kieselguhr with average particle sizes of 10 to 25 µm are used with bioprocess broths.

Filter aids are applied in two ways. As shown in Figure 13.4, filter aid can be used as a precoat on the filter medium to prevent blockage or "blinding" of the filter by solids that would otherwise wedge themselves into the pores of the cloth. Filter aid can also be added to the bioprocess broth, typically at concentrations of 1 to 5% w/w, to increase the porosity of the cake as it forms. Filter aids add to the cost of filtration; the minimum quantity needed to achieve the desired result must be established experimentally. Kieselguhr absorbs liquid; therefore, if the bioprocess product is in the liquid phase, some will be lost. Disposal of waste cell material is more difficult if it contains kieselguhr; for example, biomass cannot be used as animal feed unless the filter aid is removed. Use of filter aids is not appropriate if the cells themselves are the product of the bioprocess or when the cells require further processing after filtration for recovery of intracellular material.

13.3.2 Filtration Equipment

Plate filters are suitable for filtration of small culture batches; this type of filter gradually accumulates biomass and must be periodically opened and cleared of filter cake. Larger processes require continuous filters. *Rotary drum vacuum filters*, such as that shown in Figure 13.5, are the most widely used filtration devices in the bioprocess industry. A horizontal drum 0.5 to 3 m in diameter is covered with filter cloth and rotated slowly at 0.1 to 2 rpm. The cloth is partially immersed in an agitated reservoir containing material to be filtered. As a section of drum enters the liquid, a vacuum is applied from the interior of the drum. A cake forms on

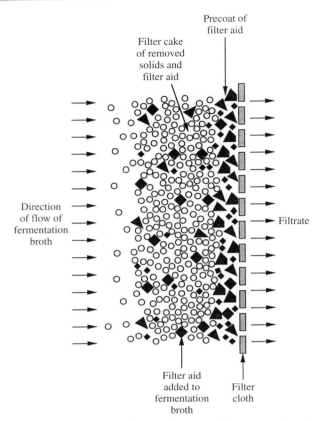

Figure 13.4 Use of filter aid in filtration of fermentation broth.

the face of the cloth while liquid is drawn through internal pipes to a collection tank. As the drum rotates out of the reservoir, the surface of the filter is sprayed with wash liquid, which is drawn through the cloth and collected in a separate holding tank. After washing, the cake is dewatered by continued application of the vacuum. The vacuum is turned off as the drum reaches the discharge zone where the cake is removed by means of a scraper, knife, or strings. Air pressure may be applied at this stage to help dislodge the filter cake from the cloth. After the cake is removed, the drum reenters the reservoir for another filtration cycle.

13.3.3 Filtration Theory

Filtration theory is used to estimate the rate of filtration. For a given pressure drop across the filter, the rate of filtration is greatest just as filtering begins. This is because the resistance to filtration is at a minimum when there are no deposited solids. The orientation of particles in the initial cake deposit is very important and can significantly influence the structure and permeability of the whole filter bed. Flow resistance due to the filter cloth can be considered

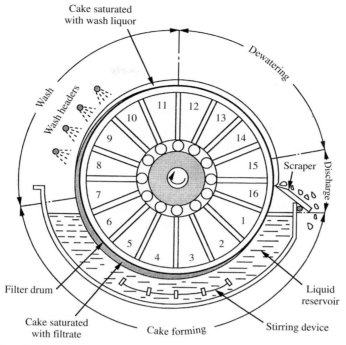

Figure 13.5 Continuous rotary drum vacuum filter. *From G.G. Brown, A.S. Foust, D.L. Katz, R. Schneidewind, R.R. White, W.P. Wood, G.M. Brown, L.E. Brownell, J.J. Martin, G.B. Williams, J.T. Banchero, and J.L. York, 1950,* Unit Operations, *John Wiley, New York.*

constant if particles do not penetrate the material; however, the resistance due to the cake increases with cake thickness.

The rate of filtration is usually measured as the rate at which liquid filtrate is collected. The filtration rate depends on the area of the filter cloth, the viscosity of the fluid, the pressure difference across the filter, and the resistance to filtration offered by the cloth and deposited solids. The rate of filtration is given by the equation:

$$\frac{1}{A}\frac{dV_f}{dt} = \frac{p}{\mu_f\left[\alpha\left(\dfrac{M_c}{A}\right) + R_m\right]} \tag{13.3}$$

where A is the filter area, V_f is the volume of filtrate, t is the filtration time, Δp is the pressure drop across the filter, μ_f is the filtrate viscosity, M_c is the total mass of solids in the cake, α is the average *specific cake resistance*, and R_m is the *filter medium resistance*. R_m includes the effect of the filter cloth. α has dimensions LM^{-1}; R_m has dimensions L^{-1}. dV_f/dt is the filtrate flow rate or *volumetric rate of filtration*. The capital cost of the filter depends on A; the bigger the area required to achieve a given filtration rate, the larger are the equipment and related investment. α is a measure of the resistance of the filter cake to flow; its value depends on the shape and size of the particles, the size of the interstitial spaces between them, and the mechanical stability of the cake. Resistance due to the filter medium is often negligible compared with the cake resistance, which is represented by the term $\alpha\ (M_c/A)$.

If the filter cake is incompressible, the specific cake resistance α does not vary with the pressure drop across the filter. However, cakes from bioprocess broths are seldom incompressible; as cell cakes compress with increasing Δp and filtration rates decline. For a compressible cake, α can be related to Δp empirically as follows:

$$\alpha = \alpha'(\Delta p)^s \tag{13.4}$$

where s is the cake *compressibility* and α' is a constant dependent largely on the size and morphology of the particles in the cake. The value of s is zero for rigid incompressible solids; for highly compressible material s is close to unity. α is also related to the average properties of the particles in the cake as follows:

$$\alpha = \frac{K_v a^2 (1 - \varepsilon)}{\varepsilon^3 \rho_p} \tag{13.5}$$

where K_v is a factor dependent on the shape of the particles, a is the particle-specific surface area:

$$a = \frac{\text{surface area of a single particle}}{\text{volume of a single particle}} \tag{13.6}$$

ε is the *porosity* of the cake:

$$\varepsilon = \frac{\text{total volume of the cake} - \text{volume of solids in the cake}}{\text{total volume of the cake}} \tag{13.7}$$

and ρ_p is the density of the particles. For compressible cakes, both ε and K_v depend on the filtration pressure drop.

It is useful to consider methods for improving the rate of filtration. A number of strategies can be deduced from the relationship between the variables in Eq. (13.3).

- *Increase the filter area A.* When all other parameters remain constant, the rate of filtration is improved if A is increased. However, this requires installation of larger filtration equipment and greater capital cost.
- *Increase the filtration pressure drop Δp.* The problem with this approach for compressible cakes is that α increases with Δp as indicated in Eq. (13.4), and higher α results in lower filtration rates. In practice, pressure drops are usually kept below about 0.5 atm to minimize cake resistance. Improving the filtration rate by increasing the pressure drop can only be achieved by reducing s, the compressibility of the cake. Addition of filter aid to the broth may reduce s to some extent.

- *Reduce the cake mass M_c.* This is achieved in continuous rotary filtration by reducing the thickness of the cake deposited per revolution of the drum and ensuring that the scraper leaves minimal cake residue on the filter cloth.
- *Reduce the liquid viscosity μ_f.* Material to be filtered is sometimes diluted if the starting viscosity is very high.
- *Reduce the specific cake resistance α.* From Eq. (13.5), possible methods of reducing α for compressible cakes are as follows:
- *Increase the porosity ε.* Cake porosity usually decreases as cells are filtered. Application of filter aid reduces this effect.
- *Reduce the shape factor of the particles Kv.* In the case of mycelial broths, it may be possible to change the morphology of the cells by manipulating the bioprocess conditions.
- *Reduce the specific surface area of the particles a.* Increasing the average size of the particles and minimizing variations in particle size reduce the value of a. Changes in bioprocess conditions and broth pretreatment are used to achieve these effects.

Integration of Eq. (13.3) allows us to calculate the time required to filter a given volume of material. Before carrying out the integration, let us substitute an expression for the mass of solids in the cake as a function of the filtrate volume:

$$M_c = c V_f \tag{13.8}$$

In Eq. (13.8), c is the mass of solids deposited per volume of filtrate; this term is related to the concentration of solids in the material to be filtered. Substituting Eq. (13.8) into Eq. (13.3), the expression for the rate of filtration becomes:

$$\frac{1}{A}\frac{dV_f}{dt} = \frac{\Delta p}{\mu_f \left[\alpha \left(\dfrac{cV_f}{A} \right) + R_m \right]} \tag{13.9}$$

Equation (13.9) can be interpreted according to the *general rate principle*, which equates the rate of a process to the ratio of the driving force and the resistance. As Δp is the driving force for filtration, the resistances offered by the filter cake and filter medium are represented by the terms summed together in the denominator of Eq. (13.9). The relative contributions of these two resistances can be estimated using the equations:

$$\text{Proportion of the total resistance due to the filter cake} = \frac{\alpha \left(\dfrac{cV_f}{A} \right)}{\alpha \left(\dfrac{cV_f}{A} \right) + R_m} \tag{13.10}$$

$$\text{Proportion of the total resistance due to the filter medium} = \frac{R_m}{\alpha \left(\dfrac{cV_f}{A} \right) + R_m} \tag{13.11}$$

A filter can be operated in two different ways. If the pressure drop across the filter is kept constant, the filtration rate will become progressively smaller as resistance due to the cake increases. Alternatively, in constant-rate filtration, the flow rate is maintained by gradually increasing the pressure drop. Filtrations are most commonly carried out at constant pressure. When this is the case, Eq. (13.9) can be integrated directly because V_f and t are the only variables: for a given filtration device and material to be filtered, each of the remaining parameters is constant.

It is convenient for integration to write Eq. (13.9) in its reciprocal form:

$$A\frac{dt}{dV_f} = \mu_f \alpha c \frac{V_f}{A\Delta p} + \frac{\mu_f R_m}{\Delta p}$$

(13.12)

Separating variables and placing constant terms outside of the integrals:

$$A\int dt = \frac{\mu_f \alpha c}{A\Delta p}\int V_f dV_f + \frac{\mu_f R_m}{\Delta p}\int dV_f$$

(13.13)

At the beginning of filtration $t = 0$ and $V_f = 0$; this is the initial condition for integration. Carrying out the integration gives:

$$At = \frac{\mu_f \alpha c}{2A\Delta p}V_f^2 + \frac{\mu_f R_m}{\Delta p}V_f$$

(13.14)

Thus, for constant-pressure filtration, Eq. (13.14) can be used to calculate either the filtrate volume V_f or the filtration time t, provided all the constants are known. However, α and R_m for a particular filtration must be evaluated beforehand.

For experimental determination of α and R_m, Eq. (13.14) is rearranged by dividing both sides of the equation by AV_f:

$$\frac{t}{V_f} = \frac{\mu_f \alpha c}{2A^2\Delta p}V_f + \frac{\mu_f R_m}{A\Delta p}$$

(13.15)

Equation (13.15) can be written more simply as:

$$\frac{t}{V_f} = K_1 V_f + K_2$$

(13.16)

where:

$$K_1 = \frac{\mu_f \alpha c}{2A^2\Delta p}$$

(13.17)

and

$$K_2 = \frac{\mu_f R_m}{A\Delta p}$$

(13.18)

K_1 and K_2 are constant during constant-pressure filtration. Therefore, from Eq. (13.16), a straight line is generated when t/V_f is plotted against V_f. The slope K_1 depends on the filtration pressure drop and the properties of the cake; the intercept K_2 also depends on the pressure drop but is independent of cake properties. α is calculated from the slope; R_m is determined from the intercept. Equation (13.16) is valid for compressible and incompressible cakes; however, K_1 for compressible cakes becomes a more complex function of Δp than is directly apparent from Eq. (13.17) because of the dependence of α on pressure.

Example 13.1 underlines the importance of laboratory testing in the design of filtration systems. Experiments are required to evaluate properties of the cake such as compressibility and specific resistance; these parameters cannot be calculated reliably from theory. It is essential that the laboratory tests be conducted with the same materials as the large-scale process. Variables such as temperature, age of the broth, and the presence of contaminants and cell debris have significant effects on filtration characteristics.

EXAMPLE 13.1 Filtration of Mycelial Broth

A 30-mL sample of broth from a penicillin bioprocess is filtered in the laboratory on a 3-cm² filter at a pressure drop of 5 psi. The filtration time is 4.5 min. Previous studies have shown that *Penicillium chrysogenum* filter cake is compressible with $s = 0.5$. If 500 L of broth from a pilot-scale bioreactor must be filtered in 1 hour, what size filter is required if the pressure drop is:

a. 10 psi?
b. 5 psi?
   The resistance due to the filter medium is negligible.

Solution

The properties of the filtrate and mycelial cake can be determined from the results of the laboratory experiment. If R_m is zero in Eq. (13.15):

$$\frac{t}{V_f} = \frac{\mu_f \alpha c}{2A^2 \Delta p} V_f$$

Substituting the expression for α for a compressible cake from Eq. (13.4):

$$\frac{t}{V_f} = \frac{\mu_f \alpha'(\Delta p)^{s-1} c}{2A^2} V_f \tag{1}$$

Rearranging gives:

$$\mu_f \alpha' c = \frac{2A^2 t}{(\Delta p)^{s-1} V_f^2}$$

Substituting values:

$$\mu_f \alpha' c = \frac{2\,(3\ \text{cm}^2)^2 (4.5\ \text{min})}{(5\ \text{psi})^{0.5-1}(30\ \text{cm}^3)^2} = 0.201\ \text{cm}^{-2}\,\text{psi}^{0.5}\,\text{min}$$

(Continued)

This value for $\mu_f \alpha' c$ is used to evaluate the area required for pilot-scale filtration. From (1):

$$A^2 = \frac{\mu_f \alpha' c (\Delta p)^{s-1}}{2t} V_f^2$$

Therefore:

$$A = \left(\frac{\mu_f \alpha' c (\Delta p)^{s-1}}{2t} \right)^{1/2} V_f$$

a. Substituting values with $\Delta p = 10\,\text{psi}$:

$$A = \left(\frac{0.201\ \text{cm}^{-2}\ \text{psi}^{0.5}\ \text{min}\ (10\ \text{psi})^{0.5-1}}{2(1\ \text{h}) \cdot \left| \dfrac{60\ \text{min}}{1\ \text{h}} \right|} \right)^{1/2} (500\ \text{L}) \cdot \left| \frac{1000\ \text{cm}^3}{1\ \text{L}} \right|$$

$$A = 1.15 \times 10^4\ \text{cm}^2 = 1.15\ \text{m}^2$$

b. When $\Delta p = 5\,\text{psi}$:

$$A = \left(\frac{0.201\ \text{cm}^{-2}\ \text{psi}^{0.5}\ \text{min}\ (5\ \text{psi})^{0.5-1}}{2(1\ \text{h}) \cdot \left| \dfrac{60\ \text{min}}{1\ \text{h}} \right|} \right)^{1/2} (500\ \text{L}) \cdot \left| \frac{1000\ \text{cm}^3}{1\ \text{L}} \right|$$

$$A = 1.37 \times 10^4\ \text{cm}^2 = 1.37\ \text{m}^2$$

Halving the pressure drop increases the area required by only 20% because at 5 psi the cake is less compressed and more porous than at 10 psi.

13.4 Centrifugation

Centrifugation is used to separate materials of different density by application of a force greater than gravity. In downstream processing, centrifugation is used to remove cells from bioprocess broths, to eliminate cell debris, to collect precipitates and crystals, and to separate phases after liquid extraction.

Steam-sterilizable centrifuges are applied when either the separated cells or bioprocess liquid is recycled back to the bioreactor or when product contamination must be prevented. Industrial centrifuges generate large amounts of heat due to friction; it is therefore necessary to have good ventilation and cooling. Aerosols created by fast-spinning centrifuges have been known to cause infections and allergic reactions in factory workers so that isolation cabinets are required for certain applications.

Centrifugation is most effective when the density difference between the particles and liquid is great, the particles are large, and the liquid viscosity is low. However, in centrifugation of biological solids such as cells, the particles are very small, the viscosity of the medium can be relatively high, and the particle density is very similar to that of the suspending fluid. These disadvantages are overcome in the laboratory with small centrifuges operated at high speed. However, problems arise in industrial centrifugation when large quantities of material must be treated.

Centrifuge capacity cannot be increased by simply increasing the size of the equipment without limit; mechanical stresses in centrifuges increase in proportion to (radius) [2] so that safe operating speeds are substantially lower in large equipment. The need for continuous through-put of material in industrial applications also restricts practical operating speeds. To overcome these difficulties, a range of centrifuges has been developed for bioprocessing. The types of centrifuges commonly used in industrial operations are described in the next section.

13.4.1 Centrifuge Equipment

Centrifuge equipment is classified according to internal structure. The *tubular bowl centrifuge* has the simplest configuration and is employed widely in the food and pharmaceutical industries. Feed enters under pressure through a nozzle at the bottom, is accelerated to rotor speed, and moves upward through the cylindrical bowl. As the bowl rotates, particles travel-ing upward are spun out and collide with the walls of the bowl as illustrated schematically in Figure 13.6. Solids are removed from the liquid if they move with sufficient velocity to reach the wall of the bowl within the residence time of the liquid in the machine. Liquid from the feed spills over a weir at the top of the bowl; solids that have collided with the walls are col-lected separately. When the thickness of sediment collecting in the bowl reaches the position

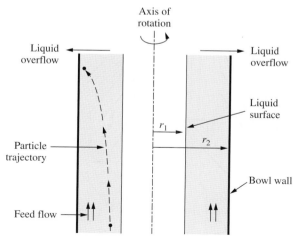

Figure 13.6 Separation of solids in a tubular bowl centrifuge.

of the liquid-overflow weir, separation efficiency declines rapidly. This limits the capacity of the centrifuge. Tubular centrifuges are applied mainly for difficult separations requiring high centrifugal forces. Solids in tubular centrifuges are accelerated by forces between 13,000 and 16,000 times the force of gravity.

A type of narrow tubular bowl centrifuge is the *ultracentrifuge*. This device is used for recovery of fine precipitates from high-density solutions, for breaking down emulsions, and for separation of colloidal particles such as ribosomes and mitochondria. It produces centrifugal forces 10^5 to 10^6 times the force of gravity. The main commercial application of ultracentrifuges has been in the production of vaccines to separate viral particles from cell debris. Typically, ultracentrifuges are operated in batch mode, so their processing capacity is restricted by the need to empty the bowl manually. However, continuous ultracentrifuges are available commercially.

An alternative to the tubular centrifuge is the *disc stack bowl centrifuge*. Disc stack centrifuges are common in bioprocessing. There are many types of disc centrifuge; the principal difference between them is the method used to discharge the accumulated solids. In simple disc centrifuges, solids must be removed periodically by hand. Continuous or intermittent discharge of solids is possible in a variety of disc centrifuges without reducing the bowl speed. Some centrifuges are equipped with peripheral nozzles for continuous solids removal; others have valves for intermittent discharge. Another method is to concentrate the solids in the periphery of the bowl and then discharge them at the top of the centrifuge using a paring device; the equipment configuration for this mode of operation is shown in Figure 13.7.

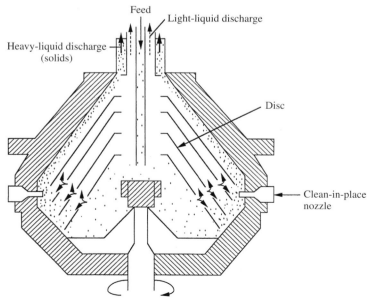

Figure 13.7 Disc stack bowl centrifuge with continuous discharge of solids.

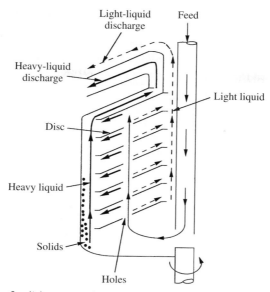

Figure 13.8 Mechanism of solids separation in a disc stack bowl centrifuge. From C.J. Geankoplis, 1983, *Transport Processes and Unit Operations*, 2nd ed., Allyn and Bacon, Boston.

Disc stack centrifuges contain conical sheets of metal called discs that are stacked one on top of the other with clearances as small as 0.3 mm. The discs rotate with the bowl and their function is to split the liquid into thin layers. As shown in Figure 13.8, feed is released near the bottom of the centrifuge and travels upward through matching holes in the discs. Between the discs, heavy components of the feed are thrown outward under the influence of centrifugal forces while lighter liquid is displaced toward the center of the bowl. As they are flung out, the solids strike the undersides of the overlying discs and slide down to the bottom edge of the bowl. At the same time, the lighter liquid flows inward over the upper surfaces of the discs to be discharged from the top of the bowl. Heavier liquid containing solids can be discharged either at the top of the centrifuge or through nozzles around the periphery of the bowl. Disc stack centrifuges used in bioprocessing typically develop forces of 5000 to 15,000 times gravity. As a guide, the minimum solid–liquid density difference for successful separation in a disc stack centrifuge is approximately 0.01 to 0.03 kg m^{-3} [2].

The performance characteristics of tubular bowl and disc stack centrifuges used for industrial separations are summarized in Table 13.2. In general, as the size of the centrifuge increases, the practical operating speed is reduced and the maximum centrifugal force decreases.

Table 13.2: Performance Characteristics of Tubular Bowl and Disc Stack Centrifuges

Centrifuge type	Bowl diameter (cm)	Speed (rpm)	Maximum centrifugal force (× gravity)	Liquid throughput (L min⁻¹)	Solids throughput (kg min⁻¹)	Motor size (kW)
Tubular bowl	10	15,000	13,200	0.4–40		1.5
	13	15,000	15,900	0.8–80		2.2
Disc stack	18	12,000	14,300	0.4–40		0.25
	33	7500	10,400	20–200		4.5
	61	4000	5500	80–800		5.6
Disc stack with nozzle discharge	25	10,000	14,200	40–150	1.5–15	15
	41	6250	8900	100–550	7–70	30
	69	4200	6750	150–1500	15–190	95
	76	3300	4600	150–1500	15–190	95

Data from *Perry's Chemical Engineers' Handbook*, 1997, 7th ed., McGraw-Hill, New York; and D.W. Green, R.H. Perry. *Perry's Chemical Engineers' Handbook*, 8th ed., McGraw-Hill: New York, Chicago, San Francisco, Lisbon, London, Madrid, Mexico City, Milan, New Delhi, San Juan, Seoul, Singapore, Sydney, Toronto, 2008, 1997, 1984, 1973, 1963, 1950, 1941, 1934.

13.4.2 Centrifugation Theory

The particle velocity achieved in a particular centrifuge compared with the settling velocity under gravity characterizes the effectiveness of centrifugation. The terminal velocity during gravity settling of a small spherical particle in dilute suspension is given by Stoke's law:

$$u_g = \frac{\rho_p - \rho_L}{18\mu} D_p^2 g \tag{13.19}$$

where u_g is the sedimentation velocity under gravity, ρ_p is the density of the particle, ρ_L is the density of the liquid, μ is the viscosity of the liquid, D_p is the particle diameter, and g is gravitational acceleration. In a centrifuge, the corresponding terminal velocity is:

$$u_c = \frac{\rho_p - \rho_L}{18\mu} D_p^2 \omega^2 r \tag{13.20}$$

where u_c is the particle velocity in the centrifuge, ω is the angular velocity of the bowl in units of rad s⁻¹, and r is the radius of the centrifuge drum. The ratio of the velocity in the centrifuge to the velocity under gravity is called the *centrifuge effect* or *g-number*, and is usually denoted Z. Therefore:

$$Z = \frac{\omega^2 r}{g} \tag{13.21}$$

The force developed in a centrifuge is Z times the force of gravity and is often expressed as so many *g*-forces. Industrial centrifuges have Z factors up to about 16,000; for small laboratory centrifuges, Z may be up to 500,000 [3].

Sedimentation occurs in a centrifuge as particles moving away from the center of rotation collide with the walls of the centrifuge bowl. Increasing the velocity of motion will improve the rate of sedimentation. From Eq. (13.20), the particle velocity in a given centrifuge can be increased by:

- Increasing the centrifuge speed, ω
- Increasing the particle diameter, D_p
- Increasing the density difference between the particle and liquid, $\rho_p - \rho_L$
- Decreasing the viscosity of the suspending fluid, μ

Whether the particles reach the walls of the bowl also depends on the time of exposure to the centrifugal force. In batch centrifuges such as those used in the laboratory, centrifuge time is increased by running the equipment longer. In continuous-flow devices such as disc stack centrifuges equipped for continuous solids discharge, the residence time is increased by decreasing the feed flow rate.

13.5 Cell Disruption

Downstream processing of culture broths usually begins with removal of the cells by filtration or centrifugation. The next step depends on the location of the desired product. For substances such as ethanol, citric acid, and antibiotics that are excreted from cells, the product is recovered from the cell-free broth using unit operations such as those described later in this chapter. In these cases, the biomass separated from the liquid is discarded or sold as a by-product. For products such as enzymes and recombinant proteins that remain in the biomass, the harvested cells must be broken open to release the desired material. A variety of methods is available to disrupt cells. Mechanical options include grinding with abrasives, high-speed agitation, high-pressure pumping, and ultrasound. Nonmechanical methods such as osmotic shock, freezing and thawing, enzymic digestion of the cell walls, and treatment with solvents and detergents can also be applied.

A widely used technique for cell disruption is high-pressure homogenization. The forces generated in this treatment are sufficient to completely disrupt many types of cells. A common apparatus for homogenization of cells is the Gaulin homogenizer, known as a French press. As indicated in Figure 13.9, this high-pressure pump incorporates an adjustable valve with restricted orifice through which cells are forced at pressures up to 550 atm. The homogenizer is of general applicability for disruption of all types of cells; however, the homogenizing valve can become blocked when used with highly filamentous organisms. Greater disruption is achieved by maintaining a small gap around the valve so that the cells strike the impact ring with high velocity. The exact mechanisms responsible for cell disruption in the Gaulin homogenizer remain a subject of debate. Cavitation, fluid shear, impact, and pressure shock may all play a role in the breakage process.

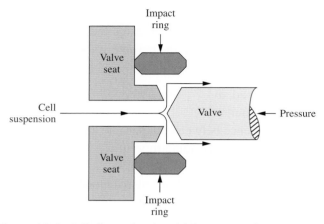

Figure 13.9 Cell disruption in a high-pressure homogenizer.

Generally, the desired degree of cell disruption and product release is not achieved during a single pass through the homogenizer, therefore multiple passes are required. Experimental results for the amount of protein released from yeast cells as a function of the number of times the suspension is passed through the equipment are shown in Figure 13.10. The extent of cell disruption in a Gaulin homogenizer is related to operating conditions by the equation [4]:

$$\ln\left(\frac{R_{max}}{R_{max} - R}\right) = k\,N\,p^{\alpha} \tag{13.22}$$

In Eq. (13.22), R_{max} is the maximum amount of protein available for release, R is the amount of protein released after N passes through the homogenizer, k is a temperature-dependent rate constant, and p is the operating pressure. Both k and α vary with cell type. The exponent α is a measure of the resistance of the cells to disruption; values of α range between 0.9 and 2.9 for bacteria and yeast [5]. However, because the exponent for a particular organism depends on the strength of the cell wall, α varies to some extent with growth conditions and phase of growth [6]. For practical purposes, α is considered independent of operating pressure, although some variation has been observed over very wide pressure ranges [7,8].

13.6 Aqueous Two-Phase Liquid Extraction

Liquid extraction is used to isolate many pharmaceutical products from animal and plant sources. In liquid extraction of bioprocess products, components dissolved in liquid are recovered by transfer into an appropriate solvent. Extraction of penicillin from aqueous broth using solvents, such as butyl acetate, amyl acetate, or methyl isobutyl ketone, and isolation of erythromycin using pentyl or amyl acetate are examples. Solvent extraction techniques are

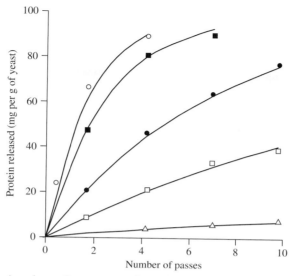

Figure 13.10 Protein release from yeast cells at 30°C as a function of operating pressure and the equivalent number of passes in recycle operation through a Gaulin homogenizer: (△) 100 kgfcm⁻², (☐) 200 kgfcm⁻², (●) 270 kgfcm⁻², (■) 400 kgfcm⁻², (○) 460 kgfcm⁻². *Adapted from P.J. Hetherington, M. Follows, P. Dunnill, and M.D. Lilly, 1971, Release of protein from baker's yeast* (Saccharomyces cerevisiae) *by disruption in an industrial homogenizer.* Trans. IChemE *49, 142–148.*

also applied for recovery of steroids, purification of vitamin B_{12} from microbial sources, and isolation of alkaloids such as morphine and codeine from raw plant material.

The simplest equipment for liquid extraction is the separating funnel used for laboratory-scale product recovery. Liquids forming two distinct phases are shaken together in the separating funnel; solute in dilute solution in one solvent transfers to the other solvent to form a more concentrated solution. The two phases are then allowed to separate and the heavy phase is withdrawn from the bottom of the funnel. The phase containing the solute in concentrated form is processed further to purify the product.

Extraction with organic solvents is a major separation technique in bioprocessing, particularly for recovery of antibiotics. However, organic solvents are unsuitable for isolation of proteins and other sensitive products. Techniques have been developed for aqueous two-phase extraction of these molecules. Aqueous solvents that form two distinct phases provide favorable conditions for separation of proteins, cell fragments, and organelles with protection of their biological activity. Two-phase aqueous systems are produced when particular polymers or a polymer and salt are dissolved together in water above certain concentrations. The liquid partitions into two phases, each containing 85 to 99% water.

Table 13.3: Examples of Aqueous Two-Phase Systems

Component 1	Component 2
Polyethylene glycol	Dextran
	Polyvinyl alcohol
	Polyvinylpyrrolidone
	Ficoll
	Potassium phosphate
	Ammonium sulfate
	Magnesium sulfate
	Sodium sulfate
Polypropylene glycol	Polyvinyl alcohol
	Polyvinylpyrrolidone
	Dextran
	Methoxypolyethylene glycol
	Potassium phosphate
Ficoll	Dextran
Methylcellulose	Dextran
	Hydroxypropyldextran
	Polyvinylpyrrolidone

From M.R. Kula, 1985, Liquid–liquid extraction of biopolymers. In: M. Moo-Young, Ed., *Comprehensive Biotechnology*, vol. 2, pp. 451–471, Pergamon Press, Oxford.

Some components used to form aqueous two-phase systems are listed in Table 13.3. When added to these mixtures, biomolecules and cell fragments partition between the phases: by selecting appropriate conditions, cell fragments can be confined to one phase while the desired protein partitions into the other phase. After partitioning, the product is removed from the extracting phase using other unit operations such as precipitation or crystallization.

The extent of differential partitioning between phases depends on the equilibrium relationship for the system. The *partition coefficient K* is defined as:

$$K = \frac{C_{Au}}{C_{Al}} \tag{13.23}$$

where C_{Au} is the equilibrium concentration of component A in the upper phase and C_{Al} is the equilibrium concentration of A in the lower phase. If $K > 1$, component A favors the upper phase; if $K < 1$, A is concentrated in the lower phase. In many aqueous systems, K is constant over a wide range of concentrations.

Partitioning is influenced by the size, electric charge, and hydrophobicity of the particles or solute molecules; biospecific affinity for one of the polymers may also play a role in some systems. The surface free energy of the phase components and the ionic composition of the liquids are of paramount importance in determining separation; K is related to both these parameters [2].

Even when the partition coefficient is low, good *product recovery* or *yield* can be achieved by using a large volume of the phase preferred by the solute. The yield of A in the upper phase, Y_{Au}, is defined as:

$$Y_{Au} = \frac{V_u C_{Au}}{V_0 C_{A0}} = \frac{V_u C_{Au}}{V_u C_{Au} + V_l C_{Al}} \tag{13.24}$$

where V_u is the volume of the upper phase, V_l is the volume of the lower phase, V_0 is the original volume of solution containing the product, and C_{A0} is the original product concentration in that solution. The yield of A in the lower phase, Y_{Al}, is defined as:

$$Y_{Al} = \frac{V_l C_{Al}}{V_0 C_{A0}} = \frac{V_l C_{Al}}{V_u C_{Au} + V_l C_{Al}} \tag{13.25}$$

The maximum possible yield for an ideal extraction stage can be evaluated using Eqs. (13.24) and (13.25) and the equilibrium partition coefficient, K. Dividing both the numerator and denominator of Eq. (13.24) by C_{Au} and recognizing that, at equilibrium, C_{Al}/C_{Au} is equal to $1/K$:

$$Y_{Au} = \frac{V_u}{V_u + \dfrac{V_l}{K}} \tag{13.26}$$

Similarly, dividing the numerator and denominator of Eq. (13.25) by C_{Al} gives:

$$Y_{Al} = \frac{V_l}{V_u K + V_l} \tag{13.27}$$

Another parameter used to characterize two-phase partitioning is the *concentration factor δ_c*, which is defined as the ratio of the product concentration in the preferred phase to the initial product concentration:

$$\delta_c = \frac{C_{Al}}{C_{A0}} \left(\text{when product partitions to the lower phase} \right) \tag{13.28}$$

$$\delta_c = \frac{C_{Au}}{C_{A0}} \left(\text{when product partitions to the upper phase} \right) \tag{13.29}$$

Aqueous extraction in polyethylene glycol–salt mixtures is an effective technique for separating proteins from cell debris. In this system, debris partitions to the lower phase while most target proteins are recovered from the upper phase. Extractions can be carried out in single-stage operations such as that depicted in Figure 13.11 using a polymer mixture that provides a suitable partition coefficient. Equilibrium is approached in extraction operations but rarely reached. It is important to consider the time taken for mass transfer and the ease of mechanical separation of the phases.

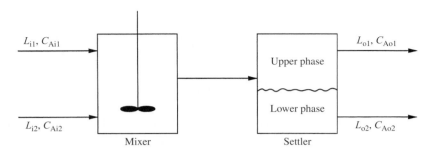

Figure 13.11 An ideal stage.

Separation of the phases is sometimes a problem because of the low interfacial tension between aqueous phases; however, very rapid large-scale extractions can be achieved by combining mixed vessels with centrifugal separators. In many cases, recovery and concentration of product with yields exceeding 90% can be achieved using a single extraction step [2]. When single-stage extraction does not give sufficient recovery, repeated extractions can be carried out in a chain or cascade of contacting and separation units.

13.7 Precipitation

Precipitation is a well-established and widely used method for recovering proteins from cell homogenates and culture broths. Precipitation is also applied in downstream processing of other bioprocess products such as citric acid and antibiotics and may be used to remove unwanted contaminants such as pigments, calcium, proteins, and nucleic acids from process streams. Typically, addition of *precipitants*, such as salts, solvents, and polymers, and changes in the pH, ionic strength, or temperature of the solution are used to reduce the solubility of the product, causing it to precipitate in the form of insoluble particles. The precipitated solids are then recovered by filtration, microfiltration, or centrifugation.

13.7.1 Protein Structure and Surface Chemistry

Precipitation is used commonly for large-scale recovery of proteins so we will focus our attention on protein applications. Proteins treated using precipitation include natural products such as enzymes and food proteins from microbial, plant, and animal sources, as well as recombinant proteins produced using genetically engineered organisms. The tendency of a protein to precipitate depends on the properties of the solvent, such as pH, ionic strength, and dielectric constant, and the size, charge, and hydrophobicity of the protein. In particular, the structure and surface chemistry of proteins play a critical role in precipitation. Proteins remain in solution only if it is energetically more favorable for them to be surrounded by solvent than to be aggregated with other protein molecules into a solid phase.

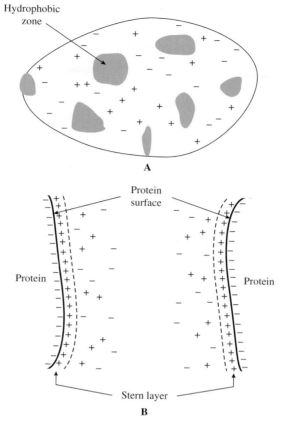

Figure 13.12 Surface chemistry of globular proteins. (A) Hydrophobic zones and nonuniform charge distribution on the surface of a protein. (B) Formation of the Stern layer of counter-ions and the electrical double layer around two neighboring soluble proteins.

Both attractive and repulsive forces exist between neighboring protein molecules in solution. The balance of these forces, that is, whether attraction overcomes repulsion or vice versa, determines whether or not the protein will precipitate. Some of the sources of attraction and repulsion between protein molecules are illustrated in Figure 13.12. In aqueous solution, water-soluble globular proteins are folded so that most of their polar, hydrophilic amino acid side chains are presented on the protein surface, while most of the hydrophobic residues, such as those associated with phenylalanine, tryptophan, leucine, valine, and methionine, are shielded within. However, as indicated schematically in Figure 13.12(A), a typical protein molecule will have some hydrophobic zones on its surface. When two protein molecules are brought together, they are attracted to each other due to interactions between opposite surface charges and any exposed hydrophobic regions.

The *isoelectric point* of a protein is the pH at which the protein carries zero net charge. Under these conditions, the positive and negative charges on the protein surface are balanced.

Although proteins can have isoelectric points anywhere between about 1 and 12, the isoelectric points of many proteins range between 4 and 6. Therefore, when suspended in electrolyte at neutral or near-neutral pH, the surfaces of these proteins have a net negative charge. As a consequence, the protein will attract positive ions from the solution to form a close layer of counter-ions over the molecule surface. This is called the *Stern layer* and is illustrated in Figure 13.12(B). Further out, a more diffuse and mobile layer of counter-ions also develops to form an *electrical double layer* surrounding the protein. When two similarly charged protein molecules are brought together, although they may be attracted by ionic and hydrophobic interactions with each other's surfaces, they are also repulsed by the interaction of their electrical double layers.

13.7.2 Precipitation Methods

In this section, we consider the principal techniques used to reduce the solubility of materials in precipitation operations, thereby inducing precipitation. For protein recovery, precipitation is aimed at separating the protein from solution without inducing major or irreversible changes to the protein structure or biological function.

Salting-Out
High salt concentrations promote the aggregation and precipitation of proteins. This phenomenon is considered to occur as a result of disruption of the hydration barriers between protein molecules, as salt causes water surrounding the protein to move into the bulk solution. The hydrophobic zones of the protein surface thus become exposed, providing sites of attraction between neighboring protein molecules. The success of salting-out therefore depends on the hydrophobicity of the protein: proteins with few exposed hydrophobic regions tend to remain in solution even at high salt concentrations.

At high ionic strengths, the *solubility* of a pure protein depends on the ionic strength of the solution according to the Cohn equation [9]:

$$\ln S = \beta - K I \qquad (13.30)$$

where S is the solubility of the protein at ionic strength I, and β and K are constants. The solubility of a protein in a particular solvent is the maximum concentration of dissolved protein that can be achieved under given conditions of temperature and pressure. The *ionic strength* of a solution is defined as:

$$I = \frac{1}{2}\Sigma z_i^2 C_i \qquad (13.31)$$

where z_i is the charge or valence of ionic component i, C_i is the molar concentration of component i in the solution, and the symbol Σ means the sum of all ionic components. Typical units for C_i and I are mol L^{-1} or mol m^{-3}. If the ionic strength of the solution is altered by changing the concentration of only one salt, Eqs. (13.30) and (13.31) can be combined and simplified to give the empirical cequation:

$$\ln S = \beta_S - K_S C_S \tag{13.32}$$

where β_S and K_S are constants and C_S is the concentration of the salt used to adjust the ionic strength of the solution. For a pure protein, the *salting-out constant* K_S depends on the properties of the protein and the salt but is relatively insensitive to pH and temperature. In contrast, the constant β_S varies strongly with pH, temperature, and the nature of the protein but is essentially independent of the salt.

Equations (13.30) and (13.31) apply only at high ionic strength. Therefore, according to the relationship represented in Eq. (13.32), a plot of $\ln S$ versus C_S at high salt concentrations should give a straight line with negative slope. Experimental data showing the change in protein solubility with salt concentration are shown in Figure 13.13. Although Eq. (13.32) holds for many pure proteins, it does not necessarily apply to protein mixtures. This is because coprecipitation can occur with the other proteins present, depending on their properties. The effect in a protein mixture is that the solubility of a particular protein will be reduced relative to that predicted using Eq. (13.32), as illustrated in Figure 13.13.

Salts with high values of K_S are more effective for salting-out than salts with low K_S. The Hofmeister series gives the relative effectiveness of different anions for protein precipitation:

$$\text{citrate} > \text{phosphate} > \text{sulphate} > \text{acetate} > \text{chloride} > \text{nitrate} > \text{thiocyanate} \tag{13.33}$$

Polyvalent cations such as calcium and magnesium depress the value of K_S, therefore, ammonium, potassium, and sodium salts are used widely. Protein precipitation at an industrial scale requires large amounts of salt; therefore, it is important that the salt used be inexpensive.

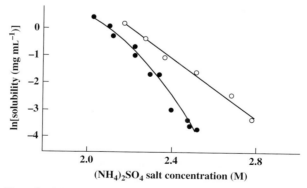

Figure 13.13 Solubility of rabbit muscle aldolase in ammonium sulphate solutions at pH 7.0: (○) pure aldolase, and (●) aldolase in a mixture with pyruvate kinase, glyceraldehyde phosphate dehydrogenase, and lactate dehydrogenase. *Adapted from R.K. Scopes, 1994,* Protein Purification: Principles and Practice, *3rd ed., Springer, New York.*

Isoelectric Precipitation

A protein at its isoelectric point has zero net charge. Under these conditions, the electrical double layer shown in Figure 13.12(B) is weakened, thereby reducing the electrostatic repulsion between neighboring protein molecules. In a solution of relatively low ionic strength, this reduction in repulsive forces may be sufficient to allow protein precipitation. As illustrated in Figure 13.14, most globular proteins have minimum solubility at or near their isoelectric point. An advantage of isoelectric precipitation compared with salting-out is that subsequent desalting of the precipitate is not required. Isoelectric precipitation may be used in conjunction with other precipitation techniques using salt, organic solvents, or polymer to achieve higher levels of product recovery.

Precipitation With Organic Solvents

Addition of a water-soluble, weakly polar solvent such as ethanol or acetone to aqueous protein solutions will generally produce a protein precipitate. As indicated in Table 13.4, organic solvents have lower *dielectric constant* or *permittivity* than water; this means that they store less electrostatic energy than water. Dielectric constants for aqueous mixtures of organic solvents are listed in the *International Critical Tables* [10]. The attraction between polar groups is stronger in solvents of low dielectric constant and weaker in solvents of high dielectric constant. Therefore, in the presence of organic solvent, oppositely charged groups of proteins experience greater attractive forces than in water, resulting in protein aggregation.

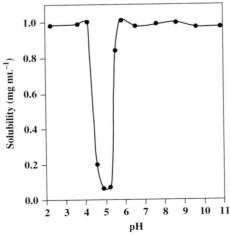

Figure 13.14 Solubility of soybean vicilin (7 S) protein as a function of pH. The protein isoelectric point is close to pH 5.0. *Adapted from Y.J. Yuan, O.D. Velev, K. Chen, B.E. Campbell, E.W. Kaler, and A.M. Lenhoff, 2002, Effect of pH and Ca2+-induced associations of soybean proteins.* J. Agric. Food Chem. 50, 4953–4958.

Table 13.4: Dielectric Constants of Water and Organic Solvents at 20°C

Solvent	Relative dielectric constant*
Acetone	21.01
Diethyl ether	4.2666
Dimethylformamide	38.25
Dioxane	2.2189
Ethanol	25.3
Methanol	33.0
2-Propanol	20.18
Water	80.1

*Ratio of the dielectric constant to the dielectric constant of a vacuum.
From *Handbook of Chemistry and Physics*, 2001, 82nd ed., CRC Press, Boca Raton, FL; and D.R. Lide, 2003, *CRC Handbook of Chemistry and Physics*. 84th ed., CRC Press, New York.

An additional effect is that organic solvents reduce the level of hydration of globular proteins by displacing water molecules into the bulk solution. This exposes interactive sites on the protein surface, thus reducing the solubility. All other properties being equal, the larger the protein, the lower is the concentration of organic solvent required for precipitation. Addition of organic solvents may be combined with salting-out and pH adjustment to improve the yield of precipitated protein.

A major concern with organic solvent precipitation is denaturation of the protein. To avoid protein structural damage, precipitation must be carried out at reduced temperatures, usually less than 10°C. Fortunately, the extent of precipitation increases at lower temperatures. Typically, addition of organic solvent to aqueous solutions causes the temperature to rise due to the heat of mixing; therefore, slow rates of addition and efficient cooling are required. In large-scale operations, the flammability of organic solvents can present safety problems. On the other hand, high volatility also assists the subsequent removal and recovery of solvent from the precipitate.

Precipitation With Polymers

Increasingly, nonionic, water-soluble polymers such as polyethylene glycol are being applied for protein precipitation. The effect of these polymers on the surface chemistry of proteins is thought to be similar to that of organic solvents. As polyethylene glycol is a dense, viscous liquid, practical concentrations for protein precipitation are limited to a maximum of about 20% w/v. Proteins of high molecular weight are precipitated at lower polyethylene glycol concentrations than proteins of low molecular weight. An advantage compared with salting-out or organic solvent precipitation is that many globular proteins are stabilized by interaction with polyethylene glycol. Residual polymer incorporated into the precipitate can be removed using unit operations such as adsorption, aqueous two-phase liquid extraction, or ultrafiltration.

13.7.3 Practical Aspects of Precipitation Operations

Precipitation depends on the nature and concentration of the protein, precipitant, and other components present, and on the temperature, pH, and ionic strength of the solution. In practice, several additional factors have a significant influence on the effectiveness of the process and the quality of precipitate formed.

Precipitates are harvested using unit operations such as filtration, microfiltration, and centrifugation. Precipitate properties such as particle size and size distribution, density, and mechanical strength determine the ease of precipitate recovery. Solid–liquid separations using filtration and centrifugation are more effective if the particle size is increased; therefore, the formation of large precipitate particles is desirable.

Most precipitation operations are carried out in batch using stirred tanks. Precipitants such as salt or organic solvent are added to the protein solution with mixing. The formation of precipitates involves a series of steps, from initial *nucleation* to particle *growth* and *aging*. Mixing within the vessel plays a critical role at all stages of the precipitation process.

- *Nucleation.* This is the primary step in precipitation, resulting in the formation of submicron-sized particles. As nucleation is often very rapid and can occur within a few seconds, good mixing is essential to ensure that the precipitant is distributed to all regions of the vessel at uniform concentration.
- *Growth.* Growth of the precipitate depends initially on diffusion processes, as solute molecules in the vicinity of the nuclei diffuse to the particles. Aggregates of size 0.1 to 1 μm are formed in this way. Subsequent growth may be promoted by mixing, which causes the particles to collide with each other and agglomerate. A balance must be struck, however, as collisions also cause particle erosion and break-up. Precipitates subjected to vigorous mixing and high levels of fluid shear during growth tend to be smaller than those produced at lower shear rates. This effect is illustrated in Figure 13.15.
- *Aging or ripening.* During this stage of precipitation, the precipitate particles continue to grow in size but also become stronger and denser. Adjustments in chemical composition may occur as some components in the precipitate resolubilize. Less intense mixing is required during the aging process to avoid particle disruption. Accordingly, the final stages of precipitation may be performed without stirring or with very low levels of agitation. As shown in Figure 13.16, aging precipitates approach a maximum stable size that is dependent on the mixing conditions. Fully aged particles have improved mechanical strength, which provides resistance to breakage during subsequent handling and processing.

After the precipitate is harvested, the protein must be separated from the precipitant and any contaminants incorporated into the particles. In most cases, for economic reasons, the bulk of

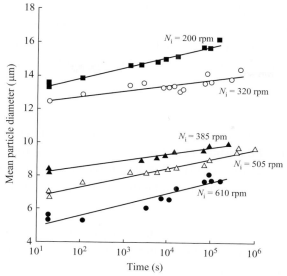

Figure 13.15 The effect of stirrer speed N_i on growth of casein precipitates. The precipitates were formed in a stirred vessel by salting-out with ammonium sulfate, then aged without stirring. *Adapted from M. Hoare, 1982, Protein precipitation and precipitate ageing. Part I: Salting-out and ageing of casein precipitates. Trans. IChemE 60, 79–87.*

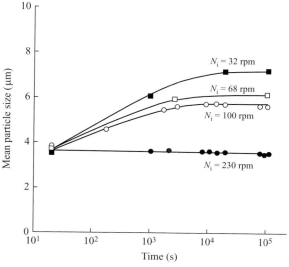

Figure 13.16 The effect of stirrer speed N_i on the size of casein precipitates during aging. *Adapted from M. Hoare, 1982, Protein precipitation and precipitate ageing. Part I: Salting-out and ageing of casein precipitates. Trans. IChemE 60, 79–87.*

the precipitant is recovered and reused. The solid precipitated particles are washed and then redissolved in a small volume of buffer to form a concentrated protein solution. Additional downstream processing using operations such as chromatography and ultrafiltration may be carried out to remove any residual precipitant and further purify the protein product.

13.8 Adsorption

Adsorption is a surface phenomenon whereby components of a gas or liquid are concentrated on the surface of solid particles or at fluid interfaces. Adsorption is the result of electrostatic, van der Waals, reactive, or other binding forces between individual atoms, ions, or molecules. Four types of adsorption can be distinguished: exchange, physical, chemical, and nonspecific.

Adsorption is used to isolate products from dilute bioprocess liquors. Several different adsorption operations are used in bioprocessing, particularly for medical and pharmaceutical products. Ion-exchange adsorption is an established practice for recovery of amino acids, proteins, antibiotics, and vitamins. Adsorption on activated charcoal is a traditional method of purification of citric acid; adsorption of organic chemicals on charcoal or porous polymer adsorbents is common in wastewater treatment.

In adsorption operations, the substance being concentrated on the surface is called the *adsorbate*; the material to which the adsorbate binds is called the *adsorbent*. Ideal adsorbent materials have a high surface area per unit volume; this can be achieved if the solid contains a network of fine internal pores that provide an extremely large internal surface area.

13.8.1 Adsorption Operations

A typical adsorption operation consists of the following stages: a *contacting* or adsorption step which loads solute on to the adsorptive resin, a *washing* step to remove residual unadsorbed material, *desorption* or *elution* of adsorbate with a suitable solvent, *washing* to remove residual eluant, and *regeneration* of the adsorption resin to its original condition. Because adsorbate is bound to the resin by physical or ionic forces, the conditions used for desorption must overcome these forces. Desorption is normally accomplished by feeding a stream of different ionic strength or pH; elution with organic solvent or reaction of the sorbed material may be necessary in some applications. Eluant containing stripped solute in concentrated form is processed to recover the adsorbate; operations for final purification include spray drying, precipitation, and crystallization. After elution, the adsorbent undergoes regenerative treatment to remove any impurities and regain adsorptive capacity. Despite regeneration, performance of the resin will decrease with use as complete removal of adsorbed material is impossible. Accordingly, after a few regenerations the adsorbent is replaced.

13.8.2 *Equilibrium Relationships for Adsorption*

Equilibrium relationships determine the extent to which material can be adsorbed on a particular surface. When an adsorbate and adsorbent are at equilibrium, there is a defined distribution of solute between the solid and fluid phases and no further net adsorption occurs. Adsorption equilibrium data are available as *adsorption isotherms*. For adsorbate A, an isotherm gives the concentration of A in the adsorbed phase versus the concentration in the unadsorbed phase at a given temperature. Adsorption isotherms are useful for selecting the most appropriate adsorbent.

Several types of equilibrium isotherm have been developed to describe adsorption relationships. However, no single model is universally applicable; all involve assumptions that may or may not be valid in particular cases. One of the simplest adsorption isotherms that accurately describes certain practical systems is the *Langmuir isotherm* shown in Figure 13.17(B). The Langmuir isotherm can be expressed as follows:

$$C_{AS}^* = \frac{C_{ASm} K_A C_A^*}{1 + K_A C_A^*} \tag{13.34}$$

In Eq. (13.34), C_{AS}^* is the equilibrium concentration or loading of A on the adsorbent in units of, for example, kg of solute per kg of solid or kg of solute per m^3 of solid. C_{ASm} is the maximum loading of adsorbate corresponding to complete monolayer coverage of all available adsorption sites, C_A^* is the equilibrium concentration of solute in the fluid phase in units of, for example, kg m^{-3}, and K_A is a constant. Because different units are used for fluid- and solid-phase concentrations, K_A usually has units such as m^3 kg^{-1} of solid.

Theoretically, Langmuir adsorption is applicable to systems where:

- Adsorbed molecules form no more than a monolayer on the surface
- Each site for adsorption is equivalent in terms of adsorption energy
- There are no interactions between adjacent adsorbed molecules

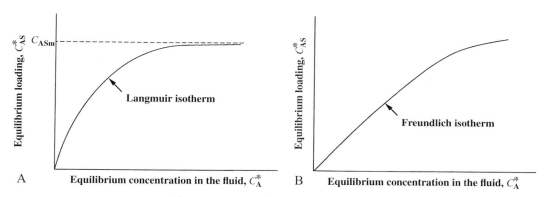

Figure 13.17 Adsorption isotherms.

In many experimental systems at least one of these conditions is not met. For example, many commercial adsorbents possess highly irregular surfaces so that adsorption is favored at particular points or "strong sites" on the surface. Accordingly, each site is not equivalent. In addition, interactions between adsorbed molecules exist for almost all real adsorption systems. Recognition of these and other factors has led to the application of other adsorption isotherms.

Of particular interest because of its widespread use in liquid–solid systems is the *Freundlich isotherm*, described by the relationship:

$$C_{AS}^* = K_F C_A^{*1/n} \qquad (13.35)$$

K_F and n are constants characteristic of the particular adsorption system; the dimensions of K_F depend on the dimensions of C_{AS}^* and C_A^* and the value of n. If adsorption is favorable, $n > 1$; if adsorption is unfavorable, $n < 1$. The form of the Freundlich isotherm is shown in Figure 13.17(B). Equation (13.35) applies to adsorption of a wide variety of antibiotics, hormones, and steroids.

There are many other forms of adsorption isotherm giving different curves for C_{AS}^* versus C_A^* [11]. Because the exact mechanisms of adsorption are not well understood, adsorption equilibrium data must be determined experimentally.

13.9 Membrane Filtration

Membrane filtration is a versatile unit operation that can be used at various stages of downstream processing to separate, concentrate, and purify products. As in conventional filtration (Section 13.3), membrane filtration relies on pressure-driven fluid flow to separate components in a mixture. In contrast with conventional filtration, separations are possible at the molecular scale between components that remain mostly in solution. Membrane separations are generally not highly selective, as different substances with similar properties may be coretained or allowed to pass with the desired product. Membrane filtration is particularly suited, however, for product concentration applications.

Membrane filtration offers several advantages compared with other unit operations used to concentrate bioproducts. Process energy requirements are low, separations can be accomplished under aseptic conditions, and exposure to harsh chemicals is not required. Either batch or continuous processing may be used and scale-up is relatively straightforward. The principal disadvantages relate to the robustness and reliability of the membranes used. For example, membranes are susceptible to fouling, have limited resistance to cleaning chemicals, solvents, and wide pH ranges, and can be damaged by fluctuations in operating pressure.

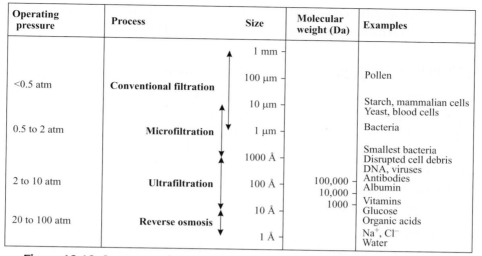

Figure 13.18 Summary of size ranges and applications of filtration processes.

Membrane filtration operations are classified according to the size of the particles or molecules retained by the membrane, the operating pressures used, and the nature of the material filtered. These features are summarized in Figure 13.18.

- *Microfiltration.* This is used to remove particulate matter such as cells and cell debris of size 0.2 to 10 μm from culture broths and cell homogenates. Typical microfiltration membranes have nominal pore diameters of 0.05 to 5 μm. *Ultrafiltration.* Membranes for ultrafiltration have pores with nominal diameter between 0.001 μm (10Å) and 0.1 μm (1000Å). Ultrafiltration is used for removing colloids, large molecules such as proteins and polysaccharides, viruses, and inclusion bodies from smaller solutes and water. Ultrafiltration membranes are capable of retaining macrosolutes with molecular weights of 10^3 to 10^6 daltons. Applications of ultrafiltration can overlap with those of ultracentrifugation (Section 13.4.1). The boundary between microfiltration and ultrafiltration is not precise, as various criteria such as pore size range, membrane structure, and type of material treated are used to define the two processes. *Reverse osmosis.* Membranes for reverse osmosis or *hyperfiltration* have nominal pore diameters of 1 to 10Å. Reverse osmosis is used to separate water from dissolved salts and other microsolutes. Reverse osmosis is not as frequently applied as microfiltration and ultrafiltration for downstream processing of biological products. Accordingly, we will not consider reverse osmosis operations in detail.

In addition to the applications already mentioned, several other membrane-based unit operations are used in bioprocessing, including dialysis, electrodialysis, and osmosis. However, these processes are not driven by a pressure drop across the membrane and rely instead on other variables such as differences in solute concentration or electromotive force. Here, we focus our attention on membrane filtration operations driven by pressure.

13.9.1 Membrane Properties

Membranes for filtration processes are manufactured from a wide variety of materials, including polysulfone, polyethersulfone, polyamide, polyacrylonitrile, polyvinyl chloride, regenerated cellulose, cellulose acetate, and ceramics. The ideal membrane has well-defined particle or solute rejection characteristics, is resistant to variations in temperature, pH, and operating pressure, allows high rates of filtration, has high mechanical strength, and is easy and economical to produce. There are two main categories of membrane structure: *symmetric* or *homogeneous* and *asymmetric* or *anisotropic*.

- *Symmetric membranes.* The pores in symmetric membranes have close to a uniform diameter throughout the depth of the membrane. This allows the entire membrane to act as a selective barrier to the passage of particles or molecules. Flow through symmetric membranes can proceed in either direction. As well as retaining material on the surface of the filter, symmetric membranes also tend to retain components of approximately the same size as the pores within the membrane itself. In this sense, the membrane acts as a *depth filter* as well as a *screen* or *surface filter*. As it is difficult to remove particles trapped within the membrane, filtration using symmetric membranes can become progressively less efficient as the membrane becomes irreversibly blocked.
- *Asymmetric membranes.* These are comprised of an ultrathin (0.1–1 μm) skin layer with very small pores supported by a thick macroporous support. The diameter of the pores therefore changes considerably through the depth of the membrane. During filtration, flow through asymmetric membranes proceeds in one direction only from the skin layer to the macroporous layer. The molecular exclusion characteristics of asymmetric membranes are determined by the size of the pores in the skin layer; however, once the solutes for passage through the membrane are selected at the upper surface, the macrovoids in the lower depths of the membrane provide high permeability and allow rapid filtration rates. Asymmetric membranes operate as screen filters by retaining material at the surface and not within the membrane itself. Accordingly, asymmetric membranes rarely block in the same way as do symmetric membranes. Cleaning is also relatively straightforward, as only the surface and not the entire filter volume requires treatment to remove residual material.

A key parameter of membrane performance is its *retention characteristics*. An ideal membrane would be capable of rejecting completely all particles or solutes above a specified size or molecular weight, while passing completely all species below that size. However, because the pores in real membranes are not all exactly the same diameter, the cut-off is imperfect. As a result, some solutes may be present after filtration in both the permeate and retentate. The selectivity of the membrane with respect to a given solute is quantified using the *rejection* or *retention coefficient*, R:

$$R = \frac{C_R - C_P}{C_R} = 1 - \frac{C_P}{C_R} \tag{13.36}$$

where C_R is the concentration of solute in the retentate and C_P is the concentration of solute in the permeate at any point during the filtration process. If the solute is rejected completely by the membrane and retained in the retentate, $C_P = 0$ and $R = 1$. Conversely, if a solute is not rejected at all by the membrane and passes through freely, its concentration on both sides of the membrane will be the same. Under these conditions, $C_R = C_P$ and $R = 0$.

Typical membrane retention characteristics are illustrated in Figure 13.19(A). Whereas an ideal membrane can accomplish sharply defined separations, real membranes have more diffuse cut-off characteristics depending on whether the membrane pore size distribution is broad or narrow. The sharp cut-off properties shown for an ideal membrane do not occur in practice. Membranes are *rated* by their manufacturers to give an indication of the minimum particle size or solute molecular weight retained [12,13]. An *absolute rating* means that the membrane can be expected to retain all particles above its rated pore size. For example, an absolute membrane rating of 0.45 μm implies that particles with size greater than 0.45 μm will not pass through under standardized operating conditions. The retention characteristics of a membrane with an absolute rating are indicated in Figure 13.19(B).

Alternatively, retention properties can be specified using a *nominal rating*. This is used mainly with ultrafiltration membranes and is expressed as a *molecular weight cut-off*. As shown in Figure 13.19(B), the molecular weight cut-off usually refers to the molecular weight of a test solute that is 90% retained by the membrane. The retention properties of a particular solute and membrane are best determined experimentally.

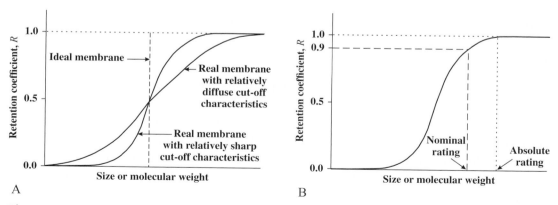

Figure 13.19 (A) Ideal and real membrane retention characteristic and (B) absolute and nominal ratings for a real membrane.

13.9.2 Membrane Filtration Theory

An important objective in the design of membrane filtration systems is estimation of the membrane area required to achieve a certain rate of filtration. The membrane area dictates the size and number of filtration units needed to achieve the desired process throughput. Alternatively, the time taken to filter a given volume of material using equipment that is already installed may be of most interest. In either case, a key parameter in analysis of membrane filtration is the *permeate flux*, *J*, which is defined as the permeate flow rate per unit area of membrane:

$$J = \frac{F_P}{A} \tag{13.37}$$

where F_P is the volumetric flow rate of permeate and A is the membrane area. Typical units for permeate flux are L m^{-2} h^{-1}, m^3 m^{-2} s^{-1}, or, more conveniently, m s^{-1}. Several theories have been developed to relate *J* to operating conditions such as the pressure exerted across the membrane, the cross-flow fluid velocity, and the properties of the material being filtered. Two models describing membrane filtration can be found in references [13–15].

13.9.3 Fouling

When micro- or ultrafiltration systems are operated, there is a progressive decline in the permeate flux over a period of time. The filtration data in Figure 13.20 illustrates this effect. The cause is *membrane fouling*. Fouling is a major limiting factor in application of membrane technology. The consequence of membrane fouling and reduced permeate flux is higher equipment costs, as larger or greater numbers of membrane filtration units are needed to achieve a given process throughput.

Fouling is irreversible under the normal range of operating conditions and cannot be alleviated by reducing the transmembrane pressure, lowering the feed solute concentration, increasing the fluid flow rate, or otherwise improving mass transfer conditions at the membrane surface. Increasing the permeate flux through a fouled membrane is usually only achieved by increasing the applied pressure drop.

Membrane fouling is considered to be a two-step process. The first step occurs within the first few minutes of contact with the feed solution and involves the adsorption of foulants onto the membrane and pore surfaces. This process is influenced by the surface charge, chemical properties, and physical morphology of the membrane. The second step, which is characterized by a relatively slow decline in permeation rate, has been ascribed to plugging of the membrane pores and/or formation of a consolidated or solidified gelatinous layer at the membrane surface. The second stage of fouling is very unpredictable, being dependent on a wide range of parameters including the membrane composition, nature of the retained material, temperature,

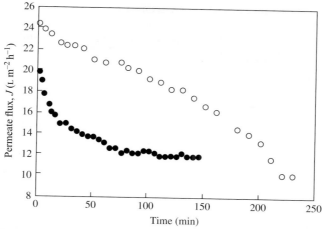

Figure 13.20 Reduction in permeate flux with operating time as a result of membrane fouling. The data were measured during ultrafiltration of (○) carrot juice, and (●) orange juice. *Data from A. Cassano, E. Drioli, G. Galaverna, R. Marchelli, G. Di Silvestro, and P. Cagnasso, 2003, Clarification and concentration of citrus and carrot juices by integrated membrane processes.* J. Food Eng. *57, 153–163.*

operating pressure, and solvent properties. However, the results can be severe, with flux reductions of up to 90% occurring within a few days.

Fouling can be incorporated into a resistances-in-series model by adding an extra resistance terms:

$$J = \frac{\Delta p}{R_M + R_C + R_F} \tag{13.38}$$

Where Δp is the pressure drop across the membrane, R_M is the resistance of the membrane, R_C is the concentration polarization resistance, and R_F is the resistance due to fouling. Because R_F reflects the physicochemical interactions between foulants and the membrane, its magnitude is relatively unaffected by operating parameters such as the transmembrane pressure and flow velocity. Table 13.5 lists values of R_M, R_C, and R_F reported in the literature for various membrane filtration applications. In most cases, the membrane resistance R_M is small relative to both the concentration polarization resistance R_C and the fouling resistance R_F. For those applications where the fouling resistance dominates, the rate of filtration can be increased most effectively by preventing fouling or improving techniques for membrane cleaning.

Cleaning and scouring procedures are applied routinely to membranes to remove adherent material. Ideally, cleaning is carried out in place to avoid having to dismantle the membrane unit. Back-flushing, pulsing, shocking, and washing the membrane using high liquid velocities and pressure drops are conventional cleaning techniques.

Table 13.5: Resistances to Membrane Filtration for Various Applications

Feed material	Membrane material	R_M (kPa L^{-1} m^2 h)	R_C (kPa L^{-1} m^2 h)	R_F (kPa L^{-1} m^2 h)
Blood, bovine	Polyvinyl alcohol	0.06	0.16	0.015
Ethanol broth	Ceramic	0.11	2.0	2.3
Ovalbumin	Polyacrylonitrile	1.2	7.9	9.4
Ovalbumin	Polysulfone	0.51	9.2	0.91
Passionfruit juice	Polysulfone	0.27	1.9	1.4
Polyvinylpyrrolidone 360				
0.1%	Polyethersulfone	0.14	2.2	0.53
2.0%	Polyethersulfone	0.14	7.9	6.7
Wheat starch effluent	Polysulfone	0.68	2.7	4.9

From M. Cheryan, 1998, *Ultrafiltration and Microfiltration Handbook*, Technomic, Lancaster, PA.

13.9.4 Membrane Filtration Equipment

Membrane filtration is carried out using modular equipment, which is available in a variety of configurations. The modules must physically support the membrane, allow a high surface area of membrane to be packed within a small volume, facilitate mass transfer between the membrane and feed stream, and minimize the risk of particle plugging or clogging.

13.10 Chromatography

Chromatography is a separation procedure for resolving mixtures into individual components. Many of the principles described for adsorption apply also to chromatography. The basis of chromatography is *differential migration*, that is, the selective retardation of solute molecules during passage through a bed of resin particles. A schematic representation of chromatography is given in Figure 13.21; this diagram shows the separation of three solutes from a mixture injected into a column. As solvent flows through the column, the solutes travel at different speeds depending on their relative affinities for the resin particles. As a result, they will be separated and appear for collection at the end of the column at different times. The pattern of solute peaks emerging from a chromatography column is called a *chromatogram*. The fluid carrying solutes through the column or used for elution is known as the *mobile phase*; the material that stays inside the column and effects the separation is called the *stationary phase*.

In *gas chromatography* (*GC*), the mobile phase is a gas. Gas chromatography is used widely as an analytical tool for separating relatively volatile components such as alcohols, ketones, aldehydes, and many other organic and inorganic compounds. However, of greater relevance to bioprocessing is *liquid chromatography*, which can take a variety of forms. Liquid chromatography finds application both as a laboratory method for sample analysis

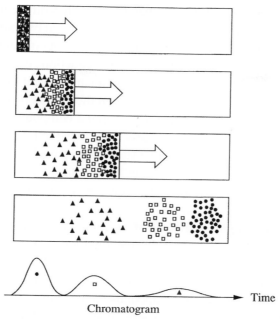

Figure 13.21 Chromatographic separation of components in a mixture. Three different solutes are shown schematically as circles, squares, and triangles. *From P.A. Belter, E.L. Cussler, and W.-S. Hu, 1988,* Bioseparations: Downstream Processing for Biotechnology, *John Wiley, New York.*

and as a preparative technique for large-scale purification of biomolecules. As a high-resolution technique, chromatography is suitable for recovery of high-purity therapeutics and pharmaceuticals.

Chromatographic methods available for purification of proteins, peptides, amino acids, nucleic acids, alkaloids, vitamins, steroids, and many other biological materials include *adsorption chromatography, partition chromatography, ion-exchange chromatography, gel chromatography*, and *affinity chromatography*. These methods differ in the principal mechanism by which molecules are retarded in the chromatography column.

- *Adsorption chromatography*. Biological molecules have varying tendencies to adsorb on polar adsorbents such as silica gel, alumina, diatomaceous earth, and charcoal. Performance of the adsorbent relies strongly on the chemical composition of the surface, which is determined by the types and concentrations of exposed atoms or groups. The order of elution of sample components depends primarily on molecule polarity. Because the mobile phase is in competition with solute for the adsorption sites, solvent properties are also important. Polarity scales for solvents are available to aid mobile-phase selection for adsorption chromatography [16].

- *Partition chromatography*. Partition chromatography relies on the unequal distribution of solute between two immiscible solvents. This is achieved by fixing one solvent (the stationary phase) to a support and passing the other solvent containing solute over it. The solvents make intimate contact, allowing multiple extractions of solute to occur. Several methods are available to chemically bond the stationary solvent to supports such as silica [17]. When the stationary phase is more polar than the mobile phase, the technique is called *normal-phase chromatography*. When nonpolar compounds are being separated, it is usual to use a stationary phase that is less polar than the mobile phase; this is called *reverse-phase chromatography*. A common stationary phase for reverse-phase chromatography is hydrocarbon with 8 or 18 carbons bonded to silica gel: these materials are called C_8 and C_{18} packings, respectively. Elution is generally in order of increasing solute hydrophobicity.

- *Ion-exchange chromatography*. The basis of separation in this procedure is electrostatic attraction between the solute and dense clusters of charged groups on the column packing. Ion-exchange chromatography can give high resolution of macromolecules and is used commercially for fractionation of antibiotics and proteins. Column packings for separation of low-molecular-weight compounds include silica, glass, and polystyrene; carboxymethyl and diethylaminoethyl groups attached to cellulose, agarose, or dextran provide suitable resins for protein chromatography. Solutes are eluted by changing the pH or ionic strength of the liquid phase; salt gradients are the most common way of eluting proteins from ion exchangers [18].

- *Gel chromatography*. This technique is also known as *molecular-sieve chromatography, exclusion chromatography, gel filtration*, and *gel-permeation chromatography*. Molecules in solution are separated using a column packed with gel particles of defined porosity. The gels most often applied are cross-linked dextran, agarose, and polyacrylamide. The speed with which components travel through the column depends on their effective molecular size. Large molecules are completely excluded from the gel matrix and move rapidly through the column to appear first in the chromatogram. Small molecules can penetrate the pores of the packing, traverse the column very slowly, and appear last in the chromatogram. Molecules of intermediate size enter the pores but spend less time there than the small solutes. Gel filtration can be used to separate proteins and lipophilic compounds. *Affinity chromatography*. This separation technique exploits the binding specificity of biomolecules. Enzymes, hormones, receptors, antibodies, antigens, binding proteins, lectins, nucleic acids, vitamins, whole cells, and other components capable of specific and reversible binding are amenable to highly selective affinity purification. The column packing is prepared by linking a binding molecule or ligand to an insoluble support; when a sample is passed through the column, only solutes with appreciable affinity for the ligand are retained. The ligand must be attached to the support in such a way that its binding properties are not seriously affected; molecules called *spacer arms* are often used to set the ligand away from the support and make it more accessible to the solute. Many ready-made support–ligand preparations are available commercially and are suitable for a wide range of proteins. The conditions for elution depend on the specific binding complex formed: elution usually involves a change in pH, ionic strength, or buffer composition. Affinity chromatography using antibody ligands is called *immunoaffinity chromatography*.

In this section, we will consider the principles of liquid chromatography for separation of biological molecules such as proteins and amino acids. The choice of stationary phase will depend to a large extent on the type of chromatography employed; however, certain basic requirements must be met. For high capacity, the solid support must be porous with high internal surface area; it must also be insoluble and chemically stable during operation and cleaning. Ideally, the particles should exhibit high mechanical strength and show little or no nonspecific binding. The low rigidity of many porous gels was initially a problem in industrial-scale chromatography; the weight of the packing material in large columns and the pressures developed during flow tended to compress the packing and impede operation. Many macroporous gels and composite materials of improved rigidity are now available for industrial use. Nevertheless, pressure may still be a limiting factor affecting column operation in some applications.

Two methods for carrying out chromatographic separations are high-performance liquid chromatography (HPLC) and fast protein liquid chromatography (FPLC). In principle, any of the types of chromatography described earlier can be executed using HPLC and FPLC techniques. Specialized equipment for HPLC and FPLC allows automated injection of samples, rapid flow of material through the column, collection of the separated fractions, and data analysis.

The differences between HPLC and FPLC lie in the flow rates and pressures used, the size of the packing material, and the resolution accomplished. In general, HPLC instruments are designed for small-scale, high-resolution analytical applications; FPLC is tailored for large-scale, preparative purifications. To achieve the high resolutions characteristic of HPLC, stationary-phase particles 2 to 5 μm in diameter are commonly used. Because the particles are so small, HPLC systems are operated under high pressure (5 to 10 MPa) to achieve flow rates of 1 to 5 mL min^{-1}. FPLC instruments are not able to develop such high pressures and are therefore operated at 1 to 2 MPa with column packings of larger size. Resolution is poorer using FPLC compared with HPLC; accordingly, it is common practice to collect only the central peak of the solute pulse emerging from the end of the column and to recycle or discard the leading and trailing edges. FPLC equipment is particularly suited to protein separations; many gels used for gel chromatography and affinity chromatography are compressible and cannot withstand the high pressures exerted in HPLC.

Chromatography is essentially a batch operation; however, industrial chromatography systems can be monitored and controlled for easy automation. Cleaning the column in place is difficult. Depending on the nature of the impurities contained in the samples, relatively harsh treatments using concentrated salt or dilute alkali solutions are required; these may affect the swelling of the gel beads and, therefore, liquid flow in the column. Regeneration in place is necessary as the repacking of large columns is laborious and time consuming. Repeated use of chromatographic columns is essential because of their high cost.

13.11 Crystallization

Crystallization is used for *polishing* or final purification of a wide range of bioprocess products. Many pharmaceuticals and fine biochemicals, including antibiotics, amino acids, β-carotene, organic acids, and vitamins, are marketed in the form of crystals. As an example, Figure 13.22 shows crystals of the amino acid glycine. Methods for bulk crystallization of proteins from bioprocess broths are also being developed. As well as for polishing, crystallization can be used within downstream processing schemes for recovery and concentration of components from the liquid phase.

The formation of solids from solution is a feature common to both precipitation (Section 13.8) and crystallization. However, an important difference is that, in contrast to precipitates, crystals often achieve very high (≥99.5%) levels of purity. In many cases, the degree of purification achieved using crystallization is so substantial that other more expensive unit operations, such as chromatography, are not required. From the original solution, a two-phase slurry or *magma* is formed containing *mother liquor* and solid crystals of varying size. The crystals are harvested from the magma using unit operations such as filtration and centrifugation then washed and dried prior to packaging. Crystallization is generally carried out at low temperatures, thus minimizing thermal damage to heat-labile products.

Figure 13.22 Glycine crystals. The scale bar represents 2 mm.
Photograph by H. Zhao.

Mass or bulk crystallization of large quantities of crystals is a complex process. Although small numbers of crystals can be produced relatively easily in the laboratory, upscaling to industrial-size equipment is a challenge. Crystallization is a two-phase, multicomponent, surface-dependent process that involves simultaneous heat and mass transfer and is sensitive to a range of conditions such as fluid hydrodynamics, particle mechanics, vessel geometry, and mode of operation. Many aspects of large-scale crystallization are closer to art than science. In addition, most of our engineering understanding of crystallization has been derived from studies of inorganic salts rather than biological molecules. Although many of the principles of crystallization are the same irrespective of the type of product, there are some features of biological systems that require special attention.

13.11.1 Crystal Properties

Crystals are characterized in terms of their *shape*, *size*, *size distribution*, and *purity*. All of these properties affect the end-use quality of the product; several also influence the ease of crystal recovery and handling.

Shape and Size

Crystals are composed of molecules, atoms, or ions arranged in an orderly, repetitive, three-dimensional array or *space lattice*. The distances between constituent units within crystals of a given substance are fixed and characteristic of that material. Under ideal growth conditions, this gives rise to reproducible and well-defined crystal shapes.

Crystals forming freely from solution appear as polyhedrons bounded by planar faces. For a given material, although the angles between the crystal faces remain constant, the space lattice allows for variation in the relative sizes of the faces. Accordingly, variation in *crystal habit* may be observed between individual crystals of a particular substance. As an example, different growth habits for hexagonal crystals are illustrated in Figure 13.23. Crystal habit is constrained by internal geometry but is subject to external factors such as the solvent used, impurities present, and conditions such as temperature and pH that affect the rate of crystal

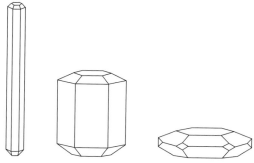

Figure 13.23 Three different crystal habits for hexagonal crystals.

growth. For some bioprocess products such as amino acids and antibiotics, it may be necessary to produce crystals with a specific habit. As solution impurities can bind to and thus selectively alter the growth of one or more crystal faces, they can strongly influence crystal habit even when present at only trace concentrations. Impurities with this effect are known as *habit modifiers*.

The shape and size of individual crystals affect important physical properties of crystalline products. Some of these are listed in Table 13.6. The ease with which crystals are recovered from magma also depends on their shape and size. For example, crystals with chunky, compact habits are easier to filter, centrifuge, wash, and dry than long, needle-like, or twinned crystals that are easily broken, or broad, flat crystals that are difficult to separate from each other.

Under ideal conditions, a crystal will maintain constant habit and geometric proportions during its growth. Such crystals are called *invariant*. The size of an invariant crystal can be expressed using only a single dimension known as the *characteristic length*, L. The value of L is often measured using sieve analysis as described further in the next section, where L is related to the sieve aperture size. The properties of individual crystals, such as surface area, volume, and mass, are defined in terms of the characteristic length and appropriate *shape factors*:

$$A_C = k_a L^2 \tag{13.39}$$

$$V_C = k_v L^3 \tag{13.40}$$

$$M_C = \rho V_C = \rho k_v L^3 \tag{13.41}$$

where A_C is the surface area of the crystal, k_a is the area shape factor, L is the characteristic length, V_C is the volume of the crystal, k_v is the volume shape factor, M_C is the mass of the crystal, and ρ is the crystal density. Methods for measuring crystal shape factors are described in the literature [19].

Table 13.6: Properties of Crystal Products Affected by Crystal Size, Shape, and Size Distribution

End-use and handling property	Processing property
Bulk density	Filtration rate
Solids flow characteristics	Centrifugation rate
Residual moisture content	Entrainment of liquid after dewatering
Caking properties in storage	
Rate of dissolution	
Dustiness	
Appearance	
Fluidization properties	
Pneumatic handling properties	

The shape and size of crystals can be affected detrimentally by *agglomeration* or the bonding together of individual crystals into larger particles. Another crystal property that particularly affects biological molecules is *polymorphism*, which is the occurrence of multiple crystal forms. The ability of biological materials to undergo polymorphic crystallization reflects the diversity of forces, including hydrogen bonding, van der Waals forces, and hydrophobic/ hydrophilic interactions, that affect the molecular conformation and packing properties of substances such as proteins, amino acids, fatty acids, and lipids. Crystallization conditions must be used to control polymorphic systems so that stable crystals of the desired shape and size are produced.

Size Distribution

Crystallization processes generate particles with a range of sizes. Crystal size distribution affects several important product and processing properties (Table 13.6). For example, difficulties with filtration are minimized, and the tendency of crystals to bind together or *cake* into large solid lumps during storage is reduced, if the crystals are of relatively uniform size. Therefore, an important objective in crystallization operations is to produce crystals with as narrow a size range as possible.

Several measurement techniques are available for measuring crystal size distribution. The simplest and most widely used is *sieving* or *screening*. Sieving is applied mostly for sizing dry particles; however, wet sieving using liquid to assist particle movement through the screens may be used to separate small particles 10 to 45 μm in size. Other methods for measuring particle size include Coulter counter, light scattering, and sedimentation techniques [19–21]. Each method gives particle size information based on particular properties of the particle; for example, sieving discriminates between particles depending on their linear dimensions whereas Coulter counters measure particle volume.

For analysis of crystal size distribution, a set of standard screens is arranged in order in a stack with the smallest-aperture screen at the bottom and the largest at the top. A pan is placed under the bottom sieve to catch the finest particles. A sample of crystals is placed on the top screen and the entire stack is shaken. The particles retained on each screen and in the pan are then collected and weighed. As illustrated in Figure 13.24, particles with a variety of shapes and sizes can pass through mesh screens to give the same size measurement in sieve analysis. Accordingly, sieving provides only an approximate indication of crystal size based on a single linear dimension of the particle.

The *mass fraction* of the crystals retained at a particular screen is calculated as:

$$\Delta w = \frac{\Delta M}{M_\mathrm{T}} \tag{13.42}$$

where Δw is the mass fraction, ΔM is the mass of particles retained on the screen, and M_T is the total sample mass. This relationship can be expended to mass density distribution by

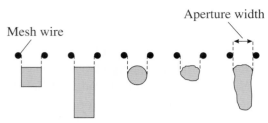

Figure 13.24 Different particle shapes and sizes giving the same size measurement by sieving.

relating m (mass density) and changes in mass fraction relative to changes in aperture size from above.

The particle size distribution developed during crystallization reflects the balance between formation of new crystals and growth of existing crystals. If tiny new crystals are constantly generated, a large number of crystals will be produced, the average particle size will be small, and the size distribution is likely to be broad. On the other hand, if few new particles are produced but the existing particles are allowed to grow in size, the number of crystals will be relatively small, the average particle size will be large, and the size distribution is likely to be narrow. Control over the relative rates of crystal formation and growth is therefore necessary to produce crystals with specified size and size distribution characteristics.

Purity

If crystals are allowed to grow slowly under constant conditions, their purity can reach very high values of 99.5 to 99.8%. However, in industrial practice, varying levels of impurities are carried over from the mother liquor into the crystalline product. The mechanisms by which impurities infiltrate crystalline materials include:

- Adsorption or deposition of impurities on crystal surfaces
- Entrapment of mother liquor between crystals forming agglomerates
- Inclusions of small quantities of mother liquor within crystals
- Substitution of impurity molecules within the crystal lattice

Impurities present on the external surfaces are removed by washing the crystals with fresh solvent during recovery by filtration or centrifugation. If agglomeration is responsible for low purity, it may be possible to reduce agglomerate formation by manipulating the crystallization conditions. Liquid inclusions within the crystals are minimized by lowering the rate of crystal growth. The substitution of impurities within crystals is a more difficult problem that arises when the impurities present have properties similar to those of the crystal-forming solute. In this case, the impurity must be removed by dissolving the crystals in a new solvent that differentially affects the solubility of the solute and impurity. If the impurity is more soluble in the new solvent than the desired product, *recrystallization* will yield a fresh population of crystals with improved purity. Recrystallization is sometimes carried out several times before crystals of the desired purity are obtained.

13.11.2 Crystal Formation and Growth

The critical phenomena in crystallization processes are the formation of new crystals and crystal growth. These processes determine the crystal size and size distribution. The driving force for both is the level of solute supersaturation in the solution.

Supersaturation

The *solubility* of a substance in a particular solvent is the maximum concentration of the solute that can be dissolved under given conditions of temperature and pressure while maintaining thermodynamic stability. A solution containing this maximum concentration is called a *saturated solution*. In many cases, it is relatively easy to generate solutions containing more dissolved solute than the saturation concentration; however, such *supersaturated solutions* are thermodynamically unstable. Some supersaturated solutions are *metastable*, meaning that the solution is capable of resisting small disturbances but becomes unstable when subjected to large perturbations.

Solubility data in the literature usually refer to pure solutes and solvents, and as impurities have a significant effect on solution properties, solubility curves for practical systems must be determined experimentally. Typical results for dissolved solute concentration as a function of solution temperature are shown in Figure 13.25. The *solubility curve* represents the condition of thermodynamic equilibrium between the solid and liquid phases. Below the solubility curve, the solution is *undersaturated* and crystallization is impossible. The broken line above the solubility curve is the *supersolubility curve*, which indicates the temperatures and solute concentrations at which spontaneous crystal formation is likely to occur. This curve is determined experimentally by slowly reducing the temperature or increasing the solute concentration in a crystal-free solution until new crystals are formed. The supersolubility curve is not as well defined as the solubility curve: its position may change with conditions such as the intensity of agitation. The zone between the solubility and supersolubility curves is known as the *metastable region* of supersaturation. In this region, although spontaneous crystal formation is unlikely, any crystals already present will grow in size. As crystallization operations are carried out mainly within the metastable region, the width of this zone has a considerable influence on the ease with which crystallization can be controlled. Above the supersolubility curve is the *unstable* or *labile region* of supersaturation. In this zone, spontaneous crystal generation occurs.

Let us consider how variation in temperature affects the solution conditions in Figure 13.25. As an undersaturated solution represented by point A is cooled, the system enters the metastable zone at point B and becomes saturated and then supersaturated. However,

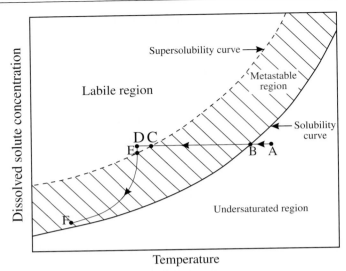

Figure 13.25 Solubility diagram showing curves for solubility and supersolubility as a function of temperature. A hypothetical pathway for batch crystallization without the use of seed crystals is indicated between points A and F.

crystal formation does not occur until the temperature reaches point C, which represents the *metastable limit*. Generation of crystals in the labile region around point D may reduce the dissolved solute concentration sufficiently so that the solution returns to the metastable zone at point E. In this region, crystallization occurs by growth of the new crystals that were formed in the labile zone. If the temperature is reduced further, the dissolved solute concentration becomes progressively lower as more and more solute leaves the solution and is incorporated into the crystals. In this example, the crystallization process is stopped at point F.

Methods other than cooling may be used to induce supersaturation. From point A in Figure 13.25, supersaturation is also achieved by increasing the dissolved solute concentration at constant temperature. Typically, this is accomplished by evaporating the solvent. In industrial practice, a combination of cooling and evaporation is often employed. Supersaturation may also be induced by adding substances that reduce the solubility of the solute or using chemical reaction to form a less soluble product. In crystallization, *salting-out* refers to the use of nonelectrolytic additives such as organic solvents as well as electrolytes such as salts to lower the solubility of solute. Addition of water-miscible solvents is often used to induce crystallization of organic compounds. For example, acetone, iso-propanol, methanol, ethanol, and ethanol–amine solutions are effective for crystallization of several antibiotics and amino acids from aqueous solution.

The level of supersaturation existing in a solution can be quantified in several different ways. The most common expressions for supersaturation are the *supersaturation driving force* ΔC, the *supersaturation ratio S*, and the *relative supersaturation* σ. These parameters are defined as:

$$\Delta C = C - C^* \tag{13.43}$$

$$S = \frac{C}{C^*} \tag{13.44}$$

and

$$\sigma = \frac{C - C^*}{C^*} \tag{13.45}$$

where C is the solute concentration and C^* is the solubility. Combining Eqs. (13.43), and (13.45) yields the relationship:

$$\sigma = \frac{\Delta C}{C^*} = S - 1 \tag{13.46}$$

Methods for measuring solubility and supersaturation in liquid solutions are described in the literature [22].

Nucleation

The formation of a new crystal from a liquid or amorphous phase is called *nucleation*. Different nucleation mechanisms have been observed, as summarized in Figure 13.26.

Primary nucleation refers to nucleation processes that occur in a previously crystal-free solution. As described in relation to Figure 13.25 primary nucleation is achieved by moving a solution into the labile region of supersaturation. Solute molecules gather together under these conditions to form clusters, which become ordered and adopt the structural geometry of the crystalline form. Primary nucleation is further classified as *homogeneous*, meaning that crystal formation occurs spontaneously in a perfectly clean solution free of impurities, or *heterogeneous*, where crystal formation is induced by the presence of inert foreign particles such as dust. Homogeneous nucleation is not relevant in industrial crystallization because

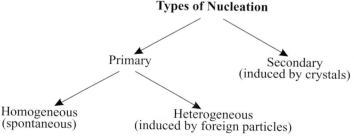

Figure 13.26 Graphic of the types of crystal nucleation.

perfectly clean solutions are virtually impossible to attain. Therefore, any primary nucleation that occurs can be assumed to be heterogeneous.

Secondary nucleation is the generation of new crystals by crystals already present in the suspension. Secondary nucleation is thought to occur as a result of several factors, including the dislodgment of extremely small crystals from the surface of larger crystals, contact due to collisions between crystals or between crystals and the vessel walls and impeller, and crystal attrition or breakage.

Generally, secondary nucleation occurs at much lower supersaturation levels than primary nucleation. To minimize the range of crystal sizes produced, excessive generation of new crystals is undesirable in most crystallization operations. Therefore, ideally, supersaturation levels are minimized and controlled, the labile region of supersaturation is avoided, and operation of the crystallizer is confined to within the metastable region. Under these conditions, supersaturation is constrained to levels considerably lower than those required for primary nucleation so that secondary nucleation dominates. This situation is typical for industrial crystallization of relatively soluble materials.

The rate at which new crystals are formed in a supersaturated solution can be represented by the empirical equation:

$$B = K_N (C - C^*)^b \tag{13.47}$$

where B is the *birth rate* of new crystals, K_N is the *nucleation rate constant*, C is the concentration of dissolved solute, C^* is the solubility of solute at the solution temperature, and b is the *order of nucleation*. K_N and b depend on the physical properties of the system, including the temperature, impurities, and operating conditions such as agitation intensity. In systems where secondary nucleation plays a key role, K_N also depends on the concentration of crystals in the suspension. B represents the number of nuclei formed per unit time and has dimensions T^{-1}; B can also be expressed on a volumetric basis with dimensions $L^{-3}T^{-1}$. The driving force for nucleation is the degree of supersaturation, $C - C^*$. Practical methods for measuring crystal nucleation rates are described in the literature [19].

Crystal Growth

Crystal growth refers to the increase in crystal size due to deposition of solute molecules in layers on the crystal surfaces. A schematic illustration of growth at two opposing crystal faces is shown in Figure 13.27. Mathematically, the rate of crystal growth can be defined in various ways. If L is the characteristic length of the crystal, the rate of solute deposition at the crystal faces perpendicular to L is expressed as:

$$G = \frac{dL}{dt} \tag{13.48}$$

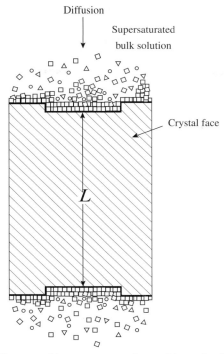

Figure 13.27 Growth of a crystal by mass transfer and layer-by-layer integration of solute molecules at opposing crystal faces.

where G is the *linear growth rate* of an individual crystal and t is time. The dimensions of G are LT^{-1}; typical units are m s^{-1} or μm h^{-1}. Table 13.7 lists some examples of linear growth rates reported for crystals of biological materials. Methods for measuring the rate of crystal growth in single- and multicrystal systems are described elsewhere [19].

Crystal growth rate can also be expressed in terms of the *mass deposition rate* or *mass growth rate*, R_G. Differentiating Eq. (13.41) with respect to time gives an expression for the rate of increase in crystal mass M_C:

$$R_G = \frac{dM_C}{dt} = 3\rho k_v L^2 \frac{dL}{dt} \tag{13.49}$$

where ρ is the crystal density and k_v is the volume shape factor. The dimensions of R_G are MT^{-1}; typical units are kg s^{-1}. Substituting Eq. (13.48) into Eq. (13.49) gives an expression relating the mass growth rate R_G to the linear growth rate G:

$$R_G = 3\rho k_v L^2 G \tag{13.50}$$

Also, from Eq. (13.39):

$$L^2 = \frac{A_C}{k_a} \tag{13.51}$$

Table 13.7: Linear Crystal Growth Rates

Substance	Temperature (°C)	Supersaturation[a]	Growth rate (m s^{-1})[b]	Reference
L-Asparagine monohydrate: (101) face	36	1.19	2.4×10^{-8}	[23]
Canavalin	20	1.5	2.6×10^{-9}	[24]
s-Carboxymethyl-L-cysteine	26	2.0	8.3×10^{-9}	[25]
Citric acid monohydrate	25	1.05	3.0×10^{-8}	[22]
	30	1.01	1.0×10^{-8}	[22]
	30	1.05	4.0×10^{-8}	[22]
Glycine	24	1.1	1.7×10^{-8}	[26]
L-Histidine, B form	20	1.13	4.2×10^{-8}	[22]
Lipase	28	2.0	5.1×10^{-9}	[27]
Lysozyme: (110) face	22	10	7.4×10^{-8}	[28]
Ovalbumin	30	3.0	2.8×10^{-10}	[29]
Sucrose	30	1.13	$b1.1 \times 10^{-8}$	[22]
	30	1.27	$b2.1 \times 10^{-8}$	[22]
	70	1.09	9.5×10^{-8}	[22]
	70	1.15	1.5×10^{-7}	[22]

[a]Supersaturation expressed as C/C^*, where C is the concentration of dissolved solute and C^* is the equilibrium solubility.
[b]Growth rate probably size dependent

where k_a is the area shape factor and A_C is the surface area of the crystal. Substituting Eq. (13.51) into Eq. (13.50) gives an alternative expression:

$$R_G = \frac{3\rho k_v A_C G}{k_a} \tag{13.52}$$

For geometrically similar crystals of the same material grown from birth in the same solution, it has been shown that when the growth rate is defined by Eq. (13.48), the rate of growth of all crystals is the same irrespective of their size [30]. This result is known as the *delta L* or ΔL *law*. The ΔL law is a reasonable generalization for many industrial crystallization processes and is often assumed in crystallization modeling. However, many crystals are known to violate the ΔL law and exhibit *size-dependent growth*, usually with large crystals growing more rapidly than small crystals. Alternatively, some crystal systems exhibit *growth dispersion*, which means that crystals of the same size grow at different rates.

A useful model of crystal growth is the *diffusion–integration theory*, which considers growth to be a two-step process. According to this model, mass transfer of solute to the crystal surface is followed by a surface reaction resulting in the molecular integration of solute into the

or

$$\Delta t = \frac{M_s}{N_c}(X_0 - X_1) \tag{13.67}$$

Equation (13.67) is used to estimate Δt, the time required to dry solids from an initial moisture content of X_0 to a final moisture content of X_1 when the drying rate is constant. From the definition of drying rate in Eq. (13.65), X_0 and X_1 are moisture contents expressed on a dry mass basis using units of, for example, kg water (kg of dry solid)$^{-1}$.

During the falling drying rate period, the drying rate N is no longer constant. Equations for drying time during this period can be developed depending on the relationship between N and X and the properties of the solid. Kinetic models for predicting the drying rate curve, including during the falling rate period when internal heat and mass transfer mechanisms are limiting, are described elsewhere [32]. Example 13.2 illustrated how to determine drying time during a constant rate of drying.

EXAMPLE 13.2 Drying Time During Constant Rate Drying

Precipitated enzyme is filtered and the filter solids washed and dried before packaging. Washed filter cake containing 10 kg of dry solids and 15% water measured on a wet basis is dried in a tray drier under constant drying conditions. The critical moisture content is 6%, dry basis. The area available for drying is 1.2 m^2. The air temperature in the drier is 35°C. At the air humidity used, the surface temperature of the wet solids is 28°C. The heat transfer coefficient is 25 J m^{-2} s^{-1} °C^{-1}. What drying time is required to reduce the moisture content to 8%, wet basis?

Solution

The initial and final moisture contents expressed on a wet basis must be converted to a dry basis:

$$15\% \text{ wet basis} = \frac{15 \text{ g water}}{100 \text{ g wet solid}} = \frac{15 \text{ g water}}{15 \text{ g water} + 85 \text{ g dry solid}}$$

$$X_0 = \frac{15 \text{ g water}}{85 \text{ g dry solid}} = 0.176$$

Similarly:

$$8\% \text{ wet basis} = \frac{8 \text{ g water}}{8 \text{ g water} + 92 \text{ g dry solid}}$$

$$X_1 = \frac{8 \text{ g water}}{92 \text{ g dry solid}} = 0.087$$

(Continued)

As X_1 is greater than the critical moisture content $X_c = 0.06$, the entire drying operation takes place with constant drying rate. Equation (13.65) is used to determine the value of N_c. The heat of vaporization Δh_v for water at 28°C, the temperature of the surface of the solids where evaporation takes place, is 2435.4 kJ kg^{-1}. Therefore:

$$N_c = \frac{25 \text{ J m}^{-2} \text{ s}^{-1} °\text{C}^{-1}(1.2 \text{ m}^2)(35-28)°\text{C}}{2435.4 \times 10^3 \text{ J kg}^{-1}} = 8.62 \times 10^{-5} \text{ kg s}^{-1}$$

Applying Eq. (13.67) to calculate the drying time:

$$\Delta t = \frac{10 \text{ kg}}{8.62 \times 10^{-5} \text{kg s}^{-1}} \cdot \left| \frac{1 \text{ h}}{3600 \text{ s}} \right| \cdot (0.176 - 0.087) = 2.87 \text{ h}$$

The time required for drying is 2.9 h.

13.12.4 Drying Equipment

A diverse range of equipment is used for drying operations [31]. including tray, screen-conveyor, screw-conveyor, rotary drum, tunnel, bin, tower, spray, fluidized bed, and flash driers. Some driers have a direct mode of heating, whereby air entering the drier is brought into contact with the wet solid. Other types of equipment apply indirect heating to the drying material through a metal wall or tray. Some drier installations use a combination of direct and indirect heating.

Most driers operate at or close to atmospheric pressure. However, tray and enclosed rotary driers may be operated under vacuum, generally with indirect heating. The advantage of using vacuum drying is that evaporation of water occurs at lower temperatures when the pressure is reduced; for example, the boiling point of water at 6 kPa or 0.06 atm is only about 36°C. This makes vacuum operation suitable for processing heat-sensitive bioprocess products. Rates of drying are also enhanced under vacuum compared with atmospheric pressure. The water vapor produced during vacuum drying is usually condensed during operation of the drier to maintain the vacuum. As an alternative to vacuum drying, flash or spray drying may be suitable for heat-labile solids because drying in these systems occurs very rapidly, usually within 0.5 to 6 seconds, so that thermal damage from prolonged exposure to heat is avoided.

Large-scale driers cannot be designed or sized using theoretical analysis alone. The drying properties of numerous batches of material must be assessed experimentally. Scale-up of drying requires appropriate laboratory- and pilot-scale testing to characterize the material being dried and the transport processes that occur. To improve energy use and cost-effectiveness, the operating efficiency of large-scale drying equipment can be improved using measures such as

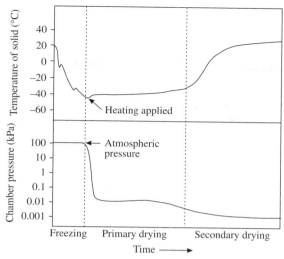

Figure 13.33 Variation of solids temperature and chamber pressure during a complete freeze-drying cycle.

preheating the inlet air with hot exhaust air, recycling some of the exhaust air, and reducing air leakage.

13.12.5 Freeze-Drying

Freeze-drying, also known as *lyophilization* or *cryodesiccation*, is used to dry unstable or heat-sensitive products at low temperature, thus protecting the material from heat damage and chemical decomposition. The material is frozen and then exposed to low pressure, which causes the frozen water within the solid to *sublimate* directly to vapor without passing through a liquid phase. As drying takes place at temperatures below 0°C, damage to biological molecules is minimized and any volatile substances are retained.

Freeze-drying is used commonly in the pharmaceutical industry; it is also used for drying some foods and for downstream processing of proteins, vaccines, and vitamins. However, the energy required for freeze-drying is substantially greater than for other drying methods, and the time required for drying is generally longer. The drying time for freeze-drying is roughly proportional to the material thickness raised to the power 1.5 to 2.0 [33,34].

The changes in operating variables during an entire freeze-drying cycle are illustrated in Figure 13.33. The duration of a complete freeze-drying cycle is typically 24 to 48 hours.

13.13 Summary of Chapter 13

At the end of Chapter 13 you should:

- Know what a *unit operation* is
- Be able to describe the five major steps involved in *downstream processing* schemes for bioprocess products
- Understand the factors affecting the choice of *cell removal operations* for treating bioprocess broths
- Understand the theory and practice of conventional *filtration*
- Understand the principles of *centrifugation*, including scale-up considerations
- Be familiar with methods used for *cell disruption*
- Be able to analyze *aqueous two-phase extractions* in terms of the *equilibrium partition coefficient, product yield*, and *concentration factor*
- Understand the mechanisms of large-scale *precipitation* for recovery of proteins from solution
- Understand the principles of *adsorption* operations and the design of *fixed-bed adsorbers*
- Know the different types of *membrane filtration* used in downstream processing, understand the importance of *concentration polarization* and *gel polarization*, and know the implications of *mass transfer control* and *fouling* for membrane filter performance
- Know the different types of *chromatography* used to separate biomolecules
- Understand the principles of *crystallization*, including characterization of the *size* and *size distribution* of crystals, the mechanisms of *crystal nucleation* and *crystal growth*, and operation of *continuous mixed suspension–mixed product removal (MSMPR)* crystallizers
- Know the role of *drying* operations in downstream processing, understand the relationship between *humidity* and the *equilibrium moisture content* of solids, and be able to describe the factors affecting *drying rate*
- Know the basic principles of *freeze-drying*

Problems

13.1. Overall product recovery

Supercoiled plasmid DNA for gene therapy applications is produced using high-yielding recombinant bacteria. After the cell culture step, the DNA is recovered using a series of five-unit operations. The processes used and the percentage product recovery for each step are as follows: microfiltration 89%, cell disruption 74%, precipitation 71%, ion-exchange chromatography 82%, gel chromatography 68%. What fraction of the plasmid DNA produced by the cells is available at the end of downstream processing?

13.2. Product yield from transgenic goats' milk

Goats are genetically engineered to produce a therapeutic protein, anti-thrombin III, in their milk. Twenty thousand liters of milk containing $10 \, g \, L^{-1}$ anti-thrombin III are treated each week to recover and purify the product. Fat is removed from the milk by skimming and centrifugation with a product yield of 75%. Caseins are then removed by salt precipitation and filtration with a product yield of 60% for these techniques. The clear whey fraction is treated further using gel chromatography and affinity chromatography; 85% yield is achieved in each of the chromatography steps.

a. What is the overall product yield?

b. How much purified anti-thrombin III is produced per week?

c. It is decided to improve the salt precipitation and filtration step to minimize product loss. If the target anti-thrombin III production rate is 80 kg week^{-1} from 20,000 L of milk, by how much must the product yield for this stage of the process be increased above the current 60%?

13.3. Laboratory algal filtration

A suspension of red *Porphyridium cruentum* microalgal cells is filtered in the laboratory using a 6-cm-diameter Büchner funnel and a vacuum pressure of 5 psi. The suspension contains 7.5 g of cells per liter of filtrate. The viscosity of the filtrate is 10^{-3} Pa s. The following data are measured.

Time (min)	1	2	3	4	5
Filtrate volume (mL)	124	191	245	289	330

a. What proportion of the total resistance to filtration is due to the cell cake after 3 minutes of filtration?

b. It is proposed to use the same Büchner funnel to filter a *P. cruentum* suspension with a higher cell concentration. What maximum cell density (mass of cells per liter of filtrate) can be handled if 200 mL of filtrate must be collected within 5 min?

13.4. Filtration of plant cells

Suspended plant cells expressing a therapeutic Fab′ antibody fragment are filtered using a laboratory filter of area 78 cm^2. A pressure drop of 9 psi is applied. The filtrate has a viscosity of 5 cP and the suspension deposits 25 g of cells per liter of filtrate. The results from the filtration experiments are correlated according to the equation:

$$\frac{t}{V_f} = K_1 V_f + K_2$$

where t is the filtration time, V_f is the filtrate volume, $K_1 = 8.3 \times 10^{-6}$ min cm^{-6}, and $K_2 = 4.2 \times 10^{-3}$ min cm^{-3}.

a. Determine the specific cake resistance and filter medium resistance.

b. The same plant cell suspension is processed using a pressure drop of 9 psi on a filter of area of 4 m^2. Estimate the time required to collect 6000 L of filtrate.

13.5. Bacterial filtration

A suspension of *Bacillus subtilis* cells is filtered under constant pressure for recovery of protease. A pilot-scale filter is used to measure the filtration properties. The filter area is 0.25 m^2, the pressure drop is 360 mmHg, and the filtrate viscosity is 4.0 cP. The cell suspension deposits 22 g of cake per liter of filtrate. The following data are measured.

Time (min)	2	3	6	10	15	20
Filtrate volume (L)	9.8	12.1	18.0	23.8	29.9	37.5

a. Determine the specific cake resistance and filter medium resistance.

b. Based on the data obtained in the pilot-scale study, what size filter is required to process 4000 L of cell suspension in 30 min at a pressure drop of 360 mmHg?

13.6. Filtration of mycelial suspensions

Pelleted and filamentous forms of *Streptomyces griseus* are filtered separately using a small laboratory filter of area 1.8 cm². The mass of wet solids per volume of filtrate is 0.25 g mL⁻¹ for the pelleted cells and 0.1 g mL⁻¹ for the filamentous culture. The viscosity of the filtrate is 1.4 cP. Five filtration experiments at different pressures are carried out with each suspension. The following results are obtained.

Filtrate volume (mL) for pelleted suspension	Pressure drop (mmHg)				
	100	250	350	550	750
Time (s)					
10	22	12	9	7	5
15	52	26	20	14	12
20	90	49	36	28	22
25	144	75	60	43	34
30	200	110	88	63	51
35	285	149	119	84	70
40	368	193	154	110	90
45	452	240	195	140	113
50	–	301	238	175	141

Filtrate volume (mL) for filamentous suspension	Pressure drop (mmHg)				
	100	250	350	550	750
Time (s)					
10	36	22	17	13	11
15	82	47	40	31	25
20	144	85	71	53	46
25	226	132	111	85	70
30	327	194	157	121	100
35	447	262	215	166	139
40	–	341	282	222	180
45	–	434	353	277	229
50	–	–	442	338	283

a. Evaluate the specific cake resistance as a function of pressure for each culture.

b. Determine the compressibility for each culture.

c. A filter press with area of 15 m^2 is used to process 20 m^3 of filamentous *S. griseus* culture. If the filtration must be completed in one hour, what pressure drop is required?

13.7. Rotary drum vacuum filtration

Continuous rotary vacuum filtration can be analyzed by considering each revolution of the drum as a stationary batch filtration. Per revolution, each cm^2 of filter cloth is used to form cake only for the period of time it spends submerged in the liquid reservoir. A rotary drum vacuum filter with drum diameter 1.5 m and filter width 1.2 m is used to filter starch from an aqueous slurry. The pressure drop is kept constant at 4.5 psi; the filter operates with 30% of the filter cloth submerged. Resistance due to the filter medium is negligible. Laboratory tests with a 5-cm^2 filter have shown that 500 mL of slurry can be filtered in 23.5 min at a pressure drop of 12 psi; the starch cake was also found to be compressible with $s = 0.57$. Use the following steps to determine the drum speed required to produce 20 m^3 of filtered liquid per hour.

a. Evaluate $\mu_f \alpha' c$ from the laboratory test data.
b. If N is the drum speed in revolutions per hour, what is the cycle time?
c. From (b), for what period of time per revolution is each cm^2 of filter cloth used for cake formation?
d. What volume of filtrate must be filtered per revolution to achieve the desired rate of 20 m^3 per hour?
e. Apply Eq. (13.15) to a single revolution of the drum to evaluate N.
f. The liquid level is raised so that the fraction of submerged filter area increases from 30 to 50%. What drum speed is required under these conditions?

13.8. Centrifugation of yeast

Yeast cells must be separated from a bioprocess broth. Assume that the cells are spherical with diameter of 5 μm and density of 1.06 g cm^{-3}. The viscosity of the culture broth is 1.36×10^{-3} N s m^{-2}. At the temperature of separation, the density of the suspending fluid is 0.997 g cm^{-3}. Five hundred liters of broth must be treated every hour.

a. Specify Σ for a suitably sized disc stack centrifuge.
b. The small size and low density of microbial cells are disadvantages in centrifugation. If instead of yeast, quartz particles of diameter 0.1 mm and specific gravity 2.0 are separated from the culture liquid, by how much is Σ reduced?

13.9. Centrifugation of food particles

Small food particles with diameter 10^{-2} mm and density 1.03 g cm^{-3} are suspended in liquid of density 1.00 g cm^{-3}. The viscosity of the liquid is 1.25 mPa s. A tubular bowl centrifuge of length 70 cm and radius 11.5 cm is used to separate the particles. If the centrifuge is operated at 10,000 rpm, estimate the feed flow rate at which the food particles are just removed from the suspension.

13.10. Cell disruption

Micrococcus bacteria are disrupted at 5°C in a Gaulin homogenizer operated at pressures between 200 and 550 kg*f*cm^{-2}. The data obtained for protein release as a function of number of passes through the homogenizer are listed in the table.

Number of passes			Pressure drop (kgf cm^{-2})			
200	300		400	500	550	
% Protein release						
1	5.0		13.5	23.3	36.0	42.0
2	9.5		23.5	40.0	58.5	66.0
3	14.0		33.5	52.5	75.0	83.7
4	18.0		43.0	66.6	82.5	88.5
5	22.0		47.5	73.0	88.5	94.5
6	26.0		55.0	79.5	91.3	–

a. How many passes are required to achieve 80% protein release at an operating pressure of 460 kgf cm^{-2}?

b. Estimate the pressure required to deliver 70% protein recovery in only two passes.

13.11. Disruption of cells cultured under different conditions

Candida utilis cells are cultivated in different bioreactors using either repeated batch or continuous culture. An experiment is conducted using a single pass in a Gaulin homogenizer to disrupt cells from the two culture systems. The following results are obtained.

	Percentage of protein released	
Type of culture	Pressure = 57 MPa	Pressure = 89 MPa
Repeated batch	66	84
Continuous	30	55

a. How many passes are required to achieve 90% protein release from the batch-culture cells at an operating pressure of 70 MPa?

b. What percentage protein is released from the continuous-culture cells using the number of passes determined in (a) and the same operating pressure?

c. What operating pressure is required to release 90% of the protein contained in the cells from continuous culture in five passes?

13.12. Enzyme purification using two-phase aqueous partitioning

Leucine dehydrogenase is recovered from 150 L of *Bacillus cereus* homogenate using an aqueous two-phase polyethylene glycol–salt system. The homogenate initially contains 3.2 units of enzyme mL^{-1}. A polyethylene glycol–salt mixture is added and two phases form. The enzyme partition coefficient is 3.5.

a. What volume ratio of upper and lower phases must be chosen to achieve 80% recovery of enzyme in a single extraction step?

b. If the volume of the lower phase is 100 L, what is the concentration factor for 80% recovery?

13.13. Recovery of viral particles

Cells of the fall armyworm *Spodoptera frugiperda* are cultured in a bioreactor to produce viral particles for insecticide production. Virus is released into the culture broth after lysis of the host cells. The initial culture volume is 5 L. An aqueous two-phase polymer solution of volume

2 L is added to this liquid; the volume of the bottom phase is 1 L. The virus partition coefficient is 10^{-2}.

a. What is the yield of virus at equilibrium?

b. Write a mass balance for the viral particles in terms of the concentrations and volumes of the phases, equating the amounts of virus present before and after addition of polymer solution.

c. Derive an equation for the concentration factor in terms of the liquid volumes and partition coefficient only.

d. Calculate the concentration factor for the viral extraction.

13.14. Enzyme salting-out

The solubility of cellulase enzyme in a buffered extract of *Trichoderma viride* cells is determined in the laboratory as a function of ammonium sulfate concentration. At $(NH_4)_2SO_4$ concentrations of 1.8, 2.2, and 2.5 M, the solubilities are 44, 0.58, and 0.07 mg L^{-1}, respectively. *T. viride* cells produced by bioprocess are homogenized to generate 800 L of buffered extract containing 25 mg L^{-1} of soluble cellulase. Cellulase is precipitated by adding $(NH_4)_2SO_4$ to the extract to give a concentration of 2.0 M and a final solution volume of 930 L.

a. Estimate the mass of enzyme recovered.

b. What is the yield or percentage recovery of cellulase in this precipitation process?

c. The residual cellulase in solution is treated for further precipitation. Assuming negligible volume change for the second increment in salt content, estimate the concentration of $(NH_4)_2SO_4$ required to recover 90% of the residual cellulase.

13.15. Precipitation of monoclonal antibody

Human IgM monoclonal antibody is produced by hybridoma cells in a stirred bioreactor. Antibody secreted into the culture medium reaches a concentration of 120 µg mL^{-1} of cell-free broth. The product is concentrated by precipitation immediately after removal of the cells. The yield of antibody recovered in the precipitate is determined as a percentage of the initial mass of antibody present in the broth. The precipitate purity or percentage of the precipitate weight that is IgM antibody is also measured. Various precipitation methods are tested and the following results are obtained.

Precipitant	Concentration of precipitant (% w/v)	pH	Antibody yield (%)	Precipitate purity (% w/w)
Ammonium sulphate	18	7.2	0.0	–
	24	7.2	49	90
	29	7.2	75	51
	31	7.2	83	18
Polyethylene glycol	8	5.5	75	95
	12	5.5	96	60
Ethanol	25	7.2	0.0	–

a. From the data provided, what are the advantages and disadvantages of using 12% w/v polyethylene glycol for this precipitation process compared with 8% w/v?

b. Using % w/v as the unit for salt concentration and $\mu g\ g^{-1}$ of cell-free broth as the unit for protein solubility, derive an empirical equation for IgM solubility as a function of ammonium sulphate concentration at pH 7.2. The density of cell-free broth can be taken as $1000\,kg\,m^{-3}$.

c. Predict the antibody yield using 27% w/v $(NH_4)_2SO_4$.

d. What mass of antibody will remain in solution if 27% w/v $(NH_4)_2SO_4$ is used to treat 100 L of cell-free broth?

e. The residual solution after salting-out with 27% w/v $(NH_4)_2SO_4$ is treated for further antibody recovery.

   (i) If a total yield of 94% is required after both precipitation steps, estimate the concentration of $(NH_4)_2SO_4$ that must be applied in the second salting-out stage.

   (ii) What mass of antibody do you expect to recover in the second precipitation step?

   (iii) When the second-stage precipitation is carried out using the $(NH_4)_2SO_4$ concentration determined in (i), it is found that the amount of antibody contained in the precipitate is somewhat more than expected. What possible reasons can you give for this?

13.16. Crystal size distribution from screen analysis

Cephalosporin C salt crystals are subjected to screen analysis for measurement of the crystal size distribution. The following results are obtained.

U.S. Sieve	Mass retained (g)
No. 30	0
No. 40	0.5
No. 50	3.05
No. 60	8.3
No. 70	10.8
No. 100	15.6
No. 140	16.8
No. 200	9.0
No. 270	6.4
Pan	4.2

a. Plot the mass density distribution as a function of average particle size.

b. What is the dominant crystal size?

13.17. Drying of benzyl penicillin

Sodium benzyl penicillin must be dried to a moisture content of 20% measured on a wet weight basis. Air at 25°C is used for the drying operation at 1 atm pressure in a fluidized bed drier.

a. What relative humidity of air is required?

b. What is the mole fraction of water in the air used for drying?

c. What is the humidity of the air?

Use the isotherm data for sodium benzyl penicillin in Figure 13.56(A) and assume that equilibrium is reached.

13.18. Drying rate for filter cake

A cake of filtered solids is dried using a screen tray drier. The dimensions of the cake are diameter of 30 cm and thickness of 5 mm. Air at 35°C is passed over both the top and bottom of the cake on the screen. The critical moisture content of the cake is 5% measured on a wet basis and the density of completely dried material is $1380 \, kg \, m^{-3}$. At the operating air flow rate, the heat transfer coefficient for gas-phase heat transfer at the surface of the solids is $45 \, W \, m^{-2} \, °C^{-1}$. At the air humidity used, the temperature of the wet cake surface is 26°C.

a. What is the rate of drying?

b. What time is required to dry the cake from a moisture content of 15% wet basis to a moisture content of 5% wet basis?

13.19. Drying time for protein crystals

Insulin crystals are dried in air in a batch drier under constant drying conditions. The moisture content associated with 20 kg dry weight of crystals is reduced from 18 g per 100 g dry solids to 8 g per 100 g dry solids in 4.5 hours. The area available for drying is $0.5 \, m^2$. The critical moisture content for this system is 5 g per 100 g dry solids. If the air temperature is 28°C and the surface temperature of the wet crystals is 20°C, what is the heat transfer coefficient for this drying operation?

13.20. Drying time for solid precipitate

Precipitated protein is centrifuged to a water content of 35% measured on a dry mass basis. It is then dried in a batch drier for 6 hours under constant drying conditions to a water content of 15%, dry basis. The critical moisture content for the solids is 8%, dry basis. How long would it take to dry a sample of the same size from a water content of 35% to 10% under the same conditions?

13.21. Drying of amino acid crystals

A batch of wet asparagine crystals containing 32 kg of dry solids and 25% moisture measured on a wet basis is dried in a tunnel drier for 3 hours using air at 28°C. The critical moisture content of the crystals is 10% dry basis. The drying area available is $12 \, m^2$. The heat transfer coefficient for gas-phase heat transfer to the surface of the crystals is $32 \, W \, m^{-2} \, °C^{-1}$. At the air humidity used, the temperature at the wet crystal surface is 24°C and the crystal equilibrium moisture content is 7.5%, dry basis.

a. What is the drying rate?

b. What is the moisture content of the dried crystals?

References

[1] K. Schügerl, Recovery of proteins and microorganisms from cultivation media by foam flotation, Adv. Biochem. Eng./Biotechnol 68 (2000) 191–233.

[2] M.-R. Kula, Recovery operations, VCH, 1985.725760

[3] H.-W. Hsu, Separations by Centrifugal Phenomena, John Wiley, 1981.

[4] P.J. Hetherington, M. Follows, P. Dunnill, M.D. Lilly, Release of protein from baker's yeast (*Saccharomyces cerevisiae*) by disruption in an industrial homogeniser, Trans. IChE. 49 (1971) 142–148.

[5] E. Keshavarz-Moore, Cell disruption: a practical approach, in: M.S. Verrall (Ed.), Downstream Processing of Natural Products, John Wiley (1996), pp. 41–52.

[6] C.R. Engler, C.W. Robinson, Effects of organism type and growth conditions on cell disruption by impingement, Biotechnol. Lett. 3 (1981) 83–88.

[7] P. Dunnill, M.D. Lilly, Protein extraction and recovery from microbial cells, in: S.R. Tannenbaum, D.I.C. Wang, (Eds.), Single-Cell Protein II, MIT Press (1975), pp. 179–207.

[8] C.R. Engler, C.W. Robinson, Disruption of *Candida utilis* cells in high pressure flow devices, Biotechnol. Bioeng. 23 (1981) 765–780.

[9]   E.J. Cohn, J.D. Ferry, Interactions of proteins with ions and dipolar ions, in: E.J. Cohn, J.T. Edsall, (Eds.), Proteins, Amino Acids and Peptides, Reinhold (1943),

[10]   International Critical Tables (1926–1930), McGraw-Hill; 1st electronic ed., International Critical Tables of Numerical Data, Physics, Chemistry and Technology, Knovel, 2003.

[11]   J.F. Richardson, J.H. Harker, (Chapters 17 and 18) fifth ed., Coulson and Richardson's Chemical Engineering, vol. 2, Butterworth-Heinemann, 2002.

[12]   M. Cheryan, Ultrafiltration and Microfiltration Handbook, Technomic (1998).

[13]   L.J. Zeman, A.L. Zydney, Microfiltration and Ultrafiltration, Marcel Dekker, 1996.

[14]   W.F. Blatt, A. Dravid, A.S. Michaels, L. Nelsen, Solute polarization and cake formation in membrane ultrafiltration: causes, consequences, and control techniques, in: J.E. Finn (Ed.), Membrane Science and Technology, Plenum (1970), pp. 47–97.

[15]   M.S. Le, J.A. Howell, Ultrafiltration, Pergamon Press, 1985.383–409.

[16]   L.R. Snyder, Classification of the solvent properties of common liquids, J. Chromatog 92 (1974) 223–230.

[17]   E.L. Johnson, R. Stevenson, Basic Liquid Chromatography, Varian Associates, 1978.

[18]   R.K. Scopes, Protein Purification: Principles and Practice, third ed., Springer-Verlag, 1994.

[19]   J. Garside, A. Mersmann, J. Nyvlt, Measurement of Crystal Growth and Nucleation Rates, second ed., Institution of Chemical Engineers, 2002.

[20]   A.D. Randolph, M.A. Larson, Theory of Particulate Processes, second ed., Academic Press, 1988.

[21]   S.J. Jančić, P.A.M. Grootscholten, Industrial Crystallization, Delft University Press, 1984.

[22]   J.W. Mullin, Crystallization, fourth ed., Butterworth-Heinemann, 2001.

[23]   S.N. Black, R.J. Davey, M. Halcrow, The kinetics of crystal growth in the presence of tailor-made additives, J. Crystal Growth 79 (1986) 765–774.

[24]   R.C. DeMattei, R.S. Feigelson, Growth rate study of canavalin single crystals, J. Crystal Growth 97 (1989) 333–336.

[25]   K. Toyokura, K. Mizukawa, M. Kurotani, Crystal growth of l-SCMC seeds in a dl-SCMC solution of pH 0.5, in: A.S. Myerson, D.A. Green, P. Meenan, (Eds.), Crystal Growth of Organic Materials, American Chemical Society (1996), pp. 72–77.

[26]   D.J. Kirwan, I.B. Feins, A.J. Mahajan, Crystal growth kinetics of complex organic compounds, in: A.S. Myerson, D.A. Green, P. Meenan, (Eds.), Crystal Growth of Organic Materials, American Chemical Society (1996), pp. 116–121.

[27]   C. Jacobsen, J. Garside, M. Hoare, Nucleation and growth of microbial lipase crystals from clarified concentrated fermentation broths, Biotechnol. Bioeng. 57 (1998) 666–675.

[28]   E. Forsythe, M.L. Pusey, The effects of temperature and NaCl concentration on tetragonal lysozyme face growth rates, J. Crystal Growth 139 (1994) 89–94.

[29]   R.A. Judge, M.R. Johns, E.T. White, Protein purification by bulk crystallization: the recovery of ovalbumin, Biotechnol. Bioeng. 48 (1995) 316–323.

[30]   W.L. McCabe, Crystal growth in aqueous solutions, parts I and II, Ind. Eng. Chem. 21 (1929) 30–33, 112–118.

[31]   Perry's Chemical Engineers' Handbook, Section 12, eighth ed., McGraw-Hill, 2008.

[32]   C.J. Geankoplis, (Chapter 9) Transport Processes and Separation Process Principles, fourth ed, Prentice Hall, 2003.

[33]   W.L. McCabe, J.C. Smith, P. Harriott, (Chapter 24) Unit Operations of Chemical Engineering, seventh ed., McGraw-Hill, 2005.

[34]   J.W. Snowman, Lyophilization, in: M.S. Verrall (Ed.), Downstream Processing of Natural Products, John Wiley (1996), pp. 275–299.

Suggestions for Further Reading

J.A. Asenjo, (Ed.), Separation Processes in Biotechnology, Marcel Dekker (1990),

B. Atkinson, F. Mavituna, Chapters 16 and 17Biochemical Engineering and Biotechnology Handbook, 2nd ed., Macmillan, 1991.

M.A. Desai, (Ed.), Downstream Processing of Proteins: Methods and Protocols, Humana Press (2010),

E. Goldberg, Handbook of Downstream Processing, Chapman and Hall, 1997.

R.G. Harrison, P. Todd, S.R. Rudge, D.P. Petrides, Bioseparations Science and Engineering, Oxford University Press, 2003.

M.R. Ladisch, Bioseparations Engineering: Principles, Practice, and Economics, John Wiley, 2001.

M.S. Verrall, Downstream Processing of Natural Products: A Practical Handbook, John Wiley, 1996.

W.L. McCabe, J.C. Smith, P. Harriott, Unit Operations of Chemical Engineering, 7th ed., McGraw-Hill, 2005.10061033

T. Oolman, T.-C. Liu, Filtration properties of mycelial microbial broths, Biotechnol. Prog. 7 (1991) 534–539.

J.F. Richardson, J.H. Harker, Chapter 7 5th ed., Coulson and Richardson's Chemical Engineering, vol. 2, Butterworth-Heinemann, 2002.

R.J. Wakeman, E.S. Tarleton, Filtration, Elsevier Science, Oxford, 1999.

H.A.C. Axelsson, Centrifugation, Pergamon Press, 1985.325346

W.W.-F. Leung, Centrifugal Separations in Biotechnology, Academic Press, 2007.

J.F. Richardson, J.H. Harker, Chapter 9 5th ed., Coulson and Richardson's Chemical Engineering, vol. 2, Butterworth-Heinemann, 2002.

Y. Chisti, M. Moo-Young, Disruption of microbial cells for intracellular products, Enzyme Microb. Technol. 8 (1986) 194–204.

C.R. Engler, Disruption of Microbial Cells, Pergamon Press, 1985.305324

M.-R. Kula, H. Schütte, Purification of proteins and the disruption of microbial cells, Biotechnol. Prog. 3 (1987) 31–42.

R. Lander, W. Manger, M. Scouloudi, A. Ku, C. Davis, A. Lee, Gaulin homogenization: a mechanistic study, Biotechnol. Prog. 16 (2000) 80–85.

A.P.J. Middelberg, Process-scale disruption of microorganisms, Biotechnol. Adv. 13 (1995) 491–551.

P.-Å. Albertsson, Partition of Cell Particles and Macromolecules, 2nd ed., John Wiley, 1971.

A.D. Diamond, J.T. Hsu, Aqueous two-phase systems for biomolecule separation, Adv. Biochem. Eng./Biotechnol 47 (1992) 89–135.

M.-R. Kula, Liquid–liquid extraction of biopolymers, Pergamon Press, 1985.451471

D.J. Bell, M. Hoare, P. Dunnill, The formation of protein precipitates and their centrifugal recovery, Adv. Biochem. Eng./Biotechnol 26 (1983) 1–71.

M.R. Ladisch, Bioseparations Engineering: Principles: Practice, and Economics, John Wiley, 2001.

M.Q. Niederauer, C.E. Glatz, Selective precipitation, Adv. Biochem. Eng./Biotechnol 47 (1992) 159–188.

F. Rothstein, R.G. Harrison, Differential precipitation of proteins, in: R.G. Harrison, (Ed.), Protein Purification Process Engineering, Marcel Dekker (1994), pp. 115–208.

F.H. Arnold, H.W. Blanch, C.R. Wilke, Analysis of affinity separations. I. Predicting the performance of affinity adsorbers, Chem. Eng. J. 30 (1985) B9–B23.

A.L. Hines, R.N. Maddox, Mass Transfer: Fundamentals and Applications, Prentice Hall, 1985.

F.L. Slejko, Adsorption Technology, in: F.L. Slejko, (Ed.), Adsorption Technology, Marcel Dekker (1985),

P.T. Cardew, M.S. Le, Membrane Processes: A Technology Guide, Royal Society of Chemistry, 1998.

M. Dosmar, D. Brose, T.H. Meltzer, M.W. Jornitz, Crossflow ultrafiltration, in: T.H. Meltzer, M.W. Jornitz, (Eds.), Filtration in the Biopharmaceutical Industry, Marcel Dekker (1998), pp. 493–532.

M.C. Porter, P.A. Schweitzer, Membrane filtration, in: P.A. Schweitzer, (Ed.), Handbook of Separation Techniques for Chemical Engineers, McGraw-Hill (1997),

R. van Reis, A.L. Zydney, M.C. Flickinger, S.W. Drew, Protein ultrafiltration, in: M.C. Flickinger, S.W. Drew, (Eds.), Encyclopedia of Bioprocess Technology: Bioprocess, Biocatalysis, and Bioseparation, John Wiley (1999), pp. 2197–2214.

Y. Chisti, M. Moo-Young, Large scale protein separations: engineering aspects of chromatography, Biotechnol. Adv. 8 (1990) 699–708.

R.A.M. Delaney, R.A. Grant, Industrial gel filtration of proteins, in: R.A. Grant, (Ed.), Applied Protein Chemistry, Applied Science (1980), pp. 233–280.

J.-C. Janson, P. Hedman, Large-scale chromatography of proteins, Adv. Biochem. Eng 25 (1982) 43–99.

M.R. Ladisch, Bioseparations Engineering: Principles: Practice, and Economics, John Wiley, 2001.

P.J. Robinson, M.A. Wheatley, J.-C. Janson, P. Dunnill, M.D. Lilly, Pilot scale affinity chromatography: purification of β-galactosidase, Biotechnol. Bioeng. 16 (1974) 1103–1112.

M.S. Verrall, Downstream Processing of Natural Products: A Practical Handbook, John Wiley, 1996.

S.M. Wheelwright, Protein Purification: Design and Scale-Up of Downstream Processing, John Wiley, 1991.

M.R. Ladisch, Bioseparations Engineering: Principles: Practice, and Economics, John Wiley, 2001.

J.F. Richardson, J.H. Harker, Chapter 15 5th ed., Coulson and Richardson's Chemical Engineering, vol. 2, Butterworth-Heinemann, 2002.

L.A. Gatlin, S.L. Nail, R.G. Harrison, Freeze drying: a practical overview, in: R.G. Harrison, (Ed.), Protein Purification Process Engineering, Marcel Dekker (1994), pp. 317–367.

R.G. Harrison, P. Todd, S.R. Rudge, D.P. Petrides, Bioseparations Science and Engineering, Oxford University Press, 2003.

PART 6

Special topics and application

CHAPTER 14

Special Topics

14.1 Metabolic Network Analysis for Metabolic Engineering

Metabolic engineering is a scientific discipline that modifies the behavior and properties of cellular reaction networks, including their structure, stoichiometry, kinetics, and control. Metabolic engineering is often guided by *in silico* analysis of metabolic networks to rationally identify which reactions in metabolism are the best to target for genetic manipulation. Common goals of metabolic engineering include:

- Increased yield and productivity of compounds synthesized by cells
- Synthesis of new products not previously produced by a particular organism
- Greater resistance to bioprocessing conditions like tolerance of hypoxia or inhibitors
- Uptake and metabolism of substrates not previously assimilated
- Enhanced biological degradation of pollutants
- Altered regulation via effector molecules including quorum-sensing molecules

Genetic manipulations to achieve these goals typically involve expressing foreign genes, changing the regulation or copy number of native genes, or deleting native genes. Metabolic engineering is often more successful when the functioning of the integrated metabolic networks is considered as opposed to a single enzyme-catalyzed reaction. Metabolic engineering now uses a systems biology approach to organism development. As an example, consider the reaction network shown in Figure 14.1. From substrate A, two major products, J and L, are formed. Commercially, L is much more valuable than J, but J and L are normally produced in about equal amounts. This is indicated by the approximately equal thickness of the arrows leading to them in Figure 14.1(A). Without a systems analysis of the reaction network, it is decided to reduce the production of J by knocking out the gene coding for the enzyme responsible for converting A to B. This is considered an appropriate strategy as it prevents the wasteful channeling of A into the pathway leading to J. The outcome of this genetic manipulation is shown in Figure 14.1(B). Eliminating the reaction A→B also prevents the reaction B→G. As indicated by the altered thickness of several of the reaction arrows in Figure 14.1(B), the cell has compensated for the loss of this direct pathway to J by adjusting the throughput of other reactions in the network. The overall result is that production of L is virtually unchanged. Such an outcome is often observed after single-step modifications to metabolic pathways.

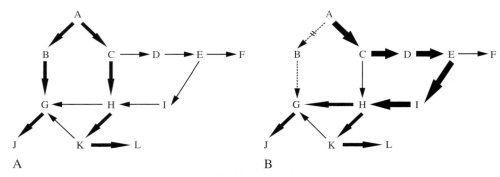

Figure 14.1 Reaction network for production of products J and L from substrate A: (A) before genetic modification of the cells, and (B) after genetic modification to prevent the conversion of A to B.

Better understanding of the interconnectedness and influence of reactions across the entire network is required to make more informed decisions about genetic engineering strategies. Typically, modification of several reactions steps is required for effective adjustment of the selectivity and yield of metabolic pathways.

Implementation of metabolic engineering strategies requires skills in engineering, computational analysis, and biology. The following sections outline a number of computational approaches that can be used to interpret phenotypic data and can be used to rationally design strains by identifying reactions, enzymes, and gene targets for modification.

14.1.1 Metabolic Flux Analysis

Figure 14.2 is a metabolic pathway diagram illustrating several active enzyme-catalyzed reactions that operate during anaerobic growth of *Saccharomyces cerevisiae*. Metabolic maps graphically present useful information about the metabolic intermediates formed and consumed during metabolism and their relationships with each other. However, they do not contain information about whether certain reactions are more or less active under specific culture conditions, or the relative contributions of specific reaction steps to the overall outcome of metabolism. For example, although the reactions required for synthesis of glycerol from glucose are represented in Figure 14.2, it is unclear if glycerol is a major or minor product of metabolism. Other information about operation of the network, such as the extent to which acetate synthesis detracts from ethanol production, and the amount of material directed into the pentose phosphate pathway relative to that processed by glycolysis, is not obvious without further analysis.

Figure 14.3 shows the results of a *metabolic flux analysis* for *S. cerevisiae* cultured under anaerobic conditions. The term *metabolic flux* denotes the rate at which material is processed through a metabolic pathway [1,2]. The flux in each branch of the reaction network is shown

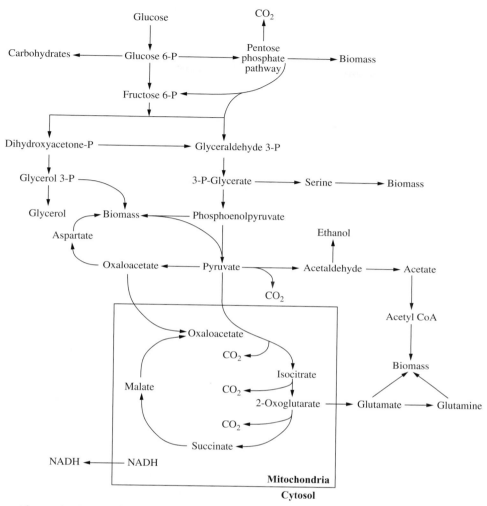

Figure 14.2 Metabolic pathways for anaerobic growth of *Saccharomyces cerevisiae*.

in Figure 14.3, where the numbers at each step indicate the flux of carbon in moles through that reaction relative to a normalized flux of 100 for glucose uptake into the cell. A negative value for a particular conversion means that the net reaction operates in the reverse direction to that shown. The quantitative information included in Figure 14.3 reflects the relative participation of all major pathways in the network. The results show, for example, that glycerol is a relatively minor product of metabolism, as the carbon flux for glycerol synthesis is only 9.43/52.76 or 18% of the flux for ethanol production. Other details of interest, for example, that acetate synthesis detracts negligibly from the carbon flow into ethanol, and that the pentose phosphate pathway draws only 8% of the carbon channeled into glycolysis, are also indicated.

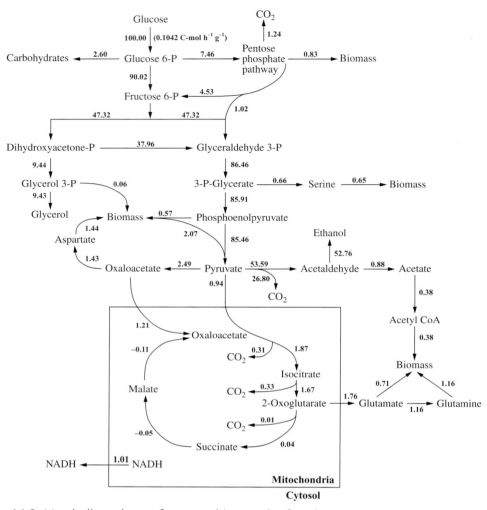

Figure 14.3 Metabolic pathways for anaerobic growth of *Saccharomyces cerevisiae* showing the fluxes of carbon in individual branches of the pathway relative to a flux of 100 for glucose uptake into the cell. The specific rate of glucose uptake was 0.1042 C-mol h^{-1} per g of biomass. *From T.L. Nissen, U. Schulze, J. Nielsen, and J. Villadsen, 1997, Flux distributions in anaerobic, glucose-limited continuous cultures of* Saccharomyces cerevisiae. Microbiology *143, 203–218. Copyright 1997. Reproduced with permission of the Society for General Microbiology.*

The greatest benefits of metabolic flux analysis are obtained when several flux maps are compared after cultures are subjected to environmental or genetic perturbation. For instance, a comparison of fluxes under culture conditions that promote low and high yeast growth rates can be used to show how metabolites are redirected from one region of the network to another to accommodate the need for faster biomass production. Flux analysis is also used to evaluate cellular responses at the molecular level for amplification or deletion of genes for specific enzymes in pathways. Principal branch points in reaction networks and the effects of enzyme and pathway regulation can be identified from flux analyses applied before and after genetic manipulation.

14.1.2 Basic Considerations of Metabolic Network Analysis

How are metabolic fluxes such as those shown in Figure 14.3 evaluated? Because the rates of reactions occurring intracellularly are quantified in flux analysis, it is reasonable to expect that experimental measurement of the intracellular metabolites and enzyme reactions would be necessary. However, a major strength of flux analysis is that it does not require kinetic information about individual reactions or the measurement of intermediate concentrations to reconstruct network properties. The complexity of metabolic networks, the difficulty of measuring metabolite levels and enzyme kinetic parameters *in vivo*, and our incomplete understanding of the reaction sequences and enzymes involved in some areas of metabolism, make this approach either impossible or very onerous in practice. Rather, flux analysis considers a limited number of experimental measurements, which could include substrate consumption, product formation, and biomass composition, in conjunction with a stoichiometric analysis of the biochemical network to estimate the internal reaction rates. In essence, highly complex, nonlinear, and intricately regulated reaction systems are reduced to a set of linear algebraic equations that can be solved with relative ease.

The methods employed in metabolic network analysis draw on those introduced in Chapter 4 to evaluate the stoichiometry of cell growth and product synthesis. Here, we extend the macroscopic description of cell metabolism developed in Section 4.6 to include reactions that occur wholly within the cell. As intracellular reactions link the consumption of substrate to the production of biomass and extracellular products, determining their fluxes provides an understanding of how the overall outcome of metabolism is achieved.

The key steps in metabolic network analysis are the construction of a model reaction network for the metabolic pathways of interest, and the application of stoichiometric constraints which capture mass balances and experimental data to evaluate intracellular fluxes. Several fundamental principles underpin flux analysis.

- *Mass balances.* Generally, flux analysis refers to the flux of carbon within metabolic pathways. Mass balances on intracellular metabolites show how carbon provided in the substrate is distributed between the different branches and reactions of the pathway. As the network model developed for flux analysis must satisfy the law of conservation of mass, mass is conserved across the entire system and at each reaction step.
- *Chemical energy balance.* Metabolic reactions generate and consume chemical energy through the interconversion of ATP, ADP, AMP, and similar compounds. Metabolic flux

analysis can be carried out using mass balancing alone; however, when energy carriers are included in the model, the amount of energy generated in one section of the metabolic network must equal the amount consumed in other reactions.

- *Redox balance.* Redox balances are mass balances on biologically active electrons. Redox carriers such as NADH, NADPH, and $FADH_2$ are involved in many cellular reactions moving energy from one metabolite to another. When redox carriers are included in flux calculations, redox generated in one part of the reaction network must be consumed in the remainder of the network.

- *Reaction thermodynamics.* Reactions in the flux model must be thermodynamically feasible. For example, if the results of flux analysis show a reverse flux through an irreversible reaction step, important thermodynamic principles have been violated. Such a result indicates an error in the analysis.

14.1.3 Network Definition and Simplification

Thousands of individual reactions are involved in cellular metabolism. Defining the network of reactions to be included in metabolic flux analysis is therefore a major undertaking. Substantial *a priori* knowledge about the structure of the pathways, the components involved, and the links between metabolic subnetworks is required. While this has been traditionally done manually, the rapid improvement in genome sequencing technologies and systems biology tools like KBase, developed at the U.S. Department of Energy, have facilitated this important step. As flux analysis relies on mass balances, the fate of all important species like carbon atoms and electrons must be defined within the metabolic model. If there are gaps or errors in the reaction network involving unknown sources or sinks of carbon and electrons, the quality of the results obtained from flux analysis will be potentially of little value.

Three types of reaction sequence are commonly found in metabolic pathways. As shown in Figure 14.4, metabolic reactions occur in *linear* cascades, *branched* or *split* pathways, and *cyclic* pathways. In linear sequences, the product of one reaction serves as the reactant for the following reaction and no alternative or additional reactions take place. Reactions in linear cascades can be lumped into a single reaction for metabolic flux analysis. This form of reaction grouping reduces the number of reactions and intermediates in the metabolic model without affecting the overall results (Figure 14.5(A)). Branched pathways can be either *diverging* or *converging*. In a diverging pathway, a metabolite functions as the reactant in at least two different conversions to form multiple products. For example, metabolite D in Figure 14.4 is a diverging pathway branch point, producing metabolites F and H that then proceed along different reaction routes. In a converging pathway, a metabolite is produced in at least two different reactions, as shown for metabolite F. Some metabolites act simultaneously as both diverging and converging branch points. In cyclic pathways, several

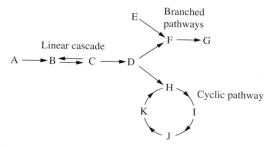

Figure 14.4 Types of reaction sequence in metabolic pathways.

metabolites serve as both reactants and products and are regenerated during each passage of the cycle.

A central tenet of metabolic network analysis generally and metabolic flux analysis specifically is that *metabolic pathways can be considered to operate under steady-state conditions*. This is a key simplifying assumption that allows us to calculate the fluxes in metabolic networks. When the system is at steady state, there can be no accumulation or depletion of intermediates and the concentrations of all metabolites remain constant. Furthermore, the total flux leading to a particular metabolite must be equal to the total flux leading away from that metabolite.

The assumption of steady state is generally accepted as valid because the rate of turnover of most metabolic intermediates is high relative to the amounts of those metabolites present in the cells at any given time. Therefore, because metabolism occurs on a relatively fast time-scale, any transients after perturbation are very rapid relative to the rate of cell growth or metabolic regulation. The steady-state assumption, which is sometimes called the *pseudo-steady-state assumption*, means that any experimental culture data used in metabolic flux analysis should be measured under conditions that can be approximated as steady-state conditions.

If the metabolic pathway in Figure 14.5(A) operates at steady state, there can be no accumulation or depletion of any of the intermediates in the pathway. If r_S and r_P are expressed in mass terms and r_1, r_2, and r_3 represent mass rates of reaction in units of, for example, g s^{-1}, we can say immediately that:

$$r_S = r_1 = r_2 = r_3 = r_P = v \tag{14.1}$$

because, at steady state, the mass rate of reaction of any intermediate in the sequence must be equal to the mass rate of reaction of all the other intermediates to avoid metabolite accumulation or depletion. Substrate uptake and product excretion must also occur at this rate to maintain constant concentrations within the cell. Accordingly, v in Eq. (14.1) can be used to

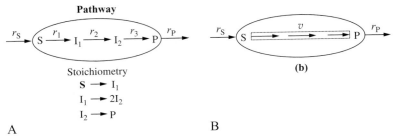

Figure 14.5 (A) Linear, unbranched reaction pathway with stoichiometry for the individual reaction steps. (B) Grouping of reactions in a linear, unbranched pathway.

represent the flux of mass through the entire pathway from S to P. Similarly, Eq. (14.1) also applies if r_S, r_P, r_1, r_2, and r_3 represent the rates at which carbon is taken up, excreted, and transferred between intermediates. As an atom, carbon remains unchanged in chemical processes and must flow through the sequence at a constant rate under steady-state conditions. In this case, v in Eq. (14.1) represents the flux of carbon (C) through the entire pathway.

The number of moles of chemical compounds is generally not conserved in reaction. Therefore, if r_S, r_P, r_1, r_2, and r_3 are molar rates in units of, for example, mol s^{-1}, information about the stoichiometry of the reactions is required to assess the relative molar fluxes through the pathway. From the individual reaction stoichiometries provided in Figure 13.5(A), the overall stoichiometry for conversion of 1 mole of S to P is S→2 P based on the stoichiometric sequence:

$$S \rightarrow I_1 \rightarrow 2\,I_2 \rightarrow 2\,P \tag{14.2}$$

As molar rates of forward reaction, r_1, r_2, and r_3 are defined as follows:

$$\begin{aligned}
r_1 &= \text{moles of S converted per unit time} \\
r_2 &= \text{moles of } I_1 \text{ converted per unit time} \\
r_3 &= \text{moles of } I_2 \text{ converted per unit time}
\end{aligned} \tag{14.3}$$

Let us perform mass balances on each of the components of the reaction sequence to find the relationships between these reaction rates. We will use the general mass balance equation derived in Chapter 4:

$$\frac{dM}{dt} = \hat{M}_i - \hat{M}_o + R_G - R_C \tag{4.26}$$

where M is mass, t is time, $\hat{M}_i$ is the mass flow rate entering the system, $\hat{M}_o$ is the mass flow rate leaving the system, R_G is the rate of generation by chemical reaction, and R_C is the rate of consumption by chemical reaction. The system in this case is the cell. Applying Eq. (4.26) to each component of the pathway in Figure 14.5(A):

$$
\frac{dM_S}{dt} = (MW)_S (r_S) - 0 + 0 - (MW)_S (r_1)
$$

$$
\frac{dM_{I1}}{dt} = 0 - 0 + (MW)_{I1} (r_1) - (MW)_{I1} (r_2)
$$

$$
\frac{dM_{I2}}{dt} = 0 - 0 + (MW)_{I2} (2r_2) - (MW)_{I2} (r_3) \tag{14.4}
$$

$$
\frac{dM_P}{dt} = 0 - (MW)_P (r_P) + (MW)_P (r_3) - 0
$$

In Eq. (14.4), MW denotes molecular weight and subscripts S, I1, I2, and P denote substrate, I_1, I_2, and product, respectively. Because the equations are mass balances, the molar reaction rates r are converted to mass rates using the molecular weights of the compounds to which they refer. The stoichiometric coefficients for each component in the reaction sequence have also been applied. For example, in the mass balance for I_1, the mass rate of generation of I_1 by chemical reaction is shown as $(MW)_{I1}(r_1)$, where r_1 is used to represent the molar rate of production of I_1.

At steady state, the derivatives with respect to time in Eq. (14.4) are zero as there can be no change in mass of any component. Applying this result allows us to cancel the molecular weight terms and simplify the equations to give:

$$
r_S = r_1
$$

$$
r_1 = r_2
$$

$$
2r_2 = r_3 \tag{14.5}
$$

$$
r_3 = r_P
$$

or

$$
r_S = r_1 = r_2 = 0.5\, r_3 = 0.5\, r_P = v \tag{14.6}
$$

Eq. (14.6) applies when the reaction rates are expressed as molar rates.

Our analysis of the reaction sequence of Figure 14.5(A) has demonstrated that the fluxes of mass, carbon atoms, and moles through a linear, unbranched metabolic pathway can be represented using a single rate v, as illustrated in Figure 14.5(B). Consistent with our definition of reaction rate, v is a forward rate, that is, v represents the rate of conversion of S. We will continue to use r_S and r_P for the rates of substrate uptake and product excretion. These rates are different from the internal flux v in that they are generally readily measurable.

The stoichiometry of the reactions must be accurate and the reaction equations must be balanced. All C atoms and electrons must be accounted for within the metabolic model. As mentioned above, metabolic flux analysis is generally carried out using fluxes expressed in moles. However, because we are interested in the distribution of carbon within pathways, flux is sometimes reported in terms of moles of C (Cmol) rather than moles of chemical compounds. The reaction equations and stoichiometry may be modified to reflect this, using chemical formulae for C-containing compounds based on 1 atom of C, and writing each reaction for conversion of 1 reactant mole of C. As an example, the reaction equation for conversion of glucose to ethanol is usually written using a basis of 1 mole of glucose:

$$C_6H_{12}O_6 \rightarrow 2\ C_2H_6O + 2\ CO_2 \tag{14.7}$$

where $C_6H_{12}O_6$ is glucose and C_2H_6O is ethanol. This equation rewritten using a basis of 1 mole of carbon is:

$$CH_2O \rightarrow 0.67\ CH_3O_{0.5} + 0.33\ CO_2 \tag{14.8}$$

where CH_2O represents glucose and $CH_3O_{0.5}$ represents ethanol. In Eq. (14.8), the formulae for all compounds are based on 1 C atom or 1 Cmol, where 1 Cmol is 6.02×10^{23} carbon atoms. The reaction of Eq. (14.8) starts with one-sixth of the mass of glucose used in Eq. (14.7); otherwise, the two reactions are the same. The advantage of writing reaction equations in Cmol is that information about the amount of carbon transferred from a reactant to a particular product is contained in the stoichiometric coefficients alone. From Eq. (14.8), we can see immediately that two-thirds of the C in glucose is transferred to ethanol. A disadvantage is that Cmol formulae for reactants and products may not be recognized as easily as their full molecular formulae.

Biomass is often included in flux analysis as a product of metabolism. Cell growth provides a sink for ATP and precursor compounds from catabolic pathways and for building-block metabolites such as amino acids, nucleotides, and lipids. A stoichiometric equation for production of biomass represented as $CH_\alpha O_\beta N_\delta$ is determined from the composition of the cells (Chapter 4) or from information about the metabolic requirements for synthesis of various biomass constituents [3]. For example, growth can be simulated by creating a lumped reaction equation that draws metabolites such as polysaccharides, proteins, DNA, RNA, and lipids from their respective biosynthetic pathways at ratios that reflect the composition of the cells. The rate of the growth reaction may then be scaled so that the flux is equal to the exponential growth rate of the organism.

Fluxes in metabolic models are normally expressed as specific rates. For example, a flux may be reported in units of mol (g cell dry weight)$^{-1}$ h^{-1} or, alternatively, as Cmol (g cdw)$^{-1}$ h^{-1} or Cmol (Cmol biomass)$^{-1}$ h^{-1}. One Cmol of biomass is represented by the chemical "formula" for dry cells normalized to one atom of carbon, as described in Chapter 4.

For ash-free cellular biomass of average C, H, O, and N composition, 1 Cmol of biomass weighs about 24.6 g.

14.1.4 Formulating the Equations Using Matrices

Consider now the branched metabolic pathway of Figure 14.6, where the pathway from S splits at intermediate I to form two products, P_1 and P_2, which are excreted at rates r_{P1} and r_{P2}, respectively. The stoichiometry of the grouped reactions for each branch is shown in Figure 14.6. We will carry out the analysis using rates expressed in units of mol $(g\ cdw)^{-1}\ h^{-1}$. To keep it simple, the pathway does not involve energy or redox conversions.

We will use this metabolic pathway to illustrate the mathematical techniques employed for metabolic flux analysis. The aim is to calculate the internal fluxes v_1, v_2, and v_3. Although the solution to this problem is trivial, it is a useful exercise to follow the same methods applied to larger, more complex systems. Using Eq. (4.26) and a basis of 1 g of cell dry weight, the unsteady-state mass balance equations for S, I, P_1, and P_2 are:

$$\frac{dM_S}{dt} = (MW)_S\ (r_S) - 0 + 0 - (MW)_S\ (v_1)$$

$$\frac{dM_I}{dt} = 0 - 0 + (MW)_I\ (0.5\ v_1) - (MW)_I\ (v_2) - (MW)_I\ (v_3)$$

$$\frac{dM_{P1}}{dt} = 0 - (MW)_{P1}\ (r_{P1}) + (MW)_{P1}\ (2v_2) - 0$$

$$\frac{dM_{P2}}{dt} = 0 - (MW)_{P2}\ (r_{P2}) + (MW)_{P2}\ (v_3) - 0$$

$$(14.9)$$

As explained in relation to Eq. (14.4), the stoichiometric coefficients 0.5 and 2 are required in the mass balance equations for I and P_1 to represent the rates of generation of these

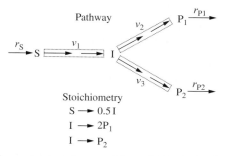

Figure 14.6 Diverging branched pathway with stoichiometry for the grouped reactions in each branch.

components, consistent with the definition of the rates as forward reaction rates. At steady state, Eq. (14.9) becomes:

$$r_S - v_1 = 0$$
$$0.5\, v_1 - v_2 - v_3 = 0$$
$$2\, v_2 - r_{P1} = 0 \tag{14.10}$$
$$v_3 - r_{P2} = 0$$

These linear algebraic equations can be arranged in matrix form. Let us define a column vector $\mathbf{v}$ containing the six rates as elements:

$$\mathbf{v} = \begin{bmatrix} r_S \\ v_1 \\ v_2 \\ v_3 \\ r_{P1} \\ r_{P2} \end{bmatrix} \tag{14.11}$$

and a matrix $\mathbf{S}$ containing the coefficients for the rates in Eq. (14.10):

$$\mathbf{S} = \begin{bmatrix} 1 & -1 & 0 & 0 & 0 & 0 \\ 0 & 0.5 & -1 & -1 & 0 & 0 \\ 0 & 0 & 2 & 0 & -1 & 0 \\ 0 & 0 & 0 & 1 & 0 & -1 \end{bmatrix} \tag{14.12}$$

Each row in $\mathbf{S}$ represents a mass balance equation so, in this example, there are four rows. Each column in $\mathbf{S}$ contains coefficients for each of the six rates that define vector $\mathbf{v}$; therefore, in this case, there are six columns. The first column in $\mathbf{S}$ contains the coefficients for r_S, the second for v_1, the third for v_2, and so on for the remainder of $\mathbf{v}$. Typically, $\mathbf{S}$ for large metabolic networks is a relatively sparse matrix with many zero elements as most metabolic reactions involve only a few components. Note that $\mathbf{S}$ can be defined in different ways and the arrangement of elements in the matrix of Eq. (14.12) is sometimes referred to in the literature as $\mathbf{S}^T$, or the transpose of the stoichiometric matrix.

Using matrix notation, Eq. (14.10) can be written succinctly as:

$$\mathbf{Sv} = 0 \tag{14.13}$$

Multiplication of $\mathbf{S}$ and $\mathbf{v}$ is carried out using the rules for matrix operations. The rules for multiplying a matrix and column vector must be used to verify that Eqs. (14.10) and (14.13) are equivalent.

14.1.5 *Underdetermined and Overdetermined Systems*

There are six rates or fluxes in Eq. (14.10) but only four mass balance equations. For any metabolic network, if there are M linear algebraic equations, one for each of M metabolites, and R reaction fluxes, the degree of freedom F in the system is:

$$F = R - M \tag{14.14}$$

Typically, the number of reactions R in a metabolic pathway is greater than the number of metabolites M. Such a system where $F > 0$ is *underdetermined*, as there is not enough information available to solve for all the rates in the mass balance equations. To solve Eq. (14.13), some of the rate elements in $\mathbf{v}$ must be measured, or additional constraints must be introduced into the equations, before the remaining rates can be determined. If F rates are measured and therefore known independently, the system becomes *determined* and a single unique solution can be found for the remaining fluxes. If more than F rates are measured, the system is *overdetermined*. An overdetermined system is not easily achieved in metabolic flux analysis but is desirable because the additional information can be used for checking purposes.

For our problem represented by Eqs. (14.10) and (14.13), let us bring the system to a determined state so that a unique solution can be found for the fluxes. According to Eq. (14.14), $F = 6 - 4 = 2$, so we need to measure two of the rates in $\mathbf{v}$. As it is much easier to observe external rather than internal fluxes, r_S, r_{P1}, and r_{P2} are obvious candidates for experimental measurement. We only need two of these, however, so let us choose to measure r_{P1} and r_{P2}. These rates can be grouped into a new vector for measured rates $\mathbf{v}_m$, while the remaining unknown rates are collected into another vector $\mathbf{v}_c$ for flux calculation. We can write Eq. (14.13) as:

$$S\mathbf{v} = S_m\mathbf{v}_m + S_c\mathbf{v}_c = 0 \tag{14.15}$$

where

$$\mathbf{v}_m = \begin{bmatrix} r_{P1} \\ r_{P2} \end{bmatrix} \tag{14.16}$$

and

$$\mathbf{v}_c = \begin{bmatrix} r_S \\ v_1 \\ v_2 \\ v_3 \end{bmatrix} \tag{14.17}$$

From the coefficients in Eq. (12.10) for r_{P1} and r_{P2}:

$$\mathbf{S_m} = \begin{bmatrix} 0 & 0 \\ 0 & 0 \\ -1 & 0 \\ 0 & -1 \end{bmatrix} \tag{14.18}$$

and

$$\mathbf{S_c} = \begin{bmatrix} 1 & -1 & 0 & 0 \\ 0 & 0.5 & -1 & -1 \\ 0 & 0 & 2 & 0 \\ 0 & 0 & 0 & 1 \end{bmatrix} \tag{14.19}$$

Both matrices $\mathbf{S_m}$ and $\mathbf{S_c}$ contain four rows, one for each mass balance equation. The elements in the two columns of $\mathbf{S_m}$ are the coefficients in the mass balance equations for the measured rates, r_{P1} and r_{P2}, respectively. The four columns of $\mathbf{S_c}$ contain the coefficients for the four rates in vector $\mathbf{v_c}$, that is, r_S, v_1, v_2, and v_3, in that order.

Solution of this problem to find the internal fluxes v amounts to solving for the unknown elements of vector $\mathbf{v_c}$. Once we have measured r_{P1} and r_{P2}, the elements of $\mathbf{v_m}$, $\mathbf{S_m}$, and $\mathbf{S_c}$ are known. We can write Eq. (14.15) as:

$$\mathbf{S_c v_c} = -\mathbf{S_m v_m} \tag{14.20}$$

or, using the rules for matrix operations:

$$\mathbf{v_c} = -\mathbf{S_c^{-1} S_m v_m} \tag{14.21}$$

The solution for $\mathbf{v_c}$ requires calculation of the inverse of matrix $\mathbf{S_c}$. The inverse of a matrix can be determined only if the matrix is square and of full rank. In our example, $\mathbf{S_c}$ is a square 4×4 matrix: this is a consequence of the system being determined so that the number of unknown rates is equal to the number of equations available to solve for them. To be of full rank, the matrix must contain no row or column that is a linear combination of other rows or columns. The determinant of a matrix of full rank is nonzero; therefore, calculating the determinant allows us to check whether $\mathbf{S_c}$ can be inverted.

For $\mathbf{S_c}$ defined in Eq. (14.19), applying appropriate software like MATLAB or Python gives det $(\mathbf{S_c}) = 1.0$. As this determinant $\neq 0$, we know that $\mathbf{S_c}$ is invertible. Further application of matrix solution software gives the inverse of $\mathbf{S_c}$ as:

$$\mathbf{S_c^{-1}} = \begin{bmatrix} 1 & 2 & 1 & 2 \\ 0 & 2 & 1 & 2 \\ 0 & 0 & 0.5 & 0 \\ 0 & 0 & 0 & 1 \end{bmatrix} \tag{14.22}$$

This result can be verified by checking that the product $\mathbf{S_c S_c^{-1}}$ is equal to the identity matrix, $\mathbf{I}$.

If matrix $\mathbf{S_c}$ is found to be singular (i.e., det $(\mathbf{Sc}) = 0$) so that $\mathbf{S_c}$ cannot be inverted, the solution cannot proceed without modifying the equations. When $\mathbf{S_c}$ is singular, it is likely that one or more of the mass balance equations used to formulate the problem is a linear combination of one or more of the other equations. In this situation, the system is under-determined. When large numbers of metabolic reactions are being analyzed, difficulties with linearly dependent elements in the equations occur with some frequency. Intracellular assays may be required to resolve the singularity created by related pathways, by providing additional information about their relative activities. Alternatively, complementary pathways might be lumped together into a single reaction. Balances on *conserved moieties*, such as ATP, ADP, and AMP, and cofactor pairs such as NAD and NADH, are linearly dependent because the total concentration of conserved moieties remains constant even though there is interchange between different members of the group. Therefore, balances on only one compound in conserved groups are included in metabolic flux analysis to avoid matrix singularity. Identifying and eliminating dependent equations by changing the metabolic network or introducing other information into the solution procedure are often required to render $\mathbf{S_c}$ invertible.

Solution of the Equations

Let us assume that the measured results for r_{P1} and r_{P2} are $r_{P1} = 0.05 \, \text{mol} \, (\text{g cdw})^{-1} \, \text{h}^{-1}$ and $r_{P2} = 0.08 \, \text{mol} \, (\text{g cdw})^{-1} \, \text{h}^{-1}$. Therefore, we can write Eq. (14.16) as:

$$\mathbf{v_m} = \begin{bmatrix} 0.05 \\ 0.08 \end{bmatrix} \, \text{mol} \, \text{g}^{-1} \, \text{h}^{-1} \tag{14.23}$$

Applying Eqs. (14.18), (14.22), and (14.23) in Eq. (14.21) gives the following equation for $\mathbf{v_c}$:

$$\mathbf{v_c} = -\begin{bmatrix} 1 & 2 & 1 & 2 \\ 0 & 2 & 1 & 2 \\ 0 & 0 & 0.5 & 0 \\ 0 & 0 & 0 & 1 \end{bmatrix} \begin{bmatrix} 0 & 0 \\ 0 & 0 \\ -1 & 0 \\ 0 & -1 \end{bmatrix} \begin{bmatrix} 0.05 \\ 0.08 \end{bmatrix} \, \text{mol} \, \text{g}^{-1} \, \text{h}^{-1} \tag{14.24}$$

Using linear algebra multiplication rules:

$$\mathbf{v}_c = - \begin{bmatrix} 1 & 2 & 1 & 2 \\ 0 & 2 & 1 & 2 \\ 0 & 0 & 0.5 & 0 \\ 0 & 0 & 0 & 1 \end{bmatrix} \begin{bmatrix} 0 \\ 0 \\ -0.05 \\ -0.08 \end{bmatrix} \text{mol g}^{-1}\,\text{h}^{-1} \tag{14.25}$$

Further multiplication gives:

$$\mathbf{v}_c = - \begin{bmatrix} -0.21 \\ -0.21 \\ -0.025 \\ -0.08 \end{bmatrix} \text{mol g}^{-1}\text{h}^{-1} \tag{14.26}$$

or

$$\mathbf{v}_c = \begin{bmatrix} 0.21 \\ 0.21 \\ 0.025 \\ 0.08 \end{bmatrix} \text{mol g}^{-1}\text{h}^{-1} \tag{14.27}$$

Equation (14.27) provides the values for the unknown fluxes. From the definition of $\mathbf{v}_c$ in Eq. (14.17), the results are $r_S = 0.21$ mol (g cdw^{-1}) h^{-1}, $v_1 = 0.21$ mol (g cdw)$^{-1}$ h^{-1}, $v_2 = 0.025$ mol (g cdw)$^{-1}$ h^{-1}, and $v_3 = 0.08$ mol (g cdw)$^{-1}$ h^{-1}. You can see that these results, together with the measured values for r_{P1} and r_{P2}, satisfy the mass balances of Eq. (14.10).

Product yields in cell culture are controlled by the *flux split ratio* or *flux partitioning ratio* Φ at critical branch points in the metabolic network. From our analysis of the diverging pathway in Figure 14.6, we can calculate the flux split ratio between the two products P$_1$ and P$_2$ at branch point I:

$$\Phi_{P1} = \frac{\text{flux to P}_1}{\text{flux to P}_1 + \text{flux to P}_2} = \frac{v_2}{v_2 + v_3} = \frac{0.025}{0.025 + 0.08} = 0.24 \tag{14.28}$$

and

$$\Phi_{P2} = \frac{\text{flux to P}_2}{\text{flux to P}_1 + \text{flux to P}_2} = \frac{v_3}{v_2 + v_3} = \frac{0.08}{0.025 + 0.08} = 0.76 \tag{14.29}$$

The problem we have just solved was very simple. Once the measured results for r_{P1} and r_{P2} were known, the algebraic equations of Eq. (14.10) could have been solved easily to arrive at the solution of Eq. (14.27) without the need to formulate the problem using matrices.

However, the procedure described here illustrates the mathematical approach required when realistic metabolic networks involving 20 to 50 fluxes are investigated. Metabolic flux analysis calculations are illustrated in Example 14.1.

EXAMPLE 14.1 Flux Analysis for Mixed Acid Fermentation

The metabolic pathways for anaerobic mixed acid fermentation have been studied extensively. Under certain culture conditions, lactic acid bacteria catabolize glucose to produce lactic acid, formic acid, acetic acid, and ethanol. The pathway for this fermentation is shown in Figure 14.7. For cells, the main purpose of the pathway is to generate ATP for growth. There is no or very little exchange of intermediates to anabolic metabolism; the pathway is also self-sufficient in reducing equivalents.

A chemostat is used to obtain experimental results for culture of *Lactococcus lactis* subsp. *lactis* under steady-state conditions. The specific rate of glucose uptake is measured as 0.01 mol $(g\ cdw)^{-1}\ h^{-1}$. The yield of lactic acid from glucose is found to be 0.49 mol lactic acid (mol glucose)$^{-1}$ and the yield of formic acid from glucose is 0.41 mol formic acid (mol glucose)$^{-1}$.

a. Draw a simplified pathway diagram suitable for metabolic flux analysis.
b. Perform a flux balance on the simplified metabolic network. Label the pathway with the flux results in units of C-mol $(g\ cdw)^{-1} h^{-1}$.
c. What is the flux split ratio at pyruvate for lactate production?
d. What is the flux split ratio at acetyl coenzyme A (acetyl CoA) for acetate and ethanol production?
e. Estimate the observed yield of ethanol from glucose in mol mol^{-1}.
f. What is the rate of ATP generation for cell growth and maintenance?

Solution

a. In this fermentation, intermediates from glycolysis are not used for biosynthesis and growth. Therefore, the pathway from glucose to pyruvate can be considered a linear sequence of reactions without branch points. The individual glycolytic reactions are grouped into a single reaction step while preserving the net cofactor requirements of glycolysis. The major branch points of the pathway are at pyruvate and acetyl CoA. The steps from acetyl CoA to acetate, and from acetyl CoA to ethanol, are also linear sequences and can be lumped into two grouped reactions. The resulting simplified pathway is shown in Figure 14.8.
b. The stoichiometric equations for the reactions in the simplified pathway can be found from biochemistry texts and are listed below. The equations are written on a mole basis.

$$glucose + 2\ ADP + 2\ NAD \rightarrow 2\ pyruvate + 2\ ATP + 2\ NADH + 2\ H_2O$$
$$pyruvate + NADH \rightarrow lactate + NAD$$
$$pyruvate + CoA + NAD \rightarrow acetylCoA + CO_2 + NADH$$
$$pyruvate + CoA \rightarrow formate + acetylCoA$$
$$acetylCoA + ADP \rightarrow acetate + CoA + ATP$$
$$acetylCoA + 2\ NADH \rightarrow ethanol + CoA + 2\ NAD$$
$$ATP \rightarrow ADP$$

The overall stoichiometry and cofactor requirements for glycolytic conversion of glucose to pyruvate are well known. The stoichiometric coefficients in the remaining reactions are unity, except that 2 NADH and 2 NAD are involved in the lumped reaction for conversion of

---

(Continued)

acetyl CoA to ethanol. The forward rates of the internal reactions are labeled v_1 to v_7 in Figure 14.8. The rates of substrate uptake and product excretion are labeled r_G, r_L, r_C, r_F, r_A, and r_E for glucose, lactate, CO_2, formate, acetate, and ethanol, respectively. Because the pathway generates a net positive amount of ATP, a reaction ATP→ADP representing ATP requirements for growth and maintenance is included so that the mass balance for ATP may be closed. Steady-state mass balances are performed for glucose and the excreted products, the internal intermediates at branch points, and the redox and energy carriers:

$$\text{glucose: } r_G - v_1 = 0$$
$$\text{pyruvate: } 2\,v_1 - v_2 - v_3 - v_4 = 0$$
$$\text{lactate: } v_2 - r_L = 0$$
$$CO_2\text{: } v_3 - r_C = 0$$
$$\text{formate: } v_4 - r_F = 0$$
$$\text{acetylCoA: } v_3 + v_4 - v_5 - v_6 = 0$$
$$\text{acetate: } v_5 - r_A = 0$$
$$\text{ethanol: } v_6 - r_E = 0$$
$$\text{NADH: } 2\,v_1 - v_2 + v_3 - 2\,v_6 = 0$$
$$\text{ATP: } 2\,v_1 + v_5 - v_7 = 0$$

Balances are performed for only one component involved in redox transfer and one component involved in energy transfer; balances for NAD and ADP are therefore not included. In total, there are 7 internal (v) and 6 external (r) fluxes, making a total of 13 unknown rates. There are 10 metabolites/cofactors giving 10 mass balance equations. From Eq. (14.14), the degree of freedom F for the system is:

$$F = 13 - 10 = 3$$

As three experimental measurements are available from chemostat cultures, the system is determined and can be solved to find a unique solution. The measured value for r_G is 0.01 mol $(\text{g cdw})^{-1}\,h^{-1}$. From the yield coefficients ($Y_{P/S}$) for lactate and formate:

$$r_L = Y_{PS}r_G = 0.49\,\text{mol mol}^{-1}(0.01\,\text{mol g}^{-1}h^{-1}) = 0.0049\,\text{mol g}^{-1}h^{-1}$$
$$r_F = Y_{PS}r_G = 0.041\,\text{mol mol}^{-1}(0.01\,\text{mol g}^{-1}h^{-1}) = 0.0041\,\text{mol g}^{-1}h^{-1}$$

The vector of measured fluxes $\mathbf{v_m}$ and the corresponding matrix $\mathbf{S_m}$ are:

$$\mathbf{v_m} = \begin{bmatrix} r_C \\ r_L \\ r_F \end{bmatrix} = \begin{bmatrix} 0.01 \\ 0.0049 \\ 0.0041 \end{bmatrix} \text{mol g}^{-1}h^{-1} \quad \mathbf{S_m} = \begin{bmatrix} 1 & 0 & 0 \\ 0 & 0 & 0 \\ 0 & -1 & 0 \\ 0 & 0 & 0 \\ 0 & 0 & -1 \\ 0 & 0 & 0 \\ 0 & 0 & 0 \\ 0 & 0 & 0 \\ 0 & 0 & 0 \\ 0 & 0 & 0 \end{bmatrix}$$

The elements of S_m are the coefficients in the mass balance equations for r_G, r_L, and r_F. Each row in S_m represents a mass balance equation; the first column in S_m contains the coefficients for r_G, the second for r_L, and the third for r_F. The remaining 10 rates define the vector of calculated fluxes, v_c. The corresponding matrix of coefficients S_c is obtained from the mass balance equations:

$$
v_c = \begin{bmatrix} v_1 \\ v_2 \\ v_3 \\ v_4 \\ v_5 \\ v_6 \\ v_7 \\ r_C \\ r_A \\ r_E \end{bmatrix} \qquad
S_c = \begin{bmatrix}
-1 & 0 & 0 & 0 & 0 & 0 & 0 & 0 & 0 & 0 \\
2 & -1 & -1 & -1 & 0 & 0 & 0 & 0 & 0 & 0 \\
0 & 1 & 0 & 0 & 0 & 0 & 0 & 0 & 0 & 0 \\
0 & 0 & 1 & 0 & 0 & 0 & 0 & -1 & 0 & 0 \\
0 & 0 & 0 & 1 & 0 & 0 & 0 & 0 & 0 & 0 \\
0 & 0 & 1 & 1 & -1 & -1 & 0 & 0 & 0 & 0 \\
0 & 0 & 0 & 0 & 1 & 0 & 0 & 0 & -1 & 0 \\
0 & 0 & 0 & 0 & 0 & 1 & 0 & 0 & 0 & -1 \\
2 & -1 & 1 & 0 & 0 & -2 & 0 & 0 & 0 & 0 \\
2 & 0 & 0 & 0 & 1 & 0 & -1 & 0 & 0 & 0
\end{bmatrix}
$$

S_c is nonsingular with $\det(S_c) = -2$. Matrix calculation software is used to determine the inverse of S_c:

$$
S_c^{-1} = \begin{bmatrix}
-1 & 0 & 0 & 0 & 0 & 0 & 0 & 0 & 0 & 0 \\
0 & 0 & 1 & 0 & 0 & 0 & 0 & 0 & 0 & 0 \\
-2 & -1 & -1 & 0 & -1 & 0 & 0 & 0 & 0 & 0 \\
0 & 0 & 0 & 0 & 1 & 0 & 0 & 0 & 0 & 0 \\
0 & -0.5 & 0 & 0 & 0.5 & -1 & 0 & 0 & 0.5 & 0 \\
-2 & -0.5 & -1 & 0 & -0.5 & 0 & 0 & 0 & -0.5 & 0 \\
-2 & -0.5 & 0 & 0 & 0.5 & -1 & 0 & 0 & 0.5 & -1 \\
-2 & -1 & -1 & -1 & -1 & 0 & 0 & 0 & 0 & 0 \\
0 & -0.5 & 0 & 0 & 0.5 & -1 & -1 & 0 & 0.5 & 0 \\
-2 & -0.5 & -1 & 0 & -0.5 & 0 & 0 & -1 & -0.5 & 0
\end{bmatrix}
$$

(*Continued*)

Applying Eq. (14.21):

$$\mathbf{v_c} = -\begin{bmatrix} -1 & 0 & 0 & 0 & 0 & 0 & 0 & 0 & 0 & 0 \\ 0 & 0 & 1 & 0 & 0 & 0 & 0 & 0 & 0 & 0 \\ -2 & -1 & -1 & 0 & -1 & 0 & 0 & 0 & 0 & 0 \\ 0 & 0 & 0 & 0 & 1 & 0 & 0 & 0 & 0 & 0 \\ 0 & -0.5 & 0 & 0 & 0.5 & -1 & 0 & 0 & 0.5 & 0 \\ -2 & -0.5 & -1 & 0 & -0.5 & 0 & 0 & 0 & -0.5 & 0 \\ -2 & -0.5 & 0 & 0 & 0.5 & -1 & 0 & 0 & 0.5 & -1 \\ -2 & -1 & -1 & -1 & -1 & 0 & 0 & 0 & 0 & 0 \\ 0 & -0.5 & 0 & 0 & 0.5 & -1 & -1 & 0 & 0.5 & 0 \\ -2 & -0.5 & -1 & 0 & -0.5 & 0 & 0 & -1 & -0.5 & 0 \end{bmatrix} \begin{bmatrix} 1 & 0 & 0 \\ 0 & 0 & 0 \\ 0 & -1 & 0 \\ 0 & 0 & 0 \\ 0 & 0 & -1 \\ 0 & 0 & 0 \\ 0 & 0 & 0 \\ 0 & 0 & 0 \\ 0 & 0 & 0 \\ 0 & 0 & 0 \end{bmatrix} \begin{bmatrix} 0.01 \\ 0.0049 \\ 0.0041 \end{bmatrix} \text{mol g}^{-1}\text{h}$$

Multiplying $\mathbf{S_m}$ with $\mathbf{v_m}$ first:

$$\mathbf{v_c} = -\begin{bmatrix} -1 & 0 & 0 & 0 & 0 & 0 & 0 & 0 & 0 & 0 \\ 0 & 0 & 1 & 0 & 0 & 0 & 0 & 0 & 0 & 0 \\ -2 & -1 & -1 & 0 & -1 & 0 & 0 & 0 & 0 & 0 \\ 0 & 0 & 0 & 0 & 1 & 0 & 0 & 0 & 0 & 0 \\ 0 & -0.5 & 0 & 0 & 0.5 & -1 & 0 & 0 & 0.5 & 0 \\ -2 & -0.5 & -1 & 0 & -0.5 & 0 & 0 & 0 & -0.5 & 0 \\ -2 & -0.5 & 0 & 0 & 0.5 & -1 & 0 & 0 & 0.5 & -1 \\ -2 & -1 & -1 & -1 & -1 & 0 & 0 & 0 & 0 & 0 \\ 0 & -0.5 & 0 & 0 & 0.5 & -1 & -1 & 0 & 0.5 & 0 \\ -2 & -0.5 & -1 & 0 & -0.5 & 0 & 0 & -1 & -0.5 & 0 \end{bmatrix} \begin{bmatrix} 0.01 \\ 0 \\ -0.0049 \\ 0 \\ -0.0041 \\ 0 \\ 0 \\ 0 \\ 0 \\ 0 \end{bmatrix} \text{mol g}^{-1}\text{h}^{-1}$$

Completing the matrix multiplication gives:

$$\mathbf{v_c} = \begin{bmatrix} 0.01 \\ 0.0049 \\ 0.0110 \\ 0.0041 \\ 0.0021 \\ 0.0131 \\ 0.0221 \\ 0.0110 \\ 0.0021 \\ 0.0131 \end{bmatrix} \text{mol g}^{-1}\text{h}^{-1} = \begin{bmatrix} v_1 \\ v_2 \\ v_3 \\ v_4 \\ v_5 \\ v_6 \\ v_7 \\ r_C \\ r_A \\ r_E \end{bmatrix}$$

All the internal and external fluxes are now known in units of mol $(g\ cdw)^{-1}\ h^{-1}$. To convert to C-mol $(g\ cdw)^{-1}\ h^{-1}$, each result except v_7 is multiplied by the number of C atoms in the molecule of the compound to which the flux refers:

$$
\begin{aligned}
\text{glucose} &= 6 \\
\text{pyruvate} &= 3 \\
\text{lactate} &= 3 \\
CO_2 &= 1 \\
\text{formate} &= 1 \\
\text{acetyl CoA} &= 2 \\
\text{acetate} &= 2 \\
\text{ethanol} &= 2
\end{aligned}
$$

The number of C atoms in acetyl CoA is taken as two, as glucose carbon is distributed only to the acetyl group with CoA acting as a conserved cofactor. The results are:

$$
\begin{bmatrix}
v_1 \\
v_2 \\
v_3 \\
v_4 \\
v_5 \\
v_6 \\
r_G \\
r_L \\
r_C \\
r_F \\
r_A \\
r_E
\end{bmatrix}
=
\begin{bmatrix}
0.06 \\
0.015 \\
0.033 \\
0.012 \\
0.004 \\
0.026 \\
0.06 \\
0.015 \\
0.011 \\
0.004 \\
0.004 \\
0.026
\end{bmatrix}
\text{C-molg}^{-1}\text{h}^{-1}
$$

Because ATP is not a component of the carbon balance, the rate of ATP conversion to ADP, v_7, is expressed in units of mol $(g\ cdw)^{-1}\ h^{-1}$ rather than C-mol $(g\ cdw)^{-1}\ h^{-1}$. The results are shown in Figure 14.9.

c. The flux split ratio at pyruvate for lactate production is:

$$
\Phi_L = \frac{\text{flux to lactate}}{\text{flux to lactate} + \text{flux to acetyl CoA}} = \frac{0.015}{0.015 + 0.033 + 0.012} = 0.25
$$

d. The flux split ratio at acetyl CoA is:

$$
\Phi_A = \frac{\text{flux to acetate}}{\text{flux to acetate} + \text{flux to ethanol}} = \frac{0.004}{0.004 + 0.026} = 0.13 \text{ for acetate production}
$$

$$
\Phi_E = \frac{\text{flux to ethanol}}{\text{flux to acetate} + \text{flux to ethanol}} = \frac{0.026}{0.004 + 0.026} = 0.87 \text{ for ethanol production}
$$

(*Continued*)

e. The observed yield of ethanol from glucose is:

$$Y'_{P/S} = \frac{r_E}{r_G} = \frac{0.0131 \text{ mol g}^{-1} \text{ h}^{-1}}{0.01 \text{ mol g}^{-1} \text{ h}^{-1}} = 1.3 \text{ mol ethanol (mol glucose)}^{-1}$$

f. The rate of ATP generation in the pathway is equal to the rate of ATP consumption for growth $= v_7 = 0.022 \text{ mol (g cdw)}^{-1} \text{ h}^{-1}$.

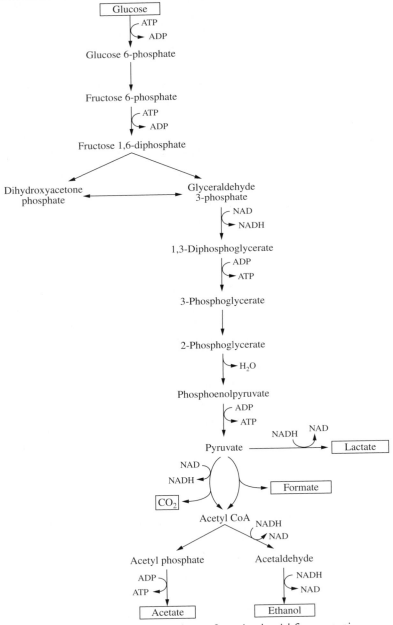

Figure 14.7 Metabolic pathway for mixed acid fermentation.

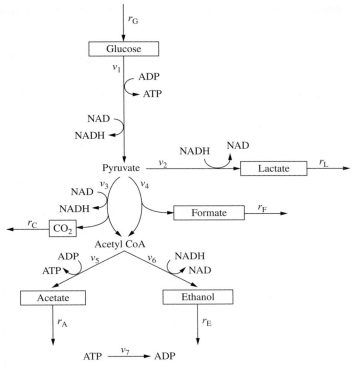

Figure 14.8 Graphic of simplified metabolic pathway for mixed acid fermentation.

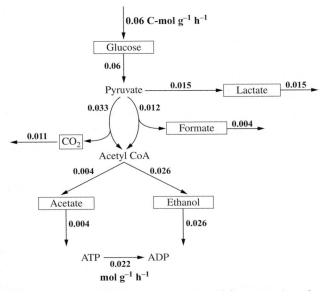

Figure 14.9 Simplified metabolic pathway for mixed acid fermentation showing carbon fluxes in C-mol g^{-1} h^{-1} and the ATP flux in mol g^{-1} h^{-1}.

14.1.6 Underdetermined Systems, Flux Balance Analysis and Elementary Flux Mode Analysis

Most network models defined for metabolic flux analysis are underdetermined. In these circumstances, a unique solution to the flux distribution cannot be found; rather, there exists an infinite number of solutions that satisfy the balance equations. Additional constraints are needed to reduce the size of the solution space and progress with the analysis. These constraints restrict the values of the fluxes to prescribed limits so that realistic solutions reflecting the biological properties of the system can be found.

When metabolic fluxes are calculated, a positive value of v indicates that the reaction operates in the forward direction whereas a negative value indicates that the reaction operates in the reverse direction. For reversible reactions, both positive and negative values may be acceptable, as both forward and reverse reactions are possible thermodynamically. In this sense, v for reversible reactions is unrestricted:

$$-\infty < v < \infty \tag{14.30}$$

In contrast, many metabolic reactions are for practical purposes irreversible, so that *thermodynamic constraints* can be imposed regarding the direction of the reaction and therefore the value of v. For example:

$$0 < v < \infty \tag{14.31}$$

may be applied as a constraint for an irreversible reaction that can proceed only in the forward direction, whereas:

$$-\infty < v < 0 \tag{14.32}$$

is valid for an irreversible reaction that can proceed only in the reverse direction. Such constraints ensure that the reactions and cyclic pathways included in the model are thermodynamically feasible. Additional *capacity constraints* reflecting restrictions on enzyme or cell function are imposed if, for example, a maximum limit on v is known:

$$0 < v < v_{max} \tag{14.33}$$

Other *regulatory constraints* relating to gene expression may also be applied. For instance, the fluxes for certain metabolic reactions may be set to zero based on transcriptional data if the gene coding for the enzyme responsible is found to be inactive.

Even after such constraints are included in the model, metabolic systems often remain underdetermined. However, a certain type of solution may be found by identifying an *objective function* and solving the equations using computational methods known as *linear programming* or *linear optimization*. The methodology of using linear optimization tools with metabolic models is commonly referred to as *Flux Balance Analysis*. Examples of objective functions include cell growth, substrate uptake, synthesis of a particular metabolite, generation or utilization of ATP, and generation or utilization of NADH. In linear programming, numerical algorithms are applied to find the set of feasible steady-state fluxes that maximizes or minimizes the objective function chosen. A common approach is to solve for the fluxes that maximize the rate of cell growth. Genome-scale stoichiometric models involving more than 2000 reactions have been developed for organisms such as *Escherichia coli* [4].

Underdetermined metabolic networks can also be analyzed using stoichiometric representations of biochemical networks and a systems biology tool called Elementary Flux Mode Analysis (EFMA). EFMA, unlike Flux Balance Analysis, does not require an objective function to identify solutions [1]. Instead, EFMA uses a type of mathematics known as convex analysis to enumerate all simplest, nondivisible metabolic pathways (elementary flux modes) residing withing a metabolic model [5]. The approach is unbiased as it does not require assumptions about what optimality principle is appropriate. The complete set EFMs can then be analyzed for desirable physiologies like the efficient conversion of substrate into ATP or biomass as a function of nutrient stress. EFMA, like FBA, is also capable of analyzing the potential impacts of metabolic engineering strategies on bioprocess hosts by predicting outcomes of gene deletions or gene additions. The references Carlson et al. (2002) and Carlson et al. (2005) examine the effect of bioplastic production in two common bioprocess hosts, *S. cerevisiae* and *E. coli* [6,7].

14.1.7 Isotope-Based Flux Analysis

Experimental and computational methods for the resolution of internal fluxes using carbon isotope labeling and analytical techniques such as mass spectrometry (MS) and nuclear magnetic resonance (NMR) spectroscopy. When cells are supplied with substrate carrying a positional label such as ^{13}C, the label is passed to metabolites derived from the substrate. The resulting pattern of molecular labeling can be used to determine the relative activities of different metabolic pathways in the cell [8]. For example, glucose taken up by *E. coli* is utilized in three metabolic pathways that operate simultaneously: the Embden–Meyerhof–Paras pathway, the Entner–Doudoroff pathway, and the pentose phosphate pathway. If the first carbon position of glucose is labeled uniformly with ^{13}C, pyruvate molecules produced in the three pathways will be labeled in different ways as illustrated in Figure 14.10. For labeled glucose entering the Embden–Meyerhof pathway, half of the resulting pyruvate is labeled with ^{13}C in the third carbon position while the remaining half is unlabeled. If labeled glucose is metabolized in the Entner–Doudoroff pathway, half of the pyruvate contains ^{13}C in the first position while the other half is unlabeled. Labeled glucose metabolized in the pentose phosphate pathway produces unlabeled pyruvate as the ^{13}C label is passed to CO_2.

In principle, therefore, the relative fluxes through the three pathways can be determined by measuring the fractions of pyruvate molecules with ^{13}C labels in the first and third positions,

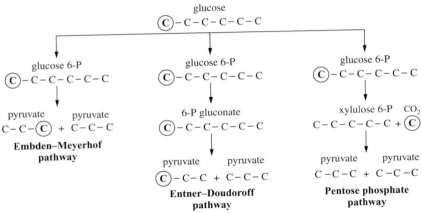

Figure 14.10 Fate of ^{13}C in labeled glucose metabolized to pyruvate. The ^{13}C label is indicated by a circle.

for example, using MS or NMR, and applying appropriate mass balance equations for the carbon atoms. Because the carbon backbone of molecules in central carbon metabolism is preserved in the amino acids, similar analysis is also possible using measurements of amino acid labeling. An advantage of this approach is that amino acids are present in much larger quantities in cells than metabolic intermediates such as pyruvate. Alternative labeling choices provide information about the relative participation of other pathways. The internal fluxes determined in this way can be used to check the reliability of calculated results from metabolic flux analysis as well as provide additional insights into pathway operation.

When extra information about pathway function is obtained from intracellular measurements, the system may reach the desirable state of being overdetermined. If this occurs, there is sufficient redundant data available to perform computational checks on all the information used in the model. This improves estimation of the unmeasured fluxes and minimizes the effects of experimental error in the measured fluxes. If experimental error can be assumed to be equally distributed among the constraints imposed, a *least-squares solution* for the fluxes may be derived. In these calculations, the redundant measurements are used to adjust the measured fluxes depending on statistical assessment of their standard deviation or prevailing level of measurement noise.

Several other computational and statistical techniques are applied routinely to test and validate the outcome of metabolic flux analysis and improve the reliability of the results. The magnitude of the experimental error associated with measured reaction rates can have a significant effect on the accuracy of calculated fluxes. The *condition number* for the system, which is calculated from the matrix of stoichiometric coefficients, is a measure of the sensitivity of the results to small changes in the parameters used in the analysis. A system is considered ill-conditioned if the condition number is too high; this indicates that the model requires restructuring to avoid such heightened sensitivity. It is important to use accurate data and, as the influence of experimental error varies for different measured rates, to choose those rates for measurement with errors that have a relatively low impact on the flux results.

14.1.8 Metabolic Control Analysis

So far in our treatment of metabolic network analysis, we have seen how the fluxes of carbon through metabolic pathways can be estimated and how this information reveals the relative contributions of reactions in the metabolic network. Metabolic flux analysis gives a snapshot of pathway function under specific culture conditions but falls short of predicting the response of metabolism to genetic, chemical, or physical change that may affect the cells. Consequently, we have not yet found a solution to the problem presented in Figure 14.1, where genetic modification to eliminate a particular metabolic reaction was ineffective for reducing the synthesis of an unwanted product.

One of the many possible explanations for the response shown in Figure 14.1(A) is that the increase in concentration of metabolite A, or the decrease in concentration of metabolite B, following knockout of A→B stimulated the activity of several enzymes, including those responsible for the reactions A→C and H→G. As a result, the impediment to production of G and therefore J was overcome. To understand this further, the relationships between metabolic fluxes, enzyme activities, and metabolite and effector concentrations must be determined. Metabolic control analysis offers a range of concepts and mathematical theorems for evaluating the control functions of metabolic pathways. Here we consider some of its basic features; further information is available in the literature [9,10].

Metabolic control analysis makes use of a set of sensitivity parameters called *control coefficients* that quantify the fractional change in a specified property of a metabolic pathway in response to a small fractional change in the activity of an enzyme. The general formula for any control coefficient ξ is:

$$\xi = \lim_{\Delta y \to 0} \frac{\dfrac{\Delta x}{x}}{\dfrac{\Delta y}{y}} \tag{14.34}$$

where $\Delta x/x$ is the fractional change in system property x in response to a fractional change $\Delta y/y$ in parameter y, evaluated in the limit as Δy approaches zero. The limit condition means that control coefficients describe responses to very small changes in system parameters.

Control coefficients are dimensionless variables that are independent of the units used to express x and y. Eq. (14.34) can be written as:

$$\xi = \frac{\dfrac{dx}{x}}{\dfrac{dy}{y}} = \frac{y}{x}\frac{dx}{dy} = \frac{d\ln x}{d\ln y} \tag{14.35}$$

Several specific control coefficients have been defined. *Flux control coefficients* quantify the effect of changes in the activities of individual enzymes on the metabolic flux through a pathway or branch of a pathway. Flux control coefficients can take either positive or negative values, as an increase in the relative activity of a specific enzyme can either enhance or diminish a particular flux. The greater the absolute value of the flux control coefficient for a particular enzyme–flux combination, the greater is the influence exerted by that enzyme on that flux. In a similar way, *concentration control coefficients* describe the sensitivity of metabolite concentrations to changes in enzyme activity. The concentration of a particular metabolite in a pathway may increase or decrease as the activity of an enzyme changes, and the magnitude of the concentration change will vary for different enzyme–metabolite combinations.

Elasticity coefficients are different from control coefficients in that they are properties of individual enzymes rather than of the metabolic system as a whole. Elasticity coefficients describe the effect of a change in the concentration of a metabolite or effector on the rate of an enzyme reaction when the concentrations of all remaining metabolites and effectors are held constant. Elasticity coefficients are positive when enzyme activity is stimulated by an increase in the concentration of a substrate or enzyme activator and negative when enzyme activity is reduced by an increased concentration of an inhibitor or reaction product. Although kinetic expressions describing the effect of substrate and inhibitor concentrations on enzyme reaction rates have been developed, these equations are considered to have limited applicability to enzymes in vivo. The presence of multiple substrates, relatively high product concentrations, and a variety of effector molecules in the intracellular environment makes kinetic models determined in vitro from initial rate data potentially unsuitable for metabolic analysis. Evaluation of the elasticities of metabolic enzymes under the conditions prevailing during *in vivo* operation of metabolic pathways is the preferred approach. Several methods have been developed to measure control and elasticity coefficients [11].

The interactions that occur between fluxes, enzymes, and metabolites and effector molecules are illustrated in Figure 14.11. Enzyme activity has a direct influence on the flux of pathways; this effect is described by the flux control coefficients. Enzyme activity also affects the concentrations of metabolites in the pathway, as represented by the concentration control coefficients. However, metabolites and other effectors exert an indirect influence on flux through their effect on enzyme activity, as measured by the elasticity coefficients. Some metabolites

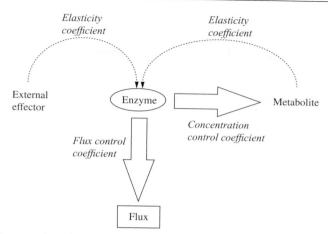

Figure 14.11 Interactions between metabolic flux, enzyme activity, and metabolites and external effectors.

are not only subjected to biochemical conversion in metabolism but are also information carriers with roles in enzyme regulation.

Control and elasticity coefficients are subject to certain constraints and relationships that can be expressed mathematically. These provide information about how fluxes are affected by enzyme regulation. An implication of the *flux control summation theorem* that has been confirmed experimentally is that flux control is shared or distributed between many enzymes in metabolic pathways. Individual enzymes generally do not have large flux control coefficients; instead, near-zero and very small coefficients are common. As a result, the presence of a single enzyme playing a dominant rate-limiting or "bottleneck" role for pathway control is relatively rare. This explains why manipulation of only one or two genes in cells often fails to influence metabolic flux, as illustrated in our hypothetical example of Figure 14.1.

To achieve significant change, coordinated modification of several enzymes is likely to be required, where the sum of the flux control coefficients is sufficiently high. An implication of the *flux control connectivity theorem*, which relates the elasticity coefficients to the flux control coefficients, is that enzymes with large elasticity coefficients, such as allosteric enzymes, do not necessarily play an important role in the control of metabolic flux. The value of the flux control coefficient must also be considered as a measure of the enzyme's influence on the pathway. There is a tendency for enzymes with relatively large elasticity coefficients to have relatively small flux control coefficients, and vice versa, so that enzymes with small elasticity coefficients have a considerable influence on flux [11,12]. This result from metabolic control analysis highlights the distinction between regulatory effects and flux control, in that enzymes subject to substantial regulation are not necessarily the best targets for genetic engineering strategies aimed at altering pathway function.

14.1.9 Metabolic Network Analysis Tools

There are several tools available for implementing the presented material. A limited list with sources is provided here:

KBase (www.kbase.us): extensive online functions including genome annotation, automated model assembly and flux balance analysis.

COBRA toolbox (https://opencobra.github.io/cobratoolbox/latest/modules/analysis/FBA/index.html): extensive online toolbox for analyzing metabolic models using flux balance and other related approaches.

COPASI (http://copasi.org/) comprehensive network simulation package that permits analysis of metabolic network models using ordinary differential equations (ODEs) with parameter sweeps, elementary flux mode analysis, and metabolic control analysis.

EFMA workshop (https://chbe.montana.edu/biochemenglab/EFMAworkshop.html) self-contained workshop for learning and applying elementary flux mode analysis to metabolic research. Content includes lecture material, step-by-step instructions, and worked examples for *E. coli*, *S. cerevisiae*, and an autotroph-heterotroph consortium.

14.2 Sustainable Bioprocessing

The environmental impacts and energy costs of industrial activities are of increasing concern to the companies involved, governmental and regulatory agencies, and the general public. Analysis of waste generation and the conservation of resources and energy are now integral to the design procedures employed for bioreactors and other components of bioprocessing. The specific areas addressed to improve the sustainability of bioprocesses include:

*   Saving energy
*   Saving water
*   Minimizing the use of materials
*   Minimizing greenhouse gas emissions
*   Minimizing the generation of effluent
*   Minimizing the environmental impact of effluent

Bioprocesses have many characteristics that render them attractive from an environmental point of view. Compared with chemical manufacturing, biological processing uses relatively nontoxic substrates and mild conditions of temperature, pressure, pH, and ionic strength. Biodegradable catalysts including enzymes and cells are employed, minimal quantities of organic solvents are required, and the conversions carried out exhibit a high level of specificity. Nevertheless, the production of cellular biomass as a waste product and the substantial energy and water requirements for maintaining aseptic operations make a significant contribution to the overall environmental burden associated with bioprocessing.

14.2.1 Sources of Waste and Pollutants in Bioprocessing

The main components of solid and liquid effluent streams produced in bioprocesses are cell biomass either before or after disruption, aqueous spent broth, and the buffers, solvents, and salt solutions used in downstream processing. Gas emissions include metabolically produced CO_2 and, in some processes, organic vapors such as ethanol and other volatile compounds. Large quantities of wastewater and dilute cleaning solutions are generated by the clean-in-place (CIP) systems used to maintain hygienic conditions. Further wastewater is produced by steam-in-place (SIP) facilities for sterilization of medium, bioreactors, piping, and other equipment such as centrifuges. The disposable plastics and polymers used for liquid and gas filtration and in the analytical laboratory for bioreactor monitoring and quality control generate additional solid waste. More broadly, gas emissions from transportation of goods and personnel can also be considered as waste products of bioprocessing.

Although effluent streams from bioprocessing generally do not contain high levels of hazardous components, large volumes of aqueous solutions containing organic and inorganic nutrients are potentially harmful in aquatic environments. Minerals and trace elements including Fe, Cu, Mo, and Mn are components of cell culture media and can become concentrated within the cellular biomass generated during cultivation. The discharge from a single 2-m^3 high-density yeast cultivation has been estimated to contain a significant quantity of around 500 g of trace metals [13]. Some compounds used in downstream processing, such as urea for solubilizing bacterial protein inclusion bodies and high-salt buffers for chromatographic separations, may present problems for waste disposal. Other chemicals including antifoams, surfactants, and fatty acids in bioreactor effluent cause foaming in waste treatment systems. Effluent from standard cleaning operations comprises dilute solutions containing bleach, phosphoric acid, and/or sodium hydroxide.

Energy is consumed in bioprocesses for heating, cooling, evaporation, pumping, agitation, aeration, and centrifugation and is provided typically in the form of electricity and steam. For mammalian cell culture and biopharmaceutical production, substantial amounts of additional energy are consumed in the distillation processes used to generate purified, pyrogen-free water (also called water-for-injection, or WFI) for medium preparation and final rinsing and flushing of equipment. Much of this energy use is essential and cannot be classified as waste, although steps may be taken to improve the efficiency of energy-consuming operations. Away from the direct requirements of the bioprocess, however, there are other areas of energy consumption in bioprocess facilities. These include operation of controlled temperature units for cold and frozen storage, and heating, ventilation, and air conditioning (HVAC) systems in clean rooms. Energy consumption in these areas can represent 50 to 75% of the total energy cost in a biopharmaceutical plant. Such facilities function continuously to preserve cell and chemical stocks and maintain cleanliness and positive air pressures even when production is not taking place. If they are over-designed or operate routinely under capacity, there may be scope to implement energy-saving measures in these areas.

14.2.2 Waste Metrics

Several parameters are used to characterize the utilization of materials and generation of waste in industrial processing. No single parameter is ideal, and none is employed universally. In principle, these metrics can be compared across different manufacturing sectors, companies, and products to gauge their relative environmental impacts.

Calculation of various indices requires definition of the system boundaries and knowledge of the mass flows into and out of the process. The *environmental-* or *E-factor* is defined as:

$$\text{E-factor} = \frac{\text{mass of waste}}{\text{mass of product}} \quad (14.36)$$

For calculation of the E-factor, waste is all the material generated by the process except the desired product. A higher E-factor means more waste and a greater negative environmental impact; the ideal E-factor is zero. An alternative index for waste generation is the *mass intensity*:

$$\text{mass intensity (MI)} = \frac{\text{mass of raw materials}}{\text{mass of product}} \quad (14.37)$$

where the ideal MI is 1. As it was originally formulated, the E-factor did not include water [14]; however, water usage is an important issue in many processes and may be represented in the calculation [15].

The E-factor and related indices do not consider the nature of the waste generated, particularly its pollutant strength and effect on human health. For example, although the E-factor for a particular process may be high, if the waste consists mainly of water and relatively benign chemicals such as inorganic salts, the effect on the environment will be lower than for a process generating relatively small quantities of toxic effluent. The *environmental quotient* has been proposed to quantify both the relative amount and type of waste produced [14]:

$$\text{environmental quotient (EQ)} = \text{E-factor} \times Q \quad (14.38)$$

where Q reflects the extent to which the waste presents an environmental problem. As an example, Q may be assigned a value of 1 for NaCl and 100 to 1000 for a toxic heavy metal. The magnitude of Q could be both volume- and location-dependent, as the ease of recycling or disposal of materials depends on those factors. An alternative approach to representing the environmental impact of waste is the *effective mass yield*:

$$\text{effective mass yield (EMY)} = \frac{\text{mass of product}}{\text{mass of non-benign raw materials}} \times 100\% \quad (14.39)$$

where substances such as water, dilute salts, dilute ethanol, dilute acetic acid, and so on, are classified as benign. A problem with indices such as those described in Eqs. (14.38) and (14.39) is that the Q value used for a particular waste component is debatable, and there is no agreed consensus about what constitutes an environmentally benign material.

Typical results from application of the E-factor to different chemical manufacturing industries are shown in Table 14.1. In the table, bulk or industrial chemicals include products such as sulfuric acid, acrylonitrile, bulk polymers, and methanol from carbon monoxide. Specialty or fine chemicals include perfumes, pesticides, coatings, inks, and sealants. Pharmaceuticals refers to small-molecule drugs such as steroids, alkaloids, antipsychotics, analgesics, and enzyme inhibitors that are produced by chemical synthesis rather than biocatalysis. The biopharmaceuticals category includes the production of therapeutic proteins using cell culture and the bioprocessing techniques that are the subject of this book. The results for biopharmaceuticals include process water. In general, E-factors become higher as the process production volume decreases and the product price increases.

Table 14.1: E-Factors for Manufacture of Chemicals and Pharmaceuticals

	Manufacturing sector			
	Bulk/industrial chemicals	Specialty/ fine chemicals	Pharmaceuticals	Biopharmaceuticals
E-factor	<1–5	5–50	25–100	2500–10,000
Annual production (tons)	10^4–10^6	10^2–10^4	10–10^3	10^{-3}–10
Annual waste production (tons)	10^4–5×10^6	5×10^2–5×10^5	250–10^5	2.5–10^5
Product price per kg	Low	Medium	High	High

Data from R.A Sheldon, 2007, The E factor: fifteen years on. *Green Chem.* 9, 1273–1283; B. Junker, 2010, Minimizing the environmental footprint of bioprocesses. Part 1: Introduction and evaluation of solid-waste disposal. *BioProcess Int.* 8(8), 62–72; and S.V. Ho, J.M. McLaughlin, B.W. Cue, and P.J. Dunn, 2010, Environmental considerations in biologics manufacturing. *Green Chem.* 12, 755–766.

Although the E-factors for pharmaceutical and biopharmaceutical production are relatively high, when the low product volumes in these sectors are considered, the total quantity of waste generated is, on average, lower than for the other types of chemical production. Nevertheless, the E-factor results draw attention to the relatively inefficient use of materials in the pharmaceuticals and biopharmaceuticals industries. This is due primarily to the use of multistep batch processing and, in the case of biopharmaceuticals, consumption of large amounts of water for equipment cleaning and steam sterilization. Compared with small-molecule drug manufacture, synthesis of therapeutic proteins using bioprocess technology has been estimated to require approximately 10 to 100 times more water per kg of product [16]. The large E-factors for pharmaceutical production can also be interpreted as reflecting the higher selling prices and profit margins in the sector and, consequently, a reduced focus on the economic benefits of lean manufacturing practices. There is considerable scope in bioprocessing for intensification of production systems, for example, using fed-batch operations at high cell concentrations, and for development of recovery and recycling programs to minimize resource consumption and waste generation. Therefore, an objective for improved bioprocess sustainability and reduced environmental impact is to identify strategies for reducing the E-factors associated with enzyme reaction and cell cultivation processes.

14.2.3 Life Cycle Analysis

The aim of life cycle analysis is to identify and evaluate the environmental impacts associated with product manufacture or process operation. Whereas the metrics described in the previous section are usually applied only to the production process itself, life cycle analysis examines the material and energy uses and releases to the environment from a much broader range of activities beyond the boundaries of the manufacturing plant. All of the "life stages" of a product may be assessed for their environmental impact, from production or extraction of the raw materials and generation of the power required in the production scheme to maintenance procedures, packaging, transportation, end use of the product, and waste treatment and disposal. This type of analysis is particularly useful for comparing the relative effects on the environment of two or more products or processes. Specialist software packages have been developed to assist the evaluation.

In principle, the all-encompassing framework of life cycle analysis reduces the risk of obtaining incorrect or misleading results for environmental impact when only some of the operations associated with a product or process are considered. Including all stages in the life cycle prevents environmental burdens from being transferred from one stage to another. In practice, carrying out a full life cycle assessment of any production scheme is a lengthy and detailed exercise involving potentially thousands of inputs, outputs, and elements of process information. Determining the system boundaries for the analysis and making decisions about the inclusion or exclusion of aspects of the process (the cut-off criteria) can influence the results substantially.

Figure 14.12 shows the elements that might be included in a life cycle analysis of a bio-process product. As well as the production process, where and how the raw materials are sourced, the use of transportation, the generation and use of electricity, and the treatment of

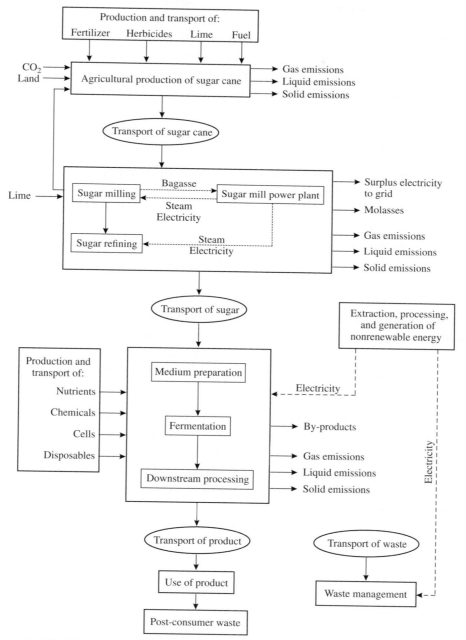

Figure 14.12 Schematic diagram of possible life cycle stages for a fermentation product.

waste are all examined. In this example, agricultural production of sugar cane and sugar milling and refining are included as life cycle stages necessary to produce substrate used in the cell cultivation. Emissions to the atmosphere and the generation of liquid and solid waste are considered for all steps in the life cycle.

Consensus-based international standards developed by the International Organization for Standardization (ISO) are used in life cycle analysis. The principles outlined in ISO 14040 provide the framework for a systematic approach to gathering and assessing the quantitative data needed to gauge environmental impact in a scientific and transparent way. The *functional unit* to be used in the analysis is determined; for a bioprocess this might be, for example, production of 1 kg of monoclonal antibody or 1 ton of valine. The functional unit serves as a reference for quantifying the input and output streams at each of the life stages in the same way as a basis was chosen for evaluating mass and energy balances in Chapters 4 and 5.

An *inventory* of the mass and energy inputs and environmental releases associated with each step in the life cycle is developed. This process is guided by process flow diagrams and uses data from industry sources, published studies, and specialized life cycle databases. Information covering areas of agriculture, energy generation, base and precious metals, transportation, bulk and specialty chemicals, biomaterials, construction materials, packaging, waste treatment, electronics, and information and communications technology is available from international life cycle database providers. Some of the data required will be location-specific, as agricultural, transport, and energy generation practices vary from country to country and region to region. For analysis of a bioprocess, the productivities, yields, and concentrations associated with operation of the bioreactor provide important information for developing the inventory of inputs and outputs for this stage of the process. Any recycling of heat or water within the production facility is considered as the inventory focuses on inputs and outputs that cross the system boundaries.

As many industrial processes produce more than one product, each product can be considered responsible for only a fraction of the resources consumed and emissions generated. *Allocation* of environmental impacts between coproducts can be a challenging aspect of life cycle analysis, as the choice of different apportioning methods can lead to different overall results. An example from bioprocessing is if the cells produced as a by-product of cultivation were sold as animal feed. An effective environmental credit would be included in the life cycle analysis to account for coproduction of this useful product.

The life cycle inventory of inputs and outputs must be translated into environmental impacts or consequences. The inventory items are aggregated into a limited set of recognized *environmental impact categories*, which are also known as "stressor" categories. There is no standard or universally agreed-on set of impact categories; however, commonly used categories include global warming, ozone depletion, acidification, eutrophication, particulate emissions, depletion of nonrenewable resources, human toxicity, ecotoxicity, and land use. As an example, if the cell cultivation and downstream processes in Figure 14.12 generate CO_2 and waste acidic buffer, the quantities involved would be assigned as contributions to the global warming and acidification impact categories, respectively. Published lists are used to guide the classification of materials to individual impact categories; some inventory items may be split between multiple categories. Within each category, some substances will be more damaging to the environment than others. Therefore, *characterization* or *equivalency factors* are applied to estimate the relative magnitude of the environmental impacts of each of the inventory components. These factors are based on scientific models of the interactions of substances with the environment. As an example, for greenhouse gases, the global warming potential of methane is 56 times greater than that of CO_2, so any methane emissions in the inventory are scored 56 times higher per unit mass than CO_2 emissions. Similarly, acidic substances in liquid effluent might be scored for their acidification potential based on how many H^+ ions they release as well as any technical factors that affect their environmental potency. Databases and life cycle analysis tools provide characterization factors for the different impact categories. Ongoing improvements in our understanding of cause and effect in environmental damage are reflected in the information available.

Typical results from a life cycle analysis are shown in Figure 14.13. In this example, the life cycle environmental impacts of two biofuels, ethanol from sugar beet and methyl ester (biodiesel) from rapeseed, are compared with those of two fossil fuels, petrol (gasoline) and diesel. The environmental effects of the fuels are quantified in terms of their contributions to the depletion of nonrenewable resources, acidification, eutrophication, ecotoxicity, and global

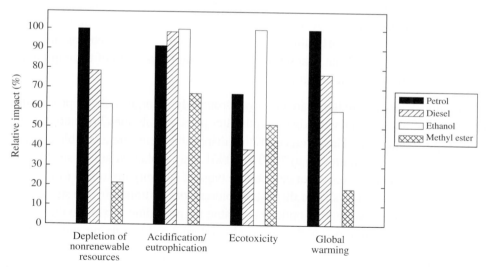

Figure 14.13 Results of life cycle analysis showing the relative environmental impacts of two fossil fuels—petrol (gasoline) and diesel—and two biofuels—ethanol from sugar beet and methyl ester from rapeseed. *Data from H. Halleux, S. Lassaux, R. Renzoni, and A. Germain, 2008, Comparative life cycle assessment of two biofuels: ethanol from sugar beet and rapeseed methyl ester.* Int. J. LCA 13, 184–190.

warming. In each environmental impact category, the fuel having the largest impact is represented as 100% and the impacts of the other three products are shown relative to that value. The life stages used in the analysis include cultivation and processing of agricultural crops in the case of the biofuels, extraction and refining in the case of the fossil fuels, ethanol fermentation, reaction of rapeseed oil with petrochemical methanol, all transportation steps, final use of the fuels in midsized cars, and application of by-products such as glycerin, rapeseed meal, and residual sugar beet pulp in the chemical industry or as animal feed. Overall, the results indicate that rapeseed methyl ester exerts the least environmental impact in three of the four categories examined, while ethanol offers improvements over the fossil fuels only in terms of nonrenewable resource consumption and global warming. More details of the analysis can be found in the original paper [17]. This example illustrates the type of environmental impact assessment that can be accomplished using life cycle methodology.

14.2.4 Disposable Bioreactors

Earlier in this chapter, practical considerations for the construction of bioreactors were described, such as the use of stainless steel for fabrication of large-scale bioreactors as pressure vessels, and the need to incorporate special design features for *in situ* steam sterilization, inoculation, sampling, and cleaning to maintain aseptic conditions. These requirements apply to most industrial bioprocesses carried out using equipment and other hardware built to last for many years. However, if a company is launching a new biotechnology product, optimizing the bioreactor design and culture performance is often of much lower priority than getting the product to market quickly. Application of ready-made, off-the-shelf disposable bioreactors can be an attractive option. A particular advantage is that none of the cleaning and sterilization facilities and procedures that account for a substantial proportion of the capital and operating costs of conventional bioprocessing is required.

Many disposable or *single-use* items are employed routinely in large-scale bioprocessing, including filter membranes and cartridges, depth filter pads, membrane chromatography capsules, and aseptic connectors. However, additional disposable items are available commercially. Plastic process bags with volumes up to 2000 to 3000 L may replace stainless steel vessels for liquid storage; disposable bioreactor or cell culture bags of capacity up to 500 to 1000 L are also used in some applications. The size of disposable bioreactors is limited by the strength of the bag material and the narrow range of options available for providing mixing to the culture contents. Bioreactor may be equipped with a disposable internal paddle mixing and sparger system.

The performance of disposable bioreactors in terms of mixing, heat transfer, and O_2 transfer is poor relative to conventional bioreactors, as intense agitation and high rates of heat and mass transfer cannot be achieved. Accordingly, their application is limited mainly to animal cell cultures where O_2 demand and heat loads are low and rapid mixing is generally not practical because of the shear sensitivity of the cells. However, as many new product developments in biotechnology involve animal cell culture, disposable bioreactors are being adopted as an alternative approach to making small-to-medium quantities of recombinant and therapeutic proteins, antibodies, viruses, and vaccines.

There may be substantial cost benefits associated with using disposable bioprocessing systems. This is because the utilities and materials required for sterilization and cleaning are minimized. If single-use components are applied everywhere throughout the manufacturing plant, a substantial amount of infrastructure normally associated with bioprocessing can be eliminated. Usually, this compensates for the costs involved in having to replace disposable items. When sterilization and cleaning are no longer required, the need for boilers, clean steam generators, steam-in-place and clean-in-place systems, and many other items of utility hardware is removed. Significant savings of energy, water, and cleaning chemicals may follow, as well as savings of time and labor.

An important concern associated with disposable bioprocessing is the environmental cost of single-use plasticware. Up to 90 disposable bags and biocontainers may be required to carry out a single 2000 L batch culture [18]. Use of disposable equipment creates plastic waste that must be incinerated as biohazardous material or disposed of as landfill after autoclaving or chemical deactivation. The potential for reusing or recycling most disposables is limited by the risk of contamination or biological carry-over and the need to separate the multiple types of plastic present in each item before recycling. Life cycle analysis has been applied to evaluate the environmental impacts associated with using 500-L stainless steel bioreactors compared with 500-L disposable bag bioreactors [19]. Despite the environmental burden associated with the generation of plastic waste, when the environmental effects of steel production are considered and depending on local conditions for electricity and steam generation and waste disposal, single-use bioreactor systems may have a lower overall impact on the environment compared with stainless steel equipment.

14.3 Summary of Chapter 14

At the end of Chapter 14, you should:

- Know the goals, assumptions, and common solution techniques for metabolic flux analysis, flux balance analysis, and elementary flux mode analysis
- Understand the importance of experimental data in flux balancing for underdetermined and overdetermined systems
- Know the basic concepts of *metabolic control analysis* for evaluating the sensitivity of fluxes to enzyme activities and metabolite concentrations
- Know the principal environmental concerns associated with resource utilization and waste generation in bioprocesses
- Understand the approaches used in *life cycle analysis* of the environmental impacts of bioprocessing

Problems

In an effort to encourage readers to become familiar with the online resources provided at the end of section 14.1, please access the following links and the specified examples, workbooks, and problems.

14.1. Stoichiometric matrices and elemental mass balances

in silico systems biology approaches including flux balance analysis, metabolic flux analysis, and elementary mode analysis require the assembly of metabolic models. These models are comprised of mass balances accounting for biochemical reactions. The balances are written succinctly as matrices known as the stoichiometric matrix. The elements of the matrices are the stoichiometric coefficients of mass balanced, biochemical reactions.

a. Create the stoichiometric matrix for the 10 reactions from glycolysis. Each reaction should have its own row and each metabolite should have its own column. If a metabolite does not participate in the reaction, its stoichiometric coefficient is 0.

1. Glucose + ATP^{4-} = Glucose-6-phosphate^{2-} + ADP^{3-} + H^+
2. Glucose-6-phosphate^{2-} → Fructose-6-phosphate^{2-}
3. Fructose-6-phosphate^{2-} + ATP4− → Fructose-1,6-bisphosphate^{4-} + ADP^{3-} + H^+
4. Fructose-1,6-bisphosphate^{4-} → Dihydroxyacetone phosphate^{2-} + Glyceraldehyde-3-phosphate^{2-}
5. Dihydroxyacetone phosphate^{2-} → Glyceraldehyde-3-phosphate^{2-}
6. Glyceraldehyde-3-phosphate^{2-} + Pi^{2-} + NAD^+ → 1,3-Bisphosphoglycerate^{4-} + NADH + H^+
7. 1,3-Bisphosphoglycerate^{4-} + ADP^{3-} → 3-Phosphoglycerate^{3-} + ATP^{4-}
8. 3-Phosphoglycerate^{3-} → 2-Phosphoglycerate^{3-}
9. 2-Phosphoglycerate^{3-} → Phosphoenolpyruvate^{3-} + H_2O
10. Phosphoenolpyruvate^{3-} + ADP^{3-} + H^+ → Pyruvate$^-$ + ATP^{4-}

b. Accurate predictions require mass-balanced reactions. Use matrix multiplication to verify that each reaction is balanced for C, H, O, N, P, and electrons. First collect the elemental equations for each metabolite being sure to account for the electrons associated with metabolite charge. Resources like Ecocyc.org or genome.jp/kegg/ are very useful for collecting data. Convert the data into a metabolite investment matrix. Each metabolite will be assigned a matrix row while C, H, O, N, P, and electrons will each be assigned a column. Multiply the stoichiometric matrix for glycolysis by the elemental investment vectors to verify each reaction is balanced for C, H, O, N, P, and electrons. This can be accomplished using MS Excel, MATLAB, Python, or similar software.

14.2. KBase (www.kbase.us) metabolic analysis tools

Using the website given, create a free account and navigate to the "Learn" tab in the tool bar (https://www.kbase.us/learn/). From here access the tutorial **"Microbial Metabolic Model Reconstruction and Analysis Tutorial"** and complete it.

14.3. CoBRA toolbox for flux balance and constraint-based analyses

(https://opencobra.github.io/cobratoolbox/latest/modules/analysis/FBA/index.html): Download and install MATLAB (https://nl.mathworks.com/help/install/). Click on **"Tutorials"** and from the "**Analysis**" section and select the tutorial for "**Computation and analysis of microbe-microbe metabolic interactions**" and complete it. Investigate other tutorials from this list as well. Additionally, from the "**Data Integration**" section, select the tutorial "**Metabotools tutorial 1**" and complete it. Alternatively, use the **Python-based** software **cobrapy** found at (https://open-cobra.github.io/cobrapy/).

14.4. COPASI (http://copasi.org/) based analysis of metabolic systems

Using the provided link, click on "**Projects**" in the toolbar to get to this website (https://basico.readthedocs.io/en/latest/). Download and install Python ((http://www.python.org/download/) and complete the "**Getting Started**" materials. From the "**Examples**" section, complete "**Accessing Models from the database**" and "**Creating and simulating a simple model**." Explore further with the tools and examples in this section.

14.5. EFMA step-by-step workshop for enumeration of elementary flux modes

(https://chbe.montana.edu/biochemenglab/EFMAworkshop.html)

Using the links at the bottom of the website, access the workshop presentation, step-by-step manual, MS Excel templates, and elementary flux mode executable (PC based). Complete the **workshop exercises** listed in the manual using the provided models and resource files. Note, solutions to all exercises can be found in the '**completed workshop exercises**' link.

References

[1]   C. Trinh, A. Wlaschin, F. Srienc, Elementary mode analysis: a useful metabolic pathway analysis tool for characterizing cellular metabolism, Appl Microbiol Biotechnol 81 (5) (2009) 813–826.

[2]   G. Stephanopoulos, J.J. Vallino, Network Rigidity and Metabolic Engineering in Metabolite Overproduction, Science 252 (5013) (1991) 1675–1681.

[3]   A.E. Beck, K.A. Hunt, R.P. Carlson, Measuring Cellular Biomass Composition for Computational Biology Applications, Processes 6 (5) (2018).

[4]   A.M. Feist, C.S. Henry, J.L. Reed, M. Krummenacker, A.R. Joyce, P.D. Karp, L.J. Broadbelt, V. Hatzimanikatis, B.Ø. Palsson, A genome-scale metabolic reconstruction for *Escherichia coli* K-12 MG1655 that accounts for 1260 ORFs and thermodynamic information, Mol. Syst. Biol. 3 (2007) 121.

[5]   S. Schuster, D.A. Fell, T. Dandekar, A general definition of metabolic pathways useful for systematic organization and analysis of complex metabolic networks, Nat Biotechnol 18 (3) (2000) 326–332.

[6]   R.P. Carlson, D.A. Fell, F. Srienc, Metabolic pathway analysis of a recombinant yeast for rational strain development, Biotechnol Bioeng 79 (2) (2002) 121–134.

[7]   R.P. Carlson, A. Wlaschin, F. Srienc, Kinetic studies and biochemical pathway analysis of anaerobic poly-(R)-3-hydroxybutyric acid synthesis in Escherichia coli, Appl Environ Microbiol 71 (2) (2005) 713–720.

[8]   Sauer, Metabolic networks in motion: ^{13}C-based flux analysis, Mol. Syst. Biol. 2 (1) (2006) 62.

[9]   G.N. Stephanopoulos, A.A. Aristidou, J. Nielsen, Metabolic Engineering: Principles and Methodologies, Academic Press, 1998.

[10] D. Fell, Understanding the Control of Metabolism, Portland Press, 1997.

[11] D.A. Fell, Metabolic control analysis: a survey of its theoretical and experimental development, Biochem. J 286 (1992) 313–330.

[12] J.R. Small, D.A. Fell, Metabolic control analysis: sensitivity of control coefficients to elasticities, Eur. J. Biochem. 191 (1990) 413–420.

[13] B. Junker, Minimizing the environmental footprint of bioprocesses. Part 1: Introduction and evaluation of solid-waste disposal, BioProcess Int 8 (8) (2010) 62–72.

[14] R.A. Sheldon, The E factor: fifteen years on, Green Chem 9 (2007) 1273–1283.

[15] M. Lancaster, Green Chemistry: An Introductory Text, Royal Society of Chemistry, 2002.

[16]   S.V. Ho, J.M. McLaughlin, B.W. Cue, P.J. Dunn, Environmental considerations in biologics manufacturing, Green Chem 12 (2010) 755–766.

[17]   H. Halleux, S. Lassaux, R. Renzoni, A. Germain, Comparative life cycle assessment of two biofuels: ethanol from sugar beet and rapeseed methyl ester, Int. J. LCA 13 (2008) 184–190.

[18]   B. Rawlings, H. Pora, A prescriptive approach to management of solid waste from single-use systems, BioProcess Int 7 (3) (2009) 40–47.

[19]   M. Mauter, Environmental life-cycle assessment of disposable bioreactors, BioProcess Int 7 (S4) (2009) 18–29.

Metabolic Network Analysis and Metabolic Engineering

H.U. Kim, T.Y. Kim, S.Y. Lee, Metabolic flux analysis and metabolic engineering of microorganisms, Mol. BioSyst 4 (2008) 113–120.

H. Shimizu, Metabolic engineering: integrating methodologies of molecular breeding and bioprocess systems engineering, J. Biosci. Bioeng 94 (2002) 563–573.

C.D. Smolke, (Ed.), Metabolic Pathway Engineering Handbook: Fundamentals, CRC Press (2010).

Conversion Factors

Entries in the same row are equivalent. For example, in Table A.1, 1 m = 3.281 ft, 1 mile = 1.609×10^3 m, etc. Exact numerical values are printed in bold type; others are given to four significant figures.

Table A.1: Length (L)

Meter (m)	Inch (in.)	Foot (ft)	Mile	Micrometer (mm)	Angstrom (Å)
1	3.9373×10^1	3.281	6.214×10^{-4}	**10^6**	**10^{10}**
2.54×10^{-2}	**1**	8.333×10^{-2}	1.578×10^{-5}	**2.54×10^4**	**2.54×10^8**
3.048×10^{-1}	**1.2×10^1**	**1**	1.894×10^{-4}	**3.048×10^5**	**3.048×10^9**
1.609×10^3	**6.336×10^4**	**5.28×10^3**	**1**	1.609×10^9	1.609×10^{13}
10^{-6}	3.937×10^{-5}	3.281×10^{-6}	6.214×10^{-10}	**1**	**10^4**
10^{-10}	3.937×10^{-9}	3.281×10^{-10}	6.214×10^{-14}	**10^{-4}**	**1**

Table A.2: Volume (L³)

Cubic meter (m³)	Liter (l or L) or cubic decimeter* (dm³)	Cubic foot (ft³)	Cubic inch (in.³)	Imperial gallon (UKgal)	U.S. gallon (USgal)
1	**10^3**	3.531×10^1	6.102×10^4	2.200×10^2	2.642×10^2
10^{-3}	**1**	3.531×10^{-2}	6.102×10^1	2.200×10^{-1}	2.642×10^{-1}
2.832×10^{-2}	2.832×10^1	**1**	**1.728×10^3**	6.229	7.481
1.639×10^{-5}	1.639×10^{-2}	5.787×10^{-4}	**1**	3.605×10^{-3}	4.329×10^{-3}
4.546×10^{-3}	4.546	1.605×10^{-1}	2.774×10^2	**1**	1.201
3.785×10^{-3}	3.785	1.337×10^{-1}	**$2.311 0^2$**	8.327×10^{-1}	**1**

*The liter was defined in 1964 as 1 dm³ exactly.

Table A.3: Mass (M)

Kilogram (kg)	Gram (g)	Pound (lb)	Ounce (oz)	Ton (t)	Imperial ton (UK ton)	Atomic mass unit* (u)
1	10^3	2.205	3.527×10^1	10^{-3}	9.842×10^{-4}	6.022×10^{-6}
10^{-3}	1	2.205×10^{-3}	3.527×10^{-2}	10^{-6}	9.842×10^{-7}	6.022×10^{-3}
4.536×10^{-1}	4.536×10^2	1	1.6×10^1	4.536×10^{-4}	4.464×10^{-4}	2.732×10^{-6}
2.835×10^{-2}	2.835×10^1	6.25×10^{-2}	1	2.835×10^{-5}	2.790×10^{-5}	1.707×10^{-5}
10^3	10^6	2.205×10^3	3.527×10^4	1	9.842×10^{-1}	6.022×10^{-9}
1.016×10^3	1.016×10^6	2.240×10^3	3.584×10^4	1.016	1	6.119×10^{-9}
1.661×10^{-27}	1.661×10^{-24}	3.661×10^{-27}	5.857×10^{-26}	1.661×10^{-30}	1.634×10^{-30}	1

*Atomic mass unit (unified); 1 u = 1/12 of the rest mass of a neutral atom of the nuclide ^{1}–C in the ground state.

Table A.4: Force (LMT^{-2})

Newton (N, kg m s^{-2})	Kilogram-force (kg$_f$)	Pound-force (lb$_f$)	Dyne (dyn, g cm s^{-2})	Poundal (pdl, lb ft s^{-2})
1	1.020×10^{-1}	2.248×10^{-1}	10^5	7.233
9.807	1	2.205	9.807×10^5	7.093×10^1
4.448	4.536×10^{-1}	1	4.448×10^5	3.217×10^1
10^{-5}	1.020×10^{-6}	2.248×10^{-6}	1	7.233×10^{-5}
1.383×10^{-1}	1.410×10^{-2}	3.108×10^{-2}	1.383×10^4	1

Table A.5: Pressure and Stress ($L^{-1}MT^{-2}$)

Pascal (Pa, N m⁻², J m⁻³, kg m⁻¹ s⁻²)	Pound-force per Inch² (psi, lb$_f$ in.⁻²)	Kilogram-force per meter² (kg$_f$ m⁻²)	Standard atmosphere (atm)	Dyne per cm² (dyn cm⁻²)	Torr (Torr, mmHg)*	Inches of water** (in. H₂O)	Bar
1	1.450×10^{-4}	1.020×10^{-1}	9.869×10^{-6}	10^{1}	7.501×10^{-3}	4.015×10^{-3}	10^{-5}
6.895×10^{3}	1	7.031×10^{2}	6.805×10^{-2}	6.895×10^{4}	5.171×10^{1}	2.768×10^{1}	6.895×10^{-2}
9.807	1.422×10^{-3}	1	9.678×10^{-5}	9.807×10^{1}	7.356×10^{-2}	3.937×10^{-2}	9.807×10^{-5}
1.013×10^{5}	1.470×10^{1}	1.033×10^{4}	1	1.013×10^{6}	$\mathbf{7.6 \times 10^{2}}$	4.068×10^{2}	1.013
10^{-1}	1.450×10^{-5}	1.020×10^{-2}	9.869×10^{-7}	1	7.501×10^{-4}	4.015×10^{-4}	10^{-6}
1.333×10^{2}	1.934×10^{-2}	1.360×10^{1}	1.316×10^{-3}	1.333×10^{3}	1	5.352×10^{-1}	1.333×10^{-3}
2.491×10^{2}	3.613×10^{-2}	2.540×10^{1}	2.458×10^{-3}	2.491×10^{3}	1.868	1	2.491×10^{-3}
10^{5}	1.450×10^{1}	1.020×10^{4}	9.869×10^{-1}	10^{6}	7.501×10^{2}	4.015×10^{2}	1

*mmHg refers to Hg at 0°C; 1 Torr = 1.00000 mmHg.

**in. H₂O refers to water at 4°C.

Table A.6: Energy, Work, and Heat ($L^2\,MT^{-2}$)

Joule (J, N m, Pa m³, W s, kg m² s⁻²)	Kilocalorie* (kcal)	British thermal unit (Btu)	Foot pound-force (ft lbf)	Liter atmosphere (L atm)	Kilowatt hour (kW h)	Erg (dyn cm)
1	2.388×10^{-4}	9.478×10^{-4}	7.376×10^{-1}	9.869×10^{-3}	2.778×10^{-7}	10^7
4.187×10^3	1	3.968	3.088×10^3	4.132×10^1	1.163×10^{-3}	4.187×10^{10}
1.055×10^3	2.520×10^{-1}	1	7.782×10^2	1.041×10^1	2.931×10^{-4}	1.055×10^{10}
1.356	3.238×10^{-4}	1.285×10^{-3}	1	1.338×10^{-2}	3.766×10^{-7}	1.356×10^7
1.013×10^2	2.420×10^{-2}	9.604×10^{-2}	7.473×10^1	1	2.815×10^{-5}	1.013×10^9
3.6×10^6	8.598×10^2	3.412×10^3	2.655×10^6	3.553×10^4	1	3.6×10^{13}
10^{-7}	2.388×10^{-11}	9.478×10^{-11}	7.376×10^{-8}	9.869×10^{-10}	2.778×10^{-14}	1

*International Table kilocalorie ($kcal_{IT}$).

Table A.7: Power (L⁻MT⁻³)

Watt (W, J s⁻¹, kg m² s⁻³)	Kilocalorie per min (kcal min⁻¹)	Foot pound-force per second (ft lb$_f$ s⁻¹)	Horsepower (British) (hp)	Metric horsepower	British thermal unit per minute (Btu min⁻¹)	Kilogram-force meter per second (kgf m s⁻¹)
1	1.433×10^{-2}	7.376×10^{-1}	1.341×10^{-3}	1.360×10^{-3}	5.687×10^{-2}	1.020×10^{-1}
$\mathbf{6.978 \times 10^{1}}$	1	5.147×10^{1}	9.358×10^{-2}	9.487×10^{-2}	3.968	7.116
1.356	1.943×10^{-2}	**1**	1.818×10^{-3}	1.843×10^{-3}	7.710×10^{-2}	1.383×10^{-1}
7.457×10^{2}	1.069×10^{1}	$\mathbf{5.5 \times 10^{2}}$	**1**	1.014	4.241×10^{1}	7.604×10^{1}
7.355×10^{2}	1.054×10^{1}	5.425×10^{2}	9.863×10^{-1}	**1**	4.183×10^{1}	$\mathbf{7.5 \times 10^{1}}$
1.758×10^{1}	2.520×10^{-1}	1.297×10^{1}	2.358×10^{-2}	2.391×10^{-2}	**1**	1.793
9.807	1.405×10^{-1}	7.233	1.315×10^{-2}	1.333×10^{-2}	5.577×10^{-1}	**1**

Table A.8: Illuminance ($L^{-2}J$)

Lux or lumen per meter^2 (lx or lm m^{-2})	Foot-candle (fc, lm ft^{-2})
1	9.290×10^{-2}
1.076×10^1	1

Table A.9: Dynamic Viscosity ($L^{-1}MT^{-2}$)

Pascal second (Pa s, N s m$^-$, kg m^{-1} s^{-1})	Poise (g cm^{-1} s^{-1}, dyn s cm$^-$)	Centipoise (cP)	kg m^{-1} h^{-1}	lb ft^{-1} h^{-1}
1	10^1	10^3	3.6×10^3	2.419×10^3
10^{-1}	1	10^2	3.6×10^2	2.419×10^2
10^{-3}	10^{-2}	1	3.6	2.419
2.778×10^{-4}	2.778×10^{-3}	2.778×10^{-1}	1	6.720×10^{-1}
4.134×10^{-4}	4.134×10^{-3}	4.134×10^{-1}	1.488	1

Ideal Gas Constant

Table B.1: Values of the Ideal Gas Constant, R

Energy unit	Temperature unit	Mole unit	R
cal	K	gmol	1.9872
J	K	gmol	8.3144
cm³ atm	K	gmol	82.057
l atm	K	gmol	0.082057
m³ atm	K	gmol	0.000082057
l mmHg	K	gmol	62.361
l bar	K	gmol	0.083144

Physical and Chemical Property Data

Table C.1: Degree of Reduction of Biological Materials

Compound	Formula	Degree of reduction γ relative to NH_3	Degree of reduction γ relative to N_2
Acetaldehyde	C_2H_4O	5.00	5.00
Acetic acid	$C_2H_4O_2$	4.00	4.00
Acetone	C_3H_6O	5.33	5.33
Adenine	$C_5H_5N_5$	2.00	5.00
Alanine	$C_3H_7O_2N$	4.00	5.00
Ammonia	NH_3	0	3.00
Arginine	$C_6H_{14}O_2N_4$	3.67	5.67
Asparagine	$C_4H_8O_3N_2$	3.00	4.50
Aspartic acid	$C_4H_7O_4N$	3.00	3.75
n-Butanol	$C_4H_{10}O$	6.00	6.00
Butyraldehyde	C_4H_8O	5.50	5.50
Butyric acid	$C_4H_8O_2$	5.00	5.00
Carbon monoxide	CO	2.00	2.00
Citric acid	$C_6H_8O_7$	3.00	3.00
Cytosine	$C_4H_5ON_3$	2.50	4.75
Ethane	C_2H_6	7.00	7.00
Ethanol	C_2H_6O	6.00	6.00
Ethene	C_2H_4	6.00	6.00
Ethylene glycol	$C_2H_6O_2$	5.00	5.00
Ethyne	C_2H_2	5.00	5.00
Formaldehyde	CH_2O	4.00	4.00
Formic acid	CH_2O_2	2.00	2.00
Fumaric acid	$C_4H_4O_4$	3.00	3.00
Glucitol	$C_6H_{14}O_6$	4.33	4.33
Gluconic acid	$C_6H_{12}O_7$	3.67	3.67
Glucose	$C_6H_{12}O_6$	4.00	4.00
Glutamic acid	$C_5H_9O_4N$	3.60	4.20
Glutamine	$C_5H_{10}O_3N_2$	3.60	4.80
Glycerol	$C_3H_8O_3$	4.67	4.67
Glycine	$C_2H_5O_2N$	3.00	4.50
Graphite	C	4.00	4.00
Guanine	$C_5H_5ON_5$	1.60	4.60
Histidine	$C_6H_9O_2N_3$	3.33	4.83
Hydrogen	H_2	2.00	2.00
Isoleucine	$C_6H_{13}O_2N$	5.00	5.50

(Continued)

Table C.1: Degree of Reduction of Biological Materials —cont'd

Compound	Formula	Degree of reduction γ relative to NH_3	Degree of reduction γ relative to N_2
Lactic acid	$C_3H_6O_3$	4.00	4.00
Leucine	$C_6H_{13}O_2N$	5.00	5.50
Lysine	$C_6H_{14}O_2N_2$	4.67	5.67
Malic acid	$C_4H_6O_5$	3.00	3.00
Methane	CH_4	8.00	8.00
Methanol	CH_4O	6.00	6.00
Oxalic acid	$C_2H_2O_4$	1.00	1.00
Palmitic acid	$C_{16}H_{32}O_2$	5.75	5.75
Pentane	C_5H_{12}	6.40	6.40
Phenylalanine	$C_9H_{11}O_2N$	4.44	4.78
Proline	$C_5H_9O_2N$	4.40	5.00
Propane	C_3H_8	6.67	6.67
iso-Propanol	C_3H_8O	6.00	6.00
Propionic acid	$C_3H_6O_2$	4.67	4.67
Pyruvic acid	$C_3H_4O_3$	3.33	3.33
Serine	$C_3H_7O_3N$	3.33	4.33
Succinic acid	$C_4H_6O_4$	3.50	3.50
Threonine	$C_4H_9O_3N$	4.00	4.75
Thymine	$C_5H_6O_2N_2$	3.20	4.40
Tryptophan	$C_{11}H_{12}O_2N_2$	4.18	4.73
Tyrosine	$C_9H_{11}O_3N$	4.22	4.56
Uracil	$C_4H_4O_2N_2$	2.50	4.00
Valeric acid	$C_5H_{10}O_2$	5.20	5.20
Valine	$C_5H_{11}O_2N$	4.80	5.40
Biomass	$CH_{1.8}O_{0.5}N_{0.2}$	4.20	4.80

Adapted from J.A. Roels, 1983, Energetics and Kinetics in Biotechnology, Elsevier, Amsterdam.

Table C.2: Mean Heat Capacities of Gases

T (°C)	Cpm (J gmol^{-1} °C^{-1})					
	Air	O_2	N_2	H_2	CO_2	H_2O
0	29.06	29.24	29.12	28.61	35.96	33.48
18	29.07	29.28	29.12	28.69	36.43	33.51
25	29.07	29.30	29.12	28.72	36.47	33.52
100	29.14	29.53	29.14	28.98	38.17	33.73
200	29.29	29.93	29.23	29.10	40.12	34.10
300	29.51	30.44	29.38	29.15	41.85	34.54

Reference state: $T_{ref} = 0$°C; $P_{ref} = 1$ atm.
Adapted from D.M. Himmelblau, 1974, Basic Principles and Calculations in Chemical Engineering, 3rd ed., Prentice Hall, Hoboken, New Jersey, U.S.

Table C.3: Heat Capacities

Compound	State	Temperature (T) unit	a	b.10²	c.10⁵	d.10⁹	Temperature range (units of T)
Acetone	l	°C	123.0	18.6			−30–60
Air	g	°C	28.94	0.4147	0.3191	−1.965	0–1500
	g	K	28.09	0.1965	0.4799	−1.965	273–1800
Ammonia	g	°C	35.15	2.954	0.4421	−6.686	0–1200
Ammonium sulphate	c	K	215.9				275–328
Calcium hydroxide	c	K	89.5				276–373
Carbon dioxide	g	°C	36.11	4.233	−2.887	7.464	0–1500
Ethanol	l	°C	103.1				0
	l	°C	158.8				100
	g	°C	61.34	15.72	−8.749	19.83	0–1200
Formaldehyde	g	°C	34.28	4.268	0.000	−8.694	0–1200
n-Hexane	l	°C	216.3				20–100
Hydrogen	g	°C	28.84	0.00765	0.3288	−0.8698	0–1500
Hydrogen chloride	g	°C	29.13	−0.1341	0.9715	−4.335	0–1200
Hydrogen sulfide	g	°C	33.51	1.547	0.3012	−3.292	0–1500
Magnesium chloride	c	K	72.4	1.58			273–991
Methane	g	°C	34.31	5.469	0.3661	−11.00	0–1200
	g	K	19.87	5.021	1.268	−11.00	273–1500
Methanol	l	°C	75.86	16.83			0–65
	g	°C	42.93	8.301	−1.87	−8.03	0–700
Nitric acid	l	°C	110.0				25
Nitrogen	g	°C	29.00	0.2199	0.5723	−2.871	0–1500
Oxygen	g	°C	29.10	1.158	−0.6076	1.311	0–1500
n-Pentane	l	°C	155.4	43.68			0–36
Sulfur (rhombic)	c	K	15.2	2.68			273–368
(monoclinic)	c	K	18.3	1.84			368–392
Sulfuric acid	l	°C	139.1	15.59			10–45
Sulfur dioxide	g	°C	38.91	3.904	−3.104	8.606	0–1500
Water	l	°C	75.4				0–100
	g	°C	33.46	0.6880	0.7604	−3.593	0–1500

C_p (J gmol⁻¹ °C⁻¹) $= a + bT + cT^2 + dT^3$ *Example*. For ammonia gas between 0°C and 1200°C: C_p (J gmol⁻¹ °C⁻¹) $= 35.15 + (2.954 \times 10^{-2})T + (0.4421 \times 10^{-5})T^2 - (6.686 \times 10^{-9})T^3$ where T is in °C. Note that some equations require T in K, as indicated. *State*: g = gas; l = liquid; c = crystal. Equations for gases apply at 1 atm.

Adapted from R.M. Felder and R.W. Rousseau, 2005, Elementary Principles of Chemical Processes, *3rd ed., John Wiley, Hoboken, NJ, USA.*

Table C.4: Heats of Combustion

Compound	Formula	Molecular weight	State	Heat of combustion Δh_c° (kJ gmol^{-1})
Acetaldehyde	C_2H_4O	44.053	l	−1166.9
			g	−1192.5
Acetic acid	$C_2H_4O_2$	60.053	l	−874.2
			g	−925.9
Acetone	C_3H_6O	58.080	l	−1789.9
			g	−1820.7
Acetylene	C_2H_2	26.038	g	−1301.1
Adenine	$C_5H_5N_5$	135.128	c	−2778.1
			g	−2886.9
Alanine (d-)	$C_3H_7O_2N$	89.094	c	−1619.7
Alanine (l-)	$C_3H_7O_2N$	89.094	c	−1576.9
			g	−1715.0
Ammonia	NH_3	17.03	g	−382.6
Ammonium ion	NH_4^+			−383
Arginine (d-)	$C_6H_{14}O_2N_4$	174.203	c	−3738.4
Asparagine (l-)	$C_4H_8O_3N_2$	132.119	c	−1928.0
Aspartic acid (l-)	$C_4H_7O_4N$	133.104	c	−1601.1
Benzaldehyde	C_7H_6O	106.124	l	−3525.1
			g	−3575.4
Butanoic acid	$C_4H_8O_2$	88.106	l	−2183.6
			g	−2241.6
1-Butanol	$C_4H_{10}O$	74.123	l	−2675.9
			g	−2728.2
2-Butanol	$C_4H_{10}O$	74.123	l	−2660.6
			g	−2710.3
Butyric acid	$C_4H_8O_2$	88.106	l	−2183.6
			g	−2241.6
Caffeine	$C_8H_{10}O_2N_4$		s	−4246.5*
Carbon	C	12.011	c	−393.5
Carbon monoxide	CO	28.010	g	−283.0
Citric acid	$C_6H_8O_7$		s	−1962.0
Codeine	$C_{18}H_{21}O_3N \cdot H_2O$		s	−9745.7*
Cytosine	$C_4H_5ON_3$	111.103	c	−2067.3
Ethane	C_2H_6	30.070	g	−1560.7
Ethanol	C_2H_6O	46.069	l	−1366.8
			g	−1409.4
Ethylene	C_2H_4	28.054	g	−1411.2
Ethylene glycol	$C_2H_6O_2$	62.068	l	−1189.2
			g	−1257.0
Formaldehyde	CH_2O	30.026	g	−570.7
Formic acid	CH_2O_2	46.026	l	−254.6
			g	−300.7
Fructose (d-)	$C_6H_{12}O_6$		s	−2813.7
Fumaric acid	$C_4H_4O_4$	116.073	c	−1334.0
Galactose (d-)	$C_6H_{12}O_6$		s	−2805.7
Glucose (d-)	$C_6H_{12}O_6$		s	−2805.0

Table C.4: Heats of Combustion—cont'd

Compound	Formula	Molecular weight	State	Heat of combustion Δh_c° (kJ gmol^{-1})
Glutamic acid (l-)	$C_5H_9O_4N$	147.131	c	−2244.1
Glutamine (l-)	$C_5H_{10}O_3N_2$	146.146	c	−2570.3
Glutaric acid	$C_5H_8O_4$	132.116	c	−2150.9
Glycerol	$C_3H_8O_3$	92.095	l	−1655.4
			g	−1741.2
Glycine	$C_2H_5O_2N$	75.067	c	−973.1
Glycogen	$(C_6H_{10}O_5)x$ per kg		s	−17,530.1*
Guanine	$C_5H_5ON_5$	151.128	c	−2498.2
Hexadecane	$C_{16}H_{34}$	226.446	l	−10,699.2
			g	−10,780.5
Hexadecanoic acid	$C_{16}H_{32}O_2$	256.429	c	−9977.9
			l	−10,031.3
			g	−10,132.3
Histidine (l-)	$C_6H_9O_2N_3$	155.157	c	−3180.6
Hydrogen	H_2	2.016	g	−285.8
Hydrogen sulfide	H_2S	34.08		−562.6
Inositol	$C_6H_{12}O_6$		s	−2772.2*
Isoleucine (l-)	$C_6H_{13}O_2N$	131.175	c	−3581.1
Isoquinoline	C_9H_7N	129.161	l	−4686.5
Lactic acid (d,l-)	$C_3H_6O_3$		l	−1368.3
Lactose	$C_{12}H_{22}O_{11}$		s	−5652.5
Leucine (d-)	$C_6H_{13}O_2N$	131.175	c	−3581.7
Leucine (l-)	$C_6H_{13}O_2N$	131.175	c	−3581.6
Lysine	$C_6H_{14}O_2N_2$	146.189	c	−3683.2
Malic acid (l-)	$C_4H_6O_5$		s	−1328.8
Malonic acid	$C_3H_4O_4$		s	−861.8
Maltose	$C_{12}H_{22}O_{11}$		s	−5649.5
Mannitol (d-)	$C_6H_{14}O_6$		s	−3046.5*
Methane	CH_4	16.043	g	−890.8
Methanol	CH_4O	32.042	l	−726.1
			g	−763.7
Morphine	$C_{17}H_{19}O_3N \cdot H_2O$		s	−8986.6*
Nicotine	$C_{10}H_{14}N_2$		l	−5977.8*
Oleic acid	$C_{18}H_{34}O_2$		l	−11,126.5
Oxalic acid	$C_2H_2O_4$	90.036	c	−251.1
Papaverine	$C_{20}H_{21}O_4N$		s	−10,375.8*
Pentane	C_5H_{12}	72.150	l	−3509.0
			g	−3535.6
Phenylalanine (l-)	$C_9H_{11}O_2N$	165.192	c	−4646.8
Phthalic acid	$C_8H_6O_4$	166.133	c	−3223.6
Proline (l-)	$C_5H_9O_2N$	115.132	c	−2741.6
Propane	C_3H_8	44.097	g	−2219.2
1-Propanol	C_3H_8O	60.096	l	−2021.3
			g	−2068.8
2-Propanol	C_3H_8O	60.096	l	−2005.8
			g	−2051.1

(Continued)

Table C.4: Heats of Combustion —cont'd

Compound	Formula	Molecular weight	State	Heat of combustion Δh_c° (kJ gmol^{-1})
Propionic acid	$C_3H_6O_2$	74.079	l	−1527.3
			g	−1584.5
1,2-Propylene glycol	$C_3H_8O_2$	76.095	l	−1838.2
			g	−1902.6
1,3-Propylene glycol	$C_3H_8O_2$	76.095	l	−1859.0
			g	−1931.8
Pyridine	C_5H_5N	79.101	l	−2782.3
			g	−2822.5
Pyrimidine	$C_4H_4N_2$	80.089	l	−2291.6
			g	−2341.6
Salicylic acid	$C_7H_6O_3$	138.123	c	−3022.2
			g	−3117.3
Serine (l-)	$C_3H_7O_3N$	105.094	c	−1448.2
Starch	$(C_6H_{10}O_5)x$ per kg		s	−17,496.6*
Succinic acid	$C_4H_6O_4$	118.089	c	−1491.0
Sucrose	$C_{12}H_{22}O_{11}$		s	−5644.9
Thebaine	$C_{19}H_{21}O_3N$		s	−10,221.7*
Threonine (l-)	$C_4H_9O_3N$	119.120	c	−2053.1
Thymine	$C_5H_6O_2N_2$	126.115	c	−2362.2
Tryptophan (l-)	$C_{11}H_{12}O_2N_2$	204.229	c	−5628.3
Tyrosine (l-)	$C_9H_{11}O_3N$	181.191	c	−4428.6
Uracil	$C_4H_4O_2N_2$	112.088	c	−1716.3
			g	−1842.8
Urea	CH_4ON_2	60.056	c	−631.6
			g	−719.4
Valine (l-)	$C_5H_{11}O_2N$	117.148	c	−2921.7
			g	−3084.5
Xanthine	$C_5H_4O_2N_4$	152.113	c	−2159.6
Xylose	$C_5H_{10}O_5$		s	−2340.5
Biomass	$CH_{1.8}O_{0.5}N_{0.2}$	24.6	s	−552

Reference conditions: 1 atm and 25°C or 20°C; values marked with an asterisk refer to 20°C. Products of combustion are taken to be CO_2 (gas), H_2O (liquid), and N_2 (gas); therefore, $\Delta h_c^\circ = 0$ for CO_2 (g), H_2O (l), and N_2 (g). *State:* g = gas; l = liquid; c = crystal; s = solid.
Adapted from Lide, D.R. (2003) CRC Handbook of Chemistry and Physics. *84th Edition, CRC Press, New York. and R.M. Felder and R.W. Rousseau, 2005,* Elementary Principles of Chemical Processes, *3rd ed., John Wiley, New York.*

Logarithms

The *natural logarithm* (ln or $\log_e$) is the inverse of the exponential function. Therefore, if:

$$y = \ln x \tag{D.1}$$

then

$$e^y = x \tag{D.2}$$

where the number e is approximately 2.71828. It also follows that:

$$\ln(e^y) = y \tag{D.3}$$

and

$$e^{\ln x} = x \tag{D.4}$$

Natural logarithms are related to *common logarithms*, or logarithms to the base 10 (written as lg, log or $\log_{10}$), as follows:

$$\ln x = \ln 10 (\log_{10} x) \tag{D.5}$$

Since ln 10 is approximately 2.30259:

$$\ln x = 2.30259 \log_{10} x \tag{D.6}$$

Zero and negative numbers do not have logarithms.

Rules for taking logarithms of products and powers are illustrated below. The logarithm of the product of two numbers is equal to the sum of the logarithms:

$$\ln(ax) = \ln a + \ln x \tag{D.7}$$

When one term of the product involves an exponential function, application of Eqs. (D.3) and (D.7) gives:

$$\ln(be^{ax}) = \ln b + ax \tag{D.8}$$

The logarithm of the quotient of two numbers is equal to the logarithm of the numerator minus the logarithm of the denominator:

$$\ln\left(\frac{a}{x}\right) = \ln a - \ln x \tag{D.9}$$

As an example of this rule, because $\ln 1 = 0$:

$$\ln\left(\frac{1}{x}\right) = -\ln x \tag{D.10}$$

The rule for taking the logarithm of a power function is:

$$\ln(x^b) = b \ln x \tag{D.11}$$

Index

Page numbers followed by *f* indicate figures; *t*, tables; *b*, boxes.